Phytochemistry in Corrosion Science
Plant Extracts and Phytochemicals as Corrosion Inhibitors

Edited by
Chandrabhan Verma, Ashish Kumar,
and Abhinay Thakur

CRC Press
Taylor & Francis Group
Boca Raton London New York

CRC Press is an imprint of the
Taylor & Francis Group, an **informa** business

Designed cover image: Shutterstock

First edition published 2024
by CRC Press
2385 NW Executive Center Drive, Suite 320, Boca Raton FL 33431

and by CRC Press
4 Park Square, Milton Park, Abingdon, Oxon, OX14 4RN

CRC Press is an imprint of Taylor & Francis Group, LLC

Names Verma, Chandrabhan, editor. | Kumar, Ashish (Chemist), editor. |
Thakur, Abhinay, editor.
Title: Phytochemistry in corrosion science : plant extracts and
phytochemicals as corrosion inhibitors / edited by Chandrabhan Verma,
Ashish Kumar, and Abhinay Thakur.
Description: First edition. | Boca Raton FL : CRC Press, 2024. |
Includes bibliographical references.
Identifiers: LCCN 2023040247 (print) | LCCN 2023040248 (ebook) |
ISBN 9781032496153 (hardback) | ISBN 9781032496160 (paperback) |
ISBN 9781003394631 (ebook)
Subjects: LCSH: Corrosion and anti-corrosives. | Botanical chemistry.
Classification: LCC TA462 .P59 2024 (print) | LCC TA462 (ebook) |
DDC 620.1/1223—dc23/eng/20231226
LC record available at https://lccn.loc.gov/2023040247
LC ebook record available at https://lccn.loc.gov/2023040248

ISBN: 978-1-032-49615-3 (hbk)
ISBN: 978-1-032-49616-0 (pbk)
ISBN: 978-1-003-39463-1 (ebk)

DOI: 10.1201/9781003394631

Typeset in Times
by codeMantra

Phytochemistry in Corrosion Science

Phytochemistry in Corrosion Science covers the use of plant extracts/phytochemicals in corrosion mitigation with industrial applications. It explores innovative and characterization approaches toward the utilization of plant extracts and their phytochemicals as potential corrosion inhibitors for several metals and their alloys.

Providing a comprehensive overview of the green aspects of plant extracts as corrosion inhibitors, this book discusses the preparation of aqueous and organic phase extracts, and their advantages, disadvantages, and use for different aggressive media. It also examines aqueous and organic extracts that have been successfully used as corrosion inhibitors for various metals and electrolyte combinations.

This book will be a useful reference for undergraduate and graduate students and academic researchers in the fields of phytochemistry, corrosion science and engineering, environmental science, chemical engineering, green chemistry, and mechanical/industrial engineering.

Contents

Chapter 21 Phytochemicals/Plant Extracts as Corrosion Inhibitors
for Copper in NaCl Solutions 445

*Elyor Berdimurodov, Khasan Berdimuradov,
Ashish Kumar, Omar Dagdag, Mohamed Rbaa,
Bhawana Jain, Oybek Mikhliev, Abduvali Kholikov,
and Khamdam Akbarov*

Chapter 22 Phytochemicals/Plant Extracts as Corrosion Inhibitors
for Zinc in NaCl Solutions 466

*Mosarrat Parveen, Mohammad Mobin, Saman Zehra,
Ruby Aslam, and Kanika Cial*

Chapter 23 Phytochemicals/Plant Extracts as Corrosion Inhibitors for
Magnesium in NaCl Solutions 475

*Xin Liu, Yifan Lv, Xuexue Xu, Sheng Wu,
and Xuerong Zheng*

Chapter 24 Sustainability/Greenness of Phytochemicals/Plant Extracts 497

*Sampad Ghosh, Nabakumar Pramanik, Rajeev Kumar,
and Vinay Kumar Chaudhary*

Chapter 25 Synergism in Anticorrosive Phenomenon
of Phytochemicals/Plant Extracts 514

Madhulata Shukla

Index.. 531

About the Editors

Dr. Chandrabhan Verma works at the Department of Chemical Engineering, Khalifa University of Science and Technology, Abu Dhabi, United Arab Emirates. Dr. Verma earned his B.Sc. and M.Sc. degrees from Udai Pratap Autonomous College, Varanasi (UP), India. He earned his Ph.D. from the Department of Chemistry, Indian Institute of Technology (Banaras Hindu University), Varanasi, under the supervision of Prof. Mumtaz A. Quraishi in Corrosion Science and Engineering. He is a member of the American Chemical Society (ACS) and a lifetime member of the World Research Council (WRC). He is a reviewer and editorial board member of various internationally recognized platforms. Dr. Verma published numerous research and review articles in different areas of science and engineering. His current research focuses on designing and developing industrially applicable corrosion inhibitors, and he has previously edited multiple books. Dr. Verma received several awards for his academic achievements, including a gold medal in M.Sc. (Organic Chemistry; 2010) and best publication awards from the Global Alumni Association of IIT-BHU (2013).

Dr. Ashish Kumar is currently Professor (Chemistry), Dean Academics and Head, Applied Science and Humanities, Nalanda College of Engineering, Bihar Engineering University, Science, Technology and Technical Education Department, Government of Bihar, India. He has been listed in the "World Ranking of Top 2% Scientists-2023" database released by Stanford University, USA. His research work is related to corrosion inhibition, material chemistry, thermodynamics, and interaction studies in liquid medium and homogeneous catalysis. He has been a reviewer of various journals. He has also published more than 100+ research papers in high-impact quality journals. He has also authored and reviewed several books and published diverse chapters (50+) in refereed books. He is a guest editor/editorial board member of several international journals and a keynote speaker and member of technical committee in various national and international conferences. He also evaluated several Ph.D. theses from India and abroad.

Dr. Abhinay Thakur is an Assistant Professor at Research and Development Cell, Lovely Professional University, Punjab, India. His research work is related to corrosion inhibition, material chemistry, thermodynamics, and interaction studies in liquid medium.

Preface

Plants are the primary source of chemicals on the earth. They can effectively convert radiant energy into carbohydrates in the presence of H_2O and CO_2. The converted carbohydrate is readily converted into phytochemicals such as saponins, terpenoids, flavonoids, tannins, and alkaloids. Owing to their natural and biological origin, these phytochemicals could be considered eco-benign, green, safe, and sustainable alternatives for various industrial and biological applications. In the last three decades, numerous aqueous and organic extracts have been successfully utilized as corrosion inhibitors for several metals and electrolyte combinations. This book emphasizes the preparation of aqueous and organic phase extracts in several aggressive media along with their pros and cons. This book will provide a comprehensive overview of the green aspects of plant extracts as potential corrosion inhibitors. Additionally, industrial case studies and future perspectives of phytochemicals/plant extracts as corrosion inhibitors will be described. The binding and adsorbance mechanism of phytochemicals on the metallic surface will be elaborated. This book will be a compilation of various chapters on phytochemicals/plant extracts in corrosion mitigation where the state of the art of technologies in the field of corrosion science and engineering, green chemistry, analytical chemistry, and industrial engineering will be reviewed and presented vividly. This book will be found very useful not only by corrosion engineering students but also by corrosion scientists and engineers in their problems in their professional capacity and those interested in corrosion. This book will cover all the recent advancements in corrosion and its inhibition by using eco-benign and sustainable plant extracts/phytochemicals for the corrosion mitigation of several metals and their alloys. Readers will find this as a compact book with the various aspects of plant extracts/phytochemicals in green chemistry mainly prepared for UG and PG students, high-level researchers, professors, and engineers in the fields of phytochemistry, corrosion science, chemical engineering, and applied fields.

Contributors

Imane Aadnane
Laboratory of Chemistry-Biology
  Applied to the Environment, Faculty
  of Sciences
Moulay Ismail University
Meknes, Morocco

Abdelmalik El Aatiaoui
Laboratory of Molecular Chemistry,
  Materials and Environment
  (LCM2E), Department of Chemistry,
  Multidisciplinary Faculty of Nador
University Mohamed I
Nador, Morocco

S.A. Abdulkareem
Department of Chemical Engineering
Federal University of Technology
Minna, Nigeria

H.L. Abubakar
Department of Chemistry
Nile University of Nigeria
Abuja, Nigeria

Y.K. Abubakar
Department of Mechanical Engineering
Federal Polytechnic
Idah, Nigeria

Khamdam Akbarov
Faculty of Industrial Viticulture and
  Food Production Technology
National University of Uzbekistan
Tashkent, Uzbekistan

A.T. Amigun
Department of Chemical and
  Geological Sciences
Al-Hikmah University
Ilorin, Nigeria

Zakia Aribou
Laboratory of Advanced Materials and
  Process Engineering, Faculty of
  Sciences
Ibn Tofaïl University
Kenitra, Morocco

Ruby Aslam
Department of Applied Chemistry,
  Corrosion Research Laboratory,
  Faculty of Engineering and
  Technology
Aligarh Muslim University
Aligarh, India

Humira Assad
Department of Chemistry, School of
  Chemical Engineering and Physical
  Sciences
Lovely Professional University
Phagwara, India

Himanshi Bairagi
Corrosion Testing Research Lab,
  Department of Chemistry
J.C. Bose University of Science and
  Technology
YMCA Faridabad, India

Priyabrata Banerjee
Electric Mobility and Tribology
    Research Group
CSIR-Central Mechanical Engineering
    Research Institute
Durgapur, India
and
Academy of Scientific and Innovative
    Research (AcSIR)
AcSIR Headquarters CSIR-HRDC
    Campus
Ghaziabad, India

Manoj K. Banjare
MATS School of Sciences
MATS University
Raipur, India

Ramesh K. Banjare
Department of Chemistry
Raipur Institute of Technology
Raipur, India

Sumayah Bashir
Department of Chemistry
Central University of Kashmir
Jammu and Kashmir, India

Kamalakanta Behera
Department of Chemistry, Faculty of
    Science
University of Allahabad
Prayagraj, India

Ghita Amine Benabdallah
Laboratory of Molecular Spectroscopy
    Modelling, Materials, Nanomaterials,
    Water and Environment, CERNE2D,
    Faculty of Sciences
Mohammed V University
Rabat, Morocco

Elyor Berdimurodov
Chemical & Materials Engineering
New Uzbekistan University
Tashkent, Uzbekistan

Medical School
Central Asian University
Tashkent, Uzbekistan
and
Faculty of Chemistry
National University of Uzbekistan
Tashkent, Uzbekistan

Khasan Berdimuradov
Faculty of Industrial Viticulture and
    Food Production Technology,
    Department of Chemistry
Shahrisabz branch of Tashkent Institute
    of Chemical Technology
Shahrisabz, Uzbekistan

Ichraq Bouhouche
Laboratory of Molecular Spectroscopy
    Modelling, Materials, Nanomaterials,
    Water and Environment, CERNE2D,
    ENSAM
Mohammed V University
Rabat, Morocco

Khalid Bouiti
Laboratory of Molecular Spectroscopy
    Modelling, Materials, Nanomaterials,
    Water and Environment, CERNE2D,
    ENSAM
Mohammed V University
Rabat, Morocco

Vinay Kumar Chaudhary
Department of Physics
Government Engineering College
Nawada, Bihar, India

Mohammed Cherkaoui
Advanced Materials and Process
  Engineering, Laboratory of
  Organic Chemistry, Catalysis and
  Environment, Faculty of Sciences
Ibn Tofaïl University
Kenitra, Morocco
and
National Higher School of Chemistry
  (NHSC)
Ibn Tofaïl University
Kenitra, Morocco

Kanika Cial
Corrosion Research Laboratory,
  Department of Applied Chemistry,
  Faculty of Engineering and
  Technology
Aligarh Muslim University
Aligarh, India

Maria Lúcia Caetano Pinto da Silva
Departamento de Engenharia Química,
  Escola de Engenharia de Lorena
Universidade de São Paulo
São Paulo, Brazil

Omar Dagdag
Department of Mechanical Engineering
Gachon University
Seongnam, Republic of Korea

Khadija Dahmani
Advanced Materials and Process
  Engineering, Laboratory of
  Organic Chemistry, Catalysis and
  Environment, Faculty of Sciences
Ibn Tofaïl University
Kenitra, Morocco

Ikram Daou
Laboratory of Chemistry-Biology
  Applied to the Environment, Faculty
  of Sciences
Moulay Ismail University
Meknes, Morocco

Walid Daoudi
Department of Chemistry, Laboratory
  of Molecular Chemistry, Materials
  and Environment (LCM2E),
  Multidisciplinary Faculty of Nador
University Mohamed I
Nador, Morocco

Pragnesh N. Dave
Department of Chemistry
Sardar Patel University
Gujarat, India

Deepti Verma
Department of Chemistry, Faculty of
  Science
University of Allahabad
Prayagraj, India

T.C. Egbosiuba
Department of Chemical Engineering
Chukwuemeka Odumegwu Ojukwu
  University
Anambra State, Nigeria

Souad El Hajjaji
Laboratory of Molecular Spectroscopy
  Modelling, Materials, Nanomaterials,
  Water and Environment, CERNE2D,
  Faculty of Sciences
Mohammed V University
Rabat, Morocco

Brahim El Ibrahimi
Department of Applied Chemistry,
  Faculty of Applied Sciences
Ibn Zohr University
Kenitra, Morocco

R. Elabor
School of the Environment
Florida Agricultural and Mechanical
    University
Tallahassee, Florida, USA

Hicham Es-soufi
Department ENSAM-Meknes
    Marjane II, Laboratory of Sciences
    and Professions of the Engineer,
    Materials and Processes
Moulay Ismail University
Meknes, Morocco

M.B. Etsuyankpa
Department of Chemistry, Faculty of
    Science
Federal University of Lafia
Nasarawa State, Nigeria

Flávia Dias Fernandes
Departamento de Engenharia Química,
    Escola de Engenharia de Lorena
Universidade de São Paulo
São Paulo, Brazil

Mouhsine Galai
Laboratory of Advanced Materials and
    Process Engineering, Faculty of
    Sciences
Ibn Tofaïl University
Kenitra, Morocco

Richika Ganjoo
Department of Chemistry, School of
    Chemical Engineering and
    Physical Sciences
Lovely Professional University
Phagwara, India

Sampad Ghosh
Department of Students Welfare
Maulana Abul Kalam Azad University
    of Technology
Nadia, India

Drishti Gupta
Department of Applied Chemistry
    (CBFS-ASAS)
Amity University
Gurugram, India

Rajesh Haldhar
School of Chemical Engineering
Yeungnam University
Gyeongsan, Republic of Korea

Bhawana Jain
Siddhachalam Laboratory
Raipur, India

Poonam Kaswan
Department of Chemistry
    (CBFS-ASAS)
Amity University
Haryana, India

Jasdeep Kaur
Department of Chemistry
Chandigarh University
Mohali, India

Babar Khan
Corrosion Research Laboratory,
    Department of Applied Chemistry,
    Faculty of Engineering and
    Technology
Aligarh Muslim University
Aligarh, India

Otmane Kharbouch
Laboratory of Organic, Organometallic
    and Theoretical Chemistry, Faculty
    of Science
Ibn Tofaïl University
Kenitra, Morocco

Abduvali Kholikov
Faculty of Industrial Viticulture and
    Food Production Technology
National University of Uzbekistan
Tashkent, Uzbekistan

Hansang Kim
Department of Mechanical Engineering
Gachon University
Seongnam, Republic of Korea

Ambrish Kumar
Department of Chemistry, Institute of
    Science
Banaras Hindu University
Varanasi, India

Ashish Kumar
Department of Science, Technology and
    Technical Education
Nalanda College of Engineering, Bihar
    Engineering University
Patna, India
and
Government of Bihar
Bihar, India

Rajeev Kumar
Department of Computer Science &
    Engineering
Nalanda College of Engineering
Nalanda, India

Rahayu Kusumastuti
Research Center for Metallurgy
National Research and Innovation
    Agency
Jakarta, Republic of Indonesia

Najoua Labjar
Laboratory of Molecular Spectroscopy
    Modelling, Materials, Nanomaterials,
    Water and Environment, CERNE2D,
    ENSAM
Mohammed V University
Rabat, Morocco

Nabil Lahrache
Laboratory of Molecular Spectroscopy
    Modelling, Materials, Nanomaterials,
    Water and Environment, CERNE2D,
    ENSAM
Mohammed V University
Rabat, Morocco

Salma Lamghafri
Laboratory of Applied Sciences,
    National School of Applied Sciences
    Al-Hoceima
Abdelmalek Essaâdi University
Tetouan, Morocco

Xin Liu
Yantai Research Institute of Harbin
    Engineering University
Harbin Engineering University
Harbin, China

Yifan Lv
Yantai Research Institute of Harbin
    Engineering University
Harbin Engineering University
Harbin, China

Pradip M. Macwan
B. N. Patel Institute of Paramedical &
    Science (Science Division), Sardar
    Patel Education Trust
Anand, India

Sadhucharan Mallick
Departments of Chemistry
Indira Gandhi National Tribal
    University (Central University)
Anuppur, India

Chandan K. Mandal
Department of Applied Chemistry
    (CBFS-ASAS)
Amity University
Gurugram, India

Bindu Mangla
Corrosion Testing Research Lab,
Department of Chemistry
J.C. Bose University of Science and
Technology
YMCA Faridabad, India

Oybek Mikhliev
Department of Chemistry
Karshi Engineering Economics Institute
Qarshi, Uzbekistan

Mohammad Mobin
Corrosion Research Laboratory,
Department of Applied Chemistry,
Faculty of Engineering and
Technology
Aligarh Muslim University
Aligarh, India

A.K. Mohammed
Department of Chemistry and
Biochemistry
North Carolina Central University
Durham, North Carolina

Hamou Moussout
Laboratory of Chemistry-Biology
Applied to the Environment, Faculty
of Sciences
Moulay Ismail University
Meknes, Morocco
and
Laboratory of Advanced Materials and
Process Engineering, Faculty of
Sciences
Ibn Tofail University
Kenitra, Morocco

M.J. Muhammad
Department of Chemistry
Federal University of Technology
Minna, Nigeria

Siti Musabikha
Research Center for Metallurgy
National Research and Innovation
Agency
Jakarta, Republic of Indonesia

S. Mustapha
Department of Chemistry
Federal University of Technology
Minna, Nigeria

Rajni Narang
Corrosion Testing Research Lab,
Department of Chemistry
J.C. Bose University of Science and
Technology
YMCA Faridabad, India

M.M. Ndamitso
Department of Chemistry
Federal University of Technology
Minna, Nigeria

Arini Nikitasari
Research Center for Metallurgy
National Research and Innovation
Agency
Jakarta, Republic of Indonesia

Moussa Ouakki
Laboratory of Organic Chemistry,
Catalysis and Environment, Faculty
of Sciences
Ibn Tofaïl University
Kenitra, Morocco
and
National Higher School of Chemistry
(NHSC)
Ibn Tofaïl University
Kenitra, Morocco

Adyl Oussaid
Laboratory of Molecular Chemistry,
    Materials and Environment
    (LCM2E), Department of Chemistry,
    Multidisciplinary Faculty of Nador
University Mohamed I
Nador, Morocco

Subhashree J. Pandya
School of Studies in Chemistry
Pt. Ravishankar Shukla University
Raipur, India

Mosarrat Parveen
Corrosion Research Laboratory,
    Department of Applied Chemistry,
    Faculty of Engineering and
    Technology
Aligarh Muslim University
Aligarh, India

Seema R. Pathak
Department of Applied Chemistry
    (CBFS-ASAS)
Amity University
Gurugram, India

Nabakumar Pramanik
Department of Chemistry
National Institute of Technology
Papum Pare, India

Siska Prifiharni
Research Center for Metallurgy
National Research and Innovation Agency
Jakarta, Republic of Indonesia

Gadang Priyotomo
Research Center for Metallurgy
National Research and Innovation Agency
Jakarta, Republic of Indonesia

Varun Rai
Department of Chemistry
University of Allahabad
Prayagraj, India

Gyandshwar K. Rao
Department of Chemistry
    (CBFS-ASAS)
Amity University
Gurugram, India

Rashmi
Corrosion Testing Research Lab,
    Department of Chemistry
J.C. Bose University of Science and
    Technology
YMCA Faridabad, India

Mohamed Rbaa
Laboratory of Organic Chemistry,
    Catalysis and Environment,
    Chemistry Department, Faculty of
    Sciences
Ibn Tofail University
Kenitra, Morocco

Issam Saber
Laboratory of Organic Chemistry,
    Catalysis and Environment, Faculty
    of Sciences
Ibn Tofaïl University
Kenitra, Morocco

Akhil Saxena
Department of Chemistry
Chandigarh University
Mohali, India

Praveen Kumar Sharma
Department of Chemistry, School of
    Chemical Engineering and Physical
    Sciences
Lovely Professional University
Phagwara, India

Shveta Sharma
Department of Chemistry, School of
    Chemical Engineering and Physical
    Sciences
Lovely Professional University
Phagwara, India

Bharti Sheokand
Department of Applied Chemistry
   (CBFS-ASAS)
Amity University
Gurugram, India

D.T. Shuaib
Department of Chemistry,
   Illinois Institute of
   Technology
Chicago, Illinois

Madhulata Shukla
Department of Chemistry,
   G.B. College
Veer Kunwar Singh University
Arrah, India

Sudhish K. Shukla
Department of Sciences, School of
   Sciences
Manav Rachna University
Faridabad, India

Ambrish Singh
Department of Chemistry
Nagaland University
Nagaland, India

Amita Somya
Department of Chemistry
Amity School of Engineering &
   Technology
Amity University
Bengaluru, India

Bousalham Srhir
National Higher School of Chemistry
   (NHSC)
Ibn Tofaïl University
Kenitra, Morocco
and
Laboratory of Advanced Materials and
   Process Engineering, Faculty of
   Sciences
Ibn Tofaïl University
Kenitra, Morocco

Abhinay Thakur
Research and Development Cell,
Lovely Professional University
Phagwara, India

J.O. Tijani
Department of Chemistry
Federal University of Technology
Minna, Nigeria

Mohamed Ebn Touhami
Laboratory of Organic, Organometallic
   and Theoretical Chemistry, Faculty
   of Science
Ibn Tofaïl University
Kenitra, Morocco

Shruti Trivedi
Department of Chemistry, Institute of
   Science
Banaras Hindu University
Varanasi, India

Burak Tuzun
Plant and Animal Production
   Department, Technical Sciences
   Vocational School of Sivas
Sivas Cumhuriyet University
Sivas, Türkiye

Priya Vashishth
Corrosion Testing Research Lab,
   Department of Chemistry
J.C. Bose University of Science and
   Technology
YMCA Faridabad, India

Monika Vats
Department of Applied Chemistry
   (CBFS-ASAS)
Amity University
Gurugram, India
and
Department of Chemistry
Dhanauri (PG) College
Dhanauri, India

Chandrabhan Verma
Department of Applied Chemistry
IITBHU
Varanasi, India

Sheng Wu
Yantai Research Institute of Harbin
    Engineering University
Harbin Engineering University
Harbin, China

Xuexue Xu
Yantai Research Institute of Harbin
    Engineering University
Harbin Engineering University
Harbin, China

Sanjukta Zamindar
Electric Mobility and Tribology
    Research Group
CSIR-Central Mechanical Engineering
    Research Institute
Durgapur, India
and
Academy of Scientific and Innovative
    Research (AcSIR)
AcSIR Headquarters CSIR-HRDC
    Campus
Ghaziabad, India

Omar Zegaoui
Laboratory of Chemistry-Biology
    Applied to the Environment, Faculty
    of Sciences
Moulay Ismail University
Meknes, Morocco

Saman Zehra
Corrosion Research Laboratory,
    Department of Applied Chemistry,
    Faculty of Engineering and
    Technology
Aligarh Muslim University
Aligarh, India

Xuerong Zheng
State Key Laboratory of Marine
    Resource Utilization in South China
    Sea, School of Materials Science &
    Engineering
Hainan University
Hainan, China

1 Green Corrosion Inhibition
Theory and Practices

*Bharti Sheokand, Seema R. Pathak,
Chandan K. Mandal, and Drishti Gupta*
Amity University

Deepti Verma and Kamalakanta Behera
University of Allahabad

Ambrish Kumar
Banaras Hindu University

Monika Vats
Amity University
Dhanauri (PG) College

Shruti Trivedi
Banaras Hindu University

1.1 INTRODUCTION

For humans, corrosion is an unquestionably serious issue that affects all aspects of daily life. Corrosion poses a serious threat to industries and moreover puts people's lives and assets at risk. Therefore, corrosion is a significant and important scientific issue [1]. Corrosion is typically understood as the loss of a metal due to corrosive substances. In a wider view, it is the deterioration of a metal as a result of exposure to its surroundings. Even while the term "corrosion" usually conjures images of metals, non-metallic materials including polymers, cement, ceramics, etc. are also susceptible to corrosion if placed in various corrosive situations.

The most common type of corrosion is the uniform or general corrosion which occurs as a result of electrochemical reactions on a totally exposed metal surface that spreads consistently to inflict maximum damage [2]. The coupling of two redox half-reactions that occur at the same pace stimulates natural metal corrosion. Electrons are released into the metal during the anodic reaction that leads to oxidation.

DOI: 10.1201/9781003394631-1

1

The cathodic reaction, which involves reducing an oxidant species like O_2, Fe^{3+}, or H^+, removes the electrons from the metal. Both of these reactions may occur concurrently on a single metal or two metals that are electrically connected to one another but are not the same [3].

The primary driving factor behind a corrosion reaction is the disparity of the potential energies between the metal ion in solid form and the product that is generated in the course of corrosion. Corrosion can therefore be regarded as the opposite of extractive metallurgy. Energy is spent on metal ores in order to produce the metals. If more energy is needed, the metal will be more thermodynamically unstable, which in turn leads to its stronger tendency to return to ore or to another oxidized form [4].

1.2 IMPACT OF CORROSION

1.2.1 MATERIAL AND ENERGY

The detrimental impacts of corrosion will cause material loss as a consequence of the equipment's deterioration [5]. Corrosion's destructive effects extend beyond metals and also include water and energy [6]. Within 90 seconds, 1 ton of steel rusts away globally. In contrast, the energy needed to produce it is comparable to the energy used by a typical family in three months [7]. By taking components of metals out of service and necessitating their replacement with a fraction of the available earthly resources, corrosion has an impact on the world's supply of metals [8]. The most frequent cause of reinforced concrete infrastructure deterioration worldwide is also the corrosion [9].

1.2.2 CONTAMINATION OF PRODUCT

Product contamination is also caused by corrosion. The so-called ancient devils of green chemistry, sulfur-containing compounds, are a significant class of air, water, and soil toxins in today's excessive environment. Sulfolane is a desirable replacement for frequently deployed industrial fluids. It has the possibility to cause the corrosion of equipment, which may lead to the spill of fluid that can contaminate groundwater [10]. The level of copper in water used for drinking may rise due to the corrosion of copper pipes, and these levels may exceed health standards and result in the metallic or unpleasant taste of water and various types of diseases [11]. Corrosion also poses a threat to food safety. Food is also contaminated because of the introduction of metallic ions into it when metallic containers are employed for packaging [12].

1.2.3 ECONOMIC

Corrosion has a significant effect on economy. Recent research by the National Association of Corrosion Engineers estimates that the entire economic loss brought on by corrosion is about 2.5 trillion US dollars. About 1,372 billion US dollars is the projected cost in the oil and gas sector caused by corrosion annually [13]. Corrosion costs industrialized countries between 3% and 4% of their GDP. Financial consequences of corrosion are divided into two categories. One is the direct cost that can be measured quantitatively. It includes the expense of replacement, the expenditure of protection, the prevention of corrosion, and development and research. The other is the indirect cost which is impossible to measure quantitatively. It includes losing items due to spills and fires, losing money from downtime, decreasing equipment effectiveness, contaminating items, polluting the environment, over-designing products to account for metal loss, and delaying that may result from lawsuits and bad will [14].

1.2.4 HUMAN LIFE AND SAFETY

Corrosion also poses a threat to human life and safety. A number of incidents are reported where due to corrosion many people have lost their lives. In terms of fatalities, injuries, and consequent health issues, the 1984 explosion at a chemical plant in Bhopal, India, was by far the deadliest corrosion-related tragedy. This project comes under India's "Green Revolution," which aimed to increase grain output by providing fertilizer and pesticides. However, a number of managerial issues and design errors came together to result in a disastrous explosion. It occurs due to the corrosion of the steel pipes through which water leaks into the tanks holding methylisocyanate; the iron corrosion products reportedly served as a catalyst for a reaction that exploded the plant, enabling methylisocyanate and some other toxic gases to evacuate, and resulted in the death of over 8,000 people. Since that day, the explosion has caused a further 15,000 deaths, and about 500,000 people are thought to be affected by ailments associated with gas [15].

On August 14, 2018, a portion of the Polcevera viaduct in Genoa, Italy, fell due to corrosion, killing 43 people. In December 1967, Silver Bridge collapsed, due to which 46 people lost their lives. The bridge had been in use for 39 years. The bridge's post-collapse analysis found that the fall was caused by the failure of a single eyebar caused by a 0.1-inch-deep corrosion-induced fault. This incident led to establishing a precedent for the USA's requirement that bridges be inspected every 2 years [16].

On June 28, 1983, a 100-foot-long piece of a bridge collapsed on a busy US Interstate highway. Three persons passed away, while three more suffered grave injuries. A support pin's rust was the reason for the collapse. On April 28, 1988, a commercial airplane traveling from Hilo to Honolulu, Hawaii, experienced a sudden cabin collapse that left one flight attendant regrettably dead and 65 passengers injured. Metal fatigue and corrosion's combined effects were the root of the issue [8].

1.3 GREEN CORROSION INHIBITORS

Green corrosion inhibitors are easily available, nontoxic, affordable, eco-friendly procedures, and environmentally acceptable goods made from renewable resources [17]. They are biodegradable and free of hazardous chemicals or heavy metals. They can prevent metal corrosion in both acidic and alkaline environments. They are

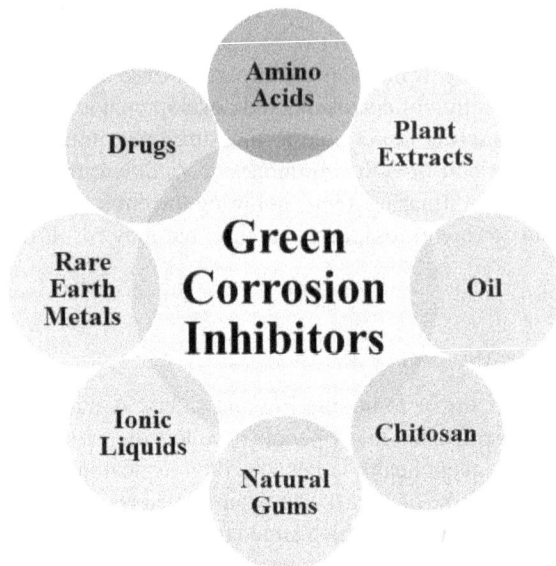

FIGURE 1.1   Various types of green corrosion inhibitors.

obtained from numerous natural sources (Figure 1.1). They include several polar atoms and electron-rich linkages, and are abundant in organic molecules. These molecules are absorbed on the surface of metal during the inhibition process and create a barrier [18]. The polar functional atoms are normally regarded as the reaction center for the adsorption mechanism [19].

1.3.1   Amino Acid–Based Green Corrosion Inhibitors

Due to properties like non-toxicity and biodegradability, amino acids occupy a special place in the category of "green corrosion inhibitors." They are simple to make in high purity at an affordable price. Majority of them have good solubility in aqueous conditions, which makes it easier to use them as corrosion inhibitors. It might reduce the need for a co-solvent for boosting solubility in the case of aqueous corrosion systems. Amino acids have minimum one carboxyl group (-COOH) and an amino group (-NH$_2$) attached to one carbon atom. Such functionalities can react with the potent acid solution, enhancing the amino acids' ability to prevent corrosion [20]. Based on the molecular composition, these can be divided into a wide variety of categories such as anionic/cationic amino acids, imino acids, W-amides, aromatic, linear/branched aliphatic, heterocyclic, sulfur-containing, and hydroxy-containing amino acids (Figure 1.2) [21]. Table 1.1 comprises amino acids and their derivatives used as corrosion inhibitors for metals and their alloys. Majority of the amino acids have heteroatoms like nitrogen, oxygen, and sulfur that offer effective adsorption centers on the surfaces of specific metals. Cysteine and other amino acids with thiol group (-SH) have been effectively used to create self-assembled monolayers on metal surfaces that can offer great defense against

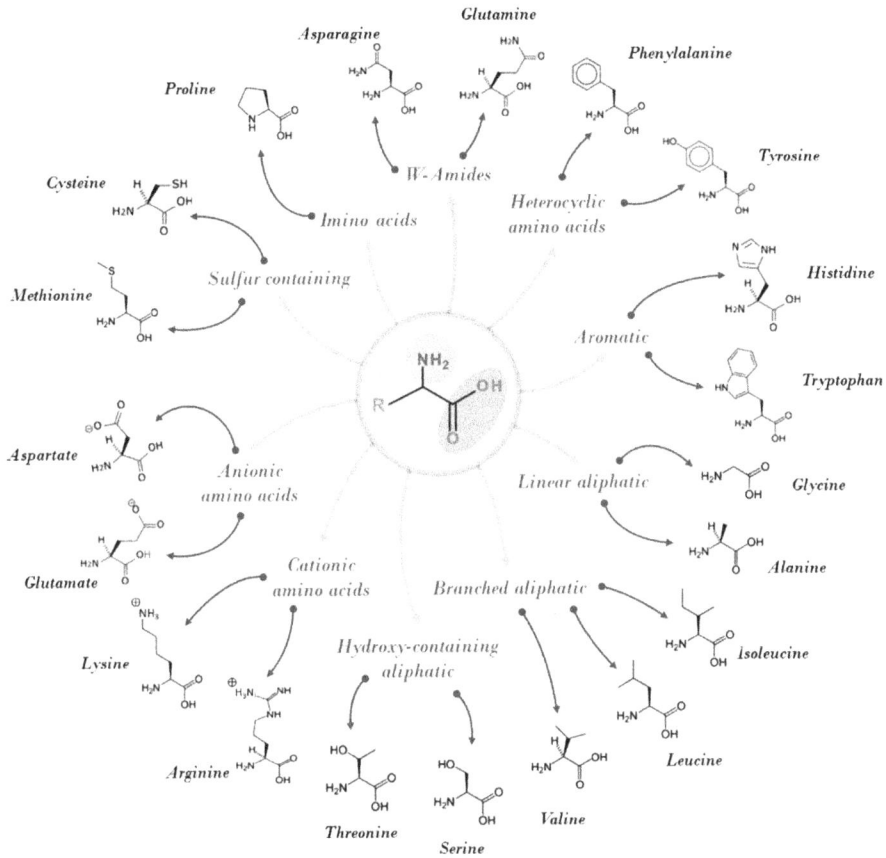

FIGURE 1.2 Structures of the 20 physiological amino acids and their chemical classification [21].

TABLE 1.1
Amino Acids and their derivatives used as Corrosion Inhibitors for Metals and its Alloys

Inhibitor	Metal	Medium	Experimental Technique	Inhibition Efficiency	References
Proline	Tin	2% NaCl	PP and EIS	65%	[22]
Cysteine	Mild steel	H$_2$SO$_4$	EM	96%	[23]
	Carbon steel	CO$_2$-saturated brine	EIS, PP, OCP,	91.8%	[24]
	Copper	solution	and WL	88%	[25]
		1 M HNO$_3$	EIS		
Methionine	Iron	HCl	EM	97.8%	[26]
	NST-44 Mild	Cassava	WL and OM	45%	[27]
	steel	1 M H$_2$SO$_4$ + 1 mM	PP	81%	[28]
	Low alloy steel	NaCl			
Aspartic acid	Copper	1 M HCl	PCM	52%	[29]

(Continued)

TABLE 1.1 (*Continued*)
Amino Acids and Their Derivatives Used as Corrosion Inhibitors for Metals and Its Alloys

Inhibitor	Metal	Medium	Experimental Technique	Inhibition Efficiency	References
Glutamate	NST-44	Cassava fluid:	WL and OM	30%	[30]
	carbon steel	hydrocyanic acid	EM	53.6%	[31]
	Copper	0.5 M HCl			
Lysine	Copper-5	0.6 M NaCl	EM	88%	[32]
Arginine	Copper	1 M HCl	EIS and PCM	63%	[33]
Threonine	Copper	1 M HCl	PP and EIS	83%	[34]
Serine	Copper	0.5 M HCl	PP and EIS	54.7%	[34]
Valine	Mild steel	1 M HCl	PP and WL	80%	[35]
Leucine	Mild steel	1 N H$_2$SO$_4$	PP and OCP	78%	[36]
Isoleucine	Copper	1 M HCl	WL, PP and EIS	94.5%	[37]
Alanine	Mild steel	1 M HCl	PP and EIS	80%	[38]
	NST-44 mild steel	Lime fluid	WL and OM	43.9%	[30]
Glycine	Steel	0.5 M HCl	EM	53.9%	[31]
	Nickel	1 N H$_2$SO$_4$	EM	72%	[39]
Tryptophan	Low carbon Steel	1 M HCl	PP and WL	91%	[40]
	Copper	0.5 M H$_2$SO$_4$	EM and GT	87%	[41]
Histidine	Copper	0.5 M HCl	EIS	74%	[42]
Tyrosine	Iron	HCl	EM	97.8%	[26]
	Low alloy steel	0.2 M ammoniated citric acid	PP and EIS	83%	[43]
Phenylalanine	Copper	0.5 M HCl	WL, PP, and EIS	72%	[44]
	Carbon Steel	Well water	WL, PP, and EIS	90%	[45]
Glutamine	Mild steel	1 M HCl	WL	96%	[46]
Asparagine	Copper	0.5 M HCl	EIS and PCM	65%	[47]

EIS, electrochemical impedance spectroscopy; EM, electrochemical methods; GT, gravimetric test; OCP, open circuit potential; OM, optical microscopy; PCM, polarization curve measurements; PP, potentiodynamic polarization; WL, weight loss method.

water corrosion. Additionally, the chemical functionalization of amino acids can be carried out using such functional groups to create improved corrosion inhibitors or to bind two amino acids altogether [48].

1.3.2 Natural Gums–Based Green Corrosion Inhibitors

Natural gums are appealing corrosion inhibitors because they are affordable, readily accessible, nontoxic, biodegradable, renewable resources, and biocompatible with the ecosystem [49]. The majority of gums include -COOH functional groups that could enhance electron and charge transfer contributions and hence make adsorption-based inhibition easier [50]. These complexes cover a substantial amount of the surface, shielding the metals from the corrosive elements. Due to the nitrogen and oxygen atoms present in arabinogalactan, polysaccharides, glucoproteins, and oligosaccharides, natural gums are thought to limit the corrosion process [51]. The abundance of oxygen atoms serves as the adsorption site. Gum's horizontal orientation at the surface of the metal caused by the concurrent adsorption of oxygen atoms increases surface covering, and as a result, protection performance increases even at low concentration levels of inhibitors [52].

Natural gums, in contrast to the majority of conventional organic corrosion inhibitors, have a variety of electron-polar functional groups that make them simple to dissolve in aqueous electrolytes. Naturally occurring gums have a variety of electron-prominent locations, allowing them to bind to atoms of metal to form extremely stable chelating complexes. Natural gums could thus offer great corrosion resistance that is even long lasting. Based on the type of electrolyte as well as metal that needs protection, natural gums include a variety of polar functional groups which can be appropriately altered. Undoubtedly, this kind of alteration is done to improve the efficiency of corrosion inhibition [53].

1.3.3 CHITOSAN-BASED GREEN CORROSION INHIBITORS

Polymers are regarded as natural corrosion inhibitors of metallic substrates because they have active sites which interact with metal ions. Chitosan has a high capability to produce films and also a stronger adherence to metallic surfaces which allows it to be coated on metals to create a protective layer against corrosion. Chitosan's versatility connected with how simple it is to chemically functionalize is a distinctive quality that offers a method to improve its adhesiveness with the surface of metal [54,55].

Chitosan has a variety of electron-rich polar or functional groups, including -OH (hydroxyl), $-NH_2$, $-CH_2OH$ (hydroxymethyl), -O- (cyclic and acyclic ether), and $-NHCOCH_3$ (amide), that help it to get adsorbed on the surface of metal and improve its solubility in solutions. It has the capacity for both chemisorption and physisorption. Typically, the process of chemisorption happens by sharing of charge, i.e., donation of an unshared pair of electrons present on the heteroatoms such as O and N through a coordinate bond. On the contrary, physisorption, also known as electrostatic adsorption, takes place when charged chitosan molecules interact with a charged metal surface through electrostatic interactions. A few of the polar chitosan substituents, such as the -OH and $-NH_2$ groups, go through protonation in aqueous systems and transform into their cationic form. However, as counter ions of the electrolyte molecules gather at the metal's positively charged surface, the metal surface gets negatively charged (Figure 1.3) [56].

Pure chitosan typically has a modest inhibitory effect. Meanwhile, functionalization might enhance the effectiveness of its protection. The inhibitory performances of the chitosan and metal nanoparticle composites are likewise superior to those of pure chitosan. This enhancement might be the result of direct contact between these nanoparticles and the metal surface. The chitosan inhibitors utilized as a supplement enhanced their ability to resist corrosion with increased immersion duration, providing longer-lasting protection [54].

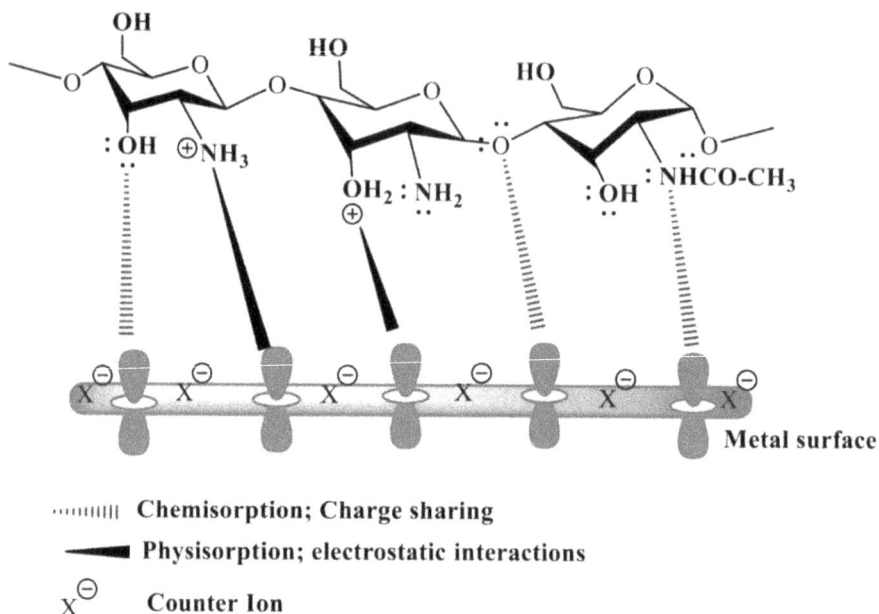

FIGURE 1.3   Diagrammatic illustration of chemisorption and physisorption mechanism of chitosan on metallic surface [56].

1.3.4   PLANT EXTRACTS–BASED GREEN CORROSION INHIBITORS

Plant extracts are considered to be an abundant source of naturally derived chemical compounds that can be extracted from several parts of plants, such as leaves, stem, root, bark, fruits, fruit peel, seeds, flowers, husks, and nuts, which contain organic compounds like alkaloids, flavonoids, heteroatoms, tannins, and nitrogen-based compounds. They are easily extracted, inexpensive, and environmentally friendly. They are also widely available and naturally biodegradable [18,57,58]. These benefits explain why plant extracts and their byproducts are used as corrosion inhibitors for metallic materials and alloys in various environments. Plant extracts are rich in polar molecules with oxygen and nitrogen as well as nonpolar compounds with aromatic rings, aliphatic chains, heterocyclic rings, and functional moieties. These are thought to promote the adsorption of inhibitors on the surfaces of metals and form a protective film on metallic surface without affecting the surroundings and hence decrease the rate of corrosion [18,59,60]. Methanol, ethanol, or an aqueous solvent is used to extract the majority of plant extracts. Based on their chemical compositions and properties, six primary types of phytochemicals and their subgroups are tabulated (Figure 1.4) [61]. According to the reports, the phytochemicals are thought to be the cause of the corrosion inhibition. So, it is crucial to identify the chemical components in plant extracts in order to forecast how well they would work in inhibiting corrosion. The phytochemical components in plant extracts can be examined and identified chemically [17]. Table 1.2 summarizes the inhibition efficiency of several plant extracts used as corrosion inhibitors for metals and their alloys.

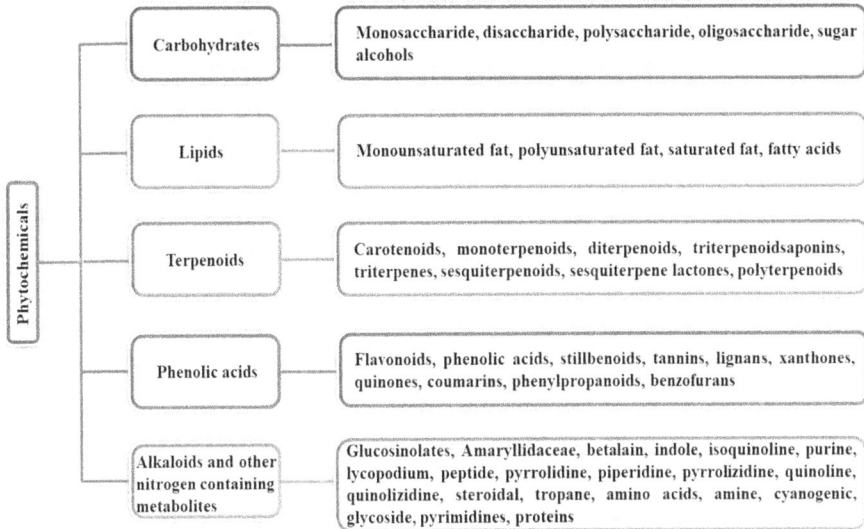

FIGURE 1.4 Categorization of phytochemicals [61].

TABLE 1.2
Several Plant Extracts used as Corrosion Inhibitors for Metals and its Alloys

Inhibitor	Metal	Medium	Experimental Technique	Inhibition Efficiency	References
Tobacco rob	N80 steel	1 M HCL	EIS	98.5%	[62]
Phyllanthus amarus	Mild steel	1 M HCL	WL, EIS, PP	95%	[63]
Justicia gendarussa	Mild steel	1 M HCL	WL, EIS, PP	93%	[57]
Piper longum	Aluminum	1 M NaOH	WL, EIS, PP	94%	[64]
Egyptian licorice	Copper	0.1 M HCl	EIS, PP	86.8%	[65]
Azadirachta indica	Copper	HNO₃	WL, EIS, PP	98%	[66]
Aquilaria subintegra	Mild steel	1 M HCl	WL, EIS, PP	93%	[59]
Red onion skin	Zinc	2 M HCl	WL	70%	[67]
Capparis spinosa	Copper	1 M HNO₃	WL, PP	82.7%	[68]
Calligonum comosum	Copper	HNO₃	Tafel, EIS	80%	[69]
Morinda tinctoria	Copper	0.5 M HCl	WL, EM	70.4%	[70]
Punica granatum	Mild steel	1 M HCl	Tafel	95%	[71]
Thyme	Mild steel	2 M HCl	WL, EIS, EM	84%	[72]
Silybum marianum	304 stainless steel	1 M HCl	WL, EIS, PP	96.4%	[73]
Ficus hispida	Mild steel	1 M HCl	WL, EIS, PP	90%	[74]
Justicia gendarussa	Mild steel	1 M HCl	WL, EIS, PP	93%	[57]
Murraya koenigii	Carbon steel	1 N HCl	WL, PP, Gasometric method	84.6%	[75]

(Continued)

TABLE 1.2 (*Continued*)
Several Plant Extracts Used as Corrosion Inhibitors for Metals and Its Alloys

Inhibitor	Metal	Medium	Experimental Technique	Inhibition Efficiency	References
Schinopsis lorentzii	Low carbon steel	1 M HCl	EIS, PP	66%	[76]
Rosmarinus offcinalis	Aluminum-magnesium alloy	3% NaCl	PP	75%	[77]
Opuntia	Aluminum	2 M HCl	WL	96%	[78]
Saraca asoca	Mild steel	0.5 M H_2SO_4	Tafel	95.5%	[79]
Eriobotrya japonica	Mild steel	0.5 M H_2SO_4	WL	96.2%	[80]
Pimenta dioica	Mild steel	1 M HCl	EIS, PP	88%	[81]

1.3.5 DRUGS-BASED GREEN CORROSION INHIBITORS

Since drugs generally have a molecular structure which is comparable to that of the traditional corrosion inhibitors, their potential to be employed as corrosion inhibitors is quite high [82]. Antifungal, antibacterial, antiemetic, and antihypertensive drugs are among those substances that have been examined as the most potential corrosion inhibitors. The efficacy of the inhibition reduced with the rising temperature but improved proportionately with the drug concentration.

The corrosion inhibition effect of penicillin (an antibacterial drug) is most likely to be connected to its basic chemical structure because of the heteroatoms found in it. Both atenolol and etilefrine have functional groups that can cause delocalization of p electron (the aromatic ring in atenolol as well as the aromatic ring and the amide group in etilefrine), which could be the cause of the adsorption [83].

Ondansetron hydrochloride occurs as either a neutral or a protonated molecule in an aqueous acidic solution. Ondansetron hydrochloride may adsorb on the metal/acid solution interface through the mechanisms listed below:

a. protonated molecules and already-adsorbed chloride ions interact electrostatically on the metal surface.
b. interactions between the unoccupied d-orbital on the surface of metal atom and the p electron of the aromatic ring.
c. interactions between the unshared electron pairs of heteroatoms and the vacant d-orbital on the surface of metal atom [84].

1.3.6 OIL-BASED GREEN CORROSION INHIBITORS

According to the method of activity, the corrosion protection properties of essential oils are often divided into three classes: (i) geometric blocking effect, (ii) active sites blocking effect, and (iii) electrocatalytic impact. By adsorbing on the surface of metal, essential oils do have the ability to prevent both anodic and cathodic processes [85].

Oil molecules from *Eucalyptus* and *Lippia alba* could also adsorb on the surface of the metal owing to the donor-acceptor interactions among the Π-electrons of aromatic ring and unoccupied d-orbitals of iron. When these essential oil molecules are adsorbed on the surface of metal, a coordinate bond could form as a result of a partial transfer of electrons from the polar atoms (O atoms) to the surface of the metal, forming a thicker protective film that covers the cathodic and anodic reaction sites and slows the process of corrosion [86].

The essential oil from the *Ammodaucus leucotrichus* fruit has oxygen atoms in functional groups (C=O) and Π-electrons in double bonds (C=C), which are in line with the characteristic properties of corrosion inhibitors. As a result, adsorption of such compounds on the surface of the metal may be responsible for the inhibitory effects of *A. leucotrichus* oil [87].

The essential oil of *Eryngium maritimum* also has oxygen atoms (O–H, C=O, C–O). Oxygenated components in *E. maritimum* oil have the ability to inhibit. The synergistic action of terpenes found in *E. maritimum* essential oil may result in inhibition effectiveness on metal surfaces [88].

1.3.7 IONIC LIQUID–BASED GREEN CORROSION INHIBITORS

Ionic liquids are made up of anions and cations (Figure 1.5) [89], at least one of which has an organic molecular structure that prevents effective close packing and has a special combination of physical-chemical characteristics [90].

Ionic liquids have many unique qualities, including low melting point (below 100°C), excellent polarity, low cytotoxicity, decreased vapor pressure, extremely high chemical and thermal stability, and less harmful impact on the environment. The biggest benefit of creating ionic liquids with unique qualities is the ability to adjust their properties to meet the demand by carefully selecting the cations and anions. That's why ionic liquids are indeed referred to as "designer chemicals." These compounds, in contrast to conventional volatile corrosion inhibitors, have an exceptionally low vapor pressure, which prevents them from evaporating and contaminating the surroundings [91].

It has been discovered that the corrosion inhibition ability of ionic liquids is superior to that of organic corrosion inhibitors because they have a high capacity for adhering to metal substrates and forming films. This protects metal from the aggressive medium and hence prevents the corrosion process. The corrosion inhibition efficacy rises because of the alkyl substituents present within ionic liquid structure. It has been observed that the temperature and concentration of ionic liquid have a significant impact on how effectively ionic liquids suppress corrosion. Lower concentration of ionic liquid is seen to have low corrosion inhibition effectiveness which can be attributed to the ionic liquids' inability to build a uniform and consistent protective coating on the surface of the metal at these concentrations. A rise in temperature, however, reduces their resistance to corrosion. It is brought on by ionic liquids' chemical breakdown, which speeds up corrosion and reduces the adsorption of ionic liquid on the steel surface [92].

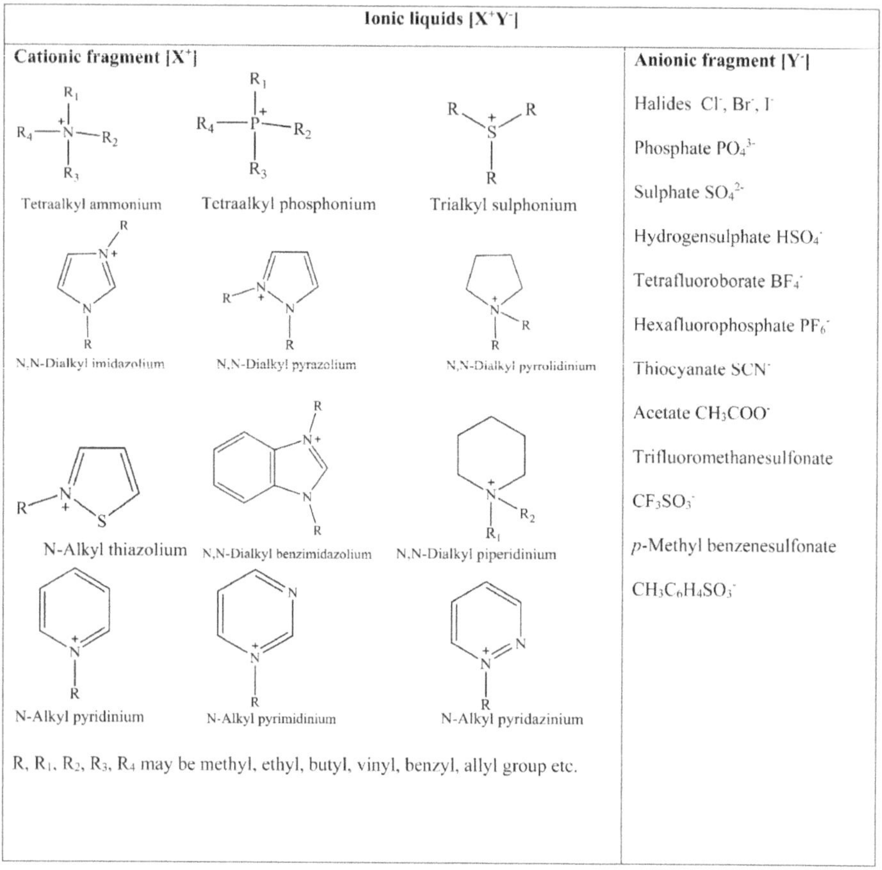

FIGURE 1.5 Structures of some commonly used cationic and anionic fragments in ionic liquids [89].

1.3.8 RARE EARTH METAL–BASED GREEN CORROSION INHIBITORS

These compounds are nontoxic; so, they work well as corrosion inhibitors [93]. In some cases, it is comparable to the inhibition offered by the chromate complexes. The employment of rare earth metal compounds in place of chromates should be the focus because chromate compounds are extremely hazardous. Because of the partial hydrolysis of the organic bonds of rare earth metals, the inhibitor system will keep certain clustered units in the form of complexes or polynuclear species. Notable inhibition is only observed in those systems in which substantial rare earth metal-organic complexes are present, indicating that when the complexes decompose into simple ions, they will no longer provide similar synergistic protection against corrosion. The dangling bonds which might be present in the solution have the ability to combine with metal ions to produce bimetallic compounds, which then accumulate on the surface of the metal to form insoluble coatings [94].

The rare earth metals work by blocking the cathodic sites of the substrate, which slows down the cathodic process [93]. Their structure contains a significant amount of hydrogen bonding that causes the monomeric units to come together to form a supramolecular polymer. These compounds are fragmented in the solution to form monomeric units, which is particularly relevant to the function of rare earth metals for protective coatings [95].

However, several of the inhibitor substances exhibited a prominent cathodic inhibition mechanism, and a greater anodic inhibition is occasionally seen. The hydrolysis of metallic ions at anodic sites produces an acidic pH, whereas the oxygen reduction at the cathodic sites results in alkaline environments, which in turn causes the precipitation of rare earth metal oxides and the subsequent creation of a protective layer. If the protective layer is not sufficiently resistant, the process of corrosion will increase the alkalinity of the solution which will eventually hydrolyze most of the rare earth metal-ligand bond, forming a thick layer of rare earth metals. These inhibitors can be employed in alkaline conditions since they perform effectively at high pH [96,97].

1.4   CONCLUSIONS

Due to its negative effects on any nation's economy and ecosystem, corrosion is a disastrous phenomenon. We must learn how to conserve resources if we have to preserve energy. Given the significant maintenance costs involved with metals and equipment, using inhibitors to reduce corrosion is a financially sound solution. Yet, current developments in corrosion inhibition point more toward the need to identify an inhibitor that is both efficient and environmentally beneficial. This is owing to the fact that an effective inhibitor may be successful at preventing corrosion, but due to the negative effects connected with its chemical formula, it may also have detrimental effects on human health and the environment. It has been shown that green corrosion inhibitors, such as those derived from plant extracts, drugs, amino acids, rare earth metals, chitosan, natural gums, ionic liquids, and oils, are the best choice to replace conventionally used, expensive, toxic synthetic inhibitors. Many ingredients included in the green inhibitors have the ability to absorb at the surface of metal and prevent its corrosion. The agenda of using green corrosion inhibitors has already begun, and it will continue to be carried out.

ACKNOWLEDGMENTS

Dr. K. Behera acknowledges the University of Allahabad for the facilities and funding, and Dr. S. Trivedi acknowledges the UGC and the Institute of Eminence (IoE)-Banaras Hindu University for the funding.

REFERENCES

[1] Liu, L.; Li, Y.; Wang, F. Electrochemical Corrosion Behavior of Nanocrystalline Materials-A Review. *J. Mater. Sci. Technol.* 2010, *26* (1), 1–14. https://doi.org/10.1016/S1005-0302(10)60001-1.

[2] Goni, L. K. M. O.; Mazumder, M. A. J. Green Corrosion Inhibitors. *Corros. Inhib.* **2019**, *30* (4), 77–92. https://doi.org/10.5772/intechopen.81376.

[3] Renock, D.; Shuller-Nickles, L. C. Predicting Geologic Corrosion with Electrodes. *Elements* **2015**, *11* (5), 331–336. https://doi.org/10.2113/gselements.11.5.331.

[4] Harsimran, S.; Santosh, K.; Rakesh, K.; Overview of Corrosion and Its Control: A Critical Review. *Proc. Eng. Sci.* **2021**, *3* (1), 13–24. https://doi.org/10.24874/PES03.01.002.

[5] Kuruvila, R.; Kumaran, S. T.; Khan, M. A.; Uthayakumar, M. A Brief Review on the Erosion-Corrosion Behavior of Engineering Materials. *Corros. Rev.* 2018, *36* (5), 435–447. https://doi.org/10.1515/corrrev-2018-0022.

[6] Raja, P. B.; Ismail, M.; Ghoreishiamiri, S.; Mirza, J.; Ismail, M. C.; Kakooei, S.; Rahim, A. A. Reviews on Corrosion Inhibitors: A Short View. *Chem. Eng. Commun.* **2016**, *203* (9), 1145–1156. https://doi.org/10.1080/00986445.2016.1172485.

[7] Javaherdashti, R. How Corrosion Affects Industry and Life. *Anti-Corrosion Methods Mater.* **2000**, *47* (1), 30–34. https://doi.org/10.1108/00035590010310003.

[8] McCafferty, E. Introduction to Corrosion Science. *Introd. Corros. Sci.* 2010, 1–575. https://doi.org/10.1007/978-1-4419-0455-3.

[9] Gharehbaghi, K.; Rahmani, F. Deterioration of Transportation Infrastructures: Corrosion of Reinforcements in Concrete Structures. *Mater. Sci. Forum* **2018**, *940 MSF*, 160–166. https://doi.org/10.4028/www.scientific.net/MSF.940.160.

[10] Bak, A.; Kozik, V.; Dybal, P.; Kus, S.; Swietlicka, A.; Jampilek, J. Sulfolane: Magic Extractor or Bad Actor? Pilot-Scale Study on Solvent Corrosion Potential. *Sustain.* **2018**, *10* (10), 3677. https://doi.org/10.3390/su10103677.

[11] Dietrich, A. M.; Glindemann, D.; Pizarro, F.; Gidi, V.; Olivares, M.; Araya, M.; Camper, A.; Duncan, S.; Dwyer, S.; Whelton, A. J.; Younos, T.; Subramanian, S.; Burlingame, G. A.; Khiari, D.; Edwards, M. Health and Aesthetic Impacts of Copper Corrosion on Drinking Water. *Water Sci. Technol.* **2004**, *49* (2), 55–62. https://doi.org/10.2166/wst.2004.0087.

[12] Rather, I. A.; Koh, W. Y.; Paek, W. K.; Lim, J. The Sources of Chemical Contaminants in Food and Their Health Implications. *Front. Pharmacol.* **2017**, *8* (NOV), 830. https://doi.org/10.3389/fphar.2017.00830.

[13] Quraishi, M. A.; Chauhan, D. S.; Ansari, F. A. Development of Environmentally Benign Corrosion Inhibitors for Organic Acid Environments for Oil-Gas Industry. *J. Mol. Liq.* **2021**, *329*, 115514. https://doi.org/10.1016/j.molliq.2021.115514.

[14] Fayomi, O. S. I.; Akande, I. G.; Odigie, S. Economic Impact of Corrosion in Oil Sectors and Prevention: An Overview. *J. Phys. Conf. Ser.* **2019**, *1378* (2), 022037. https://doi.org/10.1088/1742-6596/1378/2/022037.

[15] Hansson, C. M. The Impact of Corrosion on Society. *Metall. Mater. Trans. A Phys. Metall. Mater. Sci.* **2011**, *42* (10), 2952–2962. https://doi.org/10.1007/s11661-011-0703-2.

[16] Morgese, M.; Ansari, F.; Domaneschi, M.; Cimellaro, G. P. Post-Collapse Analysis of Morandi's Polcevera Viaduct in Genoa Italy. *J. Civ. Struct. Heal. Monit.* **2020**, *10* (1), 69–85. https://doi.org/10.1007/s13349-019-00370-7.

[17] Salleh, S. Z.; Yusoff, A. H.; Zakaria, S. K.; Taib, M. A. A.; Abu Seman, A.; Masri, M. N.; Mohamad, M.; Mamat, S.; Ahmad Sobri, S.; Ali, A.; Teo, P. T. Plant Extracts as Green Corrosion Inhibitor for Ferrous Metal Alloys: A Review. *J. Clean. Prod.* **2021**, *304*, 127030. https://doi.org/10.1016/j.jclepro.2021.127030.

[18] Hossain, N.; Asaduzzaman Chowdhury, M.; Kchaou, M. An Overview of Green Corrosion Inhibitors for Sustainable and Environment Friendly Industrial Development. *J. Adhes. Sci. Technol.* **2021**, *35* (7), 673–690. https://doi.org/10.1080/01694243.2020.1816793.

[19] de Souza, F. S.; Spinelli, A. Caffeic Acid as a Green Corrosion Inhibitor for Mild Steel. *Corros. Sci.* **2009**, *51* (3), 642–649. https://doi.org/10.1016/j.corsci.2008.12.013.

[20] Wei, H.; Heidarshenas, B.; Zhou, L.; Hussain, G.; Li, Q.; Ostrikov, K. (Ken). Green Inhibitors for Steel Corrosion in Acidic Environment: State of Art. *Mater. Today Sustain.* **2020**, *10*, 100044. https://doi.org/10.1016/j.mtsust.2020.100044.

[21] El Ibrahimi, B.; Jmiai, A.; Bazzi, L.; El Issami, S. Amino Acids and Their Derivatives as Corrosion Inhibitors for Metals and Alloys. Arab. *J. Chem.* **2020**, *13* (1), 740–771. https://doi.org/10.1016/j.arabjc.2017.07.013.

[22] El Ibrahimi, B.; Bazzi, L.; El Issami, S. The Role of PH in Corrosion Inhibition of Tin Using the Proline Amino Acid: Theoretical and Experimental Investigations. *RSC Adv.* **2020**, *10* (50), 29696–29704. https://doi.org/10.1039/d0ra04333h.

[23] Özcan, M.; Karadağ, F.; Dehri, I. Interfacial Behavior of Cysteine between Mild Steel and Sulfuric Acid as Corrosion Inhibitor. *Acta Phys. Chim. Sin.* **2008**, *24* (8), 1387–1392. https://doi.org/10.1016/S1872-1508(08)60059-5.

[24] Zhang, C.; Duan, H.; Zhao, J. Synergistic Inhibition Effect of Imidazoline Derivative and L-Cysteine on Carbon Steel Corrosion in a CO2-Saturated Brine Solution. *Corros. Sci.* **2016**, *112*, 160–169. https://doi.org/10.1016/j.corsci.2016.07.018.

[25] Milošev, I.; Pavlinac, J.; Hodošček, M.; Lesar, A. Amino Acids as Corrosion Inhibitors for Copper in Acidic Medium: Experimental and Theoretical Study. *J. Serbian Chem. Soc.* **2013**, *78* (12), 2069–2086. https://doi.org/10.2298/JSC131126146M.

[26] Zor, S.; Kandemirli, F.; Bingul, M. Inhibition Effects of Methionine and Tyrosine on Corrosion of Iron in HCl Solution: Electrochemical, FTIR, and Quantum-Chemical Study. *Prot. Met. Phys. Chem. Surfaces* **2009**, *45* (1), 46–53. https://doi.org/10.1134/S2070205109010079.

[27] Alagbe, M.; Umoru, L. E.; Afonja, A. A.; Olorunniwo, O. E. Investigation of the Effect of Different Amino-Acid Derivatives on the Inhibition of NST-44 Carbon Steel Corrosion in Cassava Fluid. *Anti-Corrosion Methods Mater.* **2009**, *56* (1), 43–50. https://doi.org/10.1108/00035590910923455.

[28] Jano, A.; Lame, A.; Kokalari, E. The Inhibition Effects of Methionine on Mild Steel in Acidic Media. *Analele Univ. "Ovidius" Constanta - Ser. Chim.* **2014**, *25* (1), 39–42. https://doi.org/10.2478/auoc-2014-0007.

[29] Gomma, G. K.; Wahdan, M. H. Effect of Temperature on the Acidic Dissolution of Copper in the Presence of Amino Acids. *Mater. Chem. Phys.* **1994**, *39* (2), 142–148. https://doi.org/10.1016/0254-0584(94)90191-0.

[30] Hamadi, L.; Mansouri, S.; Oulmi, K.; Kareche, A. The Use of Amino Acids as Corrosion Inhibitors for Metals: A Review. Egypt. *J. Pet.* **2018**, *27* (4), 1157–1165. https://doi.org/10.1016/j.ejpe.2018.04.004.

[31] Makarenko, N. V.; Kharchenko, U. V.; Zemnukhova, L. A. Effect of Amino Acids on Corrosion of Copper and Steel in Acid Medium. *Russ. J. Appl. Chem.* **2011**, *84* (8), 1362–1365. https://doi.org/10.1134/S1070427211080118.

[32] Badawy, W. A.; Ismail, K. M.; Fathi, A. M. Corrosion Control of Cu-Ni Alloys in Neutral Chloride Solutions by Amino Acids. *Electrochim. Acta* **2006**, *51* (20), 4182–4189. https://doi.org/10.1016/j.electacta.2005.11.037.

[33] Zhang, D. Q.; He, X. M.; Cai, Q. R.; Gao, L. X.; Kim, G. S. Arginine Self-Assembled Monolayers against Copper Corrosion and Synergistic Effect of Iodide Ion. *J. Appl. Electrochem.* **2009**, *39* (8), 1193–1198. https://doi.org/10.1007/s10800-009-9784-7.

[34] Zhang, D. Q.; Cai, Q. R.; Gao, L. X.; Lee, K. Y. Effect of Serine, Threonine and Glutamic Acid on the Corrosion of Copper in Aerated Hydrochloric Acid Solution. *Corros. Sci.* **2008**, *50* (12), 3615–3621. https://doi.org/10.1016/j.corsci.2008.09.007.

[35] Olivares, O.; Likhanova, N. V.; Gómez, B.; Navarrete, J.; Llanos-Serrano, M. E.; Arce, E.; Hallen, J. M. Electrochemical and XPS Studies of Decylamides of α-Amino Acids Adsorption on Carbon Steel in Acidic Environment. *Appl. Surf. Sci.* **2006**, *252* (8), 2894–2909. https://doi.org/10.1016/j.apsusc.2005.04.040.

[36] Singh, P.; Bhrara, K.; Singh, G. Adsorption and Kinetic Studies of L-Leucine as an Inhibitor on Mild Steel in Acidic Media. *Appl. Surf. Sci.* **2008**, *254* (18), 5927–5935. https://doi.org/10.1016/j.apsusc.2008.03.154.

[37] Ahmed, R. K.; Zhang, S. Bee Pollen Extract as an Eco-Friendly Corrosion Inhibitor for Pure Copper in Hydrochloric Acid. *J. Mol. Liq.* **2020**, *316*, 113849. https://doi.org/10.1016/j.molliq.2020.113849.

[38] Khaled, K. F.; Abdelshafi, N. S.; El-Maghraby, A. A.; Aouniti, A.; Al-Mobarak, N.; Hammouti, B. Alanine as Corrosion Inhibitor for Iron in Acid Medium: A Molecular Level Study. *Int. J. Electrochem. Sci.* **2012**, *7* (12), 12706–12719. ISSN: 1452-3981.

[39] Bilgiç, S.; Aksüt, A. A. Effect of Amino Acids on Corrosion of Cobalt in H2SO4. *Br. Corros. J.* **1993**, *28* (1), 59–62. https://doi.org/10.1179/000705993798268386.

[40] Fu, J. J.; Li, S. N.; Cao, L. H.; Wang, Y.; Yan, L. H.; Lu, L. De. L-Tryptophan as Green Corrosion Inhibitor for Low Carbon Steel in Hydrochloric Acid Solution. *J. Mater. Sci.* 2010, *45* (4), 979–986. https://doi.org/10.1007/s10853-009-4028-0.

[41] Moretti, G.; Guidi, F. Tryptophan as Copper Corrosion Inhibitor in 0.5 M Aerated Sulfuric Acid. *Corros. Sci.* **2002**, *44* (9), 1995–2011. https://doi.org/10.1016/S0010-938X(02)00020-3.

[42] Zhang, D. Q.; He, X. M.; Cai, Q. R.; Gao, L. X.; Kim, G. S. PH and Iodide Ion Effect on Corrosion Inhibition of Histidine Self-Assembled Monolayer on Copper. *Thin Solid Films* **2010**, *518* (10), 2745–2749. https://doi.org/10.1016/j.tsf.2009.10.150.

[43] Abdel-Fatah, H. T. M.; Kamel, M. M.; Hassan, A. A. M.; Rashwan, S. A. M.; Abd El Wahaab, S. M.; El-Sehiety, H. E. E. Adsorption and Inhibitive Properties of Tryptophan on Low Alloy Steel Corrosion in Acidic Media. Arab. *J. Chem.* **2017**, *10*, S1164–S1171. https://doi.org/10.1016/j.arabjc.2013.02.010.

[44] Zhang, D. Q.; Wu, H.; Gao, L. X. Synergistic Inhibition Effect of L-Phenylalanine and Rare Earth Ce(IV) Ion on the Corrosion of Copper in Hydrochloric Acid Solution. *Mater. Chem. Phys.* **2012**, *133* (2–3), 981–986. https://doi.org/10.1016/j.matchemphys.2012.02.001.

[45] Raja, A. S.; Rajendran, S.; Satyabama, P. Inhibition of Corrosion of Carbon Steel in Well Water by DL-Phenylalanine-Zn2+ System. *J. Chem.* **2013**, *2013*, 1–8. https://doi.org/10.1155/2013/720965.

[46] Singh, A.; Pramanik, T.; Kumar, A.; Gupta, M. Phenobarbital: A New and Effective Corrosion Inhibitor for Mild Steel in 1 M HCl Solution. *Asian J. Chem.* **2013**, *25* (17), 9808–9812. https://doi.org/10.14233/ajchem.2013.15414.

[47] Zhang, D. Q.; Cai, Q. R.; He, X. M.; Gao, L. X.; Zhou, G. D. Inhibition Effect of Some Amino Acids on Copper Corrosion in HCl Solution. *Mater. Chem. Phys.* **2008**, *112* (2), 353–358. https://doi.org/10.1016/j.matchemphys.2008.05.060.

[48] Chauhan, D. S.; Quraishi, M. A.; Srivastava, V.; Haque, J.; ibrahimi, B. E. Virgin and Chemically Functionalized Amino Acids as Green Corrosion Inhibitors: Influence of Molecular Structure through Experimental and In Silico Studies. *J. Mol. Struct.* **2021**, *1226*, 129259. https://doi.org/10.1016/j.molstruc.2020.129259.

[49] Peter, A.; Obot, I. B.; Sharma, S. K. Use of Natural Gums as Green Corrosion Inhibitors: An Overview. *Int. J. Ind. Chem.* **2015**, 6 (3), 153–164. https://doi.org/10.1007/s40090-015-0040-1.

[50] Peter, A.; Sharma, S. K.; Obot, I. B. Anticorrosive Efficacy and Adsorptive Study of Guar Gum with Mild Steel in Acidic Medium. *J. Anal. Sci. Technol.* **2016**, *7* (*1*), 1–15. https://doi.org/10.1186/s40543-016-0108-3.

[51] Eddy, N. O.; Ameh, P.; Gimba, C. E.; Ebenso, E. E. GCMS Studies on Anogessus Leocarpus (Al) Gum and Their Corrosion Inhibition Potential for Mild Steel in 0.1 M HCl. *Int. J. Electrochem. Sci.* **2011**, *6* (11), 5815–5829. ISSN: 1452-3981.

[52] Hasan, A. M. A.; Abdel-Raouf, M. E. Applications of Guar Gum and Its Derivatives in Petroleum Industry: A Review. Egypt. *J. Pet.* **2018**, *27* (4), 1043–1050. https://doi.org/10.1016/j.ejpe.2018.03.005.

[53] Verma, C.; Quraishi, M. A. Gum Arabic as an Environmentally Sustainable Polymeric Anticorrosive Material: Recent Progresses and Future Opportunities. *Int. J. Biol. Macromol.* **2021**, *184* (June), 118–134. https://doi.org/10.1016/j.ijbiomac.2021.06.050.

[54] Ashassi-Sorkhabi, H.; Kazempour, A. Chitosan, Its Derivatives and Composites with Superior Potentials for the Corrosion Protection of Steel Alloys: A Comprehensive Review. *Carbohydr. Polym.* **2020**, *237* (February), 116110. https://doi.org/10.1016/j. carbpol.2020.116110.

[55] Carneiro, J.; Tedim, J.; Ferreira, M. G. S. Chitosan as a Smart Coating for Corrosion Protection of Aluminum Alloy 2024: A Review. *Prog. Org. Coatings* **2015**, *89* (2015), 348–356. https://doi.org/10.1016/j.porgcoat.2015.03.008.

[56] Verma, C.; Quraishi, M. A. Chelation Capability of Chitosan and Chitosan Derivatives: Recent Developments in Sustainable Corrosion Inhibition and Metal Decontamination Applications. *Curr. Res. Green Sustain. Chem.* **2021**, *4* (October), 100184. https://doi. org/10.1016/j.crgsc.2021.100184.

[57] Satapathy, A. K.; Gunasekaran, G.; Sahoo, S. C.; Amit, K.; Rodrigues, P. V. Corrosion Inhibition by Justicia Gendarussa Plant Extract in Hydrochloric Acid Solution. *Corros. Sci.* **2009**, *51* (12), 2848–2856. https://doi.org/10.1016/j.corsci.2009.08.016.

[58] Chaubey, N.; Savita; Qurashi, A.; Chauhan, D. S.; Quraishi, M. A. Frontiers and Advances in Green and Sustainable Inhibitors for Corrosion Applications: A Critical Review. *J. Mol. Liq.* **2021**, *321*, 114385. https://doi.org/10.1016/j.molliq. 2020.114385.

[59] Sin, H. L. Y.; Abdul Rahim, A.; Gan, C. Y.; Saad, B.; Salleh, M. I.; Umeda, M. Aquilaria Subintergra Leaves Extracts as Sustainable Mild Steel Corrosion Inhibitors in HCl. *Meas. J. Int. Meas. Confed.* **2017**, *109*, 334–345. https://doi.org/10.1016/j. measurement.2017.05.045.

[60] Patni, N.; Agarwal, S.; Shah, P. Greener Approach towards Corrosion Inhibition. *Chinese J. Eng.* **2013**, 2013, 1–10. https://doi.org/10.1155/2013/784186.

[61] Huang, Y.; Xiao, D.; Burton-Freeman, B. M.; Edirisinghe, I. Chemical Changes of Bioactive Phytochemicals during Thermal Processing. In: *Reference Module in Food Science*; Elsevier, **2016**, *9*, 1–9. https://doi.org/10.1016/b978-0-08-100596-5.03055-9.

[62] Guo, Y.; Gao, M.; Wang, H.; Liu, Z. Tobacco Rob Extract as Green Corrosion Inhibitor for N80 Steel in HCl Solution. *Int. J. Electrochem. Sci.* **2017**, *12* (2), 1401–1420. https:// doi.org/10.20964/2017.02.25.

[63] Anupama, K. K.; Ramya, K.; Joseph, A. Electrochemical and Computational Aspects of Surface Interaction and Corrosion Inhibition of Mild Steel in Hydrochloric Acid by Phyllanthus Amarus Leaf Extract (PAE). *J. Mol. Liq.* 2016, *216*, 146–155. https://doi. org/10.1016/j.molliq.2016.01.019.

[64] Singh, A.; Ahamad, I.; Quraishi, M. A. Piper Longum Extract as Green Corrosion Inhibitor for Aluminium in NaOH Solution. Arab. *J. Chem.* **2016**, *9*, S1584–S1589. https://doi.org/10.1016/j.arabjc.2012.04.029.

[65] Deyab, M. A. Egyptian Licorice Extract as a Green Corrosion Inhibitor for Copper in Hydrochloric Acid Solution. *J. Ind. Eng. Chem.* **2015**, *22*, 384–389. https://doi. org/10.1016/j.jiec.2014.07.036.

[66] Patel, K. K.; Vashi, R. T. Azadirachta Indica Extract as Corrosion Inhibitor for Copper in Nitric Acid Medium. *Res. J. Chem. Sci.* **2015**, *5* (11), 59–66. ISSN: 2231-606X.

[67] James, A. O.; Akaranta, O.; Chemistry, I.; Harcourt, P. Corrosion Inhibition of Zinc in 2.0 M Hydrochloric Acid Solution with Acetone Extract of Red Onion Skin. *Pure Appl. Chem.* **2009**, *3* (11), 212–217. ISSN: 1996-0840.

[68] Wedian, F.; Al-Qudah, M. A.; Al-Mazaideh, G. M. Corrosion Inhibition of Copper by Capparis Spinosa L. Extract in Strong Acidic Medium: Experimental and Density Functional Theory. *Int. J. Electrochem. Sci.* **2017**, *12* (6), 4664–4676. https://doi. org/10.20964/2017.06.47.

[69] Shabani-Nooshabadi, M.; Hoseiny, F. S.; Jafari, Y. Green Approach to Corrosion Inhibition of Copper by the Extract of Calligonum Comosum in Strong Acidic Medium. *Metall. Mater. Trans. A Phys. Metall. Mater. Sci.* **2015**, *46* (1), 293–299. https://doi. org/10.1007/s11661-014-2634-1.

[70] Krishnaveni, K.; Ravichandran, J. Influence of Aqueous Extract of Leaves of Morinda Tinctoria on Copper Corrosion in HCl Medium. *J. Electroanal. Chem.* 2014, *735*, 24–31. https://doi.org/10.1016/j.jelechem.2014.09.032.

[71] Chen, G.; Zhang, M.; Pang, M.; Hou, X. Q.; Su, H.; Zhang, J. Extracts of Punica Granatum Linne Husk as Green and Eco-Friendly Corrosion Inhibitors for Mild Steel in Oil Fields. *Res. Chem. Intermed.* **2013**, *39* (8), 3545–3552. https://doi.org/10.1007/s11164-012-0861-x.

[72] Ibrahim, T.; Alayan, H.; Mowaqet, Y. A. The Effect of Thyme Leaves Extract on Corrosion of Mild Steel in HCl. *Prog. Org. Coatings* **2012**, *75* (4), 456–462. https://doi.org/10.1016/j.porgcoat.2012.06.009.

[73] Soltani, N.; Tavakkoli, N.; Khayat Kashani, M.; Mosavizadeh, A.; Oguzie, E. E.; Jalali, M. R. Silybum Marianum Extract as a N atural Source Inhibitor for 304 Stainless Steel Corrosion in 1.0M HCl. *J. Ind. Eng. Chem.* **2014**, *20* (5), 3217–3227. https://doi.org/10.1016/j.jiec.2013.12.002.

[74] Muthukrishnan, P.; Prakash, P.; Jeyaprabha, B.; Shankar, K. Stigmasterol Extracted from Ficus Hispida Leaves as a Green Inhibitor for the Mild Steel Corrosion in 1 M HCl Solution. *Arab. J. Chem.* **2019**, *12* (8), 3345–3356. https://doi.org/10.1016/j.arabjc.2015.09.005.

[75] Sharmila, A.; Prema, A. A.; Sahayaraj, P. A. Influence of Murraya Koenigii (Curry Leaves) Extract on the Corrosion Inhibition of Carbon Steel in HCL Solution. *Rasayan J. Chem.* 2010, *3* (1), 74–81.

[76] Wang, X.; Jiang, H.; Zhang, D. X.; Hou, L.; Zhou, W. J. Solanum Lasiocarpum L. Extract as Green Corrosion Inhibitor for A3 Steel in 1 M HCl Solution. *Int. J. Electrochem. Sci.* **2019**, *14* (2), 1178–1196. https://doi.org/10.20964/2019.02.06.

[77] Kliskic, M.; Radosevic, J.; Gudic, S.; Katalinic, V. Aqueous Extract of Rosmarinus officinalis L. as Inhibitor of Al±Mg Alloy Corrosion in Chloride Solution. *J. Appl. Electrochem.* **2000**, *30*, 823–830. https://doi.org/10.1023/A:1004041530105.

[78] El-Etre, A. Y. Inhibition of Aluminum Corrosion Using Opuntia Extract. *Corros. Sci.* **2003**, *45* (11), 2485–2495. https://doi.org/10.1016/S0010-938X(03)00066-0.

[79] Saxena, A.; Prasad, D.; Haldhar, R.; Singh, G.; Kumar, A. Use of Saraca Ashoka Extract as Green Corrosion Inhibitor for Mild Steel in 0.5 M H2SO4. *J. Mol. Liq.* **2018**, *258* (2017), 89–97. https://doi.org/10.1016/j.molliq.2018.02.104.

[80] Zheng, X.; Gong, M.; Li, Q.; Guo, L. Corrosion Inhibition of Mild Steel in Sulfuric Acid Solution by Loquat (Eriobotrya Japonica Lindl.) Leaves Extract. *Sci. Rep.* 2018, *8* (1), 1–15. https://doi.org/10.1038/s41598-018-27257-9.

[81] Anupama, K. K.; Ramya, K.; Shainy, K. M.; Joseph, A. Adsorption and Electrochemical Studies of Pimenta Dioica Leaf Extracts as Corrosion Inhibitor for Mild Steel in Hydrochloric Acid. *Mater. Chem. Phys.* 2015, *167*, 28–41. https://doi.org/10.1016/j.matchemphys.2015.09.013.

[82] Naseri, E.; Hajisafari, M.; Kosari, A.; Talari, M.; Hosseinpour, S.; Davoodi, A. Inhibitive Effect of Clopidogrel as a Green Corrosion Inhibitor for Mild Steel; Statistical Modeling and Quantum Monte Carlo Simulation Studies. *J. Mol. Liq.* **2018**, *269*, 193–202. https://doi.org/10.1016/j.molliq.2018.08.050.

[83] Xhanari, K.; Finšgar, M.; Knez Hrnčič, M.; Maver, U.; Knez, Ž.; Seiti, B. Green Corrosion Inhibitors for Aluminium and Its Alloys: A Review. *RSC Adv.* **2017**, *7* (44), 27299–27330. https://doi.org/10.1039/c7ra03944a.

[84] Vengatesh, G.; Karthik, G.; Sundaravadivelu, M. A Comprehensive Study of Ondansetron Hydrochloride Drug as a Green Corrosion Inhibitor for Mild Steel in 1 M HCl Medium. *Egypt. J. Pet.* **2017**, *26* (3), 705–719. https://doi.org/10.1016/j.ejpe.2016.10.011.

[85] Hossain, S. M. Z.; Razzak, S. A.; Hossain, M. M. Application of Essential Oils as Green Corrosion Inhibitors. *Arab. J. Sci. Eng.* **2020**, *45* (9), 7137–7159. https://doi.org/10.1007/s13369-019-04305-8.

[86] Gualdrón, A. F.; Becerra, E. N.; Peña, D. Y.; Gutiérrez, J. C.; Becerra, H. Q. Inhibitory Effect of Eucalyptus and Lippia alba Essential Oils on the Corrosion of Mild Steel in Hydrochloric Acid. *J. Mater. Environ. Sci.* **2013**, *4* (1), 143–158. ISSN: 2028-2508.

[87] Manssouri, M.; El Ouadi, Y.; Znini, M.; Costa, J.; Bouyanzer, A.; Desjobert, J. M.; Majidi, L. Adsorption Proprieties and Inhibition of Mild Steel Corrosion in HCl Solution by the Essential Oil from Fruit of Moroccan Ammodaucus Leucotrichus. *J. Mater. Environ. Sci.* **2015**, *6* (3), 631–646. ISSN: 2028-2508.

[88] Darriet, F.; Znini, M.; Majidi, L.; Muselli, A.; Hammouti, B.; Bouyanzer, A.; Costa, J. Evaluation of Eryngium Maritimum Essential Oil as Environmentally Friendly Corrosion Inhibitor for Mild Steel in Hydrochloric Acid Solution. *Int. J. Electrochem. Sci.* **2013**, *8* (3), 4328–4345. ISSN: 1452-3981.

[89] Gurjar, S.; Sharma, S. K.; Sharma, A.; Ratnani, S. Performance of Imidazolium Based Ionic Liquids as Corrosion Inhibitors in Acidic Medium: A Review. *Appl. Surf. Sci. Adv.* **2021**, *6*, 100170. https://doi.org/10.1016/j.apsadv.2021.100170.

[90] Styring, P. Novel Sorbent Materials for Carbon Capture. In: *Novel Materials for Carbon Dioxide Mitigation Technology*; Shi, F.; Morreale, B., (eds.); Elsevier B.V., 2015, 207–229. https://doi.org/10.1016/B978-0-444-63259-3.00007-0.

[91] Verma, C.; Ebenso, E. E.; Quraishi, M. A. Ionic Liquids as Green and Sustainable Corrosion Inhibitors for Metals and Alloys: An Overview. *J. Mol. Liq.* **2017**, *233* (2016), 403–414. https://doi.org/10.1016/j.molliq.2017.02.111.

[92] Kobzar, Y. L.; Fatyeyeva, K. Ionic Liquids as Green and Sustainable Steel Corrosion Inhibitors: Recent Developments. *Chem. Eng. J.* **2021**, *425* (July), 131480. https://doi.org/10.1016/j.cej.2021.131480.

[93] Arenas, M. A.; Bethencourt, M.; Botana, F. J.; De Damborenea, J.; Marcos, M. Inhibition of 5083 Aluminium Alloy and Galvanised Steel by Lanthanide Salts. *Corros. Sci.* **2001**, *43* (1), 157–170. https://doi.org/10.1016/S0010-938X(00)00051-2.

[94] Forsyth, M.; Markley, T.; Ho, D.; Deacon, G. B.; Junk, P.; Hinton, B.; Hughes, A. Inhibition of Corrosion on AA2024-T3 by New Environmentally Friendly Rare Earth Organophosphate Compounds. *Corrosion* **2008**, *64* (3), 191–197. https://doi.org/10.5006/1.3278465.

[95] Behrsing, T.; Deacon, G. B.; Junk, P. C. The Chemistry of Rare Earth Metals, Compounds, and Corrosion Inhibitors. In: *Rare Earth-Based Corrosion Inhibitors: A volume in Woodhead Publishing Series in Metals and Surface Engineering*; Forsyth, M.; Hinton, B. (eds.); Woodhead Publishing Limited, 2014, 1–37. https://doi.org/10.1533/9780857093585.1.

[96] Forsyth, M.; Seter, M.; Hinton, B.; Deacon, G.; Junk, P. New "Green" Corrosion Inhibitors Based on Rare Earth Compounds. *Aust. J. Chem.* **2011**, *64* (6), 812–819. https://doi.org/10.1071/CH11092.

[97] Zhao, D.; Sun, J.; Zhang, L.; Tan, Y.; Li, J. Corrosion Behavior of Rare Earth Cerium Based Conversion Coating on Aluminum Alloy. *J. Rare Earths* **2010**, *28* (SUPPL. 1), 371–374. https://doi.org/10.1016/S1002-0721(10)60338-9.

2 Phytochemicals/ Plant Extracts
Preparation, Characterization, and Properties

O. Dagdag and Hansang Kim
Gachon University

R. Haldhar
Yeungnam University

Walid Daoudi
University Mohamed I

Elyor Berdimurodov
National University of Uzbekistan
Central Asian University

2.1 INTRODUCTION

A vast variety of bioactive substances that are present in plants are referred to as phytochemicals, including pigments, phenolic and flavonoid compounds, non-flavonoid substances, alkaloids, glycosides, steroids, saponins, and terpenoids. Previous research has demonstrated that different vegetables contain various pigments such as carotenoids (*e.g.*, carotene and xanthophylls), chlorophylls (*e.g.*, chlorophyll a and chlorophyll b), betacyanins, and betaxanthins. These pigments possess strong antioxidant activities, neutralizing free radicals and protecting against diseases. Vegetables also contain phenolics and flavonoid compounds, such as phenolic acids, flavonols, flavones, flavanols, flavanones, anthocyanins, chalcones, tannins, lignans, and stilbenes [1–5].

In addition, vegetables and fruits contain different classes of alkaloids, saponins (including triterpenoidal and steroidal saponins), terpenoids (*e.g.*, monoterpenes, sesquiterpenes, and diterpenes), steroids (such as brassinosteroids and bufadienolides), and tannins (like gallotannins and ellagitannins) [6].

Phytochemicals can be classified into two categories. Primary metabolites are essential for normal growth and development of plants, including nucleic acids, carbohydrates, fatty acids, proteins, and growth regulators. Secondary metabolites are

DOI: 10.1201/9781003394631-2

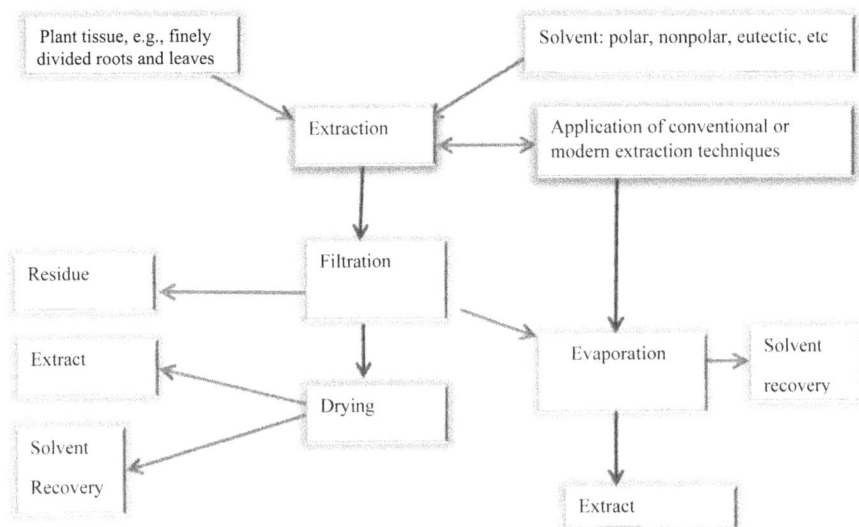

FIGURE 2.1 The process flow diagram for extracting phytochemicals from plant sources [7].

produced by plants to adapt to their environment and defend against threats. These compounds have various functions, such as insecticidal, fungicidal, antibacterial, and antiviral properties [7,8].

The extraction of bioactive compounds from plant sources involves multiple steps, which are illustrated in Figure 2.1.

Overall, the diverse functions and structural characteristics of plant phytochemicals make them of great interest to pharmacologists and biochemists. These bioactive compounds, including terpenes, terpenoids, alkaloids, and phenolic compounds, have the potential to exhibit pharmacological or toxicological effects on animals and humans [8,9].

2.2 EXTRACTION OF PHYTOCHEMICALS

2.2.1 CONVENTIONAL EXTRACTION TECHNIQUES

2.2.1.1 Maceration

Using the straightforward extraction technique of maceration, plant material is soaked in a solvent for at least 3 days while being stirred occasionally [10]. The plant material, which can be in a coarse or powder form, is immersed in the solvent at room temperature. After the extraction is completed, the mixture is strained to separate the liquid extract from the solid material through sieves or a net with tiny holes. The residue, known as the marc, is then pressed to extract any remaining liquid. The liquid extract that results is cleaned using decantation or filtration. To minimize solvent loss through evaporation, maceration is preferably carried out in a stoppered container [10]. It is crucial to remember that getting an already-concentrated extract by letting the solvent evaporate during the extraction process is not what you want. However, the product can be concentrated using vacuum evaporation [10].

The selection of an appropriate solvent in maceration is crucial as it determines the classes of phytochemicals that can be extracted from the plant material [10]. The solvent's polarity plays a significant role in extracting specific phytochemicals. For less polar flavonoids, nonionic surfactant Triton X-100, which has a hydrophilic side chain, has been used as a solvent. On the contrary, more polar flavonoids have been extracted using ethanol or a mixture of ethanol and water. Additionally, polyphenols such as anthocyanins have been effectively extracted from chokeberry using 50% ethanol as the solvent. While maceration is a simple extraction method, it has some drawbacks. It is known for its low efficiency and long extraction time. However, by optimizing the conditions, such as selecting the appropriate solvent, maceration can yield significant amounts of phenolic compounds and anthocyanins, as observed in the extraction of chokeberry. It is worth noting that maceration techniques may not be as effective for extracting certain compounds, as observed in a study on *Cajanus cajan* leaves where maceration afforded the lowest yield of flavonoids compared to other extraction methods. In summary, maceration is a straightforward extraction method that involves soaking plant material in a selected solvent for a certain period. It has its advantages and limitations, but by optimizing the conditions and solvent selection, it can yield valuable phytochemicals [11–13].

2.2.1.2 Digestion

Similar to maceration, digestion is an extractive technique that involves low heating during the extraction phase [10]. However, caution is used to prevent the temperature from changing the bioactive phytochemicals of a particular plant material. As a result, heat increases the extraction solvent's effectiveness. Most of the time, temperatures are kept between 35°C and 40°C, but for harder plant materials like barks with materials containing dismally soluble phytochemicals, they may be raised to a maximum of 50°C. Desired plant components are added to a container containing the suitable solvent that has been heated to the required temperatures during extraction. When the container is shaken periodically, the ideal temperature is maintained for a time that might be anywhere between 30 minutes and 24 hours [10].

2.2.1.2.1 Infusion and Percolation

A diluted solution of the easily soluble plant material's components is referred to as an infusion. It is an extraction method in which the plant material is submerged in a boiling solvent, in this case water, and let to stand in a stoppered container for approximately 15 minutes. At that point, the extract (tea) is drained off and separated from the marc using a filter [10]. The ideal infusion may be thought of as tea. For instance, caffeine has been drawn out of dried, crushed tea leaves from the Alokozay, Lipton, Tapal, and Tetley brands after brewing them for 2–30 minutes at temperatures between 30°C and 90°C [14]. Additionally, *Tilia cordata* fruit was used to extract phenolic chemicals at the ideal temperature of 95°C [15]. It should be mentioned that some infusions are recommended for the treatment of conditions including bronchitis, asthma, and diarrhoea. For instance, antioxidants, phenols, and flavonoids have been extracted for 10 minutes from the rhizomes of several gingers [15]. Percolation is another intriguing technique that is comparable to infusion but more effective than maceration.

The most common method for making fluid extracts like tinctures is percolation. It implies "to pass a liquid through a solid material drop by drop." The solvent, usually ethyl alcohol, gently passes through the plant material during percolation, gradually absorbing phytochemicals, and is then gradually pushed downward by a new solvent that has been placed on top [10]. Plant matter must be gently shred before being added to the percolator, ensuring the pieces are not too tiny. It will be more difficult to separate the fine particles from the extraction solvent if the particles are too small. As a result, the extract would be hazy, and the residue would collect at the percolator's base. However, it is appropriate to wet the plant matrix using the extraction solvent, allowing the plant cells to lengthen for easy phytochemical diffusion into the extraction solvent [15].

The extraction solvent is put into the percolator once the plant material has been placed inside, and it percolates through the plant material at a rate based on the type of plant material being extracted. To provide the solvent enough time to penetrate plant cells and extract the component phytochemicals, the solvent flow rate shouldn't be too high. However, the solvent percolation rate shouldn't be too low, since this would require additional solvent to accomplish the extraction. Typically, the solvent flow rate for 1 kg of plant material should be around $5\,mL\,min^{-1}$.

The chemical characteristics of the secondary metabolites to be extracted determine the extraction solvent to use. Because water hydrates plant walls and alcohol is chemically comparable to the majority of active components extracted from the plant material, water-alcohol solvent mixtures are frequently used, leading to more effective extraction. For instance, 70% ethanol was used to extract phenols, notably epicatechin, and petroleum ether was used to extract antioxidants such as phenols and flavonoids [16]. Intriguingly, percolation was utilized to extract alkaloids from wild fruits other than alcohol using an inorganic aqueous solution of hydrochloric acid [17]. Alcohol is a preservative, and thus, it also has the added benefit of protecting the extract. Leachate is the word for the extract. The plant material is compressed to collect any leftover solvent that was absorbed during the process, and the remaining solution is then added to the leachate (extract). When a colourless liquid free of phytochemicals emerges from the percolator, extraction is complete [10].

2.2.1.3   Decoction

For phytochemicals that do not break down or change with temperature, this extraction method is effective. Plant material is cooked in water for 15–60 minutes during decoction [10]. The type of plant tissues and the phytochemicals being extracted will determine how long the plant is boiled. Normal practice is to boil vulnerable plant components such as leaves, roots, blossoms, and tender stems for 15 minutes. For instance, phenols and flavonoids have been extracted from fruits, rhizomes, and leaves at 100°C using decoction and infusion [18]. Hard plant materials, such as branches and tree bark, can be boiled for an hour as an alternative. After boiling, the mixture is allowed to cool, filtered, and then the necessary amount of solution is added. Following the completion of the decoction procedure, the mixture is filtered to produce the liquid extract.

The decoction technique's extract is likely to contain a lot of unwanted byproducts. It should be emphasized that it is not the best approach for chemicals that are thermolabile. According to reports, *Syzygium cumini*'s bark extract displayed considerable antiglycation and antioxidant activity when it was extracted *via* decoction [19].

2.2.1.4   Soxhlet Extraction

Soxhlet extraction is a continuous solvent-based extraction of phytochemicals. The ground plant material is put in a thimble, which is a porous bag composed of either cellulose or hard filter paper [10]. The compartment of the Soxhlet equipment is put with the thimble filled with ground plant material. In the bottom flask, an extraction solvent such as ethanol or methanol is added. The extraction of phytochemicals occurs when the solvent is heated and turned into steam in the sample thimble, then it is condensed in the condenser on top of the equipment, and finally, it drops back. Comparing the higher yield to extraction methods based on maceration. This method has been used to extract fatty acids from hemp seeds for 8 hours at 70°C and phenolic compounds from leaves for 2 hours at 60% ethanol [20].

This extraction method is quite effective. Cynomorium, a component of traditional Chinese medicine, provided 38.21 mg g^{-1} of ursolic acid by the Soxhlet extraction technique. However, this technique runs the risk of degrading thermolabile chemicals due to the greater temperature. It's interesting to note that a research comparing the Soxhlet along with maceration extraction procedures found that the Soxhlet approach produced lower levels of extracts from both polyphenols and alkaloids [21].

2.2.2   MODERN EXTRACTION TECHNIQUES

2.2.2.1   Accelerated Solvent Extraction (ASE)

Due to its advantages, including minimal solvent requirement, large production, and relatively quick processing, this technology has grown in significance. The more favourable condition of the ASE operations is indicated by the greater solvent temperature and pressure. Compared to maceration or Soxhlet extraction, this method is a more reliable solvent extraction method. There are instances that support notable ASE performance. For instance, it was discovered that ASE outperformed supercritical fluid extraction (SFE) at recovering lipophilic and hydrophilic phytochemicals from raspberry pomace. In comparison to SFE's 15% yield, ASE's temperature along with extraction time impacts on extraction yield was noticeably less pronounced, although it still recovered 25% of the lipophilic and hydrophilic chemicals [22]. In this method, the material is packed into a stainless steel extraction cell, which is then filled with solvent and positioned between inert silica layers that are separated by cellulose filter sheets. The system is heated for a predetermined amount of time at a higher temperature and pressure; the circumstances are favourable for extraction since the viscosity is lower and the diffusion coefficient is higher. By pumping new solvent and nitrogen into the cell, the extract is collected in vials and cleaned. The inert packing material avoids the formation of aggregates in the plant matrix that might clog the system [23].

Critical ASE extraction parameters, such as temperature and pressure, along with extraction duration, are necessary to achieve improved phytochemical recovery yields. For instance, the ASE's sturdiness was tested when cocaine and benzoylecgonine were extracted from coca leaves. The ideal conditions for the extraction method were determined to be 80°C, 20 MPa, and a 10-minute extraction duration [24]. It's interesting to note that distinct classes of bioactive chemical classes are greatly affected by temperature fluctuation in different ways. For instance, it was shown that high temperatures produced the maximum production of phenolic acids, whereas low temperatures were more effective in producing high yields of flavonoids [25]. It indicates that either individually or together, the key extraction conditions may have an impact on extraction efficiency. A single parameter may sometimes have an impact on extraction. Pressure and extraction time were inconsequential, as they were when carbohydrates and phenolic compounds were recovered from barley hull, and the only factor that was discovered to affect the phytochemical extraction yield was temperature [26]. Other times, certain circumstances could show the combined effect of phytochemical extraction. For instance, it was discovered that temperature, extraction duration, and the number of cycles all had a significant impact on the phytochemical yield during the ASE of steviol glycosides from stevia leaves. An average of $91.8 \pm 3.4\%$ glycosides was recovered at the optimal conditions of 100°C, 4 minutes of extraction time, and one cycle. However, after further optimization, the experimental extraction yield increased to $100.8 \pm 3.3\%$ by lowering the particle size of plant materials to less than 0.5 mm [27]. Additionally, it was discovered that temperature and time had a substantial impact on the recovery yield of carotenoids during the extraction of those pigments from food matrices. The ideal circumstances included a temperature of 60°C and three cycles of 15 minutes of extraction time. However, it was shown that adding more solvent *via* more cycles than three did not improve the recovery of more carotenoids [28]. It may be sufficient to mention that ASE is far quicker, more effective, repeatable, environmentally friendly, more practical, and energy efficient than certain other traditional and modern extraction procedures [27].

The yield of phytochemical extraction varies depending on the kind of phytochemical. According to certain research, the extraction yield of two distinct kinds of phytochemicals may be affected in opposing ways by the extraction circumstances. For instance, it was shown that greater temperature favoured lower extraction yield of the β-glucan but higher recovery yield of phenolics throughout the extraction of β-glucans along with phenolic compounds from waxy barley using ASE. The fragmentation of the molecules at high temperatures may be the cause of the reduced extraction yield of β-glucan [29].

The solubility of the phytochemicals in the solvent has a significant impact on how effective the ASE approach is. High solubilities often increase the amount of extraction, and thus, the quantity of phytochemicals recovered. Using *n*-hexane at 80°C, 1,500 atm, static time of 10 minutes, and static cycle of 2 minutes, higher yields of relatively nonpolar compounds like butylidene dihydrophthalide, 4-hydroxy-4-methyl-2-pentanone, and 9,12-octadecanoic acid have been extracted from angelica roots [30]. A polar solute with solvent combination or a nonpolar solute with solvent combination must have properties of similar solute and solvent polarities, nevertheless, in order to provide superior yields [31]. Additionally, a solvent combination was said to produce improved yields. Surprisingly, when phenols were extracted using a combination of moderately polar solvents, they were recovered better [32], as a result of phenols' mild polarity. In fact, because of the polarity of the O–H bond, phenols are more acidic than alcohols. The presence of the benzene ring increases the polarity of phenols, weakening the O–H bond and making it easier to separate the hydrogen ion. As a result, phenols are easily produced when polar solvents are utilized.

With this method, temperature and pressure are optimized for speedy extractions that may be finished in as little as an hour. For instance, phenolic chemicals, mostly flavonoids, were extracted in under an hour utilizing the ASE approach at an ideal temperature of 80°C [33]. Similar to this, a lucerne leaf insecticide was produced, and the extract was transparent and colourless and had little residual fat [34]. In addition, four lignans were extracted from the fruits of the *Fructus schisandrae* plant at a temperature of 160°C, which is the ideal temperature, using 87% concentrated ethanol [35]. The method's versatility is demonstrated by the fact that it has also reportedly been used to extract substances from sea sponges [35].

However, we must point out that the ASE method is not necessarily preferable to Soxhlet extraction in all circumstances. For instance, the Soxhlet approach offered improved extraction robustness, and more levels of phytochemicals were recovered compared to ASE during the extraction of estragole, an essential oil from the fennel plant. Nevertheless, both approaches retrieved the same amount of estragole. The advantage of utilizing ASE over the Soxhlet method for extracting fennel's essential oils was that it used less solvent and required a shorter extraction period. The temperature of 125°C, the extraction period of 7 minutes, and the use of three cycles were ideal for ASE during the extraction of essential oils from fennel [36]. Furthermore, because Soxhlet tends to oxidize lipids, ASE has been proven to be suitable for lipid extraction [37]. Continuous heating with the hot Soxhlet extraction method causes lipid oxidation. It has been discovered that ASE works well to extract phenolic components from grains. This method is suitable since the majority of phenolic chemicals in cereals are often tightly linked to cell wall components and challenging to extract. A combination of 50% and 70% ethanol and water was used to extract the polyphenols from sorghum bran using ASE at a temperature of 150°C, respectively [38].

In summary, based on ASE's previous uses, this extraction method relies heavily on variables such as solvents, elevated extraction temperatures, and pressures and is straightforward to optimize for quick extraction. Additionally, the solvents employed in ASE are those that are often used for traditional liquid extraction methods like Soxhlet. Additionally, ASE is primarily quantitative for the extraction of polychlorinated biphenyls and polycyclic aromatic hydrocarbons, along with total hydrocarbons from reference materials. Additionally, there has been no proof or allegation of thermally labile chemical degradation during the use of ASE.

2.2.2.2 Microwave-Assisted Extraction

Microwave-assisted extraction (MAE) is a technique that utilizes microwaves and solvents to extract natural compounds from plant materials. The process involves heating the solvent and plant tissue using microwaves, which enhances the extraction kinetics. The microwaves directly impact the polar molecules, resulting in increased heating of the sample. The extraction process includes the diffusion of solvents into the sample, separation of the solute from the functional site and release of the solutes into the solvent [39,40].

MAE has been successfully used to extract various phytochemicals, such as saponins from seeds, polyphenolic antioxidants from leaves, sterols from dried mushrooms, and flavonoids from leaves. This technique is particularly effective in extracting polar compounds like flavonoids, polyphenols, and saponins. There are different methods and instruments available for MAE, including solvent-free microwave-assisted extraction (SFMAE) and pressurized microwave-assisted extraction (PMAE) [41,42].

Dried plant material contains minute traces of moisture, and when subjected to microwave heating, the moisture evaporates, creating pressure on the cell walls. This pressure causes cell wall rupture, releasing the bioactive components. Soaking the plant material in a solvent before MAE further enhances the extraction process by facilitating the hydrolysis of cellulose into soluble fractions. The increased temperature in MAE also improves solvent penetration into cell walls.

Compared to other extraction techniques like heat-reflux extraction, MAE results in complete cell wall rupture, as observed through scanning electron micrographs. In MAE, the active components are desorbed from plant material, while in heat-reflux extraction, a series of solvent penetration and solubilization processes occur. MAE is also influenced by microwave power, extraction time, temperature, and the dielectric constant of the solvent used. Lower dielectric constant solvents are advisable for extracting thermolabile compounds.

MAE offers several advantages over conventional solvent extraction techniques. It is rapid, requires less solvent, is cost-effective, and has a higher extraction rate. However, it is more suitable for smaller phenolic molecules and compounds with stability features at microwave temperature ranges. MAE can double the yield of triterpenes from *Centella asiatica* compared to the Soxhlet method. This technique is valuable for extracting phytocompounds that are lost in large quantities using traditional methods, such as flavonoids in food processing.

One significant advantage of MAE is the removal of interferences during extraction, resulting in extracts that contain analytes with minimal interferences when analysed chromatographically. This makes the final extract suitable for analysis using gas chromatography (GC) or high-performance liquid chromatography (HPLC).

In conclusion, MAE is a powerful and user-friendly technique for extracting natural compounds from plant materials. It offers advantages such as rapid extraction, lower solvent consumption, higher extraction rates, and the removal of interferences. MAE is particularly effective for extracting polar compounds and is suitable for a wide range of applications, including food processing and pharmaceutical research.

2.2.2.3 Ultrasound-Assisted Extraction, UAE (Sonication Extraction)

Ultrasound-assisted extraction (UAE) is a technique that utilizes high-frequency ultrasound waves to extract bioactive compounds from plant samples. The mechanical action of ultrasound helps to break down the plant cell walls, increasing the surface area of contact between the solvent and the sample and facilitating the release of phytochemicals. UAE has been found to be effective in extracting various compounds, such as phenolics, anthocyanins, and carotenoids, from a wide range of plant sources [43].

One of the advantages of UAE is its ability to extract compounds at lower temperatures, which helps to preserve thermolabile phytochemicals. UAE can achieve extraction efficiencies in a shorter time compared to conventional techniques like maceration, and it requires a lower quantity of solvent. This makes UAE a preferred method for many researchers [44–48].

UAE can also be combined with other extraction techniques to further enhance efficiency. For example, ultrasound-assisted hydrotropic extraction has been shown to reduce extraction time and decrease the concentration of hydrotrope required. The choice of solvent used in UAE can also impact the extraction process. Ionic liquids and eutectic solvents have been found to improve extraction yields compared to conventional organic solvents [49–57].

One of the key advantages of UAE is its environmentally friendly and sustainable nature. It has a low energy requirement, uses less solvent, and can extract green and concentrated extracts without residual solvents or impurities. UAE is also capable of preserving the chemical structure of compounds that degrade at higher temperatures, such as carotenoids, phenolics, and vitamin C.

In summary, UAE is a powerful technique for extracting bioactive compounds from plant materials. It offers advantages such as shorter extraction times, lower energy consumption, and the ability to extract compounds at lower temperatures, which is particularly beneficial for thermolabile phytochemicals. UAE can be used as a standalone method or in combination with other extraction techniques to enhance extraction efficiency.

2.2.2.4 Supercritical Fluid Extraction (SFE)

SFE technology is widely used for extracting valuable compounds from various sources on a commercial scale, including food products [58]. This technique utilizes a supercritical fluid, which exhibits properties of both a gas and a liquid when it is at its critical point, achieved by specific temperature and pressure conditions. SFE involves the solubilization and separation of extractable chemicals by flowing a solvent through a packed bed. The solvent dissolves the chemicals in the sample and then leaves the extractor. As the temperature increases and the pressure drops, the extract becomes solvent free [58].

An excellent example of a supercritical fluid is carbon dioxide, which becomes supercritical at temperatures above 31.1°C and pressures above 7,380 kPa. Carbon dioxide has been widely used in recent studies, extracting various compounds such as oil lipids, alkaloids, polyphenols, vitamins, essential oils, and phenolics from different natural sources. Supercritical CO_2 has a strong solvation power for non-polar phytochemicals, but its solubility for polar phytochemicals can be enhanced by adding small amounts of cosolvents like ethanol, methanol, water, acetone, and acetonitrile [59–63].

The use of supercritical CO_2 extraction offers several advantages, including strong solvation power, low toxicity, availability, and affordability. However, it is crucial to adjust the temperature and pressure parameters carefully to optimize yield and pre-serve the biological activities of the extracted compounds. Higher temperatures pro-mote solubility but should be considered for thermolabile molecules. In contrast, low temperatures with increased pressure can help preserve the integrity of thermolabile phytochemicals. Proper sample preparation, free from moisture, is also essential for maximum yield and product quality, as moisture can negatively affect the extraction process [64–69].

2.2.3 Structural Determination of Phytochemicals

2.2.3.1 UV–Visible Spectroscopy

UV–visible spectroscopy is compatible with identifying bioactive compounds in both isolated and mixture forms. The technique relies on the detection of maximum absorption (λ_{max}) values that correspond to specific structural features of the targeted phytochemicals. For example, total phenolic extract can be identified at 280 nm, flavones at 320 nm, phenolic acids at 360 nm, and so on. UV–visible spectroscopy offers a fast and cost-effective method of observation. Therefore, it's an excellent technique for identifying different phytochemicals present in a sample [70].

2.2.3.2 Infrared Spectroscopy (IR)

The study of IR spectroscopy focuses on the changes in molecular vibrations that occur when a molecule interacts with infrared radiation. Consequently, IR spec-troscopy is often referred to as vibrational spectroscopy. Each functional group or chemical bond in a molecule possesses unique vibrational frequencies which are determined by the bond strength and reduced masses. By analysing the absorption bands corresponding to the characteristic frequencies of the functional groups in the IR spectrum, the structure of a bioactive compound can be identified. FTIR is a powerful technique that offers high-resolution analysis of chemical constituents and aids in determining the structure of the molecule [70,71].

2.2.3.3 Nuclear Magnetic Resonance Spectroscopy

The magnetic characteristics of certain nuclei, such as ^{1}H, ^{13}C, ^{19}F, and ^{31}P, are exam-ined using NMR spectroscopy. These magnetically active nuclei generate a signal with a frequency that matches the external magnetic field when they interact with radiofrequency electromagnetic radiation, and they resonate when the oscillation frequency coincides with the intrinsic frequency. According to the intensity of the applied magnetic field, the surrounding chemical environment, and the magnetic characteristics of the involved nuclei, the measurement is often reported as a chem-ical shift. NMR spectroscopy enables the prediction of the structure of bioactive chemicals through the analysis of the frequencies (or chemical shift values) of signals and the locations of various nuclei [70,72].

2.2.3.4 Mass Spectroscopy

Mass spectroscopy is a technique used to determine the relative molecular mass of organic molecules by bombarding them with either electrons or lasers, which converts them to highly energetic charged species. This method is useful as it provides rich information required for structure determination. For instance, it can give details about places of fragmentation that can be subsequently used to predict the molecular formula of bioactive compounds. Electrospray ionization (ESI) has been found to be a highly efficient way of generating charged species from macromolecules in the structural elucidation of phenolic compounds [70,71].

2.2.4 Properties of Phytochemicals

2.2.4.1 Antimicrobial Properties

Plants contain a diverse range of phytochemicals, which have been used in traditional medicine for centuries. However, in the early 1900s, there was a shift towards synthetic medicines that were more effective and profitable. Now, there is a renewed interest in using natural chemicals for medicinal purposes due to their lower risk of side effects and cost [73,74].

Recently, the rise of antibiotic-resistant bacteria has led to increased research interest in finding new antimicrobial agents from natural sources. Experts in ethnopharmacology, botany, microbiology, and natural product chemistry are constantly studying the medicinal properties of plants and their phytochemicals. While there is still much to learn, there are already tens of thousands of known compounds in plants, with potentially hundreds of thousands more. These compounds can be classified into different categories based on their chemical structures, botanical origins, biosynthesis pathways, or biological properties. Numerous studies have been conducted both in the lab and in living organisms to investigate the effectiveness of plant phytochemicals as antimicrobial agents [75].

2.2.4.2 Antioxidant Properties

Antioxidants are compounds that can prevent or delay the oxidation of molecules. They can be natural or synthetic. However, there are concerns about the safety of synthetic antioxidants, as they have been linked to health issues such as liver damage and carcinogenicity. As a result, there is a growing interest in developing safer antioxidants from natural sources, with plants being a good option due to their traditional medicinal uses. Many medicinal plants contain phytochemicals with antioxidant properties, including popular ingredients like tamarind, cardamom, lemon grass, and galangal basil [76,77].

Preserving food and preventing bacterial and fungal contamination is a significant challenge that affects both food industries and public health. Synthetic preservatives are increasingly being viewed negatively, leading to a demand for non-toxic, natural alternatives with antioxidant or antimicrobial properties [77,78].

Herbs have long been used for their flavour and fragrance in the food industry, and some of them have antimicrobial properties as well. This has prompted a call to screen and utilize plant materials for their antioxidant and antimicrobial effects. Approximately 20% of all plant species have undergone testing for their pharmacological and biological applications, confirming their safety and advantages [79,80].

2.3 CONCLUSION

Research on natural products is getting a lot of interest globally. The extraction of bioactive components is a crucial stage in the study of natural products. The extraction procedure has put a cap on how quickly new goods can be screened. Both conventional solvent extraction techniques and contemporary, environmentally friendly extraction technologies are now used. The dependability and calibre of following analytical operations are strongly influenced by the technique of extraction that is used. The major objective of extraction is to retain the biological activity of bioactive chemicals while achieving economic feasibility, environmental friendliness, quicker extraction times, and greater yields. Modern methods have several advantages over traditional methods, including quicker extraction times, less solvent requirements, greater biological activity preservation, higher yields, and less energy usage. Longer extraction times, more solvent needs, a higher risk of bioactivity, and lower yields are characteristics of traditional procedures. The plant matrix, targeted phytochemicals, economic feasibility, and environmental effects all influence the choice of extraction method. It is difficult to draw a firm conclusion on the best extraction method due to the abundance of research.

REFERENCES

[1] U. Sarker and S. Oba, "Color attributes, betacyanin, and carotenoid profiles, bioactive components, and radical quenching capacity in selected Amaranthus gangeticus leafy vegetables," *Scientific Reports*, vol. 11, p. 11559, 2021.

[2] U. Sarker and S. Oba, "Nutritional and bioactive constituents and scavenging capacity of radicals in Amaranthus hypochondriacus," *Scientific Reports*, vol. 10, p. 19962, 2020.

[3] U. Sarker, M. N. Hossain, M. A. Iqbal, and S. Oba, "Bioactive components and radical scavenging activity in selected advance lines of salt-tolerant vegetable amaranth," *Frontiers in Nutrition*, vol. 7, p. 587257, 2020.

[4] U. Sarker and S. Oba, "Nutrients, minerals, pigments, phytochemicals, and radical scavenging activity in Amaranthus blitum leafy vegetables," *Scientific Reports*, vol. 10, p. 3868, 2020.

[5] U. Sarker and S. Oba, "Phenolic profiles and antioxidant activities in selected drought-tolerant leafy vegetable amaranth," *Scientific Reports*, vol. 10, p. 18287, 2020.

[6] M. Sharma and P. Kaushik, "Vegetable phytochemicals: an update on extraction and analysis techniques," *Biocatalysis and Agricultural Biotechnology*, vol. 36, p. 102149, 2021.

[7] C. Bitwell, I. S. Sen, C. Luke, and M. K. Kakoma, "A review of modern and conventional extraction techniques and their applications for extracting phytochemicals from plants," *Scientific African*, vol. 19, p. e01585, 2023.

[8] S. H. Taha, I. M. El-Sherbiny, A. S. Salem, M. Abdel-Hamid, A. H. Hamed, and G. A. Ahmed, "Antiviral activity of curcumin loaded milk proteins nanoparticles on potato virus Y," *Pakistan Journal of Biological Sciences: PJBS*, vol. 22, pp. 614–622, 2019.

[9] C. G. Awuchi, "The biochemistry, toxicology, and uses of the pharmacologically active phytochemicals: alkaloids, terpenes, polyphenols, and glycosides," *Journal of Food and Pharmaceutical Sciences*, vol. 7, pp. 131–150, 2019.

[10] N. Azwanida, "A review on the extraction methods use in medicinal plants, principle, strength and limitation," *Medicinal and Aromatic Plants*, vol. 4, pp. 2167–0412, 2015.

[11] N. Ćujić, K. Šavikin, T. Janković, D. Pljevljakušić, G. Zdunić, and S. Ibrić, "Optimization of polyphenols extraction from dried chokeberry using maceration as traditional technique," *Food Chemistry*, vol. 194, pp. 135–142, 2016.

[12] S. Jin, "Microwave-assisted extraction of flavonoids from Cajanus cajan leaves," *Chinese Traditional and Herbal Drugs*, vol. 42, pp. 2235–2239, 2011.

[13] V. Sharma and P. Janmeda, "Extraction, isolation and identification of flavonoid from Euphorbia neriifolia leaves," *Arabian Journal of Chemistry*, vol. 10, pp. 509–514, 2017.

[14] R. Sharif, S. W. Ahmad, H. Anjum, N. Ramzan, and S. R. Malik, "Effect of infusion time and temperature on decaffeination of tea using liquid-liquid extraction technique," *Journal of Food Process Engineering*, vol. 37, pp. 46–52, 2014.

[15] M. Cittan, E. Altuntaş, and A. Çelik, "Evaluation of antioxidant capacities and phenolic profiles in Tilia cordata fruit extracts: a comparative study to determine the efficiency of traditional hot water infusion method," *Industrial Crops and Products*, vol. 122, pp. 553–558, 2018.

[16] S. V. Chanda and M. J. Kaneria, "Optimization of conditions for the extraction of antioxidants from leaves of Syzygium cumini L. using different solvents," *Food Analytical Methods*, vol. 5, pp. 332–338, 2012.

[17] F. Zhang, B. Chen, S. Xiao, and S.-Z. Yao, "Optimization and comparison of different extraction techniques for sanguinarine and chelerythrine in fruits of Macleaya cordata (Willd) R. Br," *Separation and Purification Technology*, vol. 42, pp. 283–290, 2005.

[18] N. Mahmudati, P. Wahyono, and D. Djunaedi, "Antioxidant activity and total phenolic content of three varieties of Ginger (Zingiber officinale) in decoction and infusion extraction method," *Journal of Physics: Conference Series*, vol. 1567, p. 022028, 2020.

[19] P. Perera, S. Ekanayake, and K. Ranaweera, "Antidiabetic compounds in Syzygium cumini decoction and ready to serve herbal drink," *Evidence-Based Complementary and Alternative Medicine*, vol. 2017, p. 1083589, 2017.

[20] O. R. Alara, N. H. Abdurahman, and C. I. Ukaegbu, "Soxhlet extraction of phenolic compounds from Vernonia cinerea leaves and its antioxidant activity," *Journal of Applied Research on Medicinal and Aromatic Plants*, vol. 11, pp. 12–17, 2018.

[21] F.-S. Chin, K.-P. Chong, A. Markus, and N. K. Wong, "Tea polyphenols and alkaloids content using soxhlet and direct extraction methods," *World Journal of Agricultural Sciences*, vol. 9, pp. 266–270, 2013.

[22] N. Kryževičiūtė, P. Kraujalis, and P. R. Venskutonis, "Optimization of high pressure extraction processes for the separation of raspberry pomace into lipophilic and hydrophilic fractions," *The Journal of Supercritical Fluids*, vol. 108, pp. 61–68, 2016.

[23] W. Rahmalia, J.-F. Fabre, and Z. Mouloungui, "Effects of cyclohexane/acetone ratio on bixin extraction yield by accelerated solvent extraction method," *Procedia Chemistry*, vol. 14, pp. 455–464, 2015.

[24] A. Brachet, S. Rudaz, L. Mateus, P. Christen, and J. L. Veuthey, "Optimisation of accelerated solvent extraction of cocaine and benzoylecgonine from coca leaves," *Journal of Separation Science*, vol. 24, pp. 865–873, 2001.

[25] J. G. Figueroa, I. Borrás-Linares, J. Lozano-Sánchez, and A. Segura-Carretero, "Comprehensive identification of bioactive compounds of avocado peel by liquid chromatography coupled to ultra-high-definition accurate-mass Q-TOF," *Food Chemistry*, vol. 245, pp. 707–716, 2018.

[26] S. Sarkar, V. H. Alvarez, and M. D. Saldaña, "Relevance of ions in pressurized fluid extraction of carbohydrates and phenolics from barley hull," *The Journal of Supercritical Fluids*, vol. 93, pp. 27–37, 2014.

[27] J.-B. Jentzer, M. Alignan, C. Vaca-Garcia, L. Rigal, and G. Vilarem, "Response surface methodology to optimise accelerated solvent extraction of steviol glycosides from Stevia rebaudiana Bertoni leaves," *Food Chemistry*, vol. 166, pp. 561–567, 2015.

[28] S. Saha, S. Walia, A. Kundu, K. Sharma, and R. K. Paul, "Optimal extraction and fingerprinting of carotenoids by accelerated solvent extraction and liquid chromatography with tandem mass spectrometry," *Food Chemistry*, vol. 177, pp. 369–375, 2015.

[29] Ó. Benito-Román, V. H. Alvarez, E. Alonso, M. J. Cocero, and M. D. Saldaña, "Pressurized aqueous ethanol extraction of β-glucans and phenolic compounds from waxy barley," *Food Research International*, vol. 75, pp. 252–259, 2015.

[30] S.-K. Cho, A. Abd El-Aty, J.-H. Choi, M. Kim, and J. Shim, "Optimized conditions for the extraction of secondary volatile metabolites in Angelica roots by accelerated solvent extraction," *Journal of Pharmaceutical and Biomedical Analysis*, vol. 44, pp. 1154–1158, 2007.

[31] A. P. D. F. Machado, J. L. Pasquel-Reátegui, G. F. Barbero, and J. Martínez, "Pressurized liquid extraction of bioactive compounds from blackberry (Rubus fruticosus L.) residues: a comparison with conventional methods," *Food Research International*, vol. 77, pp. 675–683, 2015.

[32] D. T. V. Pereira, A. G. Tarone, C. B. B. Cazarin, G. F. Barbero, and J. Martínez, "Pressurized liquid extraction of bioactive compounds from grape marc," *Journal of Food Engineering*, vol. 240, pp. 105–113, 2019.

[33] S. V. Gomes, L. A. Portugal, J. P. dos Anjos, O. N. de Jesus, E. J. de Oliveira, J. P. David, et al., "Accelerated solvent extraction of phenolic compounds exploiting a Box-Behnken design and quantification of five flavonoids by HPLC-DAD in Passiflora species," *Microchemical Journal*, vol. 132, pp. 28–35, 2017.

[34] A. D. Kinross, K. J. Hageman, W. J. Doucette, and A. L. Foster, "Comparison of accelerated solvent extraction (ASE) and energized dispersive guided extraction (EDGE) for the analysis of pesticides in leaves," *Journal of Chromatography A*, vol. 1627, p. 461414, 2020.

[35] L.-C. Zhao, Y. He, X. Deng, G.-L. Yang, W. Li, J. Liang, et al., "Response surface modeling and optimization of accelerated solvent extraction of four lignans from fructus schisandrae," *Molecules*, vol. 17, pp. 3618–3629, 2012.

[36] R. Rodríguez-Solana, J. M. Salgado, J. M. Domínguez, and S. Cortés-Diéguez, "Characterization of fennel extracts and quantification of estragole: optimization and comparison of accelerated solvent extraction and Soxhlet techniques," *Industrial Crops and Products*, vol. 52, pp. 528–536, 2014.

[37] W. Chen, Y. Liu, L. Song, M. Sommerfeld, and Q. Hu, "Automated accelerated solvent extraction method for total lipid analysis of microalgae," *Algal Research*, vol. 51, p. 102080, 2020.

[38] F. Barros, L. Dykes, J. Awika, and L. Rooney, "Accelerated solvent extraction of phenolic compounds from sorghum brans," *Journal of Cereal Science*, vol. 58, pp. 305–312, 2013.

[39] S. B. Bagade and M. Patil, "Recent advances in microwave assisted extraction of bioactive compounds from complex herbal samples: a review," *Critical Reviews in Analytical Chemistry*, vol. 51, pp. 138–149, 2021.

[40] B. Kaufmann and P. Christen, "Recent extraction techniques for natural products: microwave-assisted extraction and pressurised solvent extraction," *Phytochemical Analysis: An International Journal of Plant Chemical and Biochemical Techniques*, vol. 13, pp. 105–113, 2002.

[41] S. A. Heleno, M. Prieto, L. Barros, A. Rodrigues, M. F. Barreiro, and I. C. Ferreira, "Optimization of microwave-assisted extraction of ergosterol from Agaricus bisporus L. by-products using response surface methodology," *Food and Bioproducts Processing*, vol. 100, pp. 25–35, 2016.

[42] N. W. Ismail-Suhaimy, S. S. A. Gani, U. H. Zaidan, M. I. E. Halmi, and P. Bawon, "Optimizing conditions for microwave-assisted extraction of polyphenolic content and antioxidant activity of Barleria lupulina Lindl," *Plants*, vol. 10, p. 682, 2021.

[43] A. P. D. F. Machado, B. R. Sumere, C. Mekaru, J. Martinez, R. M. N. Bezerra, and M. A. Rostagno, "Extraction of polyphenols and antioxidants from pomegranate peel using ultrasound: influence of temperature, frequency and operation mode," *International Journal of Food Science & Technology*, vol. 54, pp. 2792–2801, 2019.

[44] M. R. González-Centeno, K. Knoerzer, H. Sabarez, S. Simal, C. Rosselló, and A. Femenia, "Effect of acoustic frequency and power density on the aqueous ultrasonic-assisted extraction of grape pomace (Vitis vinifera L.)-a response surface approach," *Ultrasonics Sonochemistry*, vol. 21, pp. 2176–2184, 2014.

[45] X. Wang, Y. Wu, G. Chen, W. Yue, Q. Liang, and Q. Wu, "Optimisation of ultrasound assisted extraction of phenolic compounds from Sparganii rhizoma with response surface methodology," *Ultrasonics Sonochemistry*, vol. 20, pp. 846–854, 2013.

[46] M. Bimakr, A. Ganjloo, S. Zarringhalami, and E. Ansarian, "Ultrasound-assisted extraction of bioactive compounds from Malva sylvestris leaves and its comparison with agitated bed extraction technique," *Food Science and Biotechnology*, vol. 26, pp. 1481–1490, 2017.

[47] J. Quintero Quiroz, A. M. Naranjo Duran, M. Silva Garcia, G. L. Ciro Gomez, and J. J. Rojas Camargo, "Ultrasound-assisted extraction of bioactive compounds from annatto seeds, evaluation of their antimicrobial and antioxidant activity, and identification of main compounds by LC/ESI-MS analysis," *International Journal of Food Science*, vol. 2019, p. 3721828, 2019.

[48] M. Herrera and M. Luque de Castro, "Ultrasound-assisted extraction for the analysis of phenolic compounds in strawberries," *Analytical and Bioanalytical Chemistry*, vol. 379, pp. 1106–1112, 2004.

[49] J. Nishad, S. Saha, and C. Kaur, "Enzyme-and ultrasound-assisted extractions of polyphenols from Citrus sinensis (cv. Malta) peel: a comparative study," *Journal of Food Processing and Preservation*, vol. 43, p. e14046, 2019.

[50] K. Ghafoor, Y. H. Choi, J. Y. Jeon, and I. H. Jo, "Optimization of ultrasound-assisted extraction of phenolic compounds, antioxidants, and anthocyanins from grape (Vitis vinifera) seeds," *Journal of Agricultural and Food Chemistry*, vol. 57, pp. 4988–4994, 2009.

[51] L. Petigny, S. Périno-Issartier, J. Wajsman, and F. Chemat, "Batch and continuous ultrasound assisted extraction of boldo leaves (Peumus boldus Mol.)," *International Journal of Molecular Sciences*, vol. 14, pp. 5750–5764, 2013.

[52] L. Zhang, Y. Shan, K. Tang, and R. Putheti, "Ultrasound-assisted extraction flavonoids from Lotus (Nelumbo nuficera Gaertn) leaf and evaluation of its anti-fatigue activity," *International Journal of the Physical Sciences*, vol. 4, pp. 418–422, 2009.

[53] J. N. Del Hierro, T. Herrera, M. R. García-Risco, T. Fornari, G. Reglero, and D. Martin, "Ultrasound-assisted extraction and bioaccessibility of saponins from edible seeds: quinoa, lentil, fenugreek, soybean and lupin," *Food Research International*, vol. 109, pp. 440–447, 2018.

[54] A. Altemimi, D. G. Watson, R. Choudhary, M. R. Dasari, and D. A. Lightfoot, "Ultrasound assisted extraction of phenolic compounds from peaches and pumpkins," *PloS One*, vol. 11, p. e0148758, 2016.

[55] R. Ravanfar, M. Moein, M. Niakousari, and A. Tamaddon, "Extraction and fractionation of anthocyanins from red cabbage: ultrasonic-assisted extraction and conventional percolation method," *Journal of Food Measurement and Characterization*, vol. 12, pp. 2271–2277, 2018.

[56] J. Šic Žlabur, M. Brajer, S. Voća, A. Galić, S. Radman, S. Rimac-Brnčić, et al., "Ultrasound as a promising tool for the green extraction of specialized metabolites from some culinary spices," *Molecules*, vol. 26, p. 1866, 2021.

[57] M. R. Thakker, J. K. Parikh, and M. A. Desai, "Ultrasound assisted hydrotropic extraction: a greener approach for the isolation of geraniol from the leaves of Cymbopogon martinii," *ACS Sustainable Chemistry & Engineering*, vol. 6, pp. 3215–3224, 2018.

[58] P. Raja and A. Barron, "Basic Principles of Supercritical Fluid Chromatography and Supercrtical Fluid Extraction." Retrieved May 11, 2021, 2021.https://chem.libretexts. org/Bookshelves/Analytical_Chemistry/Physical_Methods_in_Chemistry_and_ Nano_Science_(Barron)/03%3A_Principles_of_Gas_Chromatography/3.03%3A_Basic_ Principles_of_Supercritical_Fluid_Chromatography_and_Supercrtical_Fluid_Extraction

[59] Y. N. Belo, S. Al-Hamimi, L. Chimuka, and C. Turner, "Ultrahigh-pressure supercritical fluid extraction and chromatography of Moringa oleifera and Moringa peregrina seed lipids," *Analytical and Bioanalytical Chemistry*, vol. 411, pp. 3685–3693, 2019.

[60] Y. E. Rosas-Quina and F. C. Mejía-Nova, "Supercritical fluid extraction with cosolvent of alkaloids from Lupinus mutabilis sweet and comparison with conventional method," *Journal of Food Process Engineering*, vol. 44, p. e13657, 2021.

[61] H. Zhang, Q. Li, G. Qiao, Z. Qiu, Z. Wen, and X. Wen, "Optimizing the supercritical carbon dioxide extraction of sweet cherry (Prunus avium L.) leaves and UPLC-MS/MS analysis," *Analytical Methods*, vol. 12, pp. 3004–3013, 2020.

[62] M. Khajeh, Y. Yamini, F. Sefidkon, and N. Bahramifar, "Comparison of essential oil composition of Carum copticum obtained by supercritical carbon dioxide extraction and hydrodistillation methods," *Food Chemistry*, vol. 86, pp. 587–591, 2004.

[63] M. de Andrade Lima, R. Andreou, D. Charalampopoulos, and A. Chatzifragkou, "Supercritical carbon dioxide extraction of phenolic compounds from potato (Solanum tuberosum) peels," *Applied Sciences*, vol. 11, p. 3410, 2021.

[64] E. Vági, M. Balázs, A. Komoczi, M. Mihalovits, and E. Szekely, "Fractionation of phytocannabinoids from industrial hemp residues with high-pressure technologies," *The Journal of Supercritical Fluids*, vol. 164, p. 104898, 2020.

[65] R. Goyeneche, A. Fanovich, C. R. Rodrigues, M. C. Nicolao, and K. Di Scala, "Supercritical CO2 extraction of bioactive compounds from radish leaves: yield, antioxidant capacity and cytotoxicity," *The Journal of Supercritical Fluids*, vol. 135, pp. 78–83, 2018.

[66] E. Uquiche, C. Campos, and C. Marillán, "Assessment of the bioactive capacity of extracts from Leptocarpha rivularis stalks using ethanol-modified supercritical CO2," *The Journal of Supercritical Fluids*, vol. 147, pp. 1–8, 2019.

[67] N. Hassim, M. Markom, M. I. Rosli, and S. Harun, "Scale-up approach for supercritical fluid extraction with ethanol-water modified carbon dioxide on Phyllanthus niruri for safe enriched herbal extracts," *Scientific Reports*, vol. 11, p. 15818, 2021.

[68] K. N. Prasad, B. Yang, J. Shi, C. Yu, M. Zhao, S. Xue, et al., "Enhanced antioxidant and antityrosinase activities of longan fruit pericarp by ultra-high-pressure-assisted extraction," *Journal of Pharmaceutical and Biomedical Analysis*, vol. 51, pp. 471–477, 2010.

[69] G. Ferrentino, S. Giampiccolo, K. Morozova, N. Haman, S. Spilimbergo, and M. Scampicchio, "Supercritical fluid extraction of oils from apple seeds: process optimization, chemical characterization and comparison with a conventional solvent extraction," *Innovative Food Science & Emerging Technologies*, vol. 64, p. 102428, 2020.

[70] A. Altemimi, N. Lakhssassi, A. Baharlouei, D. G. Watson, and D. A. Lightfoot, "Phytochemicals: extraction, isolation, and identification of bioactive compounds from plant extracts," *Plants*, vol. 6, p. 42, 2017.

[71] K. P. Ingle, A. G. Deshmukh, D. A. Padole, M. S. Dudhare, M. P. Moharil, and V. C. Khelurkar, "Phytochemicals: extraction methods, identification and detection of bioactive compounds from plant extracts," *Journal of Pharmacognosy and Phytochemistry*, vol. 6, pp. 32–36, 2017.

[72] M. G. Rasul, "Extraction, isolation and characterization of natural products from medicinal plants," *The International Journal of Basic Sciences and Applied Computing*, vol. 2, p. F0076122618, 2018.

[73] V. E. Tyler, "Phytomedicines: back to the future," *Journal of Natural Products*, vol. 62, pp. 1589–1592, 1999.

[74] R. Nair, T. Kalariya, and S. Chanda, "Antibacterial activity of some selected Indian medicinal flora," *Turkish Journal of Biology*, vol. 29, pp. 41–47, 2005.

[75] E. Pichersky and D. R. Gang, "Genetics and biochemistry of secondary metabolites in plants: an evolutionary perspective," *Trends in Plant Science*, vol. 5, pp. 439–445, 2000.

[76] B. Halliwell, M. A. Murcia, S. Chirico, and O. I. Aruoma, "Free radicals and antioxidants in food and in vivo: what they do and how they work," *Critical Reviews in Food Science & Nutrition*, vol. 35, pp. 7–20, 1995.

[77] J. Javanmardi, C. Stushnoff, E. Locke, and J. Vivanco, "Antioxidant activity and total phenolic content of Iranian Ocimum accessions," *Food Chemistry*, vol. 83, pp. 547–550, 2003.

[78] A. Baharlouei, G. Sharifi-Sirchi, and G. S. Bonjar, "Identification of an antifungal chitinase from a potential biocontrol agent, Streptomyces plicatus strain 101, and its new antagonistic spectrum of activity," *Philippine Agricultural Scientist*, vol. 93, pp. 439–445, 2010.

[79] A. Baharlouei, G. Sharifi-Sirchi, and G. S. Bonjar, "Biological control of Sclerotinia sclerotiorum (oilseed rape isolate) by an effective antagonist Streptomyces," *African Journal of Biotechnology*, vol. 10, pp. 5785–5794, 2011.

[80] I. Suffredini, H. S. Sader, A. G. Gonçalves, A. O. Reis, A. C. Gales, A. D. Varella, et al., "Screening of antibacterial extracts from plants native to the Brazilian Amazon Rain Forest and Atlantic Forest," *Brazilian Journal of Medical and Biological Research*, vol. 37, pp. 379–384, 2004.

3 Phytochemicals/ Plant Extracts in Corrosion Prevention
Comparison with Organic Inhibitors

Poonam Kaswan, Chandan K. Mandal, and Gyandshwar K. Rao
Amity University Haryana

Varun Rai and Kamalakanta Behera
University of Allahabad

Manoj K. Banjare
MATS University

Ramesh K. Banjare
Raipur Institute of Technology

Subhashree J. Pandya
Pt. Ravishankar Shukla University

3.1 INTRODUCTION

Metals and various materials having metals have numerous advantages, such as great mechanical power and low cost. Metal alloys are commonly used as rebars to provide strength and materials for building [1–3]. So far, due to their reactivity with environmental components, most metals and their alloys in pure form are extremely reactive and rapidly corroded [4,5]. In the purification of metallic ores, corrosion is a very critical issue. Very concentrated acidic solutions are used in these treatments to remove metallic components as well as surface contaminants such as rusts and scales. Corrosion is a natural process that weakens

DOI: 10.1201/9781003394631-3

and degrades the strength and quality of commercially important metals [6–8]. Metal corrosion is unquestionably one of the most difficult issues that organisations confront. To preserve the metal surface, corrosion inhibitors make sense. Researchers are always exploring for viable alternatives due to environmental issues and the toxicity of commercial organic corrosion inhibitors. Organic substances such as heteroatoms containing compounds, inorganic salts, plant extracts, and phytoconstituents have all been demonstrated to be efficient corrosion inhibitors for metals in a range of corrosive situations. Chemical inhibitors are widely employed in aquatic situations to prevent corrosion of metallic alloys [9]. Plant extracts are developing as economic and environmentally acceptable alternatives to hazardous chemical corrosion inhibitors. The corrosion-prevention strategies of plant extracts are physisorption, chemisorption, and retrodonation [4,8,10]. Carbohydrates, lipids, phenolic acids, terpenoids, alkaloids, and other N-containing phytochemicals cooperate in physical or chemical or retrodonation type of adsorption with the metallic surfaces [6–8,11]. The presence of such phytochemicals in plant extracts influences the mode of adsorption mechanism.

This chapter discusses about the inhibition efficiency of various plant parts of *Rauwolfia serpentina*, *Cannabis sativa*, *Neolamarkia cadamba*, *Cymbopogon citratus*, *Solanum tuberosum*, *Adhatoda vasica* leaves, *Annona squamosa*, *Pisum sativum*, *Citrus reticulate* peels, *Terminalia arjuna*, *Mangifera indica*, *Vitex negundo*, *Adhatoda vasica*, *Piper longum*, *Strychnos nuxvomica*, and *Mucuna pruriens*, as shown in Table 3.1 [13–17]. A healthy balance of hydrophilicity and hydrophobicity is required for effective corrosion prevention. Very high hydrophobicity reduces corrosion inhibition efficacy by limiting chemical solubility in polar electrolytes [18]. As a result, many exogenous substances known as corrosion inhibitors are added to acidic solutions to prevent metallic loss [19]. These supplements adsorb metals and form a protective coating around them to keep them safe from the elements. The most often used corrosion inhibitors are organic compounds, notably heterocyclics [20]. The most effective corrosion inhibitors mainly with organic compounds with aromatic rings are those with polar substituents. Organic corrosion inhibitors might be physical, chemical, or mixed type depending on how they adsorb [20,21] as shown in Figure 3.1. As a result, natural polymers, Arabic gums, polysaccharides, amino acids, and other ecologically benign corrosion inhibitors, as well as their derivatives, are receiving attention [22,23]. A search of the literature turns up several publications on the anticorrosive properties of various plant extracts [24,25]. Schiff bases containing phytochemicals reduce corrosion by utilising a diverse range of metal/electrolyte combinations [26].

3.2 CORROSION INHIBITORS FOR DIFFERENT SURFACES

3.2.1 CORROSION INHIBITORS FOR CARBON AND MILD STEEL

- The corrosion-prevention efficacy of *Kleinhovia hospita* plant extract on carbon steel specimens polarised in 1 M HCl was investigated using GLCM and SVM classification. At 3,000 mg L^{-1} of KH extract, both the hydrogen

TABLE 3.1
Name, Metal and Electrolyte, Inhibition Efficiency, and Various Extracts

S.No.	Names/Scientific Names of Plants	Metal and Electrolyte	Efficiency	References
1.	*Tunbergia fragrans*	MS/1 M HCl	81%	[37]
2.	*Crataegus oxyacantha*	Carbon steel/1 M HCl	94%	[38]
3.	*Cinnamoum tamala*	Carbon steel/0.5 M H$_2$SO$_4$	>95%	[39]
4.	*Bee* pollen extract	Cu/1 M HCl	94.5%	[40]
5.	Semi-ripe arecanut husk extract	Al-63400/HCl	–	[36]
6.	*Cinnamomum camphora* bark	LCS/0.5 M H$_2$SO$_4$	81%–89%	[41]
7.	*Dolichandra unguiscati*	MS/HCl solution	–	[42]
8.	*Calendula officinalis*	MS/1 M HCl solution	94.88%	[43]
9.	*Armoracia rusticana*	MS/0.5 M H$_2$SO$_4$	–	[44]
10.	*Portulaca grandiflora*	MS/HCl	–	[45]
11.	*Magnolia kobus* extract	MS/1 M H$_2$SO$_4$	–	[46]
12.	*Sunflower* petals extract	Carbon steel/1 M HCl solution	~75%–79%	[47]
13.	*Hibiscus sabdariffa* LE	Cu-Zn Alloy/1 M HNO$_3$	94.89%	[48]
14.	*Rosa laevigata*	Copper in 0.5 M H$_2$SO$_4$	89.8%	[49]
15.	*Punica granatum* peel extract	SS-410/15% HCl	91.79%	[50]
16.	*Euphorbia hirta* leaf extract	MS/0.5 M HCl	70%	[51]
17.	*Inula viscosa* extract	Carbon steel/1 M HCl	92%	[52]
18.	*Acanthopanax senticosus*	Q235 carbon steel/1 M HCl	98.79%	[53]
19.	*Carissa macrocarpa* leaf extract	Cu/1 M HNO$_3$	91%	[54]
20.	*Cabbage* extract	X70 steel/1 M HCl	95.87%–93.36%	[55]
21.	*Murraya koenigii* Linn	MS/5 M HCl	98.13%	[56]
22.	*Spinacia oleracea* extract	Carbon steel/1.0 M HCl	–	[57]
23.	*Xylocarpus moluccensis*	MS/1 M HCl	64%	[58]
24.	*Saussurea obvallatta*	MS/1 M HCl	90%	[59]
25.	*Turnip* peel extract	Cu/3.5 wt% NaCl	92%	[60]
26.	*Xanthium strumarium* leaves	Low-carbon steel/1 M HCl	97.02%	[61]
27.	*Physalis minima*	Low-carbon steel/1 M HCl	98%	[62]
28.	*Chrysanthemum coronarium*	Al-air battery/4 M NaOH	95.12%	[63]
29.	*Portulaca grandiflora*	Low-carbon steel in 0.5 M HCl		[64]

(Continued)

TABLE 3.1 *(Continued)*
Name, Metal and Electrolyte, Inhibition Efficiency, and Various Extracts

S.No.	Names/Scientific Names of Plants	Metal and Electrolyte	Efficiency	References
30.	*Artemisia argyi*	Carbon steel/1 M HCl	96.4%	[65]
31.	*Cytisus multiflorus*	Carbon steel/1 M H$_2$SO$_4$	95.91%	[66]
32.	*Terminalia arjuna*	Mild steel/0.2 M HCl	64.1%	[67]
33.	*Thuja occidentalis*	Al/0.5 N H$_2$SO$_4$		[68]
34.	*Sorghum vulgare*	MS/0.5 M H$_2$SO$_4$	93.6%	[69]
35.	*Reineckia carnea*	Carbon steel/1 M HCl	95.5%	[70]
36.	*Capsicum chili* extract	MS/3.5% NaCl, 2.0, 1.0 M HCl	97.1%, 94.7%, 85.2%	[28]
37.	*Eupatorium adenophora*	Cold rolled steel (CRS)/0.10 M Cl$_3$CCOOH solution	91.1%	[71]
38.	*Citrus latifolia, Citrus nobilis* L., *Euphorbia laurifolia*	API-5LX52 carbon steel	29.20%–78.21%	[72]
39.	*Papaya*	Cu/0.5 M H$_2$SO$_4$	92.9%	[73]
40.	*Lannea coromandelica*	MS/1 M H$_2$SO$_4$	88.5%	[74]
41.	*Dryopteris cochleata*	Al/1 M H$_2$SO$_4$	83.24%	[75]
42.	*Sida cordifolia*	MS/0.5 M H$_2$SO$_4$	98.83%	[76]
43.	*Litchi chinensis*	MS/0.5 M H$_2$SO$_4$	95.7%	[77]
44.	*Griffonia simplicifolia*	J55 Steel/1 M HCl	94.73%	[78]
45.	*Tamarindus indiaca* AE	MS/1 M HCl	93%	[79]
46.	*Lecaniodiscus cupaniodes*	MS/0.5 M HCl	93.46%	[80]
47.	*Zizyphus lotuse*	Cu/1 M HCl	93%	[81]
48.	*Chromolaena odorata*	MS/1 M NaCl	99.83%	[82]
49.	*Cascabela thevetia*	Carbon steel/3.5% NaCl + Na$_2$S	95%	[83]
50.	*Bagassa guianensis*	Zinc/3% NaCl	97%	[84]
51.	*Falcaria vulgaris* leaves	MS/1 M HCl	91.3%	[29]
52.	*Ananas sativum*	Al/1 M HCl	–	[32]
53.	*Azadirachta indica* leaves extract	Cu/0.5 M H$_2$SO$_4$	–	[85]
54.	*Apium graveolens L.*	Al/0.25 M NaOH	93.33%	[35]
55.	*Schinopsis lorentzii* extract	Low-carbon steel in 1 M HCl	–	[31]
56.	*Capparis spinosa* extract	Cu/1.0 M nitric acid	82%	[86]
57.	*Allium sativum*	Stainless steel/oilfield	81% ± 3%	[87]
58.	*Myrtus communis*	Cu/1 M H$_2$SO$_4$		[88]
59.	*Prosopis cineraria*	Cu/0.5 M HCl	85.85%	[89]
60.	*Vitex negundo*	Cu/Biodiesel	96%	[90]

Green corrosion inhibitors

Natural	Synthetic
1. Plant Extracts	1. Ionic Liquids
2. Chemical medicine	2. Surfactants
3. Carbohydrates	3. Synthetic polymers
4. Amino acids	4. Inhibition via MW
5. Natural polymers	5. Inhibition via US
6. Bio-surfactants	6. Inhibition via MCRs
7. Natural gases	7. PEG
8. Oleochemicals	8. Inhibitors via H_2O_2 and CO_2

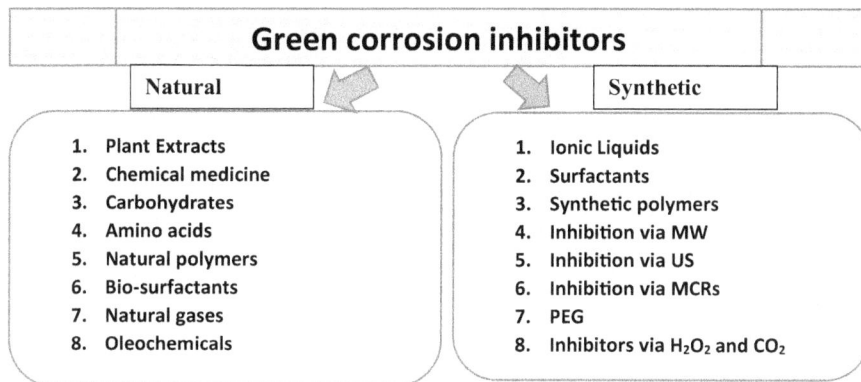

FIGURE 3.1 Green corrosion inhibitor types [12].

rate and the corrosion rate are the lowest. The inclusion of the inhibitor resulted in 99% of the maximum inhibitory effectiveness. According to the polarisation results, KH extract shows retrodonation with physisorption as the primary inhibitory mechanism [27].

- Crispy dry *Capsicum chilli* extracts, a low-cost and reusable substance, were tested for inhibitory effects on the chloric solution of mild steel using PDP and EIS. The findings showed that the coating's highest inhibitory effectiveness was 97.1%, 94.7%, and 85.2% for mild steel in 3.5% NaCl, 2.0, and 1.0 M HCl, respectively. Hydrophobic layer provided excellent protection due to the homogenous distribution [28]. Figure 3.2 shows the inhibition with different analytes.
- The research focussed on the usage of *Falcaria vulgaris* (FV) leaf extract to inhibit the mild steel corrosion process in a 1 M HCl condition. After 6 hours of immersion, EIS measurements demonstrated an inhibitory effectiveness of 91.3% at this dose [29]. Figure 3.3 shows the mechanism of phytochemicals of FV on mild steel surface.
- Husnu Gerengi and colleagues investigated the corrosion prevention of low-carbon steel in a 1 M HCl solution with varying amounts of *Schinopsis lorentzii* extract (SLE). SLE was discovered to be a mildly cathodic inhibitor, with inhibitory activity increasing as extract concentration increased [30]. Adsorption of extract molecules on the low-carbon steel surface can work as a corrosion inhibitor in a hydrochloric acid environment [31].

3.2.2 CORROSION INHIBITORS FOR AL ALLOYS

Alkaline pickling of aluminium (Al) is a serious corrosion issue in a variety of industries. We evaluated the use of natural chemicals as corrosion inhibitors for Al and its alloys in alkali environment in their utilisation for research study.

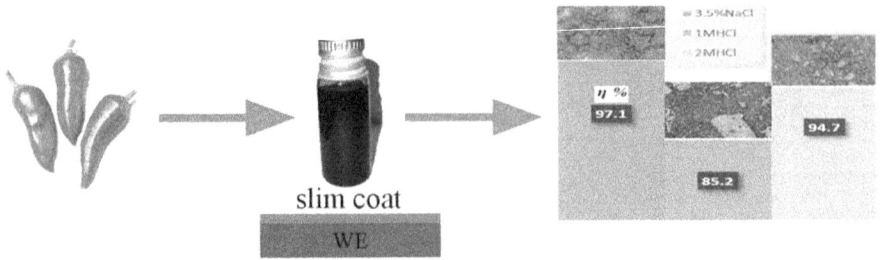

FIGURE 3.2 *Capsicum chili* extracts.

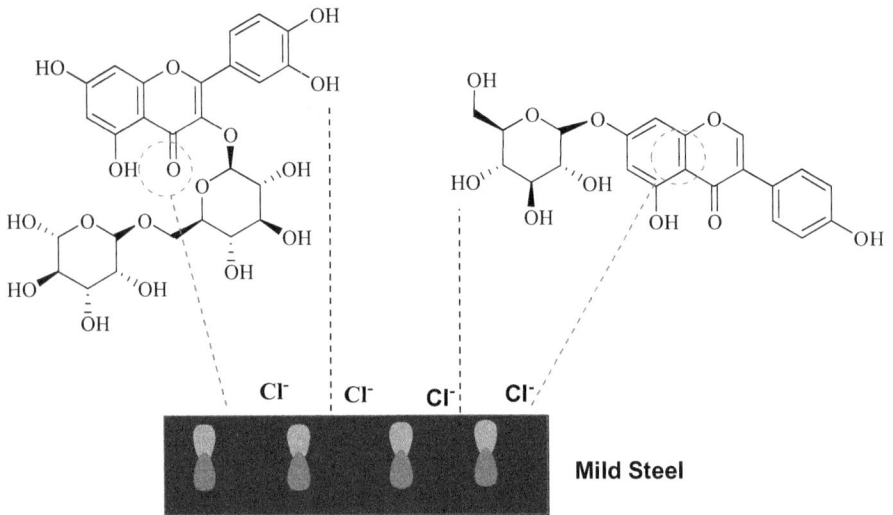

FIGURE 3.3 Mechanism of adsorption of FV leaves on mild steel.

- Rajendran et al. examined the protection from corrosion of Al using *Hibiscus rosa-sinensis* aqueous extract: 8 mL floral extract plus 50 mg L^{-1} Zn^{2+} reduced 98% of the bacteria [14].
- Ating and colleagues investigated the suppression of Al corrosion in hydrochloric acid solutions by an ethanolic extract of *Ananas sativum* leaves using weight loss and hydrogen evolution methods. The plant extract was discovered to protect metal against acid-induced corrosion. As the amount of extract and temperature were increased, the effectiveness of inhibition improved [32].
- *Sonneratia caseolaris* leaf extract (SCLE) increased steel corrosion resistance by adsorbing SCLE species to create a barrier layer in hydrochloric acid environments. In a 1 M HCl medium containing 2,500 ppm SCLE, steel showed the highest inhibitory performance of nearly 98%. When the HCl concentration was dropped from 1.00 to 0.01 M at the same SCLE

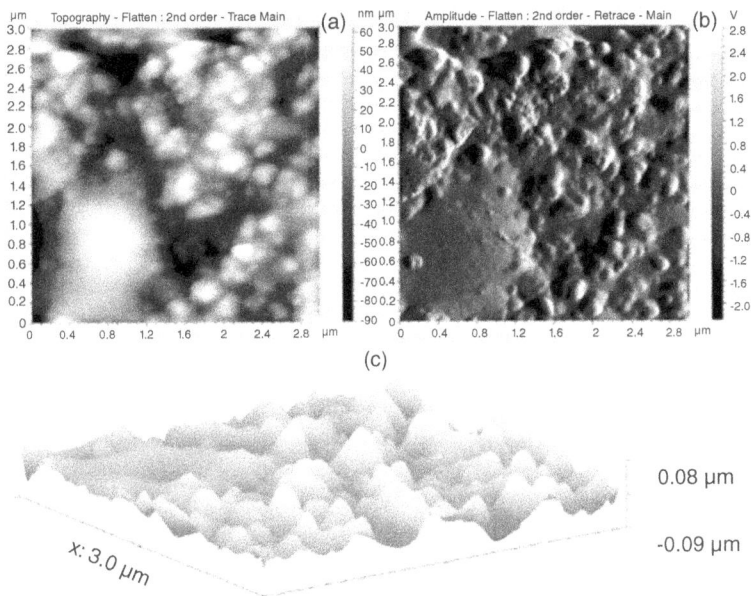

FIGURE 3.4 Atomic force microscopy (a) 2D, (b) relative Volta potential map, and (c) 3D images of the steel surface in a 1 M HCl solution containing 2,500 ppm SCLE for 24 hours. Copyright permission taken from ACS [33].

concentration, the performance decreased significantly [33]. Atomic force microscopy (AFM) 2D and 3D images as shown in Figure 3.4.

- Olusegun K. Abiola and colleagues observed inconsistent inhibitory effects that slowed both anodic and cathodic corrosion processes. The inhibitory behaviour of *Gossypium hirsutum* L. leaf and seed extracts in 2 M NaOH was evaluated using EIS, PDP, and other methods and observed a reduction in Al corrosion [34].

- Al-Moubaraki and colleagues used an aqueous extract of celery seeds to examine the corrosion of Al in NaOH solution of *Apium graveolens* L. and observed that the corrosion rate of Al in 0.25 M NaOH was reduced by increasing the dosage of AECS, and the greatest inhibition proficiency of 93.33% was attained at 1.50 g L^{-1} [35].

- Raghavendra et al. studied natural compounds obtained from semi-ripe arecanut husk extract (SAHE). The isolated SAHE was evaluated in an acid (HCl) medium as an ecologically friendly inhibitor of Al (Al-63400) corrosion [36].

3.2.3 CORROSION INHIBITION OF COPPER METAL BY PLANT EXTRACTS

- Subedi et al. showed the corrosion-prevention benefits of *Vitex negundo* leaf extract on the passivation behaviour of Al and copper (Cu) in pure biodiesel (B100) derived from waste cooking oil and its 10% blend with 90%

petrodiesel (B10). The *V. negundo* extract's highest corrosion inhibition efficiency for Al metal was about 83% in both B100 and B10, and its maximum inhibition efficiency for Cu metal was approximately 96% and 60% in B100 and B10, respectively [90].

- Bozorg et al. investigated the inhibitive actions of *Myrtus communis* extract (MCE) in sulphuric acid solution on the Cu surface. MCE was a very effective mixed-type inhibitor. MCE adsorption on the Cu surface was also observed to follow the Langmuir adsorption isotherm [88].

- *Capparis spinosa* extract was used to study Cu corrosion in 1.0 M nitric acid solution. The inhibitory efficacy increased as the inhibitor concentration in the electrolyte grew but decreased as the temperature increased, with a maximum efficiency of 82% seen at 440 mg L^{-1} of the inhibitor [86].

- *Prosopis cineraria* leaf extract was evaluated for Cu corrosion in 0.5 M HCl using WLT and EIS. It showed an efficiency of 85.85% at 1.064 g L^{-1} after 2 days of immersion [89].

- The extract of *Moringa oleifera* leaves was evaluated for Cu in 1 M nitric and phosphoric acids, and the highest efficiency of 89% was recorded after 3 hours of immersion at 0.3 mol L^{-1} at 298 K [91]. With 1 M nitric acid, a *Ceratonia siliqua* extract was tested for Cu and brass corrosion. According to the findings, the corrosion inhibitor for Cu and brass functioned well in nitric acid [92].

3.3 MECHANISMS OF INHIBITION

Physisorption, chemisorption, and mixed-mode adsorption are only a few of the processes through which phytochemicals might go through. The bulk of corrosion inhibitors were absorbed using the physisorption and chemisorption methods. They are widely used due to their high efficiency, low cost, and simplicity of synthesis, purification, and application. The principal adsorption sites in these plant extracts are polar functional groups such as heteroatoms (N, S, O) and conjugated double bonds or aromatic rings. These heterocyclic compounds' principal adsorption sites have polar functionalities with N, S, or O atoms and alternating double bonds, as well as aromatic rings in their chemical compositions. Several alternative adsorption isotherms were investigated, and Langmuir adsorption provided the most thorough explanation for the adsorption mechanism:

- Physisorption occurs through electrostatic interactions, whereas chemisorption occurs through charge sharing. It is vital to note that in an acidic medium (electrolyte), phytochemical heteroatoms are protonated and transformed into their cationic (protonated) form [93]. For example, Nnanna evaluated the corrosion inhibition of an Al alloy (AA3003) in 0.5 M HCl by extracts of different plants at 30°C and 60°C using a gravimetric approach. Extracts of *Euphorbia hirta* and *Dialum guineense* were among the plant components investigated. Researchers examined the temperature dependency of inhibition efficacy and the activation factors that drive the process to find inhibitory mechanisms. Using physical adsorption procedures, the

Langmuir adsorption isotherm was established to govern the adsorption of both plant extracts on Al alloy [16].

• Metallic surfaces, on the contrary, become negatively charged (anionic) when acidic solution counterions accumulate (e.g. chloride ions of HCl). Electrostatic attraction (ionic bonding) first drew these two oppositely charged species together. Following that, phytochemicals containing free unshared electron pairs of heteroatoms form coordination bonds with the metallic surface (chemisorption) [94,95]. Akbar Ali Samsath Begum and others studied *Spilanthes acmella* aqueous leaf extract (SA-LE) on mild steel in 1.0 M HCl solution at different temperatures. This work used WTL, TP, LPR, and EIS measurements to explore the corrosion-prevention impact. Both the Langmuir and Temkin isotherms were followed for inhibitor adsorption on the surface of mild steel [95].

• Chemisorption is commonly defined as the transfer of phytochemical non-bonding electrons into metallic atoms' d-orbitals (donation). Yet, because metals are already electron efficient, this form of donation induces inter-electronic repulsion, forcing electron transfer from metal d-orbitals to phytochemical anti-bonding molecular orbitals (retrodonation). The phenomenon of synergism occurs when both donation and retrodonation are increased [96,97]. The adsorption and corrosion inhibitory influence of *Punica granatum* (PNG) aqueous extracts on mild steel in 1 M HCl and 0.5 M H_2SO_4 at 30 (1°C) was investigated using the EIS and PDP methodologies. The results of the trials demonstrated that PNG prevented corrosion in both acid situations. Although impedance measurements show that the organic matter in the extract was adsorbed on the metal/solution interface, polarisation data indicate that the extract largely worked as a mixed-type inhibitor [98].

3.4   IMPORTANT PHYTOCHEMICALS AS CORROSION INHIBITORS

Both organic and aqueous solvents are utilised as extract mediums. Aqueous extracts usually include polar phytochemicals, while organic extracts contain non-polar phytochemicals. Triterpenes, alkaloids, saponins, phytosterol, flavonoids, phlorotannins, steroids, tannins, anthraquinones, amino acids, and phenolic compounds are among the phytochemicals implicated in corrosion inhibition. Several phytochemicals have polar functional groups that help in absorption, such as amino, hydroxyl, amide, carboxylic acid, acid chloride, ester, and others [99–101]. Aqueous extracts are chosen over organic extracts in general because they include more polar phytochemicals that allow for better connectivity with the metallic surface compared to non-polar phytochemicals found in organic extracts. These exogenous species increase the extracts' inhibitory action. Exogenous additives are used to increase the protective activity of extracts or organic inhibitors through synergism. Each extract comprises a diverse array of complex phytochemicals, some of which have polar functional groups and many linkages. These electron-rich regions promote phytochemical adsorption on metal surfaces [25,102–110]. Important phytochemicals for corrosion inhibition are shown in Charts 3.1–3.8.

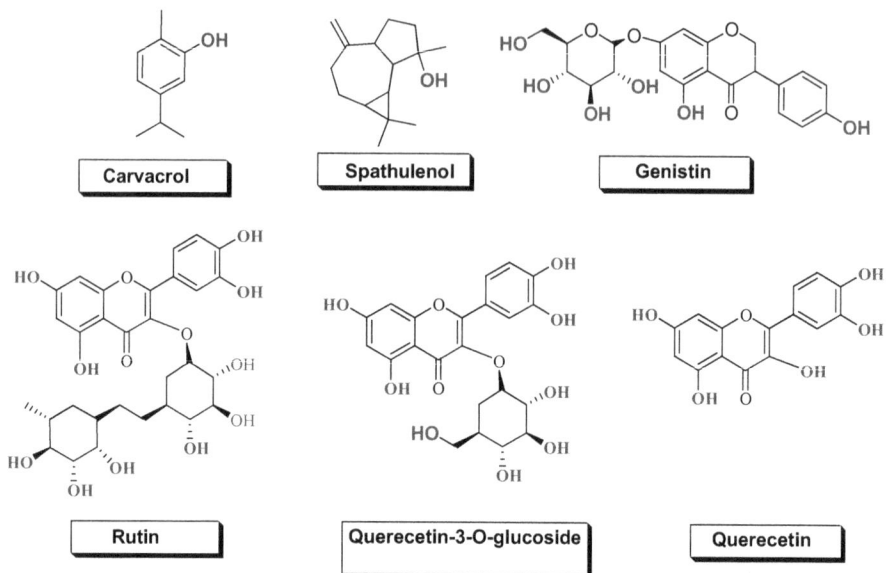

CHART 3.1   Phytochemicals for corrosion inhibition [108].

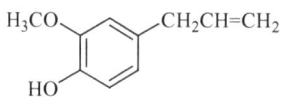

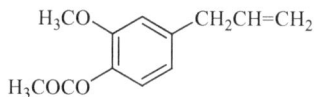

CHART 3.2   Molecular structures of chemicals extracted from *Eugenia caryophylla* [104–106].

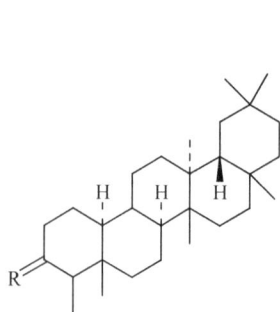

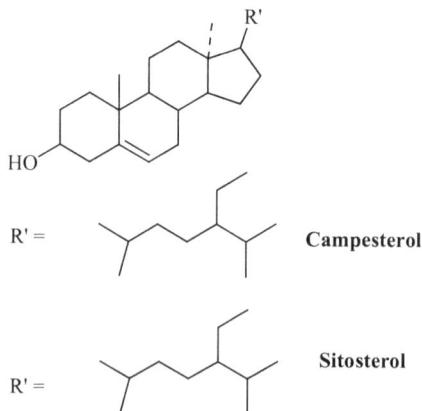

CHART 3.3   Molecular structures of chemicals extracted from *Ficus* species [103].

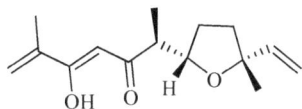

Davonone

CHART 3.4   Molecular structure of phytochemical of *Artemisia* extract [102].

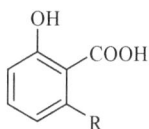 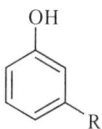 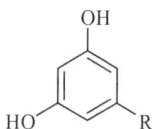 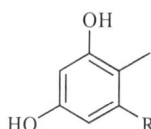

Anacardic acid | Cardanol | Cardol | 2-methyl cardol

CHART 3.5   Molecular structure of phytochemical of Cashew nut shell liquid (CNSL) [107].

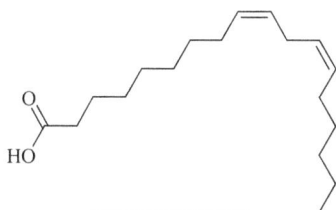 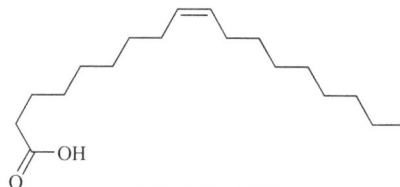

Linoleic acid | Oleic acid

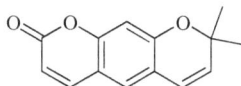

Xanthyletin

CHART 3.6   Molecular structure of phytochemical of *Citrus aurantifolia* [109].

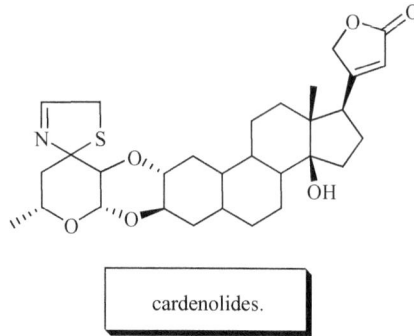

cardenolides.

CHART 3.7   Molecular structure of phytochemical of *Calotropis* extract [110].

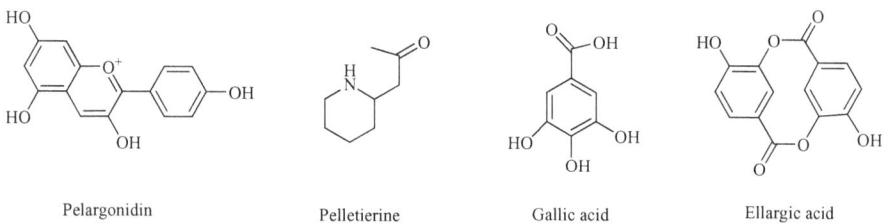

| Pelargonidin | Pelletierine | Gallic acid | Ellargic acid |

CHART 3.8   Molecular structure of phytochemical of *Punica gratum* [98].

3.5   FACTORS AFFECTING CORROSION

3.5.1   CONCENTRATION

The inhibitory proficiency increased with inhibitor concentration.

- A chemical approach was used to examine the Al corrosion in 2 M NaOH using *Phyllanthus amarus* leaf extract. The extract demonstrated outstanding inhibitory performance with 76% efficiency, which increased with the inhibitor dose. This extract acted efficiently as a corrosion inhibitor in both acidic and alkaline conditions, resulting in outstanding performance [111].
- An ethanol extract of *Acalypha torta* leaves (EAL) was tested for its ability to suppress mild steel corrosion in a 1 M HCl solution and found to follow a Langmuir isotherm. EAL is a mixed-type inhibitor, according to polarisation data. Weight loss measurements and electrochemical studies yielded similar inhibitory efficiency.
- Husnu Gerengi et al. discovered that *Schinopsis lorentzii* extract operated as a mildly cathodic inhibitor, with inhibition value increasing as extract concentration rose [31].
- The corrosion of Al in 0.25 M NaOH by *Euphorbia hirta* and *Dialum guineense* leaf extracts was investigated using WLT studies. The inhibitory activity of the tested inhibitors improved as the NaOH concentrations were lowered [112].

3.5.2 ELECTROLYTE

- Oguzie and colleagues evaluated the impact of *Gongronema latifolium* extract on Al in HCl as well as KOH. The findings indicate that perhaps the extract adsorbs on the Al surface and suppresses Al corrosion considerably in both media. In 2 M HCl and 2 M KOH, at increasing inhibitor concentration, inhibition performance improved, reaching 97.54% and 90.82%, respectively [113].
- *Sansevieria trifasciata* leaf extracts were examined for corrosion in Al with 2 M HCl as well as 2 M KOH using gasometry. After 5 hours, the Al alloy's corrosion potential decreased from 0.40 to 0.10 mmpy in 0.25 M KOH, confirming the extract's outstanding inhibitory effectiveness [114].
- When iodide was added to the corrosive electrolyte, the inhibition efficacy increased, indicating the presence of corrosion inhibition in a synergistic way for Al in extremely alkaline conditions. The corrosion-prevention efficacy of Al alloy in an alkaline environment was evaluated using *Palisota hirsute* as a corrosion inhibitor [115].

3.5.3 TEMPERATURE

- The capacity of *Ocimum basilicum* extract to inhibit corrosion in Al for both acidic and alkaline mediums was investigated, and the efficiency of inhibition rose with extract concentration but a decrement was observed with temperature rise [116].
- Cu corrosion in seawater was tested using *Xanthosoma* spp leaf extracts (XLE). At temperatures of 303, 313, 323, and 333 K, various XLE dosages (1%, 2%, 3%, 4%, and 5% v/v) were evaluated. The maximum inhibitory proficiency was reached at 300 K with a 5% v/v inhibitor concentration, whereas the lowest inhibition proficiency was seen at 333 K with a 1% v/v inhibitor concentration [117].
- The inhibitory proficiency of ZCE tested for Cu corrosion in 1 M HNO_3 solution demonstrated an increase with inhibitor dosage and a decrease with rising temperature [118].
- In 1 M HNO_3, *Trigonella stellate* extract was used for inhibition of Cu corrosion. The capacity to inhibit was shown to increase as the extract concentration increased and to decrease as the solution temperature increased [119].

3.6 ADVANTAGES

- Most previously reported inhibitors are harmful by nature because they are produced from toxic and costly solvents, chemicals, and catalysts [19].
- Moreover, the synthesis of these substances results in the formation of a variety of unwanted byproducts. As a consequence, incorporating the application of ecologically acceptable alternatives poses no economic, climate, or medical risks and provides typically effective protection at small doses [120].
- Plant extracts are naturally derived and widely available. Moreover, they are biocompatible, low cost, biodegradable, and, most importantly, non-toxic [121].

3.7 DISADVANTAGES

Plant extracts are naturally sourced and have tremendous potential as ecologically safe inhibitors. Various inhibitors, including pharmaceuticals, chemical substances, and plant extracts, have been employed in commercial applications when their non-toxicity has been demonstrated [20,122]:

- Plant extracts are high in phytochemical constituents, but very few of them have been shown to be useful for metal protection. As a consequence, it is recommended to undertake independent component separation and evaluation [123].
- The extraction processing is a source of concern because of processing times and very high temperature treatments associated with typical extraction techniques. So, other extraction procedures with tolerable processing time and appropriate operating temperature are urgently needed in future research [124,125].
- The most time-consuming operations in this step are drying and dehydration. Moreover, the solvent extraction technique makes use of solvents [126].
- In certain cases, powerful acidic or alkaline solvents are employed that are obviously labelled as hazardous to ecology and human health and will contribute to generation of hazardous waste [127,128].
- Virtually all biological compounds identified in plant extracts are capable of suppressing corrosion processes. Yet, it is unclear which molecule is responsible for the specific metal anticorrosion actions [129].

3.8 FUTURE SCOPE

1. Moreover, because most organic solvents are prohibitively costly, a wide range of aqueous and organic extracts are tested for corrosion inhibition in metals and their alloys in different electrolytes [17,124].
2. Numerous extracts have recently been explored as corrosion inhibitors at different concentrations of HCl, H_2SO_4, NaCl, and other electrolytes; however, there are few plant extracts in NaCl-based electrolytes, and therefore, their application in corrosion inhibition in such rare electrolytes should be tested [12].
3. The majority of these solvents are somewhat expensive, which may have an impact on the economics of extract preparation. Exogenous chemicals such as salts are added to extracts to improve corrosion inhibition efficacy [130,131].
4. Future studies should concentrate on the reliability of extract materials' adsorbtion/absorption over surfaces and more novel characterisation techniques based on theoretical computer modelling in order to investigate the relationship among structures with corrosion inhibition. This insight will benefit us in explaining the corrosion inhibition procedure and generating a diverse range of unique and effective corrosion inhibitors [132].

5. Separating and investigating bioactive molecules individually may be a solution; more research may be conducted to see whether such substances can reduce corrosion on their own or in combination with others *via* a synergetic effect [15].

3.9 CONCLUDING REMARKS

This chapter offered an overview of green corrosion inhibitors derived from plants. These chemicals have been shown to be a promising alternative to the currently available harmful, dangerous, and costly corrosion inhibitors. Several parts of the plant may be used to make the extracts, and a study of these green inhibitors includes substances that can adsorb on the substrate's outermost surface and produce a protective layer. Many extracts are being studied as metallic corrosion inhibitors due to their environmentally acceptable behaviour and strong inhibitory efficiency. Extracts may be made in both aqueous and organic solvents; however, aqueous extracts are preferable due to their more sustainable qualities.

ACKNOWLEDGEMENT

Dr. K. Behera acknowledges the University of Allahabad for the facilities and funding.

ABBREVIATIONS

AECS	*Apium graveolens* L.
EAL	Ethanol extract of *Acalypha torta* leaves
EIS	Electrochemical impedance spectroscopy
FV	*Falcaria vulgaris*
GLCM	Grey level co-occurrence matrix
GLE	*Gossypium hirsutum* L. leaf extracts
GSE	*Gossypium hirsutum* L. seed extracts
KH	*Kleinhovia hospita*
LPR	Linear polarisation resistance
MCE	*Myrtus communis* extract
MCISs	Modern corrosion inhibition systems
MS	Mild steel
PDP	Potentiodynamic polarisation
SB	Schiff bases
SCLE	*Sonneratia caseolaris* leaf extract
SLE	*Schinopsis lorentzii* extract
SVM	Support vector machine
TP	Tafel polarisation
VNE	*Vitex negundo* extract
WLT	Weight loss tests
XLE	*Xanthosoma* spp. leaf extracts
ZCE	*Zygophllum coccineum* L. extract

REFERENCES

[1] Dev Srivyas, P.; Charoo, M. S. Application of Hybrid Aluminum Matrix Composite in Automotive Industry. *Materials Today: Proceedings*, **2019**, *18*, 3189–3200. doi:10.1016/j.matpr.2019.07.195.

[2] Sharma, A. K.; Bhandari, R.; Aherwar, A.; Pinca-Bretotean, C. A Study of Advancement in Application Opportunities of Aluminum Metal Matrix Composites. *Materials Today: Proceedings*, **2020**, *26*, 2419–2424. doi:10.1016/j.matpr.2020.02.516.

[3] Ngo, T. D.; Kashani, A.; Imbalzano, G.; Nguyen, K. T. Q.; Hui, D. Additive Manufacturing (3D Printing): A Review of Materials, Methods, Applications and Challenges. *Composites Part B: Engineering*, **2018**, *143*, 172–196. doi:10.1016/j.compositesb.2018.02.012.

[4] Mohamad, W.; Wan, I.; Kamaruzzaman, M.; Amirah, N.; Nasir, M.; Aiman, N.; Mohd, S.; Yusof, N.; Syaizwadi, M.; Adnan, A.; et al. 25 Years of Progress on Plants as Corrosion Inhibitors through a Bibliometric Analysis Using the Scopus Database (1995–2020). *Arabian Journal of Chemistry*, **2022**, *15* (4), 103655. doi:10.1016/j.arabjc.2021.103655.

[5] Jiang, J.; Li, Y.; Liu, J.; Huang, X.; Yuan, C.; Wen, X.; Lou, D. Recent Advances in Metal Oxide-Based Electrode Architecture Design for Electrochemical Energy Storage. *Advanced Materials*, **2012**, *24*, 5166–5180. doi:10.1002/adma.201202146.

[6] Coelho, L. B.; Zhang, D.; Van Ingelgem, Y.; Steckelmacher, D.; Nowé, A.; Terryn, H. Reviewing Machine Learning of Corrosion Prediction in a Data-Oriented Perspective. *npj Materials Degradation*, **2022**, *6* (1), 8. doi:10.1038/s41529-022-00218-4.

[7] Hu, J. Y.; Zhang, S. S.; Chen, E.; Li, W. G. A Review on Corrosion Detection and Protection of Existing Reinforced Concrete (RC) Structures. *Construction and Building Materials*, **2022**, *325*, 126718. doi:10.1016/j.conbuildmat.2022.126718.

[8] Eliaz, N. Corrosion of Metallic Biomaterials: A Review. *Materials*, **2019**, *12* (3), 407. doi:10.3390/ma12030407.

[9] Ghorbani, M.; Soto Puelles, J.; Forsyth, M.; Catubig, R. A.; Ackland, L.; Machuca, L.; Terryn, H.; Somers, A. E. Corrosion Inhibition of Mild Steel by Cetrimonium Trans-4-Hydroxy Cinnamate: Entrapment and Delivery of the Anion Inhibitor through Speciation and Micellar Formation. *Journal of Physical Chemistry Letters*, **2020**, *11* (22), 9886–9892. doi:10.1021/acs.jpclett.0c02389.

[10] Royani, A.; Hanafi, M.; Manaf, A. Prospect of Plant Extracts as Eco-Friendly Biocides for Microbiologically Influenced Corrosion: A Review. *International Journal of Corrosion and Scale Inhibition*, **2022**, *11* (3), 862–888. doi:10.17675/2305-6894-2022-11-3-1.

[11] Oreko, B. U.; Samuel, B. Assessment of Inhibitive Drugs for Corrosion Inhibition Applications in Petrochemical Plants – A Review. *Saudi Journal of Engineering and Technology*, **2022**, *7* (5), 201–210. doi:10.36348/sjet.2022.v07i05.001.

[12] Alrefaee, S. H.; Rhee, K. Y.; Verma, C.; Quraishi, M. A.; Ebenso, E. E. Challenges and Advantages of Using Plant Extract as Inhibitors in Modern Corrosion Inhibition Systems: Recent Advancements. *Journal of Molecular Liquids*, **2021**, *321*, 114666. doi:10.1016/j.molliq.2020.114666.

[13] Abd-El-nabey, B. A.; Abd-El-khalek, D. E.; El-Housseiny, S.; Mohamed, M. E. Plant Extracts as Corrosion and Scale Inhibitors: A Review. *International Journal of Corrosion and Scale Inhibition*, **2020**, *9*, 1287–1328. doi:10.17675/2305-6894-2020-9-4-7.

[14] Rajendran, S.; Jeyasundari, J.; Usha, P.; Selvi, J. A.; Narayanasamy, B.; Regis, A. P. P.; Rengan, P. Corrosion Behaviour of Aluminium in the Presence of an Aqueous Extract of Hibiscus Rosa-Sinensis. *Portugaliae Electrochimica Acta*, **2009**, *27* (2), 153–164. doi:10.4152/pea.200902153.

[15] Al-Amiery, A. A.; Mohamad, A. B.; Kadhum, A. A. H.; Shaker, L. M.; Isahak, W. N. R. W.; Takriff, M. S. Experimental and Theoretical Study on the Corrosion Inhibition of Mild Steel by Nonanedioic Acid Derivative in Hydrochloric Acid Solution. *Scientific Reports*, **2022**, *12* (1), 1–21. doi:10.1038/s41598-022-08146-8.

[16] Nnanna, L. A.; Onwuagba, B. N.; Mejeha, I. M.; Okeoma, K. B. Inhibition Effects of Some Plant Extracts on the Acid Corrosion of Aluminium Alloy. *African Journal of Pure and Applied Chemistry*, **2010**, *4* (1), 11–16.

[17] Chemat, F.; Vian, M. A.; Ravi, H. K.; Khadhraoui, B.; Hilali, S.; Perino, S.; Tixier, A. S. F. Review of Alternative Solvents for Green Extraction of Food and Natural Products: Panorama, Principles, Applications and Prospects. *Molecules*, **2019**, *24* (16): 3007. doi:10.3390/molecules24163007.

[18] Umoren, S. A.; Abdullahi, M. T.; Solomon, M. M. An Overview on the Use of Corrosion Inhibitors for the Corrosion Control of Mg and Its Alloys in Diverse Media. *Journal of Materials Research and Technology*, **2022**, *20*, 2060–2093. doi:10.1016/j.jmrt.2022.08.021.

[19] Verma, C.; Ebenso, E. E.; Quraishi, M. A.; Hussain, C. M. Recent Developments in Sustainable Corrosion Inhibitors: Design, Performance and Industrial Scale Applications. *Materials Advances*, **2021**, *2* (12), 3806–3850. doi:10.1039/d0ma00681e.

[20] Rani, B. E. A.; Basu, B. B. J. Green Inhibitors for Corrosion Protection of Metals and Alloys: An Overview. *International Journal of Corrosion*, **2012**, *2012* (i), 380217. doi:10.1155/2012/380217.

[21] Raja, P. B.; Ismail, M.; Ghoreishiamiri, S.; Mirza, J.; Ismail, M. C.; Kakooei, S.; Rahim, A. A. Reviews on Corrosion Inhibitors: A Short View. *Chemical Engineering Communications*, **2016**, *203* (9), 1145–1156. doi:10.1080/00986445.2016.1172485.

[22] Verma, C.; Quraishi, M. A. Gum Arabic as an Environmentally Sustainable Polymeric Anticorrosive Material: Recent Progresses and Future Opportunities. *International Journal of Biological Macromolecules*, **2021**, *184* (April), 118–134. doi:10.1016/j.ijbiomac.2021.06.050.

[23] Bentrah, H.; Rahali, Y.; Chala, A. Gum Arabic as an Eco-Friendly Inhibitor for API 5L X42 Pipeline Steel in HCl Medium. *Corrosion Science*, **2014**, *82*, 426–431. doi:10.1016/j.corsci.2013.12.018.

[24] Popoola, L. T. Organic Green Corrosion Inhibitors (OGCIs): A Critical Review. *Corrosion Reviews*, **2019**, *37* (2), 71–102.

[25] Dar, M. A. A Review: Plant Extracts and Oils as Corrosion Inhibitors in Aggressive Media. *Industrial Lubrication and Tribology*, **2015**, *63*, 227–233. doi:10.1108/00368791111140431.

[26] Verma, C.; Quraishi, M. A. Recent Progresses in Schiff Bases as Aqueous Phase Corrosion Inhibitors: Design and Applications. *Coordination Chemistry Reviews*, **2021**, *446*, 214105. doi:10.1016/j.ccr.2021.214105.

[27] Gapsari, F.; Darmadi, D. B.; Setyarini, P. H.; Wijaya, H.; Madurani, K. A.; Juliano, H.; Sulaiman, A. M.; Hidayatullah, S.; Tanji, A.; Hermawan, H. Analysis of Corrosion Inhibition of Kleinhovia Hospita Plant Extract Aided by Quantification of Hydrogen Evolution Using a GLCM/SVM Method. *International Journal of Hydrogen Energy*, **2023**, *48*, 15392–15405. doi:10.1016/j.ijhydene.2023.01.067.

[28] Shehata, O. S.; Fatah, A. H. A.; Abdelsalm, H.; Abdel-Karim, A. M. Crispy Dry Chili Extract as an Eco-Friendly Corrosion Inhibitor for Mild Steel in Chloride Solutions: Experimental and Theoretical Studies. *Journal of Bio- and Tribo-Corrosion*, **2022**, *8* (4), 97. doi:10.1007/s40735-022-00691-z.

[29] Alimohammadi, M.; Ghaderi, M.; A, A. R. S. Falcaria Vulgaris Leaves Extract as an Eco - Friendly Corrosion Inhibitor for Mild Steel in Hydrochloric Acid Media. *Scientific Reports*, **2023**, *13*, 1–16. doi:10.1038/s41598-023-30571-6.

[30] Krishnegowda, P. M.; Venkatesha, V. T.; Krishnegowda, P. K. M.; Shivayogiraju, S. B. Acalypha Torta Leaf Extract as Green Corrosion Inhibitor for Mild Steel in Hydrochloric Acid Solution. *Industrial and Engineering Chemistry Research*, **2013**, *52* (2), 722–728. doi:10.1021/ie3018862.

[31] Ali, T. R.; Salman, T. A.; Shihab, M. S. Pomelo Leaves Extract as a Green Corrosion Inhibitor for Carbon Steel in 0.5 M HCl. *International Journal of Corrosion and Scale Inhibition*, **2021**, *10* (4), 1729–1747. doi:10.17675/2305-6894-2021-10-4-23.

[32] Ating, E. I.; Umoren, S. A.; Udousoro, I. I.; Ebenso, E. E.; Udoh, A. P. Leaves Extract of Ananas Sativum as Green Corrosion Inhibitor for Aluminium in Hydrochloric Acid Solutions. *Green Chemistry Letters and Reviews*, **2010**, *3* (2), 61–68. doi:10.1080/17518250903505253.

[33] Manh, T. D.; Huynh, T. L.; Thi, B. V.; Lee, S.; Yi, J.; Nguyen Dang, N. Corrosion Inhibition of Mild Steel in Hydrochloric Acid Environments Containing Sonneratia Caseolaris Leaf Extract. *ACS Omega*, **2022**, *7* (10), 8874–8886. doi:10.1021/acsomega.1c07237.

[34] Abiola, O. K.; Otaigbe, J. O. E.; Kio, O. J. Gossipium Hirsutum L. Extracts as Green Corrosion Inhibitor for Aluminum in NaOH Solution. *Corrosion Science*, **2009**, *51* (8), 1879–1881. doi:10.1016/j.corsci.2009.04.016.

[35] Al-Moubaraki, A. H.; Al-Howiti, A. A.; Al-Dailami, M. M.; Al-Ghamdi, E. A. Role of Aqueous Extract of Celery (Apium Graveolens L.) Seeds against the Corrosion of Aluminium/Sodium Hydroxide Systems. *Journal of Environmental Chemical Engineering*, **2017**, *5* (5), 4194–4205. doi:10.1016/j.jece.2017.08.015.

[36] Raghavendra, N.; Ishwara Bhat, J. An Investigation of Aluminum (Al-63400) Corrosion Inhibition in Hydrochloric Acid Medium by Semi-Ripe Arecanut Husk Extract: An Eco-Friendly Suitable Green Inhibitor. *Euro-Mediterranean Journal for Environmental Integration*, **2019**, *4* (1), 8. doi:10.1007/s41207-018-0094-5.

[37] Muthukumarasamy, K.; Pitchai, S.; Devarayan, K.; Nallathambi, L. Adsorption and Corrosion Inhibition Performance of Tunbergia Fragrans Extract on Mild Steel in Acid Medium. *Materials Today: Proceedings*, **2020**, *33*, 4054–4058. doi:10.1016/j.matpr.2020.06.533.

[38] Zehra, B. F.; Said, A.; Eddine, H. M.; Hamid, E.; Najat, H.; Rachid, N.; Toumert, L. I. Crataegus Oxyacantha Leaves Extract for Carbon Steel Protection against Corrosion in 1M HCl: Characterization, Electrochemical, Theoretical Research, and Surface Analysis. *Journal of Molecular Structure*, **2022**, *1259*, 132737. doi:10.1016/j.molstruc.2022.132737.

[39] Prasad, D.; Dagdag, O.; Safi, Z.; Wazzan, N.; Guo, L. Cinnamoum Tamala Leaves Extract Highly Efficient Corrosion Bio-Inhibitor for Low Carbon Steel: Applying Computational and Experimental Studies. *Journal of Molecular Liquids*, **2022**, *347*, 118218. doi:10.1016/j.molliq.2021.118218.

[40] Ahmed, R. K.; Zhang, S. Bee Pollen Extract as an Eco-Friendly Corrosion Inhibitor for Pure Copper in Hydrochloric Acid. *Journal of Molecular Liquids*, **2020**, *316*, 113849. doi:10.1016/j.molliq.2020.113849.

[41] Ezeugo, J. O.; Onukwuli, O. D.; Ikebudu, K. O. Optimization of Chrysophyllum Albidum Leaf Extract as Corrosion Inhibitor for Aluminium in 0. 5 M H$_2$SO$_4$. *World Scientific News an International Scientific Journal*, **2019**, *125* (March), 32–50.

[42] Rathod, M. R.; Rajappa, S. K.; Praveen, B. M.; Bharath, D. K. Investigation of Dolichandra Unguis-Cati Leaves Extract as a Corrosion Inhibitor for Mild Steel in Acid Medium. *Current Research in Green and Sustainable Chemistry*, **2021**, *4* (May), 100113. doi:10.1016/j.crgsc.2021.100113.

[43] El-Hashemy, M. A.; Sallam, A. The Inhibitive Action of Calendula Officinalis Flower Heads Extract for Mild Steel Corrosion in 1 M HCl Solution. *Journal of Materials Research and Technology*, **2020**, *9* (6), 13509–13523. doi:10.1016/j.jmrt.2020.09.078.

[44] Haldhar, R.; Prasad, D.; Saxena, A. Armoracia Rusticana as Sustainable and Eco-Friendly Corrosion Inhibitor for Mild Steel in 0.5M Sulphuric Acid: Experimental and Theoretical Investigations. *Journal of Environmental Chemical Engineering*, **2018**, *6* (4), 5230–5238. doi:10.1016/j.jece.2018.08.025.

[45] Fadhil, A. A.; Khadom, A. A.; Ahmed, S. K.; Liu, H.; Fu, C.; Mahood, H. B. Portulaca Grandiflora as New Green Corrosion Inhibitor for Mild Steel Protection in Hydrochloric Acid: Quantitative, Electrochemical, Surface and Spectroscopic Investigations. *Surfaces and Interfaces*, **2020**, *20*, 100595. doi:10.1016/j.surfin.2020.100595.

[46] Chung, I.-M.; Malathy, R.; Priyadharshini, R.; Hemapriya, V.; Kim, S.-H.; Prabakaran, M. Inhibition of Mild Steel Corrosion Using Magnolia Kobus Extract in Sulphuric Acid Medium. *Materials Today Communications*, **2020**, *25*, 101687. doi:10.1016/j.mtcomm.2020.101687.

[47] Khoshsang, H.; Ghaffarinejad, A. Sunflower Petals Extract as a Green, Eco-Friendly and Effective Corrosion Bioinhibitor for Carbon Steel in 1 M HCl Solution. *Chemical Data Collections*, **2022**, *37*, 100799. doi:10.1016/j.cdc.2021.100799.

[48] Shahen, S.; Abdel-Karim, A. M.; Gaber, G. A. Eco-Friendly Roselle (Hibiscus sabdariffa) Leaf Extract as Naturally Corrosion Inhibitor for Cu-Zn Alloy in 1M HNO₃. *Egyptian Journal of Chemistry*, **2022**, *65* (4), 351–361. doi:10.21608/EJCHEM.2021.92917.4392.

[49] Zhang, X.; Yang, L.; Zhang, Y.; Tan, B.; Zheng, X.; Li, W. Combined Electrochemical/Surface and Theoretical Assessments of Rosa Laevigata Extract as an Eco-Friendly Corrosion Inhibitor for Copper in Acidic Medium. *Journal of the Taiwan Institute of Chemical Engineers*, **2022**, *136*, 104408. doi:10.1016/j.jtice.2022.104408.

[50] Bhardwaj, N.; Sharma, P.; Guo, L.; Dagdag, O.; Kumar, V. Molecular Dynamic Simulation, Quantum Chemical Calculation and Electrochemical Behaviour of Punica Granatum Peel Extract as Eco-Friendly Corrosion Inhibitor for Stainless Steel (SS-410) in Acidic Medium. *Journal of Molecular Liquids*, **2022**, *346*, 118237. doi:10.1016/j.molliq.2021.118237.

[51] Aliyu, I.; Lasisi, S.; Olagunju, S. J.; Guruza, A.; Sani, H. M.; Suleiman, I. Y. Characterization of Euphorbia Hirta Leaf as Eco-Friendly Inhibitor for Protection of Mild Steel in Acidic Environment. *Journal of Materials and Environmental Science*, **2022**, *13* (02), 172–184.

[52] Kouache, A.; Khelifa, A.; Boutoumi, H.; Moulay, S.; Feghoul, A.; Idir, B.; Aoudj, S. Experimental and Theoretical Studies of Inula Viscosa Extract as a Novel Eco-Friendly Corrosion Inhibitor for Carbon Steel in 1 M HCl. *Journal of Adhesion Science and Technology*, **2022**, *36* (9), 988–1016. doi:10.1080/01694243.2021.1956215.

[53] Liao, B.; Luo, Z.; Wan, S.; Chen, L. Insight into the Anti-Corrosion Performance of Acanthopanax Senticosus Leaf Extract as Eco-Friendly Corrosion Inhibitor for Carbon Steel in Acidic Medium. *Journal of Industrial and Engineering Chemistry*, **2023**, *117*, 238–246. doi:10.1016/j.jiec.2022.10.010.

[54] El-Asri, A.; Rguiti, M. M.; Jmiai, A.; Oukhrib, R.; Bourzi, H.; Lin, Y.; Issami, S. E. Carissa Macrocarpa Extract (ECM) as a New Efficient and Ecologically Friendly Corrosion Inhibitor for Copper in Nitric Acid: Experimental and Theoretical Approach. *Journal of the Taiwan Institute of Chemical Engineers*, **2023**, *142* (November 2022), 104633. doi:10.1016/j.jtice.2022.104633.

[55] Sun, X.; Qiang, Y.; Hou, B.; Zhu, H.; Tian, H. Cabbage Extract as an Eco-Friendly Corrosion Inhibitor for X70 Steel in Hydrochloric Acid Medium. *Journal of Molecular Liquids*, **2022**, *362*, 119733. doi:10.1016/j.molliq.2022.119733.

[56] Kumar, H.; Yadav, V. Highly Efficient and Eco-Friendly Acid Corrosion Inhibitor for Mild Steel: Experimental and Theoretical Study. *Journal of Molecular Liquids*, **2021**, *335*, 116220. doi:10.1016/j.molliq.2021.116220.

[57] Hameed, R. S. A.; Aleid, G. M. S.; Mohammad, D.; Badr, M. M.; Huwaimel, B.; Suliman, M. S.; Alshammary, F.; Abdallah, M. Spinacia Oleracea Extract as Green Corrosion Inhibitor for Carbon Steel in Hydrochloric Acid Solution. *International Journal of Electrochemical Science*, **2022**, *17*, 221017. doi:10.20964/2022.10.31.

[58] Prifiharni, S.; Mashanafie, G.; Priyotomo, G.; Royani, A.; Ridhova, A.; Elya, B.; Soedarsono, J. W. Extract Sarampa Wood (Xylocarpus Moluccensis) as an Eco-Friendly Corrosion Inhibitor for Mild Steel in HCl 1M. *Journal of the Indian Chemical Society*, **2022**, *99* (7), 100520. doi:10.1016/j.jics.2022.100520.

[59] Kalkhambkar, A. G.; Rajappa, S. K.; Manjanna, J.; Malimath, G. H. Saussurea Obvallatta Leaves Extract as a Potential Eco-Friendly Corrosion Inhibitor for Mild Steel in 1 M HCl. *Inorganic Chemistry Communications*, **2022**, *143*, 109799. doi:10.1016/j.inoche.2022.109799.

[60] Fekri, M. H.; Omidali, F.; Alemnezhad, M. M.; Ghaffarinejad, A. Turnip Peel Extract as Green Corrosion Bio-Inhibitor for Copper in 3.5% NaCl Solution. *Materials Chemistry and Physics*, **2022**, *286*, 126150. doi:10.1016/j.matchemphys.2022.126150.

[61] Khadom, A. A.; Abd, A. N.; Ahmed, N. A.; Kadhim, M. M.; Fadhil, A. A. Combined Influence of Iodide Ions and Xanthium Strumarium Leaves Extract as Eco-Friendly Corrosion Inhibitor for Low-Carbon Steel in Hydrochloric Acid. *Current Research in Green and Sustainable Chemistry*, **2022**, *5* (January), 100278. doi:10.1016/j. crgsc.2022.100278.

[62] Radha, K. V; Patel, D.; Kumar, N.; Devasena, T. Investigation of Eco-Friendly Corrosion Inhibitor for Low Carbon Steel Using Extract of Physalis Minima Leaves. *Journal of Bio-and Tribo-Corrosion*, **2022**, *8* (2), 47.

[63] Huong Pham, T.; Lee, W. H.; Kim, J. G. Chrysanthemum Coronarium Leaves Extract as an Eco-Friendly Corrosion Inhibitor for Aluminum Anode in Aluminum-Air Battery. *Journal of Molecular Liquids*, **2022**, *347* (February), 118269. doi:10.1016/j. molliq.2021.118269.

[64] Khadom, A. A.; Fadhil, A. A.; Ahmed, S. K.; Mutasher, S. A. Statistical and Artificial Neural Network Analysis for Corrosion of Mild Steel in Hydrochloric Acid in Presence of Eco-Friendly Inhibitor. *IOP Conference Series: Earth and Environmental Science*, **2022**, *1055* (1), 012010. doi:10.1088/1755-1315/1055/1/012010.

[65] Wang, Q.; Liu, L.; Zhang, Q.; Wu, X.; Zheng, H.; Gao, P.; Zeng, G.; Yan, Z.; Sun, Y.; Li, Z.; et al. Insight into the Anti–Corrosion Performance of Artemisia Argyi Leaves Extract as Eco–Friendly Corrosion Inhibitor for Carbon Steel in HCl Medium. *Sustainable Chemistry and Pharmacy*, **2022**, *27*, 100710. doi:10.1016/j.scp.2022.100710.

[66] Kahlouche, A.; Ferkous, H.; Delimi, A.; Djellali, S.; Yadav, K. K.; Fallatah, A. M.; Jeon, B.-H.; Ferial, K.; Boulechfar, C.; Ben Amor, Y.; et al. Molecular Insights through the Experimental and Theoretical Study of the Anticorrosion Power of a New Eco-Friendly Cytisus Multiflorus Flowers Extract in a 1 M Sulfuric Acid. *Journal of Molecular Liquids*, **2022**, *347*, 118397. doi:10.1016/j.molliq.2021.118397.

[67] Hossain, N.; Chowdhury, M. A.; Rana, M.; Hassan, M.; Islam, S. Terminalia Arjuna Leaves Extract as Green Corrosion Inhibitor for Mild Steel in HCl Solution. *Results in Engineering*, **2022**, *14* (May), 100438. doi:10.1016/j.rineng.2022.100438.

[68] Sharma, D.; Thakur, A.; Sharma, M. K.; Sharma, R.; Kumar, S.; Sihmar, A.; ... & Om, H. Effective Corrosion Inhibition of Mild Steel Using Novel 1, 3, 4-oxadiazole-pyridine Hybrids: Synthesis, Electrochemical, Morphological, and Computational Insights. *Environmental Research*, **2023**, *234*, 116555. https://doi.org/10.1016/j. envres.2023.116555.

[69] Sharma, S.; Sharma, M.; Dheer, N.; Ujjain, S. K.; Ahuja, P.; Singh, G.; Kanojia, R. A Novel Leaf Extract of Sorghum Vulgare as an Eco-Friendly Corrosion Inhibitor for Mild Steel Corrosion in 0.5 M H$_2$SO$_4$. *Portugaliae Electrochimica Acta*, **2023**, *41* (4), 273–287. doi:10.4152/pea.2023410402.

[70] Wang, Q.; Zheng, H.; Liu, L.; Zhang, Q.; Wu, X.; Yan, Z.; Sun, Y.; Li, X. Insight into the Anti–Corrosion Behavior of Reineckia Carnea Leaves Extract as an Eco–Friendly and High–Efficiency Corrosion Inhibitor. *Industrial Crops and Products*, **2022**, *188*, 115640. doi:10.1016/j.indcrop.2022.115640.

[71] Wei, G.; Deng, S.; Li, X. Eupatorium Adenophora (Spreng.) Leaves Extract as a Highly Efficient Eco-Friendly Inhibitor for Steel Corrosion in Trichloroacetic Acid Solution. *International Journal of Electrochemical Science*, **2022**, *17*, 221182. doi:10.20964/2022.11.63.

[72] Núñez-Morales, J.; Jaramillo, L. I.; Espinoza-Montero, P. J.; Sánchez-Moreno, V. E. Evaluation of Adding Natural Gum to Pectin Extracted from Ecuadorian Citrus Peels as an Eco-Friendly Corrosion Inhibitor for Carbon Steel. *Molecules*, **2022**, *27* (7), 2111. doi:10.3390/molecules27072111.

[73] Tan, B.; Xiang, B.; Zhang, S.; Qiang, Y.; Xu, L.; Chen, S.; He, J. Papaya Leaves Extract as a Novel Eco-Friendly Corrosion Inhibitor for Cu in H_2SO_4 Medium. *Journal of Colloid and Interface Science*, **2021**, *582* (December), 918–931. doi:10.1016/j.jcis.2020.08.093.

[74] Muthukrishnan, P.; Jeyaprabha, B.; Prakash, P. Adsorption and Corrosion Inhibiting Behavior of Lannea Coromandelica Leaf Extract on Mild Steel Corrosion. *Arabian Journal of Chemistry*, **2017**, *10*, S2343–S2354. doi:10.1016/j.arabjc.2013.08.011.

[75] Nathiya, R. S.; Raj, V. Evaluation of Dryopteris Cochleata Leaf Extracts as Green Inhibitor for Corrosion of Aluminium in 1 M H_2SO_4. *Egyptian Journal of Petroleum*, **2017**, *26* (2), 313–323. doi:10.1016/j.ejpe.2016.05.002.

[76] Saxena, A.; Prasad, D.; Haldhar, R.; Singh, G.; Kumar, A. Use of Sida Cordifolia Extract as Green Corrosion Inhibitor for Mild Steel in 0.5 M H_2SO_4. *Journal of Environmental Chemical Engineering*, **2018**, *6* (1), 694–700. doi:10.1016/j.jece.2017.12.064.

[77] Ramananda Singh, M.; Gupta, P.; Gupta, K. The Litchi (Litchi Chinensis) Peels Extract as a Potential Green Inhibitor in Prevention of Corrosion of Mild Steel in 0.5 M H_2SO_4 Solution. *Arabian Journal of Chemistry*, **2019**, *12* (7), 1035–1041. doi:10.1016/j.arabjc.2015.01.002.

[78] Ituen, E.; Akaranta, O.; James, A.; Sun, S. Green and Sustainable Local Biomaterials for Oilfield Chemicals: Griffonia Simplicifolia Extract as Steel Corrosion Inhibitor in Hydrochloric Acid. *Sustainable Materials and Technologies*, **2017**, *11*, 12–18. doi:10.1016/j.susmat.2016.12.001.

[79] Jayakumar, S.; Nandakumar, T.; Vadivel, M.; Thinaharan, C.; George, R. P.; Philip, J. Corrosion Inhibition of Mild Steel in 1 M HCl Using Tamarindus Indica Extract: Electrochemical, Surface and Spectroscopic Studies. *Journal of Adhesion Science and Technology*, **2020**, *34* (7), 713–743. doi:10.1080/01694243.2019.1681156.

[80] Joseph, O. O.; Fayomi, O. S. I.; Joseph, O. O.; Adenigba, O. A. Effect of Lecaniodiscus Cupaniodes Extract in Corrosion Inhibition of Normalized and Annealed Mild Steels in 0.5 M HCl. *Energy Procedia*, **2017**, *119*, 845–851. doi:10.1016/j.egypro.2017.07.136.

[81] Jmiai, A.; El Ibrahimi, B.; Oukhrib, R.; El Issami, S.; Bazzi, L. Zizyphus Lotus as Corrosion Inhibitor for Copper in HCl Medium. *European Corrosion Congress, EUROCORR 2016*, **2016**, *3* (September), 1787–1788.

[82] Gnanasekaran, M.; Mohan, K.; Kumaravel, A.; Magibalan, S. Characterization, Corrosion Behavior, Effect of Temperature and Inhibition Studies on AA6351 Frictional Surfaced Mild Steel. *Journal of Materials Research and Technology*, **2020**, *9* (6), 16080–16092. doi:10.1016/j.jmrt.2020.11.065.

[83] Fouda, A. S.; Killa, H. M.; Farouk, A.; Salem, A. M. Calicotome Extract as a Friendly Corrosion Inhibitor of Carbon Steel in Polluted NaCl (3.5% NaCl + 16 ppm NA2S): Chemical and Electrochemical Studies. *Egyptian Journal of Chemistry*, **2019**, *62* (10), 1879–1894. doi:10.21608/EJCHEM.2019.7656.1649.

[84] Lebrini, M.; Suedile, F.; Salvin, P.; Roos, C.; Zarrouk, A.; Jama, C.; Bentiss, F. Bagassa Guianensis Ethanol Extract Used as Sustainable Eco-Friendly Inhibitor for Zinc Corrosion in 3% NaCl: Electrochemical and XPS Studies. *Surfaces and Interfaces*, **2020**, *20*, 1–31. doi:10.1016/j.surfin.2020.100588.

[85] Valek, L.; Martinez, S. Copper Corrosion Inhibition by Azadirachta Indica Leaves Extract in 0.5 M Sulphuric Acid. *Materials Letters*, **2007**, *61* (1), 148–151. doi:10.1016/j.matlet.2006.04.024.

[86] Wedian, F.; Al-Qudah, M. A.; Al-Mazaideh, G. M. Corrosion Inhibition of Copper by Capparis Spinosa L. Extract in Strong Acidic Medium: Experimental and Density Functional Theory. *International Journal of Electrochemical Science*, **2017**, *12* (6), 4664–4676. doi:10.20964/2017.06.47.

[87] Parthipan, P.; Elumalai, P.; Narenkumar, J.; Machuca, L. L.; Murugan, K.; Karthikeyan, O. P.; Rajasekar, A. Allium Sativum (Garlic Extract) as a Green Corrosion Inhibitor with Biocidal Properties for the Control of MIC in Carbon Steel and Stainless Steel in

Oilfield Environments. *International Biodeterioration and Biodegradation*, **2018**, *132* (March 2017), 66–73. doi:10.1016/j.ibiod.2018.05.005.

[88] Bozorg, M.; Farahani, T. S.; Neshati, J.; Chaghazardi, Z.; Ziarani, G. M. Myrtus Communis as Green Inhibitor of Copper Corrosion in Sulfuric Acid. *Industrial & Engineering Chemistry Research*, **2014**, *53* (11), 4295–4303.

[89] Singh Pratihar, P.; Kumar Rawat, J. Corrosion Inhibition of Copper Metal by Prosopis Cineraria Leaves as Green Corrosion Inhibitor in Acidic Medium. **2017**, *5*(6), 2412–2421.

[90] Subedi, B. N.; Amgain, K.; Bhattarai, J. Green Approach to Corrosion Inhibition Effect of Vitex Negundo Leaf Extract on Aluminum and Copper Metals in Biodiesel and Its Blend. *International Journal of Corrosion and Scale Inhibition*, **2019**, *8*, 744–759. doi:10.17675/2305-6894-2019-8-3-21.

[91] S-Fouda, A.-E.-A.; El-Dossoki, F.; El-Nadr, H.-A.; El-Hussein, A. Moringa Oleifera Plant Extract as a Copper Corrosion Inhibitor in Binary Acid Mixture ($HNO_3 + H_3PO_4$). *Zastita Materijala*, **2018**, *59* (3), 422–435. doi:10.5937/zasmat1803422f.

[92] Fouda, A. S.; Shalabi, K.; Idress, A. A. Ceratonia Siliqua Extract as a Green Corrosion Inhibitor for Copper and Brass in Nitric Acid Solutions. *Green Chemistry Letters and Reviews*, **2015**, *8* (3–4), 17–29. doi:10.1080/17518253.2015.1073797.

[93] Peme, T.; Olasunkanmi, L. O.; Bahadur, I.; Adekunle, A. S.; Kabanda, M. M.; Ebenso, E. E. Adsorption and Corrosion Inhibition Studies of Some Selected Dyes as Corrosion Inhibitors for Mild Steel in Acidic Medium: Gravimetric, Electrochemical, Quantum Chemical Studies and Synergistic Effect with Iodide Ions. *Molecules*, **2015**, *20* (9), 16004–16029. doi:10.3390/molecules200916004.

[94] Ganjoo, R.; Sharma, S.; Thakur, A.; Assad, H.; Kumar Sharma, P.; Dagdag, O.; Berisha, A.; Seydou, M.; Ebenso, E. E.; Kumar, A. Experimental and Theoretical Study of Sodium Cocoyl Glycinate as Corrosion Inhibitor for Mild Steel in Hydrochloric Acid Medium. *Journal of Molecular Liquids*, **2022**, *364*, 119988. doi:10.1016/j.molliq.2022.119988.

[95] Abdelgawad, A.; Awwad, E. M.; Khan, M. Spilanthes Acmella Leaves Extract for Corrosion Inhibition in Acid Medium. *Coatings*, **2021**, *11*, 1–23.

[96] Chen, L.; Lu, D. Organic Compounds as Corrosion Inhibitors for Carbon Steel in HCl Solution: A Comprehensive Review. *Materials (Basel)*, **2023**, *15* (6), 1–59.

[97] Verma, C.; Verma, D. K.; Ebenso, E. E. Sulfur and Phosphorus Heteroatom-Containing Compounds as Corrosion Inhibitors: An Overview. *Heteroatom Chemistry*, **2018**, *29*, e21437. doi:10.1002/hc.21437.

[98] Chidiebere, M. A.; Ogukwe, C. E.; Oguzie, K. L.; Eneh, C. N.; Oguzie, E. E. Corrosion Inhibition and Adsorption Behavior of Punica Granatum Extract on Mild Steel in Acidic Environments: Experimental and Theoretical Studies. *Industrial & Engineering Chemistry Research*, **2012**, *51*, 668–677.

[99] Proestos, C.; Zeng, M.; Elobeid, T.; Sneha, K.; Oz, F. Major Phytochemicals: Recent Advances in Health Benefits and Extraction Method. *Molecules*, **2023**, *28*, 1–41.

[100] Snehlata, K.; Sheel, R.; Kumar, B. Evaluation of Phytochemicals in Polar and Non Polar Solvent Extracts of Leaves of Aegle Marmelos (L.). *IOSR Journal of Biotechnology and Biochemistry (IOSR-JBB)*, **2018**, *4* (5), 31–38. doi:10.9790/264X-0405013138.

[101] Auwal, M. S.; Saka, S.; Mairiga, I. A.; Sanda, K. A.; Shuaibu, A.; Ibrahim, A. Preliminary Phytochemical and Elemental Analysis of Aqueous and Fractionated Pod Extracts of Acacia Nilotica (Thorn Mimosa). *Veterinery Research Forum: An International Quarterly Journal*, **2014**, *5* (2), 95–100.

[102] Suresh, J. Phytochemical and Pharmacological Properties of Artemisia. *International Journal of Pharmaceutical Sciences and Research*, **2016**, *14*, 3081–3090.

[103] Anh, H. T.; Vu, N. S. H.; Huyen, L. T.; Tran, N. Q.; Thu, H. T.; Bach, L. X.; Trinh, Q. T.; Vattikuti, S. V. P.; Nam, N. D. Ficus Racemosa Leaf Extract for Inhibiting Steel Corrosion in a Hydrochloric Acid Medium. *Alexandria Engineering Journal*, **2020**, *59*, 4449–4462. doi:10.1016/j.aej.2020.07.051.

[104] Chaieb, K.; Hajlaoui, H.; Zmantar, T.; Kahla-Nakbi, A. B. The Chemical Composition and Biological Activity of Clove Essential Oil, Eugenia Caryophyllata (Syzigium Aromaticum L. Myrtaceae): A Short Review. *Phototheraphy Research: PTR*, **2007**, *21*, 501–506. doi:10.1002/ptr.

[105] Nada, H. G.; Mohsen, R.; Zaki, M. E.; Aly, A. A. Evaluation of Chemical Composition, Antioxidant, Antibiofilm and Antibacterial Potency of Essential Oil Extracted from Gamma Irradiated Clove (Eugenia Caryophyllata) Buds. *Journal of Food Measurement and Characterization*, **2021**, *16*, 673–686. doi:10.1007/s11694-021-01196-y.

[106] Song, J. Antimicrobial Activities of Eugenia Caryophyllata Extract and Its Major Chemical Constituent Eugenol against Streptococcus Pneumoniae. *APMIS*, **2017**, 121, 1198–1206. doi:10.1111/apm.12067.

[107] Onugwu, E. Extraction, Characterization, Isolation and Identification of Pentoses from Anacardium occidentale l. (Cashew) Nut Shell. *Biomolecules*, **2022**, *9*, 465.

[108] Buchweishaija, J. Phytochemicals as Green Corrosion Inhibitors in Various Corrosive Media: A Review. *Tanzania Journal of Science*, **2008**, *35*, 77–92.

[109] Jain, A. S.; Arora, P.; Popli, H. A Comprehensive Review on Citrus Aurantifolia Essential Oil: Its Phytochemistry and Pharmacological Aspects. *Brazilian Journal of Natural Sciences*, **2020**, *3* (July), 354–364.

[110] Wadhwani, B. D. A Review on Phytochemical Constituents and Pharmacological Potential of Calotropis Procera. *RSC Advances*, **2021**, *11*, 35854–35878. doi:10.1039/d1ra06703f.

[111] Abiola, O. K.; Otaigbe, J. O. E. The Effects of Phyllanthus Amarus Extract on Corrosion and Kinetics of Corrosion Process of Aluminum in Alkaline Solution. *Corrosion Science*, **2009**, *51* (11), 2790–2793. doi:10.1016/j.corsci.2009.07.006.

[112] Nnanna, L. A.; Anozie, I. U.; Akoma, C. S.; Mejeha, I. M.; Okeoma, K. B.; Mejeh, K. I. Corrosion Control of Aluminium Alloy in Alkaline Solution Using Some Leave Extracts. *American Journal of Materials Science*, **2012**, *1* (2), 76–80. doi:10.5923/j.materials.20110102.12.

[113] Oguzie, E. E.; Onuoha, G. N.; Ejike, E. N. Effect of Gongronema Latifolium Extract on Aluminium Corrosion in Acidic and Alkaline Media. *Pigment and Resin Technology*, **2007**, *36* (1), 44–49. doi:10.1108/03699420710718751.

[114] Oguzie, E. E. Corrosion Inhibition of Aluminium in Acidic and Alkaline Media by Sansevieria Trifasciata Extract. *Corrosion Science*, **2007**, *49*, 1527–1539. doi:10.1016/j.corsci.2006.08.009.

[115] Wan, S.; Chen, H.; Zhang, T.; Liao, B.; Guo, X. Anti-Corrosion Mechanism of Parsley Extract and Synergistic Iodide as Novel Corrosion Inhibitors for Carbon Steel-Q235 in Acidic Medium by Electrochemical, XPS and DFT Methods. *Frontiers in Bioengineering and Biotechnology*, **2021**, *9* (December 2021), 1–15. doi:10.3389/fbioe.2021.815953.

[116] Authors, F. Corrosion Inhibition and Adsorption Behaviour of Ocimum Basilicum Extract on Aluminium. *Pigment & Resin Technology*, **2006**, *35*, 63–70. doi:10.1108/03699420610652340.

[117] Hart Kalada, G.; Orubite- Okorosaye, K.; James Abosede, O. Corrosion Inhibition of Copper in Seawater by Xanthosoma Spp Leaf Extract (XLE). *International Journal of Advanced Research in Chemical Science*, **2016**, *3* (12), 34–40. doi:10.20431/2349-0403.0312005.

[118] Abdallah, Y. M. Zygophllum Coccineum L. Extract as Green Corrosion Inhibitor for Copper in 1 M. *International Journal of Advanced Research*, **2015**, *2* (November 2014), 517–531.

[119] Fouda, A. S.; Mohamed, A. E.; Khalid, M. A. Trigonella Stellate Extract as Corrosion Inhibitor for Copper in 1M Nitric Acid Solution. *Journal of Chemical and Pharmaceutical Research*, **2016**, *8* (2), 86–98.

[120] Song, J. H.; Murphy, R. J.; Narayan, R.; Davies, G. B. H. Biodegradable and Compostable Alternatives to Conventional Plastics. *Philosophical Transactions of the Royal Society B: Biological Sciences*, **2009**, *364* (1526), 2127–2139. doi:10.1098/rstb.2008.0289.

[121] Zakeri, A.; Bahmani, E.; Aghdam, A. S. R. Plant Extracts as Sustainable and Green Corrosion Inhibitors for Protection of Ferrous Metals in Corrosive Media: A Mini Review. *Corrosion Communications*, **2022**, *5*, 25–38. doi:10.1016/j.corcom.2022.03.002.

[122] Miralrio, A.; Vázquez, A. E. Plant Extracts as Green Corrosion Inhibitors for Different Metal Surfaces and Corrosive Media: A Review. *Processes*, **2020**, *8* (8), 942. doi:10.3390/PR8080942.

[123] Altemimi, A.; Lakhssassi, N.; Baharlouei, A.; Watson, D. G.; Lightfoot, D. A. Phytochemicals: Extraction, Isolation, and Identification of Bioactive Compounds from Plant Extracts. *Plants*, **2017**, *6* (4), 42. doi:10.3390/plants6040042.

[124] Chaves, J. O.; de Souza, M. C.; da Silva, L. C.; Lachos-Perez, D.; Torres-Mayanga, P. C.; Machado, A. P. D. F.; Forster-Carneiro, T.; Vázquez-Espinosa, M.; González-de-Peredo, A. V.; Barbero, G. F.; et al. Extraction of Flavonoids From Natural Sources Using Modern Techniques. *Frontiers in Chemistry*, **2020**, *8* (September), 507887. doi:10.3389/fchem.2020.507887.

[125] Kharissova, O. V.; Kharisov, B. I.; González, C. M. O.; Méndez, Y. P.; López, I. Greener Synthesis of Chemical Compounds and Materials. *Royal Society Open Science*, **2019**, *6*, 191378. doi:10.1098/rsos.191378.

[126] Zhang, Q. W.; Lin, L. G.; Ye, W. C. Techniques for Extraction and Isolation of Natural Products: A Comprehensive Review. *Chinese Medicine (United Kingdom)*, **2018**, *13* (1), 1–26. doi:10.1186/s13020-018-0177-x.

[127] ACS. Guidelines for Chemical Laboratory Safety in Secondary Schools. *American Chemical Society*, **2016**, *82*, 35–45.

[128] Safe Work Australia. *Managing Risks of Hazardous Chemicals: Code of Practice.* Canberra: Safe Work Australia, **2012**.

[129] Duistermaat, J. J.; Kolk, J. A. C. Proper Actions. **2000**, *8*, 93–130. doi:10.1007/978-3-642-56936-4_2.

[130] Mrozik, W.; Rajaeifar, M. A.; Heidrich, O.; Christensen, P. Environmental Impacts, Pollution Sources and Pathways of Spent Lithium-Ion Batteries. *Energy and Environmental Science*, **2021**, *14* (12), 6099–6121. doi:10.1039/d1ee00691f.

[131] Aqeel, M.; Jamil, M.; Yusoff, I. Soil Contamination, Risk Assessment and Remediation. *Environmental Risk Assessment of Soil Contamination*, **2014**, 3–56. doi:10.5772/57287.

[132] Khnifira, M.; Boumya, W.; Attarki, J.; Mahsoune, A.; Sadiq, M.; Abdennouri, M.; Kaya, S.; Barka, N. A Combined DFT, Monte Carlo, and MD Simulations of Adsorption Study of Heavy Metals on the Carbon Graphite (111) Surface. *Chemical Physics Impact*, **2022**, *5* (August), 100121. doi:10.1016/j.chphi.2022.100121.

4 Aqueous and Organic Extracts
Advantages and Disadvantages and Their Relative Anticorrosive Performance

Babar Khan, Mohammad Mobin, and Ruby Aslam
Aligarh Muslim University

ABBREVIATIONS

AFM	Atomic force microscopy
EDS/EDX/EDAX	Energy dispersive X-ray spectroscopy
EIS	Electron impedance spectroscopy
FT-IR	Fourier transform-infrared
GC-MS	Gas chromatography-mass spectrometry
HPLC	High performance liquid chromatography
NMR	Nuclear magnetic resonance
PDP	Potentiodynamic polarization
SEM	Scanning electron microscopy
UV-VIS	Ultraviolet-visible
WL	Weight loss
XPS	X-ray photoelectron spectroscopy
XRD	X-ray diffraction

4.1 INTRODUCTION

Metals and alloys are used in various processes due to their appropriate mechanical and electrical properties [1]. Corrosion is the phenomenon that leads to the destruction or deterioration of the metals or alloys on interaction with the corrosive environments. Corrosion occurs due to the electrochemical interaction of the metals with their surrounding corrosive environment. In day-to-day life, various metals and their alloys, e.g., iron and its alloys, are used, especially in building materials [2,3]. The existing environments are highly destructive in nature for the metal and alloys and may cause their failure. Thus, to protect metals and alloys from such failure, they

DOI: 10.1201/9781003394631-4

are treated with certain compounds that will inhibit them from direct contact with the surroundings. These compounds are known as inhibitors. Corrosion inhibitors are substances that when added to the corrosive medium retard the corrosion due to the development of the thin film or layer on the metal surface by the adsorption phenomenon. Green corrosion inhibitors are of two types: inorganic and organic green corrosion inhibitors [4] (Figure 4.1).

The widely used inhibitors are inorganic inhibitors such as chromates [7], nitrites [8], phosphates [9], and molybdates [10] that provide proper inhibition from the corroding surroundings but they are not environment friendly and harm nature in one or the other way. Inorganic green corrosion inhibitors are rare earth metals [11]. Organic corrosion inhibitors may be sub-categories as surfactants [12], biopolymers [13], ionic liquids [14], drugs [15], amino acids [16], and plant extracts [17], and these categories of inhibitors have been extensively used as a green alternative of corrosion inhibitor for various metals and alloys. Due to the strict world's environmental laws, researchers are shifting to green corrosion inhibitors. To eradicate the harmful effects of corrosion inhibitors used in various industries, research is going on the green inhibitors that are obtained from the organic materials from the environment. Although these green inhibitors are not that efficient with respect to the inorganic corrosion inhibitors, they have shown appreciable inhibition efficiency [1,18].

Various methods have been reported and widely used to obtain plant extracts from various plant parts (Figure 4.2).

Various studies have shown that phytochemicals present in plant extracts derived from various parts of plants such as stem, bark, root, leave, seeds, and fruits have better inhibition efficiency at different concentrations [5]. The major phytochemicals responsible for the inhibition property are terpenoids, steroids, alkaloids, flavonoids, polyphenols, glycosides, amino acids, tannins, etc. The adsorption of the phytochemicals onto the metal surface is due to the presence of functional groups such as amino ($-NH_2$), carboxylic (-COOH), hydroxyl (-OH), and acid chloride (-COCl). [18]. There are umpteen number of literature available on the anticorrosive applications

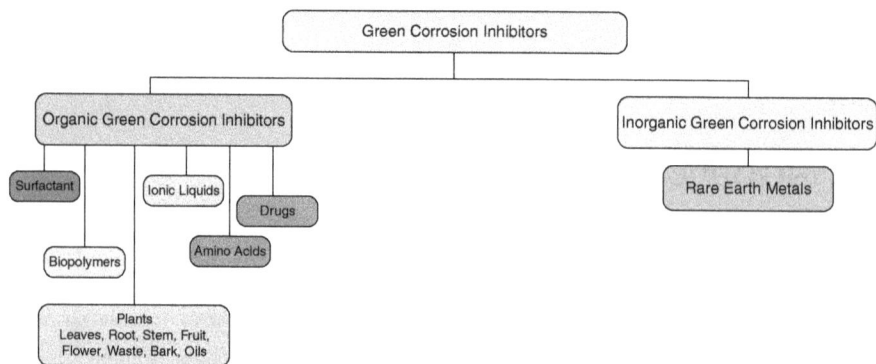

FIGURE 4.1   Green corrosion inhibitors [5,6].

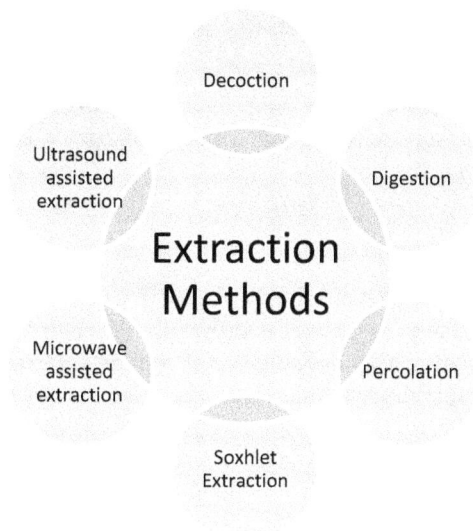

FIGURE 4.2 Methods of extraction [19].

of aqueous and non-aqueous extracts of plants for different metals. The focus of this chapter is to discuss the advantages and disadvantages associated with the use of these categories.

4.2 INHIBITION MECHANISM OF PLANT EXTRACTS

The constituents present in the plant extracts can be adsorbed on the metal surface by either the physical or the chemical phenomenon. The plant extract has phyto-chemical-containing donor sites such as O, N, and ring system that donate electrons to the vacant sites on the metal surface. The first adsorption is of the bipolar water (H_2O) and Cl^- ions. It is presumed that the water molecules from the electrolytes are adsorbed on the metal surface before the addition of the inhibitor. The inhibitor molecules replace water molecules.

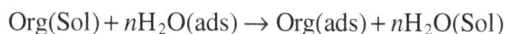

$$Org(Sol) + nH_2O(ads) \rightarrow Org(ads) + nH_2O(Sol)$$

Here, n depends upon the relative cross-sectional area of organic molecules and the water molecules. The physical adsorption is due to the electrostatic interaction between the inhibiting organic ions or dipoles and the electrically charged metal surface. The chemical adsorption is due to the charge transfer from the organic molecules present in the phytochemical to the metal surface to form the coordinate type of the bond. The preferential adsorption of organic molecules takes place because the interaction energy between water molecules and metal surface is unfavourable in comparison with the interaction energy between organic molecules and metal surface.

4.3   PLANT EXTRACT AS CORROSION INHIBITOR: A LITERATURE SURVEY

To extract phytochemicals from the plant parts, various methods are available that can produce either the aqueous extract or the organic extract depending on the type of solvent used for the extraction [20].

4.3.1   AQUEOUS PLANT EXTRACT

The aqueous extracts have shown better inhibition efficiency, and they do not affect the environment in any way as they are eco-friendly. The polar groups present in the aqueous plant extract adsorb better on the metal or alloy to form a protective film on the surface. This section discusses the inhibition efficacies of the various aqueous plant extracts.

Ali Dehghani et al. [21] observed from the FT-IR that the broad peak in the spectrum appears at 3,460 cm^{-1} and is due to the -OH bond stretching, which shows the presence of several -OH groups in the fruit shell extract of *Chinese gooseberry*. The peaks at 2,950 and 1,732 cm^{-1} in the spectrum are due to the presence of C–H and C=O stretching vibrations. The other peaks that are confirmed from the spectrum are the C=C stretching, COOH stretching, and C–N stretching at 1,640, 1,448, and 1,081 cm^{-1}, respectively (Figure 4.3).

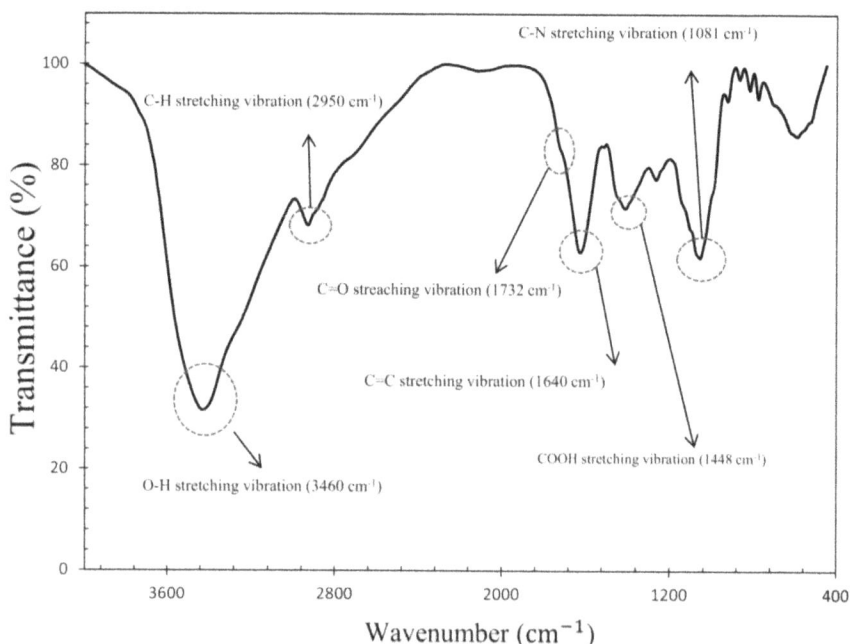

FIGURE 4.3   FT-IR spectrum of *Chinese gooseberry* fruit shell extract powder in the wave number range of 400–4,000 cm^{-1} [21].

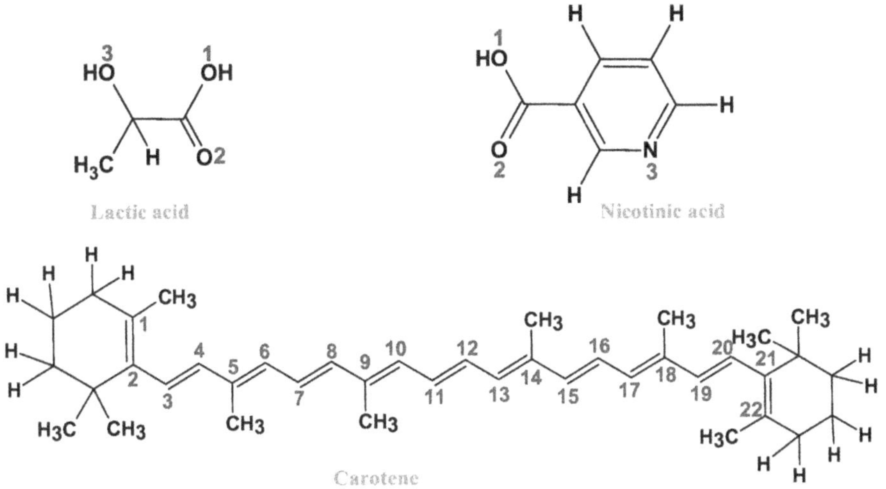

FIGURE 4.4 Chemical structures of neutral form of lactic acid, nicotinic acid, and carotene compounds existing in green corrosion inhibitor.

In another study, Ali Dehghani et al. [22] obtained the aqueous extract of *Borage* flower with components such as lactic acid, nicotinic acid, and carotene (Figure 4.4). According to the results obtained from the EIS technique, when the concentration of the *Borage* flower increased, its inhibition efficiency improved. At 800 ppm, the inhibition efficiency was found to be 91%.

The OCP results showed that the potential value becomes more negative on increasing the concentration of the extract which on reaching 20 minutes attains the stable and more negative value of the potential (Figure 4.5).

Ghasem Bahlaken et al. [23] obtained the aqueous extract of Mustard seed. The authors showed the atomic force microscopy image of the uninhibited and inhibited mild steel surface (Figure 4.6). The uninhibited sample has an average roughness of 119 nm, whereas on inhibition, the average roughness decreases to 42 nm.

Table 4.1 reports the metal, aggressive media, techniques, and inhibition efficacies of different aqueous plant extracts. The data reported showed that most of the studies were carried out in acidic media [24–43]. The reports on alkaline [41] and saline media are scarce [42,43]. These extracts exhibited more than 90% inhibition efficacies in most of the cases. However, the studies showed that the optimum inhibition efficacies are on higher concentrations of extracts.

4.3.2 Advantages and Disadvantages of Using Aqueous Plant Extracts

The polar solvent (water) is used to get the aqueous plant extract containing phytochemicals in the soluble form, whereas the non-polar solvent has phytochemicals in the organic layer. The binding to the surface of the metal is stronger in aqueous extract due to the presence of the polar phytochemicals, whereas in the organic

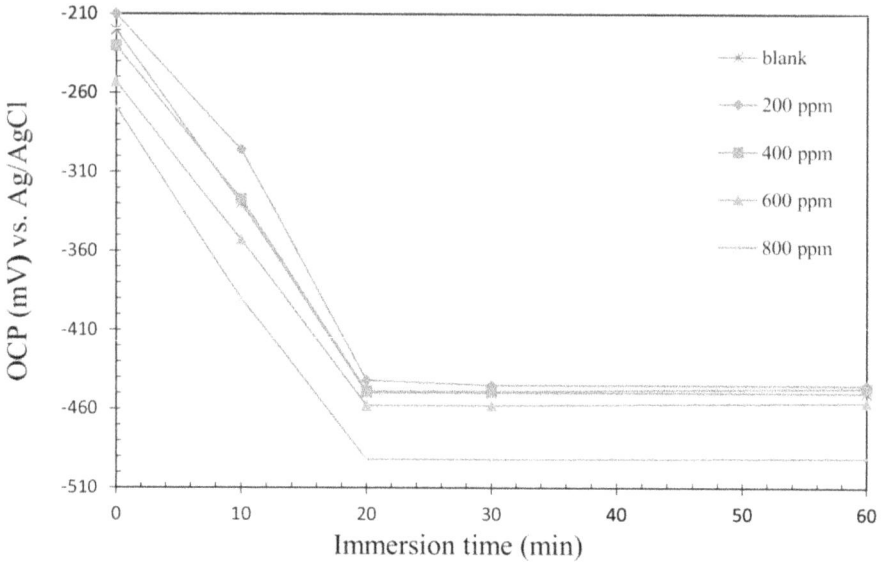

FIGURE 4.5 OCP diagram for different concentrations of *Borage* flower extract for mild steel in 1 M HCl during 1 hour immersion [22].

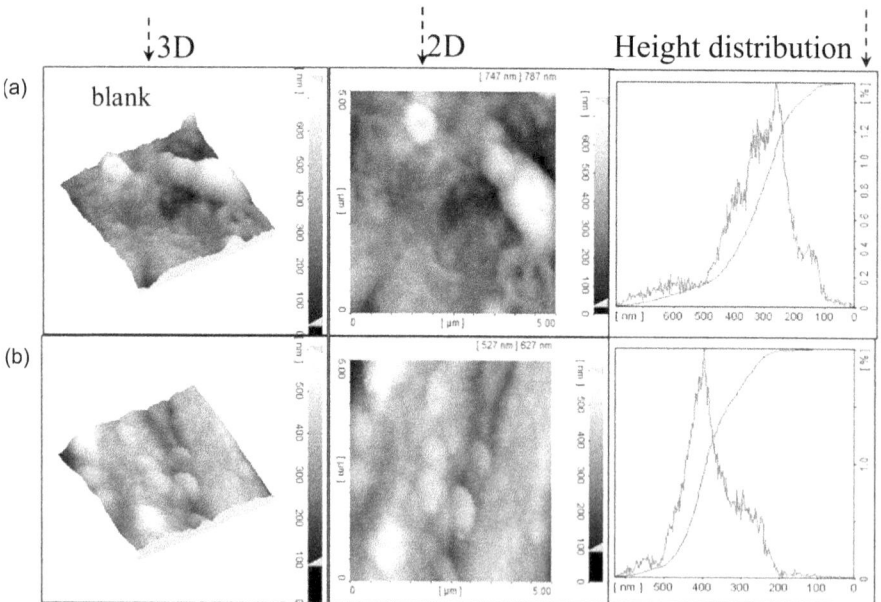

FIGURE 4.6 AFM images of corroded mild steel surface after 5 hours immersion in 1 M HCl solution with (a) 0 and (b) 200 mg·L^{-1} of *Mustard* seed extract [23].

TABLE 4.1

Anticorrosive Performance of the Aqueous Extract of Various Plants in Different Aggressive Media

S. No.	Plant	Corrosive Medium	Metal	Temperature (K)	Concentration	Corrosion Measuring Techniques	Results	References
1.	*Ananas comosus* stem extract	1 M HCl	Low-carbon steel	338	1,000 ppm	WL, PDP, EIS, UV-VIS, SEM	98%	[24]
2.	*Artemisia herba-alba* shrub extract	1 M H_3PO_4	Stainless steel	298	1 g L^{-1}	EIS, SEM-EDS, GC-MS	88%	[25]
3.	*Ziziphora* leaves extract	1 M HCl	Mild steel	–	800 ppm	FT-IR, UV-VIS, EIS	93%	[26]
4.	*Eucalyptus* leaves extract	1 M HCl	Mild steel	–	800 ppm	PDP, WL, EIS, FT-IR, UV-VIS, SEM, AFM	88%	[27]
5.	*Eucalyptus globulus* leaves extract	0.5 M H_2SO_4	Low-carbon steel	–	600 mg L^{-1}	WL, EIS, AFM, FT-IR, SEM	93%	[28]
6.	*Tinospora crispa* stem	1 M HCl	Mild steel	–	800 ppm	WL, EIS, SEM	81%	[29]
7.	*Musa paradisiac* peels (raw, ripen, over ripen) extract	1 M HCl	Mild steel	299	300 mg L^{-1}	WL, FT-IR, UV-VIS, HPLC, EIS, PDP, AFM	90%	[30]
8.	*Hyptis suaveolens* leaves extract	1 M H_2SO_4	Mild steel	–	250 mg L^{-1}	WL, FT-IR, PDP, EIS, SEM	95%	[31]
9.	*Thevetia peruviana* whole plant extract	1 M HCl	Carbon steel	318	300 ppm	WL, PDP, EIS, SEM	90%	[32]
10.	*Prunus dulcis* peels extract	0.1 M HCl	Mild steel	–	240 mg L^{-1}	FT-IR, UV-VIS, PDP, EIS, WL, SEM, EDAX, AFM	85%	[33]

(Continued)

TABLE 4.1 (Continued)
Anticorrosive Performance of the Aqueous Extract of Various Plants in Different Aggressive Media

S. No.	Plant	Corrosive Medium	Metal	Temperature (K)	Concentration	Corrosion Measuring Techniques	Results	References
11.	Brassica oleracea leaves extract	0.5 M H_2SO_4 and 1 M HCl	Q235	–	300 mg L^{-1}	WL, PDP, EIS, FT-IR, UV-VIS, XRD, XPS, SEM, EDS	92% (H_2SO_4) 94% (HCl)	[34]
12.	Garlic skin extract	0.5 M HCl	Aluminium	313	0.0250 g dm^{-3}	WL	–	[35]
13.	Hymenaea stigonocarpa fruit shell extract	0.5 M H_2SO_4	Mild steel	328	1,233.4 mg L^{-1}	PDP, EIS, FT-IR, SEM, EDS	90%	[36]
14.	Pterocarpus santalinoides leaves extract	1 Mol dm^{-3} HCl	Low-carbon steel	333	0.7 g L^{-1}	EIS, PDP, LPR, SEM-EDAX, AFM	90%	[37]
15.	Tithonia diversifolia flower extract	1 M HCl	Mild steel	–	0.7%	WL, EIS, PDP, FT-IR	95%	[38]
16.	Coriandrum sativum L seeds extract	1 M H_3PO_4	Aluminium	303	500 ppm	EIS, PDP, SEM, EDS	73%	[39]
17.	Clove seed extract	1 M HCl	Mild steel	–	800 ppm	FT-IR, UV-VIS, EIS, PDP, SEM, AFM	93%	[40]
18.	Bacopa monnieri/lawsonia stem extract	0.5 M NaOH	Low-carbon steel	–	0.8 g L^{-1}	WL, PDP	80%	[41]
19.	Aloe vera (L) Bum f. (Liliaceae) gel extract	Sea water	Carbon steel	–	4 mL	WL, PDP, EIS, FT-IR, UV-VIS, Fluorescence	98%	[42]
20.	Taraxacum officinale plant extract	Sea water	Carbon steel	328	400 mg L^{-1}	WL, PDP, EIS, SEM, FT-IR, UV-VIS	94%	[43]

extract, this type of interaction with the metal surface is not possible as the available phytochemicals are non-polar [5,44]. To prepare the aqueous extract, several steps are required, and it becomes tedious, whereas the organic extract preparation requires the use of toxic chemicals that will adversely affect the environment.

4.4 NON-AQUEOUS/ORGANIC PLANT EXTRACT

Harb et al. [45] reported the inhibition efficiency of the organic extract of *Olea europaea* L. for mild steel in NaOH (0.1 M) + NaCl (0.5 M). Figure 4.7 shows the different anodic and cathodic polarization curves for mild steel in 0.1 M NaOH + 0.5 M NaCl for the different organic (methanol, hexane, dichloromethane, ethylacetate) extracts of *Olea europaea* L. (Figure 4.7). There is a decrease in the anodic and cathodic current densities for different organic extracts, which shows the inhibition of anodic metal dissolution and cathodic reduction reaction.

Lerini et al. [46] showed the Nyquist plots for the ethanolic extract of *Bagasa guianensis* crude extract in 3% NaCl for zinc having different families of natural products (Figure 4.8). From the Nyquist plot, two capacitive loops are observed in the form of depressed semicircles. The initial loop in the plot is associated with high frequency for the charge transfer, whereas the other loop in the plot is associated with the diffusion for oxygen reduction. These results are given for the various phytochemical profiles along with the crude extract.

The anticorrosive performance of the organic extracts of various plants in an acidic medium is shown in Table 4.2. As compared to aqueous extract as we discussed earlier in Table 4.1, the inhibition efficacies offered by various non-aqueous extracts of plants are less and lie in the range of 70%–90%. Literature showed that there is no report on non-aqueous extracts of plants in other media such as H_2SO_4, NaCl, and NaOH. Moreover, there is no report on the anticorrosive application of non-aqueous extracts of plants for other metals except steel.

4.4.1 Advantages and Disadvantages of Using the Organic Plant Extracts

With organic solvents, the extraction processes are a bit easy, and it will take less time to extract the essential phytochemicals from the plant material. The inhibition efficiency of the organic extract is usually lower with respect to the aqueous extract. This behaviour is because of the reason that the adsorption is weaker due to the non-polar organic constituent present in the organic extract that does not bind efficiently with the metal surface and gives lesser protection from the corrosive environment. Moreover, the organic solvents used for the extraction process are harmful to nature in one or the other way.

4.5 CONCLUSIONS

This chapter deals with the aqueous and organic plant extracts obtained from various methods of extraction and their anticorrosive performance. The phytochemicals present in the plant parts are the essential component to protect the metal from

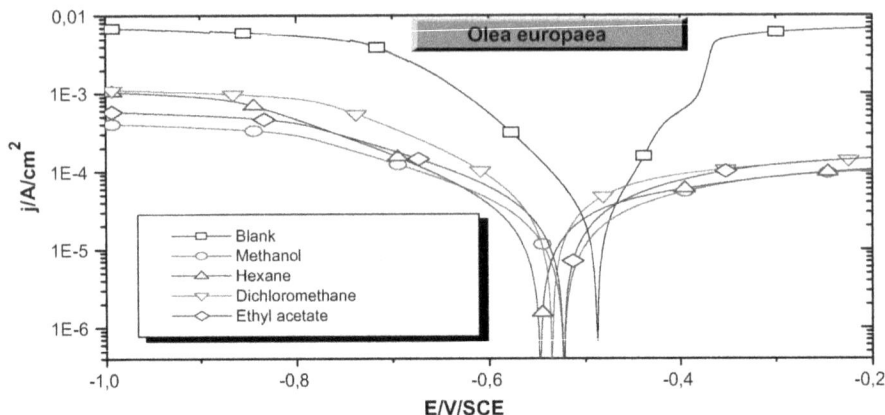

FIGURE 4.7 Polarization curves of mild steel after 1 hour of immersion in NaOH (0.1 M) + NaCl (0.5 M) in the absence and presence of different organic extracts of *Olea europaea* L. [45].

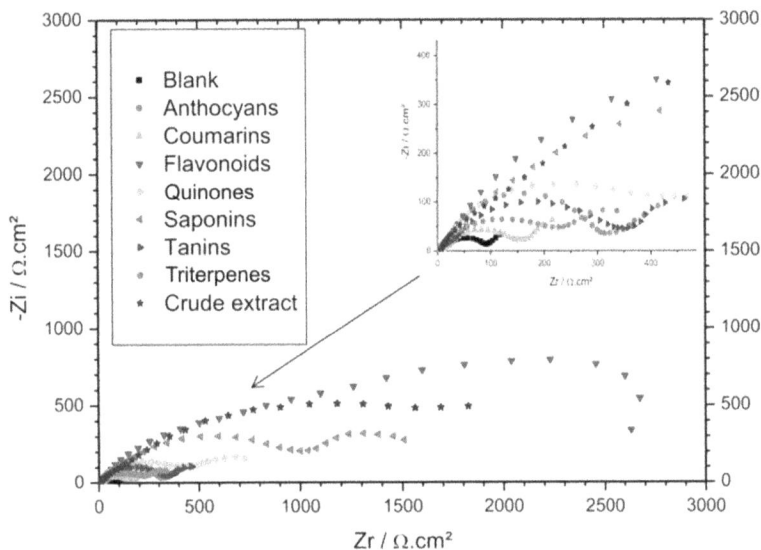

FIGURE 4.8 Nyquist plots for zinc in 3% NaCl medium containing different families and the crude extract [46].

corrosion by the barrier formation onto the metal surface. There are various methods of extraction to extract the phytochemicals from the plant parts such as digestion, decoction, percolation, Soxhlet extraction, microwave-assisted extraction, and ultra-sound-assisted extraction.

The phytochemical profile that inhibits the metal surface consists of terpenoids, alkaloids, polyphenols, glycosides, saponins, tannins, amino acids, flavonoids, and steroids. The functional groups responsible for the inhibition behaviour are the hydroxyl

TABLE 4.2
Anticorrosive Performance of the Various Plants' Organic Extracts in Acidic Medium

S. No.	Plant	Extraction Solvent	Corrosive Medium	Metal	Temperature (K)	Concentration	Corrosion Measuring Techniques	Results	References
1.	Cryptocara nigra bark extract	Hexane, dichloromethane, methanol	1 M HCl	Mild steel	–	1,000 mg L^{-1}	PDP, EIS, SEM-EDS, FT-IR, UV-VIS	52% (Hexane) 91% (CH$_2$Cl$_2$) 87% (CH$_3$OH)	[47]
2.	Ipomoea batatas extract	n-hexane	1 M HCl	Galvanized steel	318	0.7 g L^{-1}	FT-IR, EIS	64%	[48]
3.	Luffa cylindrical leaf extract	Ethanol	1 M HCl	Mild steel	333	1 g L^{-1}	WL, PDP, EIS, FT-IR	88%	[49]
4.	Zingiber officinale root extract	Methanol	1 M HCl	Mild steel	298	200 ppm	PDP, EIS, FT-IR, UV-VIS, AFM	94%	[50]
5.	Papaver sonniferum extract	Ethanol	0.2 M HCl	AISI 304 stainless steel	298	500 ppm	AFM, SEM-EDS	88%	[51]
6.	Dioscorea septemloba rhizomes extract	Ethanol	1 M HCl	Carbon steel	303.15	2.0 g L^{-1}	NMR, EIS, FT-IR, SEM	72%	[52]

(-OH), amino (-OH), carboxylic (-COOH), acid chloride (-COCl), and others. The aqueous extract has a better advantage over the organic extract as they are eco-friendly and shows higher inhibition efficiency due to the presence of polar functional groups that adsorb onto the metal surface. The organic extract takes less time to extract the essential phytochemicals from the plant part, whereas extraction in aqueous is a more tedious and time-consuming process. The organic extract is obtained by the use of organic solvents that are harmful to the environment. Polar functional groups present in the plant parts act as good corrosion inhibitors rather than non-polar functional groups as they bind strongly with the metal surface in the presence of corrosive media.

REFERENCES

[1] Ghahremani, P., Hossein, A., Dehghani, A., Bahlakeh, G., & Ramezanzadeh, B. (2022). Apple pomace extract : a potent renewable source of active biomolecules for suppressing mild steel aggression in aquatic solution. *Biomass Conversion and Biorefinery*, 0123456789. doi:10.1007/s13399-022-03581-z.

[2] Soutis, C. (2005). Carbon fiber reinforced plastics in aircraft construction. *Materials Science and Engineering: A, 412*, 171–176. doi:10.1016/j.msea.2005.08.064.

[3] Wang, M., Zhang, F., Lee, C., & Tang, Y. (2017). Low-cost metallic anode materials for high performance rechargeable batteries. *Advanced Energy Materials, 1700536*, 1–20. doi:10.1002/aenm.201700536.

[4] Alrefaee, S. H., Rhee, K. Y., Verma, C., Quraishi, M. A., & Ebenso, E. E. (2020). Challenges and advantages of using plant extract as inhibitors in modern corrosion inhibition systems: recent advancements. *Journal of Molecular Liquids, 321*, 114666. doi:10.1016/j.molliq.2020.114666.

[5] Zuliana, S., Ha, A., Koriah, S., Ali, M., Taib, A., Abu, A., Najmi, M., Mohamad, M., Mamat, S., Ahmad, S., Ali, A., & Ter, P. (2021). Plant extracts as green corrosion inhibitor for ferrous metal alloys : a review. *Journal of Cleaner Production, 304*, 127030. doi:10.1016/j.jclepro.2021.127030.

[6] Mobin, M., Shoeb, M., Aslam, R., & Banerjee, P. (2022). Synthesis, characterisation and corrosion inhibition assessment of a novel ionic liquid-graphene oxide nanohybrid. *Journal of Molecular Structure, 1262*, 133027. doi:10.1016/j.molstruc.2022.133027.

[7] Kendig, M. W., & Buchheit, R. G. (2003). Corrosion inhibition of aluminum and aluminum alloys by soluble chromates, chromate coatings, and chromate-free coatings. *Corrosion, 59*(5), 379–400. doi:10.5006/1.3277570.

[8] Ann, K. Y., Jung, H. S., Kim, H. S., Kim, S. S., & Moon, H. Y. (2006). Effect of calcium nitrite-based corrosion inhibitor in preventing corrosion of embedded steel in concrete. *Cement and Concrete Research, 36*(3), 530–535. doi:10.1016/j.cemconres.2005.09.003.

[9] Bastidas, D. M., Criado, M., La Iglesia, V. M., Fajardo, S., La Iglesia, A., & Bastidas, J. M. (2013). Comparative study of three sodium phosphates as corrosion inhibitors for steel reinforcements. *Cement and Concrete Composites, 43*, 31–38. doi:10.1016/j.cemconcomp.2013.06.005.

[10] Vukasovich, M. S., & Farr, J. P. G. (1986). Molybdate in corrosion inhibition-a review. *Polyhedron, 5*(1–2), 551–559. doi:10.1016/S0277-5387(00)84963-3.

[11] Hinton, B. R. W. (1992). Corrosion inhibition with rare earth metal salts. *Journal of Alloys and Compounds, 180*(1–2), 15–25. doi:10.1016/0925-8388(92)90359-H.

[12] Mobin, M., Aslam, R., & Aslam, J. (2017). Non toxic biodegradable cationic gemini surfactants as novel corrosion inhibitor for mild steel in hydrochloric acid medium and synergistic effect of sodium salicylate: experimental and theoretical approach. *Materials Chemistry and Physics, 191*, 151–167. doi:10.1016/j.matchemphys.2017.01.037.

[13] Mobin, M., Ahmad, I., Aslam, R., & Basik, M. (2022). Characterization and application of almond gum-silver nanocomposite as an environmentally benign corrosion inhibitor for mild steel in 1 M HCl. *Materials Chemistry and Physics*, *289*, 126491. doi:10.1016/j. matchemphys.2022.126491.

[14] Aslam, R., Mobin, M., Huda, Shoeb, M., Murmu, M., & Banerjee, P. (2021). Proline nitrate ionic liquid as high temperature acid corrosion inhibitor for mild steel: experimental and molecular-level insights. *Journal of Industrial and Engineering Chemistry*, *100*, 333–350. doi:10.1016/j.jiec.2021.05.005.

[15] Gece, G. (2011). Drugs: a review of promising novel corrosion inhibitors. *Corrosion Science*, *53*(12), 3873–3898. doi:10.1016/j.corsci.2011.08.006.

[16] El Ibrahimi, B., Jmiai, A., Bazzi, L., & El Issami, S. (2020). Amino acids and their derivatives as corrosion inhibitors for metals and alloys. *Arabian Journal of Chemistry*, *13*(1), 740–771. doi:10.1016/j.arabjc.2017.07.013.

[17] Miralrio, A., & Espinoza Vázquez, A. (2020). Plant extracts as green corrosion inhibitors for different metal surfaces and corrosive media: a review. *Processes*, *8*(8), 942. doi:10.3390/pr8080942.

[18] Shang, Z., & Zhu, J. (2021). Overview on plant extracts as green corrosion inhibitors in the oil and gas fields. *Journal of Materials Research and Technology*, *15*, 5078–5094. doi:10.1016/j.jmrt.2021.10.095.

[19] Abubakar, A., & Haque, M. (2020). Preparation of medicinal plants: basic extraction and fractionation procedures for experimental purposes. *Journal of Pharmacy and Bioallied Sciences*, *12*(1), 1. doi:10.4103/jpbs.JPBS_175_19.

[20] Alrefaee, S. H., Rhee, K. Y., Verma, C., Quraishi, M. A., & Ebenso, E. E. (2021). Challenges and advantages of using plant extract as inhibitors in modern corrosion inhibition systems: recent advancements. *Journal of Molecular Liquids*, *321*, 114666. doi:10.1016/j.molliq.2020.114666.

[21] Dehghani, A., Bahlakeh, G., & Ramezanzadeh, B. (2019b). A detailed electrochemical/ theoretical exploration of the aqueous Chinese gooseberry fruit shell extract as a green and cheap corrosion inhibitor for mild steel in acidic solution. *Journal of Molecular Liquids*, *282*, 366–384. doi:10.1016/j.molliq.2019.03.011.

[22] Dehghani, A., Bahlakeh, G., Ramezanzadeh, B., & Ramezanzadeh, M. (2019a). Potential of Borage flower aqueous extract as an environmentally sustainable corrosion inhibitor for acid corrosion of mild steel: electrochemical and theoretical studies. *Journal of Molecular Liquids*, *277*, 895–911. doi:10.1016/j.molliq.2019.01.008.

[23] Bahlakeh, G., Dehghani, A., Ramezanzadeh, B., & Ramezanzadeh, M. (2019). Highly effective mild steel corrosion inhibition in 1 M HCl solution by novel green aqueous Mustard seed extract: experimental, electronic-scale DFT and atomic-scale MC/MD explorations. *Journal of Molecular Liquids*, *293*, 111559. doi:10.1016/j. molliq.2019.111559.

[24] Mobin, M., Basik, M., & Aslam, J. (2018). Pineapple stem extract (Bromelain) as an environmental friendly novel corrosion inhibitor for low carbon steel in 1M HCl. *Measurement*, *134*, 595–605. doi:10.1016/j.measurement.2018.11.003.

[25] Boudalia, M., Tabyaoui, M., Bellaouchou, A., & Guenbour, A. (2019). Green approach to corrosion inhibition of stainless steel in phosphoric acid of Artemesia herba albamedium using plant extract. *Integrative Medicine Research*, *8*, x x, 1–11. doi:10.1016/j. jmrt.2019.09.045.

[26] Li, H., Qiang, Y., Zhao, W., & Zhang, S. (2021). A green Brassica oleracea L extract as a novel corrosion inhibitor for Q235 steel in two typical acid media. *Colloids and Surfaces A: Physicochemical and Engineering Aspects*, *616*, 126077. doi:10.1016/j. colsurfa.2020.126077.

[27] Al Mhyawi, S. R. (2014). Corrosion inhibition of aluminum in 0.5 M HCl by garlic aqueous extract. *Oriental Journal of Chemistry*, *30*(2), 541–552. doi:10.13005/ojc/300218.

[28] Policarpi, E. D. B., & Spinelli, A. (2020). Application of Hymenaea stigonocarpa fruit shell extract as eco-friendly corrosion inhibitor for steel in sulfuric acid. *Journal of the Taiwan Institute of Chemical Engineers*, 116, 215–222. doi:10.1016/j.jtice.2020.10.024.

[29] Ahanotu, C. C., Onyeachu, I. B., Solomon, M. M., Chikwe, I. S., Chikwe, O. B., & Eziukwu, C. A. (2020). Pterocarpus santalinoides leaves extract as a sustainable and potent inhibitor for low carbon steel in a simulated pickling medium. *Sustainable Chemistry and Pharmacy*, 15, 100196. doi:10.1016/j.scp.2019.100196.

[30] Divya, P., Subhashini, S., Prithiba, A., & Rajalakshmi, R. (2019). Tithonia diversifolia flower extract as green corrosion inhibitor for mild steel in acid medium. *Materials Today: Proceedings*, 18, 1581–1591. doi:10.1016/j.matpr.2019.05.252.

[31] Prabhu, D., & Rao, P. (2013). Coriandrum sativum L.-a novel green inhibitor for the corrosion inhibition of aluminium in 1.0M phosphoric acid solution. *Journal of Environmental Chemical Engineering*, 1(4), 676–683. doi:10.1016/j.jece.2013.07.004.

[32] Dehghani, A., Bahlakeh, G., Ramezanzadeh, B., & Ramezanzadeh, M. (2019b). Detailed macro-/micro-scale exploration of the excellent active corrosion inhibition of a novel environmentally friendly green inhibitor for carbon steel in acidic environments. *Journal of the Taiwan Institute of Chemical Engineers*, 100, 239–261. doi:10.1016/j.jtice.2019.04.002.

[33] Mansor, K. A., Hussein, A., & Al, J. (2019). The dual effect of stem extract of Brahmi(Bacopamonnieri) and Henna as a green corrosion inhibitor for low carbon steel in 0.5 M NaOH solution. *Case Studies in Construction Materials*, 11, e00300. doi:10.1016/j.cscm.2019.e00300.

[34] Sribharathy, V., Rajendran, S., Rengan, P., & Nagalakshmi, R. (2013). Corrosion inhibition by an aqueous extract of Aloe vera (L.) Burm F. (Liliaceae). *European Chemical Bulletin*, 2(7), 471–476.

[35] Deyab, M. A., & Guibal, E. (2020). Enhancement of corrosion resistance of the cooling systems in desalination plants by green inhibitor. *Scientific Reports*, 10(1), 4812. doi:10.1038/s41598-020-61810-9.

[36] Dehghani, A., Bahlakeh, G., Ramezanzadeh, B., & Ramezanzadeh, M. (2020). Potential role of a novel green eco-friendly inhibitor in corrosion inhibition of mild steel in HCl solution: detailed macro/micro-scale experimental and computational explorations. *Construction and Building Materials*, 245, 118464. doi:10.1016/j.conbuildmat.2020.118464.

[37] Dehghani, A., Bahlakeh, G., & Ramezanzadeh, B. (2019a). Bioelectrochemistry green eucalyptus leaf extract : a potent source of bio-active corrosion inhibitors for mild steel. *Bioelectrochemistry*, 130, 107339. doi:10.1016/j.bioelechem.2019.107339.

[38] Haldhar, R., & Prasad, D. (2020). Corrosion resistance and surface protective performance of waste material of eucalyptus globulus for low carbon steel. *Journal of Bio- and Tribo-Corrosion*, 6, 1–13. doi:10.1007/s40735-020-00340-3.

[39] Hussin, M. H., Kassim, M. J., Razali, N. N., Dahon, N. H., & Nasshorudin, D. (2011). The effect of Tinospora crispa extracts as a natural mild steel corrosion inhibitor in 1 M HCl solution. *Arabian Journal of Chemistry*, 9, S616–S624. doi:10.1016/j.arabjc.2011.07.002.

[40] Ji, G., Anjum, S., Sundaram, S., & Prakash, R. (2015). Musa paradisica peel extract as green corrosion inhibitor for mild steel in HCl solution. *Corrosion Science*, 90, 107–117. doi:10.1016/j.corsci.2014.10.002.

[41] Muthukrishnan, P., Jeyaprabha, B., & Prakash, P. (2014). Mild steel corrosion inhibition by aqueous extract of Hyptis suaveolens leaves. *International Journal of Industrial Chemistry*, 5(1), 5. doi:10.1007/s40090-014-0005-9.

[42] Fouda, A. S., Megahed, H. E., Fouad, N., & Elbahrawi, N. M. (2016). Corrosion inhibition of carbon steel in 1 M hydrochloric acid solution by aqueous extract of Thevetia peruviana. *Journal of Bio- and Tribo-Corrosion*, 2(3), 16. doi:10.1007/s40735-016-0046-z.

[43] Pal, S., Lgaz, H., Tiwari, P., Chung, I.-M., Ji, G., & Prakash, R. (2019). Experimental and theoretical investigation of aqueous and methanolic extracts of Prunus dulcis peels as green corrosion inhibitors of mild steel in aggressive chloride media. *Journal of Molecular Liquids*, *276*, 347–361. doi:10.1016/j.molliq.2018.11.099.

[44] Mobin, J. A. M., & Aslam, H. A. (2022). Corrosion inhibition performance of multi - phytoconstituents from Eucalyptus bark extract on mild steel corrosion in 5% HCl solution. *International Journal of Environmental Science and Technology*, *20*, 2441–2454. doi:10.1007/s13762-022-04152-5.

[45] Ben Harb, M., Abubshait, S., Etteyeb, N., Kamoun, M., & Dhouib, A. (2020). Olive leaf extract as a green corrosion inhibitor of reinforced concrete contaminated with seawater. *Arabian Journal of Chemistry*, *13*(3), 4846–4856. doi:10.1016/j.arabjc.2020.01.016.

[46] Lebrini, M., Suedile, F., Salvin, P., Roos, C., Zarrouk, A., Jama, C., & Bentiss, F. (2020). Bagassa guianensis ethanol extract used as sustainable eco-friendly inhibitor for zinc corrosion in 3% NaCl: electrochemical and XPS studies. *Surfaces and Interfaces*, *20*, 100588. doi:10.1016/j.surfin.2020.100588.

[47] Faiz, M., Zahari, A., Awang, K., & Hussin, H. (2020). Corrosion inhibition on mild steel in 1 M HCl solution by Cryptocarya nigra extracts and three of its constituents (alkaloids). *RSC Advances*, *10*, 6547–6562. doi:10.1039/c9ra05654h.

[48] Anyiam, C. K., Ogbobe, O., Oguzie, E. E., Madufor, I. C., Nwanonenyi, S. C., Onuegbu, G. C., Obasi, H. C., & Chidiebere, M. A. (2020). Corrosion inhibition of galvanized steel in hydrochloric acid medium by a physically modified starch. *SN Applied Sciences*, *2*(4), 520. doi:10.1007/s42452-020-2322-2.

[49] Ogunleye, O. O., Arinkoola, A. O., Eletta, O. A., Agbede, O. O., Osho, Y. A., Morakinyo, A. F., & Hamed, J. O. (2020). Heliyon Green corrosion inhibition and adsorption characteristics of Luffa cylindrica leaf extract on mild steel in hydrochloric acid environment. *Heliyon*, *6*(November 2019), e03205. doi:10.1016/j.heliyon.2020.e03205.

[50] Gadow, H. S. (2017). RSC advances investigation of the corrosion inhibition of carbon steel in hydrochloric acid solution by using ginger roots extract. *RSC Advances*, *7*, 24576–24588. doi:10.1039/C6RA28636D.

[51] Buyuksagis, A., & Dİlek, M. (2019). The use of papaver somniferum L. plant extract as corrosion inhibitor. *Protection of Metals and Physical Chemistry of Surfaces*, *55*(6), 1182–1194. doi:10.1134/S2070205119060042.

[52] Emori, W., Zhang, R.-H., Okafor, P. C., Zheng, X.-W., He, T., Wei, K., Lin, X.-Z., & Cheng, C.-R. (2020). Adsorption and corrosion inhibition performance of multi-phytoconstituents from Dioscorea septemloba on carbon steel in acidic media: characterization, experimental and theoretical studies. *Colloids and Surfaces A: Physicochemical and Engineering Aspects*, *590*, 124534. doi:10.1016/j.colsurfa.2020.124534.

5 Natural Polymers and their Derivatives as Corrosion Inhibitors

Moussa Ouakki, Zakia Aribou, Khadija Dahmani,
Issam Saber, Mouhsine Galai, Bousalham Srhir,
and Mohammed Cherkaoui
Ibn Tofaïl University

Chandrabhan Verma
IITBHU

5.1 INTRODUCTION

A significant issue in the industrial domains is corrosion inhibition of steel [1,2]. Steel has significant industry potential due to its many benefits, including low density, high resistance, and low cost [3,4]. Utilizing inhibitors is a crucial part of protecting metallic substrates against corrosion. Recently, heterocyclic organic compound derivatives were investigated as potential corrosion inhibition for steel in varying aggressive solutions (HCl, H_3PO_4, H_2SO_4, HNO_3, NaCl, etc.) [5,6]. The most employed inhibitors are heterocyclic organic compounds (quinoline, imidazole, benzodiazepine, benzimidazole, thiophene, anionic and cationic surfactant, triazepine, moxifloxacin quinoxaline, etc.) [7–13], epoxy resins [14], polymer composites [15,16], essential oil [17], and natural polymers composition [18–21]. Indeed, some natural polymers like pectin [22], Gum Arabic [23,24], chitosan [25], sodium carboxymethyl cellulose [26], and xanthan gum [27] have already been documented to demonstrate steel's potential to suppress corrosion in HCl solutions; however, the inhibitory effectiveness is still inferior to that of synthetic corrosion inhibitors.

Studies have shown that compounds with heteroatoms (oxygen, nitrogen, sulfur, and phosphorus), double and/or triple bonds, and heterocyclic aromatic rings are more effective at preventing corrosion of steel than other compounds [28,29]. These previous compounds inhibit the corrosion inhibition of steel substrates through the adsorption on steel area [30,31]. For heterocyclic organic inhibitors, previous studies displayed that the corrosion rate increases by adsorption on steel surfaces, and the inhibition efficiency follows order atoms: phosphorus > sulfur > nitrogen > oxygen [32,33]. Based on recent studies, the increases in

DOI: 10.1201/9781003394631-5

concentration of the heterocyclic organic compounds can improve the mechanical, physical, chemical, and barrier properties.

The conventional method involves employing synthetic chemicals, which pollute the environment both during production and during disposal after usage. As a corrosion inhibitor for metals, natural polymers are often non-toxic, inexpensive, widely available, environmentally acceptable, and biodegradable [34]. It is widely utilized in a variety of sectors, including the pharmaceutical, paper, textile, and food preparation industries. However, native starch has certain downsides, including poor stability when utilized at low pH levels and low water solubility [35]. Therefore, interesting alternatives to traditional product development strategies that preserve petrochemical resources and enhance certain starch qualities include physical and/or chemical alterations [36–38].

This potential chapter aims to achieve heterocyclic organic compound derivatives as potential corrosion inhibition for steel in NaCl 3% medium with high anticorrosive performance. The effect of simultaneous use of heterocyclic organic compounds' natural polymer (Starch) on the anticorrosive protection was investigated and evaluated by using polarization curves (PC), electrochemical impedance spectroscopy (EIS), scanning electron microscopy (SEM) coupled with energy-dispersive X-ray spectroscopy (EDS), X-ray diffractometer (XRD), and Fourier transform infrared (FTIR) spectroscopy. In the last step, the electrochemical behavior and corrosion detection of the steel were realized in the presence of heterocyclic organic compound at varying concentrations.

5.2 OPERATING MODE

5.2.1 Materials and Sample Preparation

The aggressive solution used in this study is 3 wt.% NaCl obtained by dissolving an appropriate amount of NaCl powder in distilled water. A range of starch concentrations between 100 and 400 ppm was prepared as well as the control solution.

Samples of mild steel with a nominal chemical composition of 0.17 wt.% C, 0.37 wt.% manganese, 0.20 wt.% Si, 0.03 wt.% S, 0.01 wt.% P, and 99.22 wt.% Fe were used for electrochemical experiments; steel samples were used with an exposed area of 1.0 cm² to the corrosive medium. Before use, the substrates were abraded with different grades of emery papers from 50 to 2,000 grit, rinsed with distilled water, degreased with ethanol, and dried in room temperature. The molecular formula of the examined inhibitor is shown in Schema 5.1.

5.2.2 Electrochemical Conditions

Electrochemical study was realized in a three-electrode electrochemical cell, composed of a mild steel plate as a working electrode (with an exposed area of 1 cm²), a platinum counter-electrode, and a saturated calomel electrode reference. An electrochemical measurement has been done using a potentiostat/galvanostat/PGZ100 controlled by the analysis software VoltaMaster4. Electrochemical experiments were

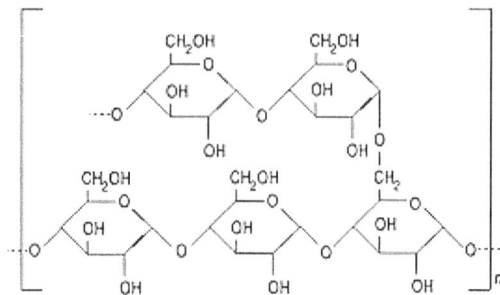

SCHEMA 5.1 Chemical structure of the starch compound.

analyzed using the electrochemical software ORIGIN PRO 6.1. Electrochemical parameters were, then, extracted using the EC-Lab V10.02 software. EIS measurements were done in the range of frequency 100 kHz to 10 mHz at an Open Circuit Potential (OCP) with 10 points per decade. The amplitude of the applied alternative current signal was 10 mV·ms^{-1}. All experiments were performed after 30 minutes immersion of mild steel in NaCl 3% medium in the absence and presence of the different concentrations of the studied inhibitors. Modeling of the impedance plots gives the charge transfer resistance (R_{ct}), the double-layer capacitance (C_{dl}), and other parameters. Inhibition efficiency ($\eta_{imp\,\%}$) is calculated by the following equation:

$$\eta_{imp}\% = \left(R_p - R_p^\circ \,/\, R_p\right) \times 100$$

where R_p° and R_{ct} are the charge transfer resistance in the absence and presence of an inhibitor, respectively.

EFM was done using a base frequency of 0.1 Hz, perturbations of 0.2 and 0.5 Hz, and sine amplitude of 10 mV. Eight cycles of sinusoidal wave form were applied for each test. All EFM measurements were performed in triplicate on parallel samples for reproductivity, and the outcomes were fitted by the in-built Gamry Echem Analyst.

The potentiodynamic polarization studies were performed on mild steel specimens by automatically changing the electrode potential from −1,200 to 300 mV ECS^{-1} versus OCP at a scan rate of 1 mV s^{-1}. The corrosion current density (i_{corr}) was calculated by extrapolating the linear Tafel segments of anodic and cathodic curves back to the corrosion potential. The inhibition efficiency was determined by using the following equation:

$$\eta_{pp}\% = \left[\left(i_{corr}^\circ - i_{corr}\right) \Big/ i_{corr}^\circ \right] \times 100$$

where i_{corr}° and i_{corr} are the corrosion current density values in the absence and presence of an inhibitor, respectively.

5.2.3 Surface and Solution Analysis

- **SEM/EDX**

    The determination of the nature of the formed film on the surface of the metal, exposed to NaCl 3% medium during 24 hours in the absence and presence of the studied inhibitor, was realized by SEM coupled with EDX analyses. These analyses were performed in UATRS-CNRST-Rabat.

- **XRD analysis**

    After 24 hours of immersion in both uninhibited and inhibited solutions, the product generated on the steel surface of the metal specimens was examined using XRD. The XRD X' PERT PRO MPD model was used to examine the surfaces of the M-steel samples. With a step size of 0.0017 and a scan rate of 7° per minute, data are captured over the range of (2) 0° to 90°.

5.3 RESULTS

5.3.1 Electrochemical Studies

5.3.1.1 Open Circuit Potential (OCP) Measurements

OCP and its evolution with time in the investigated acidic medium before and after adding various concentrations of **starch** compound after ½ hour immersion at 298 K are displayed in Figure 5.1. It can be seen that the existence of the investigated specie leads to a slight displacement of the E_{corr} toward more positive potentials compared to the blank solution.

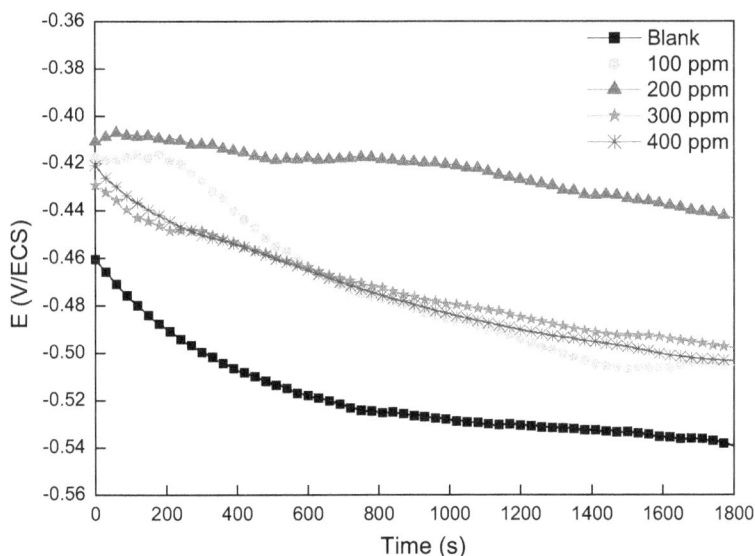

FIGURE 5.1   Open circuit potential (OCP) and its evolution with time in NaCl 3% medium before and after adding various concentrations of starch compound at 298 K.

5.3.1.2 Electrochemical Frequency Modulation (EFM)

A nondestructive corrosion measuring method called electrochemical frequency modulation (EFM) can immediately provide values of the corrosion current without the need for prior knowledge of Tafel constants. It is a small signal Ac method, similar to EIS. However, in contrast to EIS, two sine waves (of various frequencies) are simultaneously applied to the cell. The system reacts to the potential stimulation in a non-linear manner because current is a non-linear function of potential. In addition to the input frequencies, the current response also includes frequency components that are multiples, the sum, and the difference of the two input frequencies. Not at random may the two frequencies be selected. The base frequency that establishes the experiment's duration must be a tiny, integer multiple of both of them. When the two input frequencies are 2 and 5 Hz, samples of the waveform are shown in Figure 5.2.

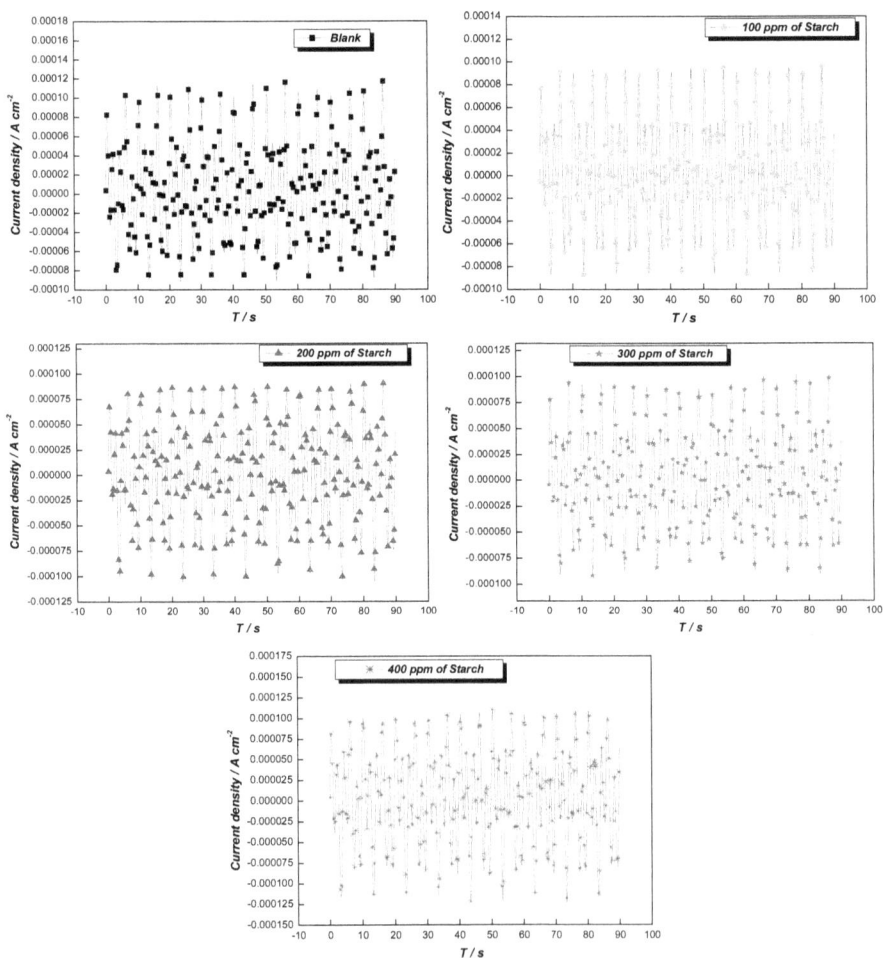

FIGURE 5.2 Waveform spectrum for the input frequencies 2 and 5 Hz for mild steel in NaCl 3% in the absence and presence of various concentrations of starch.

The greater frequency must be at least twice as great as the lesser frequency. Furthermore, the higher frequency needs to be slow enough so that charging of the double layer does not affect the current response. Usually, a good upper limit is 10 Hz. Figure 5.3 displays the intermodulation spectra derived from EFM observations. Every spectrum represents a current response in terms of frequency. The two prominent peaks, each with an amplitude of roughly 200 µA, are the result of excitation at frequencies of 2 and 5 Hz. The harmonics, sums, and differences of the two-excitation frequencies are represented by those peaks between 1 and 20 µA. The EFM140 software program uses these peaks to compute the corrosion current and the Tafel constants. It is significant to note that the current response is relatively

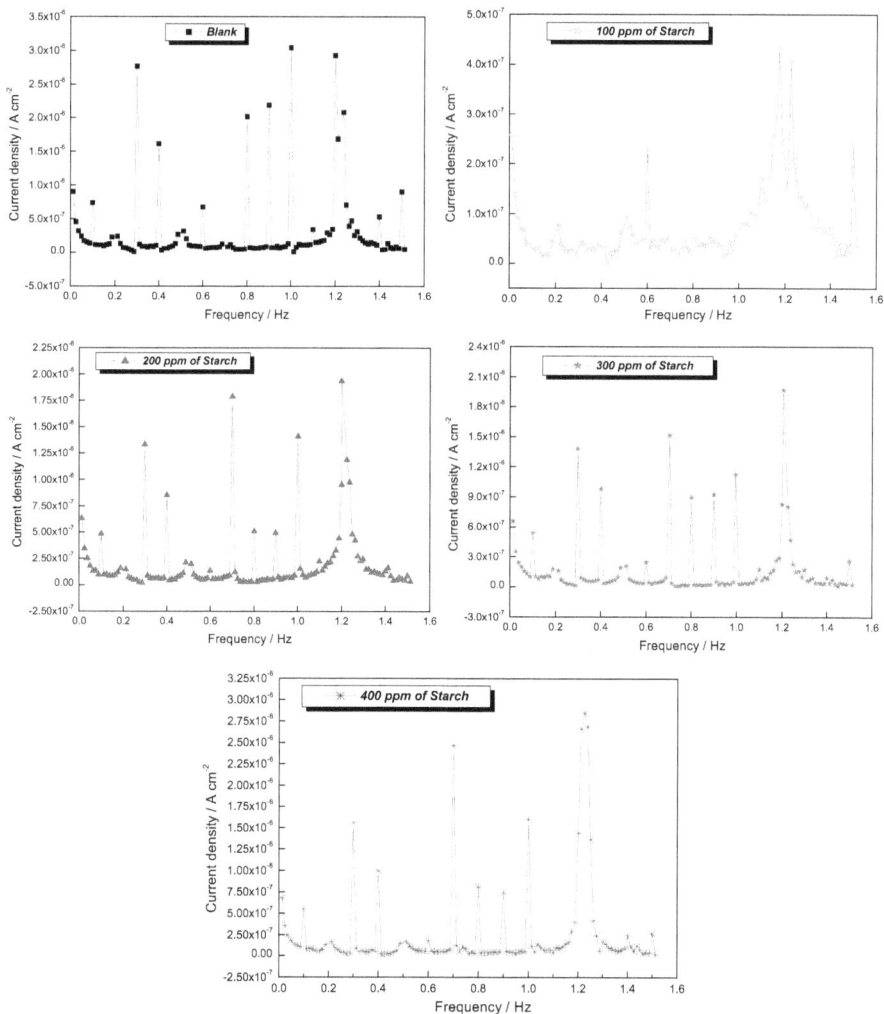

FIGURE 5.3 Intermodulation spectrum for mild steel in NaCl 3% in the absence and presence of various concentrations of starch.

tiny in the intervals between the peaks. For instance, there is a nearly no response (<100 nA) at 4.5 Hz; the peak's frequencies and amplitudes are not coincidental. They follow directly from the EFM theory [39–41].

Current responses for harmonical and intermodulation current peaks are assigned in the spectra (Figure 5.3). The larger peaks were utilized to compute the Tafel slopes (β_c and β_a), the causality factors (CF-2 and CF-3), and the corrosion current density (i_{corr}).

Table 5.1 displays the corrosion kinetic parameters at various concentrations of starch in NaCl 3% at 298 K, including inhibition efficiency ($E_{EFM}\%$), corrosion current density ($\mu A\ cm^{-2}$), Tafel constants (β_a, β_c), and causality factors (CF-2, CF-3).

The following equations were used to determine the corrosion kinetic parameters indicated in Table 5.1 using EFM techniques:

$$i_{corr} = \frac{i_w^2}{\sqrt{48\left(2i_w i_{3w} - i_{2w}\right)}}$$

$$\beta_a = \frac{i_w U_0}{2i_{2w} + 2\sqrt{3}\sqrt{2i_{3w}i_w - i_{2w}^2}}$$

$$\beta_c = \frac{i_w U_0}{2\sqrt{3}\sqrt{2i_{3w}i_w - i_{2w}^2} - 2i_{2w}}$$

$$\text{Causality factor }(2) = \frac{i_{w2 \pm w_1}}{i_{2w_1}}$$

$$\text{Causality factor }(3) = \frac{i_{2w_2 \pm w_1}}{i_{3w_1}}$$

where

U_0 is the amplitude of the sine wave distortion.

i is the instantaneous current density at the working electrode measured at frequency ω.

TABLE 5.1
Electrochemical Kinetic Parameters Obtained by EFM Technique for Mild Steel in the Absence and Presence of Starch in NaCl 3% at 298 K

	Conc. (ppm)	i_{corr} (μA cm^{-2})	β_c (mV dec^{-1})	β_a (mV dec^{-1})	CR (mpy)	$E_{EFM}\%$	C.F-2	C.F-3
NaCl 3%	–	152.87	103.1	134.0	79.59	–	2.5	3.4
Starch	100	30.59	91.34	128.6	18.89	80.0	2.0	3.7
	200	18.92	113.5	135.8	12.15	87.6	1.7	3.5
	300	24.56	117.2	131.1	14.67	83.9	1.8	3.1
	400	26.76	122.3	127.9	15.90	82.5	1.5	3.8

The inhibition efficiency ($E_{EFM}\%$) shown in Table 5.1 was obtained using the following equation:

$$E_{EFM}\% = \left(1 - \frac{i_{corr}}{i_{corr}^{\circ}}\right) \times 100$$

where i_{corr}° and i_{corr} are corrosion current densities in the absence and presence of the starch, respectively.

Table 5.1 makes clear that corrosion current densities fall as starch concentrations rise. The inhibition efficiencies rise as the starch concentration rises, reaching a maximum of 87.6% at 200 ppm. According to the EFM theory [40,42], the causality factors in Table 5.1 are extremely close to theoretical values, which should ensure the accuracy of the table slopes and corrosion current densities. The causality elements, which work as an internal review of the EFM measurement's validity, are the EFM's greatest strength [41,43,44].

5.3.1.3 Polarization Studies

Potentiodynamic polarization technique is one of the electrochemical techniques that provides information on the kinetics of the corrosion process at the metal-solution interface [45,46]. The PC of mild steel in the NaCl 3% solution in the absence and presence of different concentrations of **starch** at 298 K are depicted in Figure 5.4. It can be noticed from Figure 5.4. According to the shape of the curves obtained, we found that in the cathodic part, the curves are divided into three distinct zones [47].

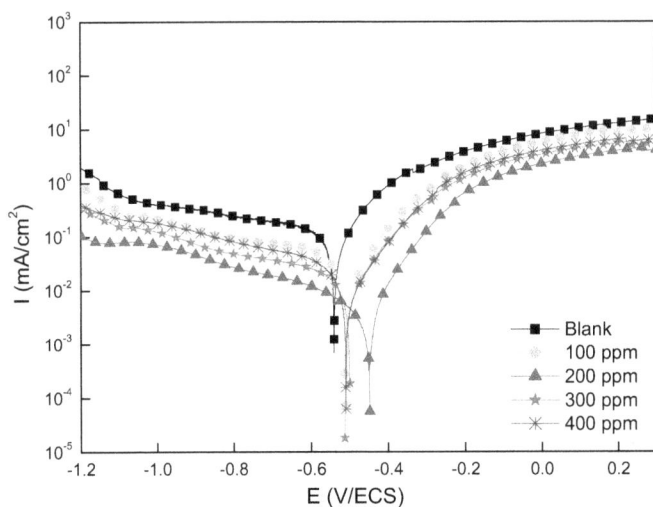

FIGURE 5.4 Potentiodynamic polarization plots of mild steel in the NaCl 3% solution in the absence and presence of starch at 298 K.

A first zone begins between the corrosion potential (E_{corr}) and the potential at the start of the plateau, which corresponds to the oxygen reduction reaction according to the equation:

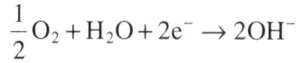

$$\frac{1}{2}O_2 + H_2O + 2e^- \rightarrow 2OH^-$$

A second zone begins at the start of plateau potential down to about $-1,000\,mV$ related to the oxygen diffusion plateau, and a third zone begins at the most negative potentials characterized by the reduction of water:

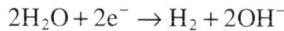

$$2H_2O + 2e^- \rightarrow H_2 + 2OH^-$$

However, the addition of starch leads to a reduction of both anodic and cathodic current densities. The decrease of the two partial anodic and cathodic currents and the large shift of the corrosion potential values toward anodic values confirm the mixed character of this inhibitor. Therefore, they can be categorized as mixed inhibitor with a slight anodic tendency to reduce the anodic dissolution of steel according to the following reaction [48]:

$$Fe \rightleftharpoons Fe^{2+} + 2e^-$$

The values of corrosion current densities (i_{corr}), corrosion potential (E_{corr}), cathodic Tafel slopes (β_c), and inhibition efficiency $\eta_{pp}\%$ in the investigated concentration range of starch in NaCl 3% solution at 298 K are presented in Table 5.2. According to Table 5.2, it may be observed that the value of i_{corr} reduces significantly with reducing the concentration of starch compound tested, achieving a minimum value of 15.3 $\mu A\,cm^{-2}$ for starch at 200 ppm. On the contrary, the inhibition effectiveness increases with inhibitor concentration to reach a maximum value of 96.7%.

The creation of a barrier film on mild steel surfaces, which tends to grow more compact and stable as the concentration of this substance increases, can be used to explain these outcomes [49,50]. Additionally, the addition of the investigated compound causes only minor changes in the cathodic slope of Tafel (β_c), indicating that

---

TABLE 5.2

Electrochemical Parameters Derived from the Tafel Plots of Mild Steel in NaCl 3% Solution before and after Adding Starch at 298 K

	Conc. (ppm)	$-E_{corr}$ (mV ECS^{-1})	i_{corr} ($\mu A\,cm^{-2}$)	$-\beta_c$ (mV dec^{-1})	β_a (mV dec^{-1})	$\eta_{pp}\%$
NaCl 3%	–	538	464	141	149	–
Starch	100	507	38.7	130	136	91.6
	200	444	15.3	129	130	96.7
	300	497	30.2	133	135	93.5
	400	505	34.0	135	138	92.6

the adsorption of this molecule onto the surface of mild steel blocks the active centers responsible for the corrosion process without altering the mechanism of the cathodic reactions. The presence of **starch** caused a change in the values of E_{corr} compared to the blank solution. According to the literature [51], if the shift of E_{corr} with inhibitor is higher than ± 85 mV compared to the uninhibited case, the inhibitor may be considered as a cathodic or anodic type inhibitor. On the contrary, if the E_{corr} variation is less than ± 85 mV, the corrosion inhibitor can be considered as a mixed-type inhibitor. In this investigation, the potential variation does not exceed $+85$ mV, which indicates that these organic products are mixed-type inhibitors with a predominant inhibition influence on the anodic reaction.

5.3.1.4 EIS Measurements

To understand the corrosion and protection mechanisms that occur on the surface of the metal in the acidic medium, in the absence and presence of **starch** at different concentrations, we have drawn electrochemical impedance diagrams with corrosion potential. The Nyquist and Bode Modulus plots for mild steel in NaCl 3% medium at 298 K in the absence and presence of various concentrations of **starch** inhibitor are given in Figures 5.5 and 5.6, respectively. These diagrams were obtained after 1/2 hour of immersion in NaCl 3% medium at 298 K.

We observed from these Nyquist diagrams (Figure 5.5) that the curves are composed of two capacitive loops: one high-frequency loop typically associated with the formation of a film on the carbon steel surface from the inhibitor molecule, iron oxides, and hydroxides, and the other low-frequency loop associated with the phenomenon of

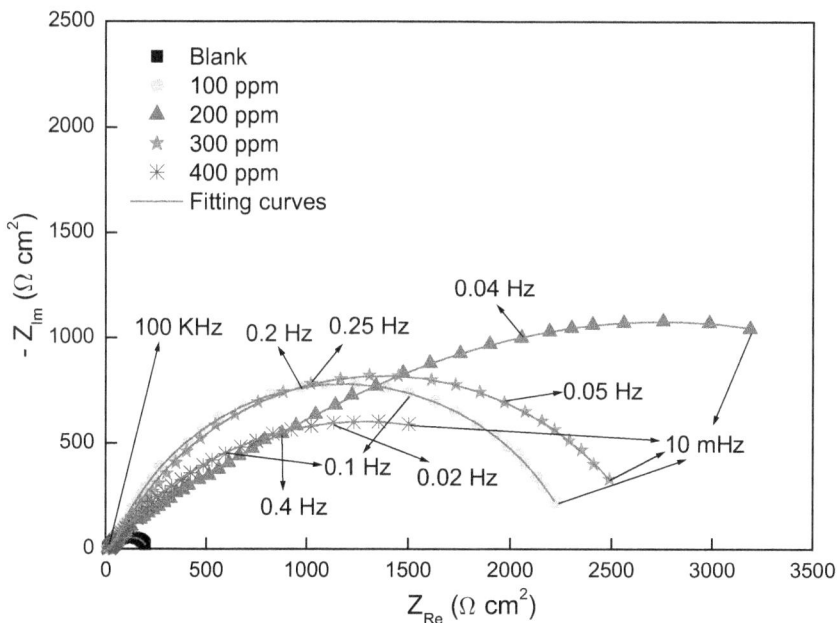

FIGURE 5.5 Nyquist impedance diagrams for mild steel obtained at 298 K in NaCl 3% medium containing different concentrations of starch.

charge transfer [47,52]. Additionally, as the diagrams virtually always have the same shape across all testing, the corrosion process has not been altered. In contrast to the blank, we noticed that the loop diameter reaches a maximum at 200 ppm for **starch**.

Figure 5.6 displays the Bode plots. The better performance of starch is indicated by the rise in absolute impedance (/Z/) at a lower frequency with increasing concentrations of starch [53]. Furthermore, the phase angle of Bode plots increases with the presence of starch. Indeed, two time constants are characterized by the existence of two relaxation processes at the surface of the mild steel due to the adsorption of starch [54,55].

Parameters that measure the electrochemical kinetics of the corrosion reaction were derived by fitting and simulating the impedance spectra using the simple equivalent circuit of the form shown in Figure 5.7. This circuit is composed of the following: R_s is the electrolyte resistance, R_{ct} is the charge transfer resistance, R_f describes the film resistance, and Q_f and Q_{ct} represent the constant phase elements used to replace film capacitance (C_f) and double-layer capacitance (C_{dl}), respectively [56,57].

The admittance (Y_{CPE}) and the impedance (Z_{CPE}) of the Constant phase element (CPE) can be represented by the following relationship:

$$Y_{CPE} = Y_0 (jw)^n$$

The C_{dl} related to the CPE parameters according to the following equation:

$$C_{dl} = \left(QR_p^{1-n}\right)^{1/n}$$

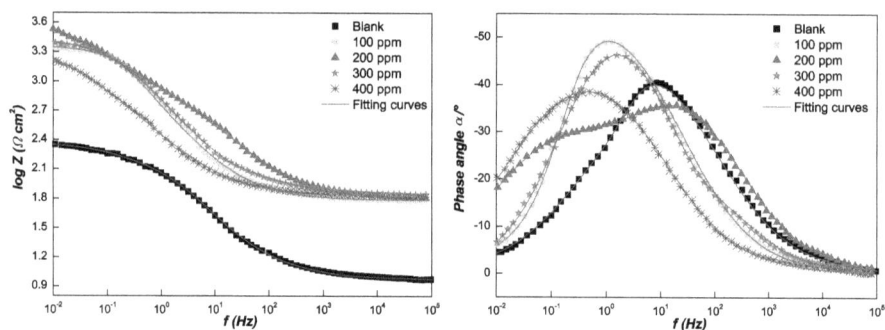

FIGURE 5.6   Bode plots of mild steel in NaCl 3% solution in the absence and presence of different concentrations of starch at 298 K.

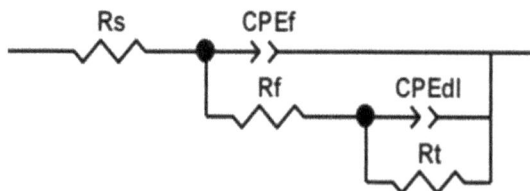

FIGURE 5.7   Equivalent circuit compatible with experimental impedance data.

where Y_0, ω, and j are, respectively, the CPE constant, the angular frequency, the imaginary number (i.e., $j^2 = -1$), and n is the phase shift (exponent) that is bound to the degree of surface inhomogeneity.

The electrochemical parameters of the impedance spectroscopy are given in Table 5.3. The values of the polarization resistance R_p of all systems are calculated using the following equation:

$$R_p = R_f + R_{tc}$$

Table 5.3 shows that the inhibitory efficacy and the charge transfer resistance (R_{tc}) reach a maximum value for the 200 ppm concentration of starch. Indeed, the capacity of the double layer reaches a minimum at the same concentration [58,59]. Moreover, the addition of 200 ppm starch also favored the increase of the film resistance R_f, with a decrease of Q_f. These evolutions indicate a thickening of the film formed and a decrease in permeability through it [60]. Consequently, the decrease in Q_f values shows that the adsorption layer formed by starch on the metal surface is stable and thick [61,62]. In the presence of 200 ppm of starch, we notice that the diameter of R_p has obviously increased from 199 to 5,095 Ω cm^2, which implies better protection of the mild steel by our starch in a corrosive environment.

We found that the "n_{ct} & n_f" index values were greater than the blank, which shows that the inhomogeneity of the steel surface decreases in the presence of starch due to the formation of an organic film on the steel surface [63,64], and the values of phase shift (n) in the presence of inhibitors are very close to unity, which suggests that electrical double layer generated due to the adsorption of starch tested at the metal-electrolyte interfaces behaves as a pseudo-capacitor type [65,66].

5.3.2 EFFECT OF TEMPERATURE AND CORROSION PROCESS ACTIVATION PARAMETERS

It is known that the temperature can modify the mechanism and action mode of an inhibitor on a metallic surface [67]. Therefore, in order to evaluate the temperature influence on inhibitor performance, the potentiodynamic PC were carried out for mild steel studied in NaCl 3% solution without and with 200 ppm of starch at the

TABLE 5.3
Electrochemical Parameters of Impedance and Corresponding Inhibition Efficiency for the Mild Steel Obtained at 298 K in NaCl 3% Medium with and without the Addition of Different Concentrations of Starch

	Conc. (ppm)	R_s (Ω cm^2)	R_f (Ω cm^2)	n_f	Q (μFS^{n-1})	R_{ct} (Ω cm^2)	n_{ct}	Q (μFS^{n-1})	R_p (Ω cm^2)	η_{imp} (%)
NaCl 3%	–	10.2	38	0.705	1,399	161.0	0.685	1,451	199.0	–
Starch	100	10.5	70	0.731	220	2,258	0.783	219	2,328	91.4
	200	14.3	995	0.602	120	4,100	0.632	132	5,095	96.0
	300	9.6	83	0.664	218	2,630	0.736	194	2,713	92.7
	400	16.2	140	0.649	203	2,425	0.901	210	2,565	92.2

temperature range from 298 to 328 K. The obtained results are shown in Figure 5.8, and their extracted parameters are provided in Table 5.4.

It is remarked that these curves exhibited the Tafel regions. It is noted also that the anodic and cathodic branches increased with increasing temperature.

The data in Table 5.4 show that current densities increase with increasing temperature, and hence, a decrease in inhibitory efficiency for the inhibitor has a concentration of 200 ppm in NaCl 3% medium. In addition, it can be noted that the inhibition efficiency decreases slightly in the presence of the inhibitor, so that the inhibitor remains effective against corrosion of the steel in NaCl 3% medium.

The thermodynamic parameters, specifically activation energy (E_a), entropy of activation (ΔS_a), and enthalpy of activation (ΔH_a) for corrosion reaction at 200 ppm for **starch**, were calculated from Arrhenius and transition state plot:

$$i_{corr} = A e^{\frac{-Ea}{RT}}$$

$$\ln\left(\frac{i_{corr}}{T}\right) = \left[\ln\left(\frac{R}{hN_a}\right) + \left(\frac{\Delta S_a}{R}\right)\right] - \frac{\Delta H_a}{RT}$$

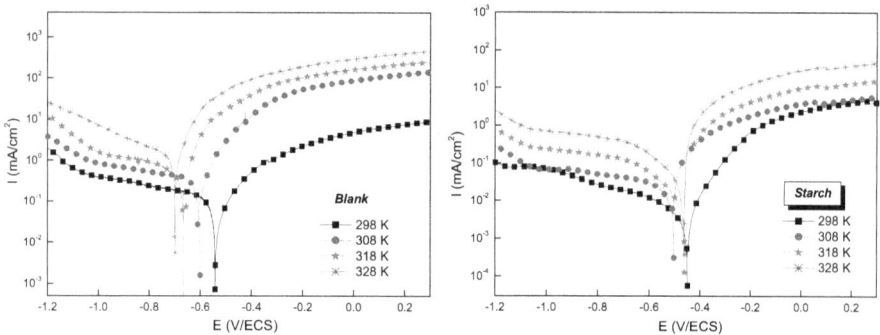

FIGURE 5.8 Effect of temperature on the polarization plots of mild steel in NaCl 3% medium for a concentration of 200 ppm of starch.

TABLE 5.4
Electrochemical and Activation Parameters, E_a, ΔH_a, and ΔS_a, of the Dissolution of Metal in NaCl 3% Medium with and without Protection

Compounds	Tempe (K)	$-E_{corr}$ (mV ECS^{-1})	i_{corr} (µA cm^{-2})	$-\beta_c$ (mV dec^{-1})	β_a (mV dec^{-1})	η_{PP}%	E_a (KJ mol^{-1})	ΔH_a (KJ mol^{-1})	ΔS_a (J mol·K^{-1})
Blank	298	538	464	141	149	–	42.3	39.7	−60.8
	308	594	758	151	154	–			
	318	664	1,353	160	158	–			
	328	696	2,170	165	161	–			
Starch	298	444	15.3	129	130	96.7	74.0	71.4	17.7
	308	498	42.9	147	150	94.3			
	318	457	109.3	155	152	91.9			
	328	458	232.4	161	156	89.3			

where R, T, A, N, and h are universal gas constant, absolute temperature, pre-exponential factor, Avogadro number, and Planck constant, respectively. The i_{corr} values were obtained from the extrapolation of Tafel plot at different temperatures with and without adding **starch** molecules. Here, i_{corr} values consider as a corrosion rate. From the Arrhenius plots, $\ln(i_{corr})$ against $1,000/T$ at an optimum concentration of **starch** is displayed in Figure 5.9.

According to Figure 5.9, a straight-line curve having a slope equal to $-E_a/RT$ and E_a was calculated from this slope. Another plot of $\ln(i_{corr}/T)$ vs. $1,000/T$ shows a straight-line curve (presented in Figure 5.9) with a slope and an intercept that are equal to $-\Delta H_a/R$ and $\ln(R/Nh) + \Delta S_a/R$, respectively.

These results enabled us to calculate E_a, ΔH_a, and ΔS_a of the mild steel in pickling medium without and with the addition of **starch** compound. The values of the thermodynamic parameters relating to this inhibitor derived from Figure 5.9 are provided in Table 5.4.

From the results in Table 5.4, we find that E_a values in the presence of the inhibitor are higher than that in the solution without the inhibitor. This increase in E_a indicates that the energy barrier of the reaction of corrosion increases in the presence of an inhibitor without changing the mechanism of dissolution [68,69]. However, the adsorption phenomenon of an organic molecule is not considered only as a chemical or physical adsorption product, but a wide range of conditions, ranging from the dominance of chemisorption or the electrostatic effects, may occur due to the nature of the complexity of the inhibition of corrosion process [70]. The positive signs of the enthalpy ΔH_a reflected the endothermic nature of the steel dissolution process [71,72]. The value of activation entropies (ΔS_a) increases and is positive in the presence of **starch** meaning that an increase in the disorder during the transformation of the reagents into activated complex [73,74].

5.3.3 SURFACE ANALYSIS

5.3.3.1 MEB/EDS

The morphology of the steel surface before and after 24 hours of immersion in the NaCl 3% medium in the absence and in the presence of a concentration of 200 ppm

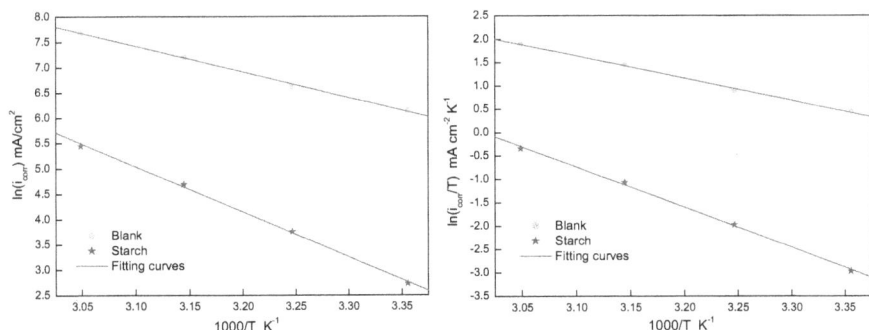

FIGURE 5.9 Arrhenius plots for mild steel corrosion in NaCl 3% medium in the absence and presence of 200 ppm of *starch*.

of **starch** inhibitor is represented in Figure 5.10. It is observed that the surface of the steel in the absence of inhibitor (Blank) is strongly corroded [75,76]. In the presence of an inhibitor in the corrosive medium, we notice a much less attacked surface. The contrasting color on the surface of the steel indicates the formation of a protective film of inhibitory **starch** molecule [77,78].

In order to get information about the composition of the surface of the mild steel before and after the formation of the protective film, energy-dispersive X-ray analysis (EDX) technique was recorded for mild steel samples immersed 24 hours in NaCl 3% medium without inhibitor (blank) and with 200 ppm of **starch**. The results of EDX spectra are shown in Figure 5.11. It can be seen that iron oxide is introduced

FIGURE 5.10   SEM micrographs of both uninhibited (blank) and inhibited MS (after the addition of 200 ppm of *starch*).

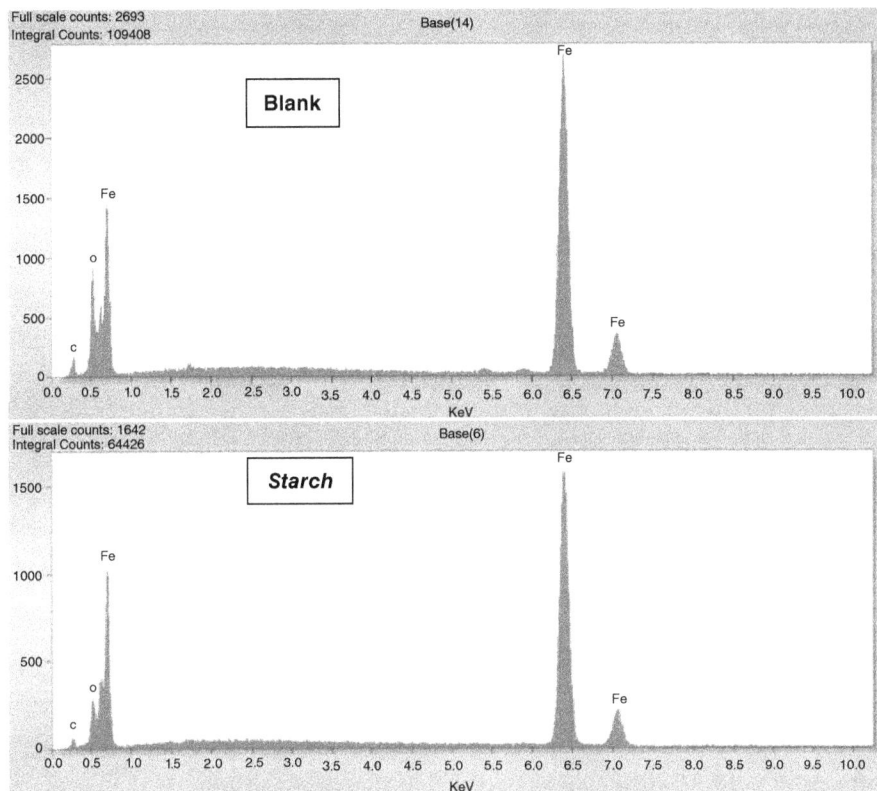

FIGURE 5.11   EDS spectra of the mild steel immersed in blank solution and inhibited solutions.

through reactions caused by the corrosion process; this is proven by the appearance of the oxygen peak in the mild steel EDX spectrum in the blank. After 24 hours of immersion, we notice a decrease in the oxygen peak in the spectra of **starch** if we compare it with that of the blank. This result confirmed the adsorption of this molecule and the formation of a protective film on the mild steel surface.

5.3.3.2   X-Ray Diffraction Analysis

The chemical composition of corrosion products generated over the metal surface in the absence and presence of the studied substance is determined using XRD patterns in this study. Figure 5.12 shows the XRD patterns of the corrosion products generated on the surface of mild steel immersed in the NaCl 3% media before and after adding starch. A modest peak near $2\theta = 29°$ consistent with magnetite (Fe_3O_4) was seen in the XRD study of the corrosion surface created in a NaCl 3% solution, in addition to diffraction peaks for iron and NaCl (Figure 5.12). The observation of iron oxide in corrosion scale generated in deaerated brines by several authors [79,80] and the potential presence of magnetite are consistent. The inclusion of starch slows the production of this oxide by forming a barrier layer that protects the surface from corrosive medium corrosive affect [81–83].

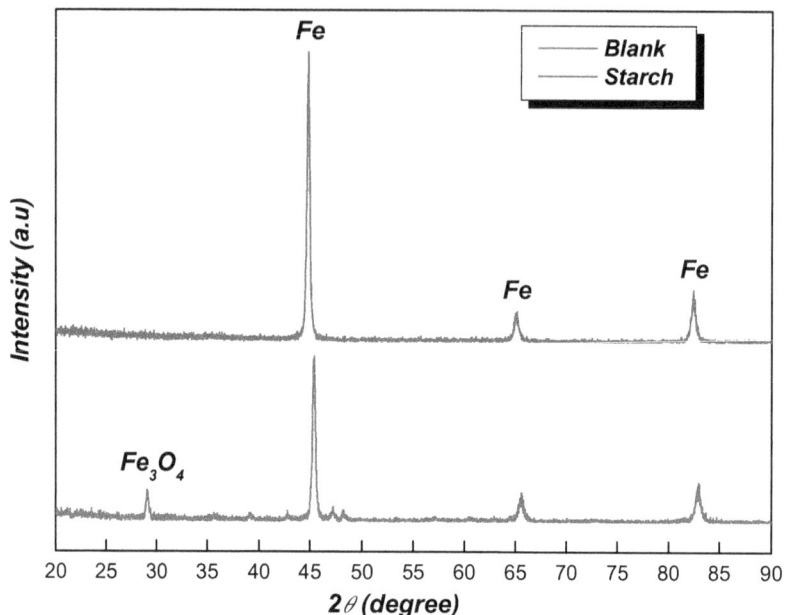

FIGURE 5.12   XRD patterns of the mild steel in the absence (blank) and the presence of *starch*.

5.4   CONCLUSION

In this study, natural polymer (starch) was evaluated as an inhibitor for mild steel in NaCl 3%, and the following conclusions were drawn:

1. EFM can be used as a rapid and nondestructive technique for corrosion rate measurements without prior knowledge of Tafel slopes.
2. Starch acts as an effective inhibitor for mild steel acidic corrosion, and its effectiveness increases with its concentration.
3. Polarization results showed that starch displayed mixed-type inhibitors with anodic behavior. The inhibition efficiency of starch followed the order: 96.7%.
4. The results of EIS indicate that the value of CPE tends to decrease, and both polarization resistance and inhibition efficiency tend to increase by the inhibitor concentration. This result can be attributed to increase the thickness of the electrical double layer.
5. The addition of the starch leads to a blockage of the corrosion of the active sites by forming a protective layer on the surface of the mild steel, which is confirmed by the surface characterization analysis utilizing SEM coupled with EDS, and XRD.

REFERENCES

[1] H. Byars, "Corrosion control in petroleum production, TPC 5," *Anti-Corrosion Methods and Materials*, 46 (1999) 350.

[2] I. Obot, D. Macdonald, and Z. Gasem, "Density functional theory (DFT) as a powerful tool for designing new organic corrosion inhibitors. Part 1: an overview," *Corrosion Science*, 99 (2015) 1–30.

[3] L. Guo, Z. S. Safi, S. Kaya, W. Shi, B. Tüzün, N. Altunay, and C. Kaya, "Anticorrosive effects of some thiophene derivatives against the corrosion of iron: a computational study," *Frontiers in Chemistry*, 6 (2018) 155.

[4] Ş. Erdoğan, Z. S. Safi, S. Kaya, D. Ö. Işın, L. Guo, and C. Kaya, "A computational study on corrosion inhibition performances of novel quinoline derivatives against the corrosion of iron," *Journal of Molecular Structure*, 1134 (2017) 751–761.

[5] K. Boumhara, H. Harhar, A. Tabyaoui, A. Bellaouchou, A. Guenbour, and A. Zarrouk, "Corrosion inhibition of mild steel in 0.5 M H_2SO_4 solution by artemisia herba-alba oil," *Journal of Bio- and Tribo-Corrosion*, 5 (2019) 1–9.

[6] S. Echihi, R. Hsissou, N. Benzbiria, M. Afrokh, M. Boudalia, A. Bellaouchou, A. Guenbour, M. Azzi, and M. Tabyaoui, "Performance of methanolic extract of artemisia herba alba as a potential green inhibitor on corrosion behavior of mild steel in hydrochloric acid solution," *Biointerface Research in Applied Chemistry*, 11 (2021) 14751–14763.

[7] O. Fergachi, F. Benhiba, M. Rbaa, R. Touir, M. Ouakki, M. Galai, B. Lakhrissi, H. Oudda, and M. E. Touhami, "Experimental and theoretical study of corrosion inhibition of mild steel in 1.0 M HCl medium by 2(-4(hloro phenyl-1H-benzo[d]imidazol)-1-yl)phenyl) methanone," *Materials Research*, 21 (2018) 1038.

[8] M. Belghiti, S. Bouazama, S. Echihi, A. Mahsoune, A. Elmelouky, A. Dafali, K. Emran, B. Hammouti, and M. Tabyaoui, "Understanding the adsorption of newly benzylidene-aniline derivatives as a corrosion inhibitor for carbon steel in hydrochloric acid solution: experimental, DFT and molecular dynamic simulation studies," *Arabian Journal of Chemistry*, 13 (2020) 1499–1519.

[9] F. Benhiba, H. Serrar, R. Hsissou, A. Guenbour, A. Bellaouchou, M. Tabyaoui, S. Boukhris, H. Oudda, I. Warad, and A. Zarrouk, "Tetrahydropyrimido-triazepine derivatives as anti-corrosion additives for acid corrosion: chemical, electrochemical, surface and theoretical studies," *Chemical Physics Letters*, 743 (2020) 137181.

[10] G. E. Badr, "The role of some thiosemicarbazide derivatives as corrosion inhibitors for C-steel in acidic media," *Corrosion Science*, 51 (2009) 2529–2536.

[11] N. Yilmaz, A. Fitoz, Ÿ. Ergun, and K. C. Emregül, "A combined electrochemical and theoretical study into the effect of 2-((thiazole-2-ylimino)methyl)phenol as a corrosion inhibitor for mild steel in a highly acidic environment," *Corrosion Science*, 111 (2016) 110–120.

[12] M. Ikpi, F. Abeng, and O. Obono, "Adsorption and thermodynamic studies for corrosion inhibition of API 5L X-52 steel in 2 M HCl solution by moxifloxacin," *World News of Natural Sciences. An International Sientific Journal*, 9 (2017) 52–61.

[13] A. Yousefi, S. Javadian, M. Sharifi, N. Dalir, and A. Motaee, "An experimental and theoretical study of biodegradable Gemini surfactants and surfactant/carbon nanotubes (CNTs) mixtures as new corrosion inhibitor," *Journal of Bio- and Tribo-Corrosion*, 5 (2019) 1–15.

[14] R. Hsissou, O. Dagdag, S. Abbout, F. Benhiba, M. Berradi, M. El Bouchti, A. Berisha, N. Hajjaji, and A. Elharfi, "Novel derivative epoxy resin TGETET as a corrosion inhibition of E24 carbon steel in 1.0 M HCl solution. Experimental and computational (DFT and MD simulations) methods," *Journal of Molecular Liquids*, 284 (2019) 182–192.

[15] R. Hssissou, B. Benzidia, N. Hajjaji, and A. Elharfi, "Elaboration, electrochemical investigation and morphological study of the coating behavior of a new polymeric poly-epoxide architecture: crosslinked and hybrid decaglycidyl of phosphorus penta methy-lene dianiline on E24 carbon steel in 3.5% NaCl," *Portugaliae Electrochimica Acta*, 37 (2019) 179–191.

[16] R. Hsissou, A. Bekhta, A. Elharfi, B. Benzidia, and N. Hajjaji, "Theoretical and elec-trochemical studies of the coating behavior of a new epoxy polymer: hexaglycidyl eth-ylene of methylene dianiline (HGEMDA) on E24 steel in 3.5% NaCl," *Portugaliae Electrochimica Acta*, 36 (2018) 101–117.

[17] K. Chatoui, S. Echihi, H. Harhar, A. Zarrouk, and M. Tabyaoui, "An investigation of carbon steel corrosion inhibition in 1 M HCl by Lepidium sativum oil as green inhibi-tor," *Journal of Materials and Environmental Sciences*, 9 (2018) 1212–1222.

[18] H. M. Abd El-Lateef, W. Albokheet, and M. Gouda, "Carboxymethyl cellulose/metal (Fe, Cu and Ni) nanocomposites as non-precious inhibitors of C-steel corrosion in HCl solutions: synthesis, characterization, electrochemical and surface morphology stud-ies," *Cellulose*, 27(14) (2020) 8039–8057.

[19] L. Huang, W. Q. Chen, S. S. Wang, Q. Zhao, H. J. Li, and Y. C. Wu, "Starch, cellu-lose and plant extracts as green inhibitors of metal corrosion: a review," *Environmental Chemistry Letters*, 20 (2022) 3235–3264. doi:10.1007/s10311-022-01400-5.

[20] M. Bello, N. Ochoa, V. Balsamo, F. López-Carrasquero, S. Coll, A. Monsalve, and G. González, "Modified cassava starches as corrosion inhibitors of carbon steel: an elec-trochemical and morphological approach," *Carbohydrate Polymers*, 82 (2010) 561–568.

[21] X. Li, S. Deng, T. Lin, X. Xie, and G. Du, "Cassava starch ternary graft copolymer as a corrosion inhibitor for steel in HCl solution," *Journal of Materials Research and Technology*, 9(2) (2020) 2196–2207.

[22] M. V. Fiori-Bimbi, P. E. Alvarez, H. Vaca, and C. A. Gervasi, "Corrosion inhibition of mild steel in HCl solution by pectin," *Corrosion Science*, 92 (2015) 192–199.

[23] H. Bentrah, Y. Rahali, and A. Chala, "Gum Arabic as an eco-friendly inhibitor for API5L X42 pipeline steel in HCl medium," *Corrosion Science*, 82 (2014) 426–431.

[24] K. Azzaoui, E. Mejdoubia, S. Jodeh, A. Lamhamdi, E. Rodriguez-Castellón, M. Algarra, A. Zarrouk, A. Errich, R. Salghi, and H. Lgaz, "Eco-friendly green inhibitor Gum Arabic (GA) for the corrosion control of mild steel in hydrochloric acid medium," *Corrosion Science*, 129 (2017) 70–81.

[25] S. A. Umoren, M. J. Banera, T. Alonso-Garcia, C. A. Gervasi, and M. V. Mirífico, "Inhibition of mild steel corrosion in HCl solution using chitosan," *Cellulose*, 20 (2013) 2529–2545.

[26] E. Bayol, A. A. Gürten, M. Dursun, and K. Kayakirilmaz, "Adsorption behaviorand inhibition corrosion effect of sodium carboxymethyl cellulose on mild steel in acidic medium," *Acta Physico-Chimica Sinica*, 24 (2008) 2236–2243.

[27] A. Biswas, P. Sagar, and G. Udayabhanu, "Experimental and theoretical studies of xan-than gum and its graft co-polymer as corrosion inhibitor for mild steel in 15% HCl," *Applied Surface Science*, 353 (2015) 173–183.

[28] K. Alaoui, R. Touir, M. Galai, H. Serrar, M. Ouakki, S. Kaya, B. Tüzün, S. Boukhris, M. EbnTouhami, and Y. El Kacimi, "Electrochemical and computational studies of some triazepine carboxylate compounds as acid corrosion inhibitors for mild steel," *Journal of Bio- and Tribo-Corrosion*, 4 (2018) 37.

[29] H. Serrar, M. Galai, F. Benhiba, M. Ouakki, Z. Benzekri, S. Boukhris, A. Hassikou, A. Souizi, H. Oudda, and M. Ebn Touhami, "Two derivatives of 7-amino-thiazolo[3,2-A]pyrimidine as inhibitors of mild steel corrosion in 1.0M HCl solution: part I: syn-thesis of inhibitors and electrochemical study," *Journal of Chemical Technology and Metallurgy*, 53(2) (2018) 324–335.

[30] M. Abdallah, M. Alfakeer, A. M. Alonazi, and S. S. Al-Juaid, "Ketamine drug as an inhibitor for the corrosion of 316 stainless steel in 2M HCl solution," *Internation Journal of Electrochemical Science*, 14 (2019) 10227–10247.

[31] F. Abeng, V. Anadebe, V. Idim, and M. Edim, "Anti-corrosion behaviour of expired tobramycin drug on carbon steel in acidic medium," *South African Journal of Chemistry*, 73 (2020) 125–130.

[32] S. M. Abd El Haleem, S. Abd El Wanees, E. E. Abd El Aal, and A. Farouk, "Factors affecting the corrosion behaviour of aluminium in acid solutions. I. Nitrogen and/or sulphur-containing organic compounds as corrosion inhibitors for Al in HCl solutions," *Corrosion Science*, 68 (2013) 1–13.

[33] A. B. Tadros, and B. Abd-el-Nabey, "Inhibition of the acid corrosion of steel by 4-amino-3-hydrazino-5-thio-1, 2, 4-triazoles," *Journal of Electroanalytical Chemistry and Interfacial Electrochemistry*, 246 (1988) 433–439.

[34] S. A. Umoren, and U. M. Eduok, "Application of carbohydrate polymers as corrosion inhibitors for metal substrates in different media: a review," *Carbohydrate Polymers*, 140 (2016), 314–341.

[35] M. M. Burrell, "Starch: the need for improved quality or quantity-an overview," *Journal Experimental Botany*, 54 (2003), 451–456.

[36] B. Balsamo, V. López-Carrasquero, F. Contreras, J. Muller, A. Laredo, E. Contó, K., and J. L. Feijoo, "Synthesis and characterization of starch-based ionic complexes," ANTEC 2010-Proceedings of the 68th Annual Technical Conference & Exhibition, Orlando, FL, May 16–20. Society of Plastics Engineers, (2010) 87–90.

[37] I. E. Rivero, V. Balsamo, and A. J. Müller, "Microwave-assisted modification of starch for compatibilizing LLDPE/starch blends," *Carbohydrate Polymers*, 75(2) (2009) 343–350.

[38] J. Singh, L. Kaur, and O. J. McCarthy, "Factors influencing the physico-chemical, morphological, thermal and rheological properties of some chemically modified starch for food application-a review," *Food Hydrocolloids*, 21 (2007), 1–22.

[39] K. F. Khaled, "Application of electrochemical frequency modulation for monitoring corrosion and corrosion inhibition of iron by some indole derivatives in molar hydrochloric acid," *Materials Chemistry and Physics*, 112 (2008) 290–300.

[40] R. W. Bosch, J. Hubrecht, W. F. Bogaerts, and B. C. Syrett, "Electrochemical frequency modulation: a new electrochemical technique for online corrosion monitoring," *Corrosion*, 57 (2001) 60.

[41] R. W. Bosch, and W. F. Bogaerts, "Instantaneous corrosion rate measurement with small-amplitude potential intermodulation techniques," *Corrosion*, 52 (1996) 204.

[42] F. Mert, C. Blawert, K. U. Kainer, and N. Hort, "Influence of cerium additions on the corrosion behaviour of high pressure die cast AM50 alloy," *Corrosion Science*, 65 (2012) 145–151.

[43] B. W. A. Sherar, P. G. Keech, and D. W. Shoesmith, "The effect of sulfide on the aerobic corrosion of carbon steel in near-neutral pH saline solutions," *Corrosion Science*, 66 (2013) 256–262.

[44] K. F. Khaled, and S. R. Al-Mhyawi, "Electrochemical and density function theory investigations of L-arginine as corrosion inhibitor for steel in 3.5% NaCl," *International Journal of Electrochemical Science*, 8 (2013) 4055–4072.

[45] O. G. Olvera, M. Rebolledo, and E. Asselin, "Atmospheric ferric sulfate leaching of chalcopyrite: thermodynamics, kinetics and electrochemistry," *Hydrometallurgy*, 165 (2016) 148–158.

[46] A. Ehsani, M. G. Mahjani, M. Hosseini, R. Safari, R. Moshrefi, and H. Mohammad Shiri, "Evaluation of Thymus vulgaris plant extract as an eco-friendly corrosion inhibitor for stainless steel 304 in acidic solution by means of electrochemical impedance spectroscopy, electrochemical noise analysis and density functional theory," *Journal of Colloid and Interface Science*, 490 (2017) 444–451.

[47] N. Ferraa, M. Ouakki, M. Cherkaoui, and M. Bennani Ziatni, "Synthesis, characterization and evaluation of apatitic tricalcium phosphate as a corrosion inhibitor for carbon steel in 3 wt% NaCl," *Journal of Bio- and Tribo-Corrosion*, 8 (2022) 23.

[48] A. Benyaich, M. Roch, J. Pagetti, and M. Troquet, "Inhibition de la corrosion du fer Armco par le bromure de triphénylbenzylphosphonium en milieu acide phosphorique," *Matériaux et Techniques*, 76 (1988) 36.

[49] F. El-Hajjaji, E. Ech-chihbi, N. Rezki, F. Benhiba, M. Taleb, D. S. Chauhan, and M. A. Quraishi, "Electrochemical and theoretical insights on the adsorption and corrosion inhibition of novel pyridinium-derived ionic liquids for mild steel in 1 M HCl," *Journal of Molecular Liquids*, 314 (2020) 113737.

[50] Y. El Kacimi, R. Touir, K. Alaoui, S. Kaya, A. Salem Abousalem, M. Ouakki, and M. EbnTouhami, "Anti-corrosion properties of 2-phenyl-4(3H)-quinazolinone-substituted compounds: electrochemical, quantum chemical, monte carlo, and molecular dynamic simulation investigation," *Journal of Bio- and Tribo-Corrosion*, 6 (2020) 47.

[51] M. Galai, M. Rbaa, M. Ouakki, A. S. Abousalem, E. Ech-Chihbi, K. Dahmani, N. Dkhireche, B. Lakhrissi, and M. EbnTouhami, "Chemically functionalized of 8-hydroxy-quinoline derivatives as efficient corrosion inhibition for steel in 1.0 M HCl solution: experimental and theoretical studies," *Surfaces and Interfaces*, 21 (2020) 100695.

[52] K. Dahmani, M. Galai, M. Cherkaoui, A. El Hasnaoui, and A. El Hessni, "Cinnamon essential oil as a novel eco-friendly corrosion inhibitor of copper in 05M sulfuric acid medium," *Journal of Materials and Environmental Sciences*, 8(5) (2017) 1676–1689.

[53] Z. Hu, Y. Meng, X. Ma, H. Zhu, J. Li, C. Li, and D. Cao, "Experimental and theoretical studies of benzothiazole derivatives as corrosion inhibitors for carbon steel in 1 M HCl," *Corrosion Science*, 112 (2016) 563–575.

[54] H. Tristijanto, M. N. Ilman, and P. T. Iswanto, "Corrosion inhibition of welded of X-52 steel pipelines by sodium molybdate in 3.5% NaCl solution," *Egyptian Journal of Petroleum*, 29(2) (2020) 155–162.

[55] B. Tan, S. Zhang, Y. Qiang, L. Guo, L. Feng, C. Liao, and S. Chen, "A combined experimental and theoretical study of the inhibition efect of three disulfde-based favouring agents for copper corrosion in 0.5 M sulfuric acid," *Journal of Colloid and Interface Science*, 526 (2018) 268–280.

[56] M. Galai, J. Ouassir, M. EbnTouhami, H. Nassali, H. Benqlilou, T. Belhaj, K. Berrami, I. Mansouri, and B. Oauki, "α-Brass and (α + β) brass degradation processes in azrou soil medium used in plumbing devices," *Journal of Bio- and Tribo-Corrosion*, 3 (2017) 30.

[57] K. Dahmani, M. Galai, M. Ouakki, A. Elgendy, R. Ez-Zriouli, R. Lachhab, S. Briche, and M. Cherkaoui, "Corrosion inhibition of copper in sulfuric acid via environmentally friendly inhibitor (Myrtus Communis): combining experimental and theoretical methods," *Journal of Molecular Liquids*, 347 (2022) 117982.

[58] R. T. Loto, C. A. Loto, A. P. Popoola, and T. Fedotova, "Electrochemical studies of the inhibition effect of 2-dimethylaminoethanol on the corrosion of austenitic stainless-steel type 304 in dilute hydrochloric acid," *Silicon*, 8(1) (2015) 145–158.

[59] C. Ogretir, B. Mihci, and G. Bereket, "Quantum chemical studies of some pyridine derivatives as corrosion inhibitors," *Journal of Molecular Structure: Theochem*, 488 (1999) 223–231.

[60] K. Rahmouni, M. Keddam, A. Srhiri, and H. Takenouti, "Corrosion of copper in 3% NaCl solution polluted by sulphide ions," *Corrosion Science*, 47 (2005) 3249–3266.

[61] G. Kardas, "The inhibition effect of 2-thiobarbituric acid on the corrosion performance of mild steel in HCl solutions," *Materials Science*, 41(3) (2005) 337–343.

[62] K. Mzioud, A. Habsaoui, M. Ouakki, M. Galai, S. El Fartah, and M. EbnTouhami, "Inhibition of copper corrosion by the essential oil of Allium sativum in 0.5M H_2SO_4 solutions," *SN Applied Sciences*, 2 (2020) 1611.

[63] A. J. Jadhav, C. R. Holkar, and D. V. Pinjari, "Anticorrosive performance of superhy-drophobic imidazole encapsulated hollow zinc phosphate nanoparticles on mild steel," *Progress in Organic Coatings*, 114 (2018) 33–39.

[64] A. Ouass, M. Galai, M. Ouakki, E. Ech-Chihbi, L. Kadiri, R. Hsissou, Y. Essaadaoui, A. Berisha, M. Cherkaoui, A. Lebkiri, and E. H. Rifi, "Poly(sodium acrylate) and poly(acrylic acid sodium) as an eco-friendly corrosion inhibitor of mild steel in normal hydrochloric acid: experimental, spectroscopic and theoretical approach," *Journal of Applied Electrochemistry*, 51(7) (2021) 1009–1032.

[65] M. Ouakki, M. Rbaa, M. Galai, B. Lakhrissi, E. H. Rifi, and M. Cherkaoui, "Experimental and quantum chemical investigation of imidazole derivatives as corrosion inhibitors on mild steel in 1.0 M hydrochloric acid," *Journal of Bio- and Tribo-Corrosion*, 4 (2018) 35.

[66] C. Verma, L. O. Olasunkanmi, T. W. Quadri, E.-S. M. Sherif, and E. E. Ebenso, "Gravimetric, electrochemical, surface morphology, DFT, and Monte Carlo simulation studies on three N-substituted 2-aminopyridine derivatives as corrosion inhibitors of mild steel in acidic medium," *The Journal of Physical Chemistry C*, 122(22) (2018) 11870–11882.

[67] M. Ouakki, M. Galai, M. Rbaa, A. S. Abousalem, B. Lakhrissi, E. H. Rifi, and M. Cherkaoui, "Investigation of imidazole derivatives as corrosion inhibitors for mild steel in sulfuric acidic environment: experimental and theoretical studies," *Ionics*, 26 (2020) 5251–5272.

[68] M. Bouklah, B. Hammouti, M. Lagrenee, and F. Bentiss, "Thermodynamic properties of 2,5-bis(4-methoxyphenyl)-1,3,4-oxadiazole as a corrosion inhibitor for mild steel in normal sulfuric acid medium," *Corrosion Science*, 48 (2006) 2831–2842.

[69] S. M. Shaban, A. A. Abd-Elaal, and S. M. Tawfik, "Gravimetric and electrochemical evaluation of three nonionic dithiol surfactants as corrosion inhibitors for mild steel in 1 M HCl solution," *Journal of Molecular Liquids*, 216 (2016) 392–400.

[70] A. Ech-chebab, M. Missioui, L. Guo, O. El Khouja, R. Lachhab, O. Kharbouch, M. Galai, M. Ouakki, A. Ejbouh, K. Dahmani, N. Dkhireche, and M. EbnTouhami, "Evaluation of quinoxaline-2(1H)-one, derivatives as corrosion inhibitors for mild steel in 1.0 M acidic media: electrochemistry, quantum calculations, dynamic simulations, and surface analysis," *Chemical Physics Letters*, 809 (2022) 140156.

[71] M. Ouakki, M. Galai, M. Cherkaoui, E.-H. Rifi, and Z. Hatim, "Inorganic compound (apatite doped by Mg and Na) as a corrosion inhibitor for mild steel in phosphoric acidic medium," *Analytical and Bioanalytical Electrochemistry*, 10(7) (2018) 943–960.

[72] R. Geethanjali, and S. Subhashini, "Thermodynamic characterization of metal dissolution and adsorption of polyvinyl alcohol-grafted poly (Acrylamide-Vinyl Sulfonate) on mild steel in hydrochloric acid," *Portugaliae Electrochimica Acta*, 33(1) (2015) 35–48.

[73] M. Elachouri, M. S. Hajji, M. Salem, S. Kertit, J. Aride, R. Coudert, and E. Essassi, "Some nonionic surfactants as inhibitors of the corrosion of iron in acid chloride solutions," *Corrosion*, 52 (1996) 103.

[74] M. Ouakki, M. Galai, M. Rbaa, A. S. Abousalem, B. Lakhrissi, E. H. Rifi, and M. Cherkaoui, "Investigation of imidazole derivatives as corrosion inhibitors for mild steel in sulfuric acidic environment: experimental and theoretical studies," *Ionics*, 26 (2020) 5251–5272.

[75] A. Dutta, S. K. Saha, U. Adhikari, P. Banerjee, and D. Sukul, "Effect of substitution on corrosion inhibition properties of 2-(substituted phenyl) benzimidazole derivatives on mild steel in 1 M HCl solution: a combined experimental and theoretical approach," *Corrosion Science*, 123 (2017) 256–266.

[76] Q. Ma, S. Qi, X. He, Y. Tang, and G. Lu, "1,2,3-Triazole derivatives as corrosion inhibitors formild steel in acidic medium: experimental and computational chemistry studies," *Corrosion Science*, 129 (2017) 91–101.

[77] N. El Hamdani, R. Fdil, M. Tourabi, C. Jama, and F. Bentiss, "Alkaloids extract of Retama monosperma (L.) Boiss. seeds used as novel eco-friendly inhibitor for carbon steel corrosion in 1 M HCl solution: electrochemical and surface studies," *Applied Surface Science*, 357 (2015) 1294–1305.

[78] S. Fouda Abdelaziz, A. Mohamed Ismail, A. Abdulraqeb Al-Khamri, and S. Ashraf Abousalem, "Experimental, quantum chemical and molecular simulation studies on the action of arylthiophene derivatives as acid corrosion inhibitors," *Journal of Molecular Liquids*, 290 (2019) 111178.

[79] Z. Wang, R. C. Moore, A. R. Felmy et al., "A study of the corrosion products of mild steel in high ionic strength brines," *Waste Management*, 21 (2001) 335.

[80] J. Jiabin Han, W. Carey, and J. Zhang, "Effect of sodium chloride on corrosion of mild steel in CO2-saturated brines," *Journal of Applied Electrochemistry*, 41 (2011) 741–749.

[81] M. Ouakki, M. Galai, Z. Benzekri, Z. Aribou, E. Ech-Chihbi, L. Guo, K. Dahmani, K. Nouneh, S. Briche, S. Boukhris, and M. Cherkaoui, "A detailed investigation on the corrosion inhibition effect of by newly synthesized pyran derivative on mild steel in 1.0 M HCl: experimental, surface morphological (SEM-EDS, DRX& AFM) and computational analysis (DFT & MD simulation). *Journal of Molecular Liquids*, 344 (2021) 117777.

[82] M. Galai, M. Rbaa, M. Ouakki, K. Dahmani, S. Kaya, N. Arrousse, N. Dkhireche, S. Briche, B. Lakhrissi, and M. EbnTouhami, "Functionalization effect on the corrosion inhibition of novel eco-friendly compounds based on 8-hydroxyquinoline derivatives: experimental, theoretical and surface treatment," *Chemical Physics Letters*, 776 (2021) 138700.

[83] H. M. Abd El-Lateef, "Corrosion inhibition characteristics of a novel salycilideneisatin hydrazine sodium sulfonate on carbon steel in HCl and a synergistic nickel ions additive: a combined experimental and theoretical perspective," *Applied Surface Science*, 501 (2020) 144237.

6 Phytochemicals/Plant Extracts as Corrosion Inhibitors for Steel Alloys in H_2SO_4 Solutions

Jasdeep Kaur and Akhil Saxena
Chandigarh University Mohali

6.1 INTRODUCTION

Many chemicals are being used as corrosion inhibitors to control corrosion, like synthetic compounds that might have significant environmental consequences and also hazardous to human health [1]. To overcome this problem, natural and biodegradable corrosion inhibitors must be employed due to their non-toxicity and affordability [2]. The use of acid solution in pickling leads to deterioration of metal surfaces. It is an electrochemical reaction that occurs basically due to interaction of metal with the environment and affected by the pH, temperature, pressure, etc. [3]. Corrosion is a destructive and continuous problem, which is also a challenging problem in front of the whole world. Complete eradication would be impractical and impossible, but prevention would be more practical and feasible [4]. The corrosion of the steel might be understood as a stepwise electrochemical procedure. Initially, the acid attack takes place at anodic sites of the surface that prompts to the migration of ferrous particles into the solution.

Discharged electrons from the anode lead their way to cathodic sites, thereby reacting with O_2 and H_2O forming hydroxyl particles. The hydroxyl particles so formed react with ferrous particles forming ferrous oxide, which on oxidation is converted into ferric oxide which is known as rust. Chemical processing plants, industries, and metal-using manufacturers spend a lot of money on corrosion. Global corrosion inhibitor demand in 2019 was $7.2 billion and will likely rise to $9.6 billion by 2026 [5]. Over $2.5 trillion have been lost due to corrosion in the financial sector every single year, and 10% of the total world's metal is damaged every year due to corrosion, according to a survey conducted by NACE in 2016 [6].

6.2 NEED FOR NATURAL CORROSION INHIBITOR

To protect the metal surface against acids, a variety of organic as well as inorganic corrosion inhibitors are being used. The high cost and non-degradable nature of synthetic inhibitors have caused researchers to look for substitutes for synthetic compounds. The damaging effects of most synthetic inhibitors prompted to adopt the natural extract as a corrosion inhibitor. The introduction of environmental friendly corrosion inhibitors is gaining popularity in the field of science. Moreover, a considerable amount of leaves fall on the ground every day, and they are most often burned, polluting Mother Nature. This waste material can be converted into a corrosion inhibitor so that we can produce a valuable extract at a low cost.

To decrease corrosion, several measures such as designing, material selection, electrochemical methods, and corrosion inhibitors are used; however, corrosion inhibitors are acknowledged as the most cost-effective and easiest to apply on surfaces [7]. Corrosion inhibitors comprise chemicals that, when introduced even in low amount to an aggressive media, slow the pace of metal corrosion [8]. Green inhibitors having heteroatoms such as nitrogen, oxygen, and sulphur will significantly decrease steel corrosion in sulphuric acid and in hydrochloric acid [9]. The heteroatoms transfer electrons from the phytochemical towards the metal, producing a coordinate connection [10,11]. The inhibitory action by the extract is mainly attributed to the phytochemical adsorption on the metallic surface, which obstructs the active centres [12]. The corrosion inhibition procedure is usually divided into two stages: the application of an eco-friendly inhibitor upon the surface of a metal, followed by the interaction of the hetero atoms of functional group on the metal. The existence of phytochemicals (as illustrated in Figure 6.1) in green extract is believed to be the reason for their effectiveness. Nature

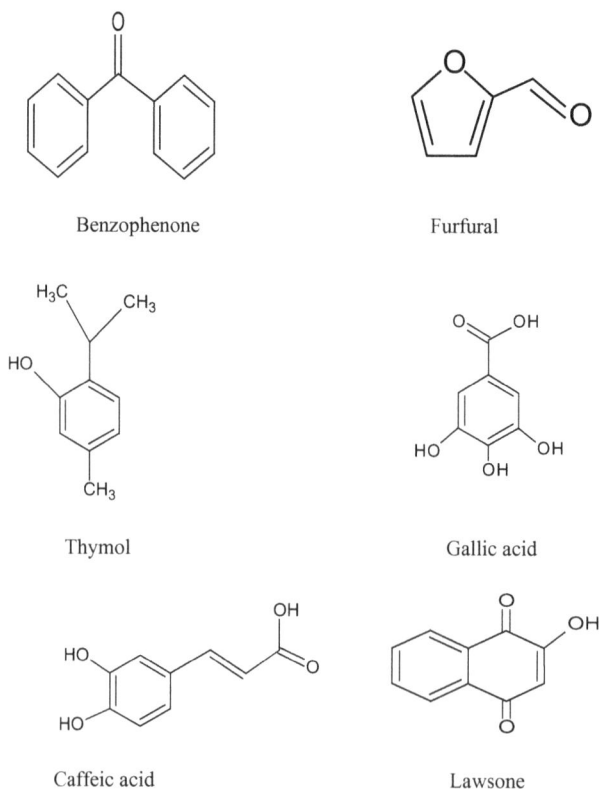

FIGURE 6.1 Phytochemical constituents in various plant extracts.

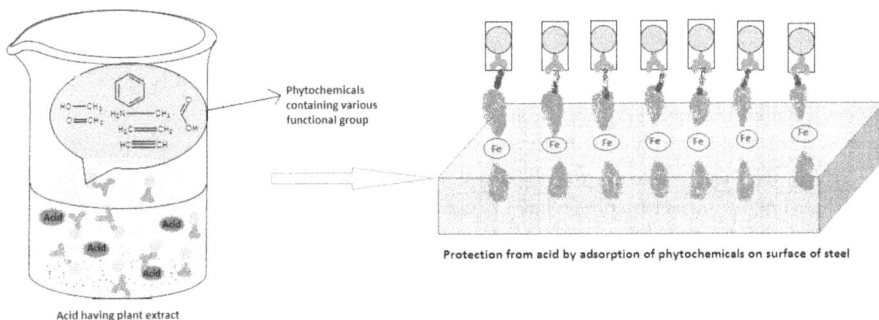

Protection from acid by adsorption of phytochemicals on surface of steel

Acid having plant extract

FIGURE 6.2 Proposed mechanism of corrosion inhibitor.

products are readily available and recyclable. Many plant components, such as roots, seeds, bark, flower, and fruits, have been utilized as best corrosion inhibitors. There seems to be a synergistic effect in corrosive control approaches that include adding halide ions to acid and plant extract solutions to boost inhibitor efficiency [13]. Natural products contain a number of phytochemicals that are rich in heteroatoms; they can be easily adsorbed on the metal surface by formation of coordination bonds. Many alkaloids like strychnine, papaverine, nicotine, and quinine were studied as inhibitors [14]. *Rauvolfia macrophylla* acts as a good corrosion inhibitor in aggressive solutions because of the involvement of perakine and tetrahydroalastonine [15]. *Geissospermum* leaves also have alkaloids with good corrosion inhibition efficiency [16].

6.3 PROPOSED MECHANISM OF CORROSION INHIBITION

Corrosion inhibition of mild steel primarily occurs in two steps: in the first step, there is a transfer of solution of acid, containing green corrosion inhibitor on the surface, which is followed by coupling of phytochemical functional group with the surface. The presence of corrosion inhibitor protective layer formed in the presence of inhibitor is shown in Figure 6.2. The existence of the inhibitor can be explained by the fact that it basically blocks the reaction sites [17,18]. Blockage is due to the formation of many results in the creation of donor-acceptor surface interactions between the inhibitor's outer electrons and the metal's vacancies [19]. Multiple bonds in organic compounds, like double bond and triple bond, are capable of adsorption on the surface of metal [20]. The existence of relatively loosely bound electrons, particularly ones found in anions, and neutral organic molecules comprising lone pair electrons favours electron transfer. In a series of similar compounds, the electron density at the functional group increases as the inhibitive efficiency increases. This is associated with rising coordinate bonding strength as electron transfer becomes easier, leading to more adsorption.

6.4 PLANT EXTRACTS AS CORROSION INHIBITOR IN H_2SO_4 MEDIUM

H_2SO_4 pickling results in corrosion of metal surface. Literature study of various plant extracts in sulphuric acid is discussed here:

Deng et al. examined the anti-corrosion effectiveness of *Ginkgo* extract on steel in two different acidic environments and compared the results in both mediums. The results demonstrated that the performance of inhibitor was extremely good even at low concentrations [21].

Bhawsar et al. conducted a study to find out the effectiveness of an extract of *Nicotiana tabacum*, having phytochemical nicotine, whose heteroatoms are responsible for the inhibition mechanism. Direct correlations exist between surface coverage and concentration but are inversely correlated to the temperature [22].

Umoren et al. explained the anti-corrosive effects of leaf and stem extracts of *Sida acuta* by using various spectroscopic techniques. The inhibitory potentials of the extracts increased as the extract concentration increased, while they dropped as the temperature increased. The leaves extract is discovered to be a more effective inhibitor than the stem extract, which may be explained by the fact that more amounts of phytochemical constituents are present in the leaves extract than the stem extracts of *S. acuta* [23].

Xianghong Li found the corrosion impact of *Dendrocalmus sinicus* leaves on steel in hydrochloride acid and sulphuric acid (H_2SO_4). Inhibitor acted as a mixed-type inhibitor. The observed inhibition efficiency was 95% in 1 M HCl and was 86% in 0.5 M H_2SO_4 [24].

Okafor et al. analysed the ability of *Nauclea latifolia*'s leaves (LV), bark (BK), and roots (RT) to suppress the steel corrosion in sulphuric acid. The phytochemicals in the extracts bind to the metal's empty orbitals and limit corrosion by obstructing the active corrosion sites. The presence of alkaloids in the root and leaf extracts is thought to be responsible for the observed inhibitory trend that is RT > LV > BK [25].

Odewunmi et al. explained the effectiveness of the watermelon rind extract (WMRE) for the steel in sulphuric acid. As the concentration of the WMRE increased, its effectiveness increased. Results of PDP show that WMRE functions as a mixed-type inhibitor [26].

Alaneme studied the anti-corrosive properties of seed husk of *Hunteria umbellata* on the mild steel in an acidic medium. Atomic absorption spectroscopy, FTIR, and SEM were performed to find the inhibition efficiency. The obtained results from various spectroscopy techniques show that the extract has various functional groups, which are responsible for the inhibitory effects [27].

Hassan et al. explained that the extract of *Citrus aurantium* acts as an efficient inhibitor in sulphuric acid. The highest inhibitor efficiency for the *C. aurantium* extract was determined to be 89% at 10 mL L^{-1} [28].

Haldhar used weight loss, electrochemical, SEM, AFM, and other theoretical studies to find the corrosion suppression characteristics of the *Myristica fragrans* fruit (MFF) extract on the steel in sulphuric acid solution. The extract of MFF reduces the pace of corrosion in sulphuric acid due to the involvement of various phytochemicals. The optimum inhibition efficiency of 87.81% was observed at 500 ppm [29].

Saxena studied the anti-corrosive behaviour of the seed extract of *Saraca ashoka* in 0.5 M sulphuric acid by using weight loss and electrochemical studies. The phytochemical epicatechin present in the seed extract was responsible for the decrease in the pace of corrosion of the steel in aggressive media [30].

Abdallah analysed the aqueous extract of various plants like parsley, curcumin, and cassia bark to determine the corrosion inhibition impact in an acidic solution. Potentiodynamic and galvanostatic studies revealed that inhibition efficiency improves as extract concentration increases. *Cassia* bark shows better efficiency than other plant extracts [31].

Saxena et al. explained the *Sida cordifolia* leaves extract as an excellent steel corrosion inhibitor in sulphuric acid. The data obtained from various techniques like Langmuir, SEM, and AFM revealed that corrosion reduction is solely based upon the adsorption process. The maximum inhibition efficiency given by plant extract is 98.96% at 500 ppm [32].

Saxena et al. investigated the *Musa acuminata* extract by using 0.5 M sulphuric acid. PDP, EIS, and weight reduction methodology were utilized to analyse the efficiency of the peel extract. Protective layer formation was proved by the SEM and AFM studies [33].

Policarpi et al. investigated the aqueous extract of *Hymenaea stigonocarpa* as an environment friendly and sustainable corrosion inhibitor. The inhibition action is proposed due to the binding of the organic molecules that block the cathodic reaction [34].

Akalezi studied the *Moringer olifera* leaves extract as a corrosion reducer in 0.5 M sulphuric acid solution. The optimum efficiency given by the extract was 93% at 1,500 ppm. According to GC/MS studies, 29 phytochemicals were examined in the extract, which are responsible for the adsorption on the surface of metal [35].

Zhou used waste feverfew root to analyse the inhibition efficiency. The resultant inhibition efficiency was 97.24% at 400 ppm. The extract has many sesquiterpene lactones that are responsible for corrosion inhibition [36].

Ikeuba explained the inhibition properties of *Gongronema latifolium* flavonoid extract for the steel. The efficiency of the extract increases as the concentration increases. Flavonoids are basically accountable for the corrosion inhibition reaction which adsorb to the surface of metal through the heteroatoms [37].

Mwakalesi investigated the aqueous extract of the *Tetradenia riparia* leaves to find the corrosion inhibition. The resultant findings of inhibition efficiency given by leaves extract was 90.6% at the concentration of 500 ppm which decreased to 4% with an increase in the temperature from 298 to 338 K [38].

More plants have been studied in literature as an efficient corrosion inhibitor in H_2SO_4 and are provided in Table 6.1.

TABLE 6.1
Plant Extracts as Corrosion Inhibitors in H_2SO_4

S. No.	Inhibitor	Efficiency/Conc. of Extract	Phytochemicals	References
1.	*Lannea coromandelica*	93% at 2,000 ppm	Tannins, flavonoids, terpenoids, and polysaccharides	[39]
2.	*Aloe vera*	96% at 300 ppm	Anthrones, chromones, carbohydrates, amino acids, proteins, and vitamins	[40]
3.	*Cryptostegia grandiflora*	87% at 500 ppm	Flavonols, phenolic compounds, cinnamic acid, and benzoic acid	[41]

(Continued)

TABLE 6.1 (*Continued*)
Plant Extracts as Corrosion Inhibitors in H$_2$SO$_4$

S. No.	Inhibitor	Efficiency/Conc. of Extract	Phytochemicals	References
4.	*Aster koraiensis*	90% at 2,000 ppm	Benzofurans, sesquiterpenoids, polyacetylenes	[42]
5.	*Armoracia rusticana*	95% at 100 ppm	Wine lactone, methoxypyrazine, and bisphenol-A	[43]
6.	*Hyptis suaveolens*	95%	Piperine, ellagic acid, tannic acid, tryptamine, caffeine, pennyroyal oil, amino acids, and caffeic acid	[44]
7.	*Acacia catechu*	93.8% at 600 ppm	(3R,4R)-3-(3,4-dihydroxyphenyl)-4-hydroxcyclohexane and (4R)-5-(1-(3,4-dihydrophenyl)-3-oxybutyl)-dihydrofuran-2(3H)-one	[45]
8.	*Adhatoda vasica*	98.9% at 3,000 ppm	Vasicinone and vasicine	[46]
9.	*Coconut*	94.3% at 500 ppm	Saponins, flavonoids, tannins, phlobatannins, polyphenols, and anthraquinones	[47]
10.	*Hemidesmus indicus*	98.05 at 4,000 ppm	Alkaloids, tannins, saponins, and steroids	[48]
11.	*Cannabis sativa*	97.31% at 200 ppm	Cannabinoids	[49]
12.	*Ficus racemosa*	90.5% at 2,500 ppm	Triterpenes, sterols, long-chain fatty acids	[50]

REFERENCES

[1] M.A. El-Hashemy, and A. Sallam, The inhibitive action of Calendula officinalis flower heads extract for mild steel corrosion in 1 M HCl solution. *J. Mater. Res. Technol.* 9 (2020) 13509–13523.

[2] S.S. Alarfaji, I.H. Ali, M.Z. Bani-Fwaz, and M.A. Bedair, Synthesis and assessment of two malonyl dihydrazide derivatives as corrosion inhibitors for carbon steel in acidic media: experimental and theoretical studies. *Molecules.* 26 (2021) 3183.

[3] N.I. Kairi, and J. Kassim, The effect of temperature on the corrosion inhibition of mild steel in 1 M HCl solution by *Curcuma longa* extract. *Int. J. Electrochem. Sci.* 8 (2013) 7138–7155.

[4] J.S. Chouhan, A. Dixit, and D. Gupta, Green inhibitors for prevention of metal and alloys corrosion: an overview. *Chem. Mat. Res.* 3(6), 16–24

[5] N. Hossain, M. Asaduzzaman Chowdhury, and M. Kchaou, An overview of green corrosion inhibitors for sustainable and environment friendly industrial development. *J. Adhes. Sci. Technol.* 35(7) (2021) 673–690.

[6] O.S. Fayomi, I.G. Akande, and S. Odigie, Economic impact of corrosion in oil sectors and prevention: an overview. *J. Phys. Conf. Ser.* 1378(2) (2019) 022037. IOP Publishing.

[7] A. Etre, *Khillah* extract as inhibitor for acid corrosion of SX 316 steel. *Appl. Surface Sci.* 252(24) (2006) 8521–8525.

[8] L. Chauhan, and G. Gunasekaran, Corrosion inhibition of mild steel by plant extract in dilute HCl medium. *Corros. Sci.* 49(3) (2007) 1143–1161.

[9] A. Satapathy, G. Gunasekaran, S. Sahoo, K. Amit, and P. Rodrigues, Corrosion inhibition by *Justicia gendarussa* plant extract in hydrochloric acid solution. *Corros. Sci.* 51(12) (2009) 2848–2856.

[10] J. Aljourani, K. Raeissi, and M. Golozar, Benzimidazole and its derivatives as corrosion inhibitors for mild steel in 1M HCl solution. *Corros. Sci.* 51(8) (2009) 1836–1843.

[11] F. Bentiss, C. Jama, B. Mernari, H. Attari, L. Kadi, M. Lebrini, M. Traisnel, and M. Lagrenée, Corrosion control of mild steel using 3,5-bis(4- methoxyphenyl)-4-amino-1,2,4-triazole in normal hydrochloric medium. *Corros. Sci.* 51(8) (2009) 1628–1635.

[12] A. Lecante, F. Robert, P.A. Blandinières, and C. Roos, Anti-corrosive properties of S. tinctoria and G. ouregou alkaloid extracts on low carbon steel. *Curr. Appl. Phys.* 11(3) (2011) 714–724.

[13] U. Eduok, S. Umoren, and A. Udoh, Synergistic inhibition effects between leaves and stem extracts of *Sida acuta* and iodide ion for mild steel corrosion in 1 M H_2SO_4 solutions. *Arabian J. Chem.* 5(3) (2012) 325–337.

[14] M. Lebrini, F. Robert, A. Lecante, and C. Roos, Corrosion inhibition of C38 steel in 1 M hydrochloric acid medium by alkaloids extract from *Oxandra asbeckii* plant. *Corros. Sci.* 53(2) (2011) 687–695.

[15] B. Ngouné, M. Pengou, A.M. Nouteza, C.P. Nanseu-Njiki, and E. Ngameni, Performances of alkaloid extract from *Rauvolfia macrophylla* Stapf toward corrosion inhibition of C38 steel in acidic media. *J. Am. Chem. Soc.* 4(5) (2019) 9081–9091.

[16] M. Faustin, A. Maciuk, P. Salvin, C. Roos, and M. Lebrini, Corrosion inhibition of C38 steel by alkaloids extract of Geissospermum laeve in 1M hydrochloric acid: electrochemical and phytochemical studies. *Corros. Sci.* 92 (2015) 287–300.

[17] A. Saxenaa, K.K. Thakura, and N. Bhardwaj, Electrochemilogycal studies and surface examination of low carbon steel by applying the extract of *Musa acuminata. Surf. Interfaces.* 18 (2020) 100436.

[18] Y. Fang, B. Suganthan, and R.P. Ramasamy, Electrochemical characterization of aromatic corrosion inhibitors from plant extracts. *J. Electroanal. Chem.* 840 (2019) 74–80.

[19] C. Akalezi, C. Ogukwe, E. Ejele, and E. Oguzie, Mild steel protection in acidic media using Mucuna pruriens seed extract. *Int. J. Corros. Scale Inhib.* 5 (2016) 132–146.

[20] N. Chaubey, Savita, A. Qurashi, D. Chauhan, and M. Quraishi, Frontiers and advances in green and sustainable inhibitors for corrosion applications: a critical review. *J. Mol. Liq.* 321 (2020) 114385.

[21] S. Deng, and X. Li, Inhibition by Ginkgo leaves extract of the corrosion of steel in HCl and H_2SO_4 solutions. *Corrosion Science.* 55 (2012) 407–415.

[22] J. Bhawsar, P.K. Jain, and P. Jain, Experimental and computational studies of *Nicotiana tabacum* leaves extract as green corrosion inhibitor for mild steel in acidic medium. *Alexandria Eng. J.* 54(3) (2015) 769–775.

[23] S.A Umoren, U.M. Eduok, M.M. Solomon, and A.P. Udoh, Corrosion inhibition by leaves and stem extracts of *Sida acuta* for mild steel in 1 M H_2SO_4 solutions investigated by chemical and spectroscopic techniques. *Arabian J. Chem.* 9 (2016) S209–S224.

[24] X. Li, S. Deng, and H. Fu, Inhibition of the corrosion of steel in HCl, H_2SO_4 solutions by *bamboo* leaf extract. *Corros. Sci.* 62 (2012) 163–175.

[25] I.E. Uwah, P.C. Okafor, and V.E. Ebiekpe, Inhibitive action of ethanol extracts from *Nauclea latifolia* on the corrosion of mild steel in H_2SO_4 solutions and their adsorption characteristics. *Arabian J. Chem.* 6(3) (2013) 285–293.

[26] N.A. Odewunmi, S.A. Umoren, and Z.M. Gasem, Utilization of *watermelon rind* extract as a green corrosion inhibitor for mild steel in acidic media. *J. Ind. Eng. Chem.* 21 (2015) 239–247.

[27] K.K. Alaneme, S.J. Olusegun, and O.T. Adelowo, Corrosion inhibition and adsorption mechanism studies of *Hunteria umbellata* seed husk extracts on mild steel immersed in acidic solutions. *Alexandria Eng. J.* 55(1) (2016) 673–681.

[28] K.H. Hassan, A.A. Khadom, and N.H. Kurshed, *Citrus aurantium* leaves extracts as a sustainable corrosion inhibitor of mild steel in sulfuric acid. *South Afr. J. Chem. Eng.* 22 (2016) 1–5.

[29] R. Haldhar, D. Prasad, and A. Saxena, *Myristica fragrans* extract as an eco-friendly corrosion inhibitor for mild steel in 0.5 M H_2SO_4 solution. *J. Env. Chem. Eng.* 6(2) (2018) 2290–2301.

[30] A. Saxena, D. Prasad, R. Haldhar, G. Singh, and A. Kumar, Use of *Saraca ashoka* extract as green corrosion inhibitor for mild steel in 0.5 M H_2SO_4. *J. Mol. Liq.* 258 (2018) 89–97.

[31] M. Abdallah, H.M. Altass, B.A. Al Jahdaly, and M.M. Salem, Some natural aqueous extracts of plants as green inhibitor for carbon steel corrosion in 0.5 M sulfuric acid. *Green Chem. Lett. Rev.* 11(3) (2018) 189–196.

[32] A. Saxena, D. Prasad, R. Haldhar, G. Singh, and A. Kumar, Use of *Sida cordifolia* extract as green corrosion inhibitor for mild steel in 0.5 M H_2SO_4. *J. Environ. Chem. Eng.* 6(1) (2018) 694–700.

[33] A. Saxena, K.K. Thakur, and N. Bhardwaj, Electrochemical studies and surface examination of low carbon steel by applying the extract of *Musa acuminata. Surf. Inter.* 18 (2020) 100436.

[34] E. de Britto Policarpi, and A. Spinelli, Application of *Hymenaea stigonocarpa* fruit shell extract as eco-friendly corrosion inhibitor for steel in sulfuric acid. *J. Taiwan Inst. Chem. Eng.* 116 (2020) 215–222.

[35] C.O. Akalezi, A.C. Maduabuchi, C.K. Enenebeaku, and E.E. Oguzie, Experimental and DFT evaluation of adsorption and inhibitive properties of *Moringa oliefera* extract on mild steel corrosion in acidic media. *Arabian J. Chem.* 13(12) (2020) 9270–9282.

[36] Z. Zhou, X. Min, S. Wan, J. Liu, B. Liao, and X. Guo, A novel green corrosion inhibitor extracted from waste feverfew root for carbon steel in H_2SO_4 solution. *Res. Eng.* 17 (2023) 100971.

[37] I. Ikeuba, B.I. Ita, P.C. Okafor, B.U. Ugi, and E.B. Kporokpo, Green corrosion inhibitors for mild steel in H_2SO_4 solution: comparative study of flavonoids extracted from *gongronema latifoliunm* with crude the extract. *Prot. Met. Phys. Chem. Surf.* 51 (2015) 1043–1049.

[38] A.J. Mwakalesi, Corrosion inhibition of mild steel in sulphuric acid solution with *tetradenia riparia* leaves aqueous extract: kinetics and thermodynamics. *Biointerface Res. Appl. Chem.* 13 (2023) 1,32

[39] P. Muthukrishnan, B. Jeyaprabha, and P. Prakash, Adsorption and corrosion inhibiting behavior of *Lannea coromandelica* leaf extract on mild steel corrosion. *Arabian J. Chem.* 10 (2013) S2343–S2354.

[40] M. Mehdipour, B. Ramezanzadeh, and S. Arman, Electrochemical noise investigation of Aloe plant extract as green inhibitor on the corrosion of stainless steel in 1 M H_2SO_4. *J. Indust. Eng. Chem.* 21 (2014) 318–327.

[41] M. Prabakaran, S. Kim, V. Hemapriya, and I. Chung, Evaluation of polyphenol composition and anti-corrosion properties of *Cryptostegia grandiflora* plant extract on mild steel in acidic medium. *J. Indust. Eng. Chem.* 37 (2016) 47–56.

[42] M. Prabakaran, S. Kim, N. Mugila, V. Hemapriya, K. Parameswari, S. Chitra, and I. Chung, *Aster koraiensis* as nontoxic corrosion inhibitor for mild steel in sulfuric acid. *J. Indust. Eng. Chem.* 52 (2017) 235–245.

[43] R. Haldhar, D. Prasad, and A. Saxena, *Armoracia rusticana* as sustainable and eco-friendly corrosion inhibitor for mild steel in 0.5M sulphuric acid: experimental and theoretical investigations. *J. Environ. Chem. Eng.* 6(4) (2018) 5230–5238.

[44] P. Muthukrishnan, B. Jeyaprabha, and P. Prakash, Mild steel corrosion inhibition by aqueous extract of *Hyptis Suaveolens* leaves. *Int. J. Ind. Chem.* 5 (2014) 5.

[45] R. Haldhar, D. Prasad, and N. Bhardwaj, Experimental and theoretical evaluation of *Acacia catechu* extract as a natural, economical and effective corrosion inhibitor for mild steel in an acidic environment. *J. Bio. Tribo. Corros.* 6 (2019) 76.

[46] M. Singh, A green approach: a corrosion inhibition of mild steel by *Adhatoda vasica* plant extract in 0.5 M H_2SO_4. *J. Mater. Environ. Sci.* 4 (2012) 119–126.

[47] S. Umoren, I. Obot, A. Israel, P. Asuquo, M. M. Solomon, U. Eduok, and A. Udoh, Inhibition of mild steel corrosion in acidic medium using coconut coir dust extracted from water and methanol as solvents. *J. Ind. Eng. Chem.* 20 (2014) 3612–3622.

[48] N. Patel, and D. Šnita, Ethanol extracts of *Hemidesmus indicus* leaves as eco-friendly inhibitor of mild steel corrosion in H_2SO_4 medium. *Chem. Pap.* 68 (2014) 1747–1754.

[49] R. Haldhar, D. Prasad, N. Mandal, F. Benhiba, I. Bahadur, and O. Dagdag, Anticorrosive properties of a green and sustainable inhibitor from leaves extract of *Cannabis sativa* plant: experimental and theoretical approach. *Colloids Surf.* 614 (2021) 126211.

[50] M. Bagga, R. Gadi, O. Yadav, R. Kumar, R. Chopra, and G. Singh, Investigation of phytochemical components and corrosion inhibition property of *Ficus racemosa* stem extract on mild steel in H_2SO_4 medium. *J. Environ. Chem. Eng.* 4 (2016) 4699–4707.

7 Composites and Nanocomposites of Polymers as Corrosion Inhibitors

Khadija Dahmani, Otmane Kharbouch,
Mouhsine Galai, Mohammed Cherkaoui,
and Mohamed Ebn Touhami
IbnTofaïl University

Imane Aadnane, Ikram Daou, and Omar Zegaoui
Moulay Ismail University

Hamou Moussout
Moulay Ismail University
University IbnTofail

7.1 INTRODUCTION

Due to its low cost, outstanding mechanical properties, and accessibility, mild steel finds use in a broad range of industries, including manufacturing, industrial cleaning, acid blasting, oil well acidizing, and other activities. However, this metal corrodes when it comes into contact with mineral acids, particularly hydrochloric acids [1,2]. As a result, steel material preservation is a recurring issue. Many researchers are now interested in this topic. Coatings and corrosion inhibitors are often used to preserve metals. Organic and inorganic inhibitors are used as a coating on metal surfaces which act as insulators for these surfaces. Additionally, because of the risks connected with damaging organic and inorganic inhibitor systems standards, which generate secondary pollutants that may cause acid rain and increase air pollution, scientists must consider the viability of certain natural materials [3,4]. In order to address this drawback, biocompatible and ecologically acceptable coatings are employed to inhibit corrosion. This is possible since green chemistry concepts are being used. In addition to being affordable and simple to handle, composites and nanocomposites are employed because they provide long-term efficiency in reducing corrosion phenomena. One way to avoid direct contact between the metal and ambient gases is to apply low-porosity composites and nanocomposites to the metal

DOI: 10.1201/9781003394631-7

surface. This reduces the redox reaction on the metal surface and prevents corrosion thanks to composites and nanocomposites [5,6]. Due to its qualities including biocompatibility, biodegradability, and functionalization potential, chitosan (CS) is a top contender in green chemistry as a coating agent to prevent metals from corroding [7]. It is mostly made up of dispersed β-D-glucosamine polysaccharides, and it is present in the shells of crustaceans including crabs, shrimp, and others. The porous structural properties of CS delay the capacity to inhibit, and this porous structure may be strengthened by the inclusion of inorganic components, leading to the production of a nanocomposite [8]. This chapter reviews and discusses the studies on chitosan, the synthesis of titanium-chitosan (7% TiO_2-CS) nanocomposite, the method of synthesis and characterization, the discussion of the results, and the inhibition mechanism of both compounds with the pertinent examples of environmentally friendly corrosion inhibitors.

Chen et al. [9] developed a brand-new anticorrosion coating using the natural materials montmorillonite (MMT) and urushiol. The titanium urushiol polymer intercalated in the montmorillonite (UTPOMMT) was created *via in situ* polymerization, and the UTPOMMT was then mixed with epoxy resin to create composite coatings (UTPOMMT/EP) (Figure 7.1). The MMT gap between the layers was increased by this method, which included adding polymerized titania urushiol to the MMT. As a result, the MMT layers were more compact and peltable in epoxy resin, improving the revêtement's barrier properties. The results of the corrosion resistance

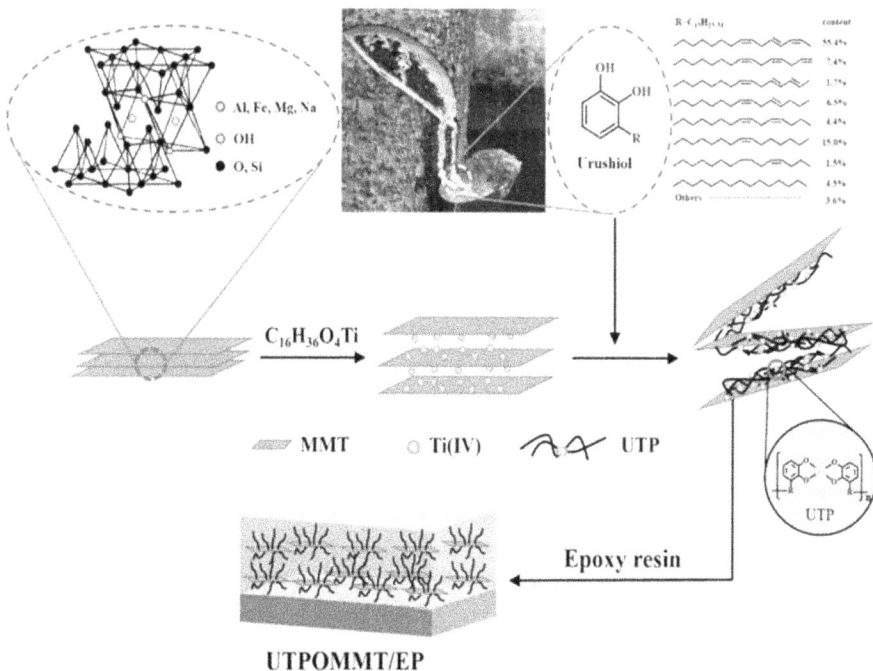

FIGURE 7.1   The preparation process of composite coating (UTPOMMT/EP).

test showed that the amount of UTPOMMT added to PE had a significant impact on the anticorrosion performance of the composite coatings. The montmorillonite and polymerized titania urushiol's synergistic effects contribute to the exceptional aging resistance of UTPOMMT/EP coatings, which are prepared with 5% of it by weight. The prepared UTPOMMT/EP composites had effective anticorrosion and anti-aging properties due to the synergistic effects of montmorillonite and polymerized titanol, and they might be used as a framework to significantly extend the lifespan of building materials. Figure 7.1 illustrates a framework for significantly extending the lifespan of anticorrosion coatings.

One research by Gharib et al. [10] examined the effectiveness of the TiO_2/Paraloid B-72 nanocomposite in preventing corrosion on copper surfaces. By filling up the gaps and valleys left by air corrosion on the copper surface, TiO_2 nanoparticles added to the Paraloid B-72 coating increase the coating's corrosion resistance, according to an examination using atomic force microscopy (AFM). Raman spectroscopy and X-ray diffraction analyses have shown that the nanocomposite coating completely covers the copper surface. This was further supported by the contact angle analysis, which revealed any water contact with the copper surface.

Another research by Ershad-Langroudi et al. [11] used two distinct alkoxysilane precursors, TEOS and TMOS, to create GPTMS-based coatings in order to better protect the aluminum substrate. When the functional group is switched from ethoxy to methoxy, or from TEOS to TMOS, the protective characteristics and polarization resistance improve (R_p). Also included in the organic-inorganic nanocomposite protective covering were aluminum oxide hydroxide (AlOOH) and titanium dioxide (TiO_2) nanoparticles, which were produced from aluminum butoxide and tetra-n-butyl titanate, respectively. According to the results of the investigation, TiO_2 may be selected over AlOOH nanoparticles for improving R_p, a rapid corrosion rate monitoring technique. However, coatings made using the sol-gel approach perform much better when both nanoparticles are added at the same time. Therefore, it can be deduced from salt spray research and potentiodynamic scans that the three primary parameters on which corrosion protection qualities are significantly dependent are silane concentration, type of silane precursors, and type of nanoparticles.

In the case of aluminum, a naturally occurring patina of oxides forms on the metal's surface, which often prevents corrosion but fails to protect the substrate in severe corrosion conditions. A corrosion inhibitor coating through Zandi-Zand et al. [12] has recently been created utilizing a two-step acid catalysis process, in which organic-inorganic hybrid sol-gel coatings are produced with bisphenol A (BPA) acting as a curing agent. Based on the investigation's findings, a silica-based coating consisting of tetramethoxysilane (TMOS) and 3-glycidoxypropyl-trimethoxysilane (GPTMS) provided superior corrosion protection for the common grade of aluminum, 1,050.

Al Jabri et al. [13] performed an experimental study for the purpose of determining a method for applying polyaniline (PANI) and titanium dioxide (TiO_2) nanocomposites to test their ability to inhibit corrosion on mild steel. Sol-gel technology was used to create TiO_2. Mild steel specimens were coated using the dip-coating procedure with polyaniline and titanium dioxide nanocomposites ($PANI/TiO_2$). Under various environmental conditions and exposure durations, stability investigations on the coated specimen were conducted. Studies on the prevention of corrosion in various processing settings employed the stable thin film-coated specimen. Infrared spectroscopy, energy dispersive X-ray analysis, X-ray diffraction, and dynamic light scattering were used to examine the properties of TiO_2 nanoparticles and PANI-TiO_2 thin films (DLS). The surface morphology and micro-structural characteristics of the thin film that had already been created were seen using scanning electron microscopy (SEM). To research the corrosion behavior of coated and uncoated materials, wet/dry testing, atmospheric tests, and dynamic potential tests were conducted. During the examination of coated and untreated specimens' corrosion behavior, the research demonstrates that polyaniline-TiO_2 composite thin films made using immersion coating technology and having a minimal film thickness may be a workable way to prevent corrosion in oil pipelines. A workable, durable, affordable, and ecologically benign method of controlling corrosion in oil pipelines with excellent film stability, as well as sustainability.

7.2 REVIEW OF CHITOSAN COMPOSITES AND NANOCOMPOSITES AS CORROSION INHIBITORS

Priya Kesari et al. [14] synthesized novel chitosan biopolymer (CS) compounds as well as chitosan-grafted acrylamide-based titanium dioxide (CS-g-PAM/TiO_2) (Figure 7.2b) and magnetite (CS-g-PAM/Fe_3O_4) hybrid nanocomposites (Figure 7.2a). These compounds were effectively verified as mild steel corrosion inhibitors in a 15% HCl solution. FTIR, APC, X-ray diffraction (XRD), and TEM were used to characterize the produced compounds. Theremogravimetrtic analysis (TGA) was used to determine the nanocomposites' thermal stability. Electrochemical analysis and gravimetric measurements were used to evaluate the anticorrosive performance, based on the electrochemical impedance spectroscopy (EIS) methodology. Using FESEM and AFM, the morphology of the metal surface was identified. By using X-ray photoelectron spectroscopy (XPS) analysis, the elemental makeup of the metal samples was established.

Zachariah et al. [15] focused on the synthesis of a chitosan/boron nitride (CS-BN) nanocomposite by a sol-gel process and incorporated into an epoxy matrix. The composite is then characterized by Fourier transform infrared spectroscopy (FTIR) and XRD. The thermal stability of the CS-BN composite was investigated by TGA. This study focuses on the synthesis of a CS-BN nanocomposite by a sol-gel process, which is then incorporated into an epoxy matrix. The composite is then characterized by Fourier transform infrared spectroscopy (FTIR) and XRD. The XRD results show that the hexagonal boron nano nitrides are finely dispersed in the chitosan matrix. The thermal stability of the CS-BN composite was studied by TGA. The study shows that boron nitride increases the thermal stability of chitosan. The corrosion protection characteristics of the biocompatible CS-BN nanocomposite on a mild steel substrate in 1 mol L^{-1} HCl solution were monitored by potentiodynamic polarization studies (Tafel), EIS, and linear polarization studies. XPS analysis was adopted to propose the corrosion mechanism and was also used to compare the corrosion inhibition behavior of BN in the composite on the mild steel surface.

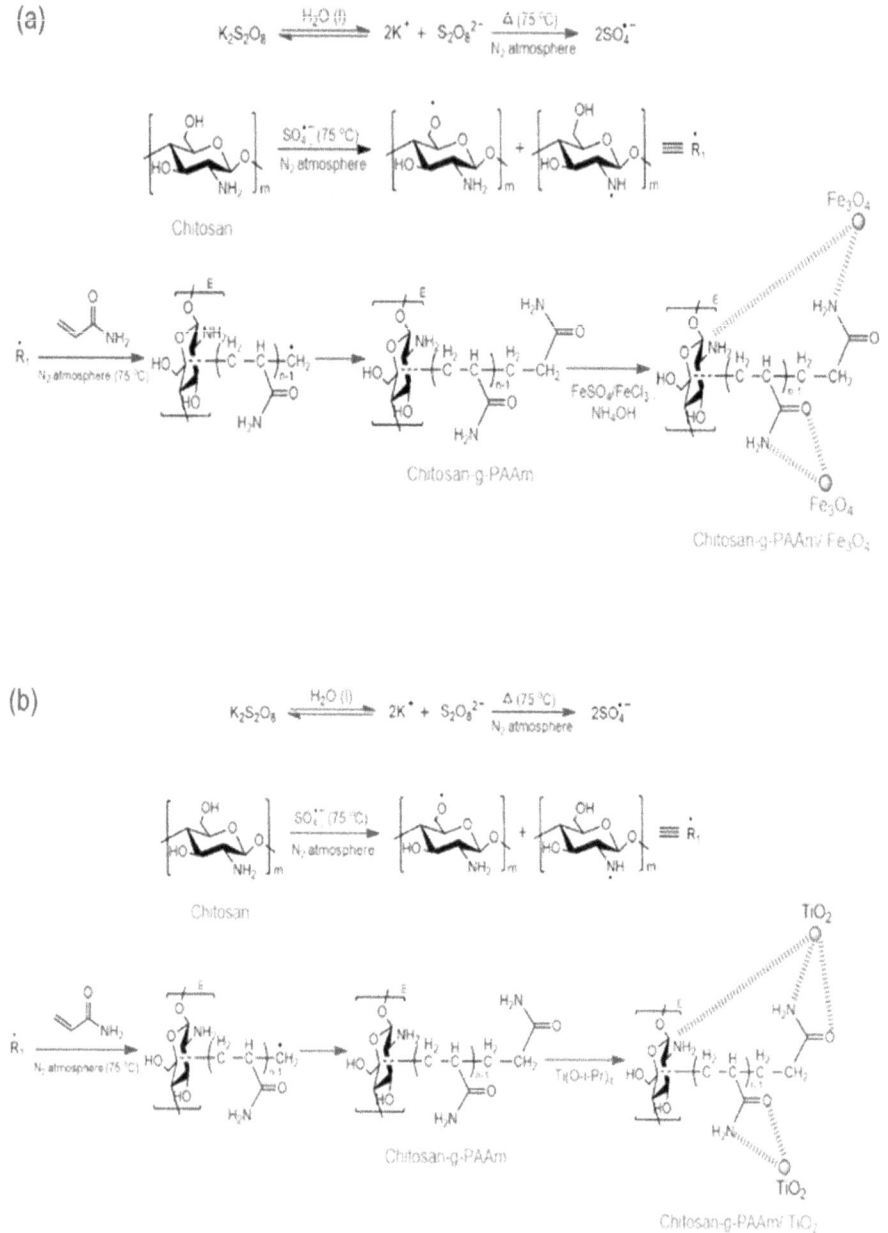

FIGURE 7.2 Plausible synthetic steps for the synthesis of the nanocomposites: (a) CS-g-PAM/Fe$_3$O$_4$ and (b) CS-g-PAM/TiO$_2$.

Zhang et al. [16] developed a unique formulation called chitosan-5-HMF (Figure 7.3) for the purpose of preventing mild steel from corroding in 1 M HCl, which is based on chitosan derivatives with varying degrees of substitution. Electrochemical experiments and theoretical research, such as molecular

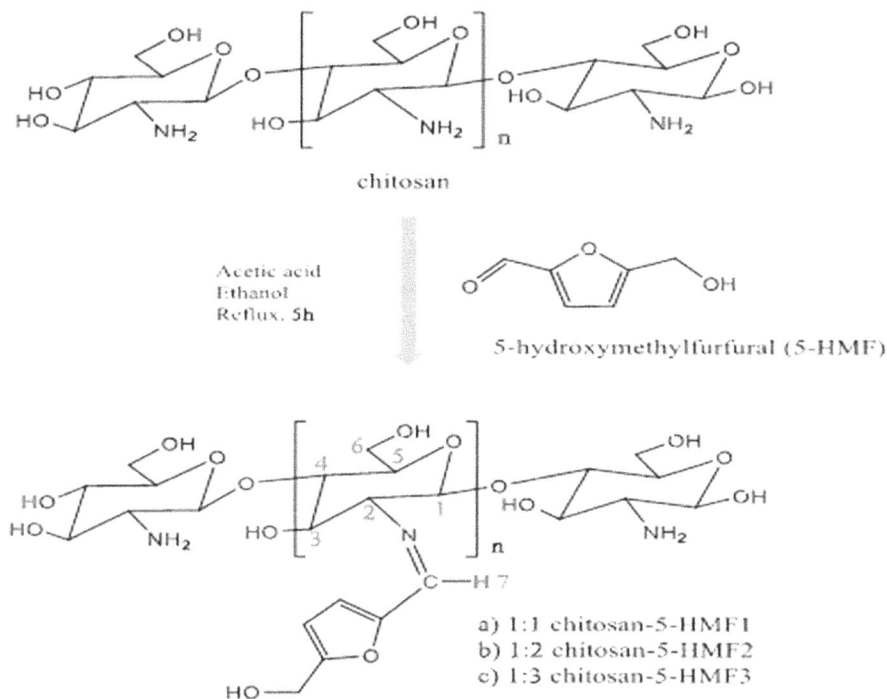

FIGURE 7.3   Synthetic routes of chitosan-5-HMF.

dynamics simulations to assess the reactivity and adsorption processes between chitosan-5-HMF and Fe, were combined to estimate the inhibitory performance of chitosan-5-HMF. SEM and contact angle are used to study the morphology and surface condition.

Wang et al. [17] created and studied four chitosan derivatives in various ways, such as weight loss, SEM, electrochemical measurements (EIS), energy dispersive spectroscopy (EDS), and molecular dynamics simulations. These derivatives were used as green corrosion inhibitors for P110 steel in 15% HCl solution. In an oil and gas well, varied flow velocities and pipeline positions were used to study the impact of fluid erosion on the corrosion inhibitor layer.

Lai et al. [18] developed a new environmentally friendly chitosan derivative and used it as a green corrosion inhibitor on aluminum alloy C3003 in 3.5wt.% NaCl solution. The CP was made by combining 4-pyridinecarboxaldehyde, chitosan, and Schiff base. The chitosan derivative with titanium dioxide (CPT) nanocomposite was then made by dispersing TiO_2 in CP. By using electrochemical methods and surface analysis, the corrosion inhibition effect of CPT on aluminum alloy C3003 at various concentrations was investigated. The creation of a chitosan-based corrosion inhibitor with a low concentration and high efficiency may be motivated by the organic/inorganic hybrid technique.

The authors started the preparation of CPT nanocomposite by combining 100mL of deionized water and 0.3g of TiO_2 nanoparticle, and the mixture was ultrasonically agitated for 1 hour. In a glass container with a circular bottom and three necks, CP (0.3g) was combined with the TiO_2 nanoparticle. The combined solution was

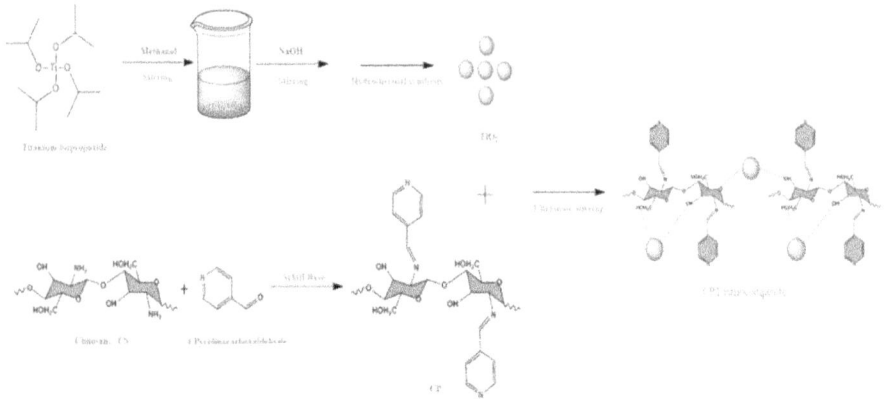

FIGURE 7.4   Synthesis routes of TiO$_2$, CP, and CPT nanocomposite.

then agitated for 3 hours at room temperature. The mixture was then repeatedly washed with ethanol and deionized water. The CPT nanocomposite was then dried for 24 hours at 60°C. The synthesis route is shown in Figure 7.4.

7.3   EXAMPLE OF A COMPLETE STUDY ON THE USE OF TWO CHITOSAN-BASED COMPOSITE AND NANOCOMPOSITE INHIBITORS

This chapter explores how two polymeric compounds based on chitosan composite (CS) and nanocomposite (7% TiO$_2$-CS) affect the control of corrosion of mild steel in an acidic medium of 1.0 M HCl.

7.3.1   WORKING METHOD

7.3.1.1   Preparation of Samples

- Preparation of chitosan (CS)

  Chitosan with a degree of deacetylation of 80% was prepared by alkaline deacetylation of chitin extracted from shrimp shells collected in Morocco. The detailed procedure of this preparation has been published in our previous paper [19].

- TiO$_2$-chitosan hybrid bionanocomposite (7% TiO$_2$-CS)

  The synthesis of the hybrid bionanocomposite (7% TiO$_2$-CS) was done using the following procedure: a well-defined amount of purified chitosan was dissolved in a (2% V/V) acetic acid solution until a colloidal gel was formed. The mixture of the titanium nanoparticles prepared previously was added dropwise into the chitosan gel slowly under a constant stirring and a temperature between 50°C and 60°C for 24 hours. The gelled solution was washed to neutral pH and centrifuged quite a few times to remove the maximum amount of water to conserve the gel before drying 72 hours until we have a dry scale.

7.3.1.2 Preparation of Electrodes and Solutions

HCl solutions were made by diluting the caustic 37% commercial laboratory grade HCl solution in distilled water. Testing for corrosion was done in a 1 M HCl solution with and without the addition of CS and 7% TiO_2-CS at various mass concentrations ranging from 0.1 to 0.4 g L^{-1} [20].

A mild steel bar was used as the source for the working electrode. Its elemental makeup has been described in previous research. Prior to usage, it was manually abraded using sandpaper of increasingly finer grits (200, 400, 800, 1,200, and 2,000). Then, distilled water and acetone were used, respectively, to clean the electrode surface. After which, it was very carefully dried with paper before being dipped into the test solution. For each test, an electrode that had recently been abraded was employed. The metal bar's proper surface area was 1 cm^2. A three-electrode cell is used, with the working electrode being made of mild steel, the counter electrode being made of platinum wire, and the reference electrode being made of silver/silver chloride (Ag/AgCl) [20,21].

7.3.1.3 Electrochemical Measurements

Electrochemical experiments (EIS and PP) were carried out using a PGZ100 potentiostat workstation and Voltamaster 4 software. The corrosion conduct of mild steel in molar HCl solutions was investigated using EIS and potentiodynamic plots in both the absence and presence of inhibitor solutions. The working electrode was submerged in the test fluid for 30 minutes to maintain a constant open circuit voltage (EOCP) [22]. The EOCP measurement was followed by the electrochemical measurements. Polarization curves in the potential range of 0.9–0.1 V were demonstrated at a scanning rate of 1 mV s^{-1}. Linear polarization resistance measurements were obtained by charting the electrode potential versus open circuit potential starting at a greater negative potential. To ensure that the electrodes were thoroughly immersed within the cylindrical Pyrex glass cell, 50 cc of diluted hydrochloric acid solution had to be used. With the exception of the temperature study, all tests were carried out outdoors. The mild steel specimen has a 1 cm^2 entire submerged surface [23,24].

When inhibitive chemicals are present or not, the corrosion current densities are, respectively, i_{corr} and $i_{corr(inhib)}$. For the examination of the effects of temperature, operating temperatures ranged from 298 to 328 K. The inhibitory effectiveness was calculated using the following equation $\eta(\%)$:

$$\eta(\%) = \frac{i_{corr} - i_{corr(inhib)}}{i_{corr}} \times 100 \tag{7.1}$$

The EIS measurements were carried out using an AC signal with a 10 mV amplitude and a frequency domain spanning from 10^5 to 10^{-1} Hz on the same workstation that was previously mentioned at open circuit potential. EIS charts have been created in both the Nyquist and Bode forms. Then, to fit charts and assess data in terms of an analog electrical circuit, Ec-Lab ver. 10.01 software was utilized. The inhibitory effectiveness obtained *via* EIS is calculated using the following equation [25,26]:

$$\eta_{EIS}(\%) = \frac{R_{CT(inhib)} - R_{CT}}{R_{CT(inhib)}} \times 100 \tag{7.2}$$

7.3.1.4 Surface Analysis

7.3.1.4.1 SEM/EDS

Using a scanning electron microscope of the Quanta FEG 450 type, the electrode surface of mild steel was examined in both the absence and presence of the researched MOFs 1 and 2 at 25°C for a day. After being gently cleaned with bi-distilled water, the samples were appropriately dried before being analyzed without further care [27].

7.3.2 RESULTS

7.3.2.1 Characterization of CS and 7% TiO$_2$-CS

7.3.2.1.1 FTIR

In the chitosan spectrum (Figure 7.5), the 3,425 cm^{-1} band is attributed to the stretching vibration of O–H and NH, the 2,894 and 2,925 cm^{-1} bands are attributed to the stretching vibration of C–H in CH$_2$ and CH$_3$, the vibration of carbonyl groups in amide I corresponds to the 1,665 cm^{-1} band, and the 1,580 cm^{-1} band corresponds to the vibration of the protonated amino groups. The C–N stretching vibration in amide (III) corresponds to 1,380 cm^{-1} band, and the 1,320 and 1,420 cm^{-1} bands are attributed to the C–H bending vibration in CH$_2$ and CH$_3$, respectively. The FTIR spectrum of TiO$_2$ shows a broadband between 420 and 750 cm^{-1} because of the overlapping of multiple bands due to the different vibrational modes of Ti-O and Ti-O-Ti in TiO$_2$. A comparison of the FTIR spectra of CS and TiO$_2$ with the spectrum of the 7% TiO$_2$-CS nanocomposite indicates the presence of distinct bands for CS and TiO$_2$ in the nanocomposite. The region 3,500–3,000 cm^{-1} is attributed to the stretching (–OH–) of water molecules. It can be seen that the 1,565 cm^{-1} band characteristic of

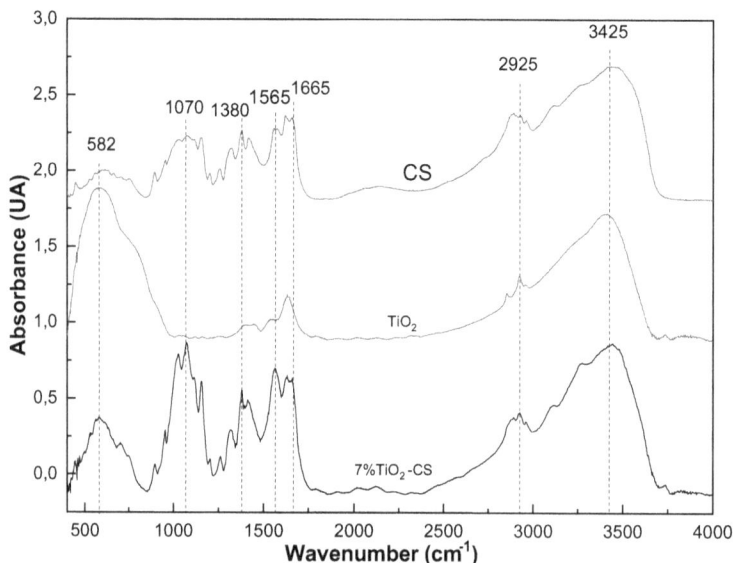

FIGURE 7.5 FTIR spectra of CS, TiO$_2$, and 7% TiO$_2$-CS.

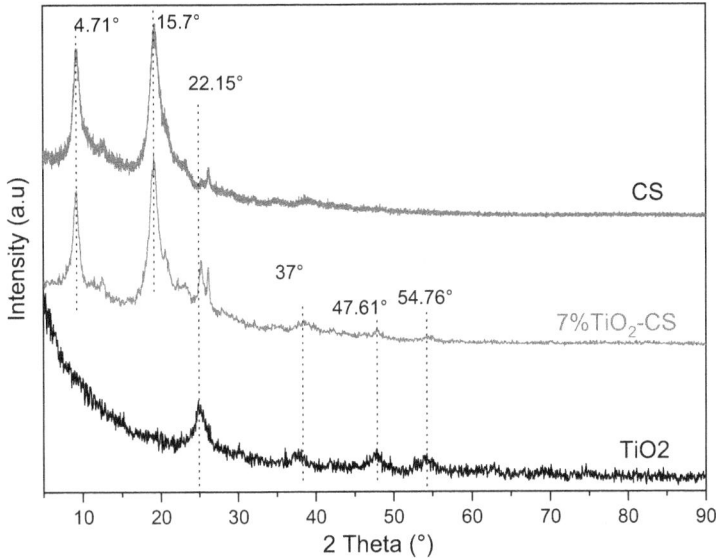

FIGURE 7.6   Typical XRD patterns of synthesized CS, TiO_2, and 7% TiO_2-CS nanocomposites.

protonated aminos has increased in intensity at the 7% TiO_2-CS spectrum indicating that the TiO_2 particles are attached to these groups, which confirms the formation of nanocomposite.

7.3.2.1.2   XRD

Figure 7.6 shows the XRD diffractograms of CS, TiO_2, and 7% TiO_2-CS. Indeed, the XRD pattern obtained for CS consists of two large distinct diffraction peaks at 2θ around 4.71° and 15.7°. These peaks are fingerprints of semi-crystalline CS, indicating the presence of -OH and -NH_2 groups attached to crystal I and crystal II in the structure of CS. The XRD pattern of TiO_2 shows diffraction peaks from 25° to 70°, indicating that TiO_2 is crystalline [28]. The TiO_2 diffractogram showed peaks at 2θ equal 22.15°, 37°, 47.61°, 54.76°, 55.01°, which is in agreement with other studies [28, 29]. These peaks confirm the presence of anatase and rutile phases [29]. They are assigned to the pure tetrahedral anatase phase of host TiO_2 with a coordination geometry of octahedral Ti^{6+} and trigonal O^{3-} planes. However, the 7% TiO_2/CS diffractogram shows the same diffraction peaks of CS as well as the peak at 22.15° of TiO_2, and this shows that there are interactions between CS and TiO_2 to form the nanocomposite.

7.3.2.1.3   TGA/DTA

Figure 7.7 presents the thermograms of CS and 7% TiO_2-CS. Both samples show continued weight loss during processing. For CS, the first loss was recorded in the temperature range from 20°C to 100°C, corresponding to a loss of humidity (10%). This can be explained by the loss of water molecules adsorbed on the polymer due to the hydrophilicity of CS. The second loss, between 200°C and 350°C accompanied by an endothermic peak at $T = 347$°C, is due to the thermal decomposition of CS.

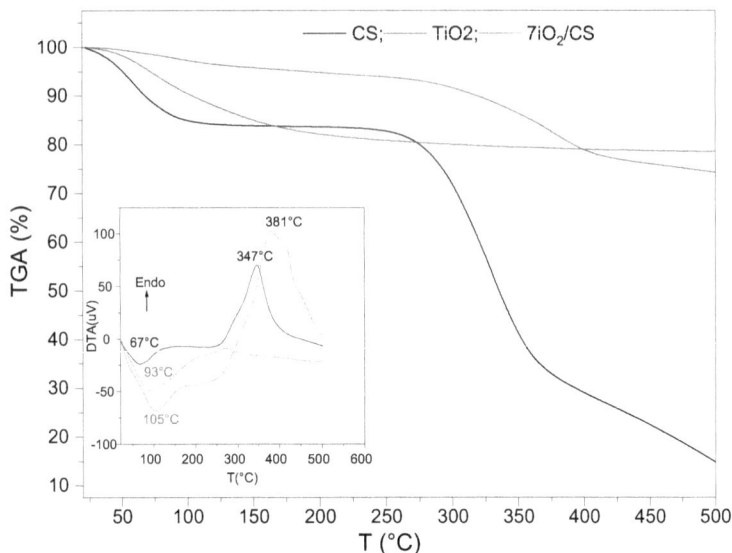

FIGURE 7.7   TGA and DTG curves of CS, TiO$_2$, and 7% TiO$_2$-CS over the temperature range of 25°C–500°C.

A third major mass loss above 350°C has been recorded due to complex dehydration of polysaccharide rings, polymerization, degradation of acetyl units, and removal of minerals in the polymer [30]. The thermogram of 7% TiO$_2$-CS nanocomposite showed a slow thermal deterioration compared to CS. The same degradation temperature was observed during the initial degradation (loss of moisture water). Indeed, a difference in loss was identified in the second degradation step due to the decomposition of CS groups in the nanocomposite with a large endothermic peak at $T = 381$°C. The third minor additional weight loss below 400°C is likely associated with complete degradation of CS. This indicates that the TiO$_2$ particles increase the thermal stability of CS in the nanocomposite since TiO$_2$ is stable in this temperature range of 20°C–500°C, and the only loss recorded is that of water absorbed on the surface of TiO$_2$ at the temperature range of 20°C–120°C.

7.3.2.1.4   TEM

Figure 7.8 shows the TEM images of CS, TiO$_2$, and 7% TiO$_2$-CS. Indeed, the TEM image of CS shows a spherical shape in morphology with their sizes ranging from 40 to 150 nm, and the average particle size value was 100 nm. The TiO$_2$ image shows that the solid is made up of TiO$_2$ grains (diameter varies between 8 and 20 nm) of spherical shape and grouped into agglomerates, as reported in this study [31]. However, the TEM image of 7% TiO$_2$-CS indicates the presence of TiO$_2$ particles dispersed on the CS matrix with no change in their average size. This image indicates that there are interactions between CS and TiO$_2$ to form the nanocomposite which is in agreement with the previous analyses.

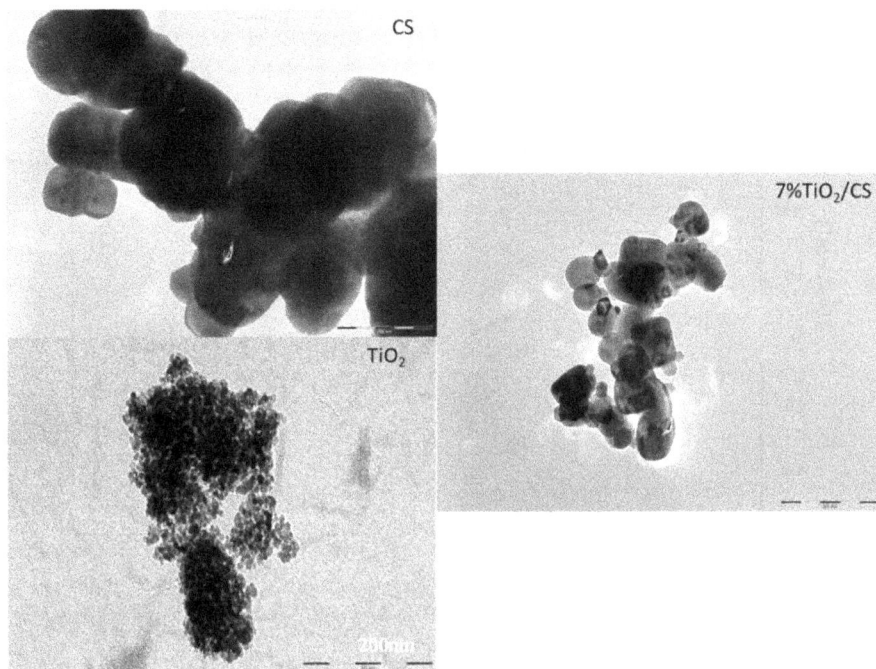

FIGURE 7.8   TEM images of CS, TiO_2, and 7% TiO_2-CS.

7.3.2.2   Electrochemical Study

7.3.2.2.1   Polarization Studies

Figure 7.9 displays the potentiodynamic polarization behavior of mild steel in 1 M HCl, both without and with various concentrations of CS and 7% TiO_2-CS. In Table 7.1, their electrochemical statistics are shown. The two examined inhibitors functioned as mixed-type inhibitors since the Ecorr does not exceed 85 mV and the anodic and cathodic curves of the Tafel plots transition to lower current densities (Figure 7.9) [32,33]. As the concentrations of the examined inhibitors are increased, the values for the corrosion current density (i_{corr}) rapidly decrease, which raises the corrosion inhibition efficiency (IE%). At a concentration of 0.4 g/L, CS and 7% TiO_2-CS had inhibitory powers of 90.7% and 95.5%, respectively. This event demonstrated the good corrosion inhibition performance of the investigated inhibitors by attaching to the active sites and forming a more durable layer on the mild steel surface. The presence of these chemicals and their ability to adsorb on the cathodic areas slows the acid-driven hydrogen evolution process. Because 7% TiO_2 was present, the IE% difference between the CS and 7% TiO_2-CS inhibitors was 4.8%. The presence of adsorbed TiO_2 has an impact on the effectiveness of the inhibition. TiO_2 has improved adsorption characteristics as a result. The anodic and cathodic slopes' respective values alter with the administration of the inhibitor (β_a and β_c). These findings suggest that the cathodic and anodic responses regulate

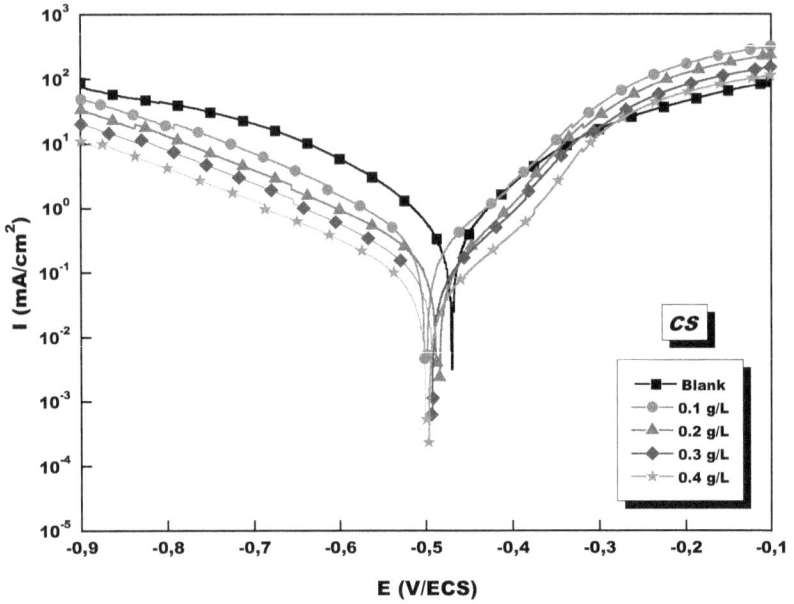

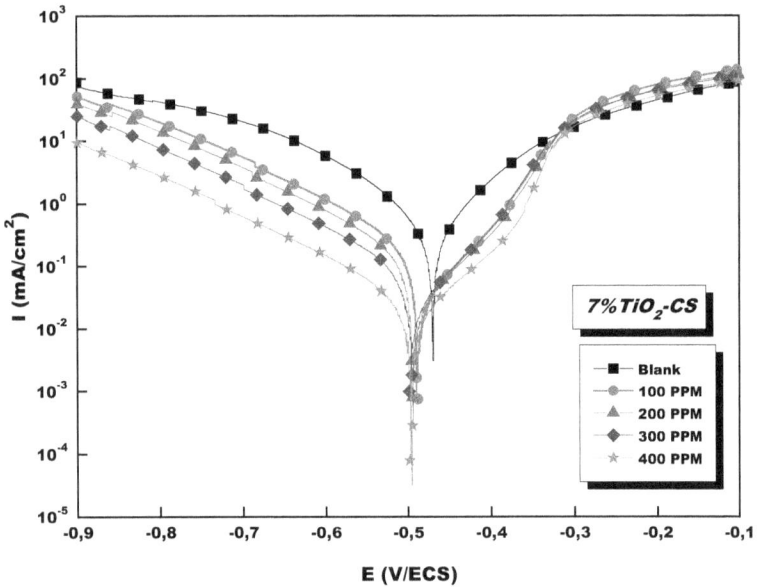

FIGURE 7.9 Polarization curves for the dissolution of mild steel in 1 M HCl at 298 K with and without the presence of CS and 7% TiO_2-CS concentrations.

TABLE 7.1

Stationary Study Parameters of Mild Steel in 1 M HCl with and without CS and 7% TiO$_2$-CS Concentrations at 298 K

Medium	Conc. (g L^{-1})	$-E_{corr}$ (mV SCE^{-1})	i_{corr} (µA cm^{-2})	$-\beta_c$ (mV dec^{-1})	β_a (mV dec^{-1})	η_{pp}%
1.0 M HCl	–	498	983	140	150	–
CS	0.1	499	255	136	145	74.0
	0.2	482	134	133	142	86.3
	0.3	489	132	135	139	86.5
	0.4	497	91	130	134	90.7
TiO$_2$-CS	0.1	486	83	138	147	91.5
	0.2	491	82	135	144	91.6
	0.3	489	73	136	138	92.5
	0.4	493	44	131	135	95.5

the inhibition [34,35]. The corrosion inhibition effectiveness (IE%) was calculated using many chemical concentrations:

$$\eta_{pp}\% = \frac{i_{corr}^{0} - i_{corr}}{i_{corr}^{0}} \times 100 \tag{7.3}$$

where, correspondingly, i_{corr} and i_{corr}^{0} represent the corrosion current densities with and without inhibitor molecules.

7.3.2.2.2 Electrochemical Impedance Spectroscopy

Stationary electrochemical methods are unable to adequately describe complex systems with several reaction stages and varying characteristics. Additionally, research has demonstrated that the essential stages of global corrosion and protection may be identified using EIS data [36]. Figure 7.10 displays mild steel Nyquist plots in 1 M HCl with and without CS or 7% TiO$_2$-CS at various concentrations. These graphs, which illustrate the heterogeneous and uneven surface of solid electrodes, show a phase shift from the true axis with a single capacitive loop in the form of a semicircle. The formation of corrosion products, the presence of contaminants, the adsorption of inhibitors, and the makeup of the layers that have formed on the electrode surface are the causes of this heterogeneity [37,38]. Table 7.2 provides the various parameters and the values of their error margins as a result of the parametric fitting utilizing a similar electrical circuit (Figure 7.10). A solution resistor, a constant phase element ($Q_{dl} = Q_2$, n_{dl}), and a charge transfer resistor were used to build this electrical circuit ($R_s = R_1$). Thus, a reasonable match was found for all experimental data in both the presence and absence of the two inhibitors. It was discovered that the impedance responses gradually alter with the addition of the inhibitor, with the molecular makeup and concentration of the two polymers having an impact on the responses' widths. Further evidence comes from the electrochemical characteristics, which show that the presence of inhibitors lowers the Q coefficient while raising the polarization resistance ($R_p = R_{ct} = R_2$). In fact, the solution's polarization resistance

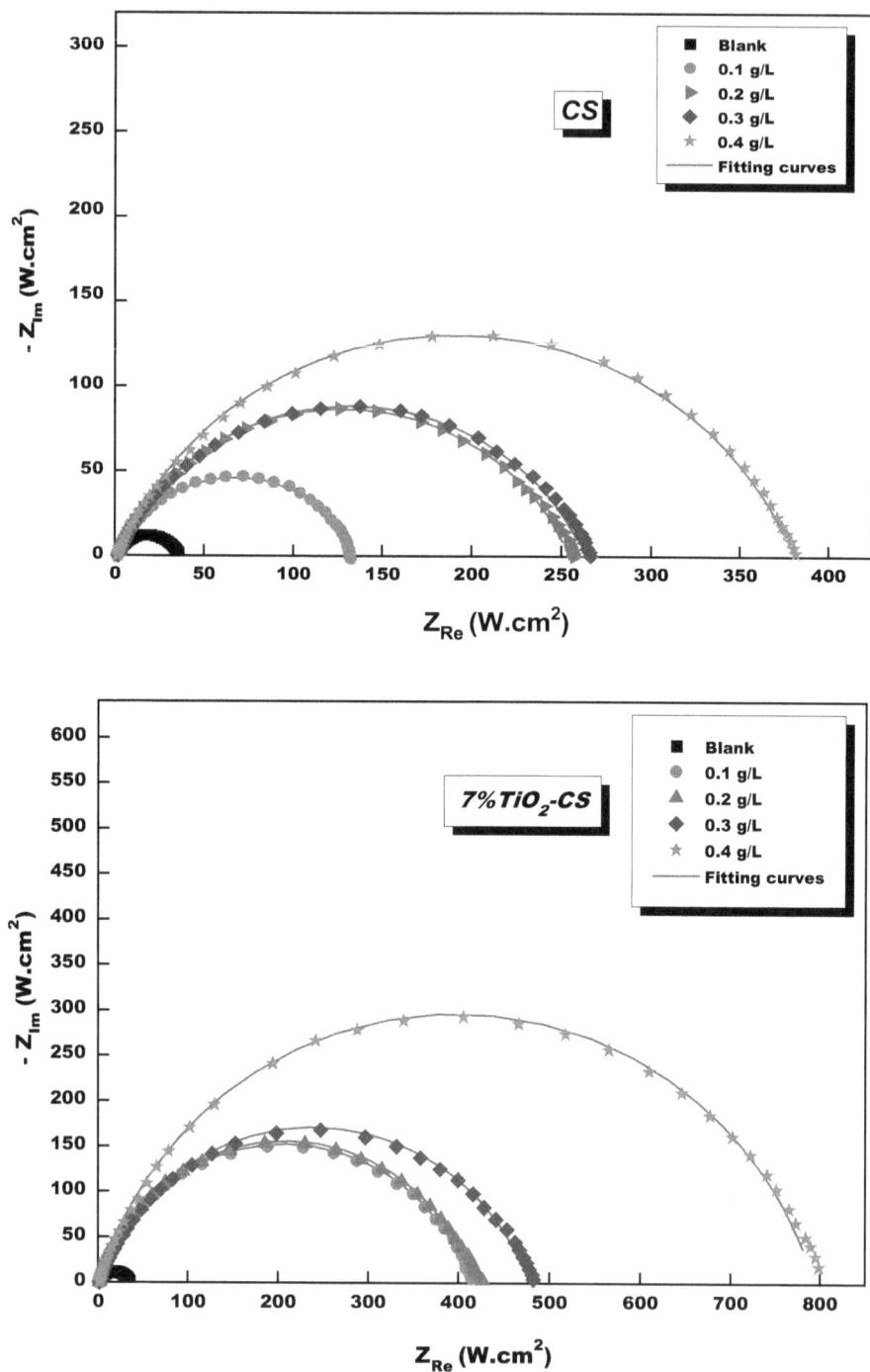

FIGURE 7.10 Nyquist plots for corrosion of mild steel in 1 M HCl in the presence and absence of CS and 7% TiO$_2$-CS concentrations at 298 K.

TABLE 7.2

Electrochemical Kinetic Parameters Obtained by EIS Technique for in 1 M HCl without and with Various Concentrations CS and 7% TiO$_2$-CS at 298 K

Medium	Conc (g L^{-1})	R_s ($\Omega \cdot$cm^2)	Q (μF$\cdot$Sn^{-1})	n	C_{dl} (μF cm^{-2})	R_{ct} ($\Omega \cdot$cm^2)	η_{imp}%	θ
1.0 M HCl	–	1,107	420	0.772	121.0	34.8	–	–
CS	0.1	1.1	328	0.7	119.0	133.4	73.9	0.739
	0.2	1.6	236	0.7	87.0	252.7	86.2	0.862
	0.3	1.6	213	0.7	96.4	261.4	86.6	0.866
	0.4	1.3	167	0.7	72.2	379.2	90.8	0.908
TiO$_2$-CS	0.1	1.4	174	0.8	95.0	410.4	91.5	0.915
	0.2	1.4	167	0.8	93.0	415.1	91.6	0.916
	0.3	1.5	166	0.7	86.5	475.8	92.6	0.926
	0.4	1.7	96	0.8	54.4	792.2	95.6	0.956

rises from 34.8 to 379.2 Ωcm^2 in the presence of 0.4 g L^{-1} of CS, and it reaches 792.2 Ωcm^2 in the presence of 0.4 g L^{-1} of 7% TiO$_2$-CS. The reduction in active sites on the mild steel surface brought on by the adherence of both inhibitors to the surface (increase in the thickness of the double layer) may be utilized to explain this outcome. The fact that the n value is less than 0.8 indicates that the mild steel surface's homogeneity has been improved by the inhibitor molecules' adsorption there. On the contrary, it can be shown that each inhibitor exhibits the same pattern of inhibitory performance at 0.4 g L^{-1}: CS (90.8%) and 7% TiO$_2$-CS (95.6%), which supports the findings of polarization investigations [39,40] (Figure 7.11). The charge transfer resistance R_{ct} values are used to compute the inhibition efficiency in accordance with the following equations:

$$\theta = \left(1 - \frac{R'_{ct}}{R_{ct}}\right) \tag{7.4}$$

$$\eta_{imp}\% = \left(1 - \frac{R'_{ct}}{R_{ct}}\right) \times 100 \tag{7.5}$$

7.3.2.2.2.1 Isotherm Adsorption   Adsorption isotherms might provide further information on how an inhibitor prevents a reaction:

$$\text{Langmuir isotherm:} \frac{\theta}{1-\theta} = K_{ads} C_{inh} \tag{7.6}$$

where K_{ads} is the equilibrium constant of the adsorption process and C_{inh} is the inhibitor concentration. The impedance information shown in Table 7.2 is used to compute the fractional coverage surface (θ). To choose the best isotherm, slope, R_2, and linear

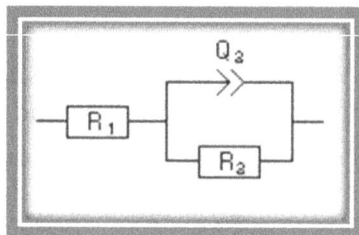

FIGURE 7.11 Equivalent circuit model for experimental data fitting.

TABLE 7.3
Thermodynamic Adsorption Properties of CS and 7%
TiO_2-CS on the Surface of Mild Steel at 298 K and 1 M HCl

Inhibitor	K_{ads} (L mol^{-1})	ΔG_{ads} (kJ mol^{-1})	R_2	Slopes
CS	32.1	−18.5	0.9993	1.03
7% TiO_2-CS	108.1	−21.5	0.99947	1.03

coefficient regression were used (Table 7.3). This equation relates the free energy of adsorption, ΔG_{ads}, to K_{ads}:

$$K_{ads} = \frac{1}{55.5} e^{-\dfrac{\Delta G_{ads}}{RT}} \tag{7.7}$$

where T is the absolute temperature, R is the universal gas constant, and 55.5 is the amount of water in the solution (mol·L^{-1}).

Figure 7.12 illustrates the Langmuir isotherm curve, with a correlation factor of around one (r 1). The Langmuir adsorption isotherm model governs the monolayer adsorption of CS and 7% TiO_2-CS on mild steel. According to this kind of isotherm, there won't be any interactions between the species that have adsorption on the electrode surface [41,42].

The computed K_{ads} and ΔG_{ads} values are shown in Table 7.3. The high K_{ads} values imply that these inhibitors stick to metal surfaces fairly strongly. Furthermore, the negative ΔG_{ads} values and the substantial interaction between the inhibitor molecules and the adsorption energy demonstrate that CS and 7% TiO_2-CS adsorb spontaneously on the mild steel surface. In reality, conventional free energy values larger than −20 kJ·mol^{-1}, which indicate physical adsorption, are connected to electrostatic interactions between charged molecules and charged metal surfaces. A chemical adsorption, or charge transfer from the inhibitor molecules to the metal surface, is indicated by values below −40 kJ·mol^{-1}. According to the literature, the findings of thermodynamic study demonstrate that the two forms of interaction engaged in the adsorption process of CS and 7% TiO_2-CS on the mild

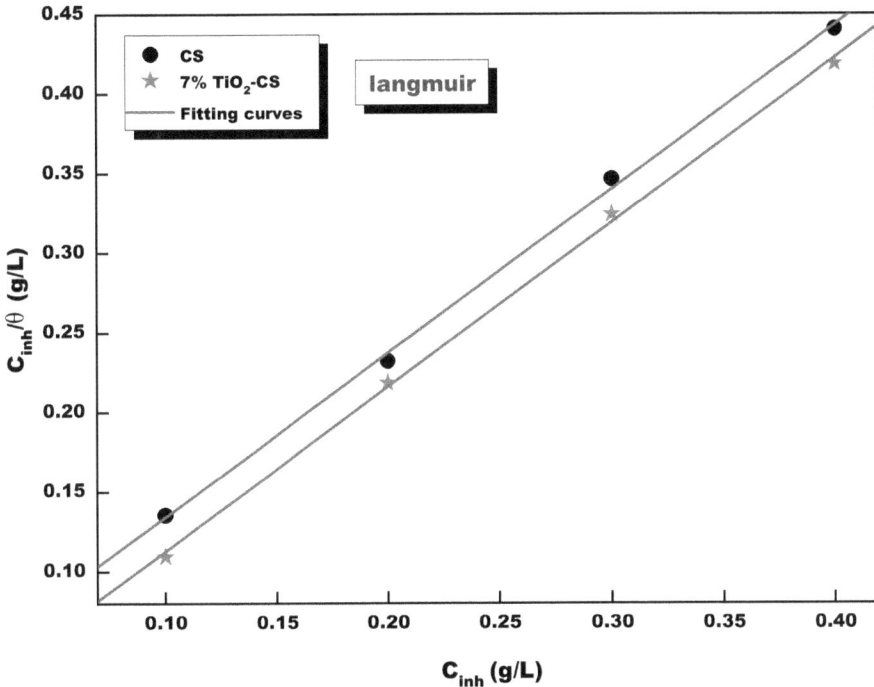

FIGURE 7.12   The Langmuir isotherm at 298 K in 1 M HCl with CS and 7% TiO$_2$-CS for mild steel.

steel surface in 1 M (molecular) HCl solution are electrostatic (ionic) adsorption and physical adsorption [43–45].

7.3.2.2.3   Effect of Temperature

Temperature is one of the most crucial factors that may determine how a metal behaves in a corrosive environment since it encourages desorption of the inhibitor and speeds up the dissolving of any produced molecule or organic complex. Electrochemical studies in potentiostatic mode were carried out in the temperature range of 298–328 K to assess how temperature impacts CS and 7% TiO$_2$-CS capacity to inhibit corrosion [46–48]. The polarization curves of mild steel before and after the addition of CS and 7% TiO$_2$-CS at a concentration of 0.4 g L^{-1} in 1 M HCl medium were thus shown after 30 minutes of immersion.

Figure 7.13 shows the effect of temperature and the presence or absence of the optimal 0.4 g L^{-1} inhibitor concentration on the polarization curves of mild steel in a 1 M HCl medium. As shown in Figure 7.13, increasing temperature increases anodic and cathodic current densities in both the absence and presence of inhibitors. This implies that the inhibitor's presence increased the corrosion kinetics of mild steel. Further evidence that temperature merely affects the rate of corrosion and not the corrosion process is provided by the curve's relative parallelism.

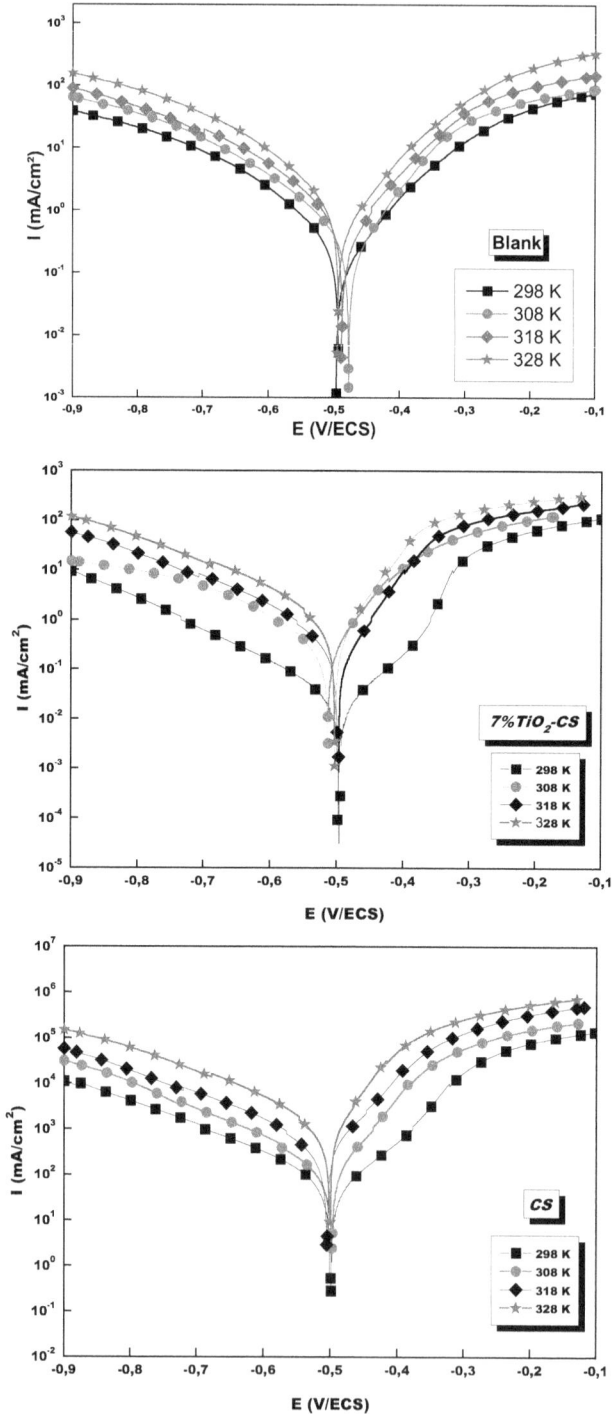

FIGURE 7.13 Temperature effects on mild steel polarization curves in 1 M HCl, both with and without the optimal concentrations of CS and 7% TiO$_2$-CS.

TABLE 7.4
Effects of Temperature on the Electrochemical Characteristics of Mild Steel in 1 M HCl with and without the Addition of the Optimal Concentrations of CS and 7% TiO$_2$-CS

Medium	T (K)	$-E_{corr}$ (mVSCE)	i_{corr} (μA cm^{-2})	Tafel Slopes (mV dec^{-1}) $-\beta_c$	β_a	η_{pp}%
1.0 M HCl	298	498	983	92	104	–
	308	491	1,200	184	112	–
	318	475	1,450	171	124	–
	328	465	2,200	161	118	–
CS	298	497	91	130	134	90.7
	308	495	134	156	114	88.8
	318	502	224	164	137	84.5
	328	498	398	147	128	81.9
7% TiO$_2$-CS	298	493	44	131	135	95.5
	308	510	99	167	138	91.7
	318	497	182	152	145	87.4
	328	500	316	157	134	85.6

Table 7.4, provides data from the potentiodynamic plots (η_{PP}, c, i_{corr}, and E_{corr}), which show that temperature increases the corrosion current density (i_{corr}) in both inhibited and uninhibited solutions. This result demonstrates that when the metal dissolving process quickens, mild steel corrosion rates increase. Additionally, the inhibitory efficiency values for CS and 7% TiO$_2$-CS decline with increasing temperature and reach 81.9% and 85.6%, respectively, at 328 K. This illustrates how the barrier has worsened [49].

7.3.2.2.3.1 The Parameters Corrosion Process Activation The calculation of thermodynamic characteristics like activation energy, activation enthalpy, and activation entropy enables the assessment and interpretation of the kind of adsorption utilized by the inhibitor when determining the efficiency of corrosion inhibition against corrosion [50]. The values of the parameters listed in Table 7.5 are calculated using the logarithm of the corrosion rate L_n (i_{corr}) function of (1,000/T) plots of Arrhenius curves (Figure 7.14). As a result, we were able to compute the corrosion process' activation energy E_a using the following equation [51,52]:

$$\ln(I_{corr}) = \ln A - \frac{E_a}{RT} \tag{7.8}$$

T is the absolute temperature, R is the perfect gas constant, and A is the Arrhenius pre-exponential constant.

The calculation of the adsorption entropy (ΔS_a) and adsorption enthalpy (ΔH_a) using the change in ($Ln(i_{corr}/T)$) vs. (1,000/T) is shown in Figure 7.14, where

TABLE 7.5

The Mild Steel Corrosion Process Activation Properties in 1 M HCl in the Presence and Absence of 0.4 g L⁻¹ CS and 7% TiO₂-CS

Medium	E_a (KJ mol⁻¹)	ΔH_a (KJ mol⁻¹)	ΔS_a (J mol·K⁻¹)
1 M HCl	21.0	18.4	−126
CS	40.0	37.4	−84.6
7% Ti-CS	53.0	50.4	−43.6

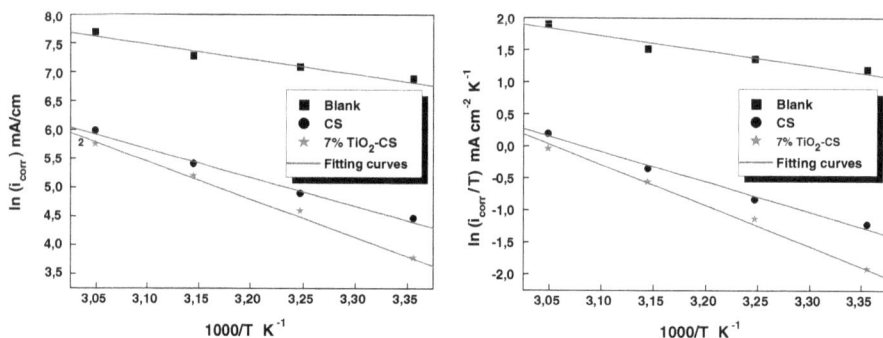

FIGURE 7.14 Arrhenius lines obtained from the mild steel corrosion current density in 1 M HCl with and without 0.4 g L⁻¹ CS and 7% TiO₂-CS.

$(-(\Delta H_a/R))$ is the slope and $[Ln(R/Nh)+((\Delta S_a/R)]$ is the point where the straight line intersects the y-axis:

$$Ln\left(\frac{I_{corr}}{T}\right)=\left(Ln\left(\frac{R}{Nh}\right)\pm\frac{\Delta S_a}{R}\right)-\frac{\Delta H_a}{RT}\qquad(7.9)$$

where N is the Avogadro number, h is the Planck constant, T is the absolute temperature, and S_a and H_a are the activation entropy and enthalpy, respectively. Grouping the results is done in Table 7.5. Figure 7.14 contrasts the mild steel's Arrhenius curves in 1 M HCl solution with those produced in the presence of 0.4 g L⁻¹ CS and 7% TiO₂-CS.

According to the literature, a certain kind of adsorption results from changing the kinetic parameters. In addition, we boost the temperature to encourage electrostatic interactions with the metal surface, and the adsorption of the protective coating is of a physical nature if the presence of our inhibitor raises the activation energy E_a in comparison to the blank. In contrast to a dip in E_a caused by an inhibitor, which is followed by an increase in the efficiency of the inhibition as the temperature rises [24,53]. There is evidence of physisorption on the surface of the mild steel in Table 7.6 data, which demonstrate an increase in activation energy E_a in the presence of the inhibitors compared to the control. The investigated inhibitor, on the contrary, irregularly blocks the cathodic sites, forming a barrier that predominantly thwarts the corrosion process. Additionally, the positive values of ΔH_a demonstrate the endothermic character of the mild steel's dissolving process and indicate that the process is endothermic.

Additionally, as shown by the negative values of entropies (ΔS_a), the activated complex in the rate-determining phase results in less disordered behavior. This shows that the step decreases disorder when the reactants pass through the activated complex by forming a connection rather than a dissociation [54–56].

7.3.2.3   Surface Characterization

7.3.2.3.1   SEM/EDS

After being immersed in 1 M HCl solution, SEM images of the mild steel are shown in Figure 7.15 with and without the addition of the inhibitors at the concentration 0.4 g L^{-1} of CS and 7% TiO$_2$-CS. Due to strong corrosive acid attack, the mild steel sample without inhibitor (Blank) suffered significant damage including corrosion deposits and cracks. In contrast, mild steel with inhibitors was very smooth and showed no corrosion, indicating a decrease in corrosion rate [57,58]. Data on the surface composition of the mild steel specimens before and after the addition of CS and 7% TiO$_2$-CS evaluated in 1 M HCl solution were collected using energy dispersive X-ray analysis (EDS). The EDS study spectra are shown in Figure 7.16. The polished mild steel sample (Blank) was examined using EDS, and the peaks it discovered match the primary elements that were present in it. However, the extra Cl peak that formed in the EDX spectra of the mild steel after immersion in 1 M HCl, which was generated by the free corrosion of the mild steel in the HCl solution (Blank). The oxygen, iron, and chlorine (Cl) peaks show a little drop and disappearance when CS and 7% TiO$_2$-CS are present, respectively. This proves that the electrode/media contact served as a site

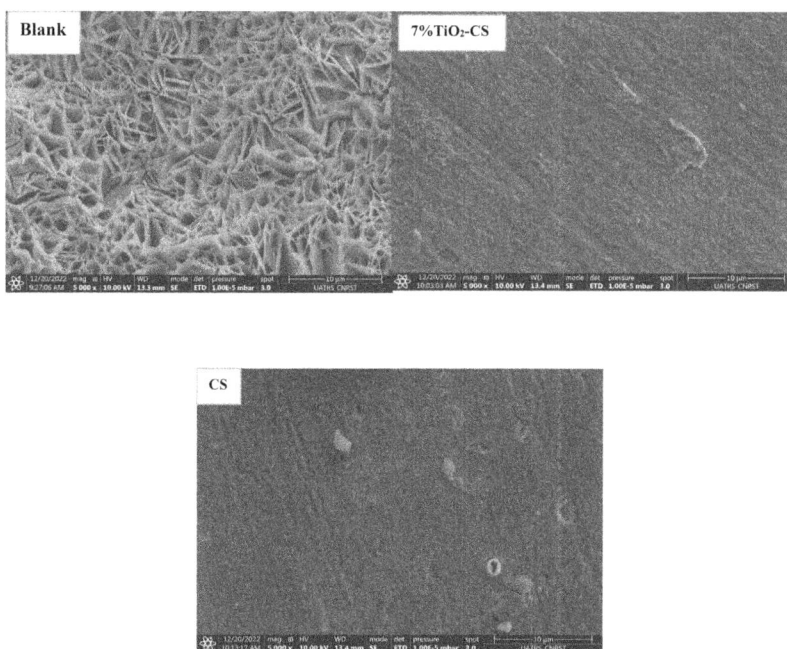

FIGURE 7.15   SEM pictures of mild steel after immersion in a 1 M HCl solution with and without 0.4 g L^{-1} additions of the inhibitors CS and TiO$_2$-CS.

of adsorptive binding for the surface active species. An examination of the principal components' weight percentages on the metal contact is provided in Table 7.6. The fact that Table 7.6 demonstrates that the fraction of atomic oxygen content decreases as atomic titanium (Ti) content emerges on the surface of the inhibited metal proves that all of the compounds studied cling to the mild steel surface [59,60].

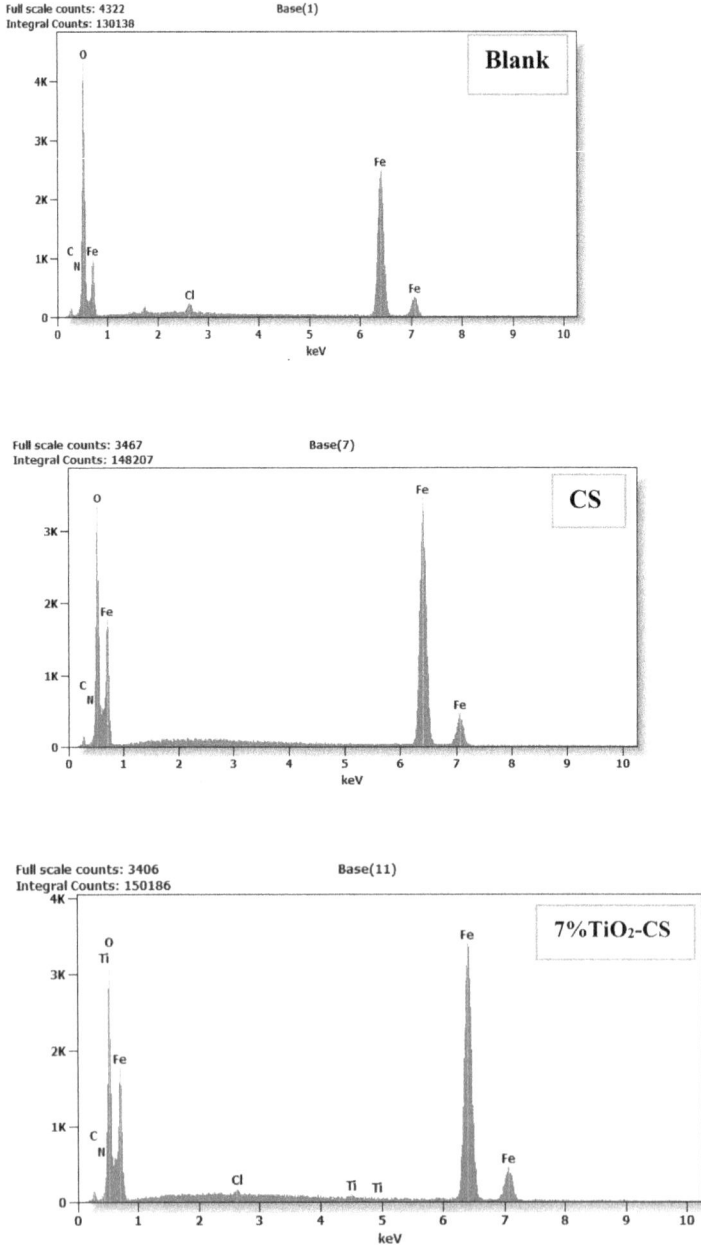

FIGURE 7.16 EDS spectra of mild steel immersed in 1 M HCl, both by themselves and with 0.4 g L^{-1} of the molecules being studied.

TABLE 7.6
EDS Investigation of the Surface
Compositions of Mild Steel

Elements	Blank	CS	7% TiO$_2$-CS
C	2.1	1.4	1.4
N	0.5	0.4	0.4
O	30.6	19.9	18.9
Cl	1.0	–	0.2
Ti	–	–	0.3
Fe	65.8	78.3	78.8
Total	100.0	100.0	100.0

7.4   CONCLUSION

This chapter compares the most recent improvements in the potency of nanocomposites and composites based on chitosan. However, using electrochemical and surface morphological (SEM-EDS) methods, two newly synthesized chitosan-based polymers (CS and 7% TiO$_2$-CS) are investigated for their inhibitory effectiveness against the corrosion of mild steel in 1 M hydrochloric acid. According to the findings, 7% TiO$_2$-CS is a far more potent inhibitor of behavior than CS: 7% TiO$_2$-CS > CS. The Langmuir technique, which was used in the experiment, demonstrated that the two items under investigation had a mixed-type character because they adsorb on the metal surface. Additionally, SEM/EDS is used to validate the development of a surface coating on the object.

REFERENCES

[1] A. Ouass et al., "Poly(sodium acrylate) and poly(acrylic acid sodium) as an eco-friendly corrosion inhibitor of mild steel in normal hydrochloric acid: experimental, spectroscopic and theoretical approach," *J. Appl. Electrochem.*, vol. 51, no. 7, pp. 1009–1032, 2021, doi:10.1007/s10800-021-01556-y.
[2] M. El Faydy et al., "Experimental and theoretical studies for steel XC38 corrosion inhibition in 1 M HCl by N-(8-hydroxyquinolin-5-yl)-methyl)-N-phenylacetamide," *J. Mater. Environ. Sci.*, vol. 7, no. 4, pp. 1406–1416, 2016.
[3] L. Kadiri et al., "Coriandrum sativum. L seeds extract as a novel green corrosion inhibitor for mild steel in 1.0 M hydrochloric and 0.5 M sulfuric solutions," *Anal. Bioanal. Electrochem.*, vol. 10, no. 2, pp. 249–268, 2018.
[4] M. Ouakki et al., "Investigation of imidazole derivatives as corrosion inhibitors for mild steel in sulfuric acidic environment: experimental and theoretical studies," *Ionics (Kiel).*, vol. 26, no. 10, pp. 5251–5272, 2020, doi:10.1007/s11581-020-03643-0.
[5] M. M. Solomon, H. Gerengi, and S. A. Umoren, "Carboxymethyl cellulose/silver nanoparticles composite: synthesis, characterization and application as a benign corrosion inhibitor for St37 steel in 15% H$_2$SO$_4$ medium," *ACS Appl. Mater. Interfaces*, vol. 9, no. 7, pp. 6376–6389, 2017, doi:10.1021/acsami.6b14153.

[6] S. Ko, A. R. Prasad, J. Garvasis, S. M. Basheer, and A. Joseph, "Stearic acid grafted chitosan/epoxy blend surface coating for prolonged protection of mild steel in saline environment," *J. Adhes. Sci. Technol.*, vol. 33, no. 20, pp. 2250–2264, 2019, doi:10.1080 /01694243.2019.1637170.

[7] L. Cordero-Arias et al., "Electrophoretic deposition of nanostructured-TiO$_2$/chitosan composite coatings on stainless steel," *RSC Adv.*, vol. 3, no. 28, pp. 11247–11254, 2013, doi:10.1039/c3ra40535d.

[8] C. Zhi, Y. Bando, C. Tan, and D. Golberg, "Effective precursor for high yield synthesis of pure BN nanotubes," *Solid State Commun.*, vol. 135, no. 1–2, pp. 67–70, 2005, doi:10.1016/j.ssc.2005.03.062.

[9] Y. Chen et al., "In-situ intercalation of montmorillonite/urushiol titanium polymer nanocomposite for anti-corrosion and anti-aging of epoxy coatings," *Prog. Org. Coatings*, vol. 165, p. 106738, 2022, doi:10.1016/J.PORGCOAT.2022.106738.

[10] A. Gharib, M. Ahmed Maher, S. Hamed Ismail, and G. Genidy Mohamed, "Effect titanium dioxide/paraloid B.72 nanocomposite coating on protection of treated Cu-Zn archaeological alloys," *Int. J. Archaeol.*, vol. 7, no. 2, p. 47, 2019, doi:10.11648/j.ija.20190702.13.

[11] A. Ershad-Langroudi, H. Abdollahi, and A. Rahimi, "Mechanical properties of sol-gel prepared nanocomposite coatings in the presence of titania and alumina-derived nanoparticles," *Plast. Rubber Compos.*, vol. 46, no. 1, pp. 25–34, 2017, doi:10.1080/146 58011.2016.1245821.

[12] R. Zandi-Zand, A. Ershad-Langroudi, and A. Rahimi, "Silica based organic-inorganic hybrid nanocomposite coatings for corrosion protection," *Prog. Org. Coatings*, vol. 53, no. 4, pp. 286–291, 2005, doi:10.1016/J.PORGCOAT.2005.03.009.

[13] H. Al Jabri, M. G. Devi, and M. A. Al-Shukaili, "Development of polyaniline - TiO$_2$ nano composite films and its application in corrosion inhibition of oil pipelines," *J. Indian Chem. Soc.*, vol. 100, no. 1, p. 100826, 2023, doi:10.1016/J.JICS.2022.100826.

[14] P. Kesari, G. Udayabhanu, A. Roy, and S. Pal, "Chitosan based titanium and iron oxide hybrid bio-polymeric nanocomposites as potential corrosion inhibitor for mild steel in acidic medium," *Int. J. Biol. Macromol.*, vol. 225, pp. 1323–1349, 2023, doi:10.1016/J. IJBIOMAC.2022.11.192.

[15] Z. P. Mathew, G. K. Shamnamol, K. P. Greeshma, and S. John, "Insight on the corrosion inhibition of nanocomposite chitosan/boron nitride integrated epoxy coating system against mild steel," *Corros. Commun.*, vol. 84, pp. 28–34, 2023, doi:10.1016/ J.CORCOM.2022.09.001.

[16] W. Zhang et al., "High-performance corrosion resistance of chemically-reinforced chitosan as ecofriendly inhibitor for mild steel," *Bioelectrochemistry*, vol. 150, p. 108330, 2023, doi:10.1016/J.BIOELECHEM.2022.108330.

[17] D. Wang, Y. Li, T. Chang, and A. Luo, "Experimental and theoretical studies of chitosan derivatives as green corrosion inhibitor for oil and gas well acid acidizing," *Colloids Surf. A Physicochem. Eng. Asp.*, vol. 628, p. 127308, 2021, doi:10.1016/J. COLSURFA.2021.127308.

[18] X. Lai, J. Hu, T. Ruan, J. Zhou, and J. Qu, "Chitosan derivative corrosion inhibitor for aluminum alloy in sodium chloride solution: a green organic/inorganic hybrid," *Carbohydr. Polym.*, vol. 265, p. 118074, 2021, doi:10.1016/J.CARBPOL.2021.118074.

[19] H. Ahlafi, H. Moussout, F. Boukhlifi, M. Echetna, M. N. Bennani, and S. My Slimane, "Kinetics of N-deacetylation of chitin extracted from shrimp shells collected from coastal area of Morocco," *Mediterr. J. Chem.*, vol. 2, no. 3, pp. 503–513, 2013, doi:10.13171/mjc.2.3.2013.22.01.20.

[20] M. Galai et al., "Synthesis, characterization and anti-corrosion properties of novel quinolinol on C-steel in a molar hydrochloric acid solution," *Port. Electrochim. Acta*, vol. 35, no. 4, pp. 233–251, 2017, doi:10.4152/pea.201704233.

[21] K. Alaoui et al., "Anti-corrosive properties of polyvinyl-alcohol for carbon steel in hydrochloric acid media: electrochemical and thermodynamic investigation," *J. Mater. Environ. Sci.*, vol. 7, no. 7, pp. 2389–2403, 2016.

[22] Y. El Kacimi et al., "Effect of silicon and phosphorus contents in steel on its corrosion inhibition in 5 M HCl solution in the presence of Cetyltrimethylammonium/KI," *J. Mater. Environ. Sci.*, vol. 7, no. 1, pp. 371–381, 2016.

[23] M. Galai et al., "α-Brass and (α + β) brass degradation processes in azrou soil medium used in plumbing devices," *J. Bio- Tribo-Corrosion*, vol. 3, no. 3, p. 30, 2017, doi:10.1007/s40735-017-0087-y.

[24] N. Dkhireche et al., "New quinoline derivatives as sulfuric acid inhibitor's for mild steel," *Anal. Bioanal. Electrochem.*, vol. 10, no. 1, pp. 111–135, 2018.

[25] M. Galai et al., "Chemically functionalized of 8-hydroxyquinoline derivatives as efficient corrosion inhibition for steel in 1.0 M HCl solution: experimental and theoretical studies," *Surf. Interfaces*, vol. 21, p. 100695, 2020, doi:10.1016/j.surfin.2020.100695.

[26] N. Errahmany et al., "Experimental, DFT calculations and MC simulations concept of novel quinazolinone derivatives as corrosion inhibitor for mild steel in 1.0 M HCl medium," *J. Mol. Liq.*, vol. 312, p. 113413, 2020, doi:10.1016/J.MOLLIQ.2020.113413.

[27] M. Ouakki et al., "Insights into corrosion inhibition mechanism of mild steel in 1 M HCl solution by quinoxaline derivatives: electrochemical, SEM/EDAX, UV-visible, FT-IR and theoretical approaches," *Colloids Surf. A Physicochem. Eng. Asp.*, vol. 611, p. 125810, 2021, doi:10.1016/J.COLSURFA.2020.125810.

[28] R. S. Dubey and S. Singh, "Investigation of structural and optical properties of pure and chromium doped TiO_2 nanoparticles prepared by solvothermal method," *Results Phys.*, vol. 7, pp. 1283–1288, 2017, doi:10.1016/J.RINP.2017.03.014.

[29] E. Filippo et al., "Enhanced photocatalytic activity of pure anatase TiO_2 and Pt-TiO_2 nanoparticles synthesized by green microwave assisted route," *Mater. Res.*, vol. 18, no. 3, pp. 473–481, 2015, doi:10.1590/1516-1439.301914.

[30] H. Moussout, H. Ahlafi, M. Aazza, and M. Bourakhouadar, "Kinetics and mechanism of the thermal degradation of biopolymers chitin and chitosan using thermogravimetric analysis," *Polym. Degrad. Stab.*, vol. 130, pp. 1–9, 2016, doi:10.1016/J.POLYMDEGRADSTAB.2016.05.016.

[31] I. Daou, R. Chfaira, O. Zegaoui, Z. Aouni, and H. Ahlafi, "Physico-chemical characterization and interfacial electrochemical properties of nanoparticles of anatase-TiO_2 prepared by the sol-gel method," *Mediterr. J. Chem.*, vol. 2, no. 4, pp. 569–582, 2013.

[32] T. A. Yousef et al., "Experimental and theoretical examinations of triazole linked saccharin derivatives as organic corrosion inhibitors for mild steel in hydrochloric acid," *J. Mol. Struct.*, vol. 1275, p. 134603, 2023, doi:10.1016/J.MOLSTRUC.2022.134603.

[33] M. Galai et al., "Effect of alkyl group position on adsorption behavior and corrosion inhibition of new naphthol based on 8-hydroxyquinoline: electrochemical, surface, quantum calculations and dynamic simulations," *J. Mol. Liq.*, vol. 335, p. 116552, 2021, doi:10.1016/j.molliq.2021.116552.

[34] S. Omanović and M. Metikoš-Huković, "The ionic conductance of barrier anodic oxide films on indium," *Solid State Ionics*, vol. 78, no. 1–2, pp. 69–78, 1995, doi:10.1016/0167-2738(95)00009-U.

[35] M. Ouakki et al., "A detailed investigation on the corrosion inhibition effect of by newly synthesized pyran derivative on mild steel in 1.0 M HCl: experimental, surface morphological (SEM-EDS, DRX & AFM) and computational analysis (DFT & MD simulation)," *J. Mol. Liq.*, vol. 344, p. 117777, 2021, doi:10.1016/j.molliq.2021.117777.

[36] M. Galai et al., "Functionalization effect on the corrosion inhibition of novel eco-friendly compounds based on 8-hydroxyquinoline derivatives: experimental, theoretical and surface treatment," *Chem. Phys. Lett.*, vol. 776, p. 138700, 2021, doi:10.1016/J.CPLETT.2021.138700.

[37] A. Popova, M. Christov, and A. Vasilev, "Mono- and dicationic benzothiazolic quaternary ammonium bromides as mild steel corrosion inhibitors: part III: influence
of the temperature on the inhibition process," *Corros. Sci.*, vol. 94, pp. 70–78, 2015,
doi:10.1016/j.corsci.2015.01.039.

[38] M. Ouakki et al., "Detailed experimental and computational explorations of pyran
derivatives as corrosion inhibitors for mild steel in 1.0 M HCl: electrochemical/sur-
face studies, DFT modeling, and MC simulation," *J. Mol. Struct.*, vol. 1261, 2022,
doi:10.1016/j.molstruc.2022.132784.

[39] M. Sahrane et al., "Experimental and theoretical studies for mild steel corrosion
inhibition in 1.0 M HCL by two stereoisomers of benzothiazinone derivatives," *Port.
Electrochim. Acta*, vol. 38, no. 1, pp. 1–17, 2020, doi:10.4152/pea.202001001.

[40] A. Leon and E. Aghion, "Effect of surface roughness on corrosion fatigue performance
of AlSi10Mg alloy produced by Selective Laser Melting (SLM)," *Mater. Charact.*, vol.
131, pp. 188–194, 2017, doi:10.1016/J.MATCHAR.2017.06.029.

[41] M. M. Abd El-Latif and M. F. Elkady, "Equilibrium isotherms for harmful ions sorption using nano zirconium vanadate ion exchanger," *Desalination*, vol. 255, no. 1–3,
pp. 21–43, 2010, doi:10.1016/J.DESAL.2010.01.020.

[42] R. Sivaraj, C. Namasivayam, and K. Kadirvelu, "Orange peel as an adsorbent in the
removal of acid violet 17 (acid dye) from aqueous solutions," *Waste Manag.*, vol. 21, no. 1,
pp. 105–110, 2001, doi:10.1016/S0956-053X(00)00076-3.

[43] M. Behpour, S. M. Ghoreishi, N. Soltani, M. Salavati-Niasari, M. Hamadanian, and A.
Gandomi, "Electrochemical and theoretical investigation on the corrosion inhibition
of mild steel by thiosalicylaldehyde derivatives in hydrochloric acid solution," *Corros.
Sci.*, vol. 50, no. 8, pp. 2172–2181, 2008, doi:10.1016/j.corsci.2008.06.020.

[44] G. Vengatesh and M. Sundaravadivelu, "Non-toxic bisacodyl as an effective corrosion inhibitor for mild steel in 1 M HCl: thermodynamic, electrochemical, SEM, EDX,
AFM, FT-IR, DFT and molecular dynamics simulation studies," *J. Mol. Liq.*, vol. 287,
p. 110906, 2019, doi:10.1016/J.MOLLIQ.2019.110906.

[45] M. Yadav, S. Kumar, R. R. Sinha, I. Bahadur, and E. E. Ebenso, "New pyrimidine
derivatives as efficient organic inhibitors on mild steel corrosion in acidic medium:
electrochemical, SEM, EDX, AFM and DFT studies," *J. Mol. Liq.*, vol. 211, pp. 135–
145, 2015, doi:10.1016/J.MOLLIQ.2015.06.063.

[46] J. Crousier, C. Antonione, Y. Massiani, and J. P. Crousier, "Effet du chrome sur la resistance a la corrosion d'alliages amorphes Fe□Ni□B□P dans H₂SO₄ 0,1 N," *Mater. Chem.*,
vol. 7, no. 5, pp. 587–604, 1982, doi:10.1016/0390-6035(82)90064-5.

[47] N. Gharda et al., "Linseed oil as a novel eco-friendly corrosion inhibitor of carbon steel
in 1 M HCl," *Surf. Rev. Lett.*, vol. 26, no. 2, 2019, doi:10.1142/S0218625X18501482.

[48] F. El-Hajjaji et al., "Experimental and quantum studies of newly synthesized pyrid-
azinium derivatives on mild steel in hydrochloric acid medium," *Mater. Today Proc.*,
vol. 13, pp. 822–831, 2019, doi:10.1016/J.MATPR.2019.04.045.

[49] A. Ech-chebab et al., "Evaluation of quinoxaline-2(1H)-one, derivatives as corrosion
inhibitors for mild steel in 1.0 M acidic media: electrochemistry, quantum calculations,
dynamic simulations, and surface analysis," *Chem. Phys. Lett.*, vol. 809, p. 140156,
2022, doi:10.1016/J.CPLETT.2022.140156.

[50] I. Ahamad, R. Prasad, E. E. Ebenso, and M. A. Quraishi, "Electrochemical and quantum chemical study of albendazole as corrosion inhibitor for mild steel in hydrochloric
acid solution," *Int. J. Electrochem. Sci.*, vol. 7, no. 4, pp. 3436–3452, 2012.

[51] U. F. Ekanem, S. A. Umoren, I. I. Udousoro, and A. P. Udoh, "Inhibition of mild steel
corrosion in HCl using pineapple leaves (Ananas comosus L.) extract," *J. Mater. Sci.*,
vol. 45, no. 20, pp. 5558–5566, 2010, doi:10.1007/s10853-010-4617-y.

[52] M. Ouakki, M. Galai, M. Cherkaoui, E. H. Rifi, and Z. Hatim, "Inorganic compound (Apatite doped by Mg and Na) as a corrosion inhibitor for mild steel in phosphoric acidic medium," *Anal. Bioanal. Electrochem.*, vol. 10, no. 7, pp. 943–960, 2018.

[53] A. K. Singh and M. A. Quraishi, "Investigation of the effect of disulfiram on corrosion of mild steel in hydrochloric acid solution," *Corros. Sci.*, vol. 53, no. 4, pp. 1288–1297, 2011, doi:10.1016/J.CORSCI.2011.01.002.

[54] M. Dahmani, A. Et-Touhami, S. S. Al-Deyab, B. Hammouti, and A. Bouyanzer, "Corrosion inhibition of c38 steel in 1 M HCl: a comparative study of black pepper extract and its isolated piperine," *Int. J. Electrochem. Sci.*, vol. 5, no. 8, pp. 1060–1069, 2010.

[55] A. Döner and G. Kardas, "N-Aminorhodanine as an effective corrosion inhibitor for mild steel in 0.5 M H_2SO_4," *Corros. Sci.*, vol. 53, no. 12, pp. 4223–4232, 2011, doi:10.1016/j.corsci.2011.08.032.

[56] A. Döner, R. Solmaz, M. Özcan, and G. Kardaş, "Experimental and theoretical studies of thiazoles as corrosion inhibitors for mild steel in sulphuric acid solution," *Corros. Sci.*, vol. 53, no. 9, pp. 2902–2913, 2011, doi:10.1016/J.CORSCI.2011.05.027.

[57] A. Dehghani, G. Bahlakeh, B. Ramezanzadeh, and M. Ramezanzadeh, "Potential of Borage flower aqueous extract as an environmentally sustainable corrosion inhibitor for acid corrosion of mild steel: electrochemical and theoretical studies," *J. Mol. Liq.*, vol. 277, pp. 895–911, 2019, doi:10.1016/J.MOLLIQ.2019.01.008.

[58] E. V. Senatore et al., "Evaluation of high shear inhibitor performance in CO_2-containing flow-induced corrosion and erosion-corrosion environments in the presence and absence of iron carbonate films," *Wear*, vol. 404–405, pp. 143–152, 2018, doi:10.1016/J.WEAR.2018.03.014.

[59] F. EL Hajjaji et al., "A detailed electronic-scale DFT modeling/MD simulation, electrochemical and surface morphological explorations of imidazolium-based ionic liquids as sustainable and non-toxic corrosion inhibitors for mild steel in 1 M HCl," *Mater. Sci. Eng. B*, vol. 289, p. 116232, 2023, doi:10.1016/J.MSEB.2022.116232.

[60] F. El-Hajjaji et al., "Electrochemical and theoretical insights on the adsorption and corrosion inhibition of novel pyridinium-derived ionic liquids for mild steel in 1 M HCl," *J. Mol. Liq.*, vol. 314, p. 113737, 2020, doi:10.1016/J.MOLLIQ.2020.113737.

8 Phytochemicals/Plant Extract as Corrosion Inhibitor for Aluminum in H$_2$SO$_4$ Solutions

Siska Prifiharni, Arini Nikitasari, Siti Musabikha,
Rahayu Kusumastuti, and Gadang Priyotomo
Research Center for Metallurgy, National
Research and Innovation Agency

8.1 INTRODUCTION

Corrosion is a major problem faced by many industries worldwide, particularly in the oil and gas industry, where equipment and pipelines are exposed to harsh environments. Aluminum is a popular choice among the various materials used in these industries due to its desirable mechanical and physical properties. However, aluminum is susceptible to corrosion in acidic environments, such as those containing sulfuric acid (H$_2$SO$_4$). The presence of H$_2$SO$_4$ leads to the formation of a passive layer on the surface of the aluminum, which can be easily damaged, leading to corrosion. Acid solutions have been applied widely in industrial applications such as acid descaling, acid cleaning, acid pickling, and mill scale removal from metal surfaces [1]. This has led to significant economic losses due to equipment failure and the need for frequent maintenance.

Various methods have been explored to address this problem to inhibit aluminum corrosion in acidic environments. Mostly, corrosion mitigation methods consist of cathodic protection, coating, corrosion inhibitors (CIs), metal selection and modified surface conditions, natural modification, plating, and design. Among those corrosion mitigation methods, CIs are more applicable in industries. In addition, using CIs is also one of the more economical approaches to minimize the potency of corrosion in each production unit. CIs are chemical substances that are added in small concentrations to corrosive media to minimize or slow down the reaction of the metal in that media. The inhibitors are used in many industrial units such as refineries, cooling systems, chemicals, oil and gas production boilers, pipelines, and other units [2].

Some CIs in industrial utilization are synthesized from low-cost raw materials with heteroatoms of aromatic or long-chain carbon systems. Unfortunately, most CIs have a high risk of toxic and hazardous potency on human health and environmental ecosystems. Some nations have a regulation to restrict the use of CIs, comprising classical oxidizing CIs such as chromate and nitrite. Because of the detrimental

DOI: 10.1201/9781003394631-8

impact of CIs, many reports of investigation related to the massive development of eco-friendly CIs [3–5]. Eco-friendly CIs or green CIs have biodegradability and no heavy metals or other toxic compounds. Natural products such as various plant extracts have advantages due to being environmentally acceptable, cheap, readily available and renewable sources of materials, and ecologically acceptable.

This chapter will review the current research on using plant extracts as CIs for aluminum in H_2SO_4 solutions. It will discuss the various mechanisms by which these plant extracts inhibit corrosion, including adsorption, film formation, and the formation of protective layers. Finally, this chapter will discuss future directions for research in this field and potential applications for using plant extracts as CIs in industry.

8.2 PRINCIPLES

Using "phytochemicals/plant extracts" as CIs for aluminum in H_2SO_4 solutions refers to the approach toward founding comprehensive efforts to chemical risk management. Plant extracts comply with the requirements of green CIs that have been declared by the Paris Commission (PARCOM) and the Registration, Evaluation, Authorization and Restriction of Chemicals (REACH), which are non-bio-accumulative, are biodegradable, and have zero or very minimal marine toxicity level [6]. Plant extracts are non-toxic, biodegradable, widely available, cost-effective, and not harmful to the environment and human health [7]. Moreover, phytochemicals present in plant extracts have been exhibited to be just as efficacious as synthetic inhibitors.

8.2.1 CORROSION IN ALUMINUM

Aluminum is classified as one of the non-ferrous metals used for all human applications globally. That metal is utilized in many products, such as architectural structures, roofing, foil insulation, windows, cladding, door, solar panels, and refrigerators. The beneficial properties of aluminum are high thermal conductivity, low density, good corrosion resistance in a certain environment, non-toxic, good machinability and forming, non-magnetic, good ductility, and so on. Aluminum is naturally categorized as a reactive metal that converts spontaneously to passive metal (aluminum oxide) due to high oxygen affinity. The following list of aluminum and alloy series are used in global applications:

1xxx: Unalloyed (pure) composition,
2xxx: Alloys in which copper is the main alloying element,
3xxx: Manganese is the main alloying element.
4xxx: Silicon is the main alloying element.
5xxx: Magnesium is the main alloying element.
6xxx: Magnesium and silicon are the primary alloying elements.
7xxx: Zinc is the primary alloying element.

Because of the passive film that develops on the metal surface, pure aluminum metal is relatively corrosion resistant compared to aluminum alloys in adding a certain element such as Fe [8,9]. However, aluminum is susceptible to corrosion which converts to its oxide. Aluminum corrosion also degrades its chemical and physical properties.

Aluminum oxide is amphoteric, which has a rapid reaction as both a base and an acid. In other words, if aluminum is exposed to acid or alkaline outside the pH range of 4–9, a corrosion attack will occur spontaneously. Spontaneous protective films of aluminum have a stable state in aqueous media with a pH range of 4.5–8.5. It is necessary to slow down the corrosion rate of aluminum in strong acids and alkaline by using CIs.

Furthermore, many investigations of corrosion behavior on aluminum and its alloys in H_2SO_4 solutions exist. Owoeye and co-workers reported that the weight loss magnitude increased with exposure time at 1 M of H_2SO_4 solution on pure aluminum [10]. In addition, adding an element to the aluminum matrix can also shift the corrosion behavior of aluminum. Because a protective thin aluminum-copper oxide film is formed, the corrosion rate of the material increases in the order Al-Cu < Al [11]. The concentration of corrosive solutions and temperature have been considered essential parameters of the corrosion process. Husaini and co-workers elucidated that the corrosion rate of pure aluminum increases with increasing H_2SO_4 concentration [12].

On the contrary, in 1 M H_2SO_4 solution, the corrosion resistance of Al-Co alloy is higher compared to Al, Al7075-T6, and Al2024-T3 alloys [13]. The role of heat treatment for aluminum alloy is considered to shift the magnitude of its corrosion rate. Compared to as-received and annealed Al 1060, quenched Al 1060 showed the strongest corrosion resistance due to its lower corrosion rate values [14]. The sulfate ion presence increased corrosion current densities while affecting the corrosion potential [15]. Because the sulfate ion is easily adsorbable onto the surface of the protective oxide film and diffuses through the pores, the aggressiveness of that anion leads to the rupture of the protective layer on the surface of Al 1060 [16]. In addition, aluminum corrodes in aqueous solutions outside of the passive range because its oxides are soluble in acid solution, producing Al^{3+} ions in the former and AlO^{2-} (aluminate) ions in the latter [17].

8.2.2 Plant Extracts as Corrosion Inhibitors

Plant extracts constitute another green alternative to inhibit aluminum's corrosion process in sulfuric acid environments because making an extract from any plant is regarded as an uncomplicated task, thus allowing more efficiency at both extraction and use of these substances for experimentation. Plant extract is a compound composed of the bioactive constituent of a plant or its parts and a specific medium acting as a solvent. The extraction yields rely on the polarity of the solvent and the techniques or methods (maceration or Soxhlet) used. Bioactive constituents in the extract give properties for a certain purpose, such as antioxidant, anti-inflammatory, antiviral, or antimicrobial effects [18].

Generally, the extracts come from the whole plant or its parts containing higher concentrations of the bioactive constituent, named phytochemicals. Phytochemicals contained in each part of the plant are described in Figure 8.1. Many studies reported that numerous plant parts extracts contain phytochemicals widely used as CIs [19–21]. For instance, phytochemicals that are promising for corrosion inhibition of aluminum in H_2SO_4 solution have been attained from leaf [22–24], stem [25–27], fruit [28,29], flower [30–32], and peel [33,34].

Although all plant components can be used as plant extracts, leaves are the most frequently used. That notion is primarily supported by secondary metabolite

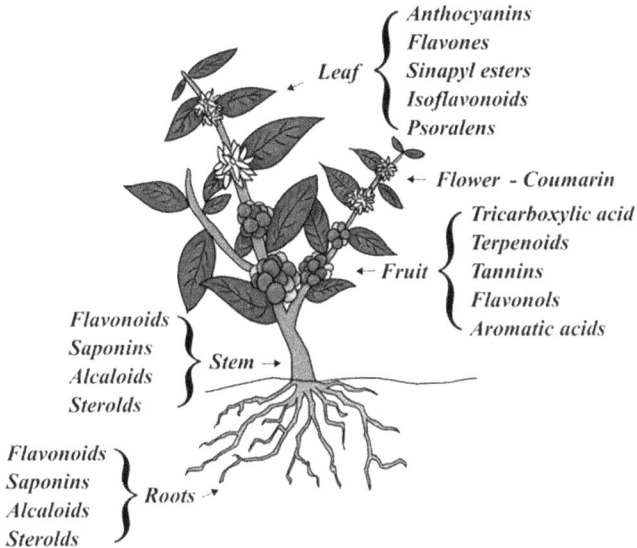

FIGURE 8.1 Phytochemicals in each part of plants.

Source: Adapted from Ref. [20].

production that always occurs in leaves [35,36]. Moreover, leaf extracts showed the best overall protective performance at low doses across various extracts. This is mainly because phytochemicals, where their synthesis occurs in the presence of radiant energy, water, and CO_2, are mostly found in leaves [19,37]. Baruku Kasuga and co-workers have investigated *Carica papaya* leaves for inhibiting aluminum corrosion in sulfuric acid [38]. *C. papaya* leaves contain phytochemicals such as alkaloids, tannins, saponins, and flavonoids, which inhibit corrosion. Green tea *Camellia sinensis* leaves also contain phytochemicals which are mainly flavonoids. Therefore, *C. sinensis* leaves can protect aluminum from corrosion in an H_2SO_4 solution [39].

Flavonoids, glycosides, alkaloids, saponins, phytosterol, tannins, anthraquinones, phenolic compounds, triterpenes, and phlorotannins are phytochemicals that have a corrosion-inhibiting action. The majority of these phytochemicals have polar functional groups that aid in absorption, such as amide ($-CONH_2$), hydroxyl ($-OH$), ester ($-COOC_2H_5$), the carboxylic acid ($-COOH$), and amino ($-NH_2$) [40,41]. Naturally occurring phytochemicals in plants are rich in electron centers and multiple bonds, contributing as adsorption centers when aluminum and inhibitors interact [42,43].

The aluminum surface might get damaged during exposure to sulfuric acid. However, the aluminum surface was changed to fine and shielded in the presence of plant extracts. These findings support using phytochemicals derived from plant extracts as a CI of aluminum in sulfuric acid. To improve their practical applicability, the impact of various parameters on the anti-corrosive capability of plant extracts has been studied. The influence of temperature on the inhibition efficiency of *Hemerocallis fulva* was analyzed [44]. Another study revealed that temperature, immersion period, and plant extract concentration impacted the anticorrosion properties of the date palm seed extracts [45].

Despite the numerous uses of plant extracts as CIs, challenges still need to be considered for future research. For example, the poor solubility of plant extracts in polar electrolytes, especially at higher concentrations, is one of the main obstacles to employing them as CIs. Practically, it has been shown that plant extracts tend to precipitate out in polar electrolytes in most cases. The preparation of plant extracts is very laborious and requires several processes, which is another drawback of employing them as CIs. Toxic solvents are frequently used in synthesizing plant extracts, which can negatively impact the environment, the soil, and aquatic life after they are released. Most of these solvents are very expensive and may negatively impact the processing of the extract [19]. Plant extracts generally have poor stability and limited shelf life. Additionally, prolonged storage of plant extracts may make them vulnerable to microbial and fungal growth, which could reduce the effectiveness of CIs [35].

8.2.3 CORROSION INHIBITION MECHANISM

Like HCl, sulfuric acid (H_2SO_4) is a staple chemical in many industries for acid pickling, acid descaling, and cleaning processes. Sulfuric acid is a highly corrosive solution during industrial processes; therefore, these practices involve the implementation of some external additives called CIs [19]. In recent years, numerous research publications have been on using plant extracts as CIs for aluminum in H_2SO_4 environments. Phytochemicals contained in plant extracts are used as CIs due to the inhibition effect via the adsorption mechanism of the inhibitor molecules on the aluminum surface [46].

The adsorption mechanism plays a prominent role in the corrosion inhibition of aluminum by plant extracts. Two types of adsorption mechanisms of CIs are physisorption and chemisorption. The process of physisorption involves the existence of an electrically charged molecule of the CI. Chemisorption requires the formation of a chemical bond between the adsorbed species (plant extracts) and the metal (aluminum) surface [42]. Figure 8.2 is the possible mechanism of the adsorption of plant extract on aluminum. The adsorption of plant extract molecules implicates O, N, and S atoms that may block active sites; hence, it can inhibit the corrosion process. The corrosion process of aluminum is inhibited since the plant extracts are adsorbed on the aluminum surface and decrease the surface area available for the cathodic and anodic reactions [47]. Heteroatoms substituted aromatic rings, and π electrons are noted to prompt the adsorption of the plant extract molecules [48–52]. Furthermore, some polar functional groups make the plant extracts more soluble in sulfuric acid solutions and tend to act as the adsorption sites of the plant extract molecule.

The charge and nature of the metal surface, electronic characteristics of the metal surface, temperature, adsorption of the solvent, ionic species, and electrochemical potential at the solution interface are the main factors that influence the adsorption of CIs [53]. The adsorption isotherms can describe the adsorption mechanism of plant extract molecules on the aluminum surface. There are various models of adsorption isotherms such as Langmuir [24,54,55], Temkin [22,30,56], Frumkin [57–59], Freundlich [60–62], and Florry Huggins [63–65] isotherms. However, many plant-based CIs in H_2SO_4 follow the Langmuir adsorption isotherm [23,25,29,31,34,38], assuming that the aluminum surface contains a fixed number of adsorption sites, and each site holds one inhibitor [26].

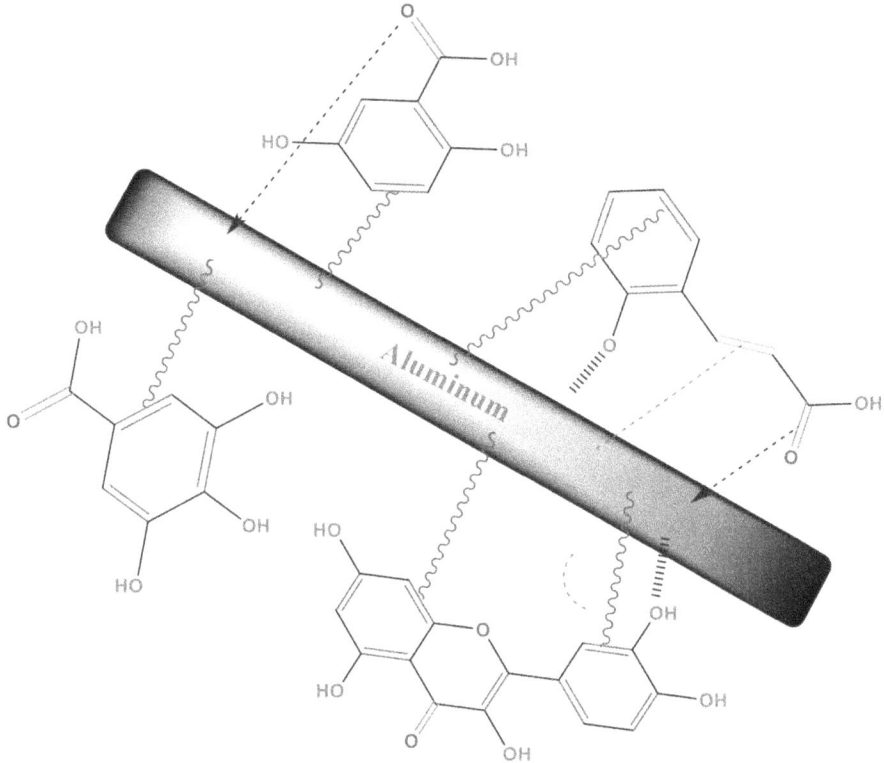

FIGURE 8.2 Mechanism of eco-friendly corrosion inhibition.

Source: Adapted from Ref. [44].

The Langmuir adsorption isotherm is an ideal isotherm for physisorption or chemisorption where there is no interaction between the plant extract molecules and the aluminum surface. The following equation can express the Langmuir adsorption [66]:

$$\frac{C_{inh}}{\theta} = \frac{1}{K_{ads}} + C_{inh} \tag{8.1}$$

where θ is the surface coverage of the inhibitor, C_{inh} is the concentration of the inhibitor, and K_{ads} is the equilibrium constant of the adsorption. Generally, a higher value of equilibrium constant of adsorption K_{ads} collateral with better inhibitor performance to adsorb on the aluminum surface [52]. The value of the equilibrium constant of the adsorption can be used to calculate the Gibbs standard free energy of adsorption (ΔG_{ads}), the standard heat of adsorption (ΔH_{ads}), and the standard entropy of adsorption (ΔS_{ads}), as follows [66]:

$$K_{ads} = \frac{1}{1,000} \exp\left(-\frac{\Delta G_{ads}}{R.T}\right) \tag{8.2}$$

$$\ln K_{ads} = \ln\frac{1}{C_{inh}} - \frac{\Delta H_{ads}}{R.T} + \frac{\Delta S_{ads}}{R} \tag{8.3}$$

The value of R at that equation is 8.314 J K^{-1} mole, and T is the absolute temperature in K. The negative value of Gibbs standard free energy ΔG_{ads} indicates the spontaneity of the adsorption process and the stability of the adsorbed plant extract on the aluminum surface. Furthermore, the magnitude of ΔG_{ads} determines the type of adsorption, whether physisorption or chemisorption. If the value of Gibbs standard free energy of adsorption is greater than −20 kJ·mol, then it is expressed as physical adsorption (physisorption), while the value of ΔG_{ads} equals or less than −40 kJ mol^{-1} is stated as chemical adsorption (chemisorption). If the ΔG_{ads} value is between −20 and −40 kJ mol^{-1}, the inhibitor is a mixture of adsorption types [67]. The positive amount of standard heat of adsorption (ΔH_{ads}) exhibits that the thermal stability of the adsorbed inhibitors increases at elevated temperatures. Moreover, the positive sign of entropy (ΔS_{ads}) reflects a rise in irregularities as a result of the desorption of H$_2$O molecules from the surface and the adsorption of the inhibitor molecules at each spot [41,52,66].

8.2.4 SYNERGISM PHENOMENON

The phenomenon of synergism refers to adding halide ions in H$_2$SO$_4$ solutions to improve the anticorrosion ability of CIs. The added halide ion is usually in the form of some organic salts such as KI, KBr, and KCl [68]. Ebenso has investigated the synergistic effect of halide ions on the corrosion inhibition of aluminum in H$_2$SO$_4$ using 2-acetyl phenothiazine [69]. According to the study's results, adding KI, KBr, and KCl improves the extract's ability to provide protection. Halide efficacy in sulfuric acid mediums followed the following rule: KI > KBr > KCl.

The mechanism of the synergism phenomenon in sulfuric acid medium is still uncertain. The iodide ions, protonated inhibitor molecule, and aluminum are stated to be combined to produce a complex layer with a thickness significantly greater than that of the individual inhibitor. Contrarily, I$^-$ were adsorbed on the steel surface and subsequently facilitated the adsorption of the protonated inhibitor molecule, leading to improved inhibition ability. Similarly, the halide anions can enhance the adsorption abilities of the CIs by creating connective bridges between the aluminum surface and the inhibitor. The iodide ion has a more significant impact on the synergism phenomenon than Cl and Br because of its larger radius, higher hydrophobicity, and lower electro-negativity [42].

8.3 METHODS

8.3.1 PREPARATION OF PLANT EXTRACTS

The literature has numerous preparation techniques for plant extracts, including maceration, decoction, heated reflux, Soxhlet, and sonication. These extraction techniques work by heating, chilling, and separating the active ingredients while the solvents are present. It is essential to select an appropriate extraction process because the inhibitory effect of plant inhibitors differs with the variation of extraction techniques. Maceration is the most popular technique to extract. The collected plant materials are washed with water to remove the dust or sand particles and cut into small pieces prior to the maceration. The plant pieces are dried at room temperature or in a hot air oven. A mechanical grinder is used to revolve the dried plants into a fine powder. The appropriate solvents, such as ethanol, methanol, and sulfuric acid, are employed to soak the dried powder of plants.

Several factors should be considered during the preparation of plant extracts. The first issue is finding a proper solvent for the extraction process. Solvents used to extract plants are chosen based on the polarity of the solute of interest. The solute will dissolve perfectly in a solvent with a similar polarity to the solute. The polarity, from most polar to least polar, of a few common solvents is as follows: Water > Methanol > Acetone > Ethyl acetate > Chloroform > Hexane [43]. According to a thorough literature review, water is the best solvent because of its unique redox-stable features, simplicity, ease of availability, non-toxic nature, non-flammability, and lack of hazards [70,71]. However, the preparation of some plant extracts needs organic solvents like ethanol and methanol.

Another important factor to consider when using plant extracts is selecting a suitable temperature for extraction. The preparation of plant extracts is significantly impacted by temperature. The solubility of phytochemicals is limited at extremely low temperatures, and their active constituents decompose at very high temperatures. To determine the best extraction yield, extraction is often done at temperatures between 60°C and 80°C [72,73]. Moreover, the temperature needed for the drying processes could be critical. The plant materials are frequently left to dry at room temperature in the shade because the high temperatures could have a negative impact. However, because this drying method takes several days or months, oven drying has also made substantial progress [74].

8.3.2 Inhibition Efficiency Measurement

The inhibition efficiency of plant-based CIs can be measured using weight loss or electrochemical methods. Weight loss measurements are conducted in a thermostatic water bath containing sulfuric acid solution with different concentrations of plant extracts; often, the effects of temperature are also applied. Aluminum specimens are weighed before being immersed in the solution. After a certain time of immersion, aluminum specimens are rinsed with distilled water, dried, and weighed. Experiments are usually performed in triplicate to ensure the repeatability of the results. The weight loss method computes the corrosion rate (v) and inhibition efficiency (IE%) according to Equations 8.4 and 8.5:

$$v = \frac{W_0 - W}{s \times t} \tag{8.4}$$

$$IE\% = \frac{v_0 - v}{v_0} \times 100 \tag{8.5}$$

where W_0 and W (mg) are the weight loss of the specimen before and after immersion, s (cm^2) is the area of the specimen, t (h) is the immersion time, and v_0 and v represent the corrosion rate in the absence and presence of CI, respectively [51,56,61].

Electrochemical measurements are carried out in a three-electrode cell by using potentiostat devices. Three-electrode cell consists of the specimen as the working electrode, platinum or graphite as a counter electrode, and a saturated calomel electrode (SCE) or Ag/AgCl electrode as the reference electrode. All potential values are recorded versus the reference electrode. Before the experiment, the working electrode is polished with sequent grades of emery paper, washed with distilled water, degreased with acetone, and dried. Electrochemical measurements have been designed for computer-assisted operation. A computer-assisted measurement system offers automatic calculation [75]. Tafel polarization and electrochemical impedance spectroscopy (EIS) are the most commonly used electrochemical measurements.

Tafel polarization curves are obtained via a variation of the potential automatically versus the reference electrode with a certain scan rate in mV s^{-1}. The corrosion current is calculated by extrapolating the anodic and cathodic Tafel lines to the point that provides (log i_{corr}) and corrosion potential (E_{corr}) for each concentration of inhibitor as well as the inhibitor-free solution [76,77]. The surface coverage (θ) and inhibition efficiency (IE%) are then calculated using (i_{corr}), as shown in the following equation:

$$IE\% = \theta \times 100 = \left[1 - \frac{i_{corr(inh)}}{i_{corr(free)}}\right] \times 100 \qquad (8.6)$$

where $i_{corr(free)}$ and $i_{corr(inh)}$ are the corrosion current densities without CI and with CI, respectively. EIS results present a Nyquist diagram, which plots the real and imaginary impedance. The main characteristics identified from the Nyquist diagram are the charge transfer resistance (R_{ct}) and the double-layer capacity (Cdl). The inhibition efficiency (IE%) obtained from the EIS is measured from the following equation:

$$IE\% = \left[1 - \left(\left(R_{ct}^0\right)/R_{ct}\right)\right] \times 100 \qquad (8.7)$$

where R_{ct}^0 and R_{ct} are the charge transfer resistance without CI and with CI, respectively. Software analysis is utilized for graphing and fitting experimental impedance data in the equivalent circuit form. Duplicate or triplicate experiments are performed to achieve the measurements' reliability and reproducibility [56].

8.3.3   SURFACE ANALYSIS

Surface analysis is usually examined *via* scanning electron microscope (SEM) and atomic force microscope (AFM) to confirm the existence of a protective inhibitor layer on the aluminum surface. The surface analysis of aluminum is investigated in a procedure that involves exposure to sulfuric acid after immersion at a certain time without and with the addition of plant extract inhibitors. After completion of the exposure, the aluminum specimens are removed, washed with distilled water, and dried. The morphologies in both the absence and presence of the plant extract inhibitors are analyzed to study the inhibition effect [78].

Generally, there are differences in the morphology of the uninhibited and inhibited aluminum surfaces. The inhibited aluminum surface has much less damage than the uninhibited Al surface. Plant extracts cause the Al surface to be smoother and less corroded, as shown in Figure 8.3. SEM characterization is often supplemented by an elemental composition analyzer and an energy dispersive X-ray (EDX), to ensure the existence of C, N, S, and O elements that play an active role in the inhibitory effect [79]. On the contrary, AFM is very significant in investigating the surface roughness of aluminum before and after immersion in sulfuric acid solution without and with the addition of plant extracts. The roughness data based on AFM clearly show that the inhibited aluminum surface appears finer due to the inhibitor adhering to the aluminum and generating the protective layer. Three-dimensional AFM micrograph of Al surface in the absence and presence of *Bassia muricata* extracts is illustrated in Figure 8.4.

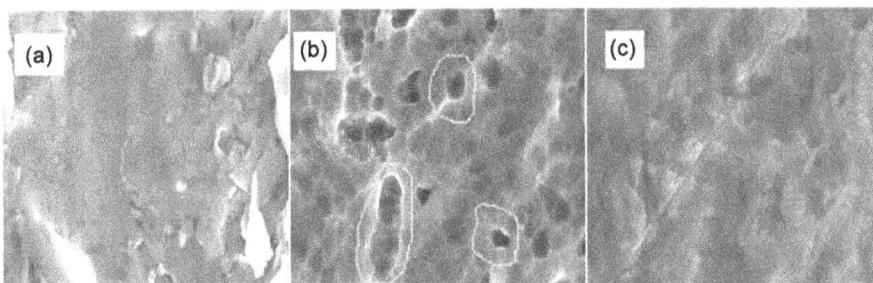

FIGURE 8.3 SEM micrograph of Al surface (a) before immersing in sulfuric acid, (b) after immersing in sulfuric acid without *Carica papaya* leaves extract, and (c) after immersing in sulfuric acid with *Carica papaya* leaves extract.

Source: Originated from Ref. [38].

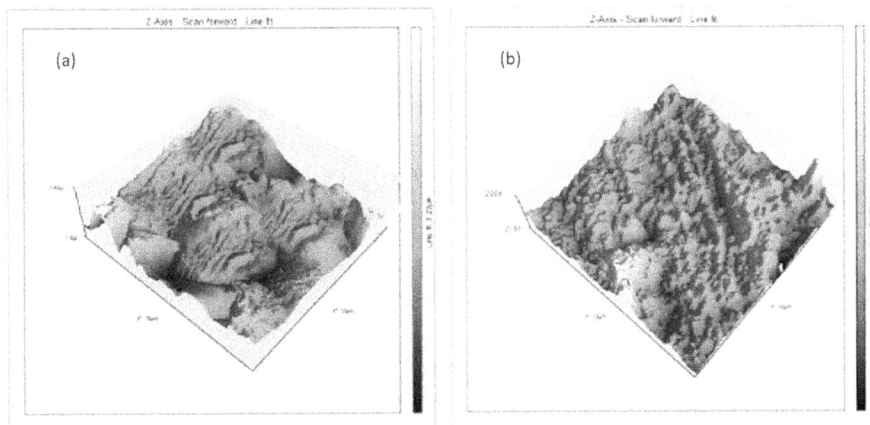

FIGURE 8.4 AFM micrograph of Al surface in the (a) absence and (b) presence of *Bassia muricata* extract in sulfuric acid.

Source: Adapted from Ref. [80].

8.4 CALCULATION COMPUTATION

8.4.1 QUANTUM CHEMICAL CALCULATION

The quantum chemical calculation is a computational method used in chemistry to calculate the behavior and properties of atoms, molecules, and solids using the principles of quantum mechanics. The primary purpose of quantum chemical calculations is to predict and explain the behavior of chemical systems and their interactions.

Many software programs are available for performing quantum chemical calculations, such as Gaussian, NWChem, and ORCA. One important output of quantum chemical calculations is the calculation of the highest occupied molecular orbital (HOMO) and lowest unoccupied molecular orbital (LUMO) energies. HOMO and LUMO are two key electronic states in a molecule, and they are important in determining its chemical and physical properties. The HOMO represents the highest energy level in which an electron is present in a stable molecule, while the LUMO represents the lowest energy level in which an electron is absent in a stable molecule [81].

The energy difference between the HOMO and LUMO is called the energy gap, and the HOMO–LUMO gap is also an important parameter in quantum chemical calculations. The energy gap provides information about the electronic properties of a molecule and can be used to predict its reactivity and electronic transitions. Molecules with a small energy gap are typically more reactive and undergo chemical reactions more easily, while molecules with a large energy gap are more stable and less reactive [82].

8.4.2 MOLECULAR DYNAMICS SIMULATION

Molecular dynamics (MD) simulation is a computational technique used to study the dynamic behavior of atoms and molecules. It involves simulating the time evolution of a molecular system by numerically solving the equations of motion for each particle in the system. The interaction between the inhibitor and the metal surface is typically modeled using a force field that includes terms for van der Waals' interactions, electrostatic interactions, and solvation effects. The inhibitor molecules are typically represented by parameters describing their size, shape, and chemical properties [83].

The effectiveness of a CI can be quantified using various parameters, such as the adsorption energy and the surface coverage. The adsorption energy is the energy required to remove the inhibitor molecule from the metal surface, and it is typically calculated by comparing the energy of the system with and without the inhibitor present. The surface coverage is the fraction of the metal surface covered by the inhibitor molecules, and it is calculated by counting the number of inhibitor molecules adsorbed on the surface [84].

The equation used to calculate the adsorption energy (ΔE_{ads}) in MD simulation is:

$$\Delta E_{ads} = E_{total} - \left(E_{metal+\,solution} + E_{inhibitor} \right) \tag{8.8}$$

where $E_{(surface+inhibitor)}$ is the total energy of the system, $E_{metal+solution}$ is the energy of the metal and the solution, and $E_{(inhibitor)}$ is the energy of the inhibitor molecules alone. A negative value of ΔE_{ads} indicates that the inhibitor is adsorbed on the metal surface and is likely to be effective in inhibiting corrosion.

8.5 CURRENT RESEARCH

Currently, plant extracts as environmentally friendly CIs are being widely investigated. The development and utilization of the extracts have been widely implemented in various materials and environments. Table 8.1 shows the research that has been done to investigate the inhibition efficiency of plant extracts as environmentally friendly CIs on aluminum in the H_2SO_4 environment.

Several researchers used water, methanol, and ethanol as solvents for the extraction, while some applied the acid solution as a solvent and used it as a CI, reporting excellent inhibitory efficacy. Methanol extract of *Dryopteris cochleata* and *Borassus flabellifer* shows better inhibition efficiency on aluminum in an H_2SO_4 medium than water extract. The phytochemical components obtained from the various solvent extracts may directly relate to these results [98,99].

Corrosion rate and inhibition efficiency were affected by sulfuric acid concentration as well. Generally, the increase in acid concentration decreases the inhibition efficiency of the extract [32], [91,92,100]. However, the inhibition efficiency of *Morinda tinctoria* extract likewise tends to increase as the acid concentration gradually increases from 0.25 to 1 M. This might be because, unlike in an HCl medium, the corrosion products and inhibitor compounds form a protective barrier on the metal surface at higher acid concentrations, resulting in higher %IE [90].

Temperature is one of the most important factors to accelerate metal corrosion. Increasing temperature accelerates the chemical reaction of anodic and cathodic inhibitors, which leads to an increase in the rate of metal corrosion. Therefore, a higher inhibitor concentration is needed at a higher temperature. In general, the efficiency of extracted inhibitor increased with increasing inhibitor concentration but decreased as the temperature increased [41]. Orie and Mathew studied the inhibition efficiency of breadfruit peel extracts using the weight loss method in 0.5 M H_2SO_4. It was revealed that inhibition efficiency increases with increasing concentration but decreases with increasing temperature [33]. A similar study was undertaken by Molina-Ocampo and co-workers using weight loss and electrochemical measurements; they noted that the inhibition efficiency of *Hibiscus sabdariffa* increases with increasing concentration but decreases with rising the testing temperature [31]. However, the ethanol extract of leaves of *Portulaca oleracea* exhibits a higher inhibition efficiency when the temperature rises [104]. In addition, Al-Bataineh et al. declared that the inhibition efficiency of *Capparis decidua* was higher when the temperature was rising [30].

Anodic, cathodic, and mixed-type CIs describe the mechanisms by which a CI interacts with a metal surface to reduce its corrosion rate. In the case of plant extract CIs for aluminum in H_2SO_4, different types can act by different mechanisms to reduce the corrosion rate. The mechanism can be categorized by potentiodynamic polarization. The results from polarization measurement show that these extracts acted as mixed-type inhibitors [44,90,93,98,99,102,105,107,110], while other extract

TABLE 8.1
Research Studies about Plant Extracts as Corrosion Inhibitors for Aluminum in H_2SO_4 Solution

Ref.	Inhibitors	Acid Concentration	Solvent Extract	Methods	Adsorption Isotherm	Inhibitor Concentration	Max. IE (%)
[85]	Striga hemonthica	0.1 M	Ethanol	Gravimetric, thermometric, weight loss	Langmuir	1.0 g L⁻¹	97.01
[86]	Newbouldia laevis	0.5 M	H_2SO_4	Weight loss	Langmuir	0.4 g L⁻¹	92
[87]	Alstonia boonei	1.0 M	Ethanol	Weight loss	Freundlich and Van't Hoff's	0.5 g L⁻¹	61
[88]	Oryza glaberrima	1.5 M	Ethanol	Wight loss, SEM		1.0 g L⁻¹	91.75
[89]	Lupine extract	0.05 M	Water	Potentiodynamic polarization, EIS	Langmuir	1,000 ppm	36.3
[90]	Morinda tinctoria	0.25–0.75 M	Water	Wight loss, potentiodynamic polarization, EIS, SEM	Freundlich	16% v/v	82.14
[80]	Bassia muricata	1.0 M	Ethanol	Weight loss, potentiodynamic polarization, EIS, SEM, AFM	Temkin adsorption isotherm	300 ppm	91.7
[32]	Calotropis	2–5 N		Weight loss	-	0.6%	82.98
[91]	Capparis decidua	0.5–2 N	Ethanol	Weight loss	-	0.4%	77.84
[92]	Citrullus colocynthis	3–5 N	Methanol	Weight loss	Langmuir	0.6%	90.01
[93]	Tecoma	1.0 M	Dichloromethane	Weight loss, polarization, EIS, EFM	Langmuir	300 ppm	90.2
[94]	Camellia sinensis	0.8 M	H_2SO_4	Weight loss, potential measurement, SEM	Langmuir	20%	41
[95]	Cassia fistula	0.5 M	H_2SO_4	Weight loss, potentiodynamic polarization	Langmuir	10 g L⁻¹	96
[96]	Tinosporacordifolia	2.0 M	Water	Potentiodynamic polarization, EIS	Langmuir	400 ppm	88.02
[44]	Hemerocallis fulva	1.0 M	Methanol	Weight loss, potentiodynamic polarization, EIS, OCP, SEM	Langmuir	600 ppm	78

(Continued)

TABLE 8.1 (Continued)
Research Studies about Plant Extracts as Corrosion Inhibitors for Aluminum in H_2SO_4 Solution

Ref.	Inhibitors	Acid Concentration	Solvent Extract	Methods	Adsorption Isotherm	Inhibitor Concentration	Max. IE (%)
[97]	Mucuna pruriens	2.0 M	Ethanol	Weight loss	Freundlich, Temkin, El-Awardy, and Adejo-Ekwenchi	0.5 g L⁻¹	62.90
[98]	Dryopteris cochleata	1.0 M	Methanol and water	Weight loss, potentiodynamic polarization, EIS, FTIR, XRD, SEM	Freundlich	2,400 ppm	95.09
[99]	Borassus flabellifer	1.0 M	Methanol and water	Weight loss, potentiodynamic polarizaion, EIS, SEM	Langmuir	0.4 g L⁻¹	65.78
[100]	Lantana camara	0.1–5 N	Ethanol	Weight loss	-	0.6%	76.36
[101]	Terminalia glaucescens Planch	0.3 M	Ethanol	Potentiodynamic polarization, EIS, FTIR, SEM	-	0.4 g L⁻¹	98.6
[102]	Carica papaya seed	2.0 M	Water	Potentiodynamic polarization, EIS, FTIR.	Langmuir	400 ppm	96.78
[103]	Caesar weed leaves extract	0.5 M	Methanol	Weight loss	Langmuir	1 g L⁻¹	82
[104]	Portulaca oleracea	2.0 M	Ethanol	Weight loss	Langmuir	0.5 g L⁻¹	45.16
[105]	Chrysophyllum albidum	0.5 M	Ethanol	Weight loss, potentiodynamic polarization, EIS, SEM	Langmuir	1.2 g L⁻¹	78.43
[106]	Spondias mombin L.	0.5 M	H_2SO_4	Weight loss, quantum chemical calculations	Langmuir	5% v/v	80.3
[107]	Pineapple crown	1.0 M	Ethanol	Weight loss, potentiodynamic polarization	-	4 g L⁻¹	68.5
[108]	Prosopis laevigata	0.5 M	Methanol	Weight loss, potentiodynamic polarization, EIS, FTIR	Langmuir	200 ppm	95.53
[31]	Hibiscus sabdariffa	0.5 M	H_2SO_4	Weight loss, potentiodynamic polarization, EIS, FTIR, SEM	Langmuir	2,000 ppm	94
[109]	Cashew	1.0 M	Ethanol	Weight loss	Langmuir, Funkin, Temkin, Flory-Huggins	1 g L⁻¹	82.5

acted as cathodic inhibitor [80,108] and anodic inhibitor [96,111]. There are two categories of inhibition mechanisms of adsorption: physical and chemical. Most plant extracts were physically adsorbed onto an aluminum surface to protect the metal [44,86,88,112]. Meanwhile, some research shows that plant extract can be chemically adsorbed to the surface [104,108].

8.6 FUTURE PREDICTIONS AND SUGGESTIONS

Based on current research and trends in corrosion inhibition, plant extracts are gaining increasing attention as potential CIs for metals such as aluminum. Some recent studies have investigated using plant extracts as CIs for aluminum in acidic environments, including H_2SO_4. For example, extracts from plants such as papaya [102,113,114], garlic [115], and neem [116] have shown promising inhibitory effects on aluminum corrosion in acidic solutions. These plant extracts contain natural compounds that can form protective films on the metal surface, thereby reducing the corrosion rate.

In terms of suggestions, further research is needed to fully understand the mechanisms behind the corrosion inhibition properties of plant extracts, as well as to optimize their effectiveness and develop practical applications for industrial use. Researchers may also explore combining plant extracts with other inhibitors to enhance their performance. Overall, using plant extracts as CIs is a promising area of research with potential practical applications in industries that rely on aluminum, such as the automotive and aerospace industries.

REFERENCES

[1] L. M. de Andrade, C. Paternoster, P. Chevallier, S. Gambaro, P. Mengucci, and D. Mantovani, "Surface processing for iron-based degradable alloys: A preliminary study on the importance of acid pickling," *Bioactive Materials*, vol. 11, pp. 166–180, May 2022, doi: 10.1016/J.BIOACTMAT.2021.09.026.

[2] A. Singh, E. E. Ebenso, and M. A. Quraishi, "Corrosion inhibition of carbon steel in HCl solution by some plant extracts," *International Journal of Corrosion*, vol. 2012, 2012, doi: 10.1155/2012/897430.

[3] N. O. Eddy and Ebenso, "Adsorption and inhibitive properties of ethanol extracts of Musa sapientum peels as a green corrosion inhibitor for mild steel in H_2SO_4," *African Journal of Pure and Applied Chemistry*, vol. 2, pp. 046–054, 2008.

[4] A. Ikeuba, B. I. Ita, R. A. Etiuma, V. M. Bassey, B. U. Ugi, and E. B. Kporokpo, "Green corrosion inhibitors for mild steel in H_2SO_4 solution: Flavonoids of Gongronema latifolium," *Chemical and Process Engineering Research*, vol. 34, pp. 1043–1049, 2015.

[5] S. S. Shivakumar and K. N. Mohana, "Centella asiatica extracts as green corrosion inhibitor for mild steel in 0.5 M sulphuric acid medium," *Pelagia Research Library*, vol. 3, no. 5, pp. 3097–3106, 2012.

[6] A. Zakeri, E. Bahmani, and A. S. R. Aghdam, "Plant extracts as sustainable and green corrosion inhibitors for protection of ferrous metals in corrosive media: A mini review," *Corrosion Communications*, vol. 5, pp. 25–38, 2022, doi: 10.1016/j.corcom.2022.03.002.

[7] G. Bahlakeh, B. Ramezanzadeh, A. Dehghani, and M. Ramezanzadeh, "Novel cost-effective and high-performance green inhibitor based on aqueous Peganum harmala seed extract for mild steel corrosion in HCl solution: Detailed experimental and electronic/atomic level computational explorations," *Journal of Molecular Liquids*, vol. 283, pp. 174–195, 2019, doi: 10.1016/j.molliq.2019.03.086.

[8] G. Priyotomo and I. N. Gede Putrayasa Astawa, "The effect of Fe-enrich phase on the pitting corrosion resistance of Al alloy in various neutral sodium chloride solutions," *International Journal of Science and Engineering*, vol. 7, no. 2, pp. 143–149, 2014, doi: 10.12777/ijse.7.2.143-149.

[9] M. G. Salama, "Corrosion inhibition of aluminum-silicon alloy in H_2SO_4 solution using some thiophene derivatives," *Zastita Materijala*, vol. 51, pp. 1–9, 2010.

[10] F. T. Owoeye, O. R. Adetunji, A. Omotosho, A. P. Azodo, and P. O. Aiyedun, "Investigation of corrosion performance of aluminum and zinc alloys in three acidic media," *Engineering Reports*, vol. 2, no. 1, pp. 1–12, 2020, doi: 10.1002/eng2.12103.

[11] A. S. Ismail, P. B. Andeng, and A. A. El-Meligi, "Investigating corrosion behaviour of aluminium and aluminium-copper alloy in H_2SO_4 as anodizing solution," *Journal of Corrosion Science and Engineering*, vol. 17, no. 2016, pp. 1–16, 2014.

[12] M. Husaini, B. Usman, and M. B. Ibrahim, "Evaluation of corrosion behaviour of aluminum in different environment," *Bayero Journal of Pure and Applied Sciences*, vol. 11, no. 1, p. 88, 2019, doi: 10.4314/bajopas.v11i1.15s.

[13] A. K. Sfikas and A. G. Lekatou, "Electrochemical behavior of Al-Al9Co2 alloys in sulfuric acid," *Corrosion and Materials Degradation*, vol. 1, no. 2, pp. 249–272, 2020, doi: 10.3390/cmd1020012.

[14] R. T. Loto, C. Loto, and E. Ayeoritsenogun Igbogbo, "Data on the corrosion resistance as-received, annealed and quenched 1060 aluminum alloy in dilute H_2SO_4 and HCl acid concentrations," *IOP Conference Series: Materials Science and Engineering*, vol. 872, no. 1, 2020, doi: 10.1088/1757-899X/872/1/012058.

[15] R. T. Loto and E. A. Igbogbo, "Corrosion behaviour of heat treated 1060 aluminium in dilute acid solutions," *Revista Tecnica De La Facultad De Ingenieria Universidad Del Zulia*, 2016, doi: 10.21311/001.39.8.01.

[16] M. Belkhaouda et al., "Effect of the heat treatment on the behaviour of the corrosion and passivation of 3003 aluminium alloy in synthetic solution," *Journal of Materials and Environmental Science*, vol. 1, no. 1, pp. 25–33, 2010.

[17] F. A. Ovat, F. O. David, and A. J. Anyandi, "Corrosion behaviour of Al (6063) alloy (as-cast and age hardened) in H_2SO_4 solution," *Journal of Materials Science Research*, vol. 1, no. 4, 2012, doi: 10.5539/jmsr.v1n4p35.

[18] A. Miralrio and A. E. Vázquez, "Plant extracts as green corrosion inhibitors for different metal surfaces and corrosive media: A review," *Processes*, vol. 8, no. 8, p. 942, Aug. 2020, doi: 10.3390/PR8080942.

[19] S. H. Alrefaee, K. Y. Rhee, C. Verma, M. A. Quraishi, and E. E. Ebenso, "Challenges and advantages of using plant extract as inhibitors in modern corrosion inhibition systems: Recent advancements," *Journal of Molecular Liquids*, vol. 321, p. 114666, 2021, doi: 10.1016/j.molliq.2020.114666.

[20] A. Thakur and A. Kumar, *Sustainable Inhibitors for Corrosion Mitigation in Aggressive Corrosive Media: A Comprehensive Study*, vol. 7, no. 2. Springer International Publishing, 2021. doi: 10.1007/s40735-021-00501-y.

[21] Y. Fang, B. Suganthan, and R. P. Ramasamy, "Electrochemical characterization of aromatic corrosion inhibitors from plant extracts," *Journal of Electroanalytical Chemistry*, vol. 840, no. March, pp. 74–83, 2019, doi: 10.1016/j.jelechem.2019.03.052.

[22] O. S. Adejo, J. A. Gbertyo, and J. U. Ahlie, "Corrosion inhibition of carbon steel in 0.5 M H_2SO_4 by," *International Journal of Modern Chemistry*, vol. 4, no. 3, pp. 137–146, 2013.

[23] S. G. Yiase, S. O. Adejo, T. G. Tyohemba, U. J. Ahile, and J. A. Gbertyo, "Thermodynamic, kinetic and adsorptive parameters of corrosion inhibition of aluminium using sorghum bicolor leaf extract in H_2SO_4," *International Journal of Advanced Research in Chemical Science*, vol. 1, no. 2, pp. 38–46, 2014.

[24] S. Adejo, S. G. Yiase, U. J. Ahile, and J. Gbertyo, "Inhibitory effect and adsorption parameters of extract of leaves of Portulaca oleracea of corrosion of aluminium. In H_2SO_4 solution DEVELOPMENT OF DYE SENSITIZED SOLAR CELL View project Corrosion studies of mild steel in sulphuric acid medium by acidime," no. August, 2013.

[25] Pushpanjali, S. A. Rao, and P. Rao, "Eco friendly green inhibitor Tinosporacordifolia (Linn.) for the corrosion control of aluminum in sulfuric acid medium," *International Journal of Innovative Research in Science, Engineering and Technology*, vol. 4, no. 2, pp. 325–333, 2015, doi: 10.15680/IJIRSET.2015.0402070.

[26] I. Y. Suleiman, M. Abdulwahab, and M. Z. Sirajo, "Anti-corrosion properties of ethanol extract of Acacia senegalensis stem on Al-Si-Fe/SiC composite in sulfuric acid medium," *Journal of Failure Analysis and Prevention*, vol. 18, no. 1, pp. 212–220, 2018, doi: 10.1007/s11668-018-0399-3.

[27] S. Prifiharni et al., "Extract sarampa wood (Xylocarpus Moluccensis) as an eco-friendly corrosion inhibitor for mild steel in HCl 1M," *Journal of the Indian Chemical Society*, vol. 99, no. 7, p. 100520, 2022, doi: 10.1016/j.jics.2022.100520.

[28] R. Chauhan, U. Garg, and R. K. Tak, "Corrosion inhibition of aluminium in acid media by citrullus colocynthis extract," *E-Journal of Chemistry*, vol. 8, no. 1, pp. 85–90, 2011, doi: 10.1155/2011/340639.

[29] N. O. Obi-Egbedi, I. B. Obot, and S. A. Umoren, "Spondias mombin L. as a green corrosion inhibitor for aluminium in sulphuric acid: Correlation between inhibitive effect and electronic properties of extracts major constituents using density functional theory," *Arabian Journal of Chemistry*, vol. 5, no. 3, pp. 361–373, 2012, doi: 10.1016/j.arabjc.2010.09.002.

[30] N. Al-Bataineh et al., "Use of Capparis decidua extract as a green inhibitor for pure aluminum corrosion in acidic media," *Corrosion Science and Technology*, vol. 21, no. 1, pp. 9–20, 2022, doi: 10.14773/cst.2022.21.1.9.

[31] L. B. Molina-Ocampo, M. G. Valladares-Cisneros, and J. G. Gonzalez-Rodriguez, "Using hibiscus sabdariffa as corrosin inhibitor for Al in 0.5 M H_2SO_4," *International Journal of Electrochemical Science*, vol. 10, pp. 388–403, 2015, [Online]. Available: www.electrochemsci.org

[32] S. Kumar and S. P. Mathur, "Corrosion inhibition and adsorption properties of ethanolic extract of calotropis for corrosion of aluminium in acidic media," *ISRN Corrosion*, vol. 2013, pp. 1–9, Mar. 2013, doi: 10.1155/2013/476170.

[33] K. J. Orie and M. Christian, "The corrosion inhibition of aluminium metal in 0.5M sulphuric acid using extract of breadfruit peels," *International Research Journal of Engineering and Technology*, vol. 2, no. 8, pp. 1–9, 2015.

[34] S. O. Adejo, J. A. Gbertyo, J. U. Ahile, and T. T. Gabriel, "Manihot esculentum root peels ethanol extract as corrosion inhibitor of aluminium in 2 M H_2SO_4," *International Journal of Scientific Engineering and Research*, vol. 4, no. 9, pp. 2308–2313, 2013.

[35] B. A. Al Jahdaly, Y. R. Maghraby, A. H. Ibrahim, K. R. Shouier, M. M. Taher, and R. M. El-Shabasy, "Role of green chemistry in sustainable corrosion inhibition: A review on recent developments," *Materials Today Sustainability*, vol. 20, p. 100242, 2022, doi: 10.1016/j.mtsust.2022.100242.

[36] M. Mehdipour, B. Ramezanzadeh, and S. Y. Arman, "Electrochemical noise investigation of Aloe plant extract as green inhibitor on the corrosion of stainless steel in 1M H_2SO_4," *Journal of Industrial and Engineering Chemistry*, vol. 21, pp. 318–327, 2015, doi: 10.1016/j.jiec.2014.02.041.

[37] M. Schreiner and S. Huyskens-Keil, "Phytochemicals in fruit and vegetables: Health promotion and postharvest elicitors," *CRC Critical Reviews in Plant Sciences*, vol. 25, no. 3, pp. 267–278, 2006, doi: 10.1080/07352680600671661.

[38] B. Kasuga, E. Park, and R. L. Machunda, "Inhibition of aluminium corrosion using Carica papaya leaves extract in sulphuric acid," *Journal of Minerals and Materials Characterization and Engineering*, vol. 06, no. 01, pp. 1–14, 2018, doi: 10.4236/jmmce.2018.61001.

[39] C. A. Loto, O. O. Joseph, R. T. Loto, and A. P. I. Popoola, "Corrosion inhibitive behaviour of camellia sinensis on aluminium alloy in H_2SO_4," *International Journal of Electrochemical Science*, vol. 9, no. 3, pp. 1221–1231, 2014.

[40] S. Perumal, S. Muthumanickam, A. Elangovan, R. Sayee Kannan, and K. K. Mothilal, "Inhibitive effect of Bauhinia tomentosa leaf extract on acid corrosion of mild steel," *International Journal of ChemTech Research*, vol. 10, no. 13, pp. 203–213, 2017.

[41] P. M. Krishnegowda, V. T. Venkatesha, P. K. M. Krishnegowda, and S. B. Shivayogiraju, "Acalypha torta leaf extract as green corrosion inhibitor for mild steel in hydrochloric acid solution," *Industrial & Engineering Chemistry Research*, vol. 52, no. 2, pp. 722–728, 2013, doi: 10.1021/ie3018862.

[42] S. Mo, H. Q. Luo, and N. B. Li, "Plant extracts as 'green' corrosion inhibitors for steel in sulphuric acid," *Chemical Papers*, vol. 70, no. 9, pp. 1131–1143, 2016, doi: 10.1515/chempap-2016-0055.

[43] A. Altemimi, N. Lakhssassi, A. Baharlouei, D. G. Watson, and D. A. Lightfoot, "Phytochemicals: Extraction, isolation, and identification of bioactive compounds from plant extracts," *Plants*, vol. 6, no. 4, 2017, doi: 10.3390/plants6040042.

[44] I. M. Chung, R. Malathy, S. H. Kim, K. Kalaiselvi, M. Prabakaran, and M. Gopiraman, "Ecofriendly green inhibitor from Hemerocallis fulva against aluminum corrosion in sulphuric acid medium," *Journal of Adhesion Science and Technology*, vol. 34, no. 14, pp. 1483–1506, Jul. 2020, doi: 10.1080/01694243.2020.1712770.

[45] S. A. Umoren, Z. M. Gasem, and I. B. Obot, "Natural products for material protection: Inhibition of mild steel corrosion by date palm seed extracts in acidic media," *Industrial & Engineering Chemistry Research*, vol. 52, no. 42, pp. 14855–14865, 2013, doi: 10.1021/ie401737u.

[46] "Plant extracts as green corrosion inhibitors for different metal surfaces and corrosive media: A review," *Processes*, vol. 8, no. 8, p. 942, 2020.

[47] P. Arora, S. Kumar, M. K. Sharma, and S. P. Mathur, "Corrosion inhibition of aluminium by Capparis decidua in acidic media," *E-Journal of Chemistry*, vol. 4, no. 4, pp. 450–456, 2007, doi: 10.1155/2007/487820.

[48] I. Pradipta, D. Kong, and J. B. L. Tan, "Natural organic antioxidants from green tea inhibit corrosion of steel reinforcing bars embedded in mortar," *Construction and Building Materials*, vol. 227, p. 117058, 2019, doi: 10.1016/j.conbuildmat.2019.117058.

[49] P. E. Alvarez, M. V. Fiori-Bimbi, A. Neske, S. A. Brandán, and C. A. Gervasi, "Rollinia occidentalis extract as green corrosion inhibitor for carbon steel in HCl solution," *Journal of Industrial and Engineering Chemistry*, vol. 58, pp. 92–99, 2018, doi: 10.1016/j.jiec.2017.09.012.

[50] R. T. Loto and C. A. Loto, "Data on the comparative evaluation of the corrosion inhibition of vanillin and vanillin admixed with rosmarinus officinalis on mild steel in dilute acid media," *Chemical Data Collections*, vol. 24, p. 100290, 2019, doi: 10.1016/j.cdc.2019.100290.

[51] W. Zhang et al., "Fructan from Polygonatum cyrtonema Hua as an eco-friendly corrosion inhibitor for mild steel in HCl media," *Carbohydrate Polymers*, vol. 238, no. March, p. 116216, 2020, doi: 10.1016/j.carbpol.2020.116216.

[52] K. H. Hassan, A. A. Khadom, and N. H. Kurshed, "Citrus aurantium leaves extracts as a sustainable corrosion inhibitor of mild steel in sulfuric acid," *South African Journal of Chemical Engineering*, vol. 22, pp. 1–5, 2016, doi: 10.1016/j.sajce.2016.07.002.

[53] I. M. Chung, R. Malathy, S. H. Kim, K. Kalaiselvi, M. Prabakaran, and M. Gopiraman, "Ecofriendly green inhibitor from Hemerocallis fulva against aluminum corrosion in sulphuric acid medium," *Journal of Adhesion Science and Technology*, vol. 34, no. 14, pp. 1483–1506, 2020, doi: 10.1080/01694243.2020.1712770.

[54] R. Khandelwal, S. K. Arora, and S. P. Mathur, "Study of plant cordia dichotoma as green corrosion inhibitor for mild steel in different acid media," *E-Journal of Chemistry*, vol. 8, no. 3, pp. 1200–1205, 2011, doi: 10.1155/2011/164589.

[55] K. M. Emran, N. M. Ahmed, B. A. Torjoman, A. A. Al-Ahmadi, and S. N. Sheekh, "Cantaloupe extracts as eco friendly corrosion inhibitors for aluminum in acidic and alkaline solutions," *Journal of Materials and Environmental Science*, vol. 5, no. 6, pp. 1940–1950, 2014.

[56] E. E. El-Katori and S. Al-Mhyawi, "Assessment of the Bassia muricata extract as a green corrosion inhibitor for aluminum in acidic solution," *Green Chemistry Letters and Reviews*, vol. 12, no. 1, pp. 31–48, 2019, doi: 10.1080/17518253.2019.1569728.

[57] J. Jacob, "Inhibitory action of Ficus carica extracts on aluminium corrosion in acidic medium," *Chemical Science Transactions*, vol. 2, no. 4, 2013, doi: 10.7598/cst2013.558.

[58] Ramirez-Arteaga, M., M. G. Valladares, and JG González Rodríguez. "Use of Prosopis laevigata as a corrosion inhibitor for Al in H$_2$SO$_4$." *International Journal of Electrochemical Science*, vol. 8, no. 5, pp. 6864–6877, 2013.

[59] M. Omotioma and O. D. Onukwuli, "Evaluation of pawpaw leaves extract as anti-corrosion agent for aluminium in hydrochloric acid medium," *Nigerian Journal of Technology*, vol. 36, no. 2, p. 496, 2017, doi: 10.4314/njt.v36i2.24.

[60] R. S. Nathiya and V. Raj, "Evaluation of Dryopteris cochleata leaf extracts as green inhibitor for corrosion of aluminium in 1 M H$_2$SO$_4$," *Egyptian Journal of Petroleum*, vol. 26, no. 2, pp. 313–323, 2017, doi: 10.1016/j.ejpe.2016.05.002.

[61] P. M. Ejikeme, S. G. Umana, and O. D. Onukwuli, "Corrosion inhibition of aluminium by Treculia Africana leaves extract in acid medium," *Portugaliae Electrochimica Acta*, vol. 30, no. 5, pp. 317–328, 2013, doi: 10.4152/pea.201205317.

[62] N. A. Madueke and N. B. Iroha, "Protecting aluminium alloy AA8011 from acid corrosion using extract from Allamanda Cathartica leaves," *International Journal of Innovative Research in Science*, vol. 7, no. 10, pp. 10251–10258, 2018, doi: 10.15680/IJIRSET.2018.0710014.

[63] G. O. Avwiri and F. O. Igho, "Inhibitive action of Vernonia amygdalina on the corrosion of aluminium alloys in acidic media," *Materials Letters*, vol. 57, no. 22-23, pp. 3705–3711, 2003, doi: 10.1016/S0167-577X(03)00167-8.

[64] H. F. Chahul, A. M. Ayuba, and S. Nyior, "Adsorptive, Kinetic, Thermodynamic and Inhibitive Properties of Cissus Populnea Stem Extract on the Corrosion of Aluminum in Acid Medium," *ChemSearch Journal*, vol. 6, no. 1, pp. 20–30, 2015, doi: 10.4314/csj.v6i1.4.

[65] A. M. Ayuba and A. Abdullateef, "Investigating the corrosion inhibition potentials of Strichnos Spinosa L. extract on aluminium in 0.3M hydrochloric acid solution," *Journal of Applied Science and Environmental Studies JASES*, vol. 4, pp. 336–348, 2021.

[66] A. Nikitasari, G. Priyotomo, A. Royani, and S. Sundjono, "Exploration of Eucheuma Seaweed algae extract as a novel green corrosion inhibitor for API 5L carbon steel in hydrochlorid acid medium," vol. 35, no. 03, pp. 596–603, 2022, doi: 10.5829/ije.2022.35.06c.13.

[67] N. Subekti, J. W. Soedarsono, R. Riastuti, and F. D. Sianipar, "Development of environmental friendly corrosion inhibitor from the extract of areca flower for mild steel in acidic media," *Eastern-European Journal of Enterprise Technologies*, vol. 2, no. 6-104, pp. 34–45, 2020, doi: 10.15587/1729-4061.2020.197875.

[68] C. Verma, E. E. Ebenso, I. Bahadur, and M. A. Quraishi, "NU," *Journal of Molecular Liquids*, 2018, doi: 10.1016/j.molliq.2018.06.110.

[69] E. E. Ebenso, "Synergistic effect of halide ions on the corrosion inhibition of aluminium in H₂SO₄ using 2-acetylphenothiazine," *Materials Chemistry and Physics*, vol. 79, no. 1, pp. 58–70, 2003, doi: 10.1016/S0254-0584(02)00446-7.

[70] D. S. Bose, L. Fatima, and H. B. Mereyala, "Green chemistry approaches to the synthesis of 5-alkoxycarbonyl-4-aryl-3,4-dihydropyrimidin-2(1H)-ones by a three-component coupling of one-pot condensation reaction: Comparison of ethanol, water, and solvent-free conditions," *Journal of Organic Chemistry*, vol. 68, no. 2, pp. 587–590, 2003, doi: 10.1021/jo0205199.

[71] R. S. Varma, "Greener and sustainable trends in synthesis of organics and nanomaterials," *ACS Sustainable Chemistry & Engineering*, vol. 4, no. 11, pp. 5866–5878, 2016, doi: 10.1021/acssuschemeng.6b01623.

[72] J. Seo, S. Lee, M. L. Elam, S. A. Johnson, J. Kang, and B. H. Arjmandi, "Study to find the best extraction solvent for use with guava leaves (Psidium guajava L.) for high antioxidant efficacy," *Food Science & Nutrition*, vol. 2, no. 2, pp. 174–180, 2014, doi: 10.1002/fsn3.91.

[73] N. A. N. Mohamad, N. A. Arham, J. Jai, and A. Hadi, "Plant extract as reducing agent in synthesis of metallic nanoparticles: A review," *Advanced Materials Research*, vol. 832, pp. 350–355, 2014, doi: 10.4028/www.scientific.net/AMR.832.350.

[74] D. S. Chauhan, M. A. Quraishi, and A. Qurashi, "Recent trends in environmentally sustainable Sweet corrosion inhibitors," *Journal of Molecular Liquids*, vol. 326, p. 115117, 2021, doi: 10.1016/j.molliq.2020.115117.

[75] G. A. Rassoul, D. M. El-Din Kassim, and B. H. EsSebbagh, "Corrosion inhibition of aluminum alloy in 50% ethylene GlYcol solution," *IICPT Iraqi Iraoi Journalof Chemical and Petroleum Engineering*, vol. 8, no. 3, pp. 53–59, 2007.

[76] F. Bentiss, M. Traisnel, and M. Lagrenee, "The substituted 1,3,4-oxadiazoles: A new class of corrosion inhibitors of mild steel in acidic media," *Corrosion Science*, vol. 42, no. 1, pp. 127–146, 2000, doi: 10.1016/S0010-938X(99)00049-9.

[77] A. E. A. S. Fouda, E. E. El-Katori, and S. Al-Mhyawi, "Methanol extract of slanum nigrum as eco-friendly corrosion inhibitor for zinc in sodium chloride polluted solutions," *International Journal of Electrochemical Science*, vol. 12, no. 10, pp. 9104–9120, 2017, doi: 10.20964/2017.10.64.

[78] L. Guo, R. Zhang, B. Tan, W. Li, H. Liu, and S. Wu, "Locust Bean Gum as a green and novel corrosion inhibitor for Q235 steel in 0.5 M H₂SO₄ medium," *Journal of Molecular Liquids*, vol. 310, p. 113239, 2020, doi: 10.1016/j.molliq.2020.113239.

[79] H. A. Mohamedien, S. M. Kamal, M. Taha, M. M. EL- Deeb, and A. G. El-Deen, "Experimental and computational evaluations of cefotaxime sodium drug as an efficient and green corrosion inhibitor for aluminum in NaOH solution," *Materials Chemistry and Physics*, vol. 290, no. April, p. 126546, 2022, doi: 10.1016/j.matchemphys.2022.126546.

[80] E. E. El-Katori and S. Al-Mhyawi, "Assessment of the Bassia muricata extract as a green corrosion inhibitor for aluminum in acidic solution," *Green Chemistry Letters and Reviews*, vol. 12, no. 1, pp. 31–48, 2019, doi: 10.1080/17518253.2019.1569728.

[81] Y. Shao et al., "Advances in molecular quantum chemistry contained in the Q-Chem 4 program package," *An International Journal at the Interface Between Chemistry and Physics*, vol. 113, no. 2, pp. 184–215, Jan. 2014, doi: 10.1080/00268976.2014.952696.

[82] M. Miar, A. Shiroudi, K. Pourshamsian, A. R. Oliaey, and F. Hatamjafari, "Theoretical investigations on the HOMO-LUMO gap and global reactivity descriptor studies, natural bond orbital, and nucleus-independent chemical shifts analyses of 3-phenylbenzo[d]thiazole-2(3H)-imine and its para-substituted derivatives: Solvent and substituent effects," *Journal of Chemical Research*, vol. 45, no. 1–2, pp. 147–158, Jan. 2021, doi: 10.1177/1747519820932091/ASSET/IMAGES/LARGE/10.1177_1747519820932091-FIG1.JPEG.

[83] A. Singh et al., "Investigation of Corrosion Inhibitors Adsorption on Metals Using Density Functional Theory and Molecular Dynamics Simulation," in Corrosion Inhibitors, A. Singh, Ed. IntechOpen, 2019, pp. 121–139. doi: 10.5772/INTECHOPEN.84126.

[84] N. I. N. Haris, S. Sobri, Y. A. Yusof, and N. K. Kassim, "An overview of molecular dynamic simulation for corrosion inhibition of ferrous metals," *Metals*, vol. 11, no. 1, p. 46, Dec. 2020, doi: 10.3390/MET11010046.

[85] S. Okeniyi, B. El-yaqub, F. Awe, P. Omale, and M. Adeyemi, "Adsorption and inhibitive properties of ethanolic extract of Striga hemonthica as a green corrosion inhibitor for Aluminium alloy in 0.1M H_2SO_4," *Journal of Chemical Society of Nigeria*, vol. 43, no. 3, pp. 635–642, 2018.

[86] L. A. Nnanna et al., "Adsorption and inhibitive properties of leaf extract of Newbouldia leavis as a green inhibitor for aluminium alloy in H_2SO_4," *American Journal of Materials Science*, vol. 1, no. 2, pp. 143–148, Aug. 2012, doi: 10.5923/j.materials.20110102.24.

[87] E. Ituen, "Adsorption and kinetic/thermodynamic characterization of aluminium corrosion inhibition in sulphuric acid by extract of Alstonia boonei," *IOSR Journal of Applied Chemistry*, vol. 3, no. 4, pp. 52–59, 2013, doi: 10.9790/5736-0345259.

[88] L. T. Popoola, "Adsorption and corrosion inhibitive properties of Oryza Glaberrima husk extract on aluminium in H_2SO_4 solution: Isotherm, kinetic and thermodynamic studies," *Studia Universitatis Babeş-Bolyai Chemia*, vol. 65, no. 4, pp. 177–201, Dec. 2020, doi: 10.24193/subbchem.2020.4.14.

[89] B. A. Abd-El-Naby, O. A. Abdullatef, H. M. El-Kshlan, E. Khamis, and M. A. Abd-El-Fatah, "Anionic effect on the acidic corrosion of aluminum and its inhibition by lupine extract," *Portugaliae Electrochimica Acta*, vol. 33, no. 5, pp. 265–274, Apr. 2016, doi: 10.4152/pea.201505265.

[90] K. Krishnaveni and J. Ravichandran, "Aqueous extract of leaves of Morinda tinctoria as a corrosion inhibitor for aluminum in sulphuric acid medium," *Journal of Adhesion Science and Technology*, vol. 29, no. 14, pp. 1465–1482, Jul. 2015, doi: 10.1080/01694243.2015.1030907.

[91] P. Arora, S. Kumar, M. K. Sharma, and S. P. Mathur, "Corrosion Inhibition of Aluminium by Capparis decidua in Acidic Media," *E-Journal of Chemistry*, vol. 4, no. 4, pp. 450–456, 2007. [Online]. Available: https://www.e

[92] R. Chauhan, U. Garg, and R. K. Tak, "Corrosion Inhibition of Aluminium in Acid Media by Citrullus Colocynthis Extract," *E-Journal of Chemistry*, vol. 8, no. 1, pp. 85–90, 2011. [Online]. Available: https://www.e

[93] A. E. A. S. Fouda, S. H. Etaiw, and M. Hammouda, "Corrosion inhibition of aluminum in 1 M H_2SO_4 by Tecoma non-aqueous extract," *Journal of Bio- and Tribo-Corrosion*, vol. 3, no. 3, Sep. 2017, doi: 10.1007/s40735-017-0090-3.

[94] W. E. Org, C. A. Loto, O. O. Joseph, R. T. Loto, and A. P. I. Popoola, "Electrochemical science corrosion inhibitive behaviour of Camellia sinensis on aluminium alloy in H_2SO_4," *International Journal of Electrochemical Science*, vol. 9, pp. 1221–1231, 2014, [Online]. Available: www.electrochemsci.org

[95] O. A. Omotosho et al., "Corrosion Resistance of Aluminium in 0.5 M H_2SO_4 in the Presence of Cassia fistula Extract," in *TMS 2018 147th Annual Meeting & Exhibition Supplemental Proceedings*, 2018, vol. Part F12, pp. 909–918. Springer International Publishing. doi: 10.1007/978-3-319-72526-0_87.

[96] S. A. Rao, P. Rao, A. Professor, and A. Professor, "Eco friendly green inhibitor Tinosporacordifolia (Linn.) for the corrosion control of aluminum in sulfuric acid medium," *International Journal of Innovative Research in Science, Engineering and Technology (An ISO)*, vol. 3297, no. 2, 2007, doi: 10.15680/IJIRSET.2015.0402070.

[97] U. J. Ahile, S. O. Adejo, J. A. Gbertyo, F. F. Idachaba, R. L. Tyohemba, and S. O. Ama, "Ethanol Stem Extract Of Mucuna Pruriens As Green Corrosion Inhibitor For Corrosion of Aluminium In H_2SO_4," *International Peer Reviewed Journal*, vol. 3, no. 5, pp. 2039–2046, 2014, [Online]. Available: www.joac.info

[98] R. S. Nathiya and V. Raj, "Evaluation of Dryopteris cochleata leaf extracts as green inhibitor for corrosion of aluminium in 1 M H₂SO₄," *Egyptian Journal of Petroleum*, vol. 26, no. 2, pp. 313–323, Jun. 2017, doi: 10.1016/j.ejpe.2016.05.002.

[99] R. S. Nathiya, S. Perumal, V. Murugesan, and V. Raj, "Evaluation of extracts of Borassus flabellifer dust as green inhibitors for aluminium corrosion in acidic media," *Materials Science in Semiconductor Processing*, vol. 104, Dec. 2019, doi: 10.1016/j.mssp.2019.104674.

[100] H. C. Sharma, S. Kumar, V. Singh, J. Sharma, and S. P. Mathur, "Green corrosion inhibitor by ethanolic extract of Lantana camara for corrosion of aluminium in acidic media," *International Journal of Pharmaceutics and Drug Analysis*, vol. 2, no. 3, pp. 341–346, 2014, [Online]. Available: www.ijpda.com

[101] O. D. Olakolegan, S. S. Owoeye, E. A. Oladimeji, and O. T. Sanya, "Green synthesis of Terminalia Glaucescens Planch (Udi plant roots) extracts as green inhibitor for aluminum (6063) alloy in acidic and marine environment," *Journal of King Saud University - Science*, vol. 32, no. 2, pp. 1278–1285, Mar. 2020, doi: 10.1016/j.jksus.2019.11.010.

[102] S. A. Rao and P. Rao, "Inhibitive effect of Carica papaya seed extract on aluminium in H₂SO₄ medium," *Journal of Materials and Environmental Science*, vol. 5, no. 2, pp. 591–598, 2014.

[103] N. Princeley Enyinnaya, A. Olubimi James, and C. Obi, "Inhibitive potential of benign Caesar-weed leaves extract (CWLE) as corrosion inhibitor of aluminium in H₂SO₄ phase," *Interdisciplinary Journal of Applied and Basic Subjects*, vol. 2, no. 1, pp. 1–13, 2022, [Online]. Available: www.visnav.in/ijabs/

[104] S. O. Adejo, S. G. Yiase, U. J. Ahile, T. G. Tyohemba, and J. A. Gbertyo, "Inhibitory effect and adsorption parameters of extract of leaves of Portulaca oleracea of corrosion of aluminium in H₂SO₄ solution," *Scholars Research Library Archives of Applied Science Research*, vol. 5, no. 1, pp. 25–32, 2013, [Online]. Available: www.scholarsresearchlibrary.com

[105] J. O. Ezeugo, O. D. Onukwuli, K. O. Ikebudu, V. C. Ezechukwu, and L. O. Nwaeto, "Optimization of Chrysophyllum albidum leaf extract as corrosion inhibitor for aluminium in 0.5 M H₂SO₄," *World Scientific News*, vol. 125, pp. 32–50, 2019, [Online]. Available: www.worldscientificnews.com

[106] N. O. Obi-Egbedi, I. B. Obot, and S. A. Umoren, "Spondias mombin L. as a green corrosion inhibitor for aluminium in sulphuric acid: Correlation between inhibitive effect and electronic properties of extracts major constituents using density functional theory," *Arabian Journal of Chemistry*, vol. 5, no. 3, pp. 361–373, Jul. 2012, doi: 10.1016/j.arabjc.2010.09.002.

[107] A. Jano, A. Lame, E. Kokalari, and E. Bicaku, "Study of corrosion inhibition of aluminum in acidic media by pineapple crown extract," *Ovidius University Annals of Chemistry*, vol. 33, no. 2, pp. 104–107, Jul. 2022, doi: 10.2478/auoc-2022-0015.

[108] J. G. Gonzalez-Rodriguez, M. Ramírez-Arteaga, M. G. Valladares, and J. G. González Rodríguez, "Use of Prosopis laevigata as a corrosion inhibitor for Al in H₂SO₄ green inhibitors for carbon steel corrosion in acid media view project use of Prosopis laevigata as a corrosion inhibitor for Al in H₂SO₄," *International Journal of Electrochemical Science*, vol. 8, pp. 6864–6877, 2013, [Online]. Available: www.electrochemsci.org

[109] M. Omotioma, O. D. Onukwuli, and I. Obiora-Okafo, "Phytochemicals and inhibitive properties of cashew extract as corrosion inhibitor of aluminium in H₂SO₄ medium," *Latin American Applied Research*, vol. 49, pp. 99–103, 2019.

[110] L. B. Molina-Ocampo, M. G. Valladares-Cisneros, and J. G. Gonzalez-Rodriguez, "Using Hibiscus sabdariffa as corrosin inhibitor for Al in 0.5 M H₂SO₄," *International Journal of Electrochemical Science*, vol. 10, no. 1, pp. 388–403, 2015.

[111] T. Yanardağ, N. Bayraktar, K. Yanardağ, Y. Özgen, and A. A. Aksüt, "The inhibitor effects of Cannabis sativa L. extracts on the corrosion of aluminium in H₂SO₄ solutions," *Communications Faculty of Sciences University of Ankara Series B*, vol. 62, no. 1, pp. 12–22, 2020, [Online]. Available: https://dergipark.org.tr/tr/pub/communb/issue/52583/749033

[112] M. Okpara, O. Nwadiuko, N. D. Ekekwe, and S. Udensi, "Inhibition by Newbouldia lea-vis leaf extract of the corrosion of aluminium in HCl and H₂SO₄ solutions," *Archives of Applied Science Research*, vol. 4, no. 1, pp. 201–217, 2012, [Online]. Available: https://www.researchgate.net/publication/267783469

[113] N. Chaubey, V. K. Singh, and M. A. Quraishi, "Papaya peel extract as potential corro-sion inhibitor for aluminium alloy in 1 M HCl: Electrochemical and quantum chemical study," *Ain Shams Engineering Journal*, vol. 9, no. 4, pp. 1131–1140, Dec. 2018, doi: 10.1016/J.ASEJ.2016.04.010.

[114] B. Kasuga, E. Park, R. L. Machunda, B. Kasuga, E. Park, and R. L. Machunda, "Inhibition of aluminium corrosion using Carica papaya leaves extract in sulphuric acid," *Journal of Minerals and Materials Characterization and Engineering*, vol. 6, no. 1, pp. 1–14, Dec. 2017, doi: 10.4236/JMMCE.2018.61001.

[115] S. R. Al-Mhyawi, "Corrosion inhibition of aluminum in 0.5 M HCL by Garlic aque-ous extract," *Oriental Journal of Chemistry*, vol. 30, no. 2, pp. 541–552, 2014, doi: 10.13005/OJC/300218.

[116] S. K. Sharma, A. Peter, and I. B. Obot, "Potential of Azadirachta indica as a green corrosion inhibitor against mild steel, aluminum, and tin: A review," *Journal of Analytical Science and Technology*, vol. 6, no. 1, pp. 1–16, Dec. 2015, doi: 10.1186/S40543-015-0067-0/FIGURES/3.

9 Phytochemicals/Plant Extracts as Corrosion Inhibitors for Different Metals/Alloys in HNO₃

Walid Daoudi, Abdelmalik El Aatiaoui, and Adyl Oussaid
University Mohamed I

Omar Dagdag
Gachon University

Salma Lamghafri
Abdelmalek Essaâdi University

Rajesh Haldhar
Yeungnam University

9.1 INTRODUCTION

Currently, one of the biggest issues in the industry field is material corrosion. Finding a clear and efficient approach to reduce these harmful impacts on metals is therefore crucial. Much effort is put to develop appropriate compounds to utilize as corrosion inhibitors (CIs). The best solution is the use of organic compounds. The choice of an inhibitor is determined by several criteria; a good inhibitor has to be eco-friendly, cheap, effective even at low concentration, chemically inert, and thermally stable. Among the most often used green CIs are plant extracts.

Plant extracts are well known for their significant biological effects, particularly their antibacterial, antifungal, anti-inflammatory, antioxidant, and antiviral properties; they can also be used in chronic degenerative illnesses like cancer and diabetes.

Most often, extracts are made from the entire plant or the parts with the highest concentrations of active constituents, named phytochemicals. These active compounds are naturally occurring in plants, and they are found in roots, leaves, stems, flowers, seeds, and fruits.

The most prevalent phytochemicals with a corrosion-inhibiting action are tannins, phenolic compounds, triterpenes, phlobatannins, alkaloids, saponins, phytosterol, and flavonoids.

DOI: 10.1201/9781003394631-9

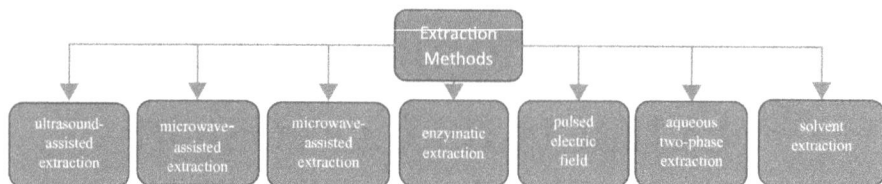

FIGURE 9.1   Extraction methods for bioactive substances.

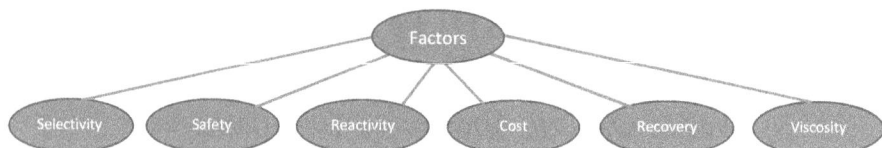

FIGURE 9.2   Schematic representation of extraction-related factors.

These bioactive substances can be extracted using a variety of methods, such as aqueous two-phase extraction and solvent extraction using organic, eutectic, and ionic liquid solvents. Figure 9.1 shows some extraction methods.

Therefore, the choice of the solvents used to extract biomolecules from plants is related to the type of plant and the part to be extracted. Also, polar solvents are used for extraction of polar compounds. In fact, solvents are influenced by the polarity of the intended solute. This latter is properly dissolved by a solvent with a similar polarity to the solute. Nonpolar solvents are destined for nonpolar compounds. The factors to be accounted for while choosing extraction solvents are shown in Figure 9.2.

Generally, plant extracts adsorb on an exposed metal/alloy surface by interaction of phytochemicals and metal surfaces through physisorption, chemisorption, and retrodonation in the presence of an acidic medium. The existence of such a phytochemical in the plant extracts determines the specific mechanism of inhibition.

The anticorrosion performance of these extracts as CIs is investigated through electrochemical techniques, namely, weight loss measurement, potentiodynamic polarization, and electrochemical impedance spectroscopy, in addition to means of surface characterizations such as scanning electron microscopy, X-ray photoelectron spectroscopy analysis, and atomic force microscopy. Moreover, a theoretical investigation is used to describe chemical reactivity and to better understand the interaction between phytochemicals and metal surface.

Nitric acid is the most commonly employed solution in the industrial field. The principal application for this acid involves pickling, industrial acid cleaning, acid descaling, and more. However, the corrosive effect of this medium has stimulated a lot of research on metal corrosion.

9.1.1   GREEN CORROSION INHIBITORS IN SEVERAL INDUSTRIES

According to Figure 9.3, inorganic and organic environmental CIs [2] are the two broad groups into which green CIs may generally be divided. The inorganic category of green-based corrosion inhibitors is heavily used in aqueous systems because of

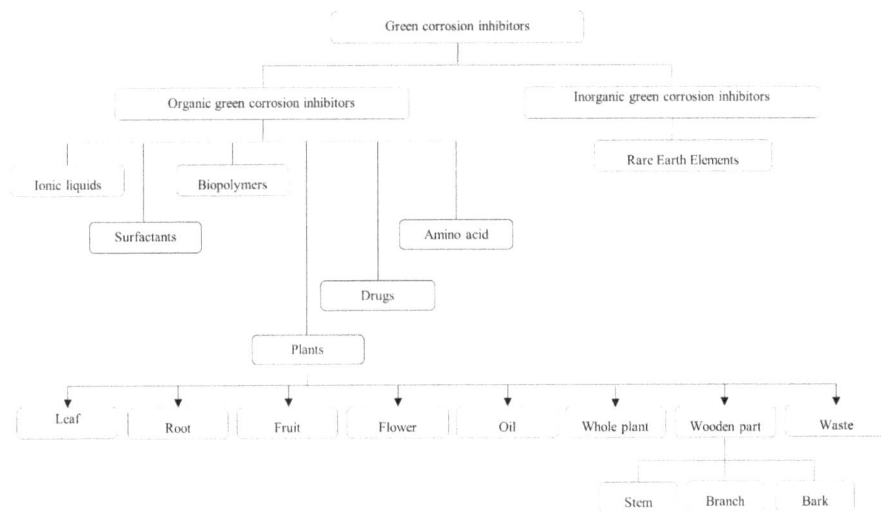

FIGURE 9.3   Groups of green corrosion inhibitors [1].

their high productivity [3], whereas the organic class of green-based CIs is made up of synthetic materials that are harmless to the environment [4]. Organic green inhibitors have advantages over inorganic inhibitors, according to Wei et al. [5]. They clarified that organic green CIs may evenly passivate the metal surface, offering the best defense against aggressive media. The brittle passive layers created by inorganic inhibitors make the metal surface vulnerable to local corrosion assaults including pitting and crevice corrosion.

A possible method to stop, slow down, or prevent corrosion is green chemistry. Despite a variety of organic compounds being widely used as CIs, their toxic nature has limited their application as inhibitors [6]. On the contrary, green CIs are made completely of natural chemical substances. These CIs are generally biodegradable, environmentally benign, affordable, renewable, and free of heavy metals or other potentially harmful substances [7]. As shown in Figure 9.4, environmentally friendly CIs are utilized extensively in a variety of industries. The phytochemicals such as flavonoids, quinine, nicotine, terpenes, alkaloids, polyphenols, carboxylic acids, and other elements like C, N, O, and S are abundant in natural product extracts. These organic extracts work well as green CIs. A variety of phytochemicals promote their adsorption by creating a thin coating (i.e., layer) on the metal surface, which protects the surface and prevents corrosion [8].

9.1.2   PLANT EXTRACTS

The use of materials made from plants in a variety of industries has become more important for a variety of reasons [10]. Primarily, the vast amounts of phytochemicals present in the crude extracts of the majority of plant species make them more useful for a variety of biological and industrial applications [11]. Additionally, plant-based crude extracts can be utilized in various electrolytic systems, alloys, and metals as

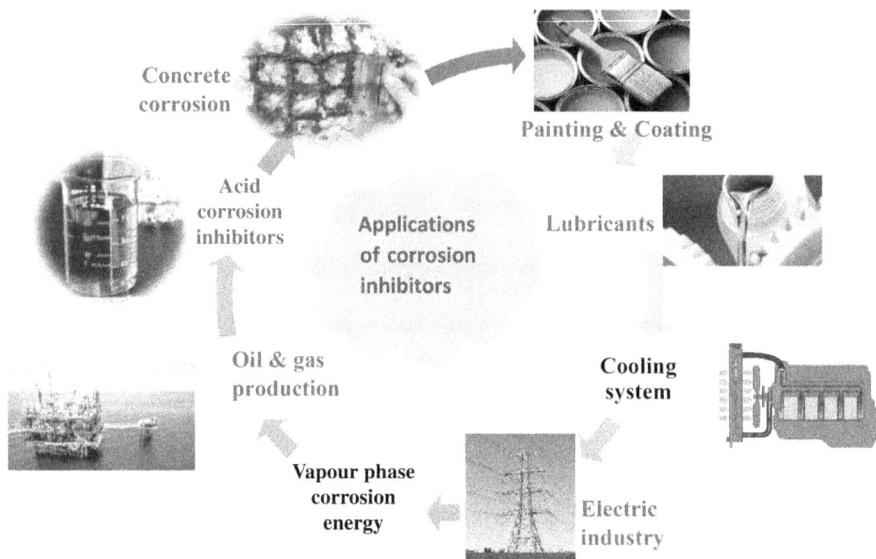

FIGURE 9.4   Green corrosion inhibitors used in a variety of sectors [9].

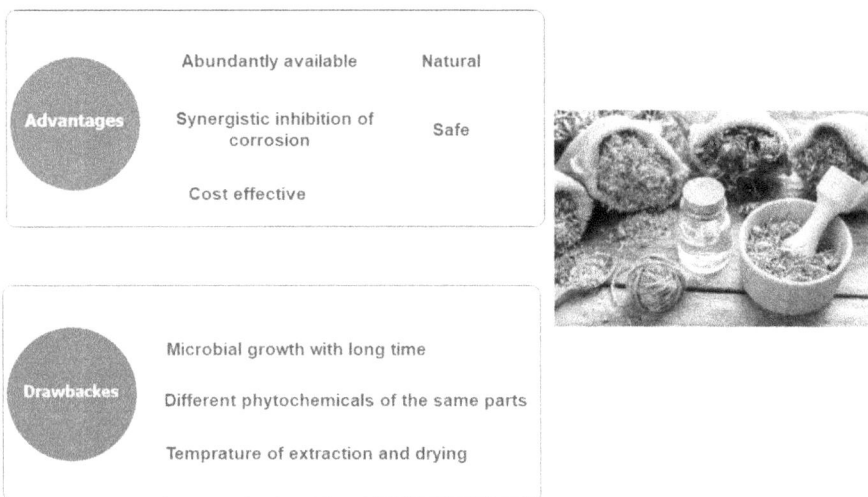

FIGURE 9.5   Application of crude extracts as corrosion inhibitors: benefits and drawbacks [9].

environmentally friendly CIs [12]. Utilizing plant-based crude extracts as environmentally friendly CIs has many benefits, including cost-effectiveness, nontoxic-
ity, high availability, time and effort savings, and low synthesis step requirements.
Additionally, various phytochemical compounds may exhibit synergistic corrosion
inhibition effects, which would help to inhibit corrosion even more [13]. The primary
benefit and drawback of employing crude extracts as CIs are shown in Figure 9.5.

However, there are some drawbacks in using plant extracts, and as a result, their use in corrosion prevention applications has come under fire. Natural plant extracts' primary drawbacks are their unstable nature and limited shelf life. Additionally, prolonged storage of plant extracts could make them vulnerable to microbial and fungal growth, which could reduce the effectiveness of CIs [10].

Additionally, the plant chemotype may exhibit some issues and result in some flaws in the performance of the inhibition. For instance, different combinations of phytochemicals taken from the same plant component or organ in other climates might exist. Finding an appropriate solvent for the extraction procedure is another crucial factor to take into account when employing natural plant extracts. Unfortunately, it is challenging to find a solvent with the desired properties, including redox stability, high availability, low toxicity, cost-effectiveness, and nonflammability. Additionally, choosing an appropriate temperature for extraction is regarded as another significant challenge because an increase in temperature runs the risk of denaturing the natural metabolites in the extracts. Additionally, the temperature needed for the drying processes may be extremely important. Some plant species require from a few hours to several days to dry out in the shade. Using high temperatures to speed up this process might have detrimental effects [10].

A variety of electron centers, including ester ($-COOC_2H_5$), ether (-O-), acid chloride (-COCl), dimethylamino ($-NMe_2$), amide ($-CONH_2$), hydroxyl (-OH), and methoxy (-OMe), are abundant in naturally occurring plant compounds. These are also abundant in multiple bonds, such as -CRN, -NQN-, -CRC-, 4CQO-, -NQO-, 4CQCo-, and 4CQN-, which serve as adsorption centers when metals and inhibitors interact [14]. On the surface of the metals, heteroatoms and electron-rich areas are what induce chemisorption, whereas the polar portions of the molecules that inhibit have been identified and discovered to produce physisorption [11]. The majority of natural compounds have complicated structures, yet they are soluble in polar electrolytes due to their interaction with polar functional groups, which have high peripheral functionalities [15]. By using inorganic salts, plant extracts' limited protective efficiency against metallic corrosion might be increased. As an example, Potassium iodide (KI) is employed because it improves the effectiveness of inhibition through a number of synergistic actions [16]. Figure 9.6 illustrates the major characteristics that are driving up the requirement for crude plant extracts as CIs. The creation of crude extracts occurs in a biphasic system with two phases: an organic phase and an aqueous phase. As shown in Figure 9.6, it should also be kept in the refrigerator at a low temperature.

In the solvent extraction process, the solvent must penetrate the plant tissue and be solubilized, and then, the compounds (phytochemicals) must be extracted [10]. Examples of several extraction solvents are shown in Figure 9.7, along with the plant active extracts that result. For various intended extracts or concentrations of phytochemicals, multiple solvents might be utilized as the extraction medium. According to Tan and Kassim [17], the ethanol extract of *Rhizophora apiculata* has an extraordinarily high amount of tannins, total flavonoids, and total phenolics compared to the acetone extract. In the other study, 70% acetone performed better as a solvent than either water or 80% methanol when it came to removing tannins from *Acacia nilotica* [18]. According to Figure 9.7, the majority of plant extracts are extracted using methanol, ethanol, or an aqueous solvent [19].

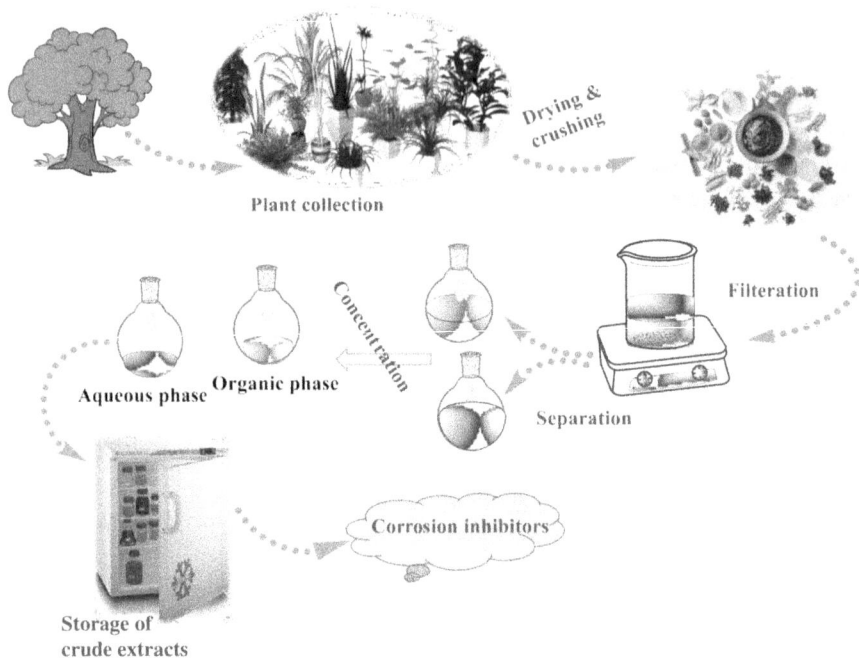

FIGURE 9.6 The scheme for gathering, drying, crushing, separating, extracting, concentrating, and storing plant extracts [9].

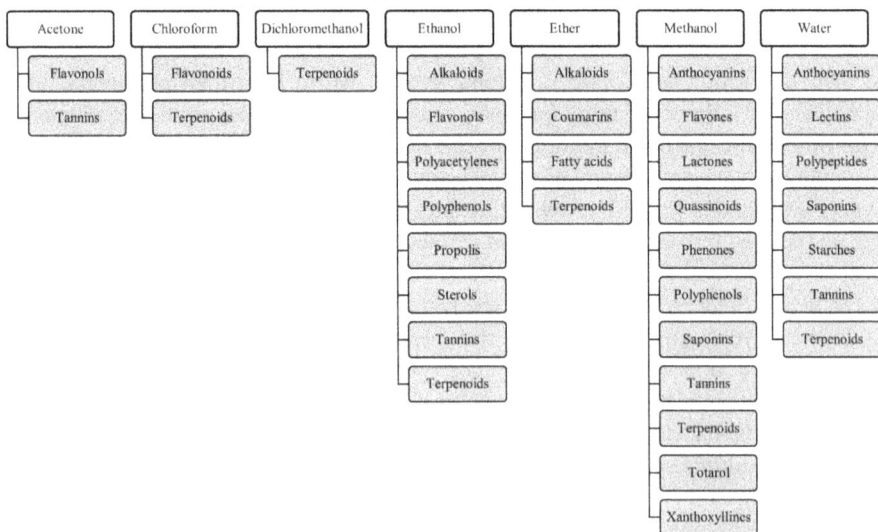

FIGURE 9.7 Phytochemicals of various types that have been extracted using various solutions [1].

Alkaloids and other nitrogen-containing metabolites	Phenolic acids	Terpenoids	Lipids	Carbohydrates
Glucosinolates, Amaryllidaceae, betalain, diterpenoid, indole, isoquinoline, lycopodium, peptide, pyrrolidine, piperidine, pyrrolizidine, quinoline, quinolizidine, steroidal, tropane, amino acids, amine, cyanogenic glycoside, purine, pyrimidines, proteins, peptides	Flavonoids, p phenolic acids, stilbenoids, tannins, lignans, xanthones, quinones, coumarins, phenylpropanoids benzofurans	• Carotenoids, • monoterpenoids, • diterpenoids, • triterpenes, • triterpenoid saponins, • sesquiterpenoids, • sesquiterpene lactones, • polyterpenoids	Monounsaturated fat, polyunsaturated fat, saturated fat and fatty acids	Monosaccharide, disaccharide, polysaccharide, oligosaccharide, sugar alcohols

FIGURE 9.8 Various phytochemicals described from plant extract [1].

9.1.3 PHYTOCHEMICALS IN PLANT EXTRACT

According to numerous studies, plant extracts are thought to be rich in naturally synthesized chemical substances such as organic compounds or bioactive chemicals that have been used as CIs in many industrial systems [1]. There have been several studies regarding the different plant parts, including the flower, leaf, root, and complete plant, that may be harvested to prevent rusting. According to Umoren et al., the phytochemical components of *Sida acuta* vary depending on the plant part, with saponins, flavonoids, tannins, alkaloids, organic acids, and anthraquinones being present in the leaves extract, while these compounds can be found only in the stem extract of other plants [20]. Researchers are looking at plant extract as a green corrosion inhibitor since different portions of the plant have varied amounts and types of phytochemicals.

Due to the existence of chemical structures that are comparable to those of typical organic molecules through previously demonstrated efficacy, plant extract can operate as a corrosion inhibitor. In accordance with their chemical makeup and properties, Figure 9.8 categorizes phytochemicals into six main groups and their subgroups. According to reports, the presence of these particular phytochemicals is what prevents rusting.

9.2 CASE STUDIES

Many studies have been performed to evaluate the performance of plant extracts as CIs for different metals and their reaction in an HNO_3 medium.

El-Dossoki et al. [21] studied the performance of *Moringa oleifera* extract in inhibiting corrosion on copper in the acid mixture ($HNO_3 + H_3PO_4$). Inhibition efficiency reached 89% at a concentration of 0.300 mol L^{-1} at 298K. The inhibitor is physically adsorbed on the Cu surface, and it follows Langmuir adsorption isotherm.

R.M. Younis [22] investigated *Carapichea ipecacuanha* extract. The inhibitor acted as a mixed type for copper corrosion in 1 M HNO_3 medium. The inhibition efficiency increased with the inhibitor concentration and reached 93%; however, it decreased with an increase in temperature. The adsorption of the inhibitor was confirmed by a decrease in double-layer capacitances, and it followed Langmuir isotherm.

Fouda et al. [23] demonstrated that *Ceratonia siliqua* is a good cathodic inhibitor against the corrosion of copper and brass. The addition of the extract increased the covering of the surface and retarded the dissolution of these metals. It was proved that the adsorption is spontaneous and physical.

Ricinus communis oil was investigated by S. Houbairi et al. The oil showed excellent inhibitory efficiency which achieved 99%. The authors reported that ricinoleic acid adsorbs locally on the copper surface without formation of monolayer. Also, the adsorbed molecules do not interact between themselves [24].

Zygophllum coccineum extract was evaluated by Fouda et al. [25]. They reported that the plant extract adsorbed to the metal surface following the Langmuir isotherm model, and the addition of this extract caused a decrease in the double-layer capacitances, which affirms that the plant extract molecules adhere to the copper surface and prevent the occurrence of corrosion.

Al Jahdaly et al. [26] proved that *Rosmarinus officinalis* extract is a mixed-type inhibitor that prevents copper corrosion in the acidic medium. Its strength is due to its ability to transfer electrons to the metal vacant orbital during the corrosion process.

Moreover, Fouda [27] has reported that the inhibitory efficiency of *Trigonela stellate* extract is related to its synergistic intermolecular effects. In fact, these molecules adsorb on the Cu surface through donor-acceptor interactions forming covalent bonds between π-electrons of the aromatic ring and vacant d-orbitals of copper surface. Also, the protonated extract adsorbs through electrostatic interactions between the positively charged molecules and the negative charge brought by Cl^- ions to the metal surface (Table 9.1).

9.3   MECHANISM OF CORROSION INHIBITION

It is generally known that the adsorption process and corrosion inhibition are closely related. As reported from the literature, there are many factors influencing adsorption process, namely, the type of metal, its charge, the organic inhibitor's chemical structure, and the type of aggressive medium.

The metal is positively charged in acidic media, while the acid anions NO^{3-} have the tendency to be adsorbed on the metal surface creating a negative charge toward the electrolyte; as a result, more protonated components will be absorbed. Usually, two adsorption modes were taken into consideration.

On the one hand, the neutral form of extract can chemically adsorb on the metal surface through bonds based on donor-acceptor interactions between electron pairs

TABLE 9.1

A Summary Report on Plant Extracts' Performance as Corrosion Inhibitors in the HNO₃ Medium

Natural Plants	Main Components	Metal	Medium	EI% (EIS)	Reference
Moringa oleifera	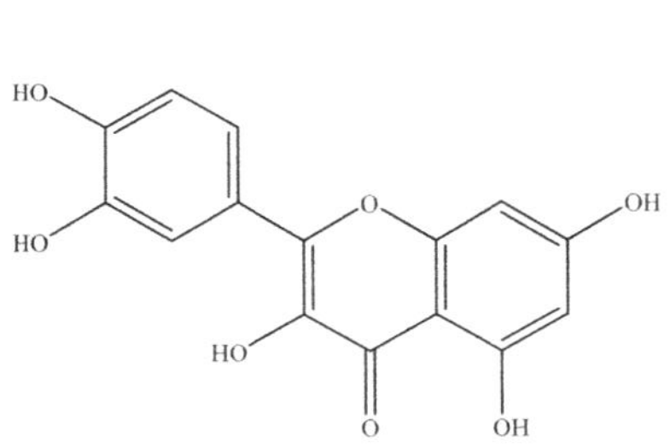 Quercetin, 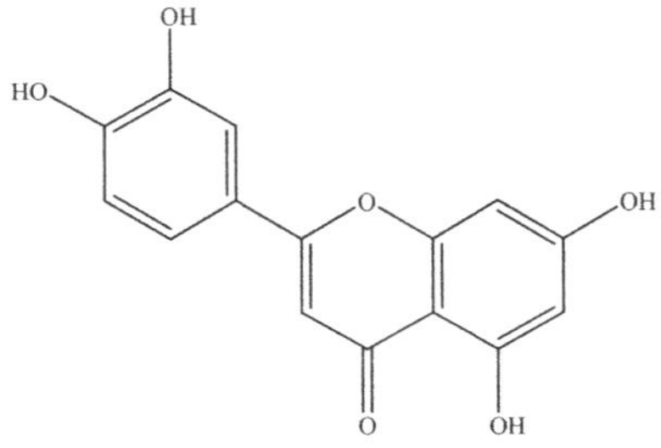 Luteolin	Copper	$HNO_3 + H_3PO_4$	89%	[21]
Carapichea ipecacuanha	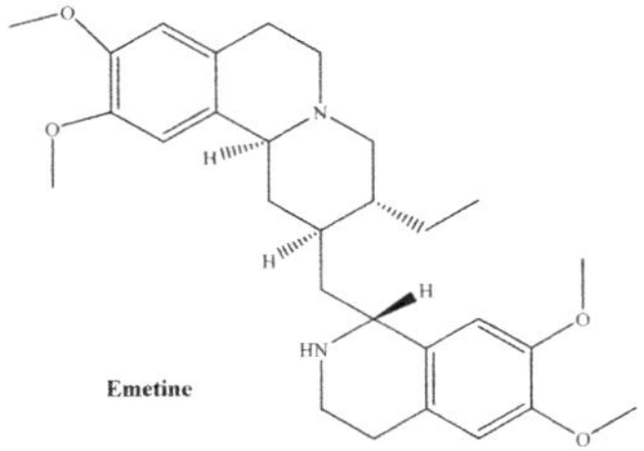 Emetine, 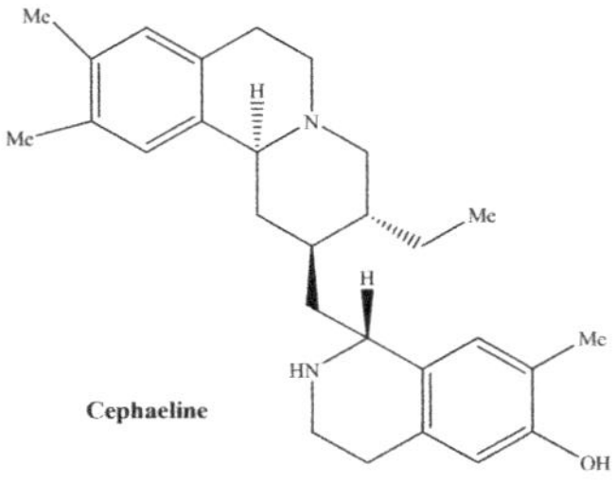 Cephaeline	Copper	1 M HNO_3	93%	[22]

(Continued)

TABLE 9.1 (Continued)
A Summary Report on Plant Extracts' Performance as Corrosion Inhibitors in the HNO₃ Medium

Natural Plants	Main Components	Metal	Medium	EI% (EIS)	Reference
Ceratonia siliqua		Copper brass	1 M HNO₃	91% 90.3%	[24]
	Tannin				
Rcicinus communis	Ricinoleic acid	Copper	2 M HNO₃	99%	[25]
Zygophllum coccineum	Quinovic acid	Copper	1 M HNO₃	91.0%	[25]

(*Continued*)

TABLE 9.1 (Continued)

A Summary Report on Plant Extracts' Performance as Corrosion Inhibitors in the HNO₃ Medium

Natural Plants	Main Components	Metal	Medium	EI% (EIS)	Reference
Rosmarinus officinalis	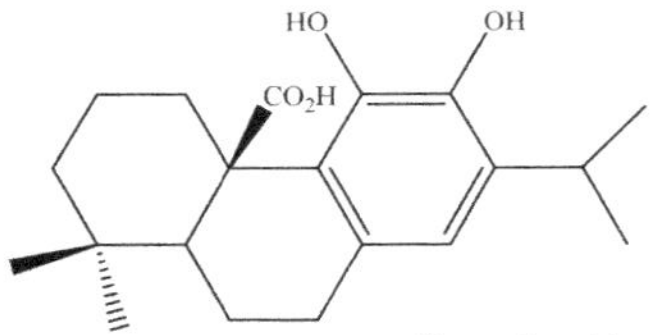 Carnosic acid 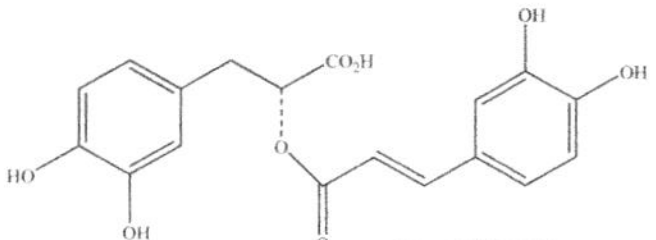 Rosmarinic acid	Copper	1 M HNO₃	77%	[26]
Trigonella stellate	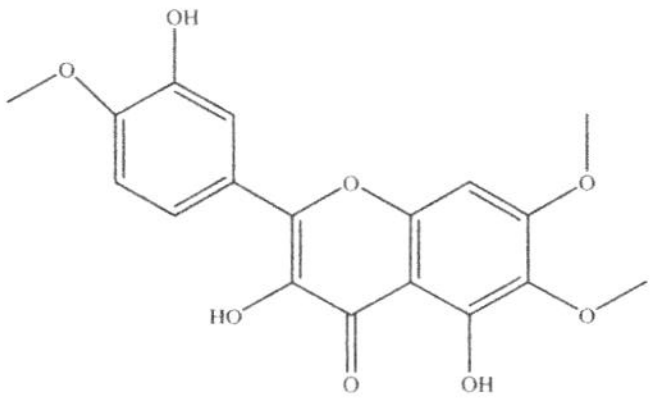	Copper	1 M HNO₃	90.4%	[27]

(Continued)

TABLE 9.1 (*Continued*)
A Summary Report on Plant Extracts' Performance as Corrosion Inhibitors in the HNO₃ Medium

Natural Plants	Main Components	Metal	Medium	EI% (EIS)	Reference
Vitex nigundo (VN)	Negundoside	Copper	3M HNO₃	98%	[28]
Saraca ashoka (SA)	Gallic acid	Copper	3M HNO₃	91%	[28]
Adhatoda vasica (AV)	Vascinone	Copper	3M HNO₃	78%	[28]

(Continued)

TABLE 9.1 (Continued)
A Summary Report on Plant Extracts' Performance as Corrosion Inhibitors in the HNO₃ Medium

Natural Plants	Main Components	Metal	Medium	EI% (EIS)	Reference
Strychnos nuxvomica (SN)	Brucine	Copper	3M HNO₃	91.6%	[29]
Piper longum (PL)	Piperine	Copper	3M HNO₃	80%	[29]
Mucuna pruriens (MP)	L-DOPA	Copper	3M HNO₃	71.6%	[29]

Structure		Side	Top
Carnosic acid			
Rosmarinic acid			
Carnosol			

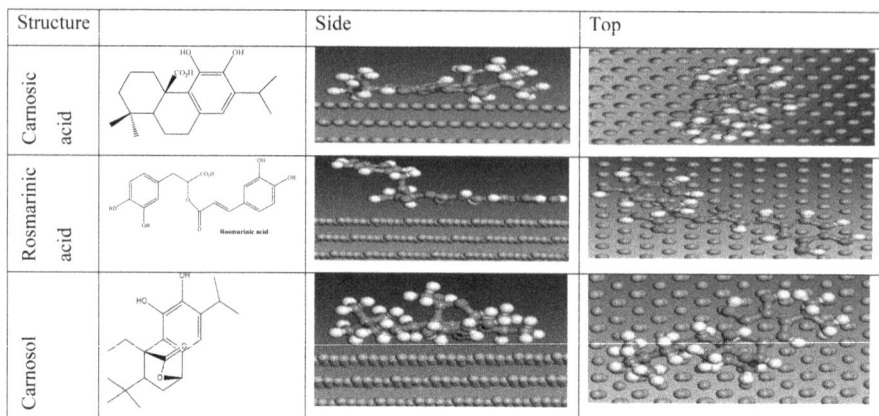

FIGURE 9.9   ROE component adsorption on the surface of copper.

of heteroatom (N,O,S), as well as π electron of aromatic ring, and the vacant d-orbital of metal atoms. On the other hand, the protonated organic molecules can interact electrostatically with the preabsorbed NO^{3-} since the Cu surface carries the positive charge in acidic medium causing electrostatic repulsion.

Furthermore, through a synergistic interaction with preadsorbed NO^{-3}, the protonated sites of the molecules can adsorb on the metal surface.

The excess negative charges that have accumulated on the metal surface can be transferred by retrodonation from the d-orbital of metal to unoccupied anti-bonding π* of extract molecules.

Adsorption of the extract is strengthened by combining all mechanisms of adsorption. As a result, a protective layer is formed that prevents the metal surface from corroding by creating a barrier between the metal and the corrosive medium [24].

Badreah A. Al Jahdaly explained how *Rosmarinus officinalis* extract (ROE) is absorbed at the copper surface in 1 M HNO_3 medium by forming coordinating bonds and transferring electrons from the adsorbed molecules to the metal surface [30]. In fact, all ROE substances possessing pairs of unshared electrons in oxygen atoms can create σ-bond with copper, and the double bonds in the compounds enable the donation of metal d-electrons to the π* orbital [31] (Figure 9.9).

Table 9.2 summarizes binding energies of different compounds of ROE. It can be noted that rosmarinic acid demonstrates the highest adsorption energy. It binds to the copper interface with a significantly higher $E_{binding}$ than carnosic acid and carnosol. As a result, the adsorption of ROE onto the copper surface is strongly influenced by rosmarinic acid indicating its performance to protect copper from corrosion in the HNO_3 medium.

The binding energy is calculated according to the following formulas:

$$E_{interaction} = E_{Tot} - (E_{Sub} + E_{int}) \tag{9.1}$$

$$E_{binding} = -E_{interaction} \tag{9.2}$$

TABLE 9.2
Estimated Energies of ROE on Cu Surface

Structures	Total Energy	Binding Energy	Rigid Adsorption Energy	Deformation Energy
Carnosic acid	−64.8866	−87.2233	−91.5315	4.3082
Rosmarinic acid	−177.2225	−104.4682	−105.0472	0.5790
Carnosol	−151.6792	−97.5361	−99.1319	1.5957

9.4 ADSORPTION ISOTHERMS

Adsorption isotherm is a mathematical formula used for interpreting and predicting the adsorption process. The equation describes the equilibrium of adsorption and the process through which a substance (adsorbate or sorbate) accumulates onto the surface of a solid. It also allows defining the interactions between the adsorbate molecules and the surface adsorption sites.

Isotherm equations confirm that the inhibitory mechanism is accurate and that the adsorption truly occurred.

Many scientists have proposed various adsorption isotherms including mathematical formulas that calculate the amount of adsorbate in the absorbent at a constant temperature, notably Langmuir, Frumkin, Flory–Huggins, Freundlich, and Temkin.

Langmuir isotherms presume that maximal adsorption occurs when the surface is entirely covered with particles that are assumed to be adsorbed as a monolayer, and each site can only accommodate one adsorbed molecule since there is no interaction between molecules on different sites.

Generally, adsorption is investigated at a specific temperature. The linear form of Langmuir isotherm model is described by the mathematic formula:

$$\text{Langmuir isotherm}: \frac{C_{inh}}{\theta} = \frac{1}{K_{ads}} + C_{inh}$$

According to Freundlich, multilayer adsorption occurs on heterogeneous surfaces with various affinities. The Freundlich isotherm equation presumes that cations and anions are simultaneously adsorbed onto the same surface. Due to this circumstance, the adsorbed cations and anions on the surface develop attractive forces.

The linear form of the Freundlich model is given by the following equation:

$$\text{Freundlich isotherm}: \ln(\theta) = \ln(K_{ads}) + z \ln(C_{inh})$$

The Temkin absorption isotherm model examines the interaction between sorbent and sorbate and predicts that the adsorption temperature will decrease linearly alongside an increase in shielding of the adsorbent surface as a result of the interaction between the sorbent and the sorbate in the adsorption phenomenon process. Adsorption is assumed to be defined by an even distribution of binding energies, with maximum binding energy. The following equation describes the linear form of the Temkin model:

$$\text{Temkin isotherm: } \theta = \frac{1}{2a}\ln\left(K_{ads}\right) - \frac{1}{2a}\ln\left(C_{inh}\right)$$

9.5   CONCLUSION

The use of phytochemicals and plant extracts as environmentally friendly CIs as an alternative to traditional CIs has been extensively described in the literature. The results showed that plant extracts can exhibit strong inhibitory efficacy in metal corrosion from a variety of plant components. The presence of heteroatoms and π-electrons in plant extracts is connected to the inhibitory effectiveness of green CIs. According to the structural investigation and molecule-surface interactions at the atomic level of detail, several experimental analysis methods have confirmed the outstanding performance of phytochemicals/plant extracts as green CIs. To study the link between the structure and the observed corrosion inhibition, advanced research should focus on using more creative characterization approaches based on theoretical simulation, like in the molecular dynamics (MD) simulations. This understanding will assist us in constructing a wide range of innovative and potentially sustainable CIs as well as in explaining the process of corrosion inhibition.

REFERENCES

[1] S. Z. Salleh, A. H. Yusoff, S. K. Zakaria, M. A. A. Taib, A. A. Seman, M. N. Masri, et al., "Plant extracts as green corrosion inhibitor for ferrous metal alloys: A review," *Journal of Cleaner Production*, vol. 304, p. 127030, 2021.
[2] B. El Ibrahimi, A. Jmiai, L. Bazzi, and S. El Issami, "Amino acids and their derivatives as corrosion inhibitors for metals and alloys," *Arabian Journal of Chemistry*, vol. 13, pp. 740–771, 2020.
[3] K. Tamalmani and H. Husin, "Review on corrosion inhibitors for oil and gas corrosion issues," *Applied Sciences*, vol. 10, p. 3389, 2020.
[4] M. Bethencourt, F. Botana, J. Calvino, M. Marcos, and M. Rodriguez-Chacon, "Lanthanide compounds as environmentally-friendly corrosion inhibitors of aluminium alloys: A review," *Corrosion Science*, vol. 40, pp. 1803–1819, 1998.
[5] H. Wei, B. Heidarshenas, L. Zhou, G. Hussain, Q. Li, and K. K. Ostrikov, "Green inhibitors for steel corrosion in acidic environment: State of art," *Materials Today Sustainability*, vol. 10, p. 100044, 2020.
[6] E. Ebenso, N. Eddy, and A. Odiongenyi, "Corrosion inhibitive properties and adsorption behaviour of ethanol extract of Piper guinensis as a green corrosion inhibitor for mild steel in H_2SO_4," *African Journal of Pure and Applied Chemistry*, vol. 2, pp. 107–115, 2008.
[7] N. O. Eddy, S. A. Odoemelam, and A. O. Odiongenyi, "Inhibitive, adsorption and synergistic studies on ethanol extract of Gnetum africana as green corrosion inhibitor for mild steel in H_2SO_4," *Green Chemistry Letters and Reviews*, vol. 2, pp. 111–119, 2009.
[8] A. Inamuddin, M. Luqman, and T. Altalhi, "Sustainable Corrosion Inhibitors," in *Materials Research Forum*, Millersville, PA, USA, 2021.

[9] B. A. Al Jahdaly, Y. R. Maghraby, A. H. Ibrahim, K. R. Shouier, A. M. Alturki, and R. M. El-Shabasy, "Role of green chemistry in sustainable corrosion inhibition: A review on recent developments," *Materials Today Sustainability*, vol. 20, p. 100242, 2022.

[10] D. S. Chauhan, M. Quraishi, and A. Qurashi, "Recent trends in environmentally sustainable Sweet corrosion inhibitors," *Journal of Molecular Liquids*, vol. 326, p. 115117, 2021.

[11] A. Miralrio and A. Espinoza Vázquez, "Plant extracts as green corrosion inhibitors for different metal surfaces and corrosive media: A review," *Processes*, vol. 8, p. 942, 2020.

[12] N. Hossain, M. Asaduzzaman Chowdhury, and M. Kchaou, "An overview of green corrosion inhibitors for sustainable and environment friendly industrial development," *Journal of Adhesion Science and Technology*, vol. 35, pp. 673–690, 2021.

[13] P. E. Alvarez, M. V. Fiori-Bimbi, A. Neske, S. A. Brandan, and C. A. Gervasi, "Rollinia occidentalis extract as green corrosion inhibitor for carbon steel in HCl solution," *Journal of Industrial and Engineering Chemistry*, vol. 58, pp. 92–99, 2018.

[14] R. Yadav and M. Agarwala, "Phytochemical analysis of some medicinal plants," *Journal of Phytology*, vol. 3, pp. 10–14, 2011.

[15] C. Verma, E. E. Ebenso, M. Quraishi, and C. M. Hussain, "Recent developments in sustainable corrosion inhibitors: Design, performance and industrial scale applications," *Materials Advances*, vol. 2, pp. 3806–3850, 2021.

[16] G. Bahlakeh, M. Ramezanzadeh, and B. Ramezanzadeh, "Experimental and theoretical studies of the synergistic inhibition effects between the plant leaves extract (PLE) and zinc salt (ZS) in corrosion control of carbon steel in chloride solution," *Journal of Molecular Liquids*, vol. 248, pp. 854–870, 2017.

[17] K. W. Tan and M. J. Kassim, "A correlation study on the phenolic profiles and corrosion inhibition properties of mangrove tannins (Rhizophora apiculata) as affected by extraction solvents," *Corrosion Science*, vol. 53, pp. 569–574, 2011.

[18] C. Y. Ishak and I. E. H. Elgailani, "Methods for extraction and characterization of tannins from some Acacia species of Sudan," *Pakistan Journal of Analytical & Environmental Chemistry*, vol. 17, p. 7, 2016.

[19] R. Haldhar, D. Prasad, A. Saxena, and P. Singh, "Valeriana wallichii root extract as a green & sustainable corrosion inhibitor for mild steel in acidic environments: Experimental and theoretical study," *Materials Chemistry Frontiers*, vol. 2, pp. 1225–1237, 2018.

[20] S. Umoren, U. Eduok, M. Solomon, and A. Udoh, "Corrosion inhibition by leaves and stem extracts of Sida acuta for mild steel in 1 M H_2SO_4 solutions investigated by chemical and spectroscopic techniques," *Arabian Journal of Chemistry*, vol. 9, pp. S209–S224, 2016.

[21] F. I. El-Dossoki, H.-A. El-Nadr, and A. El-Hussein, "Moringa oleifera plant extract as a copper corrosion inhibitor in binary acid mixture (HNO_3+ H_3PO_4)," *Zaštita materijala*, vol. 59, pp. 422–435, 2018.

[22] R. Younis, H. M. Hassan, R. Mansour, and A. El-desoky, "Corrosion inhibition of carapichea ipecacuanha extract (CIE) on copper in 1 M HNO3 solution," *International Journal of Scientific and Engineering Research*, vol. 6, pp. 761–770, 2015.

[23] A. Fouda, K. Shalabi, and A. Idress, "Ceratonia siliqua extract as a green corrosion inhibitor for copper and brass in nitric acid solutions," *Green Chemistry Letters and Reviews*, vol. 8, pp. 17–29, 2015.

[24] S. Houbairi, A. Lamiri, and M. Essahli, "Oil of ricinus communis as a green corrosion inhibitor for copper in 2 M nitric acid solution," *International Journal of Engineering Research*, vol. 3, pp. 698–707, 2014.

[25] A. Fouda, Y. Abdallah, G. Elawady, and R. Ahmed, "Zygophllum coccineum L. extract as green corrosion inhibitor for copper in 1 M HNO3 Solutions," *International Journal*, vol. 2, pp. 517–531, 2014.

[26] B. A. Al Jahdaly, "Rosmarinus officinalis extract as eco-friendly corrosion inhibitor for copper in 1 M nitric acid solution: Experimental and theoretical studies," *Arabian Journal of Chemistry*, vol. 16, p. 104411, 2023.

[27] A. Fouda, A. Mohamed, and M. Khalid, "Trigonella stellate extract as corrosion inhibitor for copper in 1M nitric acid solution," *Journal of Chemical and Pharmaceutical Research*, vol. 8, pp. 86–98, 2016.

[28] P. Mourya, N. Chaubey, V. Singh, and M. Singh, "Eco-friendly inhibitors for copper corrosion in nitric acid: Experimental and theoretical evaluation," *Metallurgical and Materials Transactions B*, vol. 47, pp. 47–57, 2016.

[29] P. Mourya, N. Chaubey, S. Kumar, V. Singh, and M. Singh, "Strychnos nuxvomica, Piper longum and Mucuna pruriens seed extracts as eco-friendly corrosion inhibitors for copper in nitric acid," *RSC Advances*, vol. 6, pp. 95644–95655, 2016.

[30] F. Benhiba, R. Hsissou, Z. Benzekri, M. Belghiti, A. Lamhamdi, A. Bellaouchou, et al., "Nitro substituent effect on the electronic behavior and inhibitory performance of two quinoxaline derivatives in relation to the corrosion of mild steel in 1M HCl," *Journal of Molecular Liquids*, vol. 312, p. 113367, 2020.

[31] E. A. Noor, "The inhibition of mild steel corrosion in phosphoric acid solutions by some N-heterocyclic compounds in the salt form," *Corrosion Science*, vol. 47, pp. 33–55, 2005.

10 Plant Extracts as Corrosion Inhibitors for Different Metals/ Alloys in HNO₃

Abhinay Thakur and Praveen Kumar Sharma
Lovely Professional University

Ashish Kumar
Bihar Engineering University
Government of Bihar

Amita Somya
Amity University

Ambrish Singh
Nagaland University

Elyor Berdimurodov
National University of Uzbekistan
Akfa University

Sumayah Bashir
Central University of Kashmir

10.1 INTRODUCTION

When metallic substances interact with their environment, corrosion inevitably takes place. As a result, the structural integrity and functional qualities of these materials deteriorate. It poses significant challenges in numerous industrial applications where metals and alloys are exposed to corrosive substances [1–4]. One such aggressive agent is nitric acid (HNO_3), known for its strong oxidizing nature and widespread use in industries such as chemical processing, metal finishing, and nuclear energy production. The corrosive nature of HNO_3 makes it crucial to develop effective corrosion inhibition strategies in order to mitigate its detrimental effects. Identifying corrosion inhibitors that are not only efficient but also environmentally

DOI: 10.1201/9781003394631-10

benign has received more attention in past decades. Conventional synthetic inhibitors have been widely employed for corrosion control. However, concerns regarding their environmental impact and potential health hazards have driven researchers to explore alternative, sustainable approaches. This has led to the emergence of plant extracts as promising candidates for corrosion inhibition. Plant extracts contain a rich variety of organic compounds, including polyphenols, flavonoids, alkaloids, and tannins, which have demonstrated inherent corrosion-inhibitive properties. These natural compounds offer several advantages over synthetic inhibitors. First, they are derived from renewable sources, making them sustainable and environmentally friendly. Additionally, plant extracts are abundant, easily accessible, and often cost-effective. Moreover, their biodegradability further contributes to their eco-friendly nature [5,6].

Extensive research has shown that plant extracts can exhibit excellent inhibitive performance on various metals and alloys in different corrosive environments, including HNO_3 solutions. This broad applicability is attributed to the diverse composition of plant extracts and the range of inhibitive mechanisms they possess. Polyphenols, for example, are known for their strong adsorption capabilities, emerging protective coatings on the metallic surface and impeding the corrosion process. Flavonoids, alkaloids, and tannins also contribute to corrosion inhibition through their ability to interact with metal ions and inhibit electrochemical reactions. The usage of plant extracts as corrosion inhibitors not only provides a sustainable substitute to conventional inhibitors but also aligns with the principles of green chemistry [7–10]. By harnessing the inherent inhibitive properties of plant extracts, researchers and engineers can develop eco-friendly corrosion inhibition strategies that contribute to the preservation of the environment. Moreover, the utilization of plant extracts offers potential cost savings and promotes the development of greener practices in various industries.

The potential of plant extracts as corrosion inhibitors for various metals and alloys in HNO_3 settings is thoroughly discussed in this chapter. It aims to highlight the significant advancements made in this field and to shed light on the mechanisms underlying corrosion inhibition by plant extracts [11–13]. Furthermore, the factors influencing inhibition efficiency, such as concentration, pH, temperature, and exposure time, are discussed in detail. This chapter also presents an analysis of recent research studies that have explored the corrosion mitigation performance of plant extracts on a range of metals and alloys exposed to HNO_3 solutions. Carbon steel, stainless steel, aluminum, and copper are among the materials investigated, as they are commonly encountered in industrial applications. The effects of plant extract composition, extraction techniques, and pre-treatment methods on their inhibitive properties are explored to offer insights into the optimization and enhancement of corrosion inhibition. Understanding the pathway of corrosion inhibition by plant extracts is crucial for their effective utilization. This chapter delves into the chemical adsorption process, film formation, and electrochemical processes involved in corrosion inhibition, elucidating the underlying principles. By comprehending these mechanisms, researchers and engineers can devise strategies to maximize the inhibitive performance of plant extracts and develop tailored approaches for specific applications.

The potential challenges and areas for improvement in utilizing plant extracts as corrosion inhibitors are discussed, along with recommendations for future research directions. Additionally, this chapter explores the scale-up and industrial implementation of plant extract-based corrosion inhibition approaches, paving the way for practical applications in various industries.

10.2 CORROSION IN HNO₃: CHALLENGES AND IMPLICATIONS

10.2.1 CORROSION MECHANISMS IN HNO₃

Corrosion mechanisms in HNO_3 environments are complex and involve a combination of chemical reactions and electrochemical processes. Understanding these mechanisms is essential for effectively combating corrosion and developing efficient corrosion inhibition strategies. This section will explore the various corrosion mechanisms observed in HNO_3 environments, including general corrosion, localized corrosion, and stress corrosion cracking. General corrosion, also referred to as uniform corrosion, is characterized by a relatively even attack on the metal substrate. In the presence of HNO_3, nitrate ions (NO_3^-) react with water to form nitric acid (HNO_3) through an equilibrium reaction. The dissociation of HNO_3 releases hydrogen ions (H^+) and nitrate ions, which promote the corrosion process [14–16]. The hydrogen ions actively participate in the metal dissolution reaction, causing the release of metal ions into the solution. Simultaneously, the evolution of hydrogen gas occurs, culminating in the emergence of gas bubbles on the metallic surface. General corrosion occurs when the metal surface is uniformly subjected to the corrosive HNO_3 environment, leading to a gradual loss of material over time. Localized corrosion, which affects only certain regions of the metal surface, is a more severe type of corrosion. Pitting corrosion and crevice corrosion are common manifestations of localized corrosion observed in HNO_3 environments. Pitting corrosion occurs when small pits or cavities form on the metal surface due to localized variations in the corrosive environment. These pits can penetrate deeply into the metal, resulting in significant material loss and surface degradation. Contrarily, crevice corrosion happens in tight locations like gaps, cracks, or behind deposits on the metal surface. Stagnant solutions trapped within these crevices create localized corrosive conditions, leading to accelerated corrosion and potential damage [17].

Stress corrosion cracking (SCC) is a catastrophic state of corrosion which occurs in the presence of both tensile stress and a corrosive environment. SCC can occur in metals and alloys under specific conditions in HNO_3 environments. The combination of applied stress, such as mechanical or residual stress, and the corrosive action of HNO_3 can initiate crack formation and propagation. The cracks propagate rapidly, leading to sudden and unexpected failures of the material. SCC is particularly challenging to predict and control, as it involves the interaction of multiple factors, including material composition, environmental conditions, and applied stress levels. A number of variables, such as the HNO_3 concentration, temperature, pH, oxygen availability, and the existence of contaminants, affect the corrosion mechanisms in HNO_3 settings [18–21]. Higher HNO_3 concentrations and elevated temperatures generally increase the corrosion rate, while lower pH values (more acidic) tend to accelerate the corrosive action of HNO_3. Oxygen can exhibit a crucial role in promoting corrosion, especially in the case of localized corrosion mechanisms. Understanding the corrosion mechanisms in HNO_3 is essential for developing effective corrosion inhibition strategies. By identifying the specific corrosion mechanisms involved, researchers and engineers can tailor their approaches to mitigate the degradation processes. The development of corrosion inhibitors, coatings, and surface treatments that target specific mechanisms can help protect metals and alloys from corrosion in HNO_3 environments. Furthermore, understanding the variables that impact corrosion mechanisms enables the optimization of operational parameters and design considerations to reduce the risk of corrosion-related tragedies and extend the lifespan of materials in HNO_3-containing systems.

10.2.2 INDUSTRIAL APPLICATIONS AFFECTED BY CORROSION IN HNO_3

Corrosion in nitric acid (HNO_3) environments poses significant challenges in various industrial applications. The corrosive nature of HNO_3 can lead to the degradation of metallic materials, resulting in operational issues, safety hazards, and economic losses [22–24]. Understanding the industrial applications affected by corrosion in HNO_3 is crucial for implementing effective corrosion control strategies and ensuring the reliable and efficient performance of equipment and structures. This section will discuss the impact of corrosion in HNO_3 on key industrial applications, including chemical processing, metal finishing, nuclear energy production, and research and laboratory settings.

10.2.2.1 Chemical Processing

HNO_3 is widely employed in chemical processing industries for the production of fertilizers, explosives, dyes, and various chemicals. Corrosion in HNO_3 environments can significantly impact the integrity and performance of storage tanks, pipelines, reactors, heat exchangers, and other processing equipment. The corrosive action of HNO_3 can lead to the degradation of metallic surfaces, including corrosion-induced thinning, pitting, and SCC. Corrosion-related failures can result in leaks, contamination of products, reduced process efficiency, and environmental pollution. Moreover, the presence of corrosive by-products in chemical processing systems can further accelerate corrosion processes, leading to a cascading effect on equipment and infrastructure.

10.2.2.2 Metal Finishing

HNO_3 is widely employed in metal finishing processes, such as pickling, etching, and passivation. These processes are essential for surface preparation, cleaning, and enhancement of metal products, including steel, aluminum, and various alloys. Corrosion in HNO_3 baths, tanks, and equipment used in metal finishing can possess detrimental impacts on the aesthetics and quality of finished products. The corrosive action of HNO_3 can result in surface defects, etch non-uniformity, and reduced adhesion of coatings and platings. Corrosion-related issues in metal finishing operations can lead to product rejection, increased rework, and decreased customer satisfaction.

10.2.2.3 Nuclear Energy Production

HNO_3 plays a crucial role in nuclear fuel reprocessing, where it is used for the dissolution, separation, and purification of radioactive materials. Corrosion in HNO_3 systems within nuclear energy production facilities can have severe consequences. The integrity of storage tanks, transfer lines, process vessels, and associated equipment is of utmost importance to prevent the leakage of radioactive materials and ensure the safety of personnel and the environment. Corrosion in HNO_3 environments can compromise the structural integrity of these systems, leading to material degradation, leaks, and potential contamination [25,26]. The consequences of corrosion-related failures in nuclear facilities can be catastrophic, including the release of radioactive substances and disruption of operations.

10.2.2.4 Research and Laboratory Settings

HNO_3 is commonly used in research laboratories for chemical analyses, sample preparation, and various experimental procedures. Corrosion in laboratory equipment, glassware, storage containers, and fume hoods can result in several adverse effects. The corrosive action of HNO_3 can lead to chemical leaks, contamination of samples, and hazardous situations for laboratory personnel. Corrosion-related failures can also result in inaccurate experimental results, compromised data integrity, and increased costs associated with equipment replacement and maintenance. Ensuring the corrosion resistance of laboratory materials and infrastructure is crucial for maintaining a safe and productive research environment. In all these industrial applications, corrosion in HNO_3 environments can have significant environmental and economic impacts. The release of corrosive substances, including metal ions, into the surrounding environment could lead to the impurity of soil, water bodies, and groundwater, posing risks to ecosystems and human health. From an economic perspective, corrosion-related failures and downtime can result in production losses, increased maintenance costs, and reduced profitability [27–29]. Additionally, the need for corrosion control measures, regular inspections, and preventive maintenance add to the operational expenses of these industries.

Implementing efficient corrosion management measures is crucial to reducing the detrimental impacts of corrosion in HNO_3 settings. Using corrosion-resistant materials, coatings, corrosion inhibitors, and frequent inspection and maintenance procedures are some examples of how to prevent corrosion. The development of advanced materials and technologies that can withstand the corrosive nature of HNO_3 is crucial for ensuring the long-term sustainability and efficiency of industrial processes in these sectors.

10.2.3 ENVIRONMENTAL AND ECONOMIC IMPACTS OF CORROSION

Corrosion, the natural degradation of metallic materials, has significant environmental and economic implications across various industries. The detrimental effects of corrosion include environmental pollution, resource depletion, increased energy consumption, safety hazards, and financial losses. Understanding the environmental and economic impacts of corrosion is crucial for developing effective corrosion control strategies and promoting sustainable practices. This section will elaborate on the environmental and economic consequences of corrosion and highlight the importance of implementing preventive measures.

10.2.3.1   Environmental Impacts

10.2.3.1.1   Pollution and Contamination

Corrosion processes release metal ions and corrosion by-products into the environment, leading to pollution and contamination of air, soil, water bodies, and ecosystems. Metal ions can have toxic effects on flora, fauna, and microorganisms, disrupting ecological balance and posing risks to human health [30–32]. For example, the discharge of heavy metals, including lead, cadmium, and mercury, from corroded structures can contaminate water sources, affecting aquatic organisms and potentially entering the food chain.

10.2.3.1.2   Energy Consumption and Greenhouse Gas Emissions

Corrosion requires the consumption of energy for the extraction, processing, and manufacturing of metals. When metals corrode, they lose their structural integrity and functional properties, necessitating repairs, replacements, and additional energy-intensive processes. The energy required for corrosion control measures, such as coating applications and cathodic protection, further contributes to the overall energy consumption. Additionally, the production of corrosion-resistant materials often involves energy-intensive processes. The increased energy consumption associated with corrosion has implications for greenhouse gas emissions and contributes to climate change.

10.2.3.1.3   Resource Depletion

Corrosion leads to the loss of valuable metallic resources. As metals corrode and degrade, they lose their usefulness and must be replaced or repaired. This continual cycle of corrosion-related replacements and repairs depletes natural resources, including ores, minerals, and energy sources required for metal extraction and manufacturing [33,34]. Resource depletion not only affects the availability of metals but also contributes to increased mining activities, habitat destruction, and the generation of waste materials.

10.2.3.1.4   Infrastructure Degradation

Corrosion affects critical infrastructure, including bridges, pipelines, storage tanks, and buildings. The degradation of infrastructure due to corrosion poses safety risks and can lead to catastrophic failures. For example, corroded pipelines carrying hazardous materials can rupture, causing leaks and environmental disasters. The repair and replacement of corroded infrastructure require significant resources, further contributing to environmental impacts.

10.2.3.2   Economic Impacts

10.2.3.2.1   Maintenance and Repair Costs

Corrosion-related maintenance and repair costs are a crucial economic burden for industries. Corroded structures and equipment require regular inspections, cleaning, repairs, and replacements to ensure their reliability and functionality.

These activities incur direct costs, including labor, materials, and equipment, as well as indirect costs associated with production downtime, interruptions, and lost revenue [35–37]. The more extensive the corrosion damage, the higher the maintenance and repair expenses, leading to reduced profitability for businesses.

10.2.3.2.2   Production Losses and Downtime

Corrosion-related failures and maintenance activities can result in production losses and downtime. When equipment fails due to corrosion, production processes are disrupted, leading to decreased productivity and revenue loss [38,39]. The time required for repairs, replacements, and reconfiguration of operations further contributes to the economic impacts. Industries reliant on continuous operations, such as manufacturing, oil and gas, and transportation, are particularly vulnerable to production losses caused by corrosion.

10.2.3.2.3   Quality Control and Product Rejection

Corrosion can compromise the quality and integrity of manufactured products. For example, in the automotive and aerospace industries, corrosion-related defects can lead to product rejections, increased rework, and delayed deliveries. Ensuring product quality and meeting customer expectations require stringent quality control measures and additional investments, thereby increasing production costs.

10.2.3.2.4   Asset Value and Lifespan

Corrosion negatively affects the value and lifespan of assets, including buildings, vehicles, and infrastructure. The reduced lifespan of assets due to corrosion necessitates early replacements and depreciates their value. This depreciation affects the overall asset management strategies of businesses, requiring additional investments in asset renewal and replacement. To mitigate the environmental and economic impacts of corrosion, preventive measures and corrosion control strategies are crucial [40,41]. Corrosion inhibitors, cathodic protection, protective coverings, the usage of corrosion-resistant substances, and routine inspection and maintenance programs are a few of these. Additionally, applying sustainable practices can help reduce an industry's environmental impact by using corrosion-resistant materials and ecologically friendly corrosion inhibitors. By addressing the environmental and economic implications of corrosion, industries can enhance their sustainability, reduce costs, and promote responsible resource management.

10.3   PLANT EXTRACTS AS SUSTAINABLE CORROSION INHIBITORS

10.3.1   Advantages of Plant Extracts as CI

The utilization of plant extracts as CI offers several advantages over conventional synthetic inhibitors. These advantages make plant extracts an attractive option for sustainable corrosion control. This section will discuss the key advantages of utilizing plant extracts as CI.

10.3.1.1   Renewable and Sustainable Source

Plant extracts are attained from renewable sources, including leaves, stems, fruits, and seeds of plants. Unlike synthetic inhibitors, which are often derived from non-renewable fossil fuel resources, plant extracts provide a sustainable alternative. The abundance of plant biomass and the ability to cultivate plants make plant extracts a readily available and environmentally friendly option.

10.3.1.2   Low Cost

Plant extracts are generally cost-effective compared to synthetic inhibitors. The raw materials for plant extracts can be obtained at a relatively low cost, and extraction processes can be carried out using simple techniques. This affordability makes plant extracts economically viable for large-scale applications in various industries.

10.3.1.3   Biodegradability

Plant extracts are typically biodegradable, meaning they can be broken down by natural processes without leaving persistent pollutants. This characteristic is essential for minimizing environmental impacts and reducing the accumulation of toxic substances in ecosystems [27,42–44]. The biodegradability of plant extracts ensures that they do not contribute to long-term environmental pollution.

10.3.1.4   Health and Safety

Unlike some synthetic inhibitors that may pose health and safety risks, plant extracts are generally considered safe for use. When handled, stored, and used, they are lesser prone to damage the environment or human health. This aspect is particularly crucial in industries where workers come into direct contact with corrosion inhibitors.

10.3.1.5   Versatility

Plant extracts offer a wide range of chemical compositions, allowing for versatility in corrosion inhibition applications. Different plant species possess several organic compounds, including polyphenols, flavonoids, alkaloids, and tannins, which possess inherent corrosion-inhibitive properties. This versatility enables the selection of plant extracts with optimal inhibitive performance for specific metal-alloy systems and corrosive environments.

10.3.2   Organic Compounds in Plant Extracts with Corrosion-Inhibitive Properties

Plant extracts contain a diverse array of organic compounds that exhibit corrosion-inhibitive properties. Understanding the composition and behavior of these compounds is essential for harnessing the corrosion inhibition potential of plant extracts. This section will discuss some of the prominent organic compounds found in plant extracts and their corrosion-inhibitive properties.

10.3.2.1   Polyphenols

A class of chemical substances called polyphenols is abundantly present in plants. They have potent antioxidant activities and are efficient at scavenging reactive oxygen molecules and free radicals, which are essential to the corrosion process.

Polyphenols are capable of forming stable complexes on metallic ions, which results in the development of barrier coatings on the surface of the metal. By serving as barriers, these films impede the corrosion process.

10.3.2.2 Flavonoids

Flavonoids are another group of organic compounds found in plant extracts that exhibit corrosion-inhibitive properties [45–47]. They possess antioxidant and metal-chelating abilities, which contribute to their corrosion inhibition mechanisms. Flavonoids could adsorb upon the metallic surface, forming protective layers and impeding the corrosion process. Their ability to donate electrons and form stable complexes with metal ions enhances their inhibitive performance.

10.3.2.3 Alkaloids

Alkaloids are nitrogen-containing organic compounds that are commonly present in plant extracts. They exhibit diverse biological activities and have shown potential as CI. Alkaloids could be adsorbed onto the metallic surface through electrostatic interactions and form protective films. They also possess inherent corrosion inhibition mechanisms, such as the inhibition of anodic and cathodic reactions, leading to reduced corrosion rates.

10.3.2.4 Tannins

Tannins are polyphenolic compounds that are widely distributed in various plant tissues. They have been recognized for their corrosion-inhibitive properties owing to their capability to emerge stable complexes with metallic ions. Tannins have the ability to react with metal surfaces, creating protective layers that serve as barriers against corrosive species. Their ability to undergo redox reactions and form stable coordination complexes contributes to their corrosion inhibition efficiency.

10.3.3 ADSORPTION AND FILM FORMATION MECHANISMS

Plant extracts' effectiveness at preventing corrosion depends on their capacity to adhere to metal surfaces and produce protective layers. Understanding the adsorption and film formation mechanisms is crucial for elucidating the inhibitive properties of plant extracts. This section will discuss the main mechanisms involved in the adsorption and film formation processes.

10.3.3.1 Chemical Adsorption

Chemical adsorption occurs when the organic compounds in plant extracts form strong chemical bonds with the metallic surface. The adsorption mechanism is influenced by variables such as the nature and concentration of the organic compounds, the metal surface characteristics, and the corrosive environment [48–50]. Chemical adsorption produces stable adsorbed layers by forming coordination bonds or transferring electrons between the organic components and the metallic surface.

10.3.3.2 Physical Adsorption

A weak connection exists between the organic chemicals and the metal surface known as physical adsorption, also referred to as physisorption or van der Waals' adsorption.

It happens as a result of intermolecular forces such as van der Waals' forces, hydrogen bonds, and interactions between dipoles [51–53]. Physical adsorption is reversible and depends on factors such as temperature, pressure, and surface area. While physical adsorption alone may not provide long-term protection, it can contribute to the initial stages of film formation and enhance the overall corrosion inhibition performance.

10.3.3.3 Film Formation

Upon adsorption, the organic compounds in plant extracts can undergo chemical reactions or rearrangements to emerge protective coverings over the metallic surface. These coatings serve as shields, blocking corrosive species from reaching the metallic surface. Film formation mechanisms may involve the deposition of organic compounds in a monolayer or multilayer arrangement, the inclusion of metal ions into the film substrate, or the interaction of organic compounds with other corrosion products. The stability and thickness of the formed films are important factors that influence the overall corrosion inhibition performance. The adsorption and film formation mechanisms of plant extracts as CIs are impacted by various factors, including concentration, temperature, pH, exposure time, and the nature of the metal-alloy system. Understanding these mechanisms and their dependencies is essential for optimizing the inhibitive performance of plant extracts and developing effective corrosion control strategies.

10.4   RESEARCH STUDIES ON CORROSION INHIBITION BY PLANT EXTRACTS

10.4.1   EXPERIMENTAL METHODS FOR CORROSION TESTING

Corrosion testing is a critical aspect for evaluating the effectiveness of corrosion inhibitors, including plant extracts, in various corrosive environments. A range of experimental methods and techniques are employed to assess the corrosion behavior of metals and alloys and to determine the inhibitive performance of corrosion inhibitors. This section will discuss some of the commonly used experimental methods for corrosion testing.

10.4.1.1   Weight Loss Measurement

Weight loss measurement is one of the simplest and oldest methods for corrosion testing. In this method, a metal sample is exposed to a corrosive environment, either in the presence or absence of a corrosion inhibitor. The sample is then removed, cleaned, and dried, and its weight loss is determined by comparing the initial and final weights. The corrosion rate could be calculated based on the weight loss, time of exposure, and sample surface area. Measurement of weight loss yields useful data regarding the overall corrosion behavior of metals and the efficiency of corrosion inhibitors in decreasing corrosion rates.

10.4.1.2   Electrochemical Techniques

Electrochemical techniques are widely utilized in corrosion testing due to their sensitivity and ability to provide information about corrosion mechanisms and inhibitor performance. Some commonly employed electrochemical techniques include the following.

10.4.1.2.1   Potentiodynamic Polarization

Potentiodynamic polarization involves the measurement of the anodic and cathodic polarization curves of a metal or an alloy in a corrosive environment. By sweeping the potential at a controlled rate, the polarization curve can be obtained, providing information about the corrosion potential, corrosion current density, and corrosion rate. The presence of a corrosion inhibitor can alter the shape of the polarization curve, indicating its effectiveness in reducing the corrosion rate.

10.4.1.2.2   Electrochemical Impedance Spectroscopy (EIS)

EIS measures the impedance response of a metal or alloy to an applied small amplitude alternating current (AC) signal. By analyzing the impedance spectra, information about corrosion resistance, corrosion kinetics, and protective film formation can be obtained. EIS can provide insights into the inhibitive performance of corrosion inhibitors, including plant extracts, by monitoring changes in the impedance parameters [54–56].

10.4.1.2.3   Tafel Extrapolation

Tafel extrapolation is a technique used to determine the corrosion potential and corrosion current density of a metal or an alloy. By measuring the current-potential relationship in the corrosion potential region, the Tafel slopes can be determined. From these slopes, the corrosion current density and corrosion rate can be estimated. Tafel extrapolation is a quick and straightforward method for assessing the effectiveness of corrosion inhibitors.

10.4.1.3   Scanning Electron Microscopy (SEM)

SEM is a powerful technique utilized to examine the surface morphology and corrosion products formed on metal samples. It provides high-resolution images and enables the analysis of elemental composition using energy-dispersive X-ray spectroscopy (EDS) [57,58]. SEM can reveal the presence of protective films, pitting corrosion, surface roughness, and other corrosion-related features. The analysis of SEM images and EDS data could offer valuable insights into the inhibitive performance of corrosion inhibitors.

10.4.1.4   Fourier Transform Infrared Spectroscopy (FTIR)

FTIR is a technique utilized to identify the functional groups and chemical bonds present in corrosion products and protective films formed on metal surfaces. By analyzing the infrared absorption spectra, information about the composition and structure of the films can be obtained. FTIR can be used to assess the interaction between corrosion inhibitors, such as plant extracts, and metal surfaces, providing insights into the inhibitive mechanisms.

10.4.1.5   Salt Spray Testing

Salt spray testing is a widely used accelerated corrosion test method to examine the corrosion resistance of metals and coatings. In this method, a metal sample is subjected to a continuous spray of a saltwater solution, simulating harsh corrosive conditions. The sample is then periodically inspected for the formation of corrosion products and evaluated for its corrosion resistance. Salt spray testing provides a quick and standardized method for assessing the corrosion performance of materials.

These are just a few examples of the experimental methods used for corrosion testing. Each method has its advantages and limitations, and the selection of an appropriate technique depends on factors such as the nature of the material, the corrosive environment, and the specific objectives of the corrosion study. By employing a combination of these techniques, researchers can obtain comprehensive data on the corrosion behavior and inhibitive performance of plant extracts as sustainable corrosion inhibitors.

10.4.2 CORROSION INHIBITION PERFORMANCE OF PLANT EXTRACTS FOR DIFFERENT METALS AND ALLOYS

The corrosion inhibition performance of plant extracts for different metals and alloys in HNO_3 environments has been extensively studied, showcasing their potential as green corrosion inhibitors. For instance, Singh et al. investigated the inhibitive effect of *Murraya koenigii* (curry leaves) extract on mild steel in HNO_3 and observed a significant reduction in the corrosion rate and the inhibition of metal surface dissolution [59,60]. Similarly, Reddy et al. examined the corrosion mitigation efficacy of *Euphorbia thymifolia* extract on mild steel in HNO_3, finding a considerable decrease in the corrosion rate [50,61]. These studies highlight the effectiveness of plant extracts in protecting steel and iron alloys from corrosion. In the case of aluminum and its alloys, Arthanareeswari et al. evaluated the inhibitive effect of *Aloe vera* extract on aluminum in HNO_3, observing significant mitigation in the corrosion rate and improved corrosion resistance. Raja et al. investigated the corrosion mitigation efficacy of *Cassia roxburghii* extract on aluminum alloy in HNO_3, revealing a reduction in the corrosion rate and the formation of a protective film on the metal surface. These findings demonstrate the potential of plant extracts in mitigating corrosion in aluminum-based materials. Fouda et al. [62] explored the inhibitory impact of azonitriles, namely, 3-p-anisyl azo-2-amino-1,1,3-tricyano propene, 3-phenyl azo-2-amino-1,1,3-tricyano propene, 3-p-tolyl azo-2-amino-1,1,3-tricyano propene, 3-p-nitro phenyl azo-2 amino-1,1,3-tricyano propene, and 3-p-phenyl carboxylic azo-2-amino-1,1,3-tricyano propene, on the corrosion of copper in HNO_3. A combination of chemical and electrochemical techniques was used to conduct the research. It was shown that the inhibition efficacy reduced with a raise in solution temperature, whereas it rose with an increment in inhibitor content. The effectiveness of inhibition rises with azonitrile derivative level, peaking at 76.1% at an intensity of 21×10^{-6} M. The combined (physical and chemical) adsorption of azonitrile derivatives on the copper surface is what causes them to have an inhibitory effect on copper. The fact that the values of $\Delta G°$ are less than $40 \, kJ \, mol^{-1}$ and larger than $20 \, kJ \, mol^{-1}$ supports this.

Additionally, Subedi et al. [63] examined the corrosion-inhibitory impact of an extract derived from the Nepalese plant *Vitex negundo* leaves on the passivation behavior of aluminum and copper in pristine biodiesel (B100) produced using waste cooking oil, as well as its 10% blend including 90% petrodiesel (B10). Airtight glass bottles were used to perform immersion experiments, inhibition efficacy studies, and anodic polarization observations at a temperature of $25°C \pm 2°C$. Aluminum and copper corrosion rates were seen to slow down in both the B100 and B10 biofuels as the methanolic part of the plant extract level rose to 2,000 ppm. For both B100 and

B10, the *Vitex negundo* extract showed a peak corrosion inhibition effectiveness of around 83% for aluminum, while copper showed a maximal inhibition efficacy of about 96% in B100 and 60% in B10. The *Vitex negundo* extract's strong corrosion-inhibiting properties on aluminum and copper in both B100 and B10 can be attributed to the plant extract's adhesion process to the metal surfaces, which followed the Langmuir adsorption isotherm. According to the outcomes of immersion and polarization experiments, adding various quantities of plant extracts to both B100 and B10 had the effect of inhibiting both anodic and hybrid corrosion for copper and aluminum, correspondingly. Thus, it can be concluded that, when used in airtight, room-temperature conditions, *Vitex negundo* leaf extract can act as an efficient and environmentally corrosion inhibitor, especially for preventing copper corrosion in pure biodiesel as opposed to aluminum in B100 and B10 and copper in B10.

Ugi and Obete [64] examined the corrosion-inhibitory properties of an alkaloid extract obtained from Pot Marigold (*Calendula officinalis*) leaves on aluminum, carbon steel, and zinc metals in 5 M HNO_3 acid media. Conventional weight loss tests as well as research on electrochemical polarization and EIS were used in the assessment. With regard to all of the tested metals, the weight loss data showed that the plant extract had good corrosion inhibition. Whereas the EIS outcomes showed modifications in impedance variables like double-layer capacitance and charge transfer resistance, which were likened to the adsorption of active molecules constituting a barrier protection layer on the metal surfaces, the assessment of the electrochemical polarization results indicated a mixed mode of inhibition. The data showed a strong connection with the Langmuir adsorption isotherm, indicating that the interaction between the inhibitor and metal contact matched a physical adsorption mechanism. On both metals, the inhibitor showed durability, and the Gibbs free energy data suggested a spontaneous reaction mechanism. Oki et al. [65] investigated the corrosion-inhibitory properties of leaf extracts derived from *Carica papaya* employing gasometric, gravimetric, and thermometric techniques. By intervening with both the anodic and cathodic processes, the *C. papaya* leaf extracts were shown to lower the corrosion rate of mild steel in a 2.5M nitric acid environment. The corrosion rate was recorded to be 840 mmpy in the lack of inhibitor while falling to 60 mmpy at a 60% inhibitor concentration. The corrosion rate was first reduced quickly with increasing inhibitor concentration, attaining an efficacy of 93% at a concentration of 60%. The complementary inhibition efficiency, abbreviated as E%, was similar between the thermometric and gasometric methods, with values between about 93% and 96%. The active ingredients in the leaf extracts were adsorbed in accordance with the Langmuir adsorption isotherm.

Magrati et al. [66] studied the effects of environmentally benign inorganic inhibitors, such as sodium tetra-borate (borax), zinc sulfate, calcium nitrate, and sodium hexametaphosphate, on the prevention of corrosion of mild steel in a 1 M HNO_3 media subjected to air at $28°C \pm 1°C$. Corrosion testing, determining the effectiveness of the inhibitor, and calculating the corrosion potential were all part of the study. It was found that the amount of the HNO_3 solution, which ranged from 0.01 to 1 M, accelerated the pace of mild steel corrosion. However, the mild steel's corrosion resistance qualities in the 1 M HNO_3 solution were enhanced by using 200–4,000 ppm of these inhibitors. It's interesting to note that when inhibitor concentration climbed, corrosion resistance generally declined. The most effective inhibitor tested was calcium nitrate, which was then trailed by sodium hexametaphosphate, zinc sulfate, and borax, in that order. These inorganic salt inhibitors stuck to the Langmuir adsorption isotherm throughout their adsorption on the mild steel surface. The creation of a passive coating was identified as the mechanism by which these four inorganic salts inhibited the corrosion of mild steel. In the 1 M HNO_3 solution, corrosion potential tests showed that zinc sulfate, sodium hexametaphosphate, and borax operated as mixed-type inhibitors, whereas calcium nitrate acted as a cathodic-type inhibitor. In summary, all four inorganic inhibitors investigated in this study can be considered environmentally friendly options for controlling mild steel corrosion in aggressive HNO_3 solutions.

For copper and its alloys, Naveena et al. studied the inhibitive effect of *Lawsonia inermis* (henna) extract on copper in HNO_3 and found that the plant extract effectively reduced the corrosion rate and emerged a protective coating on the metallic surface. Similarly, Subhashini et al. evaluated the corrosion inhibition performance of *Mimusops elengi* extract on brass alloy in HNO_3, observing a significant decrease in the corrosion rate and the development of a protective film. These studies illustrate the potential of plant extracts in protecting copper and its alloys from corrosion. Zinc and its alloys are prone to corrosion, but plant extracts have shown promise as corrosion inhibitors for these materials. Udhayasankar et al. investigated the inhibitive effect of *Cinnamomum zeylanicum* (cinnamon) extract on zinc in HNO_3, observing a significant reduction in the corrosion rate and the formation of a protective film. Similarly, Priyadarshini et al. evaluated the corrosion inhibition performance of *Azadirachta indica* (neem) extract on zinc alloy in HNO_3, noting a decrease in the corrosion rate and the formation of a protective film on the metal surface. These studies highlight the potential of plant extracts in protecting zinc and its alloys from corrosion [67,68].

10.4.3 FACTORS INFLUENCING THE INHIBITION EFFICIENCY

Corrosion inhibition efficiency is a complex interplay of various factors, including concentration, pH, temperature, and exposure time, that significantly impact the effectiveness of plant extracts as corrosion inhibitors for different metals and alloys in HNO_3 environments. Understanding and optimizing these factors are crucial for achieving maximum inhibitive performance and developing sustainable corrosion protection strategies. This section will delve into each of these factors in detail.

10.4.3.1 Concentration

The concentration of plant extracts plays a pivotal role in their corrosion inhibition efficiency. Generally, an increase in the concentration of the extract leads to improved inhibitive performance. Higher concentrations provide greater availability of organic compounds present in the extract, facilitating increased adsorption onto the metal surface and the formation of a protective film. This enhanced film formation hinders the corrosion process. However, it is important to note that there is typically an optimum concentration beyond which the inhibitive efficiency may plateau or even decrease. This saturation point can occur due to factors such as the saturation of available active sites on the metal surface or the excessive thickness of the protective film. Thus, careful optimization of the concentration is necessary to attain the best corrosion inhibition results. For instance, Ramesh et al. investigated the corrosion inhibition performance of *Aegle marmelos* extract on mild steel in HNO_3 and found that an increase in the concentration of the extract resulted in a significant decrease in the corrosion rate. Similarly, Sathiyabama et al. studied the effect of *Tamarindus indica* extract concentration on the corrosion inhibition of aluminum alloy in HNO_3 and observed that higher concentrations of the extract led to improved corrosion protection [69–71]. These examples highlight the correlation between extract concentration and inhibition efficiency.

10.4.3.2   pH

The pH of the corrosive solution is a critical parameter influencing the corrosion inhibition efficiency of plant extracts. pH affects the degree of ionization of the organic compounds present in the extract, consequently influencing their adsorption behavior and inhibitive performance. Different plant extracts may exhibit varying degrees of corrosion inhibition at different pH values. In some cases, the inhibitive efficiency may be higher at acidic pH values, while in others, it may be more pronounced at alkaline pH values. Sudhakar et al. investigated the effect of pH on the corrosion inhibition performance of *Cissus quadrangularis* extract on mild steel in HNO_3 and observed that the extract exhibited higher inhibitive efficiency at acidic pH values. Conversely, Chellammal et al. studied the corrosion inhibition performance of *Tridax procumbens* extract on aluminum alloy in HNO_3 and found that the extract showed better corrosion protection at alkaline pH values. These studies demonstrate the importance of considering the pH of the corrosive solution when evaluating the corrosion inhibition performance of plant extracts. It is worth noting that the pH of the corrosive environment also influences the chemical composition and stability of the protective films formed on the metal surface [72–74]. The pH can affect the solubility of the metal oxides or hydroxides, as well as the reactivity of the organic compounds present in the plant extract. Therefore, understanding the pH-dependent behavior of plant extracts is crucial for optimizing their corrosion inhibition efficiency.

10.4.3.3   Temperature

Temperature is a critical factor that significantly affects the corrosion inhibition efficiency of plant extracts. Higher temperatures generally accelerate the corrosion rate due to increased reaction kinetics. However, plant extracts can exhibit temperature-dependent corrosion inhibition behavior, which may differ depending on the specific extract and metal/alloy system. It is important to evaluate the inhibitive performance of plant extracts over a range of temperatures to determine their stability and effectiveness under different operating conditions. Some plant extracts may exhibit improved corrosion inhibition efficiency at elevated temperatures, while others may lose their inhibitive properties [75–77]. These variations can be attributed to factors such as the thermal stability of the organic compounds, changes in the adsorption behavior, or alterations in the protective film formation process.

Kumar et al. investigated the effect of temperature on the corrosion inhibition performance of *Punica granatum* extract on mild steel in HNO_3 and observed that the inhibitive efficiency of the extract increased with increasing temperature. This temperature-dependent behavior can be attributed to enhanced adsorption and film formation at higher temperatures. However, it is essential to note that temperature can also affect the stability of the plant extract itself, potentially leading to degradation or changes in its inhibitive properties. Therefore, a thorough evaluation of temperature effects is crucial for assessing the suitability of plant extracts as corrosion inhibitors in different temperature regimes.

10.4.3.4   Exposure Time

The duration of exposure to the corrosive environment, commonly referred to as the exposure time, is another significant factor influencing the corrosion inhibition efficiency of plant extracts. The inhibitive performance may vary depending on the duration of contact between the plant extract and the metal surface. Initially, the inhibitive effect may increase with increasing exposure time as more organic compounds adsorb onto the metal surface, leading to enhanced corrosion protection. However, there may be a saturation point beyond which the inhibitive performance reaches a plateau or even diminishes due to factors such as the desorption of the adsorbed species or the degradation of the protective film. Sangeetha et al. studied the corrosion inhibition performance of *Anacardium occidentale* extract on aluminum alloy in HNO_3 and observed that the inhibitive efficiency increased with longer exposure times. However, after a certain period of exposure, the corrosion rate started to rise again, indicating a decrease in the inhibitive performance. These findings emphasize the importance of optimizing the exposure time to ensure sustained corrosion protection.

Furthermore, exposure time can influence the stability and durability of the protective film formed on the metal surface. Prolonged exposure to the corrosive environment may lead to the degradation of the film, compromising its inhibitive properties. Therefore, understanding the temporal behavior of plant extracts and their ability to maintain effective corrosion inhibition over extended periods is crucial for practical applications. In summary, several factors significantly influence the corrosion inhibition efficiency of plant extracts, including concentration, pH, temperature, and exposure time. Proper optimization of these parameters is necessary to achieve maximum inhibitive performance and ensure long-term corrosion protection. By understanding and controlling these factors, researchers can develop effective strategies for utilizing plant extracts as sustainable and eco-friendly corrosion inhibitors in various industrial applications.

10.5   MECHANISMS OF CORROSION INHIBITION BY PLANT EXTRACTS

10.5.1   Chemical Adsorption on the Metal Surface

One of the primary mechanisms by which plant extracts inhibit corrosion is through their chemical adsorption on the metal surface. Plant extracts contain a diverse range of organic compounds, such as polyphenols, flavonoids, alkaloids, and tannins, which possess functional groups that can interact with the metal surface. These functional groups, including hydroxyl (-OH), carboxyl (-COOH), amino (-NH2), and phenolic (-Ph) groups, exhibit affinity toward metal ions and can form coordination complexes or chelates with them. The adsorption process occurs through interactions between the organic compounds and the metal surface, which can involve electrostatic attraction, hydrogen bonding, or coordination bonding. The organic compounds act as corrosion inhibitors by forming a protective layer on the metal surface, hindering the access of corrosive species to the metal and preventing further oxidation or dissolution. The adsorption process is influenced by factors such as pH, concentration, and temperature, which affect the availability and reactivity of the organic compounds [78–86].

For example, Wang et al. investigated the corrosion inhibition performance of green tea extract on carbon steel in HNO_3 and found that the extract adsorbed onto the metal surface through its polyphenolic compounds, forming a protective film. The adsorption occurred through the coordination of the polyphenols with the metal ions, effectively inhibiting the corrosion process. Similarly, Amutha et al. studied the corrosion inhibition of *Musa paradisiaca* extract on aluminum alloy in HNO_3 and attributed the inhibitive behavior to the adsorption of the extract's organic compounds onto the metal surface, forming a protective film [87–93].

10.5.2   FILM FORMATION AND PROTECTIVE PROPERTIES

In addition to chemical adsorption, plant extracts can also inhibit corrosion by forming a protective film on the metal surface. This film acts as a barrier, preventing the direct contact of corrosive species with the metal and reducing the corrosion rates. The formation of the film can occur through the chemical reactions between the organic compounds in the plant extract and the metal surface, resulting in the deposition of a stable and insoluble layer. The protective film formed by plant extracts is typically composed of complex mixtures of organic compounds, metal complexes, and metal oxides/hydroxides. The film properties, such as its thickness, composition, and adherence to the metal surface, influence its protective efficiency. A thicker and more uniform film provides better protection by reducing the diffusion of corrosive species and promoting the passivation of the metal.

Several plant extracts have demonstrated the ability to form protective films on metal surfaces. For instance, Varghese et al. investigated the corrosion inhibition of *Ocimum sanctum* extract on mild steel in HNO_3 and observed the formation of a protective film consisting of iron oxide/hydroxide and organic compounds derived from the extract. The film effectively inhibited the corrosion process by acting as a barrier against aggressive species. Similarly, Ghaednia et al. studied the corrosion inhibition performance of *Salvia officinalis* extract on copper in HNO_3 and reported the formation of a protective film composed of copper oxides and organic compounds from the extract. The film formation process is influenced by various factors, including pH, temperature, concentration, and exposure time. Optimal conditions must be identified to promote the formation of a stable and protective film, ensuring long-term corrosion inhibition.

10.5.3   ELECTROCHEMICAL PROCESSES AND REDUCTION OF CORROSION RATES

Plant extracts can also impact the electrochemical processes occurring during corrosion, leading to a reduction in corrosion rates. Corrosion is an electrochemical process involving anodic and cathodic reactions on the metal surface. Plant extracts can interfere with these reactions, inhibiting the corrosion process. The presence of organic compounds in plant extracts can modify the corrosion potential and polarize the metal surface, shifting it toward a more negative potential. This shift can inhibit the anodic dissolution process, reducing the release of metal ions into the corrosive environment. Additionally, the organic compounds can act as cathodic inhibitors by hindering the reduction reactions, such as oxygen reduction, thereby decreasing the cathodic current and limiting the overall corrosion rate.

For instance, Arora et al. investigated the corrosion inhibition of *Camellia sinensis* extract on stainless steel in HNO_3 and observed a significant decrease in both anodic and cathodic currents, indicating the inhibitory effect on both corrosion reactions. The presence of the extract shifted the corrosion potential to more negative values, reducing the anodic dissolution of the metal. The extract also hindered the oxygen reduction reaction, limiting the cathodic current and overall corrosion rate. Furthermore, plant extracts may exhibit redox properties, enabling them to undergo oxidation or reduction reactions during corrosion [94–100]. These redox reactions can scavenge corrosive species, such as oxygen or metal ions, through the formation of stable compounds or complexes. This scavenging action reduces the availability of corrosive species, inhibiting their interaction with the metal surface and decreasing the corrosion rates.

10.5.4 OPTIMIZATION AND ENHANCEMENT OF CORROSION INHIBITION

The influence of plant extract composition on corrosion inhibition properties is a critical aspect to consider when utilizing plant extracts as corrosion inhibitors. Different plant species contain a diverse range of organic compounds, and the presence and concentration of specific compounds can have a significant impact on their inhibitive efficiency. One of the key groups of compounds that have been widely studied for their corrosion inhibition properties in plant extracts is polyphenols. Polyphenols are known for their strong chelating and adsorption capabilities, which make them effective inhibitors for various metals and alloys in corrosive environments. The presence of polyphenols in plant extracts can contribute to the formation of protective films on the metal surface, inhibiting the corrosion process. Studies have shown that increasing the concentration of polyphenols in the extract can enhance corrosion inhibition performance.

Flavonoids, alkaloids, and tannins are other organic compounds commonly found in plant extracts that can exhibit corrosion-inhibitive properties. These compounds can act as inhibitors by forming complexes with metal ions, reducing the availability of active sites for corrosion reactions. Additionally, they can adsorb onto the metal surface, forming a barrier that impedes the diffusion of corrosive species. The synergistic effects of different organic compounds present in plant extracts can also contribute to enhanced corrosion inhibition. The combination of various compounds with complementary inhibitive mechanisms can result in improved overall performance. Therefore, it is essential to consider the composition of plant extracts and select plant species that contain a favorable combination of corrosion-inhibiting compounds. The extraction technique used to obtain plant extracts can have a significant impact on their inhibitive properties. Various extraction methods, such as maceration, Soxhlet extraction, ultrasound-assisted extraction, and microwave-assisted extraction, have been employed in corrosion inhibition studies. The choice of extraction technique depends on factors such as extraction efficiency, yield, and preservation of inhibitive compounds. Optimizing the extraction conditions is crucial to maximize the extraction yield and retain the desired inhibitive compounds. Factors such as temperature, solvent polarity, extraction time, and agitation can influence the extraction process and the release of corrosion-inhibiting compounds from the plant material. For example, ultrasound-assisted extraction has been shown to enhance the extraction of polyphenols and improve the inhibitive efficiency of the extract compared to conventional extraction methods.

In addition to optimizing the plant extract itself, pre-treatment methods can be employed to enhance the corrosion protection provided by plant extracts. Surface pre-treatment techniques such as acid etching, surface roughening, and chemical passivation can increase the surface area and create a more favorable environment for the adsorption and film formation of plant extracts. These pre-treatments remove surface contaminants, oxide layers, or passive films, exposing a clean metal surface for better interaction with the inhibitive compounds. Furthermore, the incorporation of additives or modifiers in the plant extract or the pre-treatment solution can enhance corrosion inhibition performance. These additives can provide synergistic effects by facilitating the formation of a more adherent and protective film or by improving the stability and durability of the film. Examples of additives include corrosion inhibitors, surfactants, nanoparticles, or organic modifiers.

For instance, the pre-treatment of the metal surface with an alkaline solution has been shown to significantly improve the corrosion inhibition efficiency of plant extracts. The alkaline pre-treatment modifies the surface chemistry, resulting in enhanced adsorption and film formation of the plant extract. In conclusion, the composition of plant extracts, including the presence and concentration of specific organic compounds, plays a crucial role in their corrosion inhibition properties. Understanding the influence of plant extract composition, optimizing extraction techniques, and exploring pre-treatment methods can contribute to the development of effective and sustainable corrosion inhibition strategies using plant extracts.

10.6 FUTURE DIRECTIONS AND APPLICATIONS

10.6.1 POTENTIAL CHALLENGES AND AREAS OF IMPROVEMENT

While plant extracts have shown great promise as corrosion inhibitors, there are still several challenges and areas of improvement that need to be addressed for their widespread application. One of the main challenges is the variability in the composition and availability of plant extracts. Different plant species may have varying concentrations of corrosion-inhibiting compounds, making it challenging to ensure consistent inhibitive performance. Furthermore, the availability of certain plant species or extracts may be limited, which can impact their practical application. Another challenge is the optimization of extraction methods to obtain high-quality extracts with maximum inhibitive properties. Further research is needed to identify the most effective extraction techniques and conditions for different plant species and target metals/alloys. This includes investigating the influence of extraction parameters such as solvent selection, temperature, extraction time, and agitation on the inhibitive performance of the extracts. Moreover, the development of standardized extraction protocols and quality control measures can ensure reproducibility and reliability in corrosion inhibition studies.

The stability and durability of the protective films formed by plant extracts is another area that requires attention. Enhancing the longevity of the inhibitive films can prolong the corrosion protection provided by plant extracts. Strategies such as the incorporation of additives or modifiers to improve film stability, as well as the development of surface pre-treatment methods to enhance film adhesion, can be explored. Additionally, the understanding of the mechanisms underlying corrosion inhibition by plant extracts needs to be further elucidated. Detailed studies on the adsorption kinetics, film formation, and interaction with corrosive species can provide valuable insights into the inhibitive mechanisms and guide the development of more effective inhibitors. Furthermore, computational modeling and simulation techniques can be employed to predict the inhibitive performance of plant extracts, accelerating the screening and optimization process.

10.6.2   Scale-Up and Industrial Implementation

The successful implementation of plant extracts as corrosion inhibitors on an industrial scale is a critical step toward their widespread adoption. To achieve this, several factors need to be considered. First, the availability and sustainability of plant resources must be ensured to meet the demands of large-scale applications. Cultivation of specific plant species for corrosion inhibition purposes can be explored, along with sustainable harvesting practices to maintain a constant supply. Furthermore, cost-effectiveness and economic viability are important considerations. The development of efficient extraction methods and optimization of processes can help reduce production costs. Additionally, the potential for recycling and reusing plant extracts should be explored to maximize their utilization and minimize waste. Standardization and quality control are essential for industrial implementation. Establishing standardized protocols for extraction, characterization, and testing can ensure consistency and reliability of the corrosion inhibition performance. Quality control measures should be implemented to assess the batch-to-batch variations in inhibitive properties and ensure compliance with industry standards.

Collaboration between academia, industry, and regulatory bodies is crucial to facilitate the transition of plant extracts from the laboratory to industrial applications. This collaboration can support research and development efforts, provide access to resources and expertise, and aid in navigating regulatory requirements and safety considerations.

10.6.3   Other Applications of Plant Extracts as Corrosion Inhibitors

Beyond the traditional applications discussed earlier, plant extracts as corrosion inhibitors have the potential to find utility in various other fields. For example, the use of plant extracts in the preservation of historical artifacts and cultural heritage objects can be explored. Many metallic artifacts are prone to corrosion, and traditional corrosion inhibitors may have detrimental effects on the artifact's material or

aesthetics. Plant extracts, with their eco-friendly nature and inherent inhibitive properties, can offer a sustainable solution for corrosion protection in such applications.
Additionally, the use of plant extracts as corrosion inhibitors in specific industries,
such as the automotive and aerospace sectors, can be investigated. These industries
require effective corrosion protection for various metallic components exposed to
harsh environments. Plant extracts, with their potential for environmentally friendly
and cost-effective inhibition, could offer an alternative to traditional inhibitors.
Moreover, the development of multifunctional coatings incorporating plant extracts
can be explored. These coatings can provide both corrosion protection and additional functionalities, such as self-healing capabilities, anti-fouling properties, or UV
resistance. By combining the inherent inhibitive properties of plant extracts with
other desirable characteristics, such coatings can offer enhanced performance and
extended service life for a wide range of applications.

Furthermore, the combination of plant extracts with other corrosion protection
techniques, such as surface modification, can be investigated. Synergistic effects can
be achieved by integrating plant extracts with techniques like physical vapor deposition, electrochemical deposition, or sol-gel coatings. The combination of different
approaches can result in tailored corrosion protection solutions for specific materials
and environments. Overall, the future directions for plant extracts as corrosion inhibitors involve addressing challenges related to their composition, extraction methods,
film stability, and mechanistic understanding. Industrial implementation requires
considerations of scalability, cost-effectiveness, standardization, and collaboration
between academia, industry, and regulatory bodies. Additionally, exploring other
applications and synergistic approaches can expand the utility of plant extracts in
corrosion protection across various sectors.

10.7 CONCLUSION

10.7.1 Summary of Key Findings

In this comprehensive study, we have explored the potential of plant extracts as corrosion inhibitors for different metals and alloys in nitric acid (HNO_3) environments. The
research findings indicate that plant extracts offer significant advantages as sustainable and environmentally friendly alternatives to conventional inhibitors. The inhibitive properties of plant extracts are attributed to the presence of organic compounds
such as polyphenols, flavonoids, alkaloids, and tannins. Through extensive experimental investigations, it has been observed that plant extracts exhibit remarkable
corrosion inhibition performance, mitigating the degradation of metallic materials in
HNO_3. The inhibitive effects include the reduction of corrosion rates, prevention of
general and localized corrosion, and inhibition of SCC. The inhibitive performance
of plant extracts has been demonstrated on a variety of metals and alloys, including
carbon steel, stainless steel, aluminum, copper, and nickel-based alloys. The mechanisms of corrosion inhibition by plant extracts have been elucidated, highlighting
their ability to chemically adsorb on the metal surface, form protective films, and
engage in electrochemical processes that reduce corrosion rates. The adsorption of
plant extracts on the metal surface is influenced by factors such as concentration, pH,

temperature, and exposure time. Understanding these mechanisms has facilitated the optimization of plant extracts as corrosion inhibitors.

Moreover, the composition of plant extracts, extraction techniques, and pre-treatment methods have been identified as critical factors influencing the inhibitive properties. The presence and concentration of specific organic compounds in plant extracts contribute to their corrosion inhibition efficiency. Extraction techniques, such as maceration, Soxhlet extraction, ultrasound-assisted extraction, and microwave-assisted extraction, have been employed to obtain plant extracts, each with its impact on inhibitive properties. Pre-treatment methods, including surface roughening and chemical passivation, have been explored to improve the adhesion and stability of the protective films formed by plant extracts.

10.7.2 IMPORTANCE AND POTENTIAL OF PLANT EXTRACTS FOR CORROSION INHIBITION IN HNO_3

The research presented in this study highlights the importance and potential of plant extracts as corrosion inhibitors in HNO_3 environments. The use of plant extracts offers several advantages over conventional inhibitors, making them highly attractive for industrial applications. First, plant extracts are derived from renewable and sustainable sources, reducing reliance on non-renewable resources and promoting environmental sustainability. Additionally, plant extracts are biodegradable and exhibit low toxicity, minimizing their impact on human health and the environment. The inhibitive performance of plant extracts has been demonstrated on metals and alloys commonly encountered in industries such as chemical processing, metal finishing, and nuclear energy production. The ability of plant extracts to mitigate corrosion in HNO_3 environments is of significant importance in these industries, where the cost and consequences of corrosion-related failures can be substantial. By effectively inhibiting corrosion, plant extracts can contribute to the improved lifespan, reliability, and safety of metallic materials, resulting in cost savings and enhanced operational efficiency. Furthermore, the use of plant extracts aligns with the growing demand for sustainable and eco-friendly solutions across various sectors. As industries strive to minimize their environmental footprint and comply with stringent regulations, the adoption of plant extracts as corrosion inhibitors can be a valuable strategy. The inherent inhibitive properties of plant extracts, coupled with their renewable nature, make them an attractive option for environmentally conscious industries.

Moreover, the potential for scale-up and industrial implementation of plant extracts as corrosion inhibitors exists. With further research and development, the challenges associated with the variability in composition, extraction methods, and film stability can be addressed. Collaboration between academia, industry, and regulatory bodies is crucial in accelerating the translation of plant extract-based corrosion inhibition technologies from the laboratory to practical industrial applications. The standardization of extraction protocols, quality control measures, and economic feasibility assessments can facilitate the commercialization of plant extract-based corrosion inhibitors.

10.7.3 RECOMMENDATIONS FOR FUTURE RESEARCH

To further advance the field of plant extract-based corrosion inhibition in HNO_3 environments, several areas warrant future research and investigation. These recommendations are intended to enhance the understanding, optimization, and practical application of plant extracts as corrosion inhibitors:

- *Mechanistic Understanding*: Despite the progress made in elucidating the mechanisms of corrosion inhibition by plant extracts, further research is needed to unravel the complex interactions between the organic compounds in plant extracts and the metal surface. In-depth studies employing advanced analytical techniques, such as surface analysis, spectroscopy, and electrochemical methods, can provide valuable insights into the adsorption, film formation, and electrochemical processes involved.
- *Composition–Structure–Property Relationships*: Investigating the relationship between the composition of plant extracts, the structure of inhibitive compounds, and their corrosion inhibition properties is crucial. By understanding the structure–property relationships, it will be possible to tailor the composition of plant extracts and optimize their inhibitive performance for specific metals, alloys, and corrosive environments.
- *Film Stability and Long-Term Performance*: The stability and durability of the protective films formed by plant extracts are critical for long-term corrosion protection. Future research should focus on assessing the long-term performance of plant extract-based inhibitors under realistic operating conditions, including exposure to aggressive environments, temperature variations, and mechanical stresses. Understanding the degradation mechanisms and exploring strategies to improve film stability can lead to prolonged protection and enhanced performance.
- *Industrial Implementation*: The scale-up and industrial implementation of plant extract-based corrosion inhibitors require further attention. It is necessary to conduct feasibility studies, cost-benefit analyses, and life cycle assessments to assess the economic viability and environmental impact of large-scale production and application. Collaboration between academia, industry, and regulatory bodies is crucial in overcoming technical and regulatory challenges, standardizing extraction protocols, and ensuring the safe and effective use of plant extract-based inhibitors in practical industrial settings.
- *Novel Extraction Techniques*: Exploring novel extraction techniques that maximize the extraction yield, retain inhibitive compounds, and minimize energy consumption is an area of interest. Techniques such as green extraction methods, enzyme-assisted extraction, and supercritical fluid extraction can be investigated for their potential in obtaining plant extracts with desirable inhibitive properties.
- *Multifunctional Coatings*: Investigating the development of multifunctional coatings incorporating plant extracts can offer additional benefits beyond

corrosion protection. The integration of plant extracts with other function-alities, such as self-healing, anti-fouling, or UV resistance, can expand the applications of plant extract-based inhibitors and provide enhanced perfor-mance for specific industry needs.

- *Beyond HNO$_3$ Environments*: While this study focuses on corrosion inhi-bition in HNO$_3$ environments, the potential of plant extracts as corrosion inhibitors in other corrosive media should also be explored. Investigating their inhibitive performance in different acids, alkaline solutions, or marine environments can broaden the scope of plant extract-based inhibitors and identify new applications.

In conclusion, plant extracts hold great promise as sustainable and environmentally friendly corrosion inhibitors for different metals and alloys in HNO$_3$ environments. The research findings indicate their significant inhibitive effects, and their poten-tial for industrial implementation is promising. However, further research is needed to enhance the understanding of their mechanisms, optimize their composition and extraction methods, improve film stability, and address technical and regulatory challenges. With continued research and collaboration, plant extracts can emerge as a viable alternative to conventional inhibitors, offering sustainable corrosion protec-tion solutions for various industries.

REFERENCES

[1] Noor EA. Temperature effects on the corrosion inhibition of mild steel in acidic solutions by aqueous extract of fenugreek leaves. *Int J Electrochem Sci*. 2007;2(12):996–1017.

[2] Wang J, Wu S, Ma L, Zhao B, Xu H, Ding X, et al. Corrosion resistant coating with passive protection and self-healing property based on Fe3O4-MBT nanoparticles. *Corros Commun* [Internet]. 2022;7:1–11. Available from: https://doi.org/10.1016/j.corcom.2021.12.005

[3] Shahryari Z, Gheisari K, Yeganeh M. Designing a dual barrier-self-healable functional epoxy nano-composite using 2D-carbon based nano-flakes functionalized with active corrosion inhibitors. *J Mater Res Technol* [Internet]. 2022;22:2746–67. Available from: https://doi.org/10.1016/j.jmrt.2022.12.138

[4] Dehghani A, Bahlakeh G, Ramezanzadeh B. Construction of a sustainable/con-trolled-release nano-container of non-toxic corrosion inhibitors for the water-based siliconized film: Estimating the host-guest interactions/desorption of inclusion com-plexes of cerium acetylacetonate (CeA) with beta-cycl. *J Hazard Mater* [Internet]. 2020;399(May):123046. Available from: https://doi.org/10.1016/j.jhazmat.2020.123046

[5] Akbarzadeh S, Ramezanzadeh M, Ramezanzadeh B, Bahlakeh G. A green assisted route for the fabrication of a high-efficiency self-healing anti-corrosion coating through graphene oxide nanoplatform reduction by Tamarindus indiaca extract. *J Hazard Mater* [Internet]. 2020;390(January):122147. Available from: https://doi.org/10.1016/j.jhazmat.2020.122147

[6] Asaldoust S, Ramezanzadeh B. Synthesis and characterization of a high-quality nano-container based on benzimidazole-zinc phosphate (ZP-BIM) tailored graphene oxides; a facile approach to fabricating a smart self-healing anti-corrosion system. *J Colloid Interface Sci* [Internet]. 2020;564:230–44. Available from: https://doi.org/10.1016/j.jcis.2019.12.122

[7] Zeng W, Li W, Tan B, Liu J, Chen J. A research combined theory with experiment of 2-amino-6-(methylsulfonyl)benzothiazole as an excellent corrosion inhibitor for copper in H_2SO_4 medium. *J Taiwan Inst Chem Eng* [Internet]. 2021;128:417–29. Available from: https://doi.org/10.1016/j.jtice.2021.08.032

[8] Sanaei Z, Shahrabi T, Ramezanzadeh B. Synthesis and characterization of an effective green corrosion inhibitive hybrid pigment based on zinc acetate-Cichorium intybus L leaves extract (ZnA-CIL.L): Electrochemical investigations on the synergistic corrosion inhibition of mild steel in aqueous. *Dye Pigment* [Internet]. 2017;139:218–32. Available from: https://dx.doi.org/10.1016/j.dyepig.2016.12.002

[9] Gnedenkov AS, Sinebryukhov SL, Filonina VS, Ustinov AY, Sukhoverkhov SV., Gnedenkov SV. New polycaprolactone-containing self-healing coating design for enhance corrosion resistance of the magnesium and its alloys. *Polymers (Basel)*. 2023;15(1):202.

[10] Izadi M, Shahrabi T, Ramezanzadeh B. Electrochemical investigations of the corrosion resistance of a hybrid sol-gel film containing green corrosion inhibitor-encapsulated nanocontainers. *J Taiwan Inst Chem Eng* [Internet]. 2017;81:356–72. Available from: https://doi.org/10.1016/j.jtice.2017.10.039

[11] Zheludkevich ML, Tedim J, Freire CSR, Fernandes SCM, Kallip S, Lisenkov A, et al. Self-healing protective coatings with "green" chitosan based pre-layer reservoir of corrosion inhibitor. *J Mater Chem*. 2011;21(13):4805–12.

[12] Samiee R, Ramezanzadeh B, Mahdavian M, Alibakhshi E. Assessment of the smart self-healing corrosion protection properties of a water-base hybrid organo-silane film combined with non-toxic organic/inorganic environmentally friendly corrosion inhibitors on mild steel. *J Clean Prod* [Internet]. 2019;220:340–56. Available from: https://doi.org/10.1016/j.jclepro.2019.02.149

[13] Hu J, Zhu Y, Hang J, Zhang Z, Ma Y, Huang H, et al. The effect of organic core-shell corrosion inhibitors on corrosion performance of the reinforcement in simulated concrete pore solution. *Constr Build Mater* [Internet]. 2021;267:121011. Available from: https://doi.org/10.1016/j.conbuildmat.2020.121011

[14] Lei Y, Qiu Z, Tan N, Du H, Li D, Liu J, et al. Polyaniline/CeO2 nanocomposites as corrosion inhibitors for improving the corrosive performance of epoxy coating on carbon steel in 3.5% NaCl solution. *Prog Org Coatings* [Internet]. 2020;139(April):105430. Available from: https://doi.org/10.1016/j.porgcoat.2019.105430

[15] Ji X, Wang W, Li W, Zhao X, Liu A, Wang X, et al. pH-responsible self-healing performance of coating with dual-action core-shell electrospun fibers. *J Taiwan Inst Chem Eng*. 2019;104:227–39.

[16] Hoai Vu NS, Hien P Van, Mathesh M, Hanh Thu VT, Nam ND. Improved corrosion resistance of steel in ethanol fuel blend by Titania nanoparticles and Aganonerion polymorphum leaf extract. *ACS Omega*. 2019;4(1):146–58.

[17] Sanaei Z, Bahlakeh G, Ramezanzadeh B. Active corrosion protection of mild steel by an epoxy ester coating reinforced with hybrid organic/inorganic green inhibitive pigment [Internet]. *J Alloys Compd*. 2017;728:1289–304. Available from: https://dx.doi.org/10.1016/j.jallcom.2017.09.095

[18] Nardeli JV, Fugivara CS, Pinto ERP, Polito WL, Messaddeq Y, Ribeiro SJL, et al. Preparation of polyurethane monolithic resins and modification with a condensed tannin-yielding self-healing property. *Polymers (Basel)*. 2019;11(11):1890.

[19] Majd MT, Asaldoust S, Bahlakeh G, Ramezanzadeh B, Ramezanzadeh M. Green method of carbon steel effective corrosion mitigation in 1 M HCl medium protected by Primula vulgaris flower aqueous extract via experimental, atomic-level MC/MD simulation and electronic-level DFT theoretical elucidation. *J Mol Liq* [Internet]. 2019;284:658–74. Available from: https://doi.org/10.1016/j.molliq.2019.04.037

[20] Cabello Mendez JA, Pérez Bueno J de J, Meas Vong Y, Portales Martínez B. Cerium compounds coating as a single self-healing layer for corrosion inhibition on aluminum 3003. *Sustain.* 2022;14(22):15056.

[21] Kardogan B, Sekercioglu K, Erşan YÇ. Compatibility and biomineralization oriented optimization of nutrient content in nitrate-reducing-biogranules-based microbial self-healing concrete. *Sustain.* 2021;13(16).

[22] Xie P, He Y, Zhong F, Zhang C, Chen C, Li H, et al. Cu-BTA complexes coated layered double hydroxide for controlled release of corrosion inhibitors in dual self-healing waterborne epoxy coatings. *Prog Org Coatings.* 2021;153(February):106164.

[23] Kaya S, Tüzün B, Kaya C, Obot IB. Determination of corrosion inhibition effects of amino acids: Quantum chemical and molecular dynamic simulation study. *J Taiwan Inst Chem Eng.* 2016;58(June):528–35.

[24] Yadav M, Gope L, Sarkar TK. Synthesized amino acid compounds as eco-friendly corrosion inhibitors for mild steel in hydrochloric acid solution: Electrochemical and quantum studies. *Res Chem Intermed.* 2016;42(3):2641–60.

[25] Satpati S, Suhasaria A, Ghosal S, Saha A, Dey S, Sukul D. Amino acid and cinnamaldehyde conjugated Schiff bases as proficient corrosion inhibitors for mild steel in 1 M HCl at higher temperature and prolonged exposure: Detailed electrochemical, adsorption and theoretical study. *J Mol Liq* [Internet]. 2021;324:115077. Available from: https://doi.org/10.1016/j.molliq.2020.115077

[26] Fawzy A, Abdallah M, Zaafarany IA, Ahmed SA, Althagafi II. Thermodynamic, kinetic and mechanistic approach to the corrosion inhibition of carbon steel by new synthesized amino acids-based surfactants as green inhibitors in neutral and alkaline aqueous media. *J Mol Liq.* 2018;265:276–91.

[27] Abd El-Lateef HM, Ismael M, Mohamed IMA. Novel Schiff base amino acid as corrosion inhibitors for carbon steel in CO_2-saturated 3.5% NaCl solution: Experimental and computational study. *Corros Rev.* 2015;33(1–2):77–97.

[28] Liu X, Okafor PC, Pan X, Njoku DI, Uwakwe KJ, Zheng Y. Corrosion inhibition and adsorption properties of cerium-amino acid complexes on mild steel in acidic media: Experimental and DFT studies. *J Adhes Sci Technol* [Internet]. 2020;34(19):2047–74. Available from: https://doi.org/10.1080/01694243.2020.1749474

[29] Alahiane M, Oukhrib R, Berisha A, Albrimi YA, Akbour RA, Oualid HA, et al. Electrochemical, thermodynamic and molecular dynamics studies of some benzoic acid derivatives on the corrosion inhibition of 316 stainless steel in HCl solutions. *J Mol Liq.* 2021;328:115413.

[30] El Ibrahimi B, Baddouh A, Oukhrib R, El Issami S, Hafidi Z, Bazzi L. Electrochemical and in silico investigations into the corrosion inhibition of cyclic amino acids on tin metal in the saline environment. *Surf Interfaces* [Internet]. 2021;23(January):100966. Available from: https://doi.org/10.1016/j.surfin.2021.100966

[31] El Ibrahimi B, El Mouaden K, Jmiai A, Baddouh A, El Issami S, Bazzi L, et al. Understanding the influence of solution's pH on the corrosion of tin in saline solution containing functional amino acids using electrochemical techniques and molecular modeling. *Surf Interfaces* [Internet]. 2019;17(May):100343. Available from: https://doi.org/10.1016/j.surfin.2019.100343

[32] Gao Y, Fan L, Ward L, Liu Z. Synthesis of polyaspartic acid derivative and evaluation of its corrosion and scale inhibition performance in seawater utilization. *Desalination* [Internet]. 2015;365:220–6. Available from: https://dx.doi.org/10.1016/j.desal.2015.03.006

[33] Haque J, Srivastava V, Verma C, Quraishi MA. Experimental and quantum chemical analysis of 2-amino-3-((4-((S)-2-amino-2-carboxyethyl)-1H-imidazol-2-yl)thio) propionic acid as new and green corrosion inhibitor for mild steel in 1 M hydrochloric acid solution. *J Mol Liq.* 2017;225:848–55.

[34] Al-Amiery AA, Mohamad AB, Kadhum AAH, Shaker LM, Isahak WNRW, Takriff MS. Experimental and theoretical study on the corrosion inhibition of mild steel by non-anedioic acid derivative in hydrochloric acid solution. *Sci Rep* [Internet]. 2022;12(1):1–21. Available from: https://doi.org/10.1038/s41598-022-08146-8

[35] Oubaaqa M, Ouakki M, Rbaa M, Abousalem AS, Maatallah M, Benhiba F, et al. Insight into the corrosion inhibition of new amino-acids as efficient inhibitors for mild steel in HCl solution: Experimental studies and theoretical calculations. *J Mol Liq* [Internet]. 2021;334:116520. Available from: https://doi.org/10.1016/j.molliq.2021.116520

[36] Srivastava V, Haque J, Verma C, Singh P, Lgaz H, Salghi R, et al. Amino acid based imidazolium zwitterions as novel and green corrosion inhibitors for mild steel: Experimental, DFT and MD studies. *J Mol Liq* [Internet]. 2017;244:340–52. Available from: https://doi.org/10.1016/j.molliq.2017.08.049

[37] Xu XT, Xu HW, Li W, Wang Y, Zhang XY. A combined quantum chemical, molcular dynamics and Monto Carlo study of three amino acids as corroison inhibitors for aluminum in NaCl solution. *J Mol Liq.* 2022;345(xxxx):117010.

[38] Thoume A, Elmakssoudi A, Left DB, Benzbiria N, Benhiba F, Dakir M, et al. Amino acid structure analog as a corrosion inhibitor of carbon steel in 0.5 M H₂SO₄: Electrochemical, synergistic effect and theoretical studies. *Chem Data Collect* [Internet]. 2020;30:100586. Available from: https://doi.org/10.1016/j.cdc.2020.100586

[39] Zhao R, Xu W, Yu Q, Niu L. Synergistic effect of SAMs of S-containing amino acids and surfactant on corrosion inhibition of 316L stainless steel in 0.5 M NaCl solution. *J Mol Liq* [Internet]. 2020;318:114322. Available from: https://doi.org/10.1016/j.molliq.2020.114322

[40] Zhang QH, Hou BS, Li YY, Zhu GY, Lei Y, Wang X, et al. Dextran derivatives as highly efficient green corrosion inhibitors for carbon steel in CO₂-saturated oilfield produced water: Experimental and theoretical approaches. *Chem Eng J.* 2021;424(April):130519.

[41] El Ibrahimi B, Jmiai A, El Mouaden K, Oukhrib R, Soumoue A, El Issami S, et al. Theoretical evaluation of some α-amino acids for corrosion inhibition of copper in acidic medium: DFT calculations, Monte Carlo simulations and QSPR studies. *J King Saud Univ - Sci* [Internet]. 2020;32(1):163–71. Available from: https://doi.org/10.1016/j.jksus.2018.04.004

[42] Fawzy A, Zaafarany IA, Ali HM, Abdallah M. New synthesized amino acids-based surfactants as efficient inhibitors for corrosion of mild steel in hydrochloric acid medium: Kinetics and thermodynamic approach. *Int J Electrochem Sci.* 2018;13(5):4575–600.

[43] Awad MI, Saad AF, Shaaban MR, Al Jahdaly BA, Hazazi OA. New insight into the mechanism of the inhibition of corrosion of mild steel by some amino acids. *Int J Electrochem Sci.* 2017;12(2):1657–69.

[44] Al-Sabagh AM, Nasser NM, El-Azabawy OE, El-Tabey AE. Corrosion inhibition behavior of new synthesized nonionic surfactants based on amino acid on carbon steel in acid media. *J Mol Liq* [Internet]. 2016;219:1078–88. Available from: https://dx. doi.org/10.1016/j.molliq.2016.03.048

[45] El Ibrahimi B, Jmiai A, El Mouaden K, Baddouh A, El Issami S, Bazzi L, et al. Effect of solution's pH and molecular structure of three linear α-amino acids on the corrosion of tin in salt solution: A combined experimental and theoretical approach. *J Mol Struct* [Internet]. 2019;1196:105–18. Available from: https://doi.org/10.1016/j.molstruc.2019.06.072

[46] El Ibrahimi B, Bazzi L, El Issami S. The role of pH in corrosion inhibition of tin using the proline amino acid: Theoretical and experimental investigations. *RSC Adv.* 2020;10(50):29696–704.

[47] Gupta NK, Verma C, Quraishi MA, Mukherjee AK. Schiff's bases derived from l-lysine and aromatic aldehydes as green corrosion inhibitors for mild steel: Experimental and theoretical studies. *J Mol Liq.* 2016;215:47–57.

[48] Vaghefinazari B, Wierzbicka E, Visser P, Posner R, Matykina E, Mohedano M, et al. Chromate-free corrosion protection strategies for magnesium alloys - a review: Part III - corrosion inhibitors and combining them with other protection strategies. *Materials (Basel)*. 2022;15:8489.

[49] Khalil S, Al-Mazaideh G, Ali N. DFT calculations on corrosion inhibition of aluminum by some carbohydrates. *Int J Biochem Res Rev*. 2016;14(2):1–7.

[50] Mobin M, Rizvi M. Polysaccharide from Plantago as a green corrosion inhibitor for carbon steel in 1 M HCl solution. *Carbohydr Polym*. 2017;160(December):172–83.

[51] Macedo RGM de A, Marques N do N, Tonholo J, Balaban R de C. Water-soluble carboxymethylchitosan used as corrosion inhibitor for carbon steel in saline medium. *Carbohydr Polym* [Internet]. 2019;205(October 2018):371–6. Available from: https://doi.org/10.1016/j.carbpol.2018.10.081

[52] B.P C, Rao P. Carbohydrate biopolymer for corrosion control of 6061 Al-alloy and 6061 Aluminum-15%(v) SiC(P) composite-green approach. *Carbohydr Polym* [Internet]. 2017;168:337–45. Available from: https://dx.doi.org/10.1016/j.carbpol.2017.03.098

[53] Fouda AEAS, El-Maksoud SAA, El-Habab AT, Ibrahim AR. Synthesis and characterization of new ethoxylated carbohydrate based surfactants for corrosion inhibition of low LCS steel in aqueous solutions. *Biointerface Res Appl Chem*. 2021;11(2):9382–404.

[54] Sánchez-Eleuterio A, Mendoza-Merlos C, Corona Sánchez R, Navarrete-López AM, Martínez Jiménez A, Ramírez-Domínguez E, et al. Experimental and theoretical studies on acid corrosion inhibition of API 5L X70 steel with novel 1-N-α-d-glucopyranosyl-1H-1,2,3-triazole xanthines. *Molecules*. 2023;28(1):460.

[55] Verma C, Olasunkanmi LO, Ebenso EE, Quraishi MA, Obot IB. Adsorption behavior of glucosamine-based, pyrimidine-fused heterocycles as green corrosion inhibitors for mild steel: Experimental and theoretical studies. *J Phys Chem C*. 2016;120(21):11598–611.

[56] Verma DK, Aslam R, Aslam J, Quraishi MA, Ebenso EE, Verma C. Computational modeling: Theoretical predictive tools for designing of potential organic corrosion inhibitors. *J Mol Struct* [Internet]. 2021;1236:130294. Available from: https://doi.org/10.1016/j.molstruc.2021.130294

[57] Verma C, Quraishi MA, Kluza K, Makowska-Janusik M, Olasunkanmi LO, Ebenso EE. Corrosion inhibition of mild steel in 1M HCl by D-glucose derivatives of dihydropyrido [2,3-d:6,5-d'] dipyrimidine-2, 4, 6, 8(1H,3H, 5H,7H)-tetraone. *Sci Rep*. 2017;7(March):1–17.

[58] Verma C, Quraishi MA. Carbohydrate polymers-modified carbon allotropes for enhanced anticorrosive activity: State-of-arts and perspective. *Chem Eng J Adv* [Internet]. 2023;13(November 2022):100428. Available from: https://doi.org/10.1016/j.ceja.2022.100428

[59] Verma C, Quraishi MA. Chelation capability of chitosan and chitosan derivatives: Recent developments in sustainable corrosion inhibition and metal decontamination applications. *Curr Res Green Sustain Chem* [Internet]. 2021;4(October):100184. Available from: https://doi.org/10.1016/j.crgsc.2021.100184

[60] Onyeachu IB, Chauhan DS, Ansari KR, Obot IB, Quraishi MA, Alamri AH. Hexamethylene-1,6-bis(N-D-glucopyranosylamine) as a novel corrosion inhibitor for oil and gas industry: Electrochemical and computational analysis. *New J Chem*. 2019;43(19):7282–93.

[61] Galai M, Rbaa M, Ouakki M, Abousalem AS, Ech-chihbi E, Dahmani K, et al. Chemically functionalized of 8-hydroxyquinoline derivatives as efficient corrosion inhibition for steel in 1.0 M HCl solution: Experimental and theoretical studies. *Surf Interfaces* [Internet]. 2020;21(September):100695. Available from: https://doi.org/10.1016/j.surfin.2020.100695

[62] Fouda AS, Fouad RR. New azonitrile derivatives as corrosion inhibitors for copper in nitric acid solution. *Cogent Chem* [Internet]. 2016;2(1):1221174. Available from: https://dx. doi.org/10.1080/23312009.2016.1221174

[63] Subedi BN, Amgain K, Joshi S, Bhattarai J. Green approach to corrosion inhibition effect of vitex negundo leaf extract on aluminum and copper metals in biodiesel and its blend. *Int J Corros Scale Inhib*. 2019;8(3):744–59.

[64] Ugi BU, Obeten ME. Corrosion inhibition performance of alkaloid extracts of calendula officinalis (pot marigold) plant on aluminium, carbon steel and zinc in 5 M nitric acid solution. *Int J Innov Res Adv Stud* [Internet]. 2016;(February 2017). Available from: www.ijiras.com

[65] Oki M, Anawe PAL, Fasakin J. Performance of mild steel in nitric acid/carica papaya leaf extracts corrosion system. *Asian J Appl Sci*. 2015;03(01):2321–893.

[66] Magrati P, Subedi DB, Pokharel DB, Bhattarai J. Appraisal of different inorganic inhibitors action on the corrosion control mechanism of mild steel in HNO3 solution. *J Nepal Chem Soc*. 2020;4:64–73.

[67] Chauhan DS, Quraishi MA, Qurashi A. Recent trends in environmentally sustainable sweet corrosion inhibitors. *J Mol Liq* [Internet]. 2021;326:115117. Available from: https://doi.org/10.1016/j.molliq.2020.115117

[68] Khamaysa OMA, Selatnia I, Lgaz H, Sid A, Lee HS, Zeghache H, et al. Hydrazone-based green corrosion inhibitors for API grade carbon steel in HCl: Insights from electrochemical, XPS, and computational studies. *Colloids Surf A Physicochem Eng Asp* [Internet]. 2021;626(April):127047. Available from: https://doi.org/10.1016/j. colsurfa.2021.127047

[69] Verma C, Quraishi MA. Carbohydrate polymer-metal nanocomposites as advanced anticorrosive materials: A perspective. *Int J Corros Scale Inhib*. 2022;11(2):507–23.

[70] Sangeetha Y, Meenakshi S, Sundaram CS. Interactions at the mild steel acid solution interface in the presence of O-fumaryl-chitosan: Electrochemical and surface studies. *Carbohydr Polym* [Internet]. 2016;136:38–45. Available from: https://dx.doi. org/10.1016/j.carbpol.2015.08.057

[71] Li X, Deng S. Synergistic inhibition effect of walnut green husk extract and potassium iodide on the corrosion of cold rolled steel in trichloroacetic acid solution. *J Mater Res Technol* [Internet]. 2020;9(6):15604–20. Available from: https://doi.org/10.1016/j. jmrt.2020.11.018

[72] Li X, Xin X, Deng S. Synergism between walnut green husk extract and sodium dodecyl benzene sulfonate on cold rolled steel in 1.0 mol/L H_2SO_4 solution. *Corros Commun* [Internet]. 2022;0–42. Available from: https://doi.org/10.1016/j.corcom.2022.05.004

[73] Feng L, Zhang S, Hao L, Du H, Pan R, Huang G, et al. Cucumber (Cucumis sativus L.) leaf extract as a green corrosion inhibitor for carbon steel in acidic solution: Electrochemical, functional and molecular analysis. *Molecules*. 2022;27(12):3826.

[74] Selles C, Benali O, Tabti B, Larabi L, Harek Y. Green corrosion inhibitor: Inhibitive action of aqueous extract of Anacyclus pyrethrum L. for the corrosion of mild steel in 0.5M H_2SO_4. *J Mater Environ Sci*. 2012;3(1):206–19.

[75] Eduok UM, Umoren SA, Udoh AP. Synergistic inhibition effects between leaves and stem extracts of Sida acuta and iodide ion for mild steel corrosion in 1M H_2SO_4 solutions. *Arab J Chem* [Internet]. 2012;5(3):325–37. Available from: https://dx.doi. org/10.1016/j.arabjc.2010.09.006

[76] Fouda AS, Shalabi K, Shaaban MS. Synergistic effect of potassium iodide on corrosion inhibition of carbon steel by Achillea santolina extract in hydrochloric acid solution. *J Bio- Tribo-Corrosion* [Internet]. 2019;5(3). Available from: https://doi.org/10.1007/ s40735-019-0260-6

[77] Ahangar M, Izadi M, Shahrabi T, Mohammadi I. The synergistic effect of zinc acetate on the protective behavior of sodium lignosulfonate for corrosion prevention of mild steel in 3.5 wt% NaCl electrolyte: Surface and electrochemical studies. *J Mol Liq* [Internet]. 2020;314:113617. Available from: https://doi.org/10.1016/j.molliq.2020.113617

[78] Bashir S, Thakur A, Lgaz H, Chung I-M, Kumar A. Computational and experimental studies on Phenylephrine as anti-corrosion substance of mild steel in acidic medium. *J Mol Liq*. 2019;293:111539.

[79] Parveen G, Bashir S, Thakur A, Saha SK, Banerjee P, Kumar A. Experimental and computational studies of imidazolium based ionic liquid 1-methyl-3-propylimidazolium iodide on mild steel corrosion in acidic solution experimental and computational studies of imidazolium based ionic liquid 1-methyl-3-propylimidazolium. *Mater Res Express*. 2020;7(1):016510.

[80] Thakur A, Kumar A. Recent trends in nanostructured carbon-based electrochemical sensors for the detection and remediation of persistent toxic substances in real-time analysis. *Mater Res Express*. 2023;10:034001.

[81] Thakur A, Savaş K, Kumar A. Recent trends in the characterization and application progress of nano-modified coatings in corrosion mitigation of metals and alloys. *Appl Sci*. 2023;13:730.

[82] Bashir S, Lgaz H, Chung IM, Kumar A. Effective green corrosion inhibition of aluminium using analgin in acidic medium: An experimental and theoretical study. *Chem Eng Commun* [Internet]. 2020;0(0):1–10. Available from: https://doi.org/10.1080/00986 445.2020.1752680

[83] Thakur A, Kumar A. Sustainable inhibitors for corrosion mitigation in aggressive corrosive media: A comprehensive study. *J Bio- Tribo-Corrosion* [Internet]. 2021;7(2): 1–48. Available from: https://doi.org/10.1007/s40735-021-00501-y

[84] Thakur A, Kumar A, Sharma S, Ganjoo R, Assad H. Materials today: Proceedings computational and experimental studies on the efficiency of Sonchus arvensis as green corrosion inhibitor for mild steel in 0. 5 M HCl solution. *Mater Today Proc* [Internet]. 2022;66:609–21. Available from: https://doi.org/10.1016/j.matpr.2022.06.479

[85] Thakur A, Kumar A. Recent advances on rapid detection and remediation of environmental pollutants utilizing nanomaterials-based (bio)sensors. *Sci Total Environ* [Internet]. 2022;834(January):155219. Available from: https://doi.org/10.1016/j.scitotenv.2022.155219

[86] Thakur A, Kaya S, Abousalem AS, Kumar A. Experimental, DFT and MC simulation analysis of Vicia Sativa weed aerial extract as sustainable and eco-benign corrosion inhibitor for mild steel in acidic environment. *Sustain Chem Pharm* [Internet]. 2022;29(July):100785. Available from: https://doi.org/10.1016/j.scp.2022.100785

[87] Bashir S, Thakur A, Lgaz H, Chung IM, Kumar A. Corrosion inhibition efficiency of bronopol on aluminium in 0.5 M HCl solution: Insights from experimental and quantum chemical studies. *Surf Interfaces* [Internet]. 2020;20(April):100542. Available from: https://doi.org/10.1016/j.surfin.2020.100542

[88] Thakur A, Sharma S, Ganjoo R, Assad H, Kumar A. Anti-corrosive potential of the sustainable corrosion inhibitors based on biomass waste: A review on preceding and perspective research. *J Phys Conf Ser*. 2022;2267(1):012079.

[89] Thakur A, Kumar A, Kaya S, Marzouki R, Zhang F, Guo L. Recent advancements in surface modification, characterization and functionalization for enhancing the biocompatibility and corrosion resistance of biomedical implants. *Coatings*. 2022;12:1459.

[90] Kaya S, Thakur A, Kumar A. The role of in silico/DFT investigations in analyzing dye molecules for enhanced solar cell efficiency and reduced toxicity. *J Mol Graph Model*. 2023;124(June):108536.

[91] Bashir S, Thakur A, Lgaz H, Chung I-M, Kumar A. Corrosion inhibition performance of acarbose on mild steel corrosion in acidic medium: An experimental and computational study. *Arab J Sci Eng* [Internet]. 2020;45(6):4773–83. Available from: https://doi.org/10.1007/s13369-020-04514-6

[92] Thakur A, Kumar A, Zhang R. Alcoholic Beverage Purification Applications of Activated Carbon. In: Verma C, Quraishi MA, editors. *Activated Carbon: Progress and Applications* [Internet]. The Royal Society of Chemistry; 2023. p. 0. Available from: https://doi.org/10.1039/BK9781839169861-00152

[93] Kumar A. Overview of the Properties, Applicability, and Recent Advancements of Some Natural Products Used as Potential Inhibitors in Various Corrosive Systems. In: *Handbook of Research on Corrosion Sciences and Engineering*, IGI Global 2003. pp. 275–310.

[94] Thakur A, Kumar A, Kaya S, Vo DVN, Sharma A. Suppressing inhibitory compounds by nanomaterials for highly efficient biofuel production: A review. *Fuel* [Internet]. 2022;312(September 2021):122934. Available from: https://doi.org/10.1016/j.fuel.2021.122934

[95] Dhonchak C, Agnihotri N. Computational Insights in the spectrophotometrically 4H-chromen-4-one complex using DFT Method. *Biointerface Res Appl Chem.* 2023;13(4):357.

[96] Verma C, Thakur A, Ganjoo R, Sharma S, Assad H. Coordination bonding and corrosion inhibition potential of nitrogen-rich heterocycles: Azoles and triazines as specific examples. *Coord Chem Rev* [Internet]. 2023;488(November 2022):215177. Available from: https://doi.org/10.1016/j.ccr.2023.215177

[97] Thakur A, Kaya S, Abousalem AS, Sharma S, Ganjoo R, Assad H, et al. Computational and experimental studies on the corrosion inhibition performance of an aerial extract of Cnicus Benedictus weed on the acidic corrosion of mild steel. *Process Saf Environ Prot* [Internet]. 2022;161:801–18. Available from: https://doi.org/10.1016/j.psep.2022.03.082

[98] Thakur A, Kaya S, Kumar A. Recent innovations in nano container-based self-healing coatings in the construction industry. *Curr Nanosci.* 2021;18(2):203–16.

[99] Sandilya S, Thakur A, Suresh Singh S, Kumar A. Recent advances, synthesis and characterization of bio-based based polymers from natural sources. *Plant Cell Biotechnol Mol Biol.* 2021;22:70–93.

[100] Sharma D, Thakur A, Sharma MK, Jakhar K, Kumar S, Sharma AK, et al. Synthesis, electrochemical, morphological, computational and corrosion inhibition studies of 3-(5-naphthalen-2-yl-[1,3,4]oxadiazol-2-yl)-pyridine against mild steel in 1 M HCl. *Asian J Chem.* 2014;35(5):1079–88.

11 Phytochemicals/Plant Extracts as Corrosion Inhibitors for Different Metals/Alloys in Phosphoric Acid

Khalid Bouiti, Nabil Lahrache, Ichraq Bouhouche, Najoua Labjar, Ghita Amine Benabdallah, and Souad El Hajjaji
Mohammed V University in Rabat

11.1 INTRODUCTION

Corrosion refers to the material or its characteristics deterioration due to physico-chemical interactions within the medium [1]. The concept recognizes that corrosion occurs naturally as a damaging process that either reduces or destroys the material's properties, rendering it ineffective in the specific function intended [2]. However, corrosion can also be a desirable process. It removes and destroys significant natural deposits. Certain manufacturing procedures involve material corrosion [3]. Aluminum anodizing, as an example, consists of the metal's surface oxidation and the formation of a protective, oxide layer on the surface [4].

Metallic corrosion is a spontaneous process whereby alloys and metals tend, through chemical agents or the atmosphere, to return to their initial oxide, sulfide, carbonate, or other more stable salt state depending on the environment [3,5].

Corrosion is a problem of significant importance, impacting the global economy and generating significant losses and damage to pollution and human safety every year [6]. It can cause dangerous and costly damage to industrial facilities, automobiles, appliances, water systems, and infrastructure [7]. Corrosion can have more significant consequences on health, safety, and the environment [7]. This destructive phenomenon has often been overlooked as a major cause of many fatal accidents due to the disappearance of evidence during the industrial incident. There are many cases of human and animal deaths due to corrosion.

Acid treatment methods are widely employed, particularly in the fields of acid cleaning and pickling, oil well stimulant treatment, and localized deposit removal [8]. Acids also have widespread applications in various chemical synthesis activities [8].

DOI: 10.1201/9781003394631-11

The aggressiveness of acids has required the employment of inhibitors to reduce the attack on the materials involved [9]. At the same time, the approaches adopted for inhibiting corrosion require an assessment of the specific performance factors of the system, since preventive measures used effectively in a particular medium can be potentially damaging under other circumstances [10].

11.2 CORROSION INHIBITION

An inhibitor is a substance that delays corrosion when added to an environment in low concentration [11]. It must not only be stable in the presence of other constituents of the environment but also not influence the stability of the species contained in this environment [12].

It must also lower the corrosion rate of the metal while maintaining the physico-chemical characteristics of the latter. The choice of corrosion inhibitors for practical purposes is based on the knowledge of their mechanism of action [13].

11.2.1 INHIBITOR MOLECULE COMPOSITION

Organic molecules have a certain development potential in terms of corrosion inhibitors: their consumption is currently preferred to that of inorganic inhibitors, mainly for reasons of ecotoxicity [14]. Organic inhibitors are generally made up of by-products from the petroleum industry [3]. They have at least one active center that can exchange electrons with metal, such as nitrogen, oxygen, phosphorus, or sulfur.

Mineral molecules are most often used in alkaline environments and more rarely in an acidic medium. The products are dissociated in the medium, ensuring inhibition processes [5]. The principal anions involved in inhibition are XO_4^{n-} type oxo anions, including phosphates, molybdates, silicates, and chromates [15]. Zn^{2+} and Ca^{2+} are the main cations, as are others forming non-soluble salts within specific anions, including OH^-.

The number of molecules in use today is becoming smaller, because most of the effective products are harmful to the environment [16].

11.2.2 ORGANIC INHIBITOR ADSORPTION

In the classification relating to the electrochemical mechanism of action, it can be distinguished as an anodic, cathodic, or mixed inhibitor [17]. The corrosion inhibitor forms a barrier layer on the metal surface, which modifies the electrochemical reactions by blocking the anodic or cathodic sites by inhibitors' adsorption onto the surface [18–20].

Adsorption is a surface phenomenon that occurs due to any surface is made up of atoms that do not have all their chemical bonds satisfied. This surface tends to fill this gap by capturing atoms and molecules in its environment [21].

There are two modes of adsorption, the physisorption or physical adsorption, which preserve the identity of the adsorbed molecules, based on three types of forces: van der Waals' forces, polar forces, and hydrogen bonds.

The force of electrostatic adsorption is a function of the difference between the charges carried by the inhibitor and the surface of the metal, which is itself a function of the difference between the corrosion potential of the metal and its potential for zero charges in the corrosive medium considered [18,22] as well as the physically adsorbed substances, which condense rapidly on the metal but are easily degraded by desorption when the temperature increases.

Chemisorption, on the other hand, is based on the pooling of electrons between the polar part of the molecule and metallic substrate [23], leading to the development of more stable chemical bonds based on higher bond energies, the electrons come mainly from the non-bonding doublets of the inhibiting molecules such as the heteroatoms. The adsorption is accompanied by a deep modification of the electronic charge distribution of the adsorbed molecules [5,23,24].

The important parameter is the electron density around the active center, which leads to the strengthening of covalent bonds between the donor atom and the metal atom. The same reason applies to cyclic amines, which are better inhibitors than aliphatic amines.

Unsaturated organic compounds are electron carriers capable of creating bonds with metal atoms. The presence of an unsaturated bond can be favorable to the inhibitory efficiency of an organic molecule in an acidic environment since it can adsorb in the same way on either a positively or negatively charged surface [3,25,26].

11.2.3 GREEN CORROSION INHIBITORS

The use of corrosion inhibitors is limited as long as they are based on substances that are unsafe for human health, such as chromium treatment [27]. Green chemistry represents the attempt to develop an approach for better management of chemical risks [28].

Ecological corrosion inhibitors are substances with characteristics that are biocompatible with nature [29], as well as low contribution and biodegradability; the classification of these inhibitors can be grouped into two categories: organic and inorganic inhibitors. Organic compounds have better inhibitory efficiency than inorganic compounds [30,31].

Green organic inhibitors include flavonoids, alkaloids, and other natural products obtained from natural resources such as plants [3,31]; for inorganic inhibitors, the inhibitory capacity depends on the molecular structure and the presence of double bonds at the heteroatoms, and alkyl chains increase the inhibition performance [29].

During the process of inhibition, molecules are adsorbed onto the surface of the metals or alloys for the formation of a barrier [32,33]. In the corrosion phase, metal ions move to the anode to transfer electrons from the anode to the cathode, with hydrogen ions, oxidizing agents, or oxygen functioning as acceptors. By delaying the oxidation or reduction reactions, the degradation of metals is minimized [34,35] (Figure 11.1).

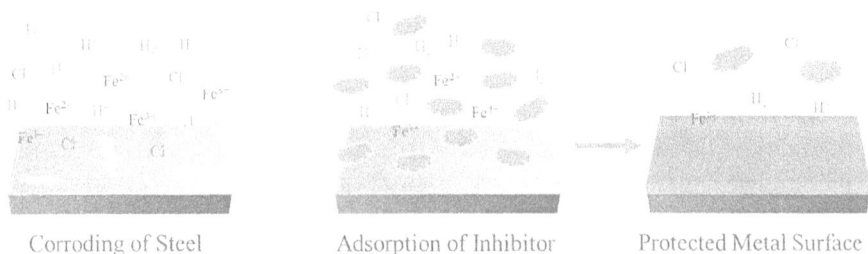

Corroding of Steel   Adsorption of Inhibitor   Protected Metal Surface

FIGURE 11.1   Process of film formation on the metal surface.

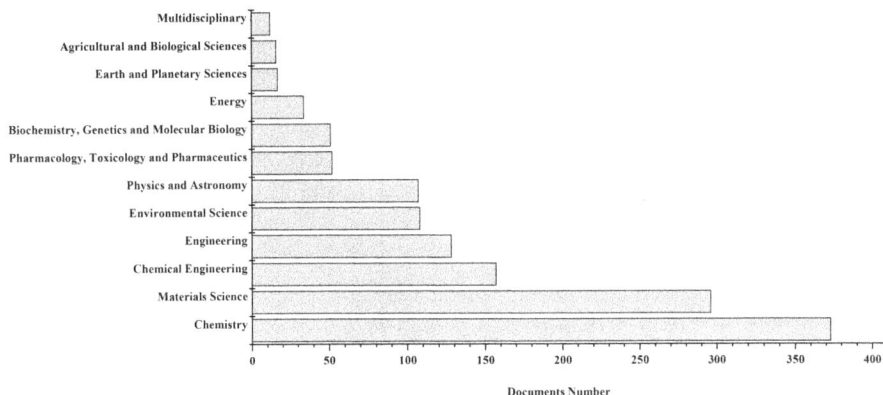

FIGURE 11.2   The fields corresponding to the publications concerning the use of plants as corrosion inhibitors.

In the literature, several works have treated the use of plant-based extracts as corrosion inhibitors, and the data were extracted from Scopus between 2010 and 2022, mostly published articles in the fields of chemistry, materials science, chemical engineering, environmental science, and others (Figure 11.2). The total number of documents found was 818 by considering the period of study, the language, and the fields related to the maximum number of publications. The results indicate 659 articles, 67 conference papers, 59 reviews, 19 chapters, and 8 conference reviews.

11.2.4   VALIDATION METHODS

11.2.4.1   Gasometric and Gravimetric Methods

The correlation between the rate of hydrogen gas evolution and the rate of corrosion has led to the recognition of the potential value of monitoring the quantity of hydrogen gas released at cathodic sites as a means of gaining insightful information about the corrosion process [36]. This monitoring can be particularly useful in studying the amount of hydrogen gas released during metallic corrosion in aggressive solutions, both with and without inhibitors [37].

Among the various techniques available for assessing the effectiveness of corrosion inhibition, the most straightforward, cost-effective, and widely utilized method is the analysis of weight loss. This method involves several steps. Initially, the metal specimen under investigation is cut into small sections of predetermined size. These sections are then subjected to degreasing, cleaning, and polishing procedures. Subsequently, the prepared samples are immersed in the corrosive medium and cleaned following established ASTM protocols [38–40] after a specific duration of exposure. By calculating the difference in mass before and after immersion, the average weight loss can be determined.

11.2.4.2 Electrochemical Impedance Spectroscopy Approach

Electrochemical impedance spectroscopy is a non-destructive method that provides valuable insights into the frequency-dependent behavior of electrochemical systems, specifically regarding energy storage and energy dissipation. This technique offers several key advantages, including its suitability for analyzing low-conductivity systems, its ability to provide mechanistic information, and its effectiveness in evaluating solution resistance [41–43].

To unravel the intricate response of the system, the collected electrochemical data are meticulously fitted to an appropriate equivalent circuit using the nonlinear least squares approach [44].

Various graphical representations, such as Nyquist, Bode, and phase angle plots, are employed to visually illustrate the data. Among these, the Nyquist plot is the most commonly used graphical representation, as it offers a concise understanding of the electrochemical behavior of the corrosive system. Moreover, it facilitates the prediction of analogous circuit elements [43,45].

By subjecting an electrochemical system to a perturbation frequency of modest amplitude, valuable information regarding the internal dynamics of the corrosion system can be obtained. Typically, the analogous circuit representing the corroding metal in an aqueous medium is a combination of resistance and capacitance, representing the corrosion surface [43,46].

The charge transfer resistance acts as a parallel resistance component that governs the corrosion rate. The capacitance at the metal/electrolyte interface is often approximated as the double-layer capacitance [47–49].

In corrosion inhibition studies, a high value of R_{ct} indicates the formation of a protective barrier that hinders the charge transfer during corrosion, resulting in a reduced corrosion rate [50]. R_{ct} signifies the discrepancy between the highest and lowest frequencies of the actual resistance component.

11.2.4.3 Polarization Measurements Technique

Potentiodynamic Polarization (PDP) is a transformative process that induces surface modifications on the electrode. The term E_{ocp} (open circuit potential) is utilized to denote the potential in the absence of cumulative current, as determined through measurements that are fitted to the potential values [51,52]. In an ideal scenario, E_{ocp} and E_{corr} would exhibit congruence. However, discrepancies between the two results may arise due to alterations occurring on the electrode surface during potential scanning.

When the potential of a working electrode is sufficiently shifted in a positive direction beyond E_{ocp}, the contribution of cathodic reduction becomes insignificant, and the measured current response primarily reflects the anodic process [51]. Conversely, at significantly negative potentials, the cathode current predominates the overall current response.

In corrosion inhibition experiments, the determination of an optimal sweep rate is crucial, as it varies depending on the specific metal environment and the type of corrosion inhibitor employed. If the scan speed is too rapid, it can lead to variations in polarization curves, potentially resulting in a misinterpretation of the polarized electrode process due to charge disturbances and insufficient time to reach a steady state [53]. On the contrary, if the measurement is conducted too slowly, it may induce changes in the interfacial structure of the polarized electrode.

11.2.4.4 Surface Analysis

SEM is widely recognized as the most commonly used technique for studying microscopic topography [54]. This powerful method capitalizes on the principles of electron-matter interactions to generate high-resolution images that unveil the surface characteristics of the material being examined [55].

By employing an electron beam that engages with the sample, valuable insights into its surface structure and composition can be obtained. This interaction between electrons and the material results in the emission of particles and radiation, which can be captured by specialized detectors. These detectors enable the collection of diverse transmitted signals, facilitating investigations into the surface topography, microstructure, and chemical composition [56].

Similarly, atomic force microscopy provides essential data pertaining to the surface geometry of metals, offering a means for comparison and topographic imaging [57–59]. For the determination of oxidizing conditions, electronic states, and equilibrium states, photoelectron X-ray spectroscopy is frequently employed [60–63].

Additionally, FTIR spectroscopy is commonly utilized to characterize corrosion inhibitors, providing crucial information regarding functional groups and vibration patterns [64]. Furthermore, UV-visible spectroscopy serves as a valuable tool for identifying functional groups, electrical transitions, and band gaps in both electrical and optical systems [65].

11.2.4.5 Theoretical Calculation Method

The current state of computer technology has revolutionized the field of corrosion research by enabling researchers to employ computational analyses and molecular simulations. These simulations allow for the generation and prediction of the performance of novel corrosion inhibitors [66]. Molecular dynamics approaches have demonstrated their efficacy in elucidating the molecular structure, electrical properties, and reactivity of potent inhibitors. Additionally, quantum chemistry simulations provide precise predictions of corrosion inhibition at the molecular level, offering the potential to reduce research costs [67,68]. This technique often combines molecular dynamics with density functional theory to explore the underlying inhibitory mechanisms at the molecular scale [69].

Notably, the energy levels of the lowest occupied and highest unoccupied orbitals play a crucial role in determining the effectiveness of corrosion inhibition in density functional theory modeling [70]. Gaussian, a widely used computational software package, can accurately determine essential properties such as the energies of the highest occupied molecular orbital and lowest unoccupied molecular orbital, energy gap, electron density, smoothness, and electron transmission altitude [71].

11.3 PHYTOCHEMICAL COMPOUND EXTRACTION

The preparation of the extracts is carried out after identifying the concerned part to proceed to the drying, grinding, and sieving process [72,73]. The drying presents

Aceton	Chloroform	Dichloromethanol	Ethanol	Ether	Methanol	Water
Flavonols	Falavonoïds	Terpenoïds	Alkaloïds	Alkaloïds	Anthocyanins	Anthocyanins
Tannins	Terpenoïds		Flavonols	Coumarins	Flavones	Lectines
			Polyacetylenes	Fatty Acids	Lactones	Polypeptides
			Polyphenols	Terpenoïds	Quassinoïds	Saponins
			Propolis		Phenones	Starches
			Sterols		Polyphenols	Tannins
			Tannins		Saponins	Terpenoïds
			Terpenoïds		Tannins	
					Terpenoïds	
					Totarol	
					Xanthoxyllines	

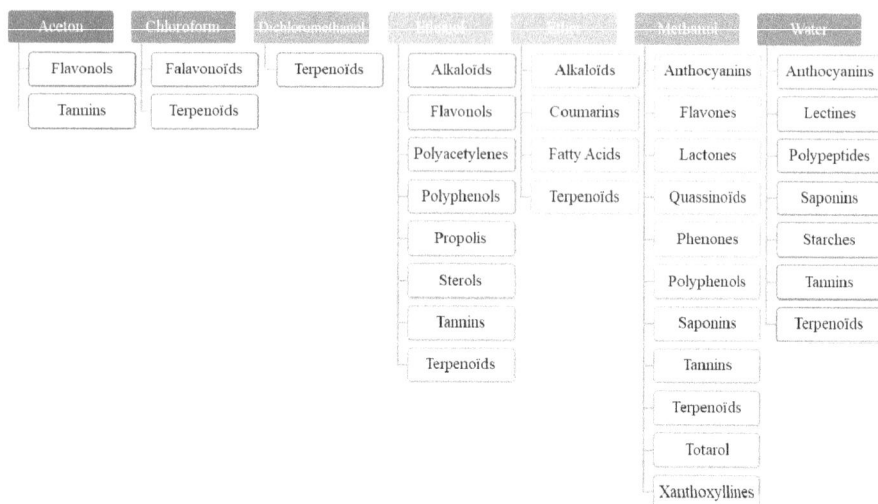

FIGURE 11.3 List of extraction solvents and the substances obtained [68].

enormous advantages as the stability of the bioactive compounds and the antioxidant capacity, as well as the fresh plants, can be under the degradation, the evaporation, or the oxidation of their components [74,75], but in certain conditions the phytochemicals can be extracted with great concentration from the one that is fresher than dried [76].

11.3.1 SOLVENT EXTRACTION

The extraction methods are based on heating, cooling, and separation of active substances. Solvent extraction is the most used, and the solvent penetrates inside the plant tissue for solubilization and extraction of substances [77,78]. Figure 11.3 lists the extraction solvents and the substances obtained. The variation of the polarity, the process time, and the temperature can affect the extract characteristics and the chemical composition [79].

The extraction procedure is affected by temperature, requiring a range of 60°C–80°C to increase mass transfer and solubility rates, thus decreasing viscosity [80]. The choice of an optimal temperature serves to ensure the non-degradability of phytochemicals. Also, the increase in the process time leads to the oxidation of phenolic compounds, which affects the inhibitory capacity [80,81].

11.3.2 MACERATION, INFUSION, DECOCTION, AND PERCOLATION

Maceration consists of soaking plant materials in a container stoppered with a solvent and letting them sit at ambient temperature for at least 3 days with frequent shaking [82]. The aim of the treatment is to soften and break down the plant's cell wall and extract the dissolved phytochemicals; the mixture is pressed or filtered. The process involves heat transfer by both conduction and convection, and the solvents are selected depending on the nature of the extracted components. Decoction and infusion employ a similar concept as maceration and are both steeped in boiled or cold water.

The period of maceration is shorter during infusion, though, and the samples are prepared by boiling in a special water volume for a defined time for decoction [82]. The decoction process is appropriate for extracting heat-stable components from harder plants and produces a higher proportion of lipid-soluble substances compared with the infusion and maceration processes.

In percolation, similar techniques are applied using a specific equipment known as a percolator. The dried powder samples are put into the device and mixed with hot water for maceration. The percolating procedure is generally carried out at a steady speed at which the process is completed and then evaporated to provide a pure extract [83].

11.3.3 Hydrodistillation and Steam Distillation

Hydrodistillation involves immersing plant material into the water in a still and heating the mixture to a boiling point. The equipment comprises a heating unit, a container, and a condenser for converting the vapors in the container into liquid, followed by a settling tank to recover the condensates and thus isolate the essential oils.

The method is recognized as particularly suitable for extracting from plant materials such as flowers and woods, although it is commonly employed for extracting natural hydrophobic plant parts characterized by a tendency to boil at a higher temperature.

The steam distillation method is the most widely applied technique. The procedure involves steaming the plants with the steam provided by a generator. Heating is the crucial parameter in determining the efficiency at which the plant material's structures decompose, burst, and reveal the aroma compounds [84].

The process consists of a bed filled with sample plants set over the steaming unit. This allows just the steam to penetrate the sample and prevents the heated water from mixing with the plant sample. As a result, the treatment involves a lower amount of steam, with less water required for distillation [85].

11.3.4 Soxhlet Extraction

Soxhlet extraction, the oldest extraction method, has proved to be the technique recommended to assess the effectiveness of various solid/liquid methods of extraction.

The raw materials in the holder are loaded onto fresh condensing solvent collected *via* a flask for distillation. When the solvent overflows, a siphon draws the liquid from the carrier and discharges it to the flask, entraining the solutes extracted in the bulk liquid. The solutes in the solvent flask are then released through distillation. The solutes remain inside the flask, while the solvent flows through the bed of solids.

The process continues until complete extraction has been achieved. Extraction by Soxhlet and thermal reflux methods are not identical. Extraction by thermal reflux involves heating the sample to boiling point inside the solvent, using a cooled container surface for condensation as the vapors rise to a boil, and then returning them to the vessel in a liquid phase, unboiled. By contrast, Soxhlet is a process employed to separate the solvent-soluble parts of an extract.

11.3.5 ULTRASONIC EXTRACTION

Ultrasonic waves are considered higher-frequency waves of sound that could be heard by humans, i.e., above 20 kHz. The propagated waves are produced by a rarefaction and expansion process. Compression leads to pressure build-up in the solution. Once the excess pressure overcomes the liquid's traction resistance, bubbles of vapor are formed. The implosive collapse is triggered by ultrasonic radiation, known as cavitation.

The bubbles implode, generating macroturbulence and interparticle collisions on a large scale as well as disturbances within the microporous biomass components. Cavitation near liquid/solid surfaces leads to liquid flowing rapidly across the cavity toward the surface. The microjets' impact leads to the detachment of the surface, particle break-up, and erosion, thus promoting bioactive or targeted component extraction through the bio-matrix. The extraction efficiency is enhanced by the increased transfer of mass through vertexing and diffusion processes.

The ultrasound's mechanical impact allows greater cellular matter penetration, thus improving the transfer of mass. Ultrasonics interfere with cellular structures for easier content extraction. Consequently, cellular disintegration and efficient transfer of mass are identified as crucial parameters in improving ultrasonic extraction. The process parameters, such as lower temperature and pressure, can be modified, compared with other methods of extraction.

Phytochemical diversity depends on several factors such as plant age, plant cycle, geographical, and climatic conditions [86]. In general, the inhibitory capacity of plant extracts derives from the chemical properties similar to the organic substances commonly used [68] (Figure 11.4).

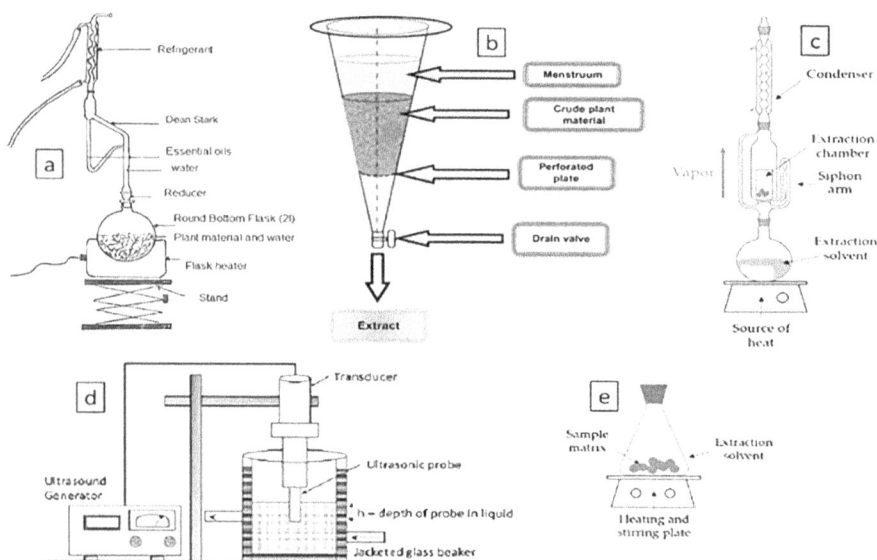

FIGURE 11.4 Extraction techniques: hydrodistillation (a), percolation (b), Soxhlet extraction (c), ultrasonic extraction (d), and maceration (e) [87–90].

11.4 PLANT EXTRACT USE FOR INHIBITION IN PHOSPHORIC MEDIUM

Phosphoric acid represents a major manufactured chemical employed as an intermediate in fertilizers [91], as an adjuvant for food processing, and for the treatment of metallic surfaces in the metallurgical industries. It is formed by sulfuric acid attacking rock phosphate. Additional techniques employ chloric acid as well as extraction by solvent [49,50].

Phosphate rock remains the main impurity source, both in suspended and dissolved form. Further contaminants, including Cl^-, can be incorporated into water treatment, particularly in saline water. The H_2SO_4 produced in the hydrometallurgical process contributes to other contaminants [92]. The specific corrosivity of impurity is affected by its properties as a chemical compound, the active component concentration, and its interactions within the specific metal surface and other acidic components [93].

Mucous and skin membranes can be irritated by high concentrations of solutions, which are corrosive. If swallowed, corrosion can affect the intestinal and gastric tracts [94]. Respiratory problems may also include nausea and vomiting. Inhaling acid mists can irritate the lungs and throat, leading to respiratory tract dysfunction [94].

Phosphoric acid penetrates the soil and is dangerous for aquatic wildlife through its acidic nature. If not diluted, it destroys vegetation. When it enters the groundwater, the phosphate left over by reducing the H_3PO_4 may promote the growth of freshwater and marine life, resulting in the proliferation of eutrophication and algae. Factory processes function in challenging environments: high and intense thermal transfer, concentrated acid, stirred and circulated solutions with eroding solids in suspension, ventilation, foaming, and corrosive acid vapors that condense onto metallic surfaces.

The acid's viscosity and density are modified by certain impurities, which form acidic sediments and sludges. On encountering metal interfaces, the deposits can affect the metal's corrosive properties, leading to local corrosion [92] (Table 11.1).

11.5 CONCLUSION

This chapter deals with the use of inhibitors prepared from plants according to different extraction methods adopted by optimizing the factors affecting the extraction process such as solvent polarity, temperature, and time. The extracted phytochemicals are considered as a substitute for synthesized inhibitors harmful to the ecosystem. For this reason, several researches have been focused on the exploitation of the anti-corrosive characteristics based on plant extracts in different media, whether acidic, basic, or neutral. This work focuses on the phosphoric medium due to its important role in different industries.

TABLE 11.1

Published Articles about Corrosion Inhibition in Phosphoric Acid by Plant Extract

Plant	Extraction Method	Solvent	Metal	Milieu	Result	References
Sargassum wightii	Soxhlet	Ethanol	Brass	0.1 N H_3PO_4	The effectiveness of inhibition increases at higher extract concentrations and reduces with temperature rise. The kinetic, thermodynamic, and activation energy data for the inhibition process were calculated.	[95]
Artemesia herba albamedium	Hydrodistillation	Water	Stainless steel	1 M H_3PO_4	The essential oil of Artemisia herba-alba proved to be a valuable inhibitor of SS corrosion in phosphoric medium 1.0 M, employing both electrochemical impedance spectroscopy and potentiodynamic polarization techniques, as well as SEM analysis. The efficiency of inhibition improved with increasing concentration, attaining 88% with 1 g L^{-1} and at 298 K. Nyquist plots provided by the impedance tests confirmed the corrosion-inhibiting properties of the plant investigated.	[96]
Coriandrum sativum	Hydrodistillation	Water	6063 Aluminum alloy	1 M H_3PO_4	The seed extract of Coriandrum sativum L. provided a valuable ecological corrosion inhibitor for aluminum alloy 6063 in 1.0 M H_3PO_4 medium, employing EIS and PDP methods at temperatures between 30°C and 50°C. SEM analysis also revealed that CSE adsorption inhibited alloy corrosion.	[97]
Ziziphus lotus	Soxhlet	Hexane	C38 steel	5.5 M H_3PO_4	Inhibition efficiency increased with Ziziphus lotus oil extract concentrations, reaching an efficiency of about 70.5% for the polarization study and about 61% for the EIS measurements at an oil concentration of 3 g L^{-1}.	[98]

(Continued)

TABLE 11.1 (*Continued*)
Published Articles about Corrosion Inhibition in Phosphoric Acid by Plant Extract

Plant	Extraction Method	Solvent	Metal	Milieu	Result	References
Lanvandula stoekas	Hydrodistillation	Water	Stainless steel	$5.5 M H_3PO_4$	The inhibiting properties of organic oil from *Lavandula stoekas* leaves on the corrosion of UB6 SS in phosphoric medium 5.5 M were investigated *via* EIS and PDP techniques. The efficiency of inhibition rises with the concentration of organic oil, reaching a max. value of 87.3% at $1.2 g L^{-1}$.	[99]
Psidium guajava	Maceration	Methanol	Mild steel	$1 M H_3PO_4$	The corrosion inhibitory and adsorption properties of *Psidium guajava* alcoholic leaves extract in $1.0 M H_3PO_4$ solution were evaluated by gravimetric, EIS, and PDP methods. An inhibitory effectiveness of up to 89% was achieved at a concentration of 800 ppm in the gravimetric tests, over an immersion time of 1 hour.	[100]
Borago officinalis	Maceration	Water	Mild steel	$1 M H_3PO_4$	Chemical hydrogen evolution and PDP were used to assess the inhibitory efficiency of *Borage* flowers aqueous extract against the corrosion of mild steel in $1.0 M$ phosphoric acid. By creating a thin coating on the metal surface, adsorption on mild steel surface was discovered to obey Langmuir and thermodynamic-kinetic isotherms. The adsorbed film was also confirmed by SEM-EDX analysis.	[101]
Acalypha indica	Soxhlet	Methanol	Mild steel	$1 N H_3PO_4$	The effect of *Acalypha indica* L. alcoholic extract on mild steel corrosion in 1 N phosphoric acid was investigated using mass loss and polarization techniques at temperatures ranging from 303 to 333 K. When compared to a blank, the corrosion rate increased with increasing temperature and decreasing inhibitor concentration. Surface analysis was also performed to determine the mechanism of corrosion inhibitor action on mild steel corrosion.	[102]

(*Continued*)

TABLE 11.1 (Continued)
Published Articles about Corrosion Inhibition in Phosphoric Acid by Plant Extract

Plant	Extraction Method	Solvent	Metal	Milieu	Result	References
Zanthoxylum alatum	Maceration	Methanol	Mild steel	H_3PO_4	Zanthoxylum alatum extract was tested for its inhibitory properties on mild steel corrosion in 20%, 50%, and 88% aqueous orthophosphoric acid by loss-weight and electrochemical impedance spectroscopy. The plant extract reduced steel corrosion significantly in 88% phosphoric. The temperature effect was investigated in the range of 50°C–80°C upon adding the plant extract to mild steel.	[103]
Fraxinus excelsior Zingiber zerumbet Isatis tinctoria	Maceration	Water	Mild steel	2 M H_3PO_4	Three plant extracts have been developed to avoid mild steel corrosion in phosphoric acid. The active ingredients used in the protective layers provide great electrochemical stability at high temperatures as well as powerful protection against acidic solutions. The interaction of organic molecules with the metal surface was studied using multi-level computational techniques. The research suggests that for materials with efficient anti-corrosion performance, conventional compounds can be substituted with natural substituted organic products.	[104]

(Continued)

TABLE 11.1 (*Continued*)
Published Articles about Corrosion Inhibition in Phosphoric Acid by Plant Extract

Plant	Extraction Method	Solvent	Metal	Milieu	Result	References
Allium sativum *Tilia cordata* *Foeniculum vulgare* *Valoniopsis* *pachynema*	Maceration Soxhlet	Methanol Ethanol	Copper Brass	8 M H$_3$PO$_4$ 0.1 N H$_3$PO$_4$	The copper electropolishing performance in orthophosphoric solution was evaluated by the galvanostatic polarization technique. Adding methanolic extracts to the electropolishing medium resulted in a reduced current limit. Polished samples were analyzed for morphological surface roughness and gloss. The high reflecting ability and inhibiting effectiveness of garlic extract were observed. A significant reflecting characteristic was reached by adding the various plant methanolic extracts to the electropolishing solution. The data on the gravimetric study are in perfect accordance with the electrochemical measures. *Valoniopsis pachynema* extract's performance in inhibiting brass corrosion in the H$_3$PO$_4$ medium was analyzed. The efficiency of inhibition indicated a favorable correlation among the gravimetric approach, potentiodynamic polarization, and electrochemical impedance spectroscopy methods. Inhibitor molecules adsorb onto the surface of brass in exothermic, spontaneous physical processes.	[105,106]
Garcinia indica	Maceration	Water	Aluminum	(0.5 M, 1.25 M, 2 M) H$_3$PO$_4$	*Garcinia indica* Choisy extract's inhibitory performance was investigated on aluminum in an H$_3$PO$_4$ medium. Analysis of the surface was performed by SEM-EDS. A regression quadratic pattern was developed and validated, before optimizing the parameters for optimum inhibitory efficacy. Optimal parameter settings were identified in terms of extract concentration and temperature (0.5 g L^{-1}, 50°F) to attain the highest efficiency of 86.19%. The predicted results revealed harmonious correlations with the experimental data.	[107]

(*Continued*)

TABLE 11.1 (Continued)
Published Articles about Corrosion Inhibition in Phosphoric Acid by Plant Extract

Plant	Extraction Method	Solvent	Metal	Milieu	Result	References
Tribulus terrestris	Soxhlet	Water	Mild steel	1 N H_3PO_4	The inhibitory performance of *Acalypha indica* L. aqueous extract against mild steel corrosion in 1.0 N H_3PO_4 medium was investigated *via* gravimetric and PDP techniques from 303 to 333 K. The corrosion rate rises with temperature increasing and inhibitor concentration decreasing relative to blank. In an H_3PO_4 media, the extract also acted as an inhibitor mixed. Surface analysis was also carried out to determine the corrosion inhibition mechanism.	[108]
Eucalyptus plant	Hydrodistillation	Water	Mild steel	0.5 M H_3PO_4 0.5 M H_2SO_4	The corrosion-inhibiting and adsorbing action of *Eucalyptus* leaf extract in sulfuric and phosphoric solutions was studied. PDP plots indicated a mixed inhibitor behavior of the extracts in both acid solutions. Based on isotherms of adsorption, in particular, the kinetic-thermodynamic, the Langmuir, the Flory Huggins, and the Temkin models, the inhibition process was analyzed. The analysis revealed a controlled activation of the corrosion mechanism.	[109]
Artemisia herba-alba	Hydrodistillation	Water	Copper	2 M H_3PO_4 with 0.3 M NaCl	*Artemisia* oil's inhibitory properties were assessed in a highly corrosive environment under various temperatures. The analysis revealed that AO is a valuable inhibitor. Natural oil effectively decreases the rate of copper corrosion. The efficiency of inhibition calculated from the gravimetric approach, EIS, and PDP plots are very similar to each other.	[110]
Argania spinosa	Hydrodistillation	Water	Copper	2 M H_3PO_4 with 3×10^{-1M} NaCl	Argan oil obtained from *Argania spinosa* has high phosphoric acid inhibition efficiency. Higher concentration improves charge transfer resistance, resulting in increased inhibition efficiency. Adsorption processes for both extracts were mostly physisorption. PDP experiments indicated that the oils acted as a mixed inhibitor.	[111]

REFERENCES

[1] Grengg, C.; Mittermayr, F.; Ukrainczyk, N.; Koraimann, G.; Kienesberger, S.; Dietzel, M., Advances in concrete materials for sewer systems affected by microbial induced concrete corrosion: A review. *Water Research* **2018**, 134, 341–352.

[2] Manam, N. S.; Harun, W. S. W.; Shri, D. N. A.; Ghani, S. A. C.; Kurniawan, T.; Ismail, M. H.; Ibrahim, M. H. I., Study of corrosion in biocompatible metals for implants: A review. *Journal of Alloys and Compounds* **2017**, 701, 698–715.

[3] Marzorati, S.; Verotta, L.; Trasatti, S., Green corrosion inhibitors from natural sources and biomass wastes. *Molecules* **2018**, 24 (1), 48.

[4] Ardelean, M.; Lascău, S.; Ardelean, E.; Josan, A., Surface treatments for aluminium alloys. *IOP Conference Series: Materials Science and Engineering* **2018**, 294, 012042.

[5] Hamdani, I.; Mokhtari, O.; Lamri, L.; Zaoui, S.; Bouknana, D.; Aouniti, A.; Berrabah, M.; Bouyanzer, A.; Hammouti, B., Bibliographic review on the problem of corrosion and their protection by green inhibitors. *Arabian Journal of Chemical and Environmental Researches* **2018**, 5, 101–123.

[6] Omer, A. M., Energy, environment and sustainable development. *Renewable and Sustainable Energy Reviews* **2008**, 12 (9), 2265–2300.

[7] Roberge, P. R., *Handbook of Corrosion Engineering.* 3rd edition; McGraw-Hill Education: New York, **2019**.

[8] Puthalath R., Murthy C.S.N., Surendranathan A.O. Reservoir formation damage during various phases of oil and gas recovery—An overview. International Journal of Earth Sciences and Engineering. 2012;5(2):224–231.

[9] Singhvi, M. S.; Zinjarde, S. S.; Gokhale, D. V., Polylactic acid: Synthesis and biomedical applications. *The Journal of Applied Microbiology* **2019**, 127 (6), 1612–1626.

[10] Sørensen, P. A.; Kiil, S.; Dam-Johansen, K.; Weinell, C. E., Anticorrosive coatings: A review. *Journal of Coatings Technology and Research* **2009**, 6 (2), 135–176.

[11] Eskişehir Osmangazi, U.; Topçu, İ. B.; Uzunömeroğlu, A., Properties of corrosion inhibitors on reinforced concrete. *Journal of Structural Engineering & Applied Mechanics* **2020**, 3 (2), 93–109.

[12] Umoren, S. A.; Solomon, M. M., Synergistic corrosion inhibition effect of metal cations and mixtures of organic compounds: A Review. *Journal of Environmental Chemical Engineering* **2017**, 5 (1), 246–273.

[13] Raja, P. B.; Ismail, M.; Ghoreishiamiri, S.; Mirza, J.; Ismail, M. C.; Kakooei, S.; Rahim, A. A., Reviews on corrosion inhibitors: A short view. *Chemical Engineering Communications* **2016**, 203 (9), 1145–1156.

[14] Verma, D. K.; Aslam, R.; Aslam, J.; Quraishi, M. A.; Ebenso, E. E.; Verma, C., Computational modeling: Theoretical predictive tools for designing of potential organic corrosion inhibitors. *Journal of Molecular Structure* **2021**, 1236, 130294.

[15] Badji, M.; Gassama, D.; Bodian, M.; Sylla-Gueye, R.; Cissé, K.; Fall, M., Use of attapulgite as a corrosion inhibitor for industrial metals and alloys: Case of aluminum, zinc and aluzinc in a 0.5 M hydrochloric acid solution. *Material Science* **2022**, 29, 195–200.

[16] Hooshmand Zaferani, S.; Sharifi, M.; Zaarei, D.; Shishesaz, M. R., Application of eco-friendly products as corrosion inhibitors for metals in acid pickling processes - A review. *Journal of Environmental Chemical Engineering* **2013**, 1 (4), 652–657.

[17] Yang, J.; Jiang, P.; Qiu, Y.; Jao, C.-Y.; Blawert, C.; Lamaka, S.; Bouali, A.; Lu, X.; Zheludkevich, M. L.; Li, W., Experimental and quantum chemical studies of carboxylates as corrosion inhibitors for AM50 alloy in pH neutral NaCl solution. *Journal of Magnesium and Alloys* **2022**, 10 (2), 555–568.

[18] Loto, R. T.; Loto, C. A.; Popoola, A. P. I., Corrosion inhibition of thiourea and thiadiazole derivatives: A review. *Journal of Materials and Environmental Science* **2012**, 3, 885–894.

[19] Patel, N. S.; Jauhariand, S.; Mehta, G. N.; Al-Deyab, S. S.; Warad, I.; Hammouti, B., Mild steel corrosion inhibition by various plant extracts in 0.5 M sulphuric acid. *International Journal of Electrochemical Science* **2013**, 8, 2635–2655.

[20] Harvey, T. J.; Walsh, F. C.; Nahlé, A. H., A review of inhibitors for the corrosion of transition metals in aqueous acids. *Journal of Molecular Liquids* **2018**, 266, 160–175.

[21] Dabrowski, A., Adsorption from theory to practice. *Advances in Colloid and Interface Science* **2001**, 93, 135–224.

[22] Obi-Egbedi, N. O.; Obot, I. B., Inhibitive properties, thermodynamic and quantum chemical studies of alloxazine on mild steel corrosion in H_2SO_4. *Corrosion Science* **2011**, 53 (1), 263–275.

[23] Agboola, O. D.; Benson, N. U., Physisorption and chemisorption mechanisms influencing micro (nano) plastics-organic chemical contaminants interactions: A review. *Frontiers in Environmental Science* **2021**, 9, 678574.

[24] Kachel, S. R.; Klein, B. P.; Morbec, J. M.; Schöniger, M.; Hutter, M.; Schmid, M.; Kratzer, P.; Meyer, B.; Tonner, R.; Gottfried, J. M., Chemisorption and physisorption at the metal/organic interface: Bond energies of naphthalene and azulene on coinage metal surfaces. *The Journal of Physical Chemistry C* **2020**, 124 (15), 8257–8268.

[25] Fouda, A. S.; Elmorsi, M. A.; Abou-Elmagd, B. S., Adsorption and inhibitive properties of methanol extract of Eeuphorbia heterophylla for the corrosion of copper in 0.5 M nitric acid solutions. *Polish Journal of Chemical Technology* **2017**, 19 (1), 95–103.

[26] Khanra, A.; Srivastava, M.; Rai, M. P.; Prakash, R., Application of unsaturated fatty acid molecules derived from microalgae toward mild steel corrosion inhibition in HCl solution: A novel approach for metal-inhibitor association. *ACS Omega* **2018**, 3 (10), 12369–12382.

[27] Jiang, S.; Chai, F.; Su, H.; Yang, C., Influence of chromium on the flow-accelerated corrosion behavior of low alloy steels in 3.5% NaCl solution. *Corrosion Science* **2017**, 123, 217–227.

[28] Hossain, N.; Asaduzzaman Chowdhury, M.; Kchaou, M., An overview of green corrosion inhibitors for sustainable and environment friendly industrial development. *Journal of Adhesion Science and Technology* **2021**, 35 (7), 673–690.

[29] Kesavan, D.; Gopiraman, M.; Sulochana, N.; Sulochana, N., Green inhibitors for corrosion of metals: A review. *Chemical Science Review and Letters* **2012**, 1, 1–8.

[30] Farahati, R.; Mousavi-Khoshdel, S. M.; Ghaffarinejad, A.; Behzadi, H., Experimental and computational study of penicillamine drug and cysteine as water-soluble green corrosion inhibitors of mild steel. *Progress in Organic Coatings* **2020**, 142, 105567.

[31] Srivastava, V.; Haque, J.; Verma, C.; Singh, P.; Lgaz, H.; Salghi, R.; Quraishi, M. A., Amino acid based imidazolium zwitterions as novel and green corrosion inhibitors for mild steel: Experimental, DFT and MD studies. *Journal of Molecular Liquids* **2017**, 244, 340–352.

[32] Bouiti, K.; aldeen Al-sharabi, H.; Bensemlali, M.; Bouhlal, F.; Abidi, B.; Labjar, N.; Laasri, S.; El Hajjaji, S., Effect of temperature on corrosion inhibition by ethanolic extract of Eriobotrya japonica seeds in chloride medium 1M. *The European Physical Journal Applied Physics* **2022**, 97, 67.

[33] Bouiti, K.; aldeen Al-sharabi, H.; Bouhlal, F.; Labjar, N.; Dahrouch, A.; Mahi, M. E.; Lotfi, E. M.; El Otmani, B.; Benabdellah, G. A.; El Hajjaji, S., Use of the ethanolic extract from Eriobotrya japonica seeds as a corrosion inhibitor of C38 in a 1 M HCl medium. *International Journal of Corrosion and Scale Inhibition* **2022**, 11 (3), 1319–1334.

[34] Feng, L.; Yang, H.; Wang, F., Experimental and theoretical studies for corrosion inhibition of carbon steel by imidazoline derivative in 5% NaCl saturated $Ca(OH)_2$ solution. *Electrochimica Acta* **2011**, 58, 427–436.

[35] Singh, P.; Srivastava, V.; Quraishi, M. A., Novel quinoline derivatives as green corrosion inhibitors for mild steel in acidic medium: Electrochemical, SEM, AFM, and XPS studies. *Journal of Molecular Liquids* **2016**, 216, 164–173.

[36] Ahmed, E. S. J.; Ganesh, G. M., A comprehensive overview on corrosion in RCC and its prevention using various green corrosion inhibitors. *Buildings* **2022**, 12 (10), 1682.

[37] El Wanees, S.; Alahmdi, M.; Alsharif, M.; Atef, Y., Mitigation of hydrogen evolution during zinc corrosion in aqueous acidic media using 5-amino-4-imidazolecarboxamide. *Egyptian Journal of Chemistry* **2018**, 62 (5), 811–825.

[38] Finšgar, M.; Jackson, J., Application of corrosion inhibitors for steels in acidic media for the oil and gas industry: A review. *Corrosion Science* **2014**, 86, 17–41.

[39] Kina, A. Y.; Ponciano, J. A. C., Inhibition of carbon steel CO_2 corrosion in high salinity solutions. *International Journal of Electrochemical Science* **2013**, 8, 12600–12612.

[40] Wade, S. A.; Lizama, Y., Clarke's solution cleaning used for corrosion product removal: Effects on carbon steel substrate. *Microscopy and Microanalysis* **2015**, 12, 170–177.

[41] Zhao, X.; Zhuang, H.; Yoon, S.-C.; Dong, Y.; Wang, W.; Zhao, W., Electrical impedance spectroscopy for quality assessment of meat and fish: A review on basic principles, measurement methods, and recent advances. *Journal of Food Quality* **2017**, 2017, 1–16.

[42] Feliu, S., Electrochemical impedance spectroscopy for the measurement of the corrosion rate of magnesium alloys: Brief review and challenges. *Metals* **2020**, 10 (6), 775.

[43] Meddings, N.; Heinrich, M.; Overney, F.; Lee, J.-S.; Ruiz, V.; Napolitano, E.; Seitz, S.; Hinds, G.; Raccichini, R.; Gaberšček, M.; Park, J., Application of electrochemical impedance spectroscopy to commercial Li-ion cells: A review. *Journal of Power Sources* **2020**, 480, 228742.

[44] Boukamp, B., Electrochemical impedance spectroscopy in solid state ionics: Recent advances. *Solid State Ionics* **2004**, 169 (1–4), 65–73.

[45] Huang, J.; Li, Z.; Liaw, B. Y.; Zhang, J., Graphical analysis of electrochemical impedance spectroscopy data in Bode and Nyquist representations. *Journal of Power Sources* **2016**, 309, 82–98.

[46] Abdel-Rehim, S. S.; Khaled, K. F.; Abd-Elshafi, N. S., Electrochemical frequency modulation as a new technique for monitoring corrosion inhibition of iron in acid media by new thiourea derivative. *Electrochimica Acta* **2006**, 51 (16), 3269–3277.

[47] Hu, J.-M.; Zhang, J.-T.; Zhang, J.-Q.; Cao, C.-N., Corrosion electrochemical characteristics of red iron oxide pigmented epoxy coatings on aluminum alloys. *Corrosion Science* **2005**, 47 (11), 2607–2618.

[48] Njoku, D. I.; Chidiebere, M. A.; Oguzie, K. L.; Ogukwe, C. E.; Oguzie, E. E., Corrosion inhibition of mild steel in hydrochloric acid solution by the leaf extract of Nicotiana tabacum. *Advances in Materials and Corrosion* **2013**, 1, 54–61.

[49] Cantrell, D. R.; Inayat, S.; Taflove, A.; Ruoff, R. S.; Troy, J. B., Incorporation of the electrode-electrolyte interface into finite-element models of metal microelectrodes. *Journal of Neural Engineering* **2008**, 5 (1), 54–67.

[50] Chauhan, D. S.; Ansari, K. R.; Sorour, A. A.; Quraishi, M. A.; Lgaz, H.; Salghi, R., Thiosemicarbazide and thiocarbohydrazide functionalized chitosan as ecofriendly corrosion inhibitors for carbon steel in hydrochloric acid solution. *International Journal of Biological Macromolecules* **2018**, 107, 1747–1757.

[51] Al-Qahtani, N.; Qi, J.; Abdullah, A. M.; Laycock, N. J.; Ryan, M. P., A review: Basics of electrochemical-thermodynamics for FeS scale formation. *International Journal of Science and Engineering Investigations* **2021**, 10 (115), 1–18.

[52] Obot, I. B.; Onyeachu, I. B.; Zeino, A.; Umoren, S. A., Electrochemical noise (EN) technique: Review of recent practical applications to corrosion electrochemistry research. *Journal of Adhesion Science and Technology* **2019**, 33 (13), 1453–1496.

[53] Scully, J. R., Polarization resistance method for determination of instantaneous corrosion rates. *Corrosion* **2000**, 56 (2), 199–218.

[54] Ponz, E.; Ladaga, J. L.; Bonetto, R. D., Measuring surface topography with scanning electron microscopy. I. EZEImage: A program to obtain 3D surface data. *Microscopy and Microanalysis* **2006**, 12 (2), 170–177.

[55] Inkson, B. J., Scanning electron microscopy (SEM) and transmission electron microscopy (TEM) for materials characterization. In *Materials Characterization Using Nondestructive Evaluation (NDE) Methods*, Woodhead Publishing: **2016**; pp. 17–43.

[56] Brydson, R., Electron energy loss spectroscopy. In *Bios in Association with the Royal Microscopical Society*, Taylor & Francis: Oxford, **2001**; p. 137. https://doi.org/10.1201/9781003076858

[57] Finšgar, M., Electrochemical, 3D topography, XPS, and ToF-SIMS analyses of 4-methyl-2-phenylimidazole as a corrosion inhibitor for brass. *Corrosion Science* **2020**, 169, 108632.

[58] Haldhar, R.; Prasad, D.; Saxena, A., Myristica fragrans extract as an eco-friendly corrosion inhibitor for mild steel in 0.5 M H_2SO_4 solution. *Journal of Environmental Chemical Engineering* **2018**, 6 (2), 2290–2301.

[59] Li, X.; Deng, S.; Fu, H., Synergistic inhibition effect of red tetrazolium and uracil on the corrosion of cold rolled steel in H_3PO_4 solution: Weight loss, electrochemical, and AFM approaches. *Materials Chemistry and Physics* **2009**, 115 (2–3), 815–824.

[60] Zarrok, H.; Zarrouk, A.; Hammouti, B.; Salghi, R.; Jama, C.; Bentiss, F., Corrosion control of carbon steel in phosphoric acid by purpald - Weight loss, electrochemical and XPS studies. *Corrosion Science* **2012**, 64, 243–252.

[61] Corrales Luna, M.; Le Manh, T.; Cabrera Sierra, R.; Medina Flores, J. V.; Lartundo Rojas, L.; Arce Estrada, E. M., Study of corrosion behavior of API 5L X52 steel in sulfuric acid in the presence of ionic liquid 1-ethyl 3-methylimidazolium thiocyanate as corrosion inhibitor. *Journal of Molecular Liquids* **2019**, 289, 111106.

[62] Bouanis, M.; Tourabi, M.; Nyassi, A.; Zarrouk, A.; Jama, C.; Bentiss, F., Corrosion inhibition performance of 2,5-bis(4-dimethylaminophenyl)-1,3,4-oxadiazole for carbon steel in HCl solution: Gravimetric, electrochemical and XPS studies. *Applied Surface Science* **2016**, 389, 952–966.

[63] Singh, A.; Ansari, K. R.; Chauhan, D. S.; Quraishi, M. A.; Kaya, S., Anti-corrosion investigation of pyrimidine derivatives as green and sustainable corrosion inhibitor for N80 steel in highly corrosive environment: Experimental and AFM/XPS study. *Sustainable Chemistry and Pharmacy* **2020**, 16, 100257.

[64] Kasaeian, M.; Ghasemi, E.; Ramezanzadeh, B.; Mahdavian, M.; Bahlakeh, G., A combined experimental and electronic-structure quantum mechanics approach for studying the kinetics and adsorption characteristics of zinc nitrate hexahydrate corrosion inhibitor on the graphene oxide nanosheets. *Applied Surface Science* **2018**, 462, 963–979.

[65] Liang, H. F.; Smith, C. T. G.; Mills, C. A.; Silva, S. R. P., The band structure of graphene oxide examined using photoluminescence spectroscopy. *Journal of Materials Chemistry C* **2015**, 3 (48), 12484–12491.

[66] Lukovits, I.; Kálmán, E.; Zucchi, F., Corrosion inhibitors-correlation between electronic structure and efficiency. *Corrosion* **2001**, 57 (1), 3–8.

[67] Guo, L.; Zhang, R.; Tan, B.; Li, W.; Liu, H.; Wu, S., Locust Bean Gum as a green and novel corrosion inhibitor for Q235 steel in 0.5 M H_2SO_4 medium. *Journal of Molecular Liquids* **2020**, 310, 113239.

[68] Salleh, S. Z.; Yusoff, A. H.; Zakaria, S. K.; Taib, M. A. A.; Abu Seman, A.; Masri, M. N.; Mohamad, M.; Mamat, S.; Ahmad Sobri, S.; Ali, A.; Teo, P. T., Plant extracts as green corrosion inhibitor for ferrous metal alloys: A review. *Journal of Cleaner Production* **2021**, 304, 127030.

[69] Tan, J.; Guo, L.; Yang, H.; Zhang, F.; El Bakri, Y., Synergistic effect of potassium iodide and sodium dodecyl sulfonate on the corrosion inhibition of carbon steel in HCl medium: A combined experimental and theoretical investigation. *RSC Advances* **2020**, 10 (26), 15163–15170.

[70] Pal, A.; Das, C., A novel use of solid waste extract from tea factory as corrosion inhibitor in acidic media on boiler quality steel. *Industrial Crops and Products* **2020**, 151, 112468.

[71] Hsissou, R.; Benhiba, F.; Khudhair, M.; Berradi, M.; Mahsoune, A.; Oudda, H.; El Harfi, A.; Obot, I. B.; Zarrouk, A., Investigation and comparative study of the quantum molecular descriptors derived from the theoretical modeling and Monte Carlo simulation of two new macromolecular polyepoxide architectures TGEEBA and HGEMDA. *Journal of King Saud University - Science* **2020**, 32 (1), 667–676.

[72] Shrestha, P. R.; Oli, H. B.; Thapa, B.; Chaudhary, Y.; Gupta, D. K.; Das, A. K.; Nakarmi, K. B.; Singh, S.; Karki, N.; Yadav, A. P., Bark extract of Lantana camara in 1M HCl as green corrosion inhibitor for mild steel. *Engineering Journal* **2019**, 23 (4), 205–211.

[73] Marsoul, A.; Ijjaali, M.; Elhajjaji, F.; Taleb, M.; Salim, R.; Boukir, A., Phytochemical screening, total phenolic and flavonoid methanolic extract of pomegranate bark (Punica granatum L): Evaluation of the inhibitory effect in acidic medium 1 M HCl. *Materials Today: Proceedings* **2020**, 27, 3193–3198.

[74] Pham, H.; Nguyen, V.; Vuong, Q.; Bowyer, M.; Scarlett, C., Effect of extraction solvents and drying methods on the physicochemical and antioxidant properties of Helicteres hirsuta Lour. leaves. *Technologies* **2015**, 3 (4), 285–301.

[75] Duan, H.; Wang, D.; Li, Y., Green chemistry for nanoparticle synthesis. *Chemical Society Reviews* **2015**, 44 (16), 5778–5792.

[76] Scavo, A.; Pandino, G.; Restuccia, A.; Mauromicale, G., Leaf extracts of cultivated cardoon as potential bioherbicide. *Scientia Horticulturae* **2020**, 261, 109024.

[77] Chauhan, D. S.; Quraishi, M. A.; Qurashi, A., Recent trends in environmentally sustainable sweet corrosion inhibitors. *Journal of Molecular Liquids* **2021**, 326, 115117.

[78] Capello, C.; Fischer, U.; Hungerbühler, K., What is a green solvent? A comprehensive framework for the environmental assessment of solvents. *Green Chemistry* **2007**, 9 (9), 927.

[79] Dai, J.; Mumper, R. J., Plant phenolics: Extraction, analysis and their antioxidant and anticancer properties. *Molecules* **2010**, 15 (10), 7313–7352.

[80] Shirmohammadli, Y.; Efhamisisi, D.; Pizzi, A., Tannins as a sustainable raw material for green chemistry: A review. *Industrial Crops and Products* **2018**, 126, 316–332.

[81] Mo, S.; Luo, H.-Q.; Li, N.-B., Plant extracts as "green" corrosion inhibitors for steel in sulphuric acid. *Chemical Papers* **2016**, 70 (9), 1131–1143.

[82] Sukhdev Swami, H.; Suman Preet Singh, K.; Gennaro, L.; Dev Dutt, R., Maceration, percolation and infusion techniques for the extraction of medicinal and aromatic plants. In *Extraction Technologies for Medicinal and Aromatic Plants*, United Nations Industrial Development Organization and the International Centre for Science and High Technology: India, **2008**; 67, 32–35.

[83] Azwanida, N. N., A review on the extraction methods use in medicinal plants, principle, strength and limitation. *Medicinal and Aromatic Plants* **2015**, 4: 196.

[84] Babu, K. G. D.; Kaul, V. K., Variation in essential oil composition of rose-scented geranium (Pelargonium sp.) distilled by different distillation techniques. *Flavour and Fragrance Journal* **2005**, 20 (2), 222–231.

[85] Masango, P., Cleaner production of essential oils by steam distillation. *Journal of Cleaner Production* **2005**, 13 (8), 833–839.

[86] Zam, W.; Bashour, G.; Abdelwahed, W.; Khayata, W., Separation and purification of proanthocyanidins extracted from pomegranate's peels (Punica Granatum). *PCI-Approved-IJPSN* **1970**, 5 (3), 1808–1813.

[87] Ouadi, Y. E.; Manssouri, M.; Bouyanzer, A.; Majidi, L.; Lahhit, N.; Bendaif, H.; Costa, J.; Chetouani, A.; Elmsellem, H.; Hammouti, B., Essential oil composition and antifungal activity of Salvia officinalis originating from North-East Morocco, against postharvest phytopathogenic fungi in apples. *Der Pharma Chemica* **2015**, 7 (9), 95–102.

[88] Mukherjee, P. K., Extraction and other downstream procedures for evaluation of herbal drugs. In *Quality Control and Evaluation of Herbal Drugs: Evaluating Natural Products and Traditional Medicine; Mukherjee, PK, Ed*, **2019**, pp. 195–236.

[89] Rojas, R.; Castro-López, C.; Sánchez-Alejo, E. J.; Niño-Medina, G.; Martínez-Ávila, G. C. G., Phenolic compound recovery from grape fruit and by- products: An overview of extraction methods. In *Grape and Wine Biotechnology*, Morata, A.; Loira, I., Eds., In Tech: **2016**.

[90] Frosi, I.; Montagna, I.; Colombo, R.; Milanese, C.; Papetti, A., Recovery of chlorogenic acids from agri-food wastes: Updates on green extraction techniques. *Molecules* **2021**, 26 (15), 4515.

[91] Cánovas, C. R.; Macías, F.; Pérez-López, R.; Basallote, M. D.; Millán-Becerro, R., Valorization of wastes from the fertilizer industry: Current status and future trends. *Journal of Cleaner Production* **2018**, 174, 678–690.

[92] Salas, B. V.; Wiener, M. S.; Martinez, J. R. S., Phosphoric acid industry: Problems and solutions. In *Phosphoric Acid Industry - Problems and Solutions*, Wiener, M. S.; Valdez, B., Eds., InTech: **2017**.

[93] Schorr, M.; Valdez, B., The phosphoric acid industry: Equipment, materials, and corrosion. *Corrosion Reviews* **2016**, 34 (1–2), 85–102.

[94] Phosphoric acid: Human health tier II assessment; 7664-38-2; IMAP Single Assessment Report: 2016/07/01/, 2016; p. 10.

[95] Kumar, R.; Chandrasekaran, V., Sargassum wightii extract as a green inhibitor for corrosion of brass in 0.1 N phosphoric acid solution. *Oriental Journal of Chemistry* **2015**, 31 (2), 939–949.

[96] Boudalia, M.; Fernández-Domene, R. M.; Tabyaoui, M.; Bellaouchou, A.; Guenbour, A.; García-Antón, J., Green approach to corrosion inhibition of stainless steel in phosphoric acid of Artemesia herba albamedium using plant extract. *Journal of Materials Research and Technology* **2019**, 8 (6), 5763–5773.

[97] Prabhu, D.; Padmalatha, R., Corrosion inhibition of 6063 aluminum alloy by Coriandrum sativum L seed extract in phosphoric acid medium. *Journal of Materials and Environmental Science* **2013**, 4 (5), 732–743.

[98] Belmaghraoui, W.; Mazkour, A.; Harhar, H.; Harir, M.; El Hajjaji, S., Investigation of corrosion inhibition of C38 steel in 5.5 M H_3PO_4 solution using *Ziziphus lotus* oil extract: An application model. *Anti-Corrosion Methods and Materials* **2019**, 66 (1), 121–126.

[99] Boudalia, M.; Guenbour, A.; Bellaouchou, A.; Laqhaili, A.; Mousaddak, M.; Hakiki, A.; Hammouti, B.; Ebenso, E. E., Corrosion inhibition of organic oil extract of leaves of Lanvandula stoekas on stainless steel in concentrated phosphoric acid solution. *International Journal of Electrochemical Science* **2013**, 8, 7414–7424.

[100] Victoria, S. N.; Prasad, R.; Manivannan, R., Psidium guajava leaf extract as green corrosion inhibitor for mild steel in phosphoric acid. *International Journal of Electrochemical Science* **2015**, 10, 2220–2238.

[101] Al-Moubaraki, A. H., Potential of borage flowers aqueous extract, *Borago officinalis* L., against the corrosion of mild steel in phosphoric acid. *Anti-Corrosion Methods and Materials* **2018**, 65 (1), 53–65.

[102] Jayanthi, K.; Sivaraju, M.; Kannan, K., Inhibiting properties of morpholine as corrosion inhibitor for mild steel in 2N sulphuric acid and phosphoric acid medium. *E-Journal of Chemistry* **2012**, 9 (4), 2213–2225.

[103] Gunasekaran, G.; Chauhan, L. R., Eco friendly inhibitor for corrosion inhibition of mild steel in phosphoric acid medium. *Electrochimica Acta* **2004**, 49 (25), 4387–4395.

[104] Al-Moubaraki, A. H.; Chaouiki, A.; Alahmari, J. M.; Al-hammadi, W. A.; Noor, E. A.; Al-Ghamdi, A. A.; Ko, Y. G., Development of natural plant extracts as sustainable inhibitors for efficient protection of mild steel: Experimental and first-principles multilevel computational methods. *Materials* **2022**, 15 (23), 8688.

[105] Taha, A. A.; Abouzeid, F. M.; Elsadek, M. M.; Habib, F. M., Effect of methanolic plant extract on copper electro-polishing in ortho-phosphoric acid. *Arabian Journal of Chemistry* **2020**, 13 (8), 6606–6625.

[106] Selva Kumar, R.; Chandrasekaran, V., Valoniopsis pachynema extract as a green inhibitor for corrosion of brass in 0.1 N phosphoric acid solution. *Metallurgical and Materials Transactions B* **2016**, 47 (2), 891–898.

[107] Prabhu, P. R.; Prabhu, D.; Rao, P., Analysis of Garcinia indica Choisy extract as eco-friendly corrosion inhibitor for aluminum in phosphoric acid using the design of experiment. *Journal of Materials Research and Technology* **2020**, 9 (3), 3622–3631.

[108] Sivaraju, M.; Kannan, K., Corrosion of mild steel in 1N Phosphoric acid with plant extract (Acalypha indica L.). *Material Science Research India* **2009**, 6, 125–136.

[109] Abdel-Gaber, A. M.; Rahal, H. T.; Beqai, F. T., Eucalyptus leaf extract as a eco-friendly corrosion inhibitor for mild steel in sulfuric and phosphoric acid solutions. *International Journal of Industrial Chemistry* **2020**, 11 (2), 123–132.

[110] Mounir, F.; Issami, S. E.; Bazzi, L.; Salghi, R.; Bammou, L.; Bazzi, L.; Eddine, A. C.; Jbara, O., Copper corrosion behavior in phosphoric acid containing chloride. *Ijrras* **2012**, 13, 574.

[111] Mounir, F.; Issami, S. E.; Bazzi, L.; Dassanayake, R.; Bazzi, L.; Chihab, A., Testing natural compound: Argan cake extract as ecofriendly inhibitor for copper corrosion in phosphoric acid. *Applied Journal of Environmental Engineering Science* **2016**, 2 (1), 2–1.

12 Plant Extracts as Corrosion Inhibitors for Copper in Various Electrolytes

Sanjukta Zamindar and Priyabrata Banerjee
Electric Mobility and Tribology Research Group, CSIR-
Central Mechanical Engineering Research Institute
Academy of Scientific and Innovative Research
(AcSIR), AcSIR Headquarters CSIR-HRDC Campus

Abhinay Thakur
Lovely Professional University

Ashish Kumar
Bihar Engineering University
Government of Bihar

12.1 INTRODUCTION

The economic impact of copper corrosion is a significant concern in various industries [1]. Corroded copper components often require costly maintenance, repairs, or even complete replacement [2]. These direct costs can be substantial, involving expenses for labour, materials, and equipment. Additionally, the downtime required for these maintenance activities can lead to production delays and decreased productivity, resulting in further financial losses. Moreover, the effects of copper corrosion extend beyond immediate economic consequences. Corrosion can compromise the integrity and functionality of structures and equipment, posing safety hazards and increasing the risk of accidents or failures. In industries such as construction, transportation, and infrastructure, the structural integrity of copper-based components is critical for ensuring the safety of individuals and the general public. Copper corrosion can also have environmental implications. Corroded copper can release metal ions into the surrounding environment, leading to contamination of soil, water, and air [3–5]. This contamination can have detrimental effects on ecosystems and human health. Additionally, the need for frequent replacement of corroded copper components contributes to resource depletion and waste generation, further impacting the environment.

DOI: 10.1201/9781003394631-12

- *Water Contamination*: Corroded copper pipes in plumbing systems can release copper ions into the water supply. This can lead to elevated levels of copper in drinking water, which poses a health risk for humans and aquatic organisms. High copper levels can cause gastrointestinal issues, liver damage, and kidney damage, and long-term exposure may even be carcinogenic [6].
- *Soil Contamination*: Copper corrosion can contaminate soil when copper-based materials, such as electrical wires or copper-based fungicides, degrade or leach into the ground. The metal ions released from corroded copper can accumulate in the soil, negatively impacting its quality and fertility. Plants may absorb these copper ions, leading to toxicity issues and reduced crop yields [7].
- *Air Pollution*: Corrosion of copper in industrial settings, such as power plants or factories, can release copper particles and dust into the air. These particles can contribute to air pollution, especially in areas with high copper corrosion rates or industrial activities. Inhalation of copper particles can cause respiratory problems, particularly in vulnerable populations such as children or individuals with pre-existing respiratory conditions [8].
- *Aquatic Ecosystems*: When copper ions leach into water bodies, such as rivers, lakes, or oceans, they can have detrimental effects on aquatic ecosystems. Copper is toxic to many aquatic organisms, including fish, invertebrates, and algae. It can disrupt the reproductive cycles of fish, impair growth and development, and harm the overall biodiversity of the aquatic ecosystem [9].
- *Contaminated Food Chain*: Copper released from corrosion can enter the food chain through various pathways. Plants grown in copper-contaminated soil may accumulate copper ions, which can be consumed by herbivores and subsequently transferred to higher trophic levels. This bioaccumulation of copper can pose risks to humans and wildlife, potentially leading to health issues and ecological imbalances [10,11].
- *Infrastructure Damage*: Copper corrosion can also have economic implications due to the degradation of infrastructure. Corroded copper pipes, electrical wiring, or other copper-based components can suffer structural damage, leading to leaks, electrical failures, or system malfunctions. Repairing or replacing corroded infrastructure can be costly, both in terms of financial resources and environmental impact [6,12].

These examples highlight how copper corrosion can extend beyond localized damage and have wide-ranging environmental implications, affecting ecosystems, human health, and overall sustainability. Additionally, here are some incidents where monuments have been damaged due to copper corrosion, listed in chronological order:

- Statue of Liberty (United States): 1984
  In 1984, it was discovered that the copper skin of the Statue of Liberty in New York City was corroded due to exposure to saltwater and air pollution.

The copper had turned green due to the formation of copper carbonate and copper sulfate. Restoration efforts were undertaken to repair and protect the monument.

- Big Ben (United Kingdom): 1990
  The iconic clock tower of Big Ben in London, England, suffered from copper corrosion in 1990. The copper roof and structures were corroded, requiring extensive restoration work to prevent further damage and ensure the monument's longevity.
- Brandenburg Gate (Germany): 2000
  The Brandenburg Gate in Berlin, Germany, underwent restoration in the year 2000 due to copper corrosion. The copper components, including the quadriga statue on top, had suffered from degradation, requiring repair and protective measures to preserve the monument.
- Eiffel Tower (France): 2005
  In 2005, it was discovered that the copper cladding on the Eiffel Tower in Paris had corroded over time. The copper had turned greenish, and the protective coating had deteriorated. Extensive refurbishment was carried out to replace the corroded copper elements and apply a new protective coating.
- St. Mark's Campanile (Italy): 2010
  St. Mark's Campanile, the bell tower in Venice, Italy, experienced copper corrosion in 2010. The copper roof and decorative elements deteriorated, leading to water leaks and structural concerns. Restoration efforts were undertaken to repair and protect the monument.

To address these issues, corrosion inhibition strategies have become a focus of research and development [13–15]. Corrosion inhibitors are substances that can be added to the corrosive environment to mitigate or prevent corrosion. They work by interfering with the electrochemical reactions occurring at the metal–electrolyte interface [16]. Corrosion inhibitors can form a protective film on the metal surface, acting as a barrier against corrosive agents. Alternatively, they can alter the electrochemical behaviour of the system, inhibiting the corrosion reactions. Corrosion inhibitors offer an effective means of reducing the economic impact of copper corrosion. By implementing corrosion inhibition strategies, industries can minimize maintenance and repair costs, prolong the service life of copper components, and enhance operational efficiency. Additionally, the use of corrosion inhibitors contributes to safer working environments and reduces the potential for accidents or failures caused by corroded copper. Furthermore, corrosion inhibitors have environmental benefits. By preventing or minimizing corrosion, the release of metal ions into the environment can be reduced, mitigating environmental contamination. Additionally, the use of corrosion inhibitors can extend the lifespan of copper components, reducing the need for frequent replacements and the associated resource consumption and waste generation.

Plant extracts have gained significant attention as potential corrosion inhibitors for copper due to their eco-friendly nature, abundance, low cost, and inherent inhibitive properties [17]. These extracts are derived from various parts of plants, such as leaves, stems, bark, and fruits, and consist of a complex mixture of organic compounds [18–23]. The use of plant extracts as corrosion inhibitors aligns with the principles of green chemistry and sustainable development [24]. Unlike traditional corrosion inhibitors that may contain toxic or environmentally harmful substances, plant extracts offer a renewable and environmentally friendly alternative. They are biodegradable and pose minimal risk to human health and the environment. Plant extracts contain a wide range of organic compounds that contribute to their inhibitive properties. These compounds include phenols, tannins, alkaloids, flavonoids, and terpenoids, among others [25,26]. Each of these compounds possesses unique characteristics and can interact with the metal surface, adsorb onto it, and form protective films that inhibit the corrosion process [27,28]. The inhibitive action of plant extracts is attributed to several mechanisms. Firstly, the organic compounds in the extracts can adsorb onto the metal surface, forming a barrier that hinders the diffusion of corrosive species and prevents their interaction with the metal [29]. Secondly, these compounds can interact with the metal ions released during the corrosion process, forming insoluble complexes that inhibit further corrosion. Additionally, plant extracts can modify the electrochemical behaviour of the metal, shifting the corrosion potential and reducing the corrosion rate [30,31]. The effectiveness of plant extracts as corrosion inhibitors for copper depends on various factors, including the composition of the extracts, concentration, exposure time, temperature, and the specific corrosive environment [32,33]. The selection of suitable plant species, optimization of extraction methods, and understanding of the structure-activity relationship of the organic compounds present in the extracts are essential for achieving optimal corrosion inhibition. It is worth noting that the inhibitive performance of plant extracts may vary depending on the specific corrosive environment. Factors such as pH, temperature, the presence of aggressive ions, and the nature of the electrolyte can influence the adsorption behaviour and film formation on the metal surface. Therefore, it is necessary to evaluate the performance of plant extract–based corrosion inhibitors under different conditions to ensure their effectiveness and reliability.

12.2 CORROSION PROCESSES IN COPPER

12.2.1 Overview of Copper Corrosion

- Copper, a widely used metal in various industries, is susceptible to corrosion when exposed to corrosive environments [34,35]. Corrosion is a complex process that involves the deterioration of metals through electrochemical reactions. Understanding the corrosion processes in copper is crucial for developing effective corrosion mitigation strategies. Copper corrosion can occur through various mechanisms, including uniform corrosion, localized corrosion (such as pitting corrosion), and galvanic corrosion [36–38]. Uniform corrosion is the most common form, where the metal corrodes uniformly over its entire surface. Localized corrosion, on the contrary, occurs in specific areas and can be more destructive, leading to the formation of pits or crevices. Galvanic corrosion arises when two dissimilar metals are in contact in the presence of an electrolyte, causing accelerated corrosion of the more reactive metal [39]. Copper corrosion is a widespread issue that affects various industries and applications worldwide. Here are some global instances that highlight the significance and impact of copper corrosion:

- *Water Distribution Systems*: Copper pipes are commonly used in water distribution systems due to their excellent corrosion resistance. However, corrosion can still occur, leading to pipe degradation and water quality issues. For example, in the United States, many older cities have experienced copper corrosion in their water supply systems, resulting in discoloration, taste changes, and elevated levels of copper in drinking water.

- *Marine Environments*: Copper is extensively used in marine applications, including shipbuilding, offshore structures, and coastal infrastructure. The corrosive nature of seawater poses a significant challenge for copper-based materials. For instance, in coastal regions such as the Arabian Gulf, where seawater salinity is high, copper corrosion can occur rapidly, impacting the structural integrity of marine installations and vessels.

- *Electrical Industry*: Copper is widely used in electrical wiring and transmission systems due to its excellent electrical conductivity. However, in areas with high humidity or aggressive industrial atmospheres, copper corrosion can occur on electrical contacts, leading to poor electrical performance, increased resistance, and potential failures. This is particularly critical in industries such as power generation, telecommunications, and electronics.

- *Industrial Processes*: Copper equipment and components used in industrial processes are susceptible to corrosion from various factors such as chemicals, high temperatures, and aggressive environments. For example, in the chemical industry, copper corrosion can occur in equipment exposed to corrosive acids or bases, impacting the efficiency and safety of the processes.

- *Architectural Applications*: Copper is a popular material in architectural designs, including roofs, facades, and decorative elements. However, atmospheric pollutants, weathering, and exposure to harsh environmental conditions can lead to copper corrosion and degradation of the aesthetic appearance. Notable instances of copper corrosion in architectural applications can be found in historic buildings and monuments around the world.

- *Heat Exchangers*: Copper is widely used in heat exchangers due to its excellent thermal conductivity. However, corrosion in heat exchangers can lead to reduced heat transfer efficiency and increased energy consumption. Instances of copper corrosion in heat exchangers can be found in industries such as HVAC systems, refrigeration, and chemical processing.

The corrosion of copper is a complex process that is influenced by several factors. Understanding these factors is crucial for effectively managing and preventing copper corrosion. The properties of the copper metal itself play a significant role in its corrosion behaviour. Copper is known for its excellent corrosion resistance due to the formation of a protective oxide layer on its surface. This oxide layer acts as a barrier, preventing further corrosion of the underlying metal. However, under certain conditions, this protective layer can be compromised, leading to corrosion. The nature of the corrosive environment also plays a crucial role in copper corrosion. The presence of certain substances in the environment can accelerate the corrosion process. One of the primary factors is the availability of dissolved oxygen. Oxygen acts as an oxidizing agent and facilitates the electrochemical reactions that lead to corrosion. In the absence of oxygen, corrosion is significantly slowed down. Aggressive ions, such as chloride ions, can also increase the corrosion rate of copper. Chloride ions can penetrate the protective oxide layer and initiate localized corrosion, such as pitting corrosion, which can lead to rapid metal degradation. The pH of the environment is another important factor influencing copper corrosion. Copper is more susceptible to corrosion in acidic and alkaline environments compared to neutral pH conditions. Acidic conditions can dissolve the protective oxide layer, exposing the underlying copper to corrosion. Alkaline conditions can lead to the formation of soluble copper complexes, which can promote corrosion. Maintaining the pH within an optimal range is crucial to prevent excessive corrosion.

External conditions such as temperature and humidity can also affect copper corrosion. Higher temperatures generally accelerate corrosion processes, as they increase the rate of chemical reactions. Additionally, high humidity levels can create a more corrosive environment by promoting the formation of electrolytes on the metal surface. Moisture can facilitate the transport of corrosive ions, leading to accelerated corrosion. To effectively manage copper corrosion, it is important to consider these factors and implement appropriate corrosion prevention strategies. This may include the use of protective coatings, corrosion inhibitors, and proper maintenance practices. Plant extracts, as discussed earlier, have shown potential as eco-friendly corrosion inhibitors for copper, providing an additional layer of protection against corrosion.

12.2.2 ELECTROCHEMICAL PRINCIPLES IN COPPER CORROSION

Understanding the electrochemical principles underlying copper corrosion is essential for comprehending the corrosion processes and designing effective corrosion inhibition strategies. Copper corrosion involves anodic and cathodic reactions that occur at the metal–electrolyte interface. Copper corrosion occurs through an electrochemical process that involves several fundamental principles. The mechanism can be explained using the following steps [40–44]:

Oxidation: Copper reacts with oxygen from the surrounding environment, resulting in the formation of copper oxide on the metal's surface. This oxidation reaction can be represented as follows:

$$2Cu + O_2 \rightarrow 2CuO$$

The copper oxide layer formed acts as a barrier, protecting the underlying copper from further corrosion.

Ionization: In the presence of moisture or water, the copper oxide layer can break down, exposing the underlying copper metal. Water molecules dissociate into hydrogen ions (H⁺) and hydroxide ions (OH⁻) through the process of ionization. The reaction can be represented as follows:

$$H_2O \rightarrow H+ \ +OH^-$$

Anodic Reaction: At localized areas on the copper surface, known as anodic sites, the copper metal undergoes oxidation, releasing copper ions (Cu^{2+}) into the surrounding solution. This anodic reaction can be represented as follows:

$$Cu \rightarrow Cu^{2+} + 2e^-$$

The copper ions are released into the surrounding environment.

Cathodic Reaction: Concurrently, at other sites on the copper surface, known as cathodic sites, reduction reactions occur. Oxygen or other oxidizing agents present in the environment react with the electrons (e−) released during the anodic reaction. The most common cathodic reaction involves the reduction of oxygen, resulting in the formation of hydroxide ions. The overall reaction can be represented as follows:

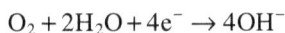

$$O_2 + 2H_2O + 4e^- \rightarrow 4OH^-$$

The hydroxide ions formed can react with the copper ions to form copper hydroxide.

Formation of Corrosion Products: Copper hydroxide ($Cu(OH)_2$) formed combines with carbon dioxide from the air to produce copper carbonate ($CuCO_3$). Copper carbonate often appears as greenish corrosion products commonly known as copper patina or verdigris.

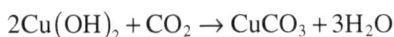

$$2Cu(OH)_2 + CO_2 \rightarrow CuCO_3 + 3H_2O$$

Continuation of the Process: The electrochemical process of copper corrosion continues as long as moisture, oxygen, and an electrolyte (such as water) are present. The cycle repeats with the anodic and cathodic reactions occurring at different sites on the copper surface, leading to the gradual degradation and corrosion of the metal. The electrochemical reactions occurring in copper corrosion are governed by principles such as Faraday's laws of electrolysis and the Nernst equation, which describe the relationship between the current, charge, and electrochemical potentials. Understanding these principles allows for the quantification and prediction of corrosion rates, as well as the determination of the effectiveness of corrosion inhibition strategies.

12.2.3 Importance of Corrosion Inhibition Strategies

Given the economic and environmental consequences of copper corrosion, the development of corrosion inhibition strategies is of great importance. Corrosion inhibitors are substances that can be added to the corrosive environment to suppress or retard the corrosion process. They function by interfering with the electrochemical reactions occurring at the metal–electrolyte interface, either by forming protective films on the metal surface or by altering the electrochemical behaviour. Corrosion inhibitors offer several benefits. They can extend the service life of copper components, reducing the need for frequent maintenance, repairs, and replacements. This results in cost savings for industries that utilize copper-based equipment or structures. Moreover, corrosion inhibitors can minimize production downtime caused by corrosion-related failures, improving operational efficiency and productivity. Corrosion inhibition strategies are of significant importance worldwide in various industries and applications. Here are some global instances that highlight the importance of corrosion inhibition strategies:

- *Oil and Gas Industry*: The oil and gas industry is highly susceptible to corrosion due to the aggressive nature of the environments involved, such as offshore platforms, pipelines, and refining facilities. Corrosion in this industry can lead to equipment failures, leaks, and environmental hazards. The implementation of corrosion inhibition strategies, including the use of inhibitors such as corrosion inhibitors, coatings, and cathodic protection, is crucial to ensure the safety, integrity, and longevity of the infrastructure and to prevent costly shutdowns and accidents [45,46].
- *Transportation Infrastructure*: Corrosion in transportation infrastructure, such as bridges, highways, and tunnels, can have significant economic and safety implications. For example, in coastal areas, the exposure of steel-reinforced concrete structures to chloride ions can lead to corrosion of the reinforcing steel, compromising the structural integrity of the infrastructure. Corrosion inhibition strategies, such as the application of corrosion inhibitors and protective coatings, are essential for mitigating corrosion and extending the service life of these critical assets [47,48].
- *Aerospace Industry*: The aerospace industry relies heavily on corrosion inhibition strategies to ensure the safety and reliability of aircraft. Corrosion can occur due to environmental exposure, including moisture, high-altitude conditions, and the presence of aggressive chemicals in the atmosphere. Effective corrosion inhibition measures, including protective coatings, surface treatments, and the use of corrosion inhibitors, are vital to prevent corrosion-related failures and maintain the structural integrity and performance of aircraft components [49,50].
- *Power Generation*: Corrosion is a significant concern in power generation facilities, including fossil fuel power plants, nuclear power plants, and renewable energy infrastructure. Corrosion can affect the efficiency and reliability of equipment such as boilers, heat exchangers, and turbines. Corrosion inhibition strategies, such as the use of corrosion inhibitors, protective coatings, and water treatment programmes, are critical for minimizing corrosion-related damages, optimizing energy production, and reducing maintenance and replacement costs [51].

- *Manufacturing and Industrial Processes*: Corrosion can impact various manufacturing and industrial processes, leading to reduced productivity, equipment failures, and increased maintenance costs. Industries such as chemical processing, metal fabrication, and electronics manufacturing rely on corrosion inhibition strategies to protect equipment and maintain production efficiency. The implementation of corrosion inhibitors, surface treatments, and corrosion-resistant materials is essential to prevent corrosion-related disruptions and ensure smooth operations [52].
- *Infrastructure in Coastal Environments*: Coastal environments pose particular challenges due to the corrosive nature of saltwater and the presence of chloride ions. Infrastructure such as ports, coastal buildings, and offshore structures is highly susceptible to corrosion. Corrosion inhibition strategies, including the use of protective coatings, cathodic protection systems, and corrosion-resistant materials, are crucial for preserving the integrity and durability of coastal infrastructure and minimizing maintenance and repair costs [53,54].

The choice of a corrosion inhibitor depends on factors such as the specific corrosive environment, the type of corrosion process occurring, and the desired inhibitive mechanism. Various types of corrosion inhibitors are available, including organic compounds, inorganic compounds, and naturally derived substances like plant extracts. Each type of inhibitor has its advantages, limitations, and compatibility with different corrosive environments. Developing effective corrosion inhibition strategies requires a thorough understanding of the corrosion mechanisms, the properties of the corrosive environment, and the behaviour of the inhibitors. The performance of corrosion inhibitors should be evaluated through laboratory experiments and field trials to ensure their effectiveness, compatibility, and long-term stability.

12.3   PLANT EXTRACTS AS CORROSION INHIBITORS FOR COPPER

12.3.1   Advantages of Plant Extracts as Corrosion Inhibitors

The utilization of plant extracts as corrosion inhibitors for copper offers several advantages compared to traditional inhibitors. These advantages contribute to their increasing popularity and the exploration of their potential in corrosion protection applications. One significant advantage of plant extracts is their eco-friendliness and sustainability [55,56]. Plant extracts are derived from renewable sources, making them an environmentally conscious choice. Unlike synthetic inhibitors that may contain toxic or harmful substances, plant extracts are generally considered safe and biodegradable. This aligns with the principles of green chemistry and sustainable development, making plant extracts a more desirable option for corrosion inhibition. Another advantage of plant extracts is their abundance and cost-effectiveness. Plants are widely available in nature, and their extracts can be obtained from various plant parts, such as leaves, stems, bark, and fruits. This accessibility makes plant extracts a cost-effective alternative to synthetic inhibitors, which may require complex synthesis processes or expensive raw materials.

Plant extracts also exhibit inherent inhibitive properties that make them suitable for corrosion protection. They contain a diverse range of organic compounds, such as phenols, tannins, alkaloids, flavonoids, and terpenoids, which contribute to their inhibitive action. These compounds can interact with the metal surface, adsorb onto it, and form protective films that inhibit corrosion. Additionally, plant extracts may possess antioxidant properties, which help in scavenging reactive species and reducing oxidative damage to the metal. Furthermore, plant extracts offer versatility in terms of their composition and properties. Different plant species contain varying compositions of organic compounds, allowing for a wide range of inhibitive properties. This versatility enables the selection and customization of plant extracts based on the specific corrosive environment and desired inhibitive mechanisms. It also provides opportunities for exploring synergistic effects by combining multiple plant extracts or combining them with other corrosion inhibitors.

12.3.2 CHALLENGES IN IMPLEMENTING PLANT EXTRACTS

While plant extracts show promise as corrosion inhibitors for copper, there are several challenges associated with their implementation in practical applications [24,57–60]. These challenges must be addressed to maximize the effectiveness and reliability of plant extract–based corrosion inhibition strategies. One significant challenge is the variability in the composition of plant extracts. The composition of plant extracts can vary depending on factors such as plant species, geographical location, extraction methods, and even seasonal variations. This variability can affect the consistency and reproducibility of their inhibitive performance. Standardization of extraction protocols, quality control measures, and characterization techniques are crucial for ensuring the reliability of plant extract–based corrosion inhibitors. Another challenge is the optimization of extraction methods to obtain maximum inhibitive properties. Different extraction techniques, such as solvent extraction, maceration, Soxhlet extraction, and ultrasound-assisted extraction, can yield extracts with varying compositions and inhibitive efficiencies. The selection of the appropriate extraction method and optimization of extraction parameters, such as solvent type, extraction time, temperature, and plant-to-solvent ratio, are essential for obtaining extracts with optimal inhibitive properties.

The stability and durability of plant extract–based corrosion inhibitors are also important considerations. Factors such as exposure time, temperature fluctuations, and the presence of aggressive species in the corrosive environment can impact the performance and effectiveness of these inhibitors over extended periods. Stability tests and optimization of formulation and application methods are crucial for enhancing the longevity and performance of plant extract–based corrosion inhibitors. Furthermore, compatibility issues may arise when incorporating plant extracts into practical corrosion protection systems. The compatibility of plant extracts with coating materials, additives, and other components of the corrosion protection system should be evaluated to ensure their synergistic interaction and long-term stability.

12.3.3 Extraction Methods and Characterization Techniques

The extraction of plant extracts for corrosion inhibition applications is a multi-step process that involves the isolation of active compounds from the plant material [61–63]. Various extraction methods can be employed, each offering distinct advantages and limitations. The selection of an appropriate extraction method depends on several factors, including the target compounds, the plant material, the desired yield, and the inhibitive properties required. One of the most commonly used extraction methods is solvent extraction, which involves the use of organic solvents to dissolve and extract the active compounds from the plant material. The choice of solvent is crucial and depends on factors such as the polarity of the target compounds and the solubility of the plant constituents. Several techniques can be employed in solvent extraction, including maceration, reflux, Soxhlet extraction, and ultrasound-assisted extraction. Each technique has its benefits in terms of efficiency, selectivity, and environmental impact. In addition to solvent extraction, other extraction techniques are also employed in the extraction of plant extracts for corrosion inhibition. These techniques include steam distillation, supercritical fluid extraction, microwave-assisted extraction, and solid-phase extraction. Steam distillation is commonly used for extracting essential oils from aromatic plants. Supercritical fluid extraction utilizes supercritical fluids such as carbon dioxide to extract compounds from plant material. Microwave-assisted extraction uses microwave irradiation to enhance the extraction process. Solid-phase extraction involves the use of solid sorbents to selectively retain and release target compounds.

The selection of the appropriate extraction method should consider various factors such as the desired yield of the target compounds, the specific compounds being targeted, the availability of equipment, and the scalability of the process. It is crucial to optimize the extraction conditions to ensure maximum extraction efficiency and the preservation of the inhibitive properties of the plant extracts. Once the plant extracts are obtained, their characterization becomes essential to understand their composition, properties, and inhibitive potential. Various characterization techniques can be employed to analyse the chemical composition and structure of plant extracts [64–66]. Spectroscopic methods such as Fourier-transform infrared spectroscopy (FTIR), nuclear magnetic resonance (NMR) spectroscopy, and mass spectrometry (MS) are commonly used to identify and analyse the functional groups and chemical constituents present in the extracts. Chromatographic techniques, such as high-performance liquid chromatography (HPLC) and gas chromatography-mass spectrometry (GC-MS), play a crucial role in the separation and identification of individual compounds present in the plant extracts. These techniques allow researchers to determine the concentration and purity of the active compounds and understand their distribution within the extract. In addition to spectroscopic and chromatographic techniques, imaging techniques such as scanning electron microscopy (SEM) and atomic force microscopy (AFM) can provide valuable insights into the surface morphology and film formation properties of the plant extracts. SEM allows for visualizing the surface topography and the presence of any protective films formed by the extracts. AFM provides high-resolution images of the surface, allowing for the observation of nanoscale features and film thickness [16].

Furthermore, the inhibitive performance of plant extracts can be evaluated using electrochemical techniques. Electrochemical methods such as potentiodynamic polarization, electrochemical impedance spectroscopy (EIS), and cyclic voltammetry are commonly employed to assess the inhibitive efficiency, corrosion rates, and electrochemical behaviour of copper in the presence of plant extract–based inhibitors. These techniques provide valuable information about the effectiveness of the extracts in suppressing the corrosion process and forming protective barriers on the copper surface. Overall, the extraction and characterization of plant extracts for corrosion inhibition applications require a multidisciplinary approach combining knowledge from chemistry, materials science, and electrochemistry. The selection of the appropriate extraction method and the utilization of characterization techniques enable researchers to identify and understand the active compounds present in the extracts and evaluate their inhibitive potential. These studies contribute to the development of sustainable and eco-friendly corrosion inhibition strategies using plant extracts.

12.4 EVALUATION OF PLANT EXTRACTS AS CORROSION INHIBITORS FOR COPPER

Plant extracts have emerged as promising alternatives for corrosion inhibition of copper due to their eco-friendly nature, abundance, and cost-effectiveness. Several studies have investigated the inhibitive properties of plant extracts from various regions around the world, showcasing their potential as corrosion inhibitors. Here are a few global examples:

- Neem (*Azadirachta indica*) extract: Neem is a widely distributed tree native to the Indian subcontinent. Studies have demonstrated the corrosion-inhibiting effect of neem extracts on copper in different corrosive environments. For example, researchers in India found that neem leaf extract effectively reduced the corrosion rate of copper in acidic media. The presence of active compounds such as alkaloids, flavonoids, and tannins in neem extract contributed to its inhibitive properties [67].
- Green tea (*Camellia sinensis*) extract: Green tea is popular worldwide and known for its antioxidant properties. Researchers investigated the inhibitive effect of green tea extract on copper alloy corrosion in 0.1 M Na_2SO_4. The study found that the extract significantly reduced the corrosion rate of copper and formed a protective film on the metal surface. The high content of polyphenols, particularly catechins, in green tea extract, contributed to its corrosion inhibition capabilities [68].
- Olive (*Olea europaea*) leaf extract: Olive trees are extensively cultivated in Mediterranean regions, and their leaves have been used for various medicinal purposes. Researchers in Greece evaluated the corrosion inhibition properties of olive leaf extract on copper in a chloride-containing environment [69]. The study found that the extract effectively reduced the corrosion rate and exhibited good adsorption behaviour on the copper surface. The presence of phenolic compounds, including oleuropein and hydroxytyrosol, in olive leaf extract played a crucial role in corrosion inhibition [70].

- Aloe vera (*Aloe barbadensis*) gel extract: *Aloe vera* is a succulent plant widely cultivated in different parts of the world. Researchers in Morocco investigated the inhibitive effect of *Aloe vera* gel extract on copper corrosion in chloride-containing solutions [71]. The study revealed that the gel extract significantly reduced the corrosion rate and improved the corrosion resistance of copper. The presence of bioactive compounds such as polysaccharides, flavonoids, and phenolic acids in the gel extract contributed to its inhibitive properties.
- *Xanthosoma* sp. leaf extract: Investigation on inhibition efficiency of *Xanthosoma* sp. leaf extracts (XLE) was made for copper corrosion in saltwater. At various temperatures of 303, 313, 323, and 333 K, different dosages of XLE (1%, 2%, 3%, 4%, and 5% v/v) were examined. At 300 K with an inhibitor concentration of 5% v/v, the maximum inhibition efficiency of 85% was achieved, whereas at 333 K with an inhibitor concentration of 1%v/v, the lowest inhibition proficiency of 18.11% was noted [72].

These examples highlight the global exploration of plant extracts as corrosion inhibitors for copper. They demonstrate the potential of plant extracts from different regions in providing effective corrosion protection for copper in various corrosive environments. The diverse composition of plant extracts from different plant species offers opportunities for the development of customized corrosion inhibition strategies based on regional plant resources. Additionally, several researchers such as El-Asri et al. utilized *Carissa macrocarpa* extract (ECM) for the protection of copper in $0.5\,M\,HNO_3$ medium [73]. ECM is mainly comprised of 31% of pentadecanal, 11% of tetradecanal, and 10% of tetradecane. The corrosion inhibition efficiency of ECM has been strongly demonstrated with the help of gravimetric analysis (*viz.* weight loss experiment) and electrochemical analysis (*viz.* EIS and PDP). The temperature effect on corrosion inhibition has been investigated. The corrosion current density (I_{corr}) value dramatically increases with an increase in temperature for a particular solution. The desorption process that is the lack of adsorption of the molecules is supposed to be responsible for the falling of inhibition efficiency of ECM at higher temperature, *vide.* From the PDP experiment, the I_{corr} for $0.5\,M\,HNO_3$ solution is $25.75\,\mu A\,cm^{-2}$ while the value of I_{corr} for the highest concentration of ECM ($3\,g\,L^{-1}$) is $2.31\,\mu A\,cm^{-2}$. It has been observed that with an increase in the concentration of ECM, I_{corr} value significantly decreases.

The theoretical investigations *like* density functional theory (DFT) and Monte Carlo (MC) simulation corroborate the experimental results. The highly negative adsorption energies of the three main components *viz.* pentadecanal, tetradecanal, and tetradecane and their perpendicular interaction correspond to better adsorption that is better surface coverage. The equilibrium adsorption images obtained from MC simulation have been represented in Figure 12.1.

The excellent inhibitory activity of *Atriplex leucoclada* extract (ALE) in 1 M HCl media has been investigated by Ahmed et al. [74]; 91.5% of inhibition efficiency was achieved with a mixed type of inhibition. Besides the electrochemical analysis, morphological analyses *like* SEM, AFM, and X-ray photoelectron spectroscopy (XPS) were carried out to understand the corrosion inhibition phenomena of ALE.

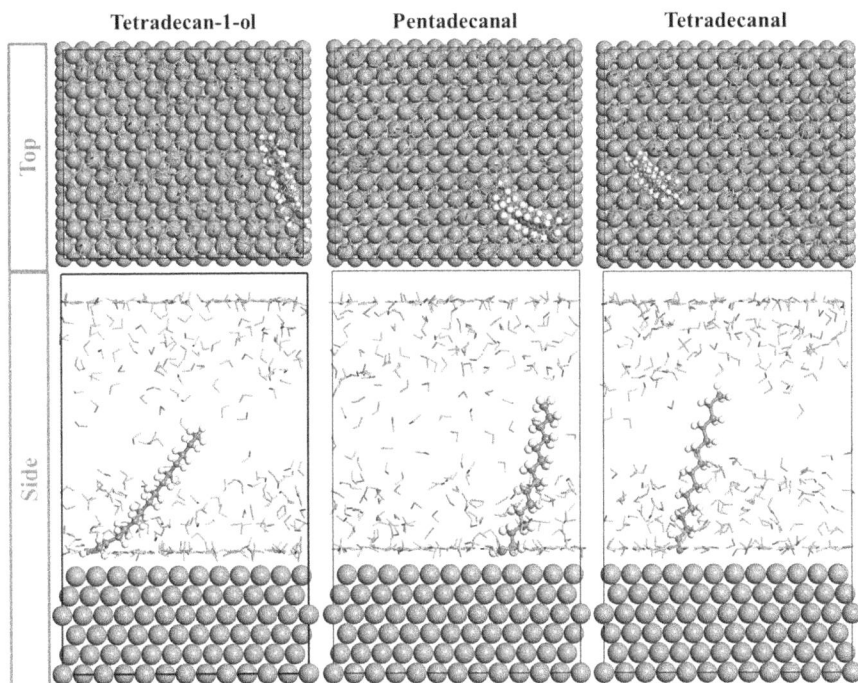

FIGURE 12.1 Equilibrium adsorption images obtained from MC simulation. Reproduced with permission from Ref. [73].

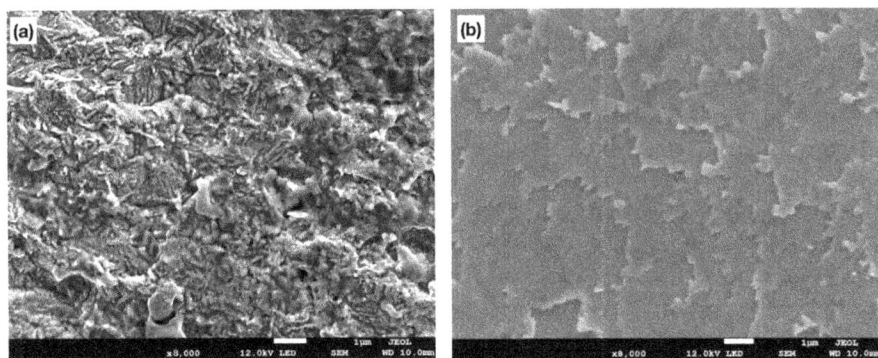

FIGURE 12.2 SEM images of copper surface after immersion in 1 M HCl for 24 hours (a) without ALE and (b) with 8 g L^{-1} ALE. Reproduced with permission from Ref. [74].

In Figure 12.2, it has been observed that the SEM image of the copper surface immersed within 1 M HCl for 24 hours without ALE shows several deteriorations, while the SEM image of the copper surface immersed within 1 M HCl for 24 hours with 8 g L^{-1} of ALE is significantly less deteriorated, that is, less corrosion occurs.

The AFM images of the copper surface immersed within 1 M HCl for 8 hours without ALE show several pits while the pits become shallow and small with the

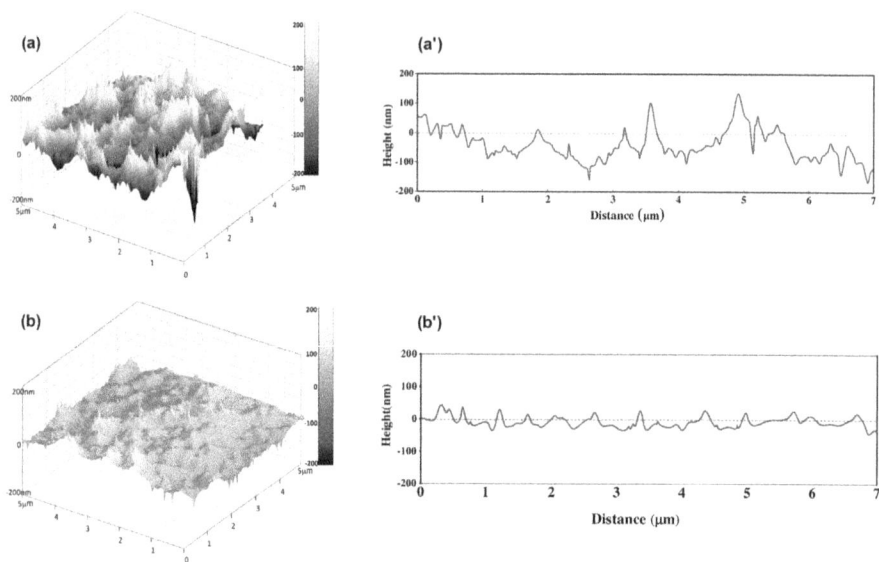

FIGURE 12.3   3D surface topography images of copper surface after immersion in 1 M HCl for 6 hours (a) without ALE and (b) with 8 g L^{-1} ALE and their corresponding height profiles (a') without ALE and (b') with 8 g L^{-1} ALE. Reproduced with permission from Ref. [74].

addition of 8 g L^{-1} ALE (*vide*, Figure 12.3a and b). The corresponding height profiles exhibit lots of roughness for the copper surface immersed within 1 M HCl for 8 hours without ALE (*vide*, Figure 12.3a'), while the roughness significantly diminished for the copper surface immersed within 1 M HCl for 8 hours with 8 g L^{-1} ALE (*vide*, Figure 12.3b').

Tan et al. reported papaya leave extract (PLE) as an efficient green corrosion inhibitor for the protection of Cu in the presence of 0.5 mol L^{-1} H$_2$SO$_4$ environment [75]. This extract acts as a mixed type of inhibitor at 298, 303, and 308 K temperature. The corrosion inhibition efficiency was evaluated by means of electrochemical investigation such as EIS and potentiodynamic polarization (PDP). The Nyquist plots without and with four different concentrations of PLE at three different temperatures have been presented in Figure 12.4. The charge transfer resistance (R_{ct}) and the electric double-layer capacitance (C_{dl}) lead in formation of one capacitive loop in the high-frequency region of the Nyquist plot. While in the low-frequency region, the Warburg impedance is brought on by either the dissolved oxygen (DO) in the bulk solution diffusing to the copper electrode surface or the Cu^{2+} of the copper electrode surface diffusing into the H$_2$SO$_4$ solution. With the addition of PLE, the Warburg impedance become extinct in the low frequency range, indicating that the PLE creates a compact protective film at the metal–electrolyte interface that prevents Cu^{2+} ions from diffusing into the bulk solution. It is also important to point out that the diameter of capacitive loops increases with the addition of PLE. This is explained by the fact that PLE forms a compact protective film on the Cu electrode surface which facilitates the displacement of H$_2$O molecules from the working electrode surface.

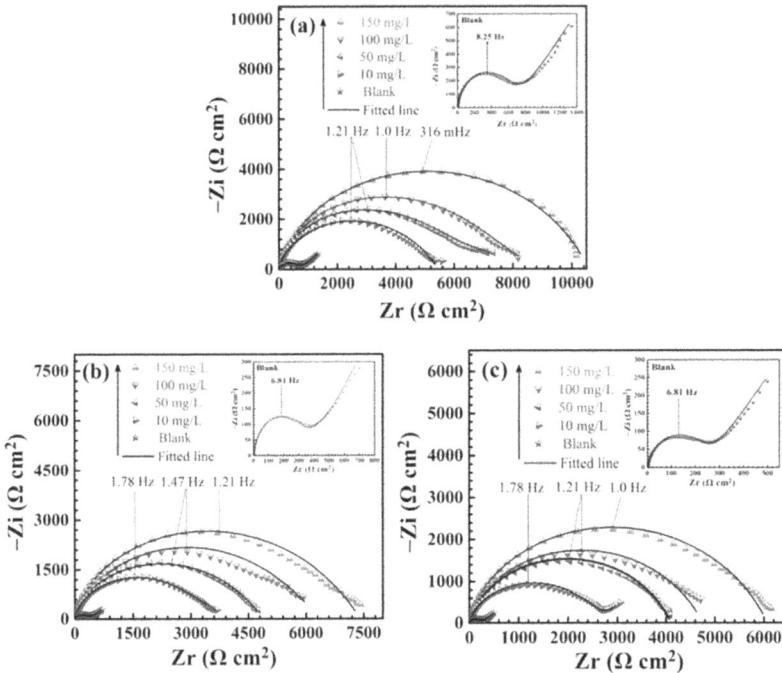

FIGURE 12.4 Nyquist plots for copper immersed in aerated 0.5 M H$_2$SO$_4$ solution without and with varying concentrations of PLE at (a) 298 K, (b) 298 K, 303 K, and (c) 308 K temperature. Reproduced with permission from Ref. [75].

Consequently, the process of charge transfer becomes more challenging and copper corrosion can be efficiently prevented.

The corresponding electrical equivalent circuits used for fitting the Nyquist plots have been represented in Figure 12.5. In the equivalent circuit, R_f, R_s, and W are the abbreviated forms of film resistance, solution resistance, and Warburg constant. CPE_f and CPE_{dl} denote constant phase angle components for film capacitance (C_f) and C_{dl}.

The utilization of plant extracts as corrosion inhibitors not only addresses the economic and environmental challenges associated with copper corrosion but also promotes sustainable and eco-friendly corrosion protection practices.

12.5   MECHANISMS OF PLANT EXTRACTS IN CORROSION INHIBITION

12.5.1   Adsorption Behaviour of Plant Extracts on Copper Surfaces

The adsorption behaviour of plant extracts on copper surfaces is a critical aspect of their corrosion inhibition mechanisms [76–78]. By understanding the adsorption process, researchers can gain valuable insights into the interaction between plant extracts and metal surfaces, leading to the formation of a protective layer that hinders corrosion. The adsorption behaviour is influenced by various factors, including the

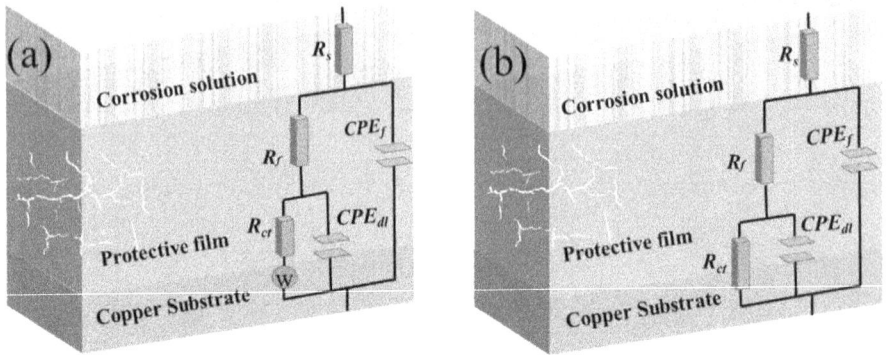

FIGURE 12.5 The corresponding electrical equivalent circuits used for fitting the Nyquist plots (a) with Warburg impedance and (b) without Warburg impedance. Reproduced with permission from Ref. [75].

chemical composition of the plant extract, the surface properties of the metal, and the characteristics of the corrosive environment. Plant extracts consist of a diverse array of organic compounds, such as phenols, alkaloids, flavonoids, and tannins, which possess functional groups capable of interacting with metal surfaces. These functional groups, such as hydroxyl (-OH), carboxyl (-COOH), and amino (-NH$_2$) groups, can form coordination bonds with copper atoms, facilitating the adsorption process [24,79]. In addition to the chemical composition of the plant extract, external factors play a role in the adsorption behaviour. The pH of the solution, temperature, concentration of the extract, and contact time all influence the adsorption process. For example, changes in pH can affect the ionization state of the functional groups in the plant extract, thereby influencing their interaction with the copper surface. Temperature variations can alter the kinetics of the adsorption process, while the concentration and contact time determine the amount of plant extract available for adsorption.

Studies have shown that the adsorption of plant extracts on copper surfaces can follow Langmuir or Freundlich isotherms, which describe the adsorption behaviour in terms of monolayer or multilayer adsorption, respectively [39,80]. The Langmuir isotherm suggests the formation of a monolayer of adsorbed molecules on the metal surface, where further adsorption is not possible once the surface is fully covered. On the other hand, the Freundlich isotherm suggests the formation of multilayer adsorption, indicating that adsorbed molecules can accumulate on top of each other, forming a thicker protective layer. To investigate the adsorption behaviour of plant extracts on copper surfaces, various surface analysis techniques are employed. Fourier-transform infrared spectroscopy (FTIR) is commonly used to examine the functional groups involved in the adsorption process. By analysing the characteristic absorption peaks in the infrared spectrum, researchers can identify the specific functional groups present in the plant extract and study their interaction with the metal surface. XPS is another valuable technique used to investigate the adsorption behaviour. It provides information about the chemical composition of the surface and can determine the presence of adsorbed species on the copper surface. XPS allows researchers to identify the elements and their oxidation states, providing insights into the coordination bonds formed between the plant extract and the metal surface.

The adsorbed plant extract molecules form a protective layer on the copper surface, effectively hindering the access of corrosive species and inhibiting corrosion. The protective layer acts as a physical barrier, preventing the diffusion of aggressive ions and reducing the electrochemical reactions that lead to corrosion. The functional groups present in the plant extract can also interact with the metal surface, forming a passive film that enhances the corrosion resistance of copper.

12.5.2 FILM FORMATION AND PROTECTIVE BARRIER PROPERTIES

Plant extracts can contribute to the formation of protective films on copper surfaces, which act as barriers against corrosive species. The film formation process involves the adsorption of plant extract molecules on the metal surface, followed by their aggregation and organization into a protective layer. The film acts as a physical barrier, preventing the diffusion of corrosive species to the metal surface and reducing the corrosion rate [81]. The film-forming ability of plant extracts depends on various factors, including the concentration of active compounds, the presence of functional groups, and environmental conditions. Plant extracts rich in polyphenolic compounds, such as tannins and flavonoids, have been found to exhibit excellent film-forming properties. These compounds can form complexes with copper ions, leading to the formation of insoluble precipitates or coordination compounds that contribute to film formation. The protective barrier properties of the film formed by plant extracts are assessed using techniques such as EIS and SEM. EIS provides information about the film's resistance to ion transport and its capacitance, which are indicative of the barrier properties. SEM allows the visualization of the morphology and thickness of the film, providing insights into its protective ability. The film formed by plant extracts acts as a physical barrier, reducing the diffusion of corrosive species to the metal surface. It also hinders the access of oxygen and moisture, which are essential for the corrosion process. The presence of polyphenolic compounds in plant extracts contributes to the film's stability and its ability to repair itself in the event of damage.

12.5.3 SURFACE PASSIVATION BY PLANT EXTRACTS

Surface passivation refers to the formation of a passive layer on the metal surface, which effectively inhibits further corrosion [65,82]. Plant extracts can promote the passivation of copper surfaces by facilitating the formation of a stable and protective oxide layer. Passivation occurs through the interaction between the active compounds in the plant extract and the metal surface, leading to the formation of a passive film. The passive film formed by plant extracts consists mainly of copper oxide and copper hydroxide. The active compounds in the plant extract can act as complexing agents, facilitating the formation of stable oxide/hydroxide compounds. The passive film acts as a physical barrier, preventing the diffusion of corrosive species and providing long-term corrosion protection. Techniques such as XPS and Raman spectroscopy are employed to analyse the passive film formed on copper surfaces treated with plant extracts. These techniques provide information about the chemical composition and structure of the passive layer, confirming the presence of copper oxide/hydroxide compounds. In addition to the passive film formation, plant extracts can also influence the electrochemical behaviour of copper surfaces, leading to a shift in the corrosion potential and the suppression of anodic and cathodic reactions [32,83]. The active compounds in the plant extract can act as inhibitors by adsorbing onto active sites on the metal surface, inhibiting the corrosion reactions. Understanding the mechanisms of film formation and surface passivation by plant extracts provides insights into their long-term corrosion inhibition capabilities. The combination of film formation, barrier properties, and surface passivation contributes to the overall corrosion protection provided by plant extract–based inhibitors, making them effective and sustainable alternatives for copper corrosion control.

12.6 FUTURE PROSPECTS AND CHALLENGES

12.6.1 Emerging Trends in Plant Extract-Based Corrosion Inhibitors

The use of plant extracts as corrosion inhibitors for copper continues to evolve, with several emerging trends shaping the field. These trends reflect advancements in research, technology, and the growing demand for sustainable corrosion protection strategies. Some of the key emerging trends in plant extract–based corrosion inhibitors are as follows.

12.6.1.1 Synergistic Combinations

Synergistic combinations of plant extracts have gained significant attention in the field of corrosion inhibition. Researchers are exploring the potential of combining extracts with complementary inhibitive properties to achieve synergistic effects, which can result in improved corrosion protection compared to using individual plant extracts. The rationale behind synergistic combinations lies in the diverse chemical compositions and modes of action of different plant extracts. Each plant extract contains a unique set of organic compounds with specific functional groups that can interact with metal surfaces and inhibit corrosion. By combining plant extracts with different compositions, it is possible to enhance the inhibitive performance by targeting multiple corrosion mechanisms simultaneously.

Several studies have demonstrated the benefits of synergistic combinations in corrosion inhibition. For example, a combination of plant extracts rich in phenolic compounds and those containing alkaloids or flavonoids have shown enhanced inhibitive efficiency compared to using individual extracts. This is because phenolic compounds can form protective layers on the metal surface through their ability to chelate metal ions and scavenge free radicals, while alkaloids and flavonoids exhibit passivation effects and can inhibit specific corrosion reactions. The synergistic effects observed in these combinations can be attributed to various factors. Firstly, the different compounds in the plant extracts can interact with the metal surface through different mechanisms, leading to more comprehensive and effective corrosion inhibition. The combined extracts can act on multiple corrosion processes, such as metal dissolution, reduction of corrosive species, or formation of protective films, resulting in improved overall performance. Additionally, the combination of plant extracts may enhance the adsorption behaviour on the metal surface. Different compounds in the extracts may exhibit complementary adsorption characteristics, such as different adsorption sites or affinities for the metal surface. This can lead to the formation of a more stable and dense protective layer, effectively reducing the access of corrosive species to the metal surface.

The synergistic combinations of plant extracts can also result in improved stability and durability of the protective layer. The presence of multiple compounds can enhance the film-forming properties, cohesion, and adhesion of the protective layer, making it more resistant to mechanical stress, temperature variations, and other environmental factors. This ensures prolonged corrosion protection and extends the service life of the metal. Moreover, synergistic combinations offer the advantage of utilizing multiple plant sources, expanding the range of available natural inhibitors. This can be particularly beneficial in regions where specific plants or extracts are limited or not easily accessible. By combining extracts from various plant sources, researchers can harness the inhibitive properties of a wider array of natural compounds, increasing the potential for effective corrosion inhibition.

However, it is important to note that the selection and optimization of synergistic combinations require careful consideration. The choice of plant extracts, their ratios, and the compatibility of their chemical compositions should be thoroughly evaluated. The concentration and compatibility of the combined extracts, as well as their stability over time, need to be taken into account to ensure the desired synergistic effects are achieved.

12.6.1.2 Nanotechnology

The integration of nanotechnology with plant extract–based corrosion inhibitors has emerged as a promising approach for achieving improved performance in corrosion inhibition. Nanotechnology offers unique opportunities to enhance the inhibitive properties of plant extracts by utilizing nanostructures, including nanoparticles, nanocomposites, and nanocoatings. One of the key advantages of nanotechnology in corrosion inhibition is the ability to synthesize nanoparticles using plant extracts as reducing or stabilizing agents. Plant extracts contain bioactive compounds, such as phenols, flavonoids, and tannins, which can serve as reducing agents to convert metal ions into nanoparticles. These nanoparticles can possess high surface area-to-volume ratios, allowing for increased contact with the metal surface and improved adsorption behaviour. Nanoparticles synthesized from plant extracts can exhibit enhanced adsorption properties due to their small size and large surface area. The nanoparticles can attach to the metal surface more effectively, forming a dense and uniform protective layer. This layer acts as a physical barrier, inhibiting the diffusion of corrosive species and reducing the corrosion rate of the metal.

Furthermore, nanocomposites can be formed by incorporating plant extract nanoparticles into matrices such as polymers or ceramics. The combination of plant extract nanoparticles with other materials can lead to synergistic effects, where the properties of both components are combined to provide enhanced corrosion inhibition. For example, the incorporation of plant extract nanoparticles into a polymer matrix can improve the mechanical strength, stability, and adhesion of the coating, resulting in prolonged corrosion protection. Nanocoatings, on the other hand, involve the deposition of thin layers of plant extract nanoparticles or nanocomposites onto the metal surface. These coatings can offer superior corrosion resistance by providing a uniform and continuous barrier against corrosive environments. The nanoscale thickness of the coatings ensures minimal interference with the substrate's properties while effectively preventing corrosion processes. The use of nanotechnology in combination with plant extract–based corrosion inhibitors also enables the development of smart and responsive coatings. By incorporating stimuli-responsive nanoparticles or nanocomposites into the coatings, their properties can be modulated in response to specific environmental conditions. For instance, pH-responsive nanoparticles can release inhibitive compounds in acidic environments, where corrosion is more likely to occur, thereby enhancing corrosion inhibition efficiency.

Characterization techniques such as transmission electron microscopy (TEM), SEM, X-ray diffraction (XRD), and Fourier-transform infrared spectroscopy (FTIR) are commonly employed to study the morphology, composition, and structure of plant extract–based nanomaterials. These techniques provide valuable insights into the size, shape, distribution, and stability of the nanoparticles, as well as their interaction with the metal surface. While the integration of nanotechnology with plant extract–based corrosion inhibitors offers significant advantages, several challenges need to be addressed. The scalability of the synthesis methods, the stability of the nanoparticles or nanocomposites over time, and the potential toxicity of the nanomaterials require careful consideration. Additionally, the compatibility between the nanomaterials and the coating or matrix materials should be evaluated to ensure long-term performance and durability.

12.6.1.3   Green Extraction Methods

The development of green extraction methods for plant extracts has emerged as a significant focus in the field of corrosion inhibition. Conventional extraction methods typically involve the use of organic solvents, which can have adverse environmental impacts due to their toxicity, flammability, and potential for environmental pollution. In response to these concerns, researchers have been exploring alternative extraction techniques that are more sustainable and environmentally friendly. One of the green extraction methods gaining attention is ultrasound-assisted extraction (UAE). UAE utilizes high-frequency ultrasound waves to enhance the extraction process. The ultrasound waves create cavitation bubbles in the solvent, leading to the disruption of plant cell walls and facilitating the release of active compounds. UAE offers several advantages, including reduced extraction time, lower solvent consumption, and improved extraction efficiency. The process is also relatively energy-efficient and can be easily scaled up for industrial applications. Microwave-assisted extraction (MAE) is another green extraction method that has shown promise in obtaining plant extracts for corrosion inhibition applications. In MAE, microwave radiation is used to generate heat, promoting the extraction of active compounds from the plant material. The selective heating effect of microwaves on polar molecules allows for efficient extraction while minimizing the degradation of thermally sensitive compounds.

MAE offers rapid extraction, reduced solvent usage, and enhanced extraction yields compared to conventional methods. Supercritical fluid extraction (SFE) is a green extraction technique that utilizes supercritical fluids, such as carbon dioxide (CO_2), as the extraction solvent. Under specific temperature and pressure conditions, CO_2 reaches a supercritical state where it exhibits the properties of both a gas and a liquid. In this state, CO_2 can penetrate the plant material and extract the desired compounds effectively. SFE offers several advantages, including the use of a non-toxic and non-flammable solvent, high selectivity, and the ability to obtain extracts free from residual solvents.

Green extraction methods not only contribute to environmental sustainability but also help to preserve the integrity and activity of the active compounds present in plant extracts. Conventional extraction methods may subject the plant material to high temperatures and prolonged exposure to solvents, which can degrade or modify the desired compounds. In contrast, green extraction methods operate under milder conditions, minimizing the degradation and alteration of the active compounds. This ensures that the extracted plant extracts retain their inhibitive properties, providing reliable and effective corrosion protection. The selection of an appropriate green extraction method depends on several factors, including the plant material, target compounds, desired yield, and inhibitive properties required. Each method has its advantages and limitations, and the choice should be based on the specific needs of the corrosion inhibition application. It is important to consider factors such as extraction efficiency, solvent consumption, energy requirements, scalability, and the potential impact on the environment. To evaluate the effectiveness of green extraction methods, various analytical techniques can be employed to characterize the extracted plant extracts. Spectroscopic methods, such as Fourier-transform infrared spectroscopy (FTIR) and NMR spectroscopy, can be used to identify and quantify the chemical composition of the extracts. Chromatographic techniques, including HPLC and gas chromatography-mass spectrometry (GC-MS), can separate and analyse individual compounds present in the extracts. These characterization techniques provide valuable insights into the composition, structure, and potential inhibitive properties of the plant extracts obtained through green extraction methods.

12.6.1.4 Waste Biomass Utilization

The utilization of waste biomass as a source for plant extracts in corrosion inhibition applications has gained significant attention in recent years. Waste biomass refers to organic materials that are generated as by-products of various industries, including agriculture, forestry, and food processing. Instead of being discarded as waste, these biomass residues can be repurposed and transformed into valuable plant extracts, offering a sustainable and cost-effective solution for corrosion inhibition. Agricultural residues, such as straws, husks, stems, and shells, are abundant sources of waste biomass. These residues are typically left unused after harvesting or processing crops and can contribute to environmental pollution if not properly managed. However, researchers have recognized the potential of agricultural residues as renewable and readily available feedstock for extracting corrosion inhibitors. By converting these residues into plant extracts, their value is maximized, and their environmental impact is minimized. Food processing by-products, such as fruit peels, pomace, and seed residues, also present valuable opportunities for waste biomass utilization. These by-products are generated during the processing of fruits, vegetables, grains, and other food commodities. Instead of being discarded, they can be processed to extract active compounds that exhibit corrosion inhibition properties. This approach not only reduces waste generation but also adds value to the food processing industry by creating additional revenue streams and reducing disposal costs.

The conversion of waste biomass into plant extracts involves sustainable processes that align with the principles of the circular economy. Various extraction techniques, such as solvent extraction, UAE, and MAE, can be employed to obtain the desired compounds from biomass residues. The choice of extraction method depends on factors such as the characteristics of the biomass, the target compounds, and the desired yield. Green extraction methods, as discussed earlier, are particularly suitable for waste biomass utilization as they minimize solvent usage, energy consumption, and environmental impact. The utilization of waste biomass for corrosion inhibition not only contributes to environmental sustainability but also offers economic benefits. By repurposing waste materials, the cost of acquiring feedstock for plant extract production is reduced, as waste biomass is often available at lower or no cost. This can lead to cost savings in corrosion inhibition applications, making them more economically viable for various industries. Furthermore, the utilization of waste biomass aligns with the concept of resource efficiency. Instead of relying solely on virgin plant materials, which may require significant land, water, and energy resources for cultivation and processing, waste biomass provides an alternative and sustainable source. It reduces the pressure on natural resources and promotes the efficient use of existing biomass. To ensure the quality and effectiveness of plant extracts derived from waste biomass, thorough characterization and quality control measures are necessary. Analytical techniques, such as chromatography, spectroscopy, and microscopy, can be employed to determine the chemical composition, structure, and inhibitive properties of the extracts. This characterization process helps in understanding the potential corrosion inhibition mechanisms and optimizing the extraction parameters for enhanced performance.

12.6.2 CHALLENGES IN IMPLEMENTING PLANT EXTRACTS FOR COPPER CORROSION INHIBITION

While plant extracts show great promise as corrosion inhibitors for copper, several challenges need to be addressed for their successful implementation. These challenges include:

12.6.2.1 Standardization and Reproducibility

One of the challenges in utilizing plant extracts for corrosion inhibition is the variation in their composition due to various factors. Different plant species, geographical locations, extraction methods, and seasonal variations can all contribute to differences in the chemical composition of plant extracts. This variability can lead to inconsistent inhibitive performance across different batches of extracts, making it difficult to ensure reproducibility and reliability. To address this challenge, standardization of extraction protocols, characterization techniques, and evaluation methods is crucial. Developing standardized extraction procedures that outline specific parameters such as solvent type, extraction time, temperature, and plant-to-solvent ratio can help achieve consistent results. Characterization techniques such as spectroscopy, chromatography, and microscopy should also be standardized to ensure accurate and reliable analysis of the extract composition. Furthermore, standardized evaluation methods, such as electrochemical techniques, should be established to assess the inhibitive performance of plant extract–based inhibitors in a consistent and reproducible manner. By establishing these standards, researchers and industry practitioners can compare results across studies and batches, enabling better understanding and optimization of plant extract–based corrosion inhibition strategies.

12.6.2.2   Stability and Shelf Life

The stability and shelf life of plant extract–based inhibitors are important considerations for their practical application. Plant extracts are susceptible to degradation over time, which can lead to a decline in their inhibitive properties. Factors such as temperature, light exposure, and air oxidation can accelerate the degradation process and compromise the effectiveness of the inhibitors. To address stability concerns, it is essential to develop strategies that enhance the stability and prolong the shelf life of plant extract–based inhibitors. Proper storage conditions, such as cool and dark environments, can help minimize degradation. Packaging materials that protect against light and moisture can also contribute to the stability of the extracts. Additionally, formulation techniques such as encapsulation or incorporation into matrices can improve the stability and controlled release of the active compounds. Research efforts should focus on identifying stabilizing agents or additives that can prevent or slow down the degradation of plant extracts, ensuring their long-term effectiveness as corrosion inhibitors.

12.6.2.3   Compatibility with Coatings and Substrates

In many corrosion protection applications, coatings or surface treatments are applied to the metal substrate to provide an additional barrier against corrosion. When integrating plant extract–based inhibitors with existing coatings or substrates, compatibility issues may arise. Factors such as adhesion, chemical interactions, and long-term effectiveness need to be carefully considered to ensure the compatibility and synergy between the plant extract–based inhibitors and the coating systems. Compatibility studies should evaluate the compatibility of the plant extracts with different coating materials, including paints, polymers, or other protective films. Adhesion tests, such as pull-off tests or cross-cut adhesion tests, can assess the bonding strength between the coating and the substrate. Compatibility can also be evaluated by examining the interfacial interactions between the plant extract–based inhibitors and the coating materials through techniques like spectroscopy or microscopy. Moreover, long-term studies should be conducted to determine the stability and effectiveness of the combined system under realistic operating conditions, including exposure to corrosive environments.

12.6.2.4 Cost-Effectiveness

While plant extracts are generally considered cost-effective compared to synthetic inhibitors, the overall cost of implementing plant extract–based corrosion inhibition strategies should be evaluated. Factors such as the cost of extraction methods, availability and sustainability of plant resources, scalability of production, and compatibility with existing industrial processes can impact the economic feasibility of large-scale applications. To improve cost-effectiveness, optimization of extraction methods can help reduce the cost associated with obtaining plant extracts. Finding alternative sources of plant materials or utilizing waste biomass, as discussed earlier, can also contribute to cost savings. Developing efficient extraction techniques that minimize solvent usage, energy consumption, and processing time can further enhance cost-effectiveness. Additionally, exploring opportunities for integration within existing industrial processes or synergies with other applications, such as utilizing waste heat or by-products, can provide additional economic benefits. Furthermore, life cycle assessments (LCAs) can be conducted to evaluate the overall environmental and economic impacts of plant extract–based corrosion inhibition strategies. By considering the entire life cycle of the inhibitors, including raw material production, extraction, formulation, application, and disposal, a comprehensive analysis can be performed to assess their cost-effectiveness and sustainability compared to conventional corrosion inhibition methods.

12.6.3 STRATEGIES FOR SUSTAINABLE CORROSION PROTECTION

To overcome the challenges and promote sustainable corrosion protection, several strategies can be employed.

12.6.3.1 Collaboration and Knowledge Sharing

Collaboration and knowledge sharing are crucial for advancing the field of plant extract–based corrosion inhibitors. Researchers, industry professionals, and policymakers need to work together to share their expertise, resources, and best practices. Collaborative efforts can accelerate research, development, and implementation efforts by pooling knowledge and conducting joint projects. Collaborative networks and partnerships can be established to facilitate communication and collaboration among stakeholders. Platforms such as conferences, workshops, and forums can provide opportunities for researchers and industry professionals to exchange ideas, present their findings, and discuss challenges and potential solutions. Furthermore, the collaboration between academia, industry, and regulatory bodies can help bridge the gap between research and practical applications, ensuring that the developed inhibitors meet the requirements and standards of the industry.

12.6.3.2 Optimization of Extraction Processes

To maximize the potential of plant extract–based corrosion inhibitors, optimization of extraction processes is crucial. Traditional extraction methods may not always yield the desired quantity, purity, and stability of plant extracts. Therefore, exploring innovative extraction techniques, alternative solvents, and process parameters can lead to more efficient and sustainable extraction methods. Techniques such as

UAE, MAE, and SFE offer greener and more effective alternatives to conventional extraction methods. Optimization studies should focus on identifying the optimal conditions for extraction, including solvent selection, extraction time, temperature, and plant-to-solvent ratio. Additionally, the use of emerging technologies, such as green chemistry principles and process intensification, can further enhance extraction efficiency while minimizing environmental impacts.

12.6.3.3   Performance Enhancement

Continuous research and development efforts are needed to improve the inhibitive performance of plant extract–based inhibitors. This involves exploring new plant sources, investigating novel extraction techniques, understanding the mechanisms of inhibition, and developing strategies to enhance adsorption, film formation, and surface passivation properties. Screening studies can be conducted to identify plants with high inhibitive potential and investigate their active compounds. By understanding the structure-activity relationship of these compounds, researchers can optimize the composition of plant extracts for enhanced corrosion inhibition. Moreover, studying the interaction between plant extract–based inhibitors and metal surfaces using advanced characterization techniques can provide insights into the adsorption mechanisms and film formation properties. This knowledge can guide the design and optimization of plant extract–based inhibitors for specific corrosion environments and applications.

12.6.3.4   Field Testing and Validation

Conducting field testing and real-world evaluations of plant extract–based inhibitors is crucial to validate their performance in practical applications. Laboratory-scale studies provide valuable preliminary data, but field testing allows for the assessment of inhibitors under realistic operating conditions. Long-term monitoring and performance assessments are necessary to evaluate the durability, efficacy, and cost-effectiveness of plant extract–based inhibitors. Field trials should involve exposure to corrosive environments relevant to the target industries, such as marine, oil and gas, or infrastructure applications. Data collected from field testing can provide insights into the long-term performance of the inhibitors, including their effectiveness in preventing corrosion, durability under exposure to environmental factors, and compatibility with existing systems. These data can inform further optimization and refinement of plant extract–based corrosion inhibition strategies.

12.6.3.5   Integration with Existing Corrosion Protection Strategies

Plant extract–based inhibitors can be integrated with existing corrosion protection strategies, such as coatings, inhibitors, and surface treatments. Synergistic combinations and hybrid systems can leverage the strengths of different approaches, providing comprehensive and long-lasting corrosion protection. By combining plant extract–based inhibitors with existing protective coatings or surface treatments, synergistic effects can be achieved, leading to improved corrosion resistance. Compatibility studies should be conducted to evaluate the interaction between plant extract–based inhibitors and existing systems. Techniques such as adhesion testing, surface analysis, and electrochemical evaluation can assess the compatibility, adhesion, and long-term effectiveness of the combined systems. Additionally, integrating plant extract–based inhibitors into existing industrial processes or infrastructure projects can be facilitated by considering factors such as scalability, ease of application, and cost-effectiveness.

By addressing these challenges and implementing sustainable strategies, plant extract–based corrosion inhibitors have the potential to offer effective, eco-friendly, and economically viable solutions for copper corrosion protection in various industries. Through collaboration, optimization, performance enhancement, field testing, and integration with existing strategies, the field can advance toward the widespread adoption and implementation of plant extract–based corrosion inhibition approaches.

12.7 CONCLUSION

In conclusion, the utilization of plant extracts as corrosion inhibitors for copper holds significant promise in addressing the economic, environmental, and safety challenges posed by copper corrosion. The economic impact of copper corrosion is substantial, with direct costs associated with repairs, maintenance, and component replacement, as well as indirect costs resulting from production delays and equipment failures. Moreover, the integrity and functionality of structures can be compromised by copper corrosion, leading to safety hazards and environmental contamination. Corrosion inhibition strategies have gained attention as effective means to mitigate copper corrosion. Plant extracts offer several advantages as corrosion inhibitors, making them an attractive alternative to traditional approaches. Firstly, plant extracts are abundant, widely available, and cost-effective compared to synthetic inhibitors. They possess diverse bioactive compounds, such as alkaloids, phenols, flavonoids, and organic acids, which contribute to their inhibitive properties. These compounds can interact with copper surfaces and modify the electrochemical reactions occurring at the metal–electrolyte interface, leading to corrosion inhibition. The adsorption behaviour of plant extracts on copper surfaces is a crucial mechanism for corrosion inhibition. The active compounds in the extracts can adsorb onto the metal surface, forming a protective layer that acts as a barrier against corrosive species. This layer prevents direct contact between the metal and the electrolyte, inhibiting corrosion reactions. Additionally, the formation of a passive film by plant extracts further enhances the corrosion resistance of copper. The passive film acts as a physical and chemical barrier, reducing the corrosion rate and extending the lifespan of copper materials.

Plant extracts also exhibit surface passivation properties, promoting the formation of stable and protective oxide/hydroxide layers on copper surfaces. These layers contribute to the reduction of metal dissolution and the inhibition of corrosion propagation. The chemical composition and structure of the passive film can be analysed using advanced techniques such as XPS and Raman spectroscopy, providing insights into the protective nature of the film formed by plant extracts. While the potential of plant extracts as corrosion inhibitors is promising, some challenges need to be addressed for their successful implementation. Standardization and reproducibility of the inhibitive performance across different batches of plant extracts are crucial to ensure consistent corrosion protection. The stability and shelf life of plant extract–based inhibitors also need to be improved to maintain their inhibitive properties over time. Compatibility with coatings and substrates, as well as the cost-effectiveness of large-scale applications, are additional challenges that require attention. Looking towards the future, emerging trends in plant extract–based corrosion inhibitors offer exciting prospects. Synergistic combinations of different plant extracts, integration of nanotechnology to enhance inhibitive properties, the development of green extraction methods, and the utilization of waste biomass as sources of plant extracts are some of the trends shaping the field. Collaborative efforts among researchers, industry professionals, and policymakers are essential to advance the understanding and application of plant extract–based corrosion inhibitors. By embracing emerging trends and adopting sustainable strategies, plant extract–based corrosion inhibitors have the potential to contribute to enhanced corrosion control, reduced economic burden, and a more environmentally conscious approach to corrosion protection for copper materials.

REFERENCES

[1] A. Fateh, M. Aliofkhazraei, A.R. Rezvanian, Review of corrosive environments for copper and its corrosion inhibitors, *Arab. J. Chem.* 13 (2020) 481–544. https://doi.org/10.1016/j.arabjc.2017.05.021.

[2] S.B. Adeloju, H.C. Hughes, The corrosion of copper pipes in high chloride-low carbonate mains water, *Corros. Sci.* 26 (1986) 851–870. https://doi.org/10.1016/0010-938X(86)90068-5.

[3] J.D. Hochmuth, J. Asselman, K.A.C. De Schamphelaere, Are interactive effects of harmful algal blooms and copper pollution a concern for water quality management?, *Water Res.* 60 (2014) 41–53. https://doi.org/10.1016/j.watres.2014.03.041.

[4] Y. Zhu, X. Zhu, Q. Xu, Y. Qian, Water quality criteria and ecological risk assessment for copper in Liaodong Bay, China, *Mar. Pollut. Bull.* 185 (2022) 114164. https://doi.org/10.1016/j.marpolbul.2022.114164.

[5] S. Cui, K. Zhou, R. Ding, J. Wang, Y. Cheng, G. Jiang, Monitoring the soil copper pollution degree based on the reflectance spectrum of an arid desert plant, Spectrochim. *Acta Part A Mol. Biomol. Spectrosc.* 263 (2021) 120186. https://doi.org/10.1016/j.saa.2021.120186.

[6] I. Vargas, D. Fischer, M. Alsina, J. Pavissich, P. Pastén, G. Pizarro, Copper corrosion and biocorrosion events in premise plumbing, *Materials (Basel).* 10 (2017) 1036. https://doi.org/10.3390/ma10091036.

[7] A. Srivastava, R. Balasubramaniam, Microstructural characterization of copper corrosion in aqueous and soil environments, *Mater. Charact.* 55 (2005) 127–135. https://doi.org/10.1016/j.matchar.2005.04.004.

[8] J. Briffa, E. Sinagra, R. Blundell, Heavy metal pollution in the environment and their toxicological effects on humans, *Heliyon.* 6 (2020) e04691. https://doi.org/10.1016/j.heliyon.2020.e04691.

[9] C.F. Carolin, P.S. Kumar, A. Saravanan, G.J. Joshiba, M. Naushad, Efficient techniques for the removal of toxic heavy metals from aquatic environment: A review, *J. Environ. Chem. Eng.* 5 (2017) 2782–2799. https://doi.org/10.1016/j.jece.2017.05.029.

[10] V. Kumar, S. Pandita, G.P. Singh Sidhu, A. Sharma, K. Khanna, P. Kaur, A.S. Bali, R. Setia, Copper bioavailability, uptake, toxicity and tolerance in plants: A comprehensive review, *Chemosphere.* 262 (2021) 127810. https://doi.org/10.1016/j.chemosphere.2020.127810.

[11] I. Yruela, Copper in plants: Acquisition, transport and interactions, *Funct. Plant Biol.* 36 (2009) 409. https://doi.org/10.1071/FP08288.

[12] R. Bender, D. Féron, D. Mills, S. Ritter, R. Bäßler, D. Bettge, I. De Graeve, A. Dugstad, S. Grassini, T. Hack, M. Halama, E. Han, T. Harder, G. Hinds, J. Kittel, R. Krieg, C. Leygraf, L. Martinelli, A. Mol, D. Neff, J. Nilsson, I. Odnevall, S. Paterson, S. Paul, T. Prošek, M. Raupach, R.I. Revilla, F. Ropital, H. Schweigart, E. Szala, H. Terryn, J. Tidblad, S. Virtanen, P. Volovitch, D. Watkinson, M. Wilms, G. Winning, M. Zheludkevich, Corrosion challenges towards a sustainable society, *Mater. Corros.* 73 (2022) 1730–1751. https://doi.org/10.1002/maco.202213140.

[13] B.A. Abd-El-Nabey, A.M. Abdel-Gaber, M.E.S. Ali, E. Khamis, S. El-Housseiny, Inhibitive action of cannabis plant extract on the corrosion of copper in 0.5 M H_2SO_4, *Int. J. Electrochem. Sci.* 8 (2013) 5851–5865. https://doi.org/10.1016/S1452-3981(23)14727-4.

[14] M. Bozorg, T. Shahrabi Farahani, J. Neshati, Z. Chaghazardi, G. Mohammadi Ziarani, Myrtus communis as green inhibitor of copper corrosion in sulfuric acid, *Ind. Eng. Chem. Res.* 53 (2014) 4295–4303. https://doi.org/10.1021/ie404056w.

[15] S.F.L.A. da Costa, S.M.L. Agostinho, J.C. Rubim, Spectroelectrochemical study of passive films formed on brass electrodes in 0.5 m H_2SO_4 aqueous solutions containing benzotriazole (BTAH), *J. Electroanal. Chem. Interfacial Electrochem.* 295 (1990) 203–214. https://doi.org/10.1016/0022-0728(90)85016-X.

[16] M.A. Deyab, Egyptian licorice extract as a green corrosion inhibitor for copper in hydrochloric acid solution, *J. Ind. Eng. Chem.* 22 (2015) 384–389. https://doi.org/10.1016/j.jiec.2014.07.036.

[17] N. Chaubey, Savita, A. Qurashi, D.S. Chauhan, M.A. Quraishi, Frontiers and advances in green and sustainable inhibitors for corrosion applications: A critical review, *J. Mol. Liq.* 321 (2021) 114385. https://doi.org/10.1016/j.molliq.2020.114385.

[18] V. Kola, I.S. Carvalho, Plant extracts as additives in biodegradable films and coatings in active food packaging, *Food Biosci.* 54 (2023) 102860. https://doi.org/10.1016/j.fbio.2023.102860.

[19] W. Schwab, R. Davidovich-Rikanati, E. Lewinsohn, Biosynthesis of plant-derived flavor compounds, *Plant J.* 54 (2008) 712–732. https://doi.org/10.1111/j.1365-313X.2008.03446.x.

[20] F. Chassagne, T. Samarakoon, G. Porras, J.T. Lyles, M. Dettweiler, L. Marquez, A.M. Salam, S. Shabih, D.R. Farrokhi, C.L. Quave, A systematic review of plants with antibacterial activities: A taxonomic and phylogenetic perspective, *Front. Pharmacol.* 11 (2021). https://doi.org/10.3389/fphar.2020.586548.

[21] C. Bitwell, S. Sen Indra, C. Luke, M.K. Kakoma, A review of modern and conventional extraction techniques and their applications for extracting phytochemicals from plants, *Sci. African.* 19 (2023) e01585. https://doi.org/10.1016/j.sciaf.2023.e01585.

[22] T. Wu, W. Zhu, L. Chen, T. Jiang, Y. Dong, L. Wang, X. Tong, H. Zhou, X. Yu, Y. Peng, L. Wang, Y. Xiao, T. Zhong, A review of natural plant extracts in beverages: Extraction process, nutritional function, and safety evaluation, *Food Res. Int.* 172 (2023) 113185. https://doi.org/10.1016/j.foodres.2023.113185.

[23] M.A. Deyab, Corrosion inhibition of aluminum in biodiesel by ethanol extracts of Rosemary leaves, *J. Taiwan Inst. Chem. Eng.* 58 (2016) 536–541. https://doi.org/10.1016/j.jtice.2015.06.021.

[24] S.H. Alrefaee, K.Y. Rhee, C. Verma, M.A. Quraishi, E.E. Ebenso, Challenges and advantages of using plant extract as inhibitors in modern corrosion inhibition systems: Recent advancements, *J. Mol. Liq.* 321 (2021) 114666. https://doi.org/10.1016/j.molliq.2020.114666.

[25] N. Palaniappan, I. Cole, F. Caballero-Briones, S. Manickam, K.R. Justin Thomas, D. Santos, Experimental and DFT studies on the ultrasonic energy-assisted extraction of the phytochemicals of Catharanthus roseus as green corrosion inhibitors for mild steel in NaCl medium, *RSC Adv.* 10 (2020) 5399–5411. https://doi.org/10.1039/C9RA08971C.

[26] Alkaloids as green and environmental benign corrosion inhibitors: An overview, *Int. J. Corros. Scale Inhib.* 8 (2019). https://doi.org/10.17675/2305-6894-2019-8-3-3.

[27] M. Mobin, I. Ahmad, M. Murmu, P. Banerjee, R. Aslam, Corrosion inhibiting properties of polysaccharide extracted from Lepidium meyenii root for mild steel in acidic medium: Experimental, density functional theory, and Monte Carlo simulation studies, *J. Phys. Chem. Solids.* 179 (2023) 111411. https://doi.org/10.1016/j.jpcs.2023.111411.

[28] A. Ishak, F.V. Adams, J.O. Madu, I.V. Joseph, P.A. Olubambi, Corrosion inhibition of mild steel in 1M hydrochloric acid using Haematostaphis barteri leaves extract, *Procedia Manuf.* 35 (2019) 1279–1285. https://doi.org/10.1016/j.promfg.2019.06.088.

[29] M. Mobin, Huda, S. Zamindar, P. Banerjee, Mechanistic insight into adsorption and anti-corrosion capability of a novel surfactant-derived ionic liquid for mild steel in HCl medium, *J. Mol. Liq.* 385 (2023) 122403. https://doi.org/10.1016/j.molliq.2023.122403.

[30] H. Li, S. Zhang, Y. Qiang, Corrosion retardation effect of a green cauliflower extract on copper in H_2SO_4 solution: Electrochemical and theoretical explorations, *J. Mol. Liq.* 321 (2021) 114450. https://doi.org/10.1016/j.molliq.2020.114450.

[31] R.K. Ahmed, S. Zhang, Bee pollen extract as an eco-friendly corrosion inhibitor for pure copper in hydrochloric acid, *J. Mol. Liq.* 316 (2020) 113849. https://doi.org/10.1016/j.molliq.2020.113849.

[32] X. Zhang, W. Li, G. Yu, X. Zuo, W. Luo, J. Zhang, B. Tan, A. Fu, S. Zhang, Evaluation of Idesia polycarpa Maxim fruits extract as a natural green corrosion inhibitor for copper in 0.5 M sulfuric acid solution, *J. Mol. Liq.* 318 (2020) 114080. https://doi.org/10.1016/j.molliq.2020.114080.

[33] A. Jmiai, A. Tara, S. El Issami, M. Hilali, O. Jbara, L. Bazzi, A new trend in corrosion protection of copper in acidic medium by using Jujube shell extract as an effective green and environmentally safe corrosion inhibitor: Experimental, quantum chemistry approach and Monte Carlo simulation study, *J. Mol. Liq.* 322 (2021) 114509. https://doi.org/10.1016/j.molliq.2020.114509.

[34] M.A. Amin, K.F. Khaled, Copper corrosion inhibition in O2-saturated H_2SO_4 solutions, *Corros. Sci.* 52 (2010) 1194–1204. https://doi.org/10.1016/j.corsci.2009.12.035.

[35] R. del P.B. Hernández, I.V. Aoki, B. Tribollet, H.G. de Melo, Electrochemical impedance spectroscopy investigation of the electrochemical behaviour of copper coated with artificial patina layers and submitted to wet and dry cycles, *Electrochim. Acta.* 56 (2011) 2801–2814. https://doi.org/10.1016/j.electacta.2010.12.059.

[36] A.A. Attia, E.M. Elmelegy, M. El-Batouti, A.-M.M. Ahmed, Anodic corrosion inhibition in presence of protic solvents, *Asian J. Chem.* 28 (2016) 267–272. https://doi.org/10.14233/ajchem.2016.18978.

[37] K. Habib, In-situ monitoring of pitting corrosion of copper alloys by holographic interferometry, *Corros. Sci.* 40 (1998) 1435–1440. https://doi.org/10.1016/S0010-938X(98)00049-3.

[38] R.M. Souto, M.P. Sánchez, M. Barrera, S. González, R.C. Salvarezza, A.J. Arvia, The kinetics of pitting corrosion of copper in alkaline solutions containing sodium perchlorate, *Electrochim. Acta.* 37 (1992) 1437–1443. https://doi.org/10.1016/0013-4686(92)87019-V.

[39] G. Kear, B.D. Barker, F.C. Walsh, Electrochemical corrosion of unalloyed copper in chloride media--A critical review, *Corros. Sci.* 46 (2004) 109–135. https://doi.org/10.1016/S0010-938X(02)00257-3.

[40] H. Li, S. Zhang, B. Tan, Y. Qiang, W. Li, S. Chen, L. Guo, Investigation of Losartan Potassium as an eco-friendly corrosion inhibitor for copper in 0.5 M H_2SO_4, *J. Mol. Liq.* 305 (2020) 112789. https://doi.org/10.1016/j.molliq.2020.112789.

[41] S. Ren, M. Cui, X. Chen, S. Mei, Y. Qiang, Comparative study on corrosion inhibition of N doped and N,S codoped carbon dots for carbon steel in strong acidic solution, *J. Colloid Interface Sci.* 628 (2022) 384–397. https://doi.org/10.1016/j.jcis.2022.08.070.

[42] H. Huang, Z. Wang, Y. Gong, F. Gao, Z. Luo, S. Zhang, H. Li, Water soluble corrosion inhibitors for copper in 3.5 wt% sodium chloride solution, *Corros. Sci.* 123 (2017) 339–350. https://doi.org/10.1016/j.corsci.2017.05.009.

[43] Z. Tao, G. Liu, Y. Li, R. Zhang, H. Su, S. Li, Electrochemical investigation of tetrazo-lium violet as a novel copper corrosion inhibitor in an acid environment, *ACS Omega*. 5 (2020) 4415–4423. https://doi.org/10.1021/acsomega.9b03475.

[44] R. Solmaz, E. Altunbaş Şahin, A. Döner, G. Kardaş, The investigation of synergis-tic inhibition effect of rhodanine and iodide ion on the corrosion of copper in sul-phuric acid solution, *Corros. Sci.* 53 (2011) 3231–3240. https://doi.org/10.1016/j.corsci.2011.05.067.

[45] T.L. Skovhus, R.B. Eckert, E. Rodrigues, Management and control of microbiologically influenced corrosion (MIC) in the oil and gas industry-overview and a North Sea case study, *J. Biotechnol.* 256 (2017) 31–45. https://doi.org/10.1016/j.jbiotec.2017.07.003.

[46] A.A. Olajire, Corrosion inhibition of offshore oil and gas production facilities using organic compound inhibitors - A review, *J. Mol. Liq.* 248 (2017) 775–808. https://doi.org/10.1016/j.molliq.2017.10.097.

[47] S. Nolan, M. Rossini, C. Knight, A. Nanni, New directions for reinforced concrete coastal structures, *J. Infrastruct. Preserv. Resil.* 2 (2021) 1. https://doi.org/10.1186/s43065-021-00015-4.

[48] W. Raczkiewicz, M. Bacharz, K. Bacharz, M. Teodorczyk, Reinforcement corrosion testing in concrete and fiber reinforced concrete specimens exposed to aggressive exter-nal factors, *Materials (Basel).* 16 (2023) 1174. https://doi.org/10.3390/ma16031174.

[49] L. Li, M. Chakik, R. Prakash, A review of corrosion in aircraft structures and graphene-based sensors for advanced corrosion monitoring, *Sensors.* 21 (2021) 2908. https://doi.org/10.3390/s21092908.

[50] N. Faisal, Ö.N. Cora, M.L. Bekci, R.E. Śliwa, Y. Sternberg, S. Pant, R. Degenhardt, A. Prathuru, Defect Types, in: M.G.R. Sause, E. Jasiūnienė (eds) *Structural Health Monitoring Damage Detection Systems for Aerospace*, Springer, 2021, pp. 15–72. https://doi.org/10.1007/978-3-030-72192-3_3.

[51] K. Li, Y. Zeng, Corrosion of heat exchanger materials in co-combustion thermal power plants, *Renew. Sustain. Energy Rev.* 161 (2022) 112328. https://doi.org/10.1016/j.rser.2022.112328.

[52] H. Cockings, High Temperature Corrosion, in: *Encyclopedia of Materials: Metals and Alloys*, Elsevier, 2022, pp. 464–475. https://doi.org/10.1016/B978-0-12-819726-4.00064-8.

[53] B. Tansel, K. Zhang, Effects of saltwater intrusion and sea level rise on aging and corrosion rates of iron pipes in water distribution and wastewater collection systems in coastal areas, *J. Environ. Manage.* 315 (2022) 115153. https://doi.org/10.1016/j.jenvman.2022.115153.

[54] Y. Zhang, B.M. Ayyub, J.F. Fung, Projections of corrosion and deterioration of infra-structure in United States coasts under a changing climate, *Resilient Cities Struct.* 1 (2022) 98–109. https://doi.org/10.1016/j.rcns.2022.04.004.

[55] H.S. Gadow, M.M. Motawea, H.M. Elabbasy, Investigation of myrrh extract as a new corrosion inhibitor for α-brass in 3.5% NaCl solution polluted by 16 ppm sulfide, *RSC Adv.* 7 (2017) 29883–29898. https://doi.org/10.1039/C7RA04271J.

[56] F.A. Ayeni, S. Alawode, D. Joseph, P. Sukop, V. Olawuyi, T.E. Alonge, O.O. Alabi, O. Oluwabunmi, F.I. Alo, Investigation of *Sida acuta* (wire weed) plant extract as cor-rosion inhibitor for aluminium-copper-magnessium alloy in acidic medium, *J. Miner. Mater. Charact. Eng.* 02 (2014) 286–291. https://doi.org/10.4236/jmmce.2014.24033.

[57] P.C. Okafor, V.I. Osabor, E.E. Ebenso, Eco-friendly corrosion inhibitors: Inhibitive action of ethanol extracts of Garcinia kola for the corrosion of mild steel in H_2SO_4 solutions, *Pigment Resin Technol.* 36 (2007) 299–305. https://doi.org/10.1108/03699420710820414.

[58] E.E. Oguzie, Evaluation of the inhibitive effect of some plant extracts on the acid corrosion of mild steel, *Corros. Sci.* 50 (2008) 2993–2998. https://doi.org/10.1016/j.corsci.2008.08.004.

[59] U.M. Eduok, S.A. Umoren, A.P. Udoh, Synergistic inhibition effects between leaves and stem extracts of Sida acuta and iodide ion for mild steel corrosion in 1M H_2SO_4 solutions, *Arab. J. Chem.* 5 (2012) 325–337. https://doi.org/10.1016/j.arabjc.2010.09.006.

[60] A.S. Fouda, K. Shalabi, M.S. Shaaban, Synergistic effect of potassium iodide on corrosion inhibition of carbon steel by Achillea santolina extract in hydrochloric acid solution, *J. Bio- Tribo-Corrosion.* 5 (2019) 71. https://doi.org/10.1007/s40735-019-0260-6.

[61] G. Ji, S.K. Shukla, P. Dwivedi, S. Sundaram, R. Prakash, Inhibitive effect of Argemone mexicana plant extract on acid corrosion of mild steel, *Ind. Eng. Chem. Res.* 50 (2011) 11954–11959. https://doi.org/10.1021/ie201450d.

[62] P.M. Krishnegowda, V.T. Venkatesha, P.K.M. Krishnegowda, S.B. Shivayogiraju, Acalypha torta leaf extract as green corrosion inhibitor for mild steel in hydrochloric acid solution, *Ind. Eng. Chem. Res.* 52 (2013) 722–728. https://doi.org/10.1021/ie3018862.

[63] S.A. Umoren, Z.M. Gasem, I.B. Obot, Natural products for material protection: Inhibition of mild steel corrosion by date palm seed extracts in acidic media, *Ind. Eng. Chem. Res.* 52 (2013) 14855–14865. https://doi.org/10.1021/ie401737u.

[64] Y. Qiang, S. Zhang, B. Tan, S. Chen, Evaluation of Ginkgo leaf extract as an eco-friendly corrosion inhibitor of X70 steel in HCl solution, *Corros. Sci.* 133 (2018) 6–16. https://doi.org/10.1016/j.corsci.2018.01.008.

[65] A. Jmiai, B. El Ibrahimi, A. Tara, M. Chadili, S. El Issami, O. Jbara, A. Khallaayoun, L. Bazzi, Application of Zizyphus Lotuse - pulp of Jujube extract as green and promising corrosion inhibitor for copper in acidic medium, *J. Mol. Liq.* 268 (2018) 102–113. https://doi.org/10.1016/j.molliq.2018.06.091.

[66] C. Verma, E.E. Ebenso, I. Bahadur, M.A. Quraishi, An overview on plant extracts as environmental sustainable and green corrosion inhibitors for metals and alloys in aggressive corrosive media, *J. Mol. Liq.* 266 (2018) 577–590. https://doi.org/10.1016/j.molliq.2018.06.110.

[67] B.S. Swaroop, S.N. Victoria, R. Manivannan, Azadirachta indica leaves extract as inhibitor for microbial corrosion of copper by Arthrobacter sulfureus in neutral pH conditions-A remedy to blue green water problem, *J. Taiwan Inst. Chem. Eng.* 64 (2016) 269–278. https://doi.org/10.1016/j.jtice.2016.04.007.

[68] T. Ramde, S. Rossi, C. Zanella, Inhibition of the Cu65/Zn35 brass corrosion by natural extract of Camellia sinensis, *Appl. Surf. Sci.* 307 (2014) 209–216. https://doi.org/10.1016/j.apsusc.2014.04.016.

[69] A.M. Abdel-Gaber, B.A. Abd-El-Nabey, E. Khamis, D.E. Abd-El-Khalek, A natural extract as scale and corrosion inhibitor for steel surface in brine solution, *Desalination.* 278 (2011) 337–342. https://doi.org/10.1016/j.desal.2011.05.048.

[70] M.A. Deyab, Q. Mohsen, E. Bloise, M.R. Lazzoi, G. Mele, Experimental and theoretical evaluations on Oleuropein as a natural origin corrosion inhibitor for copper in acidic environment, *Sci. Rep.* 12 (2022) 7579. https://doi.org/10.1038/s41598-022-11598-7.

[71] B. Benzidia, M. Barbouchi, R. Hsissou, M. Zouarhi, H. Erramli, N. Hajjaji, A combined experimental and theoretical study of green corrosion inhibition of bronze B66 in 3% NaCl solution by Aloe saponaria (syn. Aloe maculata) tannin extract, *Curr. Res. Green Sustain. Chem.* 5 (2022) 100299. https://doi.org/10.1016/j.crgsc.2022.100299.

[72] K. G. Hart, K. Orubite-Okorosaye, A. O. James, Corrosion inhibition of copper in seawater by Xanthosoma Spp Leaf Extract (XLE), *Int. J. Adv. Res. Chem. Sci.* 3 (2016). https://doi.org/10.20431/2349-0403.0312005.

[73] A. El-Asri, M. Rguiti, A. Jmiai, R. Oukhrib, H. Bourzi, Y. Lin, S. El Issami, Carissa macrocarpa extract (ECM) as a new efficient and ecologically friendly corrosion inhibitor for copper in nitric acid: Experimental and theoretical approach, *J. Taiwan Inst. Chem. Eng.* 142 (2023) 104633. https://doi.org/10.1016/j.jtice.2022.104633.

[74] R.K. Ahmed, S. Zhang, Atriplex leucoclada extract: A promising eco-friendly anticorrosive agent for copper in aqueous media, *J. Ind. Eng. Chem.* 99 (2021) 334–343. https://doi.org/10.1016/j.jiec.2021.04.042.

[75] B. Tan, B. Xiang, S. Zhang, Y. Qiang, L. Xu, S. Chen, J. He, Papaya leaves extract as a novel eco-friendly corrosion inhibitor for Cu in H_2SO_4 medium, *J. Colloid Interface Sci.* 582 (2021) 918–931. https://doi.org/10.1016/j.jcis.2020.08.093.

[76] P.B. Raja, M. Fadaeinasab, A.K. Qureshi, A.A. Rahim, H. Osman, M. Litaudon, K. Awang, Evaluation of green corrosion inhibition by alkaloid extracts of Ochrosia oppositifolia and isoreserpiline against mild steel in 1 M HCl medium, *Ind. Eng. Chem. Res.* 52 (2013) 10582–10593. https://doi.org/10.1021/ie401387s.

[77] E.E. Oguzie, K.L. Oguzie, C.O. Akalezi, I.O. Udeze, J.N. Ogbulie, V.O. Njoku, Natural products for materials protection: Corrosion and microbial growth inhibition using Capsicum frutescens biomass extracts, *ACS Sustain. Chem. Eng.* 1 (2013) 214–225. https://doi.org/10.1021/sc300145k.

[78] M.S. Al-Otaibi, A.M. Al-Mayouf, M. Khan, A.A. Mousa, S.A. Al-Mazroa, H.Z. Alkhathlan, Corrosion inhibitory action of some plant extracts on the corrosion of mild steel in acidic media, *Arab. J. Chem.* 7 (2014) 340–346. https://doi.org/10.1016/j.arabjc.2012.01.015.

[79] A. Zakeri, E. Bahmani, A.S.R. Aghdam, Plant extracts as sustainable and green corrosion inhibitors for protection of ferrous metals in corrosive media: A mini review, *Corros. Commun.* 5 (2022) 25–38. https://doi.org/10.1016/j.corcom.2022.03.002.

[80] A. Yurt, B. Duran, H. Dal, An experimental and theoretical investigation on adsorption properties of some diphenolic Schiff bases as corrosion inhibitors at acidic solution/mild steel interface, *Arab. J. Chem.* 7 (2014) 732–740. https://doi.org/10.1016/j.arabjc.2010.12.010.

[81] A. Liu, X. Ren, J. Zhang, C. Wang, P. Yang, J. Zhang, M. An, D. Higgins, Q. Li, G. Wu, Theoretical and experimental studies of the corrosion inhibition effect of nitrotetrazolium blue chloride on copper in 0.1 M H_2SO_4, *RSC Adv.* 4 (2014) 40606–40616. https://doi.org/10.1039/C4RA05274A.

[82] R.S. Nathiya, V. Raj, Evaluation of Dryopteris cochleata leaf extracts as green inhibitor for corrosion of aluminium in 1 M H_2SO_4, *Egypt. J. Pet.* 26 (2017) 313–323. https://doi.org/10.1016/j.ejpe.2016.05.002.

[83] B.A. Al Jahdaly, Rosmarinus officinalis extract as eco-friendly corrosion inhibitor for copper in 1 M nitric acid solution: Experimental and theoretical studies, *Arab. J. Chem.* 16 (2023) 104411. https://doi.org/10.1016/j.arabjc.2022.104411.

13 Phytochemicals/Plant Extracts as Corrosion Inhibitors for Zinc in Various Electrolytes

S. Mustapha
Federal University of Technology

R. Elabor
Florida Agricultural and Mechanical University

A. T. Amigun
Al-Hikmah University

M. B. Etsuyankpa
Federal University of Lafia

T. C. Egbosiuba
Chukwuemeka Odumegwu Ojukwu University

D. T. Shuaib
Illinois Institute of Technology

H. L. Abubakar
Nile University of Nigeria

Y. K. Abubakar
Federal Polytechnic

M. J. Muhammad, J. O. Tijani,
S. A. Abdulkareem, and M. M. Ndamitso
Federal University of Technology

A. K. Mohammed
North Carolina Central University

DOI: 10.1201/9781003394631-13

13.1   INTRODUCTION

About 70% of the earth's surface is covered by water, which helps with global trade in the transportation of goods. As a result, the marine industry has emerged as one of the most crucial pillars of economic growth. However, metabolites influence the materials collectively to produce corrosion, making the marine environment a very harsh corrosive environment for metallic materials used in the ocean sector (Li and Ning, 2019). Atmospheric corrosion is a natural and spontaneous breakdown of exposed metals and alloys, resulting in aesthetic and structural damage and substantial associated expenses (Bernardi et al., 2020).

The primary cause of infrastructure and industrial equipment degradation and abandonment in the marine environment is always corrosion of the materials. Corrosion is characterized by rough surface texture and colourful byproducts (Khayatazad et al., 2020). The cumulative losses from corrosion outweigh those from other natural disasters, and this is widely acknowledged. An international issue is the result of corrosive action. Safety is often put at risk, and technology development is slowed down by corrosion. Metal tools, outdoor furniture, charcoal grills, and car body panels are a few examples of the many items that suffer from corrosion's detrimental effects on property life. The cost of maintaining degraded equipment and structures, the overdesign of structures to allow for corrosion, the waste of valuable resources, the shutdown of equipment due to corrosion failure, product contamination, the reduction and loss of efficiency and useful products and resources, and the replacement of degraded equipment and structures are some of the economic effects. Around billion dollars are spent annually on global corrosion costs (Akpanyung and Loto, 2019). Because corrosion can negatively affect safety, material conservation, and the economy in various engineering applications, it should be prevented appropriately. According to the literature, failure to avoid corrosion led to expensive corrosion-related problems, including the need for upkeep, rehabilitation, and replacement of harmed structures. However, research on the causes of material corrosion in the marine environment and corrosion prevention techniques is necessary (Zuliana et al., 2021).

The oxidation of the metal surface is a natural mechanism by chemical and electrochemical processes and the accumulation of microorganisms. Because the oxidation reaction of the anode charge occurs on the metal surface more quickly than that of the cathode charge, corrosion starts in the anode charge. The corrosion process requires specific aiding substances, such as chemical or natural components, known as inhibitors (Kokilaramani et al., 2021). Therefore, much research has developed novel chemicals acceptable for corrosion inhibitors. Inorganic and organic substances applied sparingly to corrosive conditions successfully slow the corrosion rate. Effective organic inhibitors have N, S, or O atoms or electronegative functional groups and electrons in triple or conjugated double bonds as part of their structures. The inhibitory effects of these chemical compounds are due to adsorbate-metal surface interactions. However, the successful use of plant extracts as green corrosion inhibitors must be pursued, as many organic inhibitors are poisonous and insufficiently cost-effective (Alvarez et al., 2017).

There are a lot of plant biomaterials in nature that are accessible, affordable, non-toxic, and biodegradable. They are candidates for metal corrosion inhibitors because they contain heteroatoms and/or electrons. Numerous studies have been conducted using plant biomaterials as metal corrosion inhibitors in various corrosive conditions. The factors that affect corrosion and corrosion inhibitors are thoroughly examined in this chapter, focusing on plant extract as a corrosion inhibitor and its mechanism. A basic understanding of the variables influencing plant extracts as corrosion inhibitors was also clarified.

13.2 CORROSION

Corrosion is a natural and undesirable process that causes the deterioration and degradation of metallic materials. It occurs when metals react with their surrounding environment, leading to the formation of corrosion products and the weakening of the material's structural properties. Corrosion is influenced by various factors, including moisture, oxygen, temperature, pH, corrosive substances (such as salts or acids), and the type of metal or alloy involved (Abdeen et al., 2019). Different forms of corrosion (see Figure 13.1) can occur, including uniform corrosion, localized corrosion (such as pitting or crevice corrosion), galvanic corrosion, and stress corrosion cracking.

The deterioration of mechanical properties such as ductility and strength usually follows the degradation of metals due to corrosion. This leads to the depletion of materials and thicknesses and eventual failure. These severe outcomes of the corrosion process have begun to be a burden to the global oil and gas industry (Alamri, 2020).

13.2.1 Factors That Enhance Corrosion

Corrosion is a complex electrochemical process that occurs when metals interact with their environment. Several factors can enhance or accelerate the corrosion process. It is important to note that the specific corrosion factors depend on the metal, the environment, and the specific conditions involved. Different metals have varying degrees of corrosion resistance, and the severity of corrosion can vary significantly in different environments. Here are some common factors that contribute to enhanced corrosion:

13.2.1.1 Moisture

Moisture, such as humidity, rain, or condensation, significantly accelerates corrosion. Water acts as an electrolyte, facilitating the flow of ions and promoting corrosion reactions. Moisture plays a crucial role in the corrosion process and is often one of the most significant factors that enhance corrosion. When metals are exposed to moisture, such as high humidity, rain, or condensation, it creates an environment conducive to corrosion (Soufeiani et al., 2020).

Moisture acts as an electrolyte, facilitating the flow of ions between the metal surface and the surrounding environment. This promotes electrochemical reactions that lead to corrosion. Water provides the necessary medium for the transportation

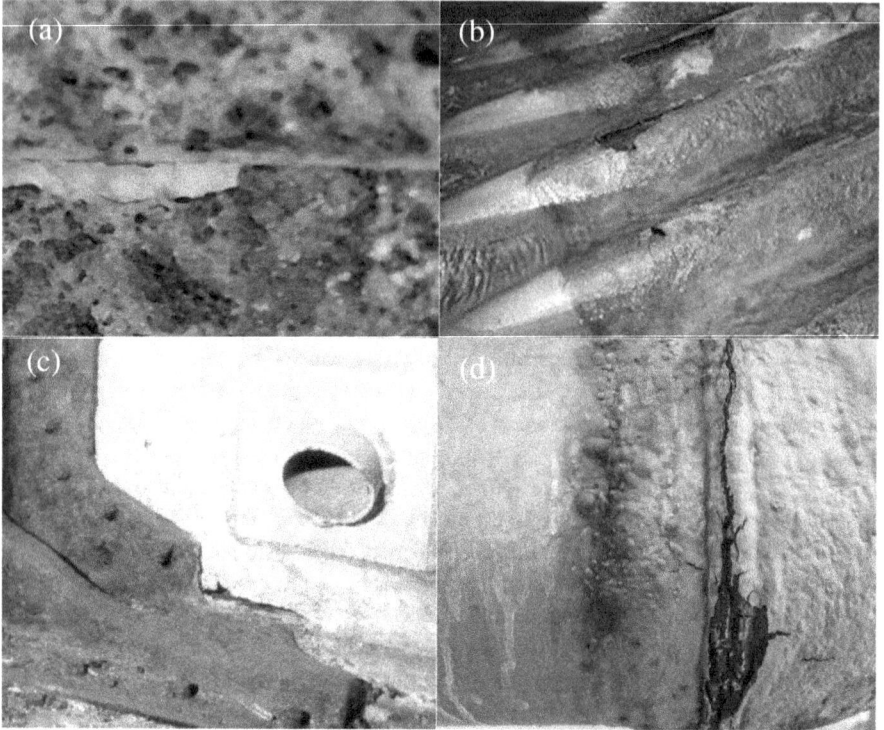

FIGURE 13.1   The images of (a) uniform corrosion, (b) localized corrosion, (c) galvanic corrosion, and (d) stress corrosion cracking.

of ions, enabling oxidation and reduction reactions to occur (Koushik et al., 2021). Moisture brings oxygen into contact with the metal surface, which is essential for the corrosion process. Oxygen reacts with the metal, initiating oxidation reactions and forming metal oxides or hydroxides. These compounds are often the corrosion products that weaken the metal's integrity. Moisture helps establish a corrosion cell, which is a localized electrochemical system. The metal surface acts as an anode, where oxidation occurs, while dissolved oxygen and moisture act as the cathode, facilitating reduction reactions. This galvanic cell leads to accelerated corrosion. Moisture can carry hygroscopic salts, such as chlorides, sulphates, or carbonates, into the environment. These salts have a high affinity for water and can absorb atmospheric moisture onto the metal surface. Hygroscopic salts significantly increase the corrosion rate by providing additional electrolytes and creating a more corrosive environment. Moisture can influence the pH of the surrounding environment. When moisture reacts with atmospheric gases or dissolved substances, it creates acidic or alkaline conditions. These pH changes directly affect the corrosion rate by altering the chemical reactions occurring on the metal surface (Gong et al., 2020). Moisture forms thin water films on metal surfaces, creating a microenvironment that promotes corrosion. Water films can trap corrosive substances, such as salts, pollutants, or acids, against the metal surface. They concentrate ions and promote localized corrosion, such as pitting or crevice corrosion. Cyclic exposure to moisture and subsequent drying, therefore, accelerates corrosion. During wet conditions, the metal corrodes, and when the surface dries, corrosion products can be left behind, creating an uneven surface. This repeated wet-dry cycling leads to corrosion pits, cracks, or fissures, further enhancing corrosion propagation.

The severity and specific mechanisms of corrosion in the presence of moisture can vary depending on factors such as the type of metal, the composition of the environment, and the exposure conditions. Proper corrosion prevention measures, such as protective coatings, surface treatments, and moisture control, are crucial to mitigate the detrimental effects of moisture on metal corrosion.

13.2.1.2 Oxygen

The availability of oxygen is essential for the corrosion process to occur. Oxygen reacts with the metal surface to form metal oxides, often corrosion products (Harsimran et al., 2021). Oxygen influences the rate and extent of corrosion which plays a complex role in corrosion, affecting various electrochemical reactions and contributing to the degradation of metals. Oxygen reacts with the metal surface, initiating oxidation reactions that form metal oxides or hydroxides. This process is often the initial step in corrosion, where metal atoms lose electrons and transform into metal ions. The resulting metal oxides/hydroxides may not provide adequate protection against further corrosion, and they can be porous or loosely adherent, exposing the metal to further degradation (Palanisamy, 2019).

Oxygen participates in electrochemical reactions within the corrosion cell. It acts as the cathodic reactant, consuming electrons and facilitating reduction reactions. Oxygen reduction occurs when dissolved oxygen reacts with water and electrons at the cathode. This reaction complements the anodic oxidation process, completing the corrosion cell and allowing the corrosion process to continue. Oxygen influences the corrosion potential of metal, thus establishing an equilibrium potential at which the anodic and cathodic reactions are balanced. The corrosion potential determines the tendency of a metal to corrode in a particular environment. Oxygen shifts the corrosion potential, making the metal more susceptible to corrosion by creating a more favourable environment for oxidation reactions (Yadav et al., 2021).

In some cases, metals can form a protective oxide layer on their surface that provides resistance against corrosion. This passive film is formed by reacting the metal with oxygen in the environment. However, if the passive film is damaged or oxygen is limited, the protective layer may break down, leading to localized corrosion or general corrosion (Xia et al., 2023). Oxygen contributes to pitting corrosion, localized corrosion characterized by small pits or holes on the metal surface. Pitting occurs when oxygen becomes depleted within a crevice or pit, creating an environment conducive to the anodic dissolution of the metal (Wang et al., 2022b). Oxygen concentration cells can form within these regions, exacerbating corrosion and forming deeper pits. Differences in oxygen concentration across a metal surface lead to the formation of oxygen-concentration cells, and these cells create localized variations in the corrosion rate, with the areas of higher oxygen concentration exhibiting lower corrosion rates and the areas of lower oxygen concentration experiencing accelerated corrosion. These concentration cells occur due to differences in oxygen diffusion or barriers restricting oxygen access.

The specific effect of oxygen on corrosion can vary depending on factors such as the type of metal, the nature of the corrosive environment, and other corrosive agents. Understanding and controlling oxygen availability and its interaction with other factors is crucial in corrosion prevention and mitigation strategies.

13.2.1.3 Salts and Electrolytes

Corrosion is typically enhanced in the presence of salts and other electrolytes. These substances increase the conductivity of the electrolyte, allowing more rapid corrosion reactions to occur. Salts and electrolytes significantly enhance corrosion processes and are often considered critical factors in accelerating metal degradation. Their presence in the environment can have several effects on corrosion.

Salts and electrolytes, when dissolved in water or moisture, increase the ionic conductivity of the solution. This enhanced conductivity facilitates the movement of ions, including metal ions, to and from the metal surface. It promotes electrochemical reactions by enabling the transfer of electrons and ions, thereby accelerating the corrosion process. Salts dissociate into their constituent ions in water, creating an electrolyte solution. This formation of ions provides the necessary conductive medium for electrochemical reactions to occur on the metal surface. Ions present in the electrolyte can participate in redox reactions, facilitating both anodic and cathodic processes. Salts and electrolytes can contribute to galvanic corrosion, which occurs when two dissimilar metals are in contact with an electrolyte (Dillon, 2019). The difference in electrochemical potential between the metals and the electrolyte creates a galvanic cell. This cell leads to accelerated corrosion of the less noble (more active) metal, which acts as the anode, while the more noble metal acts as the cathode.

Salts and electrolytes can create concentration cells on the metal surface, leading to localized variations in the corrosion rate. These concentration cells can occur due to differences in electrolyte concentration across the metal surface, resulting from evaporation, temperature gradients, or stagnant areas. The areas of higher electrolyte concentration exhibit lower corrosion rates, while regions of more insufficient concentration experience accelerated corrosion (Liu et al., 2020). Certain ions present in salts and electrolytes can directly contribute to corrosion. For example, chloride ions are well-known corrosion promoters, particularly for metals such as steel or stainless steel. Chloride ions can break down passive oxide layers, initiate pitting corrosion, and promote the anodic dissolution of metal ions (Parangusan et al., 2021). Depending on the metal and specific conditions, sulphate and bromide ions can also exhibit corrosive behaviour (Kumar et al., 2022).

Salts present in the environment can deposit onto the metal surface. This deposition can lead to the formation of salt layers or crusts that trap moisture, creating localized corrosion cells. Additionally, some salts are hygroscopic, meaning they can absorb atmospheric moisture. These hygroscopic salts can deliquesce, forming concentrated solutions that enhance the corrosive environment around the metal (Katona et al., 2023). Some salts and electrolytes can alter the pH of the surrounding environment. For example, acidic salts such as sulphates or nitrates can create an acidic environment, while basic salts such as carbonates or hydroxides can induce alkaline conditions. Significant pH changes directly affect the corrosion rate and the electro-chemical reactions on the metal surface.

The specific impact of salts and electrolytes on corrosion depends on factors such as the type of metal, the composition and concentration of the electrolyte, tempera-ture, and other environmental factors. Understanding the role of salts and electrolytes in corrosion is crucial for designing corrosion prevention strategies and selecting appropriate materials in corrosive environments.

13.2.1.4 Acids and Bases

Strong acids or bases can corrode metals by attacking the metal surface and pro-moting dissolution or chemical reactions. Acidic or alkaline environments can be particularly corrosive to certain metals. Acids and bases can significantly affect the corrosion process, depending on the type and concentration of the corrosive sub-stances involved. Strong acids or bases can accelerate corrosion by directly attacking the metal surface or altering the surrounding environment.

Strong acids react with metals, leading to their dissolution or chemical reactions with the metal surface. Acids can remove the protective oxide layer on the metal, exposing it to further corrosion (Anh et al., 2020). Similarly, strong bases can react with metals to form metal hydroxides, which may not provide effective protection against corrosion. These direct attacks result in the rapid degradation of the metal. Acids and bases change the pH of the surrounding environment. Acids lower the pH, creating an acidic environment, while bases raise the pH, resulting in alkaline con-ditions. Extreme pH values accelerate corrosion by altering the chemical reactions occurring on the metal surface. For example, low pH increases the rate of hydro-gen evolution, leading to more aggressive corrosion. Certain metals, such as iron, are particularly susceptible to acidic corrosion. In strong acids, the metal undergoes accelerated dissolution, forming corrosion products. Acidic corrosion can lead to the loss of metal mass and the formation of pits, grooves, or irregular surface features (Chen et al., 2020). While alkaline environments are generally less corrosive than acidic ones, certain metals can still undergo corrosion in alkaline conditions. For example, aluminium can corrode strongly alkaline solutions, such as concentrated sodium hydroxide (caustic soda), forming aluminium hydroxide. Alkaline corro-sion results in surface etching, pitting, or the formation of alkaline batteries on the metal surface. Cleaning agents that contain acids or bases can cause corrosion if used improperly or if they come into contact with sensitive metals. For example, strong acid-based cleaners remove protective layers on metal surfaces, increasing corrosion susceptibility (Yang, 2021). Therefore, following appropriate guidelines and recom-mendations when using cleaning agents to prevent corrosion damage is essential.

Acid rain, precipitation with a low pH due to atmospheric pollutants, can sig-nificantly enhance corrosion. When acid rain contacts metal surfaces, it can dissolve or react with the protective oxide layer, leaving the metal vulnerable to corrosion. Acid rain damages metal structures like steel, copper, or aluminium. Acidic or alkaline environments in industrial processes involving acids or bases can expose metal equipment and structures to corrosive conditions. Examples include chemical manufacturing, mining, wastewater treatment, and battery production. The corrosive effects depend on the specific acids or bases used, their concentration, temperature, and exposure duration. It is essential to consider the compatibility of metals with acids and bases and implement appropriate corrosion prevention measures, such as material selection, protective coatings, and proper maintenance, to mitigate the effects of acids and bases on corrosion.

13.2.1.5 Pollution and Contaminants

Environmental pollutants, such as sulphur compounds, industrial emissions, and airborne particles, can accelerate corrosion. These contaminants may react with the metal surface, creating aggressive environments and promoting corrosion (Wang et al., 2022b). Environmental pollution and contaminants from industrial emissions, atmospheric pollutants, chemical spills, or other sources significantly impact corrosion processes by introducing additional corrosive agents or accelerating existing corrosion mechanisms. For example, sulphur compounds, such as sulphur dioxide and hydrogen sulphide, can react with metal surfaces to form metal sulphides, which are often less protective and more susceptible to corrosion (Becker et al., 2022). Acidic pollutants directly attack metal surfaces, causing corrosion through acid etching or dissolution. Pollution and contaminants can create more aggressive environments for corrosion. For instance, acidic pollutants lower the pH of the surrounding environment, increasing the corrosion rate (Soufeiani et al., 2020). Similarly, pollutants rich in chlorides or other corrosive ions can enhance the conductivity of the electrolyte, promoting more rapid corrosion.

Airborne pollutants, such as sulphur compounds, nitrogen, or particulate matter, can deposit on metal surfaces and contribute to corrosion. These pollutants act as catalysts, accelerators, or initiators of corrosion reactions (Nascimento and Furtado, 2022). Additionally, particulate matter traps moisture and creates localized areas of high humidity, promoting corrosion. Certain gases in the atmosphere, such as sulphur dioxide, hydrogen sulphide, chlorine, or ammonia, can directly react with metal surfaces and initiate corrosion. These gases penetrate protective coatings or penetrate through metal porosity, leading to localized corrosion or general corrosion.

Emissions from industrial processes, such as sulphur dioxide from combustion or acid fumes from chemical production, can contribute to corrosion. These emissions settle on metal surfaces and corrode the metal directly or react with environmental moisture to form corrosive substances. Pollutants and contaminants promote the accumulation of corrosion products on metal surfaces (Xiao et al., 2021). These chemically aggressive deposits create localized corrosion cells, crevices, or galvanic couples that accelerate corrosion. Corrosion product accumulation also traps moisture and other corrosive agents, leading to prolonged exposure and increased corrosion rates.

Some pollutants and contaminants can foster microbial activity on metal surfaces, leading to microbiologically influenced corrosion. Microorganisms like bacteria or fungi produce corrosive byproducts or create localized microenvironments that accelerate corrosion. Pollution and contaminants compromise the integrity of protective coatings, such as paint or corrosion inhibitors (Rao and Mulky, 2023), and thus degrade or penetrate the coatings, exposing the metal surface to corrosive agents and facilitating corrosion. The effects of pollution and contaminants on corrosion depend on the type of metal, the specific pollutants involved, and the environmental conditions. Mitigating pollution-induced corrosion requires effective control measures, regular maintenance, and appropriate corrosion prevention strategies.

13.2.1.6   Temperature

Elevated temperatures can increase the rate of corrosion by accelerating chemical reactions. Higher temperatures cause thermal cycling, leading to the expansion and contraction of metals, which facilitate crack formation and corrosion (Dacillo and Zarrouk, 2023). The effect of temperature on corrosion is multifaceted and can impact corrosion rates, the types of corrosion mechanisms, and the overall degradation of metals.

Higher temperatures generally lead to increased reaction rates in chemical processes, including corrosion. Elevated temperatures provide greater thermal energy to reactants, enabling more rapid chemical reactions at the metal surface and in the corrosive environment (Ma et al., 2021). This results in accelerated corrosion rates and more severe degradation of metals. Corrosion reactions often require a certain amount of energy to occur, known as activation energy. Higher temperatures provide the necessary energy to overcome the activation barriers, facilitating the corrosion process.

Consequently, increasing the temperature can lower the activation energy, leading to more frequent and efficient corrosion reactions. Temperature affects the mobility of reactants and products involved in corrosion reactions. The diffusion of ions, dissolved gases, and other reactive species is enhanced at higher temperatures. This increased mobility promotes the transport of corrosive agents to the metal surface and facilitates the removal of corrosion products, further accelerating the corrosion process (Xu et al., 2022). Temperature can alter the properties of the corrosive solution, including its pH, conductivity, and solubility. Higher temperatures increase the ionization of dissolved species, affect the dissociation of acids or bases, and modify the solubility of salts. These changes impact the corrosivity of the environment and influence the types and rates of corrosion reactions.

Fluctuating temperatures, such as in thermal cycling conditions, can contribute to corrosion. Alternating between high and low temperatures induces expansion and contraction of the metal, leading to stress accumulation and the initiation of cracks or fissures. Subsequent exposure to corrosive agents during thermal cycling exacerbates corrosion by providing pathways for corrosion propagation. Some metals form protective passive films that help resist corrosion. However, elevated temperatures can destabilize these films and reduce their protective properties. Thermal energy disrupts the passive film's structure, causing it to break down or become less adherent, leading to increased vulnerability to corrosion (Sadeghi et al., 2019). When dissimilar metals are in contact, temperature changes can create differential thermal expansion. This generates mechanical stress at the metal interface, damaging protective films and promoting localized corrosion, such as galvanic or crevice corrosion.

Certain metals and alloys are susceptible to specific types of corrosion at high temperatures. For example, oxidation, sulphidation, carburization, or nitridation occurs in industrial environments with elevated temperatures, such as combustion systems, chemical plants, or power generation facilities. High-temperature corrosion mechanisms involve complex reactions between metals and reactive species at elevated temperatures (Bell et al., 2019). The impact of temperature on corrosion depends on factors such as the type of metal, the corrosive environment, and other contributing factors. Proper temperature control, material selection, and thermal insulation are crucial considerations in corrosion prevention and mitigation strategies, particularly in high-temperature applications.

13.2.1.7 Galvanic Corrosion

When two dissimilar metals are in contact with an electrolyte, galvanic corrosion can occur. This type of corrosion is driven by the difference in electrochemical potential between the two metals, leading to accelerated corrosion of the less noble (more active) metal. Galvanic corrosion, or bimetallic corrosion, occurs when two different metals or alloys come into contact with an electrolyte (Francis et al., 2020). This contact creates a galvanic cell, accelerating the corrosion of the less noble (more active) metal.

Galvanic corrosion arises from the difference in electrochemical potentials between two dissimilar metals or alloys. A galvanic cell is established when they are electrically connected through an electrolyte. The more active metal acts as the anode, undergoing oxidation and releasing electrons, while the more noble metal acts as the cathode, undergoing reduction and consuming electrons. This electrochemical process promotes corrosion at the anode. The anode, the less noble metal, experiences accelerated corrosion due to galvanic corrosion (Falconer et al., 2023). The galvanic current flowing from the anode to the cathode increases the rate of oxidation reactions, leading to faster metal dissolution and corrosion. The anode becomes the sacrificial metal, sacrificing itself to protect the more noble metal in the galvanic couple.

Galvanic corrosion provides cathodic protection to the more noble metal in the galvanic couple (Nergis et al., 2019). As the anode corrodes, it releases electrons consumed by the cathode, preventing or reducing its corrosion. This principle is used in cathodic protection systems, where sacrificial anodes (usually made of more active metals) are placed to protect the protected structure or component. An electrolyte, such as moisture or a corrosive solution, is essential for galvanic corrosion. The electrolyte enables the flow of ions between the anode and cathode, facilitating electrochemical reactions. The electrolyte provides the medium for ion transport, leading to metal dissolution at the anode and reduction reactions at the cathode. The extent of galvanic corrosion depends on the position of the metals in the galvanic series (Harsimran et al., 2021). Metals farther apart in the galvanic series exhibit a larger potential difference, resulting in more severe galvanic corrosion. It is important to consider the compatibility of metals when designing structures or using different metals in contact to minimize galvanic corrosion.

Galvanic corrosion contributes to localized corrosion phenomena, such as pitting or crevice. Differences in oxygen concentration or pH between the anode and cathode regions create localized cells, leading to accelerated corrosion and the initiation of localized corrosion attacks (Xu and Tan, 2019). The galvanic corrosion potential of different metal combinations and implementing appropriate measures, such as electrical insulation, coatings, or the use of compatible metals, are crucial to prevent or mitigate galvanic corrosion in practical applications.

13.2.1.8 Mechanical Stress

Applied stress or mechanical factors, such as vibrations, cyclic loading, or friction, can cause localized damage to the protective oxide layer on metals, making them more susceptible to corrosion. Mechanical stress, such as applied loads, vibrations, cyclic loading, or friction, can significantly impact the corrosion behaviour of metals (Abdollahzadeh et al., 2021). The combination of mechanical stress and corrosion can lead to accelerated degradation and failure of metal structures.

Mechanical stress promotes the initiation and propagation of stress corrosion cracking. Stress corrosion cracking (SCC) is a phenomenon where the combined action of tensile stress and a corrosive environment leads to the formation and propagation of cracks in the metal (Basukumar and Arun, 2022). The presence of corrosive species, such as moisture or specific chemicals, along with mechanical stress, significantly reduces the threshold stress required for crack initiation and accelerates crack propagation. Mechanical stress interacts with corrosion in a phenomenon known as fatigue corrosion. Fatigue occurs when a metal is subjected to cyclic loading, leading to the initiation and propagation of cracks. In the presence of a corrosive environment, the fatigue life of a metal is significantly reduced due to the combined effects of cyclic loading and corrosion.

Mechanical stress can increase the wear rate of metals when combined with corrosion. Corrosive substances act as abrasives, accelerating the wear of the metal surface. The mechanical stress exacerbates this process by promoting the removal of the protective oxide layer and exposing fresh metal surfaces to corrosion and further wear. Mechanical stress contributes to the formation and propagation of crevices, leading to crevice corrosion (Gong et al., 2022). Gaps or crevices form at interfaces or joints when metal is under stress. These crevices can trap corrosive substances and restrict oxygen and ion diffusion, creating localized environments promoting crevice corrosion. Mechanical stress affects galvanic corrosion, which occurs when two dissimilar metals are in contact with the presence of an electrolyte. The applied stress induces deformation and micro-galvanic effects at the metal interfaces, leading to accelerated galvanic corrosion (Chen et al., 2021).

Mechanical stress, such as scratching, grinding, or deformation, can cause surface damage to the metal. This damage removes the protective oxide layer or creates stress concentration sites, making the metal more vulnerable to corrosion initiation and propagation. This promotes corrosion fatigue, the combined effect of cyclic loading and corrosion. The cyclic loading causes localized deformation and strain accumulation, while the corrosive environment weakens the metal's resistance to fatigue failure (Zhao et al., 2022). This combination results in accelerated crack initiation and growth, leading to premature failure of the metal under cyclic loading conditions.

The effects of mechanical stress on corrosion vary depending on factors such as the type of metal, the applied stress level, the corrosive environment, and the presence of other factors like temperature or moisture. Mitigating the effects of mechanical stress on corrosion involves implementing proper design practices, selecting appropriate materials, applying protective coatings, and employing stress-relief measures.

13.2.1.9 Surface Condition

Surface roughness, scratches, or imperfections can create sites for corrosion initiation. Irregularities on the metal surface disrupt the protective oxide layer, exposing the underlying metal to corrosion agents. The condition of the metal surface is a significant factor that can influence the occurrence and progression of corrosion. The surface condition refers to the metal surface's quality, smoothness, cleanliness, and integrity.

Many metals naturally form a thin oxide layer on their surface, which acts as a protective barrier against corrosion. A well-maintained, intact oxide layer inhibits the penetration of corrosive agents and slows down the corrosion process. However, if the surface is compromised or damaged, the protective oxide layer is disrupted, making the metal more susceptible to corrosion (Unune et al., 2022). Rough surfaces, characterized by irregularities, scratches, or pits, can enhance corrosion initiation and propagation. Corrosion tends to occur preferentially in areas with higher surface roughness, as these regions have larger surface areas, greater vulnerability to corrosive agents, and increased possibilities for forming crevices or localized corrosion sites. Surface contaminants, such as dirt, oil, salts, or pollutants, act as corrosion initiators or accelerators (Tukhlievich and Verma, 2023). These contaminants create localized galvanic cells, provide sites for the adsorption of corrosive species, or introduce corrosive chemicals onto the metal surface. Additionally, deposits trap moisture and corrosive agents against the metal, promoting corrosion.

13.2.1.10 Coating and Protective Layer Degradation

Coatings and protective layers are vital in mitigating corrosion by providing a barrier between the metal surface and the corrosive environment (Lawal et al., 2023). Protective coatings, such as paint, plating, or corrosion inhibitors, can deteriorate over time. If the coating is compromised, the underlying metal becomes vulnerable to corrosion.

Coatings and protective layers are a physical barrier separating the metal from the corrosive environment. When these layers degrade or are damaged, they lose their ability to prevent direct contact between the metal and corrosive agents such as moisture, oxygen, acids, or salts (Rao and Mulky, 2023). This leads to an increased susceptibility of the metal to corrosion. Coating degradation exposes the metal surface to a corrosive environment. Without the protective layer, the metal becomes vulnerable to attack from aggressive substances present in the environment, resulting in accelerated corrosion rates.

Coating defects, such as pinholes, cracks, or delamination, create localized exposed metal areas. These defects serve as initiation sites for corrosion, allowing corrosive agents to penetrate and attack the metal surface (Qureshi et al., 2022). Localized corrosion mechanisms are worsened by coating degradation, leading to galvanic cells forming between the exposed metal and the intact coated regions. This occurs when dissimilar metals or areas with different coating integrity are in contact with the presence of an electrolyte. Galvanic corrosion accelerates the corrosion of the less noble metal, promoting localized corrosion and further compromising the protective system.

When a coating degrades or becomes permeable, it traps corrosive substances against the metal surface. Moisture, salts, acids, or other corrosive agents accumulate at the interface between the coating and the metal, leading to underfilm corrosion. The trapped corrosive substances cause localized corrosion, blistering, or the formation of corrosive pits beneath the degraded coating (Peltier and Thierry, 2022). The entrapment of corrosion products hinders the ability to inspect and maintain the coating system, making it challenging to identify and repair coating defects. Coating degradation reduces the effective service life of the protective system. A degraded or compromised coating requires frequent maintenance, repair, or replacement to ensure continued corrosion protection. Premature coating failure can lead to increased maintenance costs, downtime, and potential damage to the underlying structure or equipment. The proper coating selection, application, and regular inspection to maintain the integrity of protective layers. Timely repair or reapplication of coatings is essential to ensure continued corrosion protection and extend the lifespan of the metal substrate.

13.2.2 FACTORS THAT AFFECT CORROSION INHIBITORS

Corrosion inhibitors are substances used to mitigate or slow down the corrosion process. While corrosion inhibitors can effectively protect metals, several factors influence their performance.

13.2.2.1 Environment

The corrosive environment in which the metal is exposed plays a crucial role in the performance of corrosion inhibitors. Factors such as temperature, humidity, pH, the presence of aggressive chemicals, and the composition of the electrolyte can all impact the effectiveness of inhibitors. Inhibitors may perform differently in acidic, alkaline, or saline environments, and their efficiency can vary accordingly (Al-Moubaraki and Obot, 2021). The environmental pH and acidity can influence corrosion inhibitors' performance. Inhibitors have optimal pH ranges in which they are most effective. Highly acidic or alkaline environments alter the chemical properties of the inhibitor and affect its ability to form and maintain a protective film on the metal surface (Abdel-Karim and El-Shamy, 2022). Temperature plays a critical role in the effectiveness of corrosion inhibitors. Elevated temperatures can increase the corrosion rate and may require inhibitors with higher thermal stability and effectiveness. Additionally, the temperature can affect inhibitors' solubility and diffusion rate, influencing their ability to reach and protect the metal surface. The presence of oxygen impacts the performance of corrosion inhibitors (Verma et al., 2019). Inhibitors may function by reducing oxygen availability at the metal surface, preventing or slowing down oxidation reactions. Higher oxygen concentrations challenge inhibitors, requiring higher or more potent inhibitor formulations.

Moisture and humidity levels in the environment affect the performance of corrosion inhibitors. Some inhibitors need a certain moisture level to form a protective film, while excessive moisture can dilute or wash away the inhibitor from the metal surface. Humidity promotes moisture absorption onto the metal, potentially impacting the inhibitor's ability to adhere and protect the surface (Al-Amiery et al., 2023).

13.2.2.2 Inhibitor Concentration

The concentration of the corrosion inhibitor is a critical factor that affects its performance. In general, higher inhibitor concentrations provide better corrosion protection. However, there may be an optimal concentration range beyond which increasing the inhibitor concentration may not significantly improve the protection. It is essential to determine the appropriate dosage or concentration of the inhibitor based on the specific application and requirements.

The concentration of the corrosion inhibitor directly affects its inhibition efficiency (Masum et al., 2019). The higher concentrations of inhibitors lead to greater corrosion protection. A higher concentration increases the probability of inhibitor molecules adsorbing onto the metal surface and forming a protective layer. The adsorbed inhibitor molecules act as a barrier, reducing the access of corrosive agents to the metal and inhibiting corrosion reactions (de Souza Morais et al., 2023). Corrosion inhibitors may reach a saturation point where increasing the concentration beyond a certain threshold does not provide additional benefits. This occurs when the inhibitor molecules form a complete monolayer on the metal surface, and further inhibitor molecules cannot effectively adsorb. Increasing the concentration of the inhibitor will not result in a proportional increase in inhibition efficiency. Each corrosion inhibitor has an optimal concentration range where it exhibits maximum efficiency. This concentration range depends on various factors, including the type of inhibitor, the specific metal and environment, and the desired level of corrosion protection. Deviating from the optimal concentration can lead to diminished inhibition efficiency. The solubility of the corrosion inhibitor is another important consideration. Insoluble inhibitors may require higher concentrations to achieve effective inhibition, as they must be dissolved or dispersed in the system to interact with the metal surface (Liu et al., 2023). Conversely, highly soluble inhibitors may be effective at lower concentrations, as they readily dissociate and adsorb onto the metal.

13.2.2.3 Inhibitor Type and Composition

The choice of corrosion inhibitor and its chemical composition can significantly influence its effectiveness. Different inhibitors have varying mechanisms of action and are suitable for specific corrosive environments. Some common types of inhibitors include organic compounds, inorganic compounds, passivation, film-forming inhibitors, and volatile corrosion inhibitors (VCIs). Selecting the most appropriate inhibitor for a particular metal and environment is crucial for adequate corrosion protection (Abdel-Karim and El-Shamy, 2022).

Corrosion inhibitors can be classified into several types, including organic inhibitors, inorganic inhibitors, mixed inhibitors, and volatile inhibitors. Organic compounds containing nitrogen, phosphorus, or oxygen atoms form a protective film on the metal surface, acting as a barrier against corrosive agents (Abdelwedoud et al., 2022). Organic inhibitors are generally effective in aqueous environments and provide temporary and long-term protection. Inorganic inhibitors include substances such as chromates, molybdates, and phosphates. They form insoluble compounds on the metal surface, creating a protective layer. Inorganic inhibitors are effective in specific environments and can provide reasonable protection against certain types of corrosion (Al-Amiery et al., 2023).

Mixed inhibitors combine organic and inorganic components to provide enhanced corrosion protection. Combining different inhibitor types improves the performance and effectiveness of corrosion inhibition. Mixed inhibitors are designed to target specific corrosion mechanisms or environments. Volatile inhibitors, also known as vapour-phase inhibitors (VPIs), release vapour-phase corrosion inhibitors that protect the metal surface (Karki, 2020). These inhibitors work by adsorbing onto the metal and forming a protective layer. VPIs are particularly useful for protecting enclosed or hard-to-reach areas.

The composition of corrosion inhibitors influences their ability to prevent or reduce corrosion. The active ingredients in corrosion inhibitors determine their mode of action and affinity for metal surfaces. These ingredients include organic or inorganic compounds, salts, surfactants, or polymers. Selecting suitable active ingredients depends on the type of corrosion, the metal being protected, and the specific environment. The concentration of the inhibitor affects its effectiveness. A higher concentration of inhibitors may provide more vital corrosion protection; however, an excessive concentration can lead to other issues, such as fouling or reduced efficiency (Khan et al., 2022). Finding the optimal concentration is crucial to achieving the desired corrosion inhibition. The surface condition and characteristics of the metal can impact the performance of corrosion inhibitors. Factors such as surface roughness, cleanliness, pre-existing corrosion products, and the nature of the metal's passive film can affect the inhibitor's adsorption and protective properties.

13.2.2.4   Inhibitor Adsorption

Corrosion inhibitors typically work by adsorbing onto the metal surface and forming a protective barrier or film. The ability of the inhibitor to adsorb onto the metal surface depends on factors such as molecular structure, surface charge, intermolecular interactions, and the nature of the metal surface. Strong adsorption ensures better coverage and protection against corrosive species (Zhang et al., 2021).

Corrosion inhibitors can adsorb onto the metal surface, forming a protective film. This film acts as a barrier between the metal and the corrosive environment, preventing or slowing down corrosion. The adsorption of inhibitors alters the electrochemical reactions occurring at the metal surface, reducing the rate of anodic or cathodic reactions (Kaya et al., 2023a). Corrosion inhibitors that adsorb onto the metal surface can inhibit specific corrosion reactions. They hinder the oxidation or reduction reactions involved in the corrosion process. By blocking or reducing the access of corrosive species to the metal surface, inhibitors can effectively reduce the corrosion rate.

Corrosion inhibitors often form insoluble complexes with metal ions in a corrosive environment. These complexes can precipitate or deposit onto the metal surface, forming a protective layer. The formation of insoluble complexes reduces the availability of metal ions for corrosion reactions, thus suppressing the corrosion process (Vasyliev et al., 2020). Inhibitors that adsorb onto the metal surface can react with corrosion products, such as metal oxides or hydroxides. They dissolve the existing corrosion products or inhibit their formation and growth. This interaction prevents the accumulation of corrosive products, which can further enhance the protection of the metal surface.

The extent of inhibitor adsorption and surface coverage is essential to corrosion inhibition. A higher degree of adsorption and surface coverage leads to better protection against corrosion. However, excessive adsorption can form thick and less protective inhibitor films, which may hinder the diffusion of reactants and slow the corrosion inhibition process (Thakur and Kumar, 2021). The reversibility of inhibitor adsorption is crucial for the long-term effectiveness of corrosion inhibitors. Ideally, inhibitors should reversibly adsorb onto the metal surface, allowing them to re-establish a protective layer if the original film is damaged or depleted. Reversible adsorption ensures continuous corrosion protection and enhances the durability of the inhibitor's performance.

13.2.2.5 Temperature and Time

Temperature and exposure time can influence the effectiveness of corrosion inhibitors. Higher temperatures may increase the rate of inhibitor degradation or reduce its adsorption capacity (Kobzar and Fatyeyeva, 2021). The duration of exposure to the corrosive environment can affect the inhibitor's longevity and efficiency. Long-term exposure or prolonged immersion leads to inhibitor depletion or breakdown, requiring reapplication or maintenance.

Elevated temperatures can accelerate the degradation of corrosion inhibitors. Higher temperatures enhance chemical reactions, increase the rate of inhibitor decomposition, and reduce the inhibitor's effectiveness. The increased kinetic energy of molecules at higher temperatures leads to faster chemical reactions, reducing the inhibitor's longevity. Some corrosion inhibitors may become more volatile at higher temperatures, leading to their evaporation or depletion from the system (Calegari et al., 2022). This loss of inhibitor concentration can diminish the protective effects and increase the susceptibility of the metal to corrosion. Corrosion inhibitors should possess good thermal stability to withstand higher temperatures without significant degradation. Inhibitors that decompose or undergo chemical changes at elevated temperatures may lose their inhibiting properties, rendering them ineffective.

Corrosion inhibitors should maintain their effectiveness over an extended period. Time is a critical factor in assessing the long-term stability and performance of inhibitors. Some inhibitors may gradually degrade or lose their inhibiting properties over time, which results in diminished corrosion protection. Depending on the system and exposure conditions, corrosion inhibitors may deplete over time due to reactions with corrosive species, surface adsorption, or chemical transformations. As the inhibitor concentration decreases, its protective effect may diminish, and the corrosion rate can increase. The corrosive environment may change over time, potentially affecting the performance of corrosion inhibitors (Shang and Zhu, 2021). For example, the composition of the electrolyte, pH, or presence of contaminants may evolve, altering the effectiveness of the inhibitor or introducing new corrosion mechanisms. However, to effectively address the impact of temperature and time on corrosion inhibitors, it is essential to select inhibitors that are stable and retain their inhibiting properties, conduct thorough testing and evaluation of the inhibitors' performance, monitor inhibitor concentrations, and regularly assess the corrosion rate and condition of the protected metal under the anticipated temperature and exposure condition.

13.2.2.6   Flow or Dynamic Conditions

Inhibitor performance can vary under dynamic conditions, such as in flowing fluids or environments with frequent changes in fluid composition. The flow rate, turbulence, and the presence of solid particles can affect the inhibitor's distribution, adsorption, and renewal on the metal surface. The corrosive medium's flow in dynamic conditions affects the corrosion inhibitor's transport to the metal surface (Zhu et al., 2021). The inhibitor molecules must be efficiently transported to the corroding sites for protection. Flow can enhance the mass transport of the inhibitor, ensuring a more uniform and continuous coverage of the metal surface, which can enhance the inhibiting effect.

Flow conditions lead to the dilution of the inhibitor concentration in the system. High flow rates cause increased turbulence and mixing, reducing the inhibitor's concentration near the metal surface (Obot et al., 2019). This dilution effect decreases the effectiveness of the corrosion inhibitor, leading to reduced corrosion protection. Flow conditions can influence the kinetics of inhibitor adsorption, reducing the residence time of the inhibitor at the metal surface and limiting the adsorption process. This could result in incomplete inhibitor coverage or weaker inhibitor-metal interactions, compromising the effectiveness of the corrosion inhibitor.

In systems with high flow rates, mechanical erosion and corrosion, known as erosion-corrosion, can occur (Senatore et al., 2021). Flow-induced turbulence and high-velocity fluid streams cause accelerated wear and damage to the protective film formed by the corrosion inhibitor. This could lead to the exposure of bare metal surfaces, making them more susceptible to corrosion. Some corrosion inhibitors rely on passivation, where a protective oxide layer forms on the metal surface, to provide corrosion protection. Flow conditions, particularly high-velocity flow, could disrupt or strip away the passivating layer, diminishing the inhibiting effect of the corrosion inhibitor. These affect the stability and longevity of corrosion inhibitors. Aggressive flow, high temperatures, or exposure to certain chemicals degrades or destabilizes the inhibitor molecules, thus reducing their inhibiting effectiveness.

13.2.2.7   Compatibility and Stability

Corrosion inhibitors should be compatible with the protected metal and any existing surface treatments or coatings. Incompatibility could lead to adverse reactions, reduced protection, or accelerated corrosion. Corrosion inhibitors should be compatible with the metal or alloy they are intended to protect. Compatibility refers to the ability of the inhibitor to interact with the metal surface and form a stable and adherent protective layer (Kaya et al., 2023c). If the inhibitor is incompatible, it may not effectively adsorb or react with the metal surface, reducing corrosion inhibition efficiency. The inhibitors incompatible with the environment may undergo chemical reactions, precipitation, or degradation, which can diminish their ability to provide corrosion protection (Alamri, 2020). Corrosion inhibitors should possess good stability in the corrosive environment over a desired period. Stability refers to the ability of the inhibitor to resist degradation or breakdown when exposed to corrosive conditions. If the inhibitor is unstable, it may decompose or react prematurely, reducing its effectiveness and limiting its long-term protection capabilities.

The stability of corrosion inhibitors directly impacts their inhibition efficiency over time (Askari et al., 2021). Inhibitors that maintain stability and effectiveness for extended periods provide long-lasting corrosion protection. However, if an inhibitor's stability diminishes with time, it may lose its inhibitive properties, necessitating frequent reapplication or replacement. The compatibility and stability of corrosion inhibitors also have implications for environmental considerations. Inhibitors that are not environmentally friendly or degrade into harmful byproducts can harm ecosystems or human health. Therefore, the compatibility and stability of inhibitors should be assessed not only in terms of corrosion protection but also in terms of their environmental impact.

13.3   PLANT EXTRACTS AS CORROSION INHIBITORS

Plant extracts have gained significant attention as potential corrosion inhibitors due to their environmentally friendly nature and abundance in nature. Using plant extracts as corrosion inhibitors offers environmental benefits compared to conventional chemical inhibitors (Kumari, 2022). Plant extracts derived from renewable resources are biodegradable and have low toxicity. They provide a sustainable alternative to synthetic corrosion inhibitors, reducing the environmental impact and potential health risks associated with traditional inhibitor chemicals.

13.3.1   INHIBITORY MECHANISMS

Plant extracts contain various phytochemical compounds, such as phenols, flavonoids, tannins, alkaloids, and organic acids, which can act as corrosion inhibitors. These compounds can scavenge free radicals and inhibit the oxidation reactions that lead to metal corrosion (Abdel-Karim and El-Shamy, 2022). These compounds (see Figure 13.2) possess inhibitory properties by forming a protective film on the metal surface, reducing the anodic or cathodic reactions, or inhibiting corrosion product deposition. They may also act as mixed-type inhibitors, affecting both the anodic and cathodic processes simultaneously.

Some plant extracts can promote the formation of protective passive films on the metal surface. These films act as a barrier, preventing the access of corrosive agents to the metal substrate. Plant extracts containing compounds like tannins and polyphenols facilitate the formation and stabilization of passive films, thereby inhibiting corrosion (Wamba-Tchio et al., 2022). The inhibitory action of plant extracts is often attributed to their ability to adsorb onto the metal surface, forming a protective layer. The adsorbed plant extracts can modify the electrochemical properties of the metal, reduce the availability of corrosive agents, and impede corrosion reactions. The adsorption process is typically influenced by factors such as pH, temperature, concentration, and the chemical composition of the extract.

Certain plant extracts possess film-forming properties, allowing them to form a physical barrier on the metal surface. These extracts create a protective layer that acts as a sacrificial coating, shielding the metal from the corrosive environment. Plant extracts also exhibit synergistic effects when combined with other inhibitors or corrosion-resistant materials such as zinc and silver ions (Abdel Hameed et al.,

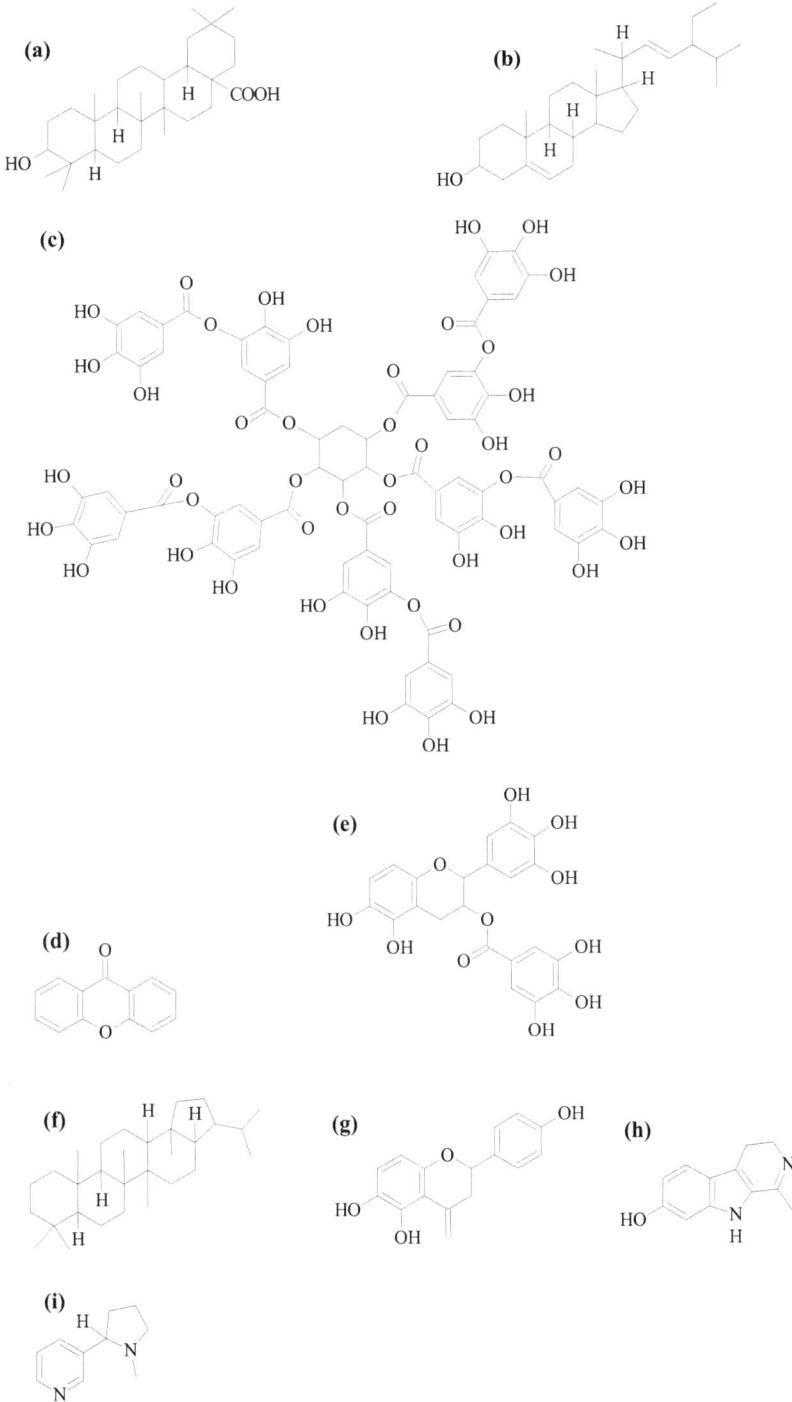

FIGURE 13.2   The structure of (a) saponins, (b) phytosterol, (c) tannins, (d) anthraquinones, (e) phenolic compound, (f) triterpene, (g) flavonoid, (h) glycoside, and (i) alkaloid.

2021; Jothi et al., 2022). These combinations enhance the overall corrosion inhibition efficiency by providing multiple modes of action and improving the stability of the protective layer on the metal surface.

13.3.2 Film Formation

Many plant extracts contain organic compounds that can adsorb onto the metal surface, forming a protective film. This film acts as a physical barrier, preventing the corrosive agents from accessing the metal surface (Kaya et al., 2023a). The film may also be chemically reactive, forming complexes with metal ions and stabilizing the surface.

Plant extracts contain diverse natural compounds with inherent antioxidant and anti-corrosive properties, making them suitable for forming protective films on metal surfaces (Ong et al., 2021). The active compounds in plant extracts inhibit corrosion by effectively blocking the access of corrosive agents to the metal surface. They act as corrosion inhibitors by suppressing anodic or cathodic reactions or forming passive films that impede the corrosion process, forming a protective film on metal surfaces through various mechanisms. These include adsorption onto the metal surface, chelation with metal ions, formation of complexes with metal oxide/hydroxide layers, or polymerization to create a barrier layer. Plant extracts are combined with other compounds or polymers to enhance their corrosion protection efficiency (Go et al., 2019; Verma et al., 2022). Synergistic effects can be achieved by combining plant extracts with additives such as nanoparticles, polymers, or other natural compounds to improve the film-forming properties and overall performance of the coating.

Some plant extracts exhibit self-healing properties, where the film formed on the metal surface can reseal or regenerate in the presence of damage or defects. This ability to heal and reform the protective film enhances the long-term effectiveness of plant extract-based coatings for corrosion protection. However, using plant extracts as film-forming materials for corrosion protection poses some challenges. These include the variability in the composition and effectiveness of plant extracts due to factors like plant species, extraction methods, and environmental conditions. Additionally, the durability and long-term stability of the plant extract-based films need to be evaluated to ensure their reliability under different environmental conditions.

13.3.3 Passivation Enhancement

Plant extracts have been found to enhance the passivation of metals, promoting the formation of a stable and protective oxide layer. The compounds in the extracts facilitate the growth and stability of the passive film, reducing the metal's susceptibility to corrosion. Plant extracts have been reported to enhance the passivation of metals by facilitating the formation of a stable and protective oxide layer on the metal surface (Oguike et al., 2020; Kaya et al., 2023b). The organic compounds in the plant extracts interact with the metal ions and accelerate the formation of a passivating oxide film. This passivation layer is a barrier against corrosion, reducing the metal's susceptibility to degradation.

13.3.4 Environmental Friendliness

One of the significant advantages of using plant extracts as corrosion inhibitors is their eco-friendliness compared to traditional inhibitors. Plant extracts are biodegradable from renewable resources, reducing the environmental impact associated with corrosion inhibition. Plant extracts are renewable resources, making them attractive for environmentally friendly corrosion inhibition (Verma et al., 2023). Using plant extracts can reduce the need for synthetic and potentially harmful chemical inhibitors, promoting sustainability and reducing environmental impact. Plant extracts contain many bioactive compounds, such as phenols, flavonoids, tannins, alkaloids, and terpenoids. These compounds possess inherent antioxidant and antimicrobial properties, contributing to corrosion inhibition by scavenging free radicals, inhibiting metal oxidation, and preventing microbial-induced corrosion.

Extracting and utilizing plant compounds for corrosion inhibition often involves green synthesis methods (Al Shibli et al., 2022). These methods generally involve simple extraction techniques using environmentally friendly solvents, such as water or ethanol, and do not require complex chemical processes or high-energy consumption. Plant extracts are applied as coatings or incorporated into protective films to inhibit corrosion. These application methods are typically eco-friendly, as they do not involve using hazardous solvents or releasing harmful byproducts. Plant extract-based coatings are easily removed or degraded, minimizing their environmental impact. Plant extracts are biodegradable and pose minimal risks to the environment. They are broken down into non-toxic byproducts, reducing their environmental persistence and potential ecological effects. Also, some plant extracts possess anti-fouling properties, preventing the attachment and growth of marine organisms on metal surfaces. This additional benefit can be valuable in environments where biofouling can accelerate corrosion.

13.3.5 Cost-Effectiveness

Plant extracts are generally cost-effective compared to synthetic inhibitors (Bhardwaj et al., 2021). They can be obtained from readily available plant sources, and the extraction processes are relatively simple and low-cost, making them attractive for industrial applications. Plant extracts are often readily available and accessible compared to synthetic corrosion inhibitors. Plants containing the desired extract are easily sourced, cultivated, or even found in local regions, making them cost-effective options for corrosion control. The extraction of plant compounds can be achieved using relatively simple techniques, such as solvent extraction or maceration. These processes are generally less expensive than the production of synthetic inhibitors. Additionally, the low-cost production of plant extracts can contribute to their cost-effectiveness. The sustainable nature of plant extracts aligns with the growing emphasis on eco-friendly and sustainable solutions. Renewable resources offer long-term cost benefits and reduce dependency on non-renewable resources.

13.3.6 LIMITATIONS

While plant extracts offer potential as corrosion inhibitors, their performance can vary depending on the plant species, extraction method, and specific environmental conditions. Standardization of the extraction process and a detailed understanding of the active compounds in the extracts are necessary to ensure consistent and reliable inhibition performance. Plant extracts contain numerous chemical compounds based on plant species, growth conditions, extraction methods, and storage conditions. These variabilities make it challenging to standardize and optimize the inhibition efficiency of plant extracts across different environments and metal systems. Plant extracts may exhibit varying degrees of inhibition effectiveness on various metal substrates. Some extracts perform well in inhibiting corrosion on certain metals while showing limited or no effect on others. Achieving broad-spectrum inhibition across different metals can be challenging, requiring the selection of specific plant extracts tailored to each metal system.

The stability of plant extracts can be a concern for long-term corrosion protection. Many plant compounds are prone to degradation under environmental conditions such as temperature, light, and oxygen exposure. This can lead to a loss of inhibition efficiency over time and a reduced lifespan of the protective coating formed by the plant extract. The plant extracts often require the formation of a protective film on the metal surface to inhibit corrosion. However, achieving good film formation and adhesion can be challenging. The film may be porous, uneven, or easily disrupted, compromising corrosion protection and initiating localized corrosion.

The production of consistent plant extracts with reliable inhibition performance can be challenging to achieve on a large scale. Plant sourcing, extraction methods, and processing variations can result in inconsistent inhibition efficiency. The scalability of plant extract-based corrosion inhibitors may also be limited due to the availability and sustainability of plant resources. In some cases, the compatibility of plant extracts with other coatings, surface treatments, or protective materials can be an issue. Plant extracts may interact unfavourably with additional corrosion protection systems, resulting in reduced effectiveness or undesired interactions that compromise the overall performance.

13.4 MECHANISM OF PLANT EXTRACT AS CORROSION INHIBITORS

Cathodic, anodic, or mixed-type inhibitors are the several categories of corrosion inhibitors based on how they prevent corrosion. Inhibitors of cathodic corrosion reduce the corrosion potential to lower values by blocking cathodic reactions, including oxygen reduction and hydrogen evolution. Anodic corrosion inhibitors interact with the reactive sites on the metal surface to passivate them and raise the corrosion potential's values. Inhibitors of a mixed kind fall outside either the cathodic or anodic categories. These inhibitors can shield the metal surface through physisorption, chemisorption, or film formation. Chemisorption is caused by donor–acceptor interactions between unoccupied orbitals on the metal surface and free electron pairs in the inhibitor, whereas electrostatic contacts between inhibitor molecules and the metal surface drive physical sorption. The film creates a physical barrier between the metal surface and the corrosive media, preventing corrosive attacks (Zhao et al., 2021).

An organic molecule can exist as a neutral or protonated species depending on the pH of a system. Unquestionably, negative anions in a system with a positively charged surface will draw the anions to it. The strength of the inhibitor species' adsorption will depend on how well the adsorbed anions on the metal surface recharge if the inhibitor species are in the system as protonated species. As seen in Figure 13.3, the so-called physisorption mechanism, the inhibitor species are electrostatically adsorbed on the metal surface in this scenario.

According to claims made about plant extracts used as metal corrosion inhibitors, their form of adsorption is physisorption. In this scenario, electrostatic forces hold the extract molecules to the metal surfaces, and a rise in system temperature often weakens the adsorption link, decreasing the inhibition's effectiveness. In a corrosive environment, neutral inhibitor molecules could donate their electron pairs to an open orbital in a metal substrate, forming a covalent or co-ordinate bond-chemisorption mechanism. This sort of adsorption benefits from higher system temperatures; specifically, inhibition efficiency rises as the temperature rises. The protonated inhibitor species can be neutralized on a metal surface, allowing the liberated heteroatoms to form a covalent or co-ordinated connection with the metal. On the same metal surface, physisorption and chemisorption mechanisms coexist. The best technique to identify the dominant adsorption mechanism on a metal surface is to calculate the standard enthalpy of adsorption (Umoren et al., 2019).

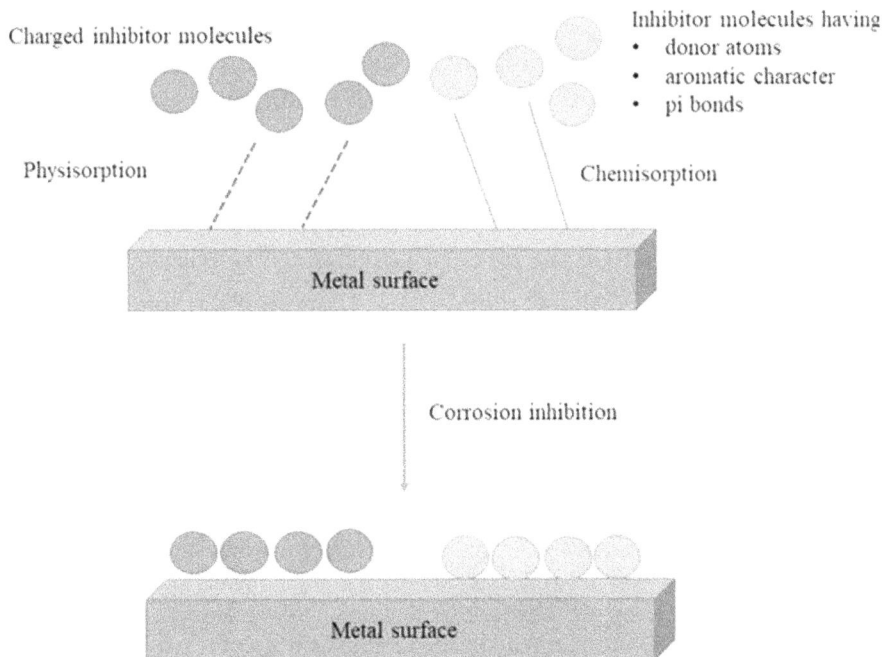

FIGURE 13.3 The mechanism of corrosion inhibitor (Eddy et al., 2022).

13.5 PLANT EXTRACTS COMBINED WITH ZINC IONS AGAINST CORROSION

Plant extracts have gained considerable attention due to their abundant availability, biodegradability, and low toxicity. The synergistic effect of plant extracts and zinc ions offers a promising avenue for combating corrosion while minimizing the environmental impact. Zinc is a well-known metal sacrificial anode to protect other metals from corrosion. The release of zinc ions in the corrosive environment forms a protective layer on the metal surface, hindering corrosion. The addition of zinc ions to various coatings and inhibitors has demonstrated enhanced corrosion protection. Combining plant extracts and zinc ions is a promising approach to enhance corrosion protection. Plant extracts act as organic inhibitors, and when combined with zinc ions, they create a more robust and effective protective layer on the metal surface. The synergistic effect not only improves the inhibition efficiency but also reduces the consumption of zinc, making it a sustainable and economically viable solution. Various studies have explored the application of plant extract and zinc ion-based inhibitors on different metal substrates, including steel, aluminium, and copper. The review analyses the performance of these inhibitors in other corrosive environments, such as acidic, alkaline, and saline solutions. Table 13.1 summarizes the combination of plant extracts with zinc ions as a compelling approach for mitigating corrosion in an environmentally friendly and sustainable manner.

13.6 CONCLUSION

Plant extracts have shown considerable potential as corrosion inhibitors for various metals and alloys. They possess organic compounds, such as phenols, alkaloids, flavonoids, and tannins, which can form protective films on the metal surface, inhibiting corrosion reactions. The use of plant extracts as corrosion inhibitors aligns with the principles of green chemistry and sustainable practices. They are renewable, readily available, and often obtained as byproducts from agricultural processes. Their utilization as corrosion inhibitors reduces reliance on synthetic chemicals and minimizes environmental impact. The corrosion inhibition mechanisms of plant extracts are multifaceted. They function through adsorption onto the metal surface, forming protective films, scavenging free radicals, inhibiting corrosion-related reactions, or combining these mechanisms. The specific mode of action varies depending on the extract's composition and the metal system. While plant extracts show promise, challenges remain regarding their stability, consistency, and compatibility with different metal systems and environments. Extract variability, extraction methods, and formulation techniques must be optimized to ensure consistent and reliable performance as corrosion inhibitors. Compared to conventional corrosion inhibitors, plant extracts can offer cost advantages, mainly from abundant and inexpensive plant sources. Combining plant extracts with zinc ions presents a compelling approach for mitigating corrosion in an environmentally friendly and sustainable manner. The synergistic effect of these two components enhances the inhibition efficiency, providing a viable alternative to conventional corrosion inhibitors. However, economic considerations, scalability, and the development of efficient extraction and formulation processes need further exploration to make plant extract-based inhibitors commercially viable.

TABLE 13.1
The Synergistic Effect of Plant Extract and Zinc Ions on Corrosion Inhibition

Plant Extract	Finding	References
Garcinia gummi-gutta leaf	The concentration of *Garcinia gummi-gutta* leaf extract (GGLE) varied from 100 to 6,000 ppm, and the result indicates that corrosion inhibition efficiency was amplified by raising the inhibitor concentration. The maximum inhibition efficiency was 82.2% at 6,000 ppm concentration. EIS results show the development of a protective layer of inhibitor molecule over the metal surface, and potentiodynamic polarization (PDP) demonstrates that the inhibitor operates as a mixed-type inhibitor. Scanning electron and atomic force microscopies were executed to assess the surface morphology and roughness, respectively.	Shamnamol et al. (2023)
Nettle leaves	The NLE and zinc nitrate combination effectively inhibited mild steel corrosion in a chloride solution. The mixture formed a barrier film on the anodic and cathodic regions, with the best corrosion inhibition efficiency achieved using equal amounts of NLE and zinc nitrate. The deposited film increased barrier properties and corrosion resistance over long exposure times. Organic-inorganic complexes like histamine-zinc-serotonin, histamine-zinc-quercetin, and serotonin-zinc-quercetin can be deposited on the steel surface through electron donation and chelation. The theoretical results suggest stronger interfacial adhesion of serotonin-zinc-quercetin complex inhibitors, indicating higher inhibition effects.	Ramezanzadeh et al. (2019a)
Chicory leaves	The CLE extract contains chicoric and caffeic acid, which interact with metal cations to form insoluble complexes on steel surfaces. These organic compounds have high corrosion inhibition efficiency (96%), retarding anodic and cathodic reactions. The effective chelation between these compounds reduces steel corrosion rate. The protective film formed from complexes of Zn^{2+}-CLE and Fe^{2+}-CLE was confirmed by SEM/EDS analysis. MC/MD simulations confirmed the organic-inorganic inhibitors' strong surface binding and higher inhibition effects.	Sanaei et al. (2019)
Eucalyptus leaves	UV-Vis analysis showed the successful formation of complexes and chelates in the ELE:Zn sample. Surface characterizations showed extensive degradation without additives. FT-IR and GIXRD spectra showed complexation and chelation, leading to Fe^{2+}-Zn^{2+}-ELE complexes formation. The highest hydrophobicity film was formed on the 200 ppm ELE + 600 ppm Zn sample. The inhibition mechanism in ELE:Zn samples was mainly based on mixed behaviour, while the adsorption mechanism was donor-acceptor.	Bahlakeh et al. (2019)

(Continued)

TABLE 13.1 (*Continued*)
The Synergistic Effect of Plant Extract and Zinc Ions on Corrosion Inhibition

Plant Extract	Finding	References
Primrose flower	This study investigated the synergistic impact of PPE and metal cations on the surface of steel panels. FT-IR spectroscopy showed O–Zn bonds, while SEM/EDS showed Fe-PPE and zinc hydroxide combined with corrosion products. The simultaneous use of PPE and Zn cations resulted in protective films without corrosion products. GIXRD patterns showed poly[di-*m*-pentanoato-zinc(II)] and poly(1,1′-ferrocenylene) compounds. AFM micrographs confirmed the smoother PPE:Zn sample. Higher Zn and lower PPE amounts resulted in more water repellency. The best performance was observed in the 300 ppm PPE + 700 ppm Zn sample, with IE values of 95.3% and 94.1%, respectively. Green molecules could potentially adsorb on the iron-based adsorbent, creating an anticorrosive barrier over the surface.	Akbarzadeh et al. (2020)
Lemon balm (LB)	The mixed inhibitors showed more corrosion resistance, with the addition of 200 ppm LB to 600 ppm Zn^{2+} providing the highest corrosion inhibition efficiency of approximately 93%. The mechanism involved precipitating zinc oxide/hydroxide on the cathodic zone and constructing complexes from LB organic compounds. The lowest corrosion current density was found for specimens exposed to 200 ppm LB to 600 ppm Zn^{2+} solution. Lower corrosion products (iron oxide/hydroxide) were formed on the surface, possibly due to the protective inhibitive film generation. The molecular-scale computational results confirmed the adsorption of organic-inorganic Zn-LB complexes on the MS surface.	Asadi et al. (2020)
Ervatamia divaricata	A weight loss study found that a formulation containing 2.5 cm³ of EDE and 25 ppm Zn^{2+} has 98% inhibition efficiency in controlling carbon steel corrosion in groundwater. The synergistic effect between EDE and Zn^{2+} is evident. A protective film, consisting of the Fe^{2+}-Corynanthean complex and $Zn(OH)_2$, controls the anodic and cathodic reactions. The protective film's hardness is lower than polished metal but higher than corroded metal. Electrochemical studies show that inhibitors increase linear polarization resistance, corrosion current, charge transfer resistance, double-layer capacitance, impedance, and phase angle. Adsorption isotherm studies show that inhibitors adsorb on the metal surface according to the Langmuir adsorption isotherm.	Shanthy et al. (2021)

(Continued)

TABLE 13.1 (*Continued*)
The Synergistic Effect of Plant Extract and Zinc Ions on Corrosion Inhibition

Plant Extract	Finding	References
Valerian	The valerian extract effectively reduces the corrosion rate in 1.0 M HCl by adsorbing molecules to the metal surface. Its inhibition efficiency increases with concentration and synergistically affects zinc sulphate. The valerian extract's adsorption obeys the Temkin adsorption isotherm model, acting as a mixed-type inhibitor. The extract can potentially create protective layers on the metal surface.	El-Katori et al. (2019)
Watermelon	The survey showed the impact of WME:Zn complex on MS anti-corrosion properties. Results showed a smooth protective layer, increased inhibition efficiency, and reduced surface roughness. Mixed oxygen reduction and Fe-dissolution reactions mitigated corrosion. WME and Zn interaction was successful, with protected samples displaying semi-hydrophilic and hydrophobic behaviours.	Mofidabadi et al. (2022)
Green tannic acid	The study used electrochemical measurements and surface characterizations to investigate the synergistic inhibition of tannic acid (TA) and Zn(II) in saline solution. Results showed that 250 ppm TA + 750 ppm Zn was the most efficient corrosion inhibitor hybrid. The mixed-type inhibition mechanism was concluded. No visible corrosion products were detected, and the TA-Zn film made the surface more hydrophobic.	Kaghazchi et al. (2021)
Ginger, pomegranate, and celery	Ginger, pomegranate, and celery extracts are effective zinc electroplating mild steel additives. The crystal structure of the electroplated steel changes with extract concentration and plating time. Ginger additives have the brightest micrographic image and are more corrosion-resistant than control steel. The electroplated steel with ginger extract has the lowest corrosion rate, while unplated steel has meagre resistance.	Loto and Akinyele (2020)
Mangifera indica	The study investigated the promotion of *Mangifera indica* L. extract inhibition properties by zinc cations. Electrochemical tests showed that a combination of 400 mg L^{-1} Zn^{2+} and 400 mg L^{-1} *Mangifera indica* L. extract yielded the highest corrosion inhibition efficiency (91%). A protective film with organic active molecules and zinc cations effectively covered steel surfaces, promoting metal corrosion resistance.	Ramezanzadeh et al. (2019a)
Terminalia catappa	At room temperature, XRD and SEM analyses show no corrosion in zinc and carbon steel 1020 plates. Biodiesel's increased corrosivity is due to O$_2$, H$_2$O, CO$_2$, and ROO radicals. The tropical almond leaf extract additive is an environmentally and economically friendly alternative to synthetic additives. It effectively inhibits zinc and carbon steel corrosion 1020 when interacting with soybean oil biodiesel.	Fernandes et al. (2021)

DECLARATION OF COMPETING INTEREST

The authors confirm that they have no financial or personal relationships that could influence the work described in this chapter.

REFERENCES

Abdeen, D. H., Hachach, M. El, & Koc, M. (2019). A review on the corrosion behaviour of nanocoatings on metallic substrates. *Materials (Basel, Switzerland)*, *12*, 210.

Abdel Hameed, R. S., Aljuhani, E. H., Felaly, R., & Munshi, A. M. (2021). Effect of expired paracetamol-Zn^{2+} system and its synergistic effect towards iron dissolution inhibition and green inhibition performance. *Journal of Adhesion Science and Technology*, *35*(8), 838–855.

Abdel-Karim, A. M., & El-Shamy, A. M. (2022). A review on green corrosion inhibitors for protection of archeological metal artifacts. *Journal of Bio-and Tribo-Corrosion*, *8*(2), 35.

Abdelwedoud, B. O., Damej, M., Tassaoui, K., Berisha, A., Tachallait, H., Bougrin, K., Mehmeti, V., & Benmessaoud, M. (2022). Inhibition effect of N-propargyl saccharin as corrosion inhibitor of C38 steel in 1 M HCl, experimental and theoretical study. *Journal of Molecular Liquids*, *354*, 118784.

Abdollahzadeh, A., Bagheri, B., Abbasi, M., Sharifi, F., & Moghaddam, A. O. (2021). Mechanical, wear and corrosion behaviors of AZ91/SiC composite layer fabricated by friction stir vibration processing. *Surface Topography: Metrology and Properties*, *9*(3), 035038.

Akbarzadeh, S., Ramezanzadeh, M., Bahlakeh, G., & Ramezanzadeh, B. (2020). A detailed investigation of the chloride-induced corrosion of mild steel in the presence of combined green organic molecules of Primrose flower and zinc cations. *Journal of Molecular Liquids*, *297*, 111862.

Akpanyung, K. V., & Loto, R. T. (2019). Pitting corrosion evaluation: a review. *Journal of Physics: Conference Series*, *1378*, 022088.

Al Shibli, F. S. Z. S., Bose, S., Kumar, P. S., Rajasimman, M., Rajamohan, N., & Vo, D. V. N. (2022). Green technology for sustainable surface protection of steel from corrosion: a review. *Environmental Chemistry Letters*, *20*, 1–19.

Al-Amiery, A. A., Yousif, E., Isahak, W. N. R. W., & Al-Azzawi, W. K. (2023). A review of inorganic corrosion inhibitors: types, mechanisms, and applications. *Tribology in Industry*, *44*(2), 313.

Alamri, A. H. (2020). Localized corrosion and mitigation approach of steel materials used in oil and gas pipelines-an overview. *Engineering Failure Analysis*, *116*, 104735.

Al-Moubaraki, A. H., & Obot, I. B. (2021). Corrosion challenges in petroleum refinery operations: sources, mechanisms, mitigation, and future outlook. *Journal of Saudi Chemical Society*, *25*(12), 101370.

Alvarez, A. P. E., Fiori-Bimbi, M. V., Neske, A., Brand, S. A., Gervasi, C. A., & Brand, S. A. (2017). Rollinia occidentalis extract as green corrosion inhibitor for carbon steel in HCl solution. *Journal of Industrial and Engineering Chemistry*, *58*, 92–99.

Anh, H. T., Vu, N. S. H., Huyen, L.T., Tran, N. Q., Thu, H. T., Bach, L. X., Trinh, Q. T., Vattikuti, S. P., & Nam, N. D. (2020). Ficus racemosa leaf extract for inhibiting steel corrosion in a hydrochloric acid medium. *Alexandria Engineering Journal*, *59*(6), 4449–4462.

Asadi, N., Ramezanzadeh, M., Bahlakeh, G., & Ramezanzadeh, B. (2020). Theoretical MD/DFT computer explorations and surface-electrochemical investigations of the zinc/iron metal cations interactions with highly active molecules from Lemon balm extract toward the steel corrosion retardation in saline solution. *Journal of Molecular Liquids*, *310*, 113220.

Askari, M., Aliofkhazraei, M., Jafari, R., Hamghalam, P., & Hajizadeh, A. (2021). Downhole corrosion inhibitors for oil and gas production-a review. *Applied Surface Science Advances, 6,* 100128.

Bahlakeh, G., Dehghani, A., Ramezanzadeh, B., & Ramezanzadeh, M. (2019). Combined molecular simulation, DFT computation and electrochemical studies of the mild steel corrosion protection against NaCl solution using aqueous Eucalyptus leaves extract molecules linked with zinc ions. *Journal of Molecular Liquids, 294,* 111550.

Basukumar, H. K., & Arun, K. V. (2022). Tensile behavior of pre-stress corroded and post hydrogen embrittled spring steel. *Materials Today: Proceedings, 49,* 2000–2006.

Becker, J., Pellé, J., Rioual, S., Lescop, B., Le Bozec, N., & Thierry, D. (2022). Atmospheric corrosion of silver, copper and nickel exposed to hydrogen sulphide: a multi-analytical investigation approach. *Corrosion Science, 209,* 110726.

Bell, S., Steinberg, T., & Will, G. (2019). Corrosion mechanisms in molten salt thermal energy storage for concentrating solar power. *Renewable and Sustainable Energy Reviews, 114,* 109328.

Bernardi, E., Vassura, I., Raffo, S., Nobili, L., Passarini, F., De, D., & Morcillo, M. (2020). Influence of inorganic anions from atmospheric depositions on weathering steel corrosion and metal release. *Construction and Building Materials, 236,* 117515.

Bhardwaj, N., Sharma, P., & Kumar, V. (2021). Phytochemicals as steel corrosion inhibitor: an insight into mechanism. *Corrosion Reviews, 39*(1), 27–41.

Calegari, F., Sousa, I., Ferreira, M. G., Berton, M. A., Marino, C. E., & Tedim, J. (2022). Influence of the operating conditions on the release of corrosion inhibitors from spray-dried carboxymethylcellulose microspheres. *Applied Sciences, 12*(4), 1800.

Chen, X., Gussev, M., Balonis, M., Bauchy, M., & Sant, G. (2021). Emergence of micro-galvanic corrosion in plastically deformed austenitic stainless steels. *Materials & Design, 203,* 109614.

Chen, X., Yang, L., Dai, H., & Shi, S. (2020). Exploring factors controlling pre-corrosion fatigue of 316L austenitic stainless steel in hydrofluoric acid. *Engineering Failure Analysis, 113,* 104556.

Dacillo, K. B., & Zarrouk, S. J. (2023). Stress corrosion cracking in metal alloys exposed to geothermal fluids with high non-condensable gas content. *Geothermics, 111,* 102724.

de Souza Morais, W. R., da Silva, J. S., Queiroz, N. M. P., de Paiva e Silva Zanta, C. L., Ribeiro, A. S., & Tonholo, J. (2023). Green corrosion inhibitors based on plant extracts for metals and alloys in corrosive environment: a technological and scientific prospection. *Applied Sciences, 13*(13), 7482.

Dillon, P. B. (2019). Preliminary study of galvanic corrosion on veneer anchors. In *Proceedings of 13th North American Masonry Conference,* Salt Lake City, Utah (pp. 1782–1794).

Eddy, N. O., Ibok, U. J., Garg, R., Garg, R., Iqbal, A., Amin, M., Mustafa, F., Egilmez, M., & Galal, A. M. (2022). A brief review on fruit and vegetable extracts as corrosion inhibitors in acidic environments. *Molecules, 27*(9), 2991.

El-Katori, E. E., Fouda, A. S., & Mohamed, R. R. (2019). The synergistic impact of the aqueous valerian extract and zinc ions for the corrosion protection of mild steel in acidic environment. *Zeitschrift für Physikalische Chemie, 233*(12), 1713–1739.

Falconer, C., Wang, Y., Sridharan, K., & Couet, A. (2023). Activity gradient mass transport accelerated corrosion of stainless steel in molten salt. *Applied Materials Today, 32,* 101850.

Fernandes, F. D., Ferreira, L. M., & da Silva, M. L. C. P. (2021). Evaluation of the corrosion inhibitory effect of the ecofriendly additive of Terminalia catappa leaf extract added to soybean oil biodiesel in contact with zinc and carbon steel 1020. *Journal of Cleaner Production, 321,* 128863.

Francis, R., Turnbull, A., & Hinds, G. (2020). Bimetallic Corrosion (Guides for Good Practice in Corrosion Control No. 5).

Go, L. C., Holmes, W., Depan, D., & Hernandez, R. (2019). Evaluation of extracellular poly-meric substances extracted from waste activated sludge as a renewable corrosion inhibi-tor. *PeerJ, 7*, e7193.

Gong, K., Wu, M., Liu, X., & Liu, G. (2022). Nucleation and propagation of stress corrosion cracks: modeling by cellular automata and finite element analysis. *Materials Today Communications, 33*, 104886.

Gong, K., Wu, M., Xie, F., Liu, G., & Sun, D. (2020). Effect of dry/wet ratio and pH on the stress corrosion cracking behavior of rusted X100 steel in an alternating dry/wet envi-ronment. *Construction and Building Materials, 260*, 120478.

Harsimran, S., Santosh, K., & Rakesh, K. (2021). Overview of corrosion and its control: a critical review. *Proceedings on Engineering Sciences, 3*(1), 13–24.

Jothi, K. J., Balachandran, S., & Palanivelu, K. (2022). Synergistic combination of Phyllanthus niruri/silver nanoparticles for anticorrosive application. *Materials Chemistry and Physics, 279*, 125794.

Kaghazchi, L., Naderi, R., & Ramezanzadeh, B. (2020). Construction of a high-performance anti-corrosion film based on the green tannic acid molecules and zinc cations on steel: elec-trochemical/surface investigations. *Construction and Building Materials, 262*, 120861.

Kaghazchi, L., Naderi, R., & Ramezanzadeh, B. (2021). Synergistic mild steel corrosion miti-gation in sodium chloride-containing solution utilizing various mixtures of phytic acid molecules and Zn^{2+} ions. *Journal of Molecular Liquids, 323*, 114589.

Karki, N. (2020). Development of green corrosion inhibitor from natural products of Nepal (Doctoral dissertation, Department of Chemistry).

Katona, R. M., Knight, A. W., Maguire, M., Bryan, C. R., & Schaller, R. F. (2023). Considerations for realistic atmospheric environments: an application to corrosion test-ing. *Science of the Total Environment, 885*, 163751.

Kaya, F., Solmaz, R., & Geçibesler, İ. H. (2023a). Investigation of adsorption, corrosion inhi-bition, synergistic inhibition effect and stability studies of Rheum ribes leaf extract on mild steel in 1 M HCl solution. *Journal of the Taiwan Institute of Chemical Engineers, 143*, 104712.

Kaya, F., Solmaz, R., & Gecibesler, I. H. (2023b). Adsorption and corrosion inhibition capa-bility of Rheum ribes root extract (Işgın) for mild steel protection in acidic medium: a comprehensive electrochemical, surface characterization, synergistic inhibition effect, and stability study. *Journal of Molecular Liquids, 372*, 121219.

Kaya, F., Solmaz, R., & Geçibesler, İ. H. (2023c). The use of methanol extract of Rheum ribes (Işgın) flower as a natural and promising corrosion inhibitor for mild steel protection in 1 M HCl solution. *Journal of Industrial and Engineering Chemistry, 122*, 102–117.

Khan, M. A. A., Irfan, O. M., Djavanroodi, F., & Asad, M. (2022). Development of sustainable inhibitors for corrosion control. *Sustainability, 14*(15), 9502.

Khayatazad, M., Pue, L. D., & Waele, W. D. (2020). Developments in the built environ-ment detection of corrosion on steel structures using automated image processing. *Developments in the Built Environment, 3*(April), 100022.

Kobzar, Y. L., & Fatyeyeva, K. (2021). Ionic liquids as green and sustainable steel corrosion inhibitors: recent developments. *Chemical Engineering Journal, 425*, 131480.

Kokilaramani, S., Al-Ansari, M. M., Rajasekar, A., Al-Khattaf, F. S., Hussain, A., & Govarthanan, M. (2021). Microbial influenced corrosion of processing industry by re-circulating waste water and its control measures-a review. *Chemosphere, 265*, 129075.

Koushik, B. G., Van den Steen, N., Mamme, M. H., Van Ingelgem, Y., & Terryn, H. (2021). Review on modelling of corrosion under droplet electrolyte for predicting atmospheric corrosion rate. *Journal of Materials Science & Technology, 62*, 254–267.

Kumar, S., Katyal, P., Chaudhary, R. N., & Singh, V. (2022). Assessment of factors influ-encing bio-corrosion of magnesium-based alloy implants: a review. *Materials Today: Proceedings, 56*, 2680–2689.

Kumari, P. (2022). Plant extracts as corrosion inhibitors for aluminum alloy in NaCl environment-recent review. *Journal of the Chilean Chemical Society, 67*(2), 5490–5495.

Lawal, S. L., Afolalu, S. A., Jen, T. C., & Akinlabi, E. T. (2023). Effects of microstructural variables on corrosion protective layer of steel. In *2023 International Conference on Science, Engineering and Business for Sustainable Development Goals (SEB-SDG)* (Vol. 1, pp. 1–9). IEEE.

Li, Y., & Ning, C. (2019). Bioactive materials latest research progress of marine microbiological corrosion and bio-fouling, and new approaches of marine anti-corrosion and anti-fouling. *Bioactive Materials, 4*, 189–195.

Liu, L., Xu, Y., Zhu, Y., Wang, X., & Huang, Y. (2020). The roles of fluid hydrodynamics, mass transfer, rust layer and macro-cell current on flow accelerated corrosion of carbon steel in oxygen containing electrolyte. *Journal of the Electrochemical Society, 167*(14), 141510.

Liu, Z., Fan, B., Zhao, J., Yang, B., & Zheng, X. (2023). Benzothiazole derivatives-based supramolecular assemblies as efficient corrosion inhibitors for copper in artificial seawater: formation, interfacial release and protective mechanisms. *Corrosion Science, 212*, 110957.

Loto, C. A., & Akinyele, M. (2020). Effect of ginger, pomegranate and celery extracts on zinc electrodeposition, surface morphology and corrosion inhibition of mild steel. *Alexandria Engineering Journal, 59*(2), 933–941.

Ma, B., Shin, D., & Banerjee, D. (2021). One-step synthesis of molten salt nanofluid for thermal energy storage application-a comprehensive analysis on thermophysical property, corrosion behavior, and economic benefit. *Journal of Energy Storage, 35*, 102278.

Masum, M. M. I., Siddiqa, M. M., Ali, K. A., Zhang, Y., Abdallah, Y., Ibrahim, E., Qiu, W., Yan, C., & Li, B. (2019). Biogenic synthesis of silver nanoparticles using Phyllanthus emblica fruit extract and its inhibitory action against the pathogen Acidovorax oryzae strain RS-2 of rice bacterial brown stripe. *Frontiers in Microbiology, 10*, 820.

Mofidabadi, A. H. J., Dehghani, A., & Ramezanzadeh, B. (2022). Investigating the effectiveness of watermelon extract-zinc ions for steel alloy corrosion mitigation in sodium chloride solution. *Journal of Molecular Liquids, 346*, 117086.

Nascimento, R. C., & Furtado, L. B. (2022). Carbon dots as corrosion inhibitors: synthesis, molecular structures and corrosion inhibition. *Anticorrosive Nanomaterials: Future Perspectives, 56*, 122.

Nergis, D. P. B., Nejneru, C., Nergis, D. D. B., Savin, C., Sandu, A. V., Toma, S. L., & Bejinariu, C. (2019). The galvanic corrosion behavior of phosphated carbon steel used at carabiners manufacturing. *Revista de Chimie (Bucharest), 70*(1), 215.

Obot, I. B., Onyeachu, I. B., & Umoren, S. A. (2019). Alternative corrosion inhibitor formulation for carbon steel in CO_2-saturated brine solution under high turbulent flow condition for use in oil and gas transportation pipelines. *Corrosion Science, 159*, 108140.

Oguike, R. S., Oni, O., Barambu, A. U., Balarak, D., Buba, T., Okeke, C. U., Momoh, L. S., Onimisi, S., & Nwada, W. J. (2020). Computational stimulation and experimental study on corrosion inhibition qualities of Emilia sonchifolia leaf extract for copper (CU131729) in hydrochloric acid. *Computational Chemistry, 9*(01), 18.

Ong, G., Kasi, R., & Subramaniam, R. (2021). A review on plant extracts as natural additives in coating applications. *Progress in Organic Coatings, 151*, 106091.

Palanisamy, G. (2019). Corrosion inhibitors. In *Corrosion Inhibitors*, 1–24.

Parangusan, H., Bhadra, J., & Al-Thani, N. (2021). A review of passivity breakdown on metal surfaces: influence of chloride-and sulfide-ion concentrations, temperature, and pH. *Emergent Materials, 4*(5), 1187–1203.

Peltier, F., & Thierry, D. (2022). Review of Cr-free coatings for the corrosion protection of aluminum aerospace alloys. *Coatings, 12*(4), 518.

Qureshi, T., Wang, G., Mukherjee, S., Islam, M. A., Filleter, T., Singh, C. V., & Panesar, D. K. (2022). Graphene-based anti-corrosive coating on steel for reinforced concrete infrastructure applications: challenges and potential. *Construction and Building Materials, 351*, 128947.

Ramezanzadeh, M., Bahlakeh, G., & Ramezanzadeh, B. (2019). Study of the synergistic effect of Mangifera indica leaves extract and zinc ions on the mild steel corrosion inhibition in simulated seawater: computational and electrochemical studies. *Journal of Molecular Liquids, 292*, 111387.

Ramezanzadeh, M., Bahlakeh, G., Ramezanzadeh, B., & Sanaei, Z. (2019). Adsorption mechanism and synergistic corrosion-inhibiting effect between the green Nettle leaves extract and Zn^{2+} cations on carbon steel. *Journal of Industrial and Engineering Chemistry, 77*, 323–343.

Rao, P., & Mulky, L. (2023). Microbially influenced corrosion and its control measures: a critical review. *Journal of Bio-and Tribo-Corrosion, 9*(3), 57.

Sadeghi, E., Markocsan, N., & Joshi, S. (2019). Advances in corrosion-resistant thermal spray coatings for renewable energy power plants. Part I: effect of composition and microstructure. *Journal of Thermal Spray Technology, 28*, 1749–1788.

Sanaei, Z., Bahlakeh, G., Ramezanzadeh, B., & Ramezanzadeh, M. (2019). Application of green molecules from Chicory aqueous extract for steel corrosion mitigation against chloride ions attack; the experimental examinations and electronic/atomic level computational studies. *Journal of Molecular Liquids, 290*, 111176.

Senatore, E. V., Pinto, M. C. P., Souza, E. A., Barker, R., Neville, A., & Gomes, J. P. (2021). Effects of pre-filmed $FeCO_3$ on flow-induced corrosion and erosion-corrosion in the absence and presence of corrosion inhibitor at 60°C. *Wear, 480*, 203927.

Shamnamol, G. K., Rugma, P., John, S., & Jacob, J. M. (2023). Unraveling the synergistic effect of cationic and anionic salt on the corrosion inhibition performance of Garcinia gummigutta leaf extract against mild steel in HCl medium. *Results in Chemistry, 5*, 100728.

Shang, Z., & Zhu, J. (2021). Overview on plant extracts as green corrosion inhibitors in the oil and gas fields. *Journal of Materials Research and Technology, 15*, 5078–5094.

Shanthy, P., Thangakani, J. A., Karthika, S., Joycee, S. C., Rajendran, S., & Jeyasundari, J. (2021). Corrosion inhibition by an aqueous extract of ervatamia divaricata. *International Journal of Corrosion and Scale Inhibition, 10*(1), 331–348.

Soufeiani, L., Foliente, G., Nguyen, K. T., & San Nicolas, R. (2020). Corrosion protection of steel elements in façade systems-a review. *Journal of Building Engineering, 32*, 101759.

Thakur, A., & Kumar, A. (2021). Sustainable inhibitors for corrosion mitigation in aggressive corrosive media: a comprehensive study. *Journal of Bio-and Tribo-Corrosion, 7*, 1–48.

Tukhlievich, B. E., & Verma, C. (Eds.). (2023). *Corrosion Prevention Nanoscience: Nanoengineering Materials and Technologies*. Walter de Gruyter GmbH & Co KG.

Umoren, S. A., Solomon, M. M., Obot, I. B., & Suleiman, R. K. (2019). A critical review on the recent studies on plant biomaterials as corrosion inhibitors for industrial metals. *Journal of Industrial and Engineering Chemistry, 76*, 91–115.

Unune, D. R., Brown, G. R., & Reilly, G. C. (2022). Thermal based surface modification techniques for enhancing the corrosion and wear resistance of metallic implants: a review. *Vacuum, 203*, 111298.

Vasyliev, G., Vorobyova, V., & Zhuk, T. (2020). Raphanus sativus L. extract as a scale and corrosion inhibitor for mild steel in tap water. *Journal of Chemistry, 2020*, 1–9.

Verma, C., Alfantazi, A., Quraishi, M. A., & Rhee, K. Y. (2023). Are extracts really green substitutes for traditional toxic corrosion inhibitors? Challenges beyond origin and availability. *Sustainable Chemistry and Pharmacy, 31*, 100943.

Verma, C., Ebenso, E. E., & Quraishi, M. A. (2019). Alkaloids as green and environmental benign corrosion inhibitors: an overview. *International Journal of Corrosion and Scale Inhibition, 8*(3), 512–528.

Verma, C., Hussain, C. M., Quraishi, M. A., & Alfantazi, A. (2022). Green surfactants for corrosion control: design, performance and applications. *Advances in Colloid and Interface Science, 311*, 102822.

Wamba-Tchio, O. R., Pengou, M., Teillout, A. L., Baumier, C., Mbomekallé, I. M., De Oliveira, P., Nanseu-Njiki, C. P., & Ngameni, E. (2022). Electrochemical study and experimental simulation of the synergistic effect of a formulation based on Ficus pumila Linn. leaves extract and zinc sulfate on the XC38 steel corrosion inhibition in NaCl solution. *Journal of Electroanalytical Chemistry, 919*, 116553.

Wang, J., Du, M., Li, G., & Shi, P. (2022). Research progress on microbiological inhibition of corrosion: a review. *Journal of Cleaner Production, 373*, 133658.

Wang, Z., Huang, Y., Wang, X., Xu, Y., & Cai, F. (2022). Effects of oyster as macrofouling organism on corrosion mechanisms of a high-strength low-alloy steel. *Corrosion Science, 207*, 110580.

Xia, D. H., Ji, Y., Zhang, R., Mao, Y., Behnamian, Y., Hu, W., & Birbilis, N. (2023). On the localized corrosion of AA5083 in a simulated dynamic seawater/air interface-part 1: corrosion initiation mechanism. *Corrosion Science, 213*, 110985.

Xiao, K., Li, Z., Song, J., Bai, Z., Xue, W., Wu, J., & Dong, C. (2021). Effect of concentrations of Fe^{2+} and Fe^{3+} on the corrosion behavior of carbon steel in Cl^- and SO_4^{2-} aqueous environments. *Metals and Materials International, 27*, 2623–2633.

Xu, L., Guan, F., Ma, Y., Zhang, R., Zhang, Y., Zhai, X., Dong, X., Wang, Y., Duan, J., & Hou, B. (2022). Inadequate dosing of THPS treatment increases microbially influenced corrosion of pipeline steel by inducing biofilm growth of Desulfovibrio hontreensis SY-21. *Bioelectrochemistry, 145*, 108048.

Xu, Y., & Tan, M. Y. (2019). Probing the initiation and propagation processes of flow accelerated corrosion and erosion corrosion under simulated turbulent flow conditions. *Corrosion Science, 151*, 163–174.

Yadav, A., Kumar, R., Pandey, U. P., & Sahoo, B. (2021). Role of oxygen functionalities of GO in corrosion protection of metallic Fe. *Carbon, 173*, 350–363.

Yang, H. M. (2021). Role of organic and eco-friendly inhibitors on the corrosion mitigation of steel in acidic environments-a state-of-art review. *Molecules, 26*(11), 3473.

Zhang, R., Yu, X., Yang, Q., Cui, G., & Li, Z. (2021). The role of graphene in anti-corrosion coatings: a review. *Construction and Building Materials, 294*, 123613.

Zhao, D., Han, C., Peng, B., Cheng, T., Fan, J., Yang, L., Chen, L., & Wei, Q. (2022). Corrosion fatigue behavior and anti-fatigue mechanisms of an additively manufactured biodegradable zinc-magnesium gyroid scaffold. *Acta Biomaterialia, 153*, 614–629.

Zhao, J., Tu, Z., & Hwa, S. (2021). Carbon corrosion mechanism and mitigation strategies in a proton exchange membrane fuel cell (PEMFC): a review. *Journal of Power Sources, 488*, 229434.

Zhu, Y., Sun, Q., Wang, Y., Tang, J., Wang, Y., & Wang, H. (2021). Molecular dynamic simulation and experimental investigation on the synergistic mechanism and synergistic effect of oleic acid imidazoline and L-cysteine corrosion inhibitors. *Corrosion Science, 185*, 109414.

Zuliana, S., Ha, A., Koriah, S., Ali, M., Taib, A., Abu, A., Najmi, M., Mohamad, M., Mamat, S., Ahmad, S., Ali, A., & Ter, P. (2021). Plant extracts as green corrosion inhibitor for ferrous metal alloys: a review. *Journal of Cleaner Production, 304*, 127030.

14 Phytochemicals/Plant Extracts as Corrosion Inhibitors for Brass in Various Electrolytes

Pragnesh N. Dave
Sardar Patel University

Pradip M. Macwan
B. N. Patel Institute of Paramedical & Science

14.1 INTRODUCTION

Recent years have seen a surge in the development of green corrosion inhibitors. In keeping with green chemistry, researchers are dedicated to creating green corrosion inhibitors (Huang et al., 2022). Natural corrosion inhibitors accordingly arise at the perfect moment. Green corrosion inhibitors are typically found in abundance in natural plants with low toxicity, low cost, moderate performance, and mild environmental impact (Ismail, 2016; Verma et al., 2017). Nevertheless, not entirely of them may be utilised as primal matter. In reality, plant extracts and polysaccharides are utilised as metal-protective agents since they are very active components. Natural substances such polysaccharides, polyphenol polymers, tannin, flavonoids, anthraquinones, saponins, and alkaloids are abundant in plant extracts, as we are all aware. Starch, cellulose, pectin, lentinan, and the polysaccharide found in *Lycium barbarum* are the primary plant polysaccharides. N, O, and bonding as well as aromatic heterocycles are included in these natural products and are the primary dynamic ingredients for effective corrosion prevention. The dynamic groups can relate with the metal exterior and create an adsorption-protecting coating that prevents metal corrosion (Verma et al., 2020).

Starch and cellulose are two of the more prevalent polysaccharides found in plants and are more prevalent than the other ones. They primarily satisfy the prerequisites for industrial corrosion inhibitors, which include high yield and ease of access. Researchers have also discovered that starch and cellulose have good corrosion-inhibiting effects on a variety of metals (Popoola, 2019). Other extracts that have been taken out of leaves, stems, flowers, and seeds undoubtedly have significant utility as well. Several plant extracts, including chondroitin sulphate, *Tagetes erecta*, *Moringa oleifera*, and *Lycium shawii*, have been employed as green corrosion inhibitors. Inhibitors of corrosion are thought to be investigated in less than 1% of the

DOI: 10.1201/9781003394631-14

world's plant species (Abdel-Gaber et al., 2021; Chen et al., 2021). This implies that there is quiet much room for improvement in plant application.

In general, using starch, cellulose, and additional plant extracts as metal corrosion inhibitors promotes a green environment. There are several different corrosion inhibitors available today, including organic carbon steel copper corrosion inhibitor BWF-XZ530, polymer stainless steel/copper corrosion inhibitor WDZ-02, organic carbon steel copper corrosion inhibitor HO-660, and organic corrosion inhibitor LS-01. Nevertheless, unlike conventional corrosion inhibitors, there are no natural corrosion inhibitors that may be used in industrial production.

In order to be employed for a specific purpose, a substance or activity called a plant extract is taken from the plant's tissue, typically by treating it with a solvent (Abd-El-nabey et al., 2020). Solid or liquid extraction is the first step in the purifying process for plants. The action of separating one or more parts present in a solid item by solubilising it in a fluid is so defined. This substance, commonly recognised as a solvent, can be either a liquid or a gas (water vapour or supercritical fluids). Plant extracts can be categorised into the following categories: water-soluble plant extracts, oil-soluble plant extracts, essential oils, spray-dried powder, enzyme-hydrolysed vegetable protein powder, the pure active ingredient, hulled fruit essential powder, liposome compressed microcapsules, polysaccharides or other porous polymer-encapsulated microcapsules and microspheres immersed extract. Freshly made fruit or vegetable juice has also been utilised in the home or in a professional beauty shop. The active components in a plant extract will vary depending on which plant parts were utilised to make it, such as the roots, branches, leaves, bark, flowers, garlands, fruits, seeds, shoots, and so on. The producer of some plant extracts commonly uses this categorisation technique since it is more useful in the production, usage, storage, and transportation of the product.

Metal and its alloys used in diverse industries undergo an unavoidable but regulated process of corrosion (Pais & Rao, 2019). After iron, aluminium, and copper, zinc is presently the fourth most commonly consumed metal worldwide. It possesses effective anticorrosive qualities. Zinc is mostly used in the process of galvanisation, which involves coating iron or steel with thin zinc coatings to stop corrosion. Materials used in autos, electrical components, and home furnishings are created by combining zinc with other metals, including copper (to create brass), iron, and aluminium. It also has a variety of uses in medicines, including a number of pigments, activators, catalysts, batteries, toys, and toys (Hebbar et al., 2015).

14.2   CORROSION (GAYA ET AL., 2019)

The degradation of a material or its qualities as a result of interactions between the substance and its environment is known as corrosion (Adejoro et al., 2015; Chigondo & Chigondo, 2016). Corrosion may seriously harm metal and alloy structures, which can have an economic impact on product losses, safety, environmental pollution, and repair and replacement costs. Corrosion is an unwanted phenomenon that needs to be avoided because of these negative impacts (Patni et al., 2013). According to reports, a number of elements have a significant impact on metallic corrosion. Some of the elements include the following: Temperature, non-homogeneous soil,

oxygen concentration, stressed metallic section, concrete/soil contact, high voltage direct current (HVDC) electric transmission, DC transit systems, damaged or
scratched surface, moist/dry electrolyte, and different alloys or metals are additional
considerations.

14.3   CORROSION INHIBITORS (GAYA ET AL., 2019)

Adsorption of their particles onto the corroding metal exterior is how corrosion is
inhibited, and how well it is inhibited relies on the mechanical, structural, and chemical properties of the adsorption layers that are produced under certain conditions
(Eddy et al., 2010; Salim et al., 2019). One of the authorised methods for reducing
and/or preventing corrosion is the usage of inhibitors to regulate corrosion in metals
and blends exposed to harsh environments. A corrosion inhibitor is a substance that,
when given to an environment in modest concentrations, effectively slows down the
pace at which a metal exposed to that environment corrodes (Patni et al., 2013). Due
to their toxicity and disposal challenges, particularly in the oceanic industry where
sea life is at risk, several inorganic inhibitors, especially those comprising phosphate,
chromate, and additional dense metals, are now gradually being limited or forbidden
by numerous environmental rules (Roy et al., 2014). Corrosion inhibitors fall into
one of two categories: those that increase the production of a defensive oxide coating
by an oxidising action and those that prevent corrosion by selectively adhering to the
metal exterior and forming a wall that blocks the entry of corrosive substances (Patni
et al., 2013).

14.4   ACTIVE CONSTITUENTS PRESENT IN GREEN
INHIBITOR (KAUR ET AL., 2022)

The inclusion of phytochemicals contributes to the effectiveness of green inhibitors as inhibitors (as shown in Figure 14.1). Green inhibitors, such as plant extract,
which contain heteroatoms like O, N, and S-containing combinations (Chauhan &
Gunasekaran, 2007; El-Etre, 2006; El-Etre et al., 2005; Oguzie, 2006; Satapathy
et al., 2009), as well as phytochemicals like *Glycyrrhiza glabra*, whose chief components are flavonoids, glycyrrhizin, isoflavonoids, and chalcone (Alibakhshi et al.,
2018), can effectively control the corrosion of mild steel in sulphuric acid and hydrochloric acid. These green inhibitors slow down or even stop mild steel's corrosion
response.

14.5   PLANT EXTRACTS AS CORROSION
INHIBITORS (GAYA ET AL., 2019)

Natural substances, some of which have difficult molecular arrangements and dif-
fering chemical, biotic, and bodily characteristics, are found in plants. The main
reason why naturally occurring substances are utilised so frequently is because they
are affordable, cost-effective, and widely available. These benefits explain the usage
of plant extracts and plant-derived compounds as corrosion inhibitors for metals and

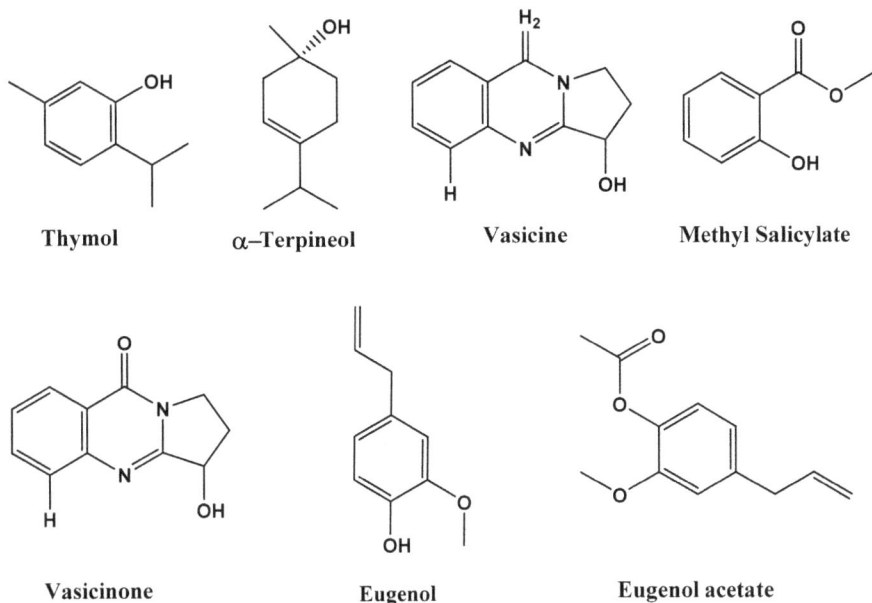

Thymol   α–Terpineol   Vasicine   Methyl Salicylate

Vasicinone   Eugenol   Eugenol acetate

FIGURE 14.1   Various phytochemical components in the plant extract.

blends in several situations (Patni et al., 2013). Given their effectiveness in works as inhibitors, plant extracts have the potential to substitute synthetic organic and inorganic inhibitors. The structure of the energetic constituent governs how green inhibitors work, and as a result, several ideas have been proposed by researchers to date to explain this phenomenon (Bouanis et al., 2016; Chigondo & Chigondo, 2016). As corrosion inhibitors, or "green corrosion inhibitors," a variety of plant extracts can be utilised. The extracts of *Camellia sinensis*, *Datura stramonium*, *Moringa oleifera*, *Cocos nucifera*, *Gongronema latifolium*, *Acalypha torta*, and *Curcuma longa* have been studied in the current chapter as corrosion inhibitors.

14.5.1 EXTRACTS OF *CAMELLIA SINENSIS* (GREEN AND BLACK TEA)

Using gravimetric and spectroscopic techniques, Yahaya et al. (2017) examined the inhibitory activity of *Green tea extracts (GTE)* and *Black tea extracts (BTE)* on slight steel in an acid environment. According to the results, both extracts' ability to suppress growth increases as extract concentration increases and decreases as temperature rises. At a concentration of $0.25 \, \text{g L}^{-1}$ of inhibitor, maximum inhibition efficiencies of 83.1% and 81.7% for GTE and BTE, respectively, were achieved. Both inhibitors adsorption behaviour on the mild steel surface followed the Langmuir adsorption isotherm, and their adsorption-based inhibition of mild steel corrosion entailed a spontaneous, exothermic, and physisorptive adsorption process. The development of the active Compound-Fe complex was also shown by the FTIR spectra. Hence, GTE and BTE from *C. sinensis* are prospective, low-cost, and environmentally friendly mild steel corrosion inhibitors.

At room temperature, the impact of *Carica papaya* (pawpaw) leaves and *C. sinensis* (tea) extracts as an organic green inhibitor on the corrosion of (duplex) brass (65%–35% Cu–Zn alloy) in 1 M HNO_3 (nitric acid) (Loto et al., 2011) were investigated. Techniques for measuring potential and the rate of weight loss/corrosion were employed in the experimental study. The green tea leaves were used to make the tea extract. The outcomes demonstrated that the extracts effectively inhibited corrosion on the brass exam samples in the 1 M nitric acid employed.

14.5.2 *Datura stramonium* Leaf Extract (DSLE)

Atomic absorption spectroscopy (AAS), gravimetric (mass loss), electrochemical (Tafel and potentiodynamic polarisation) and phytochemical screening techniques were used to evaluate *D. stramonium* leaf extract as a corrosion inhibitor in 1 M HCl acid solution (Oke et al., 2018). At room temperature, it was discovered that the extract efficiently repressed the corrosion process. The effectiveness of the inhibition grew with increasing extract content (by over 400% increases). At a concentration of 0.5 g L^{-1}, DSLE in 1 M HCl was found to have a maximum inhibitory effectiveness of 98.69%. The findings of the potentiodynamic polarisation showed that DSLE worked as a mixed-type inhibitor, functioning on together cathodic and anodic spots. The AAS analysis demonstrates that when the attentiveness of the extract increases, the concentration of Fe^{2+} in the electrolyte drops. Studies on the extract's adsorption on metal surfaces revealed that a physisorption process occurs as attentiveness rises and that chemisorption is also common. The Temkin and Freundlich isotherm models' presumptions were satisfied by the adsorption investigations.

14.5.3 *Cocos nucifera* Linn Extract

The weight loss approach, which is thought to be additional enlightening than other laboratory procedures, has been used to evaluate the corrosion-inhibitive impact of coconut water as an eco-friendly inhibitor for the corrosion control of mild steel in 0.5 M solution of H_2SO_4 acid (Adzor, 2016); 30–110 mL of coconut water was used in the tests. At intervals of between 24 and 192 hours, the test coupons were fully submerged in corroding liquid containing varied inhibitor concentrations. The outcomes demonstrated that the test coupons were affected differently depending on the inhibitor concentration in the corrodent. As the inhibitor concentration was raised, it was discovered that the corrosion rate decreased while the inhibitor efficiency rose. The study found maximal inhibition efficiencies of 89.07% and 81.57% for 24 and 48 hours of immersion duration, respectively, at concentrations of 90 and 110 mL. According to the study, coconut water has inhibitory qualities that can lessen mild steel's corrosion in an acidic environment.

14.5.4 *Moringa oleifera* Leaf Extract

Using gravimetric, gasometric, and potentiodynamic polarisation methods, the inhibitory impact of *M. oleifera* leaf extract on the corrosion of a strengthened steel block in a 2 M solution of HCl was investigated (Odusote et al., 2015). In all three of the investigational circumstances, the study found that as extract concentration rises, so does inhibition efficiency. After 120 hours of exposure, 1.0 g L⁻¹ reached an efficacy of 92.31% during gravimetric measurement. During the gasometric test, the volume of hydrogen gas evolved decreases as exposure duration increases. It was observed that the extract decreased both the rate of corrosion and the evolution of hydrogen gas. The degree at which hydrogen gas is produced, which depends on the concentration of the extract, is decreased by the creation of an adsorption layer on the metal's surface. The anodic breakdown and cathodic hydrogen evolution mechanisms are modified by the *M. oleifera* leaf extract, according to the results of potentiodynamic polarisation. Also, it was shown that when extract concentration rises, the density of the corrosive currents reduces. This reduction in corrosion is brought on by the leaf extract's enhanced adsorption on the metal surface.

14.5.5 *Gongronema latifolium* Extract

Using gasometric techniques at 303, 313, and 323 K, Onwumelu and colleagues (2018) investigated the corrosion prevention effect of methanol extract of *Gongronema latifolium* slight steel in HCl solution. The attentiveness of extracts augmented from 0.1% to 0.5% w/v, and it was discovered that the efficacy of the inhibition reduced as the temperature rose. At 303 K, it was discovered that the greatest effectiveness of inhibition was 77.17%. The inhibited system's activation energy values were higher than those found for the uncontrolled system. The fashion of the inhibition efficacy with temperature and the measured standards of E_a, d_s, and ΔG_{ads} led to the physical adsorption mechanism being postulated for the adsorption of the inhibitor. The application of *G. latifolium* extract to the exterior of minor steel tracked the isotherms for Langmuir, Freundlich, and El-Awardy.

14.5.6 *Acalypha torta* and *Curcuma longa* Leaf Extract

Weighing loss, potentiodynamic polarisation, electrochemical impedance spectroscopy, chronoamperometric measurements, and scanning electron microscopic interpretations were used to investigate the inhibitory effect of ethanol extract of *Acalypha torta* leaves (EAL) on corrosion of minor steel in 1 M HCl solution (Krishnegowda et al., 2013). The activation factors influencing the adsorption process were computed and analysed. The adsorption of EAL on mild steel trails a Langmuir adsorption isotherm. According to polarisation measurements, the EAL functions as a mixed-type inhibitor. The inhibitory efficiency determined by electrochemical testing and weight loss assays was in good agreement.

14.6 APPLICATION OF PLANT EXTRACTS FOR METALS PROTECTION

14.6.1 Plant Extract as an Inhibitor in HCl Stock

Acids are used in industries to clean the metal exterior from erosion, discoloration, and other contaminants (Kaur et al., 2022). Several plant extracts can be utilised as efficient inhibitors in the HCl medium to lessen low-carbon steel's dissolving.

Lebrini et al. (2011) used potentiodynamic polarisation and electrochemical impedance spectroscopy to explain the effective corrosion prevention action of *Oxandra asbeckii* plant extract on C38 steel in 1 M HCl. The alkaloids extract from this plant is a mixed-type inhibitor, according to polarisation curves. *O. asbeckii* plant extract adsorption stuck to the Langmuir adsorption isotherm. External analysis is also done in 1 M HCl through and deprived of the addition of plant extract to help explain the effectiveness of the inhibition. When *O. asbeckii* extract is combined with HCl, protective coating forms on the surface of the metal due to the presence of liriodenine, azafluorenone, and other terpenoids, according to phytochemical analysis of the extract.

On the basis of the constituents found in the extract, such as ellagic acid and tannic acid, which are the primary phytochemicals, Behpour et al. (2012) calculated the inhibitory impact of *Punica granatum* peel on mild steel in 2 M HCl. It has been demonstrated to be a mixed kind of inhibitor using polarisation curves. The authors also looked at how temperature affected the extract's ability to block steel. Many variables including concentration, the presence of various phytochemicals, functional group interactions with little carbon steel, temperature, and the presence of phytochemicals all affect how well a particular corrosion inhibitor inhibits corrosion. It is clear from the literature that a successful study has been done to identify mild steel's green inhibitors.

14.6.2 Plant Extract as Inhibitor in H_2SO_4 Medium

H_2SO_4 is also a frequently employed pickling media, much like HCl (Kaur et al., 2022). As this acid is also potent, it dissolves metal surfaces when applied to them for cleaning purposes. Bhawsar et al. (2015) discussed the minor steel corrosion inhibition efficiency of *Nicotiana tabacum* leaves extract, which functions as an inhibitory molecule in 2 M H_2SO_4 with an efficacy of 94% at 1,000 ppm. Temperature has an inverse relationship with inhibition efficacy but a direct relationship with concentration. The development of a protective layer on the metal's surface also complies with Langmuir's isotherm. Abdallah et al. (2018) looked at the effectiveness of aqueous extracts of cassia, curcumin, and parsley bark in inhibiting the corrosion of carbon steel. Efficiency is explained by a variety of methods, including potentiodynamic anodic polarisation and weight reduction approaches. During their analysis, the scientists also came to the conclusion that curcumin extract exhibits the lowest corrosion inhibition efficacy among the three, whereas cassia bark exhibits greater inhibition efficiency than parsley extract. The polarisation curve provides an explanation for these mixed inhibitors.

14.7 MECHANISM OF INHIBITION BY PLANT EXTRACT (UMOREN ET AL., 2019)

As was previously mentioned, several organic bioactive compounds were present in plant extracts. The centre of interaction with metal surfaces is thought to be heteroatoms and/or p-electrons found in the bioactive compounds (Honarmand et al., 2017; Khadraoui et al., 2014; Stephen & Adebayo, 2018). The pH of the corrosive media, the actual state of the particle in the medium, the anions present in the medium, the temperature of the medium, and the charge on the substrate exterior are the main determinants of the kind of adsorption between an organic molecule and a metal surface. By calculating the difference (d) between the corrosion potential (E_{corr}) and the potential of zero charge (PZC) ($E_{q=0}$), one may calculate the charge on a metal exterior (Gerengi et al., 2016; Khadom & Yaro, 2011; Saadouni et al., 2016). The substrate exterior is predicted to pick up positive charges if $d = E_{corr}E_{iq=0}$ is negative (Gerengi et al., 2016). The metal surface should be negatively charged if $d = E_{corr}E_{q=0}$ is positive (Gerengi et al., 2016). An organic molecule can occur as a neutral or protonated specie reliant on the pH of a system. Unquestionably, negative anions in a system with a positively charged surface will draw the anions to it. The intensity of the inhibitor species' adsorption will depend on how well the adsorbed anions on the metal surface recharge if the inhibitor species are present in the system as protonated species. The so-called physisorption process occurs in this scenario, when the inhibitor species are electrostatically adsorbed on the metal exterior.

Several revisions on plant extracts as anti-corrosion agents for metals reported that physisorption was the method of adsorption (El Bribri et al., 2013; Srivastava et al., 2017). In this scenario, electrostatic forces hold the extract molecules to the metal surfaces, and a rise in system temperature typically weakens the adsorption link, resulting in a drop in inhibitory efficacy. A covalent or co-ordinate kind of bond-chemisorption mechanism can be formed when neutral inhibitor molecules in a corrosive solution give their electron pairs to empty orbitals of a metal substrate. This sort of adsorption benefits from higher system temperatures; specifically, inhibition efficiency rose as temperature rose. It should be noted that on a metal surface, protonated inhibitor species can be neutralised, allowing the liberated heteroatoms to form a covalent or coordination bond with the metal surface.

14.8 STARCH AND CELLULOSE AS METAL CORROSION INHIBITORS (HUANG ET AL., 2022)

Plant polysaccharides and their derivatives have received a lot of interest due to the high number of active adsorption sites they contain, which improves their ability to stop corrosion (Vaidya et al., 2022). Starch, cellulose, pectin, lentinan, ginseng polysaccharide, astragalus polysaccharide, and other plant-based polysaccharides are among the most common plant sources of plant polysaccharides because they are inexpensive, easy to regenerate, and readily degraded. These polymers can contain more active groups because of the heteroatoms in their macromolecular structure (Bhardwaj et al., 2021). They may make strong bonds with surface metal ions to provide a thick layer of protection that envelops the metal surface like a thick blanket. Plant polysaccharide polymers have different inhibitory effects depending on the individual molecular type, which are primarily influenced by molecular mass, the presence of polar groups, the number of rings, and the quantity of adsorption centres. These elements work together to define the adsorbed layer's stability, compactness, and thickness on the metal surface.

Since polysaccharides have a large number of active adsorption sites, they can form stronger bonds with metal ions and have a greater effect on inhibiting corrosion. This is not the case, though. The majority of experimental findings (Shahini et al., 2020; Suhaimi et al., 2019) indicate that plant polysaccharides' ability to control corrosion on metals is less effective than anticipated. Owing to weak solubility, limited corrosion inhibition efficacy, insufficient alteration procedures for single polysaccharides, and a lack of corrosion inhibition mechanisms, no plant polysaccharides have been effectively exploited in the current conserving business. As plant polysaccharide corrosion inhibitors, starch and cellulose are now the subjects of extensive research. The second-most reproducible biological macromolecules in nature are starch polysaccharides, which have long been used as a source of sustenance (Aljeaban et al., 2020; Umoren et al., 2016). They are extensively utilised in the food, biomedical, textile, and other industries. Moreover, cellulose polysaccharide, which is the utmost prevalent and plentiful in nature, is the highest in yield (Jiang et al., 2019). Like starch, cellulose is used in a variety of industrial processes. The two plant polysaccharides that are most prevalent are starch and cellulose. They are thus utilised to explain the drawbacks of plant polysaccharides as inhibitors of metal corrosion and the accompanying fixes. This offers useful direction for advancing the manufacturing of green polymer corrosion inhibitors that are affordable, effective, and environmentally benign.

14.8.1 NATURAL STARCH

The primary nutrition for human, animal, and plant survival is starch, which is the most significant carbohydrate in plants. Starch is a high molecular weight carbohydrate molecule that has the chemical formula $(C_6H_{10}O_5)_n$. Amylose and amylopectin are the two molecular types. Polyglucose molecule chains connected by 2-1.4 glycosides make up amylose. A molecular weight of 32,000–160,000 and 200–980 glucose groups is found in an amylose molecule. The molecular assembly of amylopectin is dissimilar. There are 2-1.6 side chains in addition to 2-1.4 glycosidic linkages. A single amylopectin molecule has a molecular weight of 100,000–1,000,000 and 600–6,000 glucose groups. A natural polymer called starch was once used to prevent metal from corroding. Molecules include many of -OH groups. In particular, chemical adsorption and the formation of a coordination link between the O atom's lone pair electrons and iron's vacant d orbital are compatible with the properties of corrosion inhibitors. In acid and neutral salt solutions, starch and its derivatives are frequently employed as corrosion inhibitors for steel and aluminium. Metal corrosion is impacted in a specific way by natural starch. According to Rosliza and Wan Nik (2010), cassava starch was 93% (1,000 mg L^{-1}) effective at inhibiting the oxidation of aluminium alloy in saltwater. In a different experiment, Charitha and Rao (2017) found that starch had a 63.44% inhibitory effectiveness for the 6061 aluminium alloy in 0.1 M HCl. It is obvious that different plant starches may have varying levels of inhibitory effectiveness. Hence, it appears that starch does not have a very high metal-inhibiting effectiveness. The majority of plant starches have an inhibitory effectiveness of around 70%. Starches offer no benefit over current organic or inorganic corrosion inhibitors when it comes to extracts' ability to stop corrosion, much alone compared to them. Thus, natural starch cannot be employed as a corrosion inhibitor in industrial settings.

14.8.1.1 Reason for This

Several sugar rings are present in starch molecules, making them more water soluble. Yet that is not the case. Both amylose and amylopectin have extremely low solubility at room temperature; however, amylose is somewhat more solvable than amylopectin. Moreover, due to its simple swelling, amylopectin is solvable in hot water. Amylose, in contrast, does not enlarge in hot water, which results in a same solubility in both boiling and room temperature water. Each glucose unit in starch has many hydroxyl groups, and the grape lining units that link them form a polymer complex. Amylose molecules, on the contrary, are spiral-shaped rather than linear. It is insoluble in cold water because its -OH groups generate intramolecular hydrogen bonds inside the spiral, stopping it from rebonding with water molecules. Amylose dissolves in water when it is heated because molecular motion breaks the intermolecular hydrogen bonds. Because of a thin coating on the surface, amylopectin is insoluble in water, preventing the hydroxyl group from coming into touch with the liquid. Although starch has a lot of lively adsorption spots, only few of the open free hydroxyl groups are bound to vacant metal orbitals. Starch cannot, by itself, completely shield the metal surface from corrosion by forming a protective coating there. It makes no difference whatever plant the starch originates from, whether it is maize, pea, mung bean, or sweet potato starch. When directly applied to industry, natural plant starches don't seem to have the same inhibitory capacity.

14.8.2 Non-natural Starch

Furthermore, having limited solubility, natural starch often contains extra than 700 mg L^{-1} of corrosion inhibitor. This is comparable to how much plant extract there is. The dose for the majority of conventional corrosion inhibitors, however, is 50 mg L^{-1} or less. Undoubtedly, there is a lot of starch present, which makes it impossible to use even a tiny amount of industrial corrosion inhibitors. Starch is a commonly utilised raw ingredient in the food, aquaculture, textile, and industrial industries because it is readily available and inexpensive. Regrettably, the anti-corrosion business has not used starch. As natural starch cannot satisfy industrial standards, more research is required to determine how to enhance natural starch's corrosion prevention capabilities.

Just the structure and physicochemical characteristics of starch are altered during modification; chemical reagents are not used. High starch raw materials are processed physically using techniques including solvent extraction, mechanical extrusion, and pre-gelatinisation. Using sweet potato tuber as the raw material, Anyiam et al. (2020) improved natural starch by extrusion. At 700 mg L^{-1}, the greatest corrosion inhibition effectiveness was 64.26%. When *Dioscorea hispida* starch was processed with dimethyl sulfone by Othman et al. (2018). They discovered that it has a wide variety of potential uses as a green corrosion inhibitor for carbon steel in the 0.6 M series. Clearly, physical alteration of starch cannot have the desired impact on corrosion inhibition; the effectiveness of corrosion inhibition is still only approximately 70%. It might not be significantly better than natural starch for inhibiting corrosion, even if there is more starch present, because physical alteration has not fundamentally altered the bonding mechanism of active groups. Fewer hydroxyl groups are still available to interact with the metal atoms.

The issue of starch's insufficient ability to properly suppress corrosion is still not sufficiently resolved by physical modification. The researchers then attempt chemical modifications to starch. Starch has a significant number of glycosidic linkages and -OH groups in its chemical structure. By adding groups with corrosion inhibition function by chemical change caused by the breaking of glycosidic bonds or the activity of -OH, the corrosion inhibition performance can be increased. A novel type of polymer material called starch graft copolymer is created by chemically altering starch. Hydrophilic semi-rigid chains that make up the framework of the starch macromolecular chain are grafted onto these molecules to add various functional groups. This not only keeps the starch's original properties while enhancing the function of the functional groups but also significantly increases the polymer's overall efficacy at inhibiting corrosion. The more typical starch copolymers are binary ones. For instance, in a solution of $1.0\,mol\,L^{-1}$ HCl, the cassava starch-acrylamide binary graft copolymer effectively inhibited corrosion on cold-rolled steel. Chemically modified cassava starch has a substantially higher corrosion prevention efficacy than normal cassava starch. Yet there are issues with environmental contamination while making the cassava starch-acrylamide binary graft copolymer. In a different testing, microwave radiation was used to create oxidation-phosphate potato starch in order to investigate ecologically acceptable methods (Xiang et al., 2014).

The corrosion inhibitor effectively prevents corrosion while having no negative effects on the environment or public health. Based on research on binary starch graft copolymers, a third material containing corrosion inhibitor groups is added to create ternary graft copolymers, which have a greater molecular volume and more adsorption sites. On cold steel, a ternary graft copolymer of cassava starch and sodium allylsulfonate with acrylamide had a high corrosion prevention efficacy of 97.2%. Similar to this, the synthesis process uses hazardous chemicals and convoluted reaction pathways.

14.8.3   CELLULOSE

Another macromolecular polysaccharide made of glucose is cellulose. There are three -OH groups present in each structural monomer. The glucose unit is eventually produced from the cellulose after acid hydrolysis (Biotech et al., 2021). In reality, cellulose is a polymer created when a single glucose is dehydrated. Why then do cellulose and glucose exhibit such disparate physical and chemical characteristics? The intramolecular hydrogen bond responsible for starch's low water solubility also causes this. The fibre molecule has a linear shape and between 3,000 and 5,000 glucose units thanks to the 1-4-glycosidic link. Many hydrogen bonds and van der Waals forces hold the long-chain linear macromolecules together, preventing them from dissolving in water. Many O atoms in cellulose can join in coordination bonds with metal atoms to stop corrosive media from coming into touch with metal.

14.9 NEED OF GREEN INHIBITORS FOR CORROSION CONTROL (PAIS & RAO, 2019)

Effective inhibitors are synthetic molecules with numerous linkages and hetero-atoms. Unfortunately, several of the organic inhibitors are allegedly dangerous or expensive. Both the application process and the preparation process might result in toxicity. Occasionally, the inhibitor's toxicity might have harmful impacts on people or the environment. In addition, certain inorganic substances, such as chromates, have been claimed to have excellent anticorrosive properties but are exceedingly harmful to both humans and the environment. The protection of the environment and human health need to come first. In this context, there is a persistent call for environmentally acceptable inhibitors to take the place of hazardous inhibitors (Chigondo & Chigondo, 2016). The last ten years have seen an increase in research on ecologically safe, eco-friendly inhibitors. They're referred to as "green inhibitors." Green corrosion inhibitors are eco-friendly substances free of heavy metals and other toxic substances. Due to their adsorptive qualities, they are often referred to as site-blocking components or adsorption site blockers. Green inhibitors, also known as eco-friendly inhibitors, are chemicals that are biocompatible with the environment (Sharma et al., 2015). Moreover, there is a growing body of study on the use of natural materials such as plant extracts, essential oils, surfactants, and biopolymers as efficient corrosion inhibitors that are also benign to the environment (Znini et al., 2012). Plant materials have proven to be excellent and efficient corrosion inhibitors among the numerous kinds of green inhibitors (Prabhu & Rao, 2013). Plant extracts are thought of as biologically derived green inhibitors. They have been displacing synthetic organic and inorganic inhibitors throughout the past few decades. These extracts are rich sources of organic compounds such as tannins, alkaloids, and flavonoids and contain significant amounts of these substances.

Although there are a few review articles on the usage of various plant yields as corrosion inhibitors for zinc, there isn't a substantial body of literature on the use of plant products as corrosion inhibitors for copper, mild steel, stainless steel, aluminium, and its alloys. As it is a significant nonferrous metal, zinc has many uses. This concise overview provides a vivid picture of the use of plant products to prevent zinc from corroding in a variety of acidic, alkaline, and neutral conditions. This enables the early literature, its contents, and contemporary revisions that contribute to sustainable and green production to be easily accessed by corrosion engineers and material scientists.

14.10 CORROSION INHIBITION OF ZINC WITH PLANT PRODUCTS

Many ideas have been proposed to explain the mechanism of action of green inhibitors, which is dependent on the assembly of the active component (Bouanis et al., 2016). Natural inhibitors' active ingredients vary among plant classes, but they all have a proportion of structural resemblances with their synthetic equals. For example, pepper contains an alkaloid piperine (Quraishi et al., 2015), garlic encompasses allyl propyl disulphide (Rani & Basu, 2012), fennel seeds encompass limonene

(20.8%) and pinene (17.8%) trailed by myrcene (15%) and fenchone (12.5%), mustard seeds contain an alkaloid berberine (Lahhit et al., 2011), soya bean encloses tannins, pectins, flavonoids, steroids, and glycosides, carrot encloses pyrrolidine, and castor seed encloses the alkaloid ricinine (Rani & Basu, 2012). In eucalyptus oil, the active constituent is monomtrene-1,8-cineole. Volatile monoterpenes, hexuronic acid, neutral sugar remains, canaric and triterpene acids are only a few of the active components found in gum exudate. Both reducing and non-reducing sugars are also present. Unsaturated fatty acids, primary and secondary amines, bioflavonoids, and other compounds are all present in *Garcinia kola* seed extract. Ascorbic acid, pigments, amino acids, flavonoids, and carotene are all present in the calyx extract (Chigondo & Chigondo, 2016). Plant extracts include polar or hetero atoms like S, N, O, P, and aromatic rings, which favour the adsorption of electrons from donor atoms like S, N, O, P, and aromatic rings of inhibitors onto the unoccupied d-orbitals of metal surface atoms through a donor-acceptor interaction. The formation of a protective coating that shields the metal from the hostile environment is caused by the adsorption of inhibitor molecules on the metal surface (Emran et al., 2018). Corrosion is therefore prevented. A description of green inhibitors that are used completely to control the corrosion of zinc is provided below.

14.10.1 *ACHILLEA FRAGRANTISSIMA* (LAVENDER COTTON)

The *Asteraceae* family includes the little perennial plant *Achillea fragrantissima*. Using the use of an *A. fragrantissima* aqueous extract as an inhibitor, Ali et al. examined the corrosion of zinc in 0.5 M HCl. Measurements of weight loss, hydrogen evolution, and polarisation were used to compute the corrosion rates. The effectiveness of the inhibition was shown to rise with concentration and fall with temperature. *A. fragrantissima* adhered to a zinc surface in accordance with the Langmuir adsorption isotherm. 800 ppm of the inhibitor was found to have the highest effectiveness, 82%. The primary components were identified as α-thujone (60.9%), β-thujone (9.1%), sabinene (4.1%), and camphor (3.7%) (Ali et al., 2014) (Figure 14.2). The extract's components bind to the metal surface and serve as an inhibitor. Hence, one might draw the conclusion that the main ingredient, thujone, may be to blame for the extract's inhibitive effect.

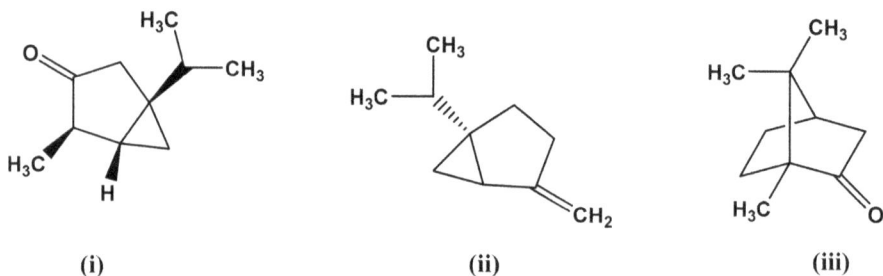

(i)   (ii)   (iii)

FIGURE 14.2 Chief components of *Achillea fragrantissima* extract: (a) α-Thujone, (b) sabinene, and (c) camphor.

14.10.2 *ALOE VERA*

Using a weight loss approach, Abiola and James investigated the impact of *Aloe vera* leaf extract on the corrosion of zinc in a 2 M HCl solution. Zinc corrosion was prevented by *A. vera* extract in a 2 M HCl solution; the effectiveness of the inhibition improved with aloe vera extract content but decreased with temperature. At 30°C, a concentration of 10 v/v inhibitor produced a 67.1% efficiency. The inhibitor molecules adhered to the surface in accordance with the Langmuir adsorption isotherm. Aloin A, aloin B, and aloe emodins (Abiola & James, 2010) are a few of the physiologically active substances found in *A. vera*. Aloe emodins display heterocyclic chemical molecules with nitrogen or oxygen atoms, and the molecular assembly of aloin indicates an anthraquinone connection to a pentose with five oxygen atoms (Figure 14.3). These chemical molecules may be the cause of *A. vera*'s inhibitory effects because they are present in the extract.

14.10.3 *OCIMUM TENUIFLORUM* (TULSI)

As a promising green zinc corrosion inhibitor in H_2SO_4, Sanjay et al. (2015) investigated the corrosion inhibition capabilities of *Ocimum tenuiflorum* leaf extract. The findings demonstrated that zinc corrosion was suppressed at various concentrations of the *O. tenuiflorum* extract and that the effectiveness of the inhibition varied with extract concentration and temperature in the H_2SO_4 medium at 0.5, 1.0, and 2.0 N at 30°C and 60°C. At 333 K, an efficiency maximum of 86.2% was noted. It was discovered that the concentrations of the principal ingredients, which changed as the plant developed, were bisabolene (13%–20%), 1,8-cineole (9%–33%), and methyl chavicol (2%–12%).

14.10.4 *TRIGONELLA FOENUM GRAECUM* (FENUGREEK)

At 30°C, 35°C, 40°C, and 45°C, Abdel-Gaber (2016) investigated the inhibitory effects of fenugreek seed extract on the corrosion of zinc. Fenugreek seed extract was employed as an efficient inhibitor for the corrosion of zinc in sulphuric acid media.

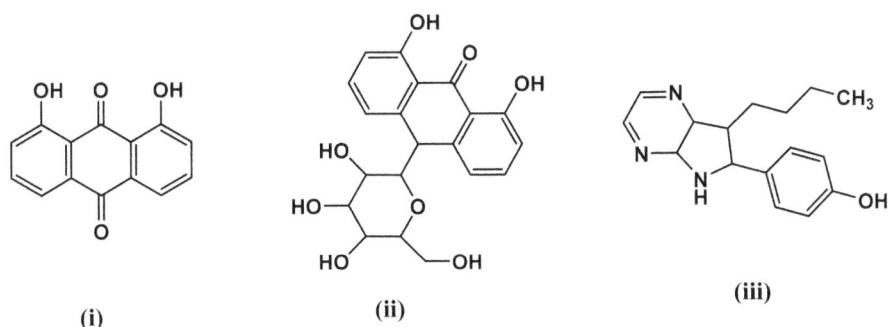

FIGURE 14.3 Chief components of *Aloe vera* extract: (a) Aloe emodin, (b) aloin, and (c) aloe emodin.

It functions as an anodic-type inhibitor. It was discovered that inhibition increases with higher extract concentrations but diminishes with higher temperatures. The production and adsorption of organozinc complex onto the metal surface give the metal its corrosion resistance. Lysine and L-tryptophan (alkaloids), diosgenin, and 4-hydroxyisoleucine are the chemical components that prevent rusting (Figure 14.4). In a sulphuric acid environment, fenugreek seed extract successfully prevented the corrosion of zinc by acting as an anodic type inhibitor.

14.10.5 CANNABIS (HEMP)

El-Housseiny used *Cannabis* plant extract to study how the corrosion of metal in 0.5 M sulphuric acid was inhibited. It was discovered that when inhibitor concentration rose, so did inhibition effectiveness. For the trials, *Cannabis* extract with 20% ethanol was employed. With 10 ppm of the inhibitor in 0.5 M sulphuric acid, the highest level of inhibition efficacy of 90% was discovered. The inhibitor compounds' adsorption complied with the Flory-Huggins isotherm (El-Housseiny, 2017).

14.10.6 MORINGA OLEIFERA (DRUMSTICK TREE)

With the use of *Moringa oleifera* leaf extract, Dass et al. examined the corrosion of zinc in a 0.5 M hydrochloric acid media. When the concentration of the extract grew, the effectiveness of the inhibition increased as well. It was discovered that the greatest inhibitory effectiveness was 67.50%. The chemical molecules in the leaf

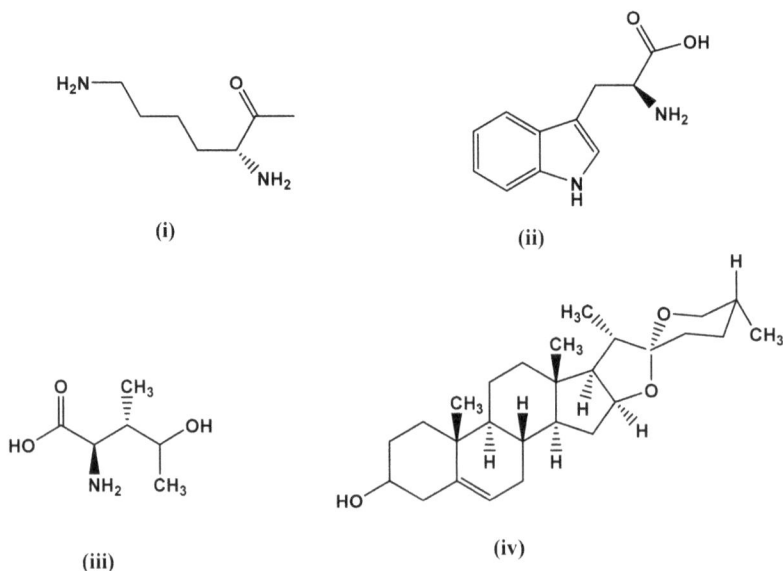

FIGURE 14.4   Major constituents of *Trigonella foenum graecum* extract: (a) Lysine, (b) L-tryptophan, (c) 4-hydroxyisoleucine, and (d) diosgenin.

extracts that include functional groups such $-C=O$, $C=N$, $C-OH$, $C=C$, and $Ph-OH$ were mostly responsible for the inhibitory effect. The corrosion inhibition action was caused by the presence of organic substances as alkaloids, saponins, tannins, and phenols (Onen et al., 2015).

14.10.7 *MANSOA ALLIACEA* (GARLIC VINE)

Using an ethanol extract of *Mansoa alliacea*, Suedile et al. investigated the suppression of corrosion of zinc in 3% NaCl. Electrochemical impedance spectroscopy and linear polarisation were used for the studies. According to potentiodynamic polarisation curves, the plant extract acts as a mixed-type inhibitor. The Langmuir adsorption isotherm was followed by the adsorption behaviour. Almost 90% of the inhibition was produced by the extract. The occurrence of flavonoids in the crude extract was what caused the inhibitory activity (Suedile et al., 2014). In Figure 14.5, numerous active ingredients are shown schematically.

14.10.8 *NYPA FRUTICANS* WURMB (MANGROVE)

The effects of extract from *Nypa fruticans* Wurmb on the suppression of zinc corrosion in hydrochloric acid were explored by Orubite et al. When *N. fruticans* Wurmb extract was present, zinc's optimal inhibitory effectiveness was 36.43%. For the investigations, fresh *N. fruticans* Wurmb leaves were extracted using methanol. It was proposed that tannin or bulky nitrogen-containing compounds were to blame

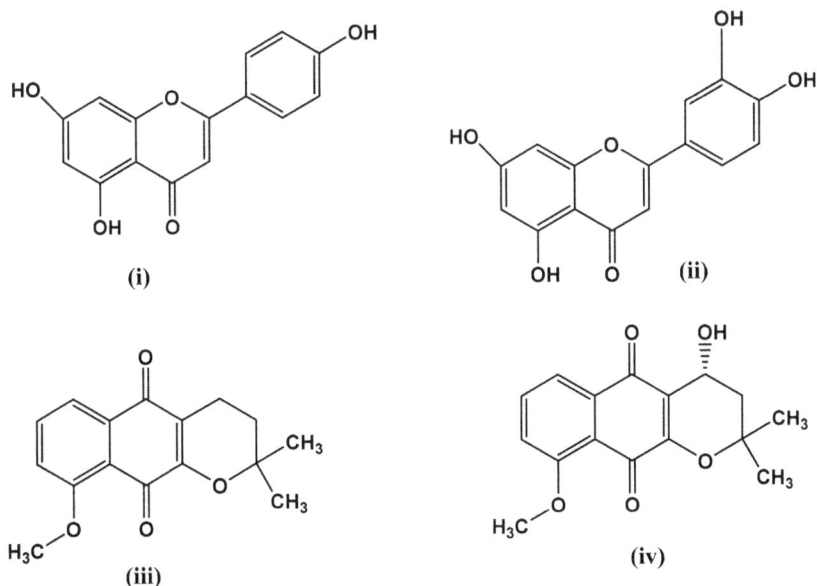

FIGURE 14.5 Major constituents of *Mansoa alliacea* extract: (a) Apigenin, (b) luteolin, (c) 9-methoxy-alpha-lapachone, and (d) 4-hydroxy-9-methoxyalpha-lapachone.

for the inhibition. *N. fruticans* constituents adhered to the zinc surface in accordance with the Langmuir adsorption isotherm (Orubite-Okorosaye & Oforka, 2005).

14.10.9 *Mangifera indica* (Mango)

In order to limit the corrosion of zinc sheets in a solution of 5 M sulphuric acid, Ugi et al. examined the adsorption properties and the inhibitory effectiveness of flavonoids, alkaloids, and tannin extracts of *Mangifera indica* leaves. At 30°C, the extract provided inhibitory efficiencies for alkaloids, flavonoids, and tannins of 96.2%, 85.3%, and 70.6%, respectively. The chemical molecules in the leaf extracts that include functional groups such as –C=O, C=N, C–OH, C=C, and Ph–OH were mostly responsible for the inhibitory effect. The corrosion of zinc in 0.5 M hydrochloric acid solution was researched using *M. indica* leaf extract. With an increase in extract concentration, the inhibitory effectiveness grew. A maximum inhibitory efficacy of 58.75% was discovered. The existence of organic compounds like alkaloids, saponins, tannins, and phenols was accountable for the corrosion inhibition activity (Onen et al., 2015; Ugi et al., 2015).

14.10.10 *Allium cepa* (Red Onion)

James and Akaranta (2011) used an acetone extract of red onion skin to study the corrosion of zinc in a 2 M hydrochloric acid solution (*Allium cepa*). One of the flavonoid substances present in red onion skin is quercetin. It is a conjugated molecule that has electron-rich heteroatoms and carbonyl groups that can act as excellent adsorption sites on the metal surface, preventing the corrosion of the zinc. The extract's degree of inhibition improved with temperature and concentration.

14.10.11 *Lupinus* (Lupin)

Abd-El-Naby et al. (2014) used lupine seed (*Lupinus*) extracts in a 0.5 M sodium chloride medium at 30°C to study the mitigation of zinc corrosion. With a 40 ppm concentration of the inhibitor, a maximum efficacy of 89.1% was attained. The inhibitor molecules adhered to the surface in accordance with the Langmuir adsorption isotherm. Up to 5% of quinolizidine alkaloids are found in lupine seeds; the key chemical components that cause the inhibition are lupanine, multiflorane, and sparteine; a schematic illustration is given in Figure 14.6.

14.10.12 *Damsissa*

Abd-El-Naby et al. used *Damsissa* extracts in a 0.5 M sodium hydroxide and 0.5 M sodium chloride medium at 30°C to study the corrosion activity of zinc. For *Damsissa*, a maximum efficiency of 1,000 and 15 ppm in 0.5 M sodium hydroxide and 0.5 M sodium chloride medium, respectively, was found. The inhibitor compounds' adsorption complied with the Flory-Huggins isotherm. Lactones, damsin, ambrosin, and coumarins are some of the essential chemical components of *Damsissa* (Figure 14.7) (Abd-El-Naby et al., 2012, 2014).

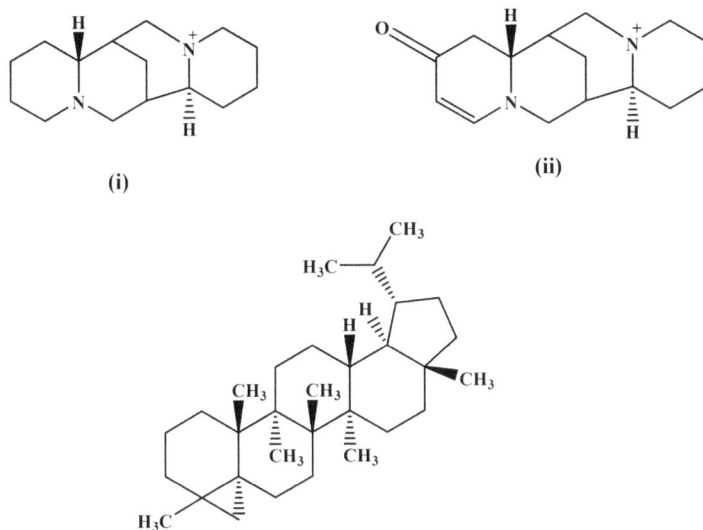

FIGURE 14.6 Major components of *Lupinus* extract: (a) Sparteine, (b) multiflorane, (c) lupanine.

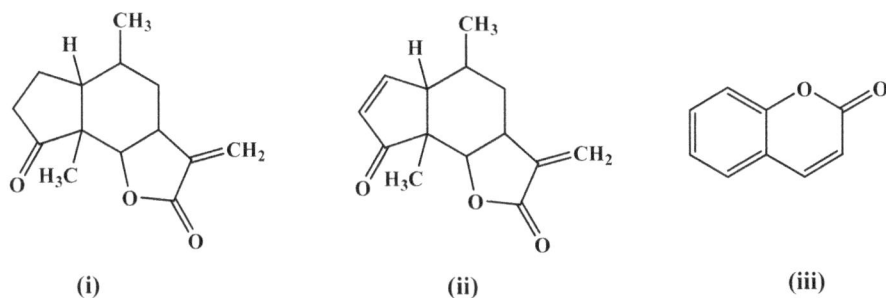

FIGURE 14.7 Main constituents of *Damsissa*: (a) Ambrosin, (b) damsia, and (c) coumarin.

14.10.13 *HLFABAR*

Abd-El-Naby et al. (2012, 2014) used *Hlfabar* extracts in a 0.5 M sodium hydroxide and 0.5 M sodium chloride media at 30°C to study the corrosion activity of zinc. Hlfabar showed maximum effectiveness at 1,400 and 40 ppm in 0.5 M sodium hydroxide and 0.5 M sodium chloride medium, respectively, of 60.2% and 94.7%. The adsorption of the inhibitor compounds was compatible with Langmuir adsorption isotherm. Hydroxyl-eudesmol derivatives are a key component responsible for the inhibitory action.

14.10.14 *CORIANDRUM SATIVUM* (*CORIANDER*)

Coriander seed extracts and the addition of zinc oxide nanoparticles were used to assess the corrosion inhibition tests of zinc in 0.5 M HCl. The outcomes showed that Np-natural inhibitor was superior to natural inhibitor in terms of effectiveness.

The extract included a huge number of organic components that were in charge of preventing corrosion. Np-natural inhibitor and natural inhibitor both achieved maximum efficiencies of 93.01% and 89.41%, respectively.

14.11 PLANT EXTRACTS AS INHIBITORS FOR CORROSION OF CARBON STEEL (ABD-EL-NABEY ET AL., 2020)

Since carbon steel is inexpensive and has strong ductility, toughness, and mechanical qualities, it may be employed in a variety of applications (Khorrami et al., 2014). Pipelines, structural forms, vehicle body parts, ships, bridges, tin cans, and other items are examples of typical uses (Tuaweri et al., 2015). Yet, in many applications, corrosion causes steel to be destroyed and degraded, reducing its lifetime. Using inhibitors is one of the most efficient and cost-effective ways to shelter metals alongside corrosion.

14.11.1 PLANT EXTRACTS AS INHIBITORS FOR CORROSION OF CARBON STEEL IN ACID MEDIA

Using weight loss and electrochemical impedance spectroscopy techniques, Gunasekaran and Chauhan (2004) investigated the inhibitory impact of *Zanthoxylum alatum* plant extract on the corrosion of mild steel in 20%, 50%, and 88% aqueous phosphoric acid. In 88% phosphoric acid, the extract was able to decrease steel corrosion more successfully than in 20% phosphoric acid. According to the results, this extract is efficient up to 70°C. The mechanism of mild steel's corrosion prevention in phosphoric acid medium was examined using surface analysis (XPS) and (FT-IR). Using weight loss and hydrogen gas evolution techniques, Orubite and Oforka (2004) investigated the inhibitory effect of the extract of *N. fruticans* Wurmb leaves on the corrosion of mild steel in hydrochloric acid solution. The highest percentage, 75.18%, was noted. Using colorimetric, weight loss, AC impedance, and Tafel polarisation methods, Krishnveni et al. (2013) investigated the inhibitory capabilities of the *Morinda tinctoria* plant leaves extract in the corrosion of steel in acid media. Charge transfer is used to stop the process as the extract functions as a mixed-type inhibitor.

Okoronkwo et al. (2015) used weight loss and thermometric techniques to investigate the inhibitory effect of chitosan derived from *Archachatina marginata* shells on the corrosion of ordinary carbon steel in acid environments. Fourier transform infrared spectroscopy analysis was used to characterise the produced chitosan. Chitosan has a good inhibitory effectiveness of 93.2%, according to the data. The effects of lemon balm extract on the inhibition of mild steel corrosion in 1 M HCl solution using electrochemical and theoretical methods were investigated and the results of electrochemical impedance spectroscopy indicate that a solution containing 800 ppm of extract had a maximum inhibitory effectiveness of 95%. The adsorption of active inhibitive chemicals such as caryophyllene, germacrene, citral, luteolin, chlorogenic acid, and rosmarinic acid on the surface of mild steel was a contributing factor in the extract's outstanding corrosion inhibition action on mild steel in HCl solution.

14.11.2 PLANT EXTRACTS AS INHIBITORS FOR CORROSION OF CARBON STEEL IN NEUTRAL AND ALKALINE MEDIA

Using potentiodynamic polarisation methods, *Silene marmarica* was investigated as an ecologically friendly steel corrosion inhibitor in $0.5\,M\ H_2SO_4$ by Abd-El-Naby et al. (2012). The findings demonstrated that *S. marmarica* prevented steel from corroding in $0.5\,M\ H_2SO_4$, and iodide ions had a synergistic impact. *S. marmarica*'s adsorption behaviour was created to suit the Langmuir isotherm and the Kinetic-thermodynamic model in both the absence and presence of iodide ions. Akbarzadeh et al. (2011) examined the inhibition efficiency of Kraft lignin (KL) and Soda lignin (SL) of the corrosion of mild steel in 3.5% NaCl solution at pH's 6 and 8, employing weight loss, electrochemical methods, and surface analysis for inhibitor concentrations of 50–800 ppm at 25°C. The greatest inhibition efficiencies for pH 6 and 8 for KL and SL extracts, respectively, were 95% and 92%, and 97% and 95%, respectively, for SL and pH 6 and 8, respectively, at 800 ppm of inhibitor concentration. *Phyllanthus amarus* extract (PAE)-Zn^{2+} system for preventing corrosion of carbon steel in aqueous solution containing 60 ppm Cl^- ion was explored by Sangeetha et al. (2011). The results showed that the formulation comprising 2 mL of PAE and 25 ppm Zn^{2+} had a 98% inhibitory efficiency. A protective film is generated on the steel surface, as shown by AC impedance spectra; polarisation studies revealed that this system operates as a mixed-type inhibitor, and FTIR spectra revealed that the protective film is composed of the Fe^{2+}-Phyllanthus complex.

14.12 PLANT EXTRACTS AS INHIBITORS FOR CORROSION OF ALUMINIUM AND ITS ALLOYS

Aluminium and its alloys have a wide range of technical applications as a result of the combination of the following qualities: light weight, mechanical strength, strong corrosion resistance, and nontoxic quality (Alam & Husain Ansari, 2017). For many important applications in the fields of aircraft, the food industry, construction, heat exchange, and electrical transmission, aluminium and its alloys are the material of choice. An organic, protective oxide coating gives aluminium corrosion resistance in various situations. Because of its amphoteric nature, this film can disintegrate when exposed to either acidic (media with a pH below 5) or alkaline (media with a pH above 9) conditions. Due to their numerous uses, aluminium and its alloys are exposed to acids and bases during pickling, descaling, and electrochemical etching (Abd-El-Nabey et al., 2012).

14.12.1 PLANT EXTRACTS AS INHIBITORS FOR CORROSION OF ALUMINIUM AND ITS ALLOYS IN ACID MEDIA

El-Hosary et al. (1972) investigated the inhibitory effect of *Hibiscus subdariffa* (karkade) extract as green corrosion inhibitor for aluminium and zinc in 2 N HCl solution. They also investigated how molasses prevented the corrosion of steel, copper, and aluminium. Using the weight loss approach, Avwiri and Igho (2003) investigated the effects of *Vernonia amygdalina* (biter leaf) on the acidic corrosion of 25 and 3RS aluminium alloys. The outcomes indicated that the maximum inhibitory efficiencies were achieved with 0.1 M HCl and 0.1 M HNO$_3$, respectively, at 49.5% and 72.5%. Obot and Obi-Egbedi (2010) investigated the effects of *Chlomolaena odorata* L. leaf extracts on the suppression of the corrosion of aluminium in 2 M HCl using gasometric and thermometric methods. The results showed that this extract was a highly effective inhibitor of aluminium corrosion in acidic conditions. The extract's constituents adhere to the Langmuir isotherm when they adsorb on aluminium.

14.12.2 PLANT EXTRACTS AS INHIBITORS FOR CORROSION OF ALUMINIUM AND ITS ALLOYS IN NEUTRAL AND ALKALINE MEDIA

Oguzie (2007) investigated the inhibitory effect of the *Sansevieria trifasciata* leaf extract on the corrosion of aluminium in 2 M HCl and 2 M KOH solutions. The outcomes showed that this extract worked well as an inhibitor in both settings. In the presence of halide ions, synergistic effects improved the effectiveness of inhibition. The inhibitory behaviour was explained by a physical adsorption mechanism, and the adsorption process followed the Freundlich isotherm. Using chemical and electrochemical methods, Abdel-Gaber et al. (2008) investigated the suppression of aluminium corrosion in 2 M NaOH by *Damsissa* (*Ambrosia maritime* L.) extract. The extract exhibited mixed-type inhibitory behaviour. The findings suggested that *Damsissa* extract would work well as an inhibitor of aluminium corrosion in alkaline conditions. The impedance data confirmed that extracts were remarkably stable during storage for up to 35 days. Rosliza and Wan Nik (2010) used gasometry, potentiodynamic polarisation, linear polarisation resistance, and electrochemical impedance spectroscopy experiments to investigate the usage of tapioca starch for improving corrosion resistance of AA 6061 alloy in sea water. The findings showed that tapioca starch reduced corrosion rates, corrosion current densities, and double-layer capacitance and increased the polarisation resistance in seawater. The experimental data are nicely suited by the Langmuir adsorption isotherm.

14.13 PLANT EXTRACTS AS INHIBITORS FOR CORROSION OF COPPER AND ITS ALLOYS

Due to its strong electrical and thermal conductivities, copper is one of the most significant metals in industry. As a result, it is utilised in several industrial applications, including shipbuilding, sea water desalination, pipelines, heat conductors, and electrical power lines (Khaled, 2008). Controlling copper corrosion is one of the biggest issues in copper applications because copper metal corrodes in different ways depending on the environment. Sangeetha and Fredimoses (2011) examined the effectiveness of the acid extract of *Azadirachta indica* seed as corrosion inhibitor for mild copper metal in 1, 2, and 3 N HNO$_3$ solution. The findings showed that *A. indica* seed had a 95% efficacy at 1% inhibitor concentration throughout a range of time periods as an excellent copper corrosion inhibitor in nitric acid. Shah et al. (2013) used weight loss, potentiodynamic polarisation, electrochemical impedance spectroscopy, scanning electron microscopy, with energy dispersive X-ray (EDX), AAS, and ion chromatography to investigate the inhibitory effect of Mangure tannin as a green inhibitor for the corrosion of copper in 0.5 M HCl solution (IC). At 3 g L^{-1}, the maximum inhibition efficiency was 82%. According to the findings, mangrove tannin behaves as a cathodic inhibitor and its adsorption process follows the Langmuir isotherm.

Using potentiodynamic polarisation and electrochemical impedance spectroscopy methods, Shabani-Nooshabadi et al. (2015) examined the corrosion inhibition effect of *Calligonum comosum* extract on copper in 2 M HCl solution. A maximum inhibitory efficacy of 80.06% was attained in 0.8 g L^{-1} of the extract, according to the polarisation experiments, which demonstrated that this extract functions as a mixed-type inhibitor. The Langmuir adsorption isotherm was used to discuss the extract's inhibitory effects. Deivanayagam et al. (2015) used weight loss measures to examine the inhibitory impact of an alcoholic extract of *Mimusops elengi* leaves on copper corrosion in natural sea water. After 120 hours of immersion, the maximal inhibition effectiveness was 86.84%. The chemical adsorption process was used to adhere the inhibitor to the metal surface.

Wedian et al. (2017) employed measures of weight loss and polarisation to determine the effectiveness of a *Capparis spinosa* L. extract as a green inhibitor for copper corrosion in high acid environments. According to the findings, this extract effectively inhibits copper corrosion in strong acid media, with a maximum inhibition efficacy of 82.7% at 440 ppm of the extract. The potentiodynamic, chemical, and weight loss estimates all agreed rather well. Oukhrib et al. (2017) investigated *Ziziphus lotus* (wild jujube) as a corrosion inhibitor for copper in seawater utilising weight loss and polarisation techniques. After being submerged in inhibited and uninhibited electrolytes, the morphology of the copper surface was examined using a scanning electron microscope. The findings showed that the extract works effectively to prevent copper corrosion in seawater, with a 93% effectiveness level at 5 g L^{-1}.

14.14 PLANT EXTRACTS AS INHIBITORS FOR CORROSION OF NICKEL AND ITS ALLOYS

Because nickel and its alloys are resistant to oxidative corrosion and aqueous corrosion at high temperatures, they are widely used in the industrial, aviation, marine, chemical, petrochemical, oil and gas, nuclear, and conventional power-producing industries, as well as the textile, desalination, and food processing industries. The need for nickel has increased over the past 40 years, which has resulted in higher production rates, since nickel is an essential component of super alloys, stainless steel, and other alloys as well as a protective coating for materials with lesser corrosion resistance. Nickel makes an excellent foundation for alloys that need strength at high temperatures because it has good resistance to many corrosive solutions, including mildly acidic and alkaline solutions. In the electrochemical series E_0 Ni^{2+}/Ni $= -0.25$ V, nickel held a middle place and is more noble than Sn, Pb, and Cu (Walia & Singh, 2005). New machines have several components composed of pure nickel. Electrodes made of nickel are crucial components for petrochemical technology and energy conversion equipment. While nickel corrosion and passivation do occur in acid solutions, nickel electrodes are mostly coated with electrochemically active $Ni(OH)_2$ layers in aqueous conditions (Zinola & Luna, 1995). El-Etre et al. (2005) investigated the aqueous extract of henna (*Lawsonia*) leaves as a corrosion inhibitor for C-steel, nickel, and zinc in acid, neutral, and alkaline solution. In all tested medium, it was discovered that the extract worked well as a corrosion inhibitor for the three metals. The extract functioned as a mixed-type inhibitor, and the foundation for its inhibitory activity was the adsorption of its component molecules on the surface of the metal.

14.15   EXTRACTION METHODS (MIRALRIO & VÁZQUEZ, 2020)

A wide range of extraction techniques are used nowadays. Nonetheless, a succinct analysis of extraction techniques is provided below. The next step is to choose the region of the plant that has the highest concentration of the required active compounds. The leaves, flowers, seeds, fruits, roots, and stems as well as all other parts of the plant are used to create the extracts. In essence, extraction methods operate by heating, cooling, and separating the active components while the solvent is present. Percolation, digestion, infusion decoction, and maceration are a few of the traditional extraction methods (Gharby et al., 2022). In general, depending on the outcomes required, the extract method's format can be applied. During maceration, materials that have been crushed, mashed, or sliced may also have previously dried, and they are immersed in the extraction solvent for at least three days while being vigorously agitated. The target substance's solubilisation of the active molecules by the solvent's dilution opens the door to possible extraction. In the final combination, filtration can separate the suspended particles. The advantage of this method is that the complete essence is removed without being changed, and the active components are easily soluble.

By using a short maceration with hot water during the infusion process, the extract is created. The solubilisation of the components that are most easily soluble results in the production of the extract. In a manner similar to this, the decoction procedure involves cooking the crude drug in a certain volume of water for a specific period of time. In order to improve the extraction solvent's solubility and prevent the active chemicals from decomposing, the digesting process suggests macerating the raw materials in the presence of a somewhat heated solvent. The input material is placed in a conical tank called a percolator that has an adjustable closure as part of the percolation filtering procedure, which operates at room temperature. After that, the percolator has to be covered and filled with solvent so that the extract may be obtained drop by drop (Zhang et al., 2018). The benefits of percolation are found in the excellent performance of the active ingredients, the speed with which they may be produced, and the affordability of the raw materials.

Hot continuous extraction and ultrasound or sonication are more complex techniques (Azmir et al., 2013). The first one makes use of a glass body with a hot flask, a syphon arm, a thimble, an extraction chamber, and a condenser (Azmir et al., 2013). Briefly put, the solvent-filled boiling flask is heated, and the resulting vapour is condensed. The remaining liquid drops into the thimble containing the raw materials, and the extract fills the extraction chamber to start the syphon arm, which then dumps the remaining liquid back into the boiling flask. The reflux process needs to be stopped for the required amount of extraction to be reached. Last but not least, sonication is a technique that increases the permeability of cell walls by employing high-intensity ultrasounds to produce cavitation, which damages cellular membranes (Nn, 2015). Cells that have been subjected to sonication burst, releasing their contents for further extraction. The resulting liquids are clarified by decantation or filtration after being created using the aforementioned methods.

Since it solubilises the active compounds as it moves through the plant tissues and enables their extraction, the solvent is crucial to the extraction procedures. It has been established that extraction solvents affect the physical, chemical, and antioxidant characteristics of the produced extracts because the concentration of flavonoids, saponins, phenolic compounds, and other chemicals contained in plant extracts differs depending on the extraction solvent (Neffati et al., 2017; Pham et al., 2015; Seal, 2016). Several solvents have been used in order to extract plant extracts with the required concentration of active compounds. The solvent used will consequently determine how well active compounds are extracted; the most common solvents are water, methanol, ethyl acetate, dichloromethane, and hexane. Given that it is easily available, non-toxic, inert, and affordable, water may be the most useful extraction solvent (Duan et al., 2015; Varma, 2016). While not all plant extracts are offered as aqueous extracts, it is feasible to experiment with other solvents. To achieve the optimum result, a variety of options must be tried since solvents are selective.

14.16 CHARACTERISATION TECHNIQUES

Many experimental approaches are therefore available for this aim in order to properly characterise the extracts offered as corrosion inhibitors. For assessing a metal's susceptibility to localised corrosion, such as pitting and crevice, cyclic potentiodynamic polarisation (PP) is utilised (Esmailzadeh et al., 2018). The measurement and analysis of the current generated by a changing voltage in a working electrode is the basis of polarisation tests like PP. Electrochemical impedance spectroscopy (EIS) is another frequently used method to examine anti-corrosion effectiveness in very short testing intervals. This method is employed to establish a system's impedance in terms of the frequency of a variable potential. The most common graphical representations of the findings of EIS analysis are Nyquist plots (Wang et al., 2021), which are based on models with similar electrical circuits. EIS displays additional information, such as the system's operation and various levels of resistance. By figuring out the link between electrochemical potential and produced currents on charged electrodes, the linear polarisation resistance (LPR) approach may be utilised to calculate the corrosion rate (Mansfeld, 2009). The weight loss (WL) method, which is based on the mass lost by corrosion and directly monitored to determine the corrosion rate, is less complicated and time-consuming. To determine the effectiveness of the corrosion inhibitor, some parameters that can be determined in both the presence and absence of the material can be utilised, such as corrosion current density measured by PP.

The most frequent methods used to characterise surfaces are spectroscopy and microscopy. The metal surface with and without a corrosion inhibitor may be clearly compared using a scanning electron microscope (SEM), which also yields additional morphological data (Saxena et al., 2019; Vengatesh et al., 2019). Similar to this, the atomic force microscope (AFM) collects data on the metal surface's shape for topographical imaging and comparisons (Finšgar, 2020; Haldhar et al., 2018). To determine oxidation states, stoichiometry, and electronic states, X-ray photoelectron spectroscopy (XPS) is frequently utilised (Bouanis et al., 2016).

Fourier transform infrared spectroscopy (FT-IR) is typically used for complementary characterisations in order to learn more about the functional groups and vibrational modes of corrosion inhibitors. Similarly, functional groups, electronic transitions, and optical band gaps may all be better understood via ultraviolet-visible spectroscopy (UV-VIS).

14.17   DRAWBACKS AND FUTURE PERSPECTIVE

According to the various study findings discussed in this chapter, plant extracts are trustworthy, affordable, and environmentally friendly inhibitors of metal corrosion. Regarding the use of plant extracts as inhibitors of metal corrosion, there are still several unsolved problems. Is the inhibitive property caused by all bioactive compounds in plant extract? Does inhibition result from the simultaneous action of bioactive molecules? Are there any elements present that could lessen others' capacity to inhibit? What precisely does plant extract inhibition involve? The fact that plant extracts may limit activity through a mechanism other than simple adsorption, as is frequently claimed, is evident. Thus, future studies on plant extracts as corrosion inhibitors should concentrate on isolating the exact active component that is accountable for the anticorrosive characteristic of extract. The question of whether such an active molecule exerts its inhibitory characteristic on its own or in concert with other molecules should be a key focus of research. The mechanism by which plant extracts reduce corrosion will be better understood as a result of such a discovery. Also, it is seen in peer-reviewed scientific studies that plant extracts do not function well in hostile conditions and cannot resist extremely high temperatures.

14.18   CONCLUSIONS

The biggest issue in industries is metal corrosion. Green inhibitors are shown to be effective when taking into account environmental and ecological factors. The aforementioned discussion demonstrates how effective natural plant extracts are at preventing corrosion in mild steel, and how we can use green inhibitors in place of other chemicals that are used to control corrosion. Green inhibitors are eco-friendly and also very affordable, and they have the added benefit of allowing us to use plant extracts as corrosion inhibitors to safeguard the environment.

The potential for a broad range of operating environments and efficiency in corrosion prevention and control utilising green inhibitors is great. Successful research suggests that environmentally friendly corrosion inhibitors are a more dependable and long-lasting alternative than conventional corrosion inhibitors. Biodegradable components were given by natural biomass sources, which were effective and environmentally beneficial when employed as corrosion inhibitors.

The focus of contemporary research is on substances of a natural origin, such as extracts from plant components and natural polymers. Plant extracts are widespread; they are cheap, nontoxic, biodegradable, and biocompatible, and they include hetero-atoms and/or π-electrons, which make them appropriate as metal corrosion inhibitors. They have been evaluated under a variety of corrosive conditions, including HCl and H_2SO_4. Plant extracts have been found to be effective in each of these mediums. Adsorptive mechanisms, such as physisorption, chemisorption, or both, are the basis of the inhibitory mechanism induced by plant extracts.

Future studies should consequently concentrate on this area. For researchers who study corrosion as well as students of materials science and allied fields, this chapter will stand out as a masterpiece.

REFERENCES

Abd-El-Nabey, B. A., Abd-El-khalek, D. E., El-Housseiny, S., & Mohamed, M. E. (2020). Plant extracts as corrosion and scale inhibitors: A review. *International Journal of Corrosion and Scale Inhibition*, 9(4), 1287–1328. https://doi.org/10.17675/2305-6894-2020-9-4-7.

Abd-El-Nabey, B. A., Abdel-Gaber, A. M., Elawady, G. Y., & El-Housseiny, S. (2012). Inhibitive action of some plant extracts on the alkaline corrosion of aluminum. *International Journal of Electrochemical Science*, 7(9), 7823–7839.

Abd-El-Naby, B. A., Abdullatef, O. A., Abd-El-Gaber, A. M., Shaker, M. A., & Esmail, G. (2012). Effect of some natural extracts on the corrosion of zinc in 0.5 M NaCl. *International Journal of Electrochemical Science*, 7(7), 5864–5879.

Abd-El-Naby, B. A., Abdullatef, O. A., Abd-El-Gaber, A. M., Shaker, M. A., & Esmail, G. (2014). Electrochemical studies on the inhibitive action of Damssisa and Halfabar on the alkaline corrosion of zinc. *International Journal of Electrochemical Science*, 9(3), 1163–1178.

Abdallah, M., Altass, H. M., Al Jahdaly, B. A., & Salem, M. M. (2018). Some natural aqueous extracts of plants as green inhibitor for carbon steel corrosion in 0.5 M sulfuric acid. *Green Chemistry Letters and Reviews*, 11(3), 189–196. https://doi.org/10.1080/17518253.2018.1458161.

Abdel-Gaber, A. M. (2016). Effect of immersion time and temperature on the inhibition of the acid corrosion of zinc by fenugreek seeds extract. *International Journal of Applied Chemistry*, 3, 231.

Abdel-Gaber, A. M., Khamis, E., Abo-ElDahab, H., & Adeel, S. (2008). Inhibition of aluminium corrosion in alkaline solutions using natural compound. *Materials Chemistry and Physics*, 109(2–3), 297–305. https://doi.org/10.1016/j.matchemphys.2007.11.038.

Abdel-Gaber, A. M., Rahal, H. T., & El-Rifai, M. S. (2021). Green approach towards corrosion inhibition in hydrochloric acid solutions. *Biointerface Research in Applied Chemistry*, 11(6), 14185–14195. https://doi.org/10.33263/BRIAC116.1418514195.

Abiola, O. K., & James, A. O. (2010). The effects of Aloe vera extract on corrosion and kinetics of corrosion process of zinc in HCl solution. *Corrosion Science*, 52, 661–664. https://doi.org/10.1016/j.corsci.2009.10.026.

Adejoro, I. A., Ojo, F. K., & Obafemi, S. K. (2015). Corrosion inhibition potentials of ampicillin for mild steel in hydrochloric acid solution. *Journal of Taibah University for Science*, 9(2), 196–202. https://doi.org/10.1016/j.jtusci.2014.10.002.

Adzor S. A., & Udoye. B. (2016). Corrosion inhibitive effects of coconut (Cocos nucifera Linn) water for mild steel in acidic medium. *European Journal of Material Sciences*, 3(2), 1–12.

Akbarzadeh, E., Ibrahim, M. N. M., & Rahim, A. A. (2011). Corrosion inhibition of mild steel in near neutral solution by Kraft and Soda lignins extracted from oil palm empty fruit bunch. *International Journal of Electrochemical Science*, *6*(11), 5396–5416.

Alam, T., & Husain Ansari, A. (2017). Review on aluminium and its alloys for automotive applications polymer matrix nanocomposites view project metal matrix reinforced ceramic composites view project. May. https://www.researchgate.net/publication/317075488.

Ali, A. I., Megahed, H. A., & El-Etre, M. (2014). Zinc corrosion in HCl in the presence of aqueous extract of Achillea fragrantissima. January *5*, 923–930.

Alibakhshi, E., Ramezanzadeh, M., Bahlakeh, G., Ramezanzadeh, B., Mahdavian, M., & Motamedi, M. (2018). Glycyrrhiza glabra leaves extract as a green corrosion inhibitor for mild steel in 1 M hydrochloric acid solution: Experimental, molecular dynamics, Monte Carlo and quantum mechanics study. *Journal of Molecular Liquids*, *255*, 185–198. https://doi.org/10.1016/j.molliq.2018.01.144.

Aljeaban, N. A., Goni, L. K. M. O., Alharbi, B. G., Jafar, M. A., Ali, S. A., Chen, T., Quraishi, M. A., & Al-Muallem, H. A. (2020). Polymers decorated with functional motifs for mitigation of steel corrosion: An overview. *International Journal of Polymer Science*, *2020*, 9512680. https://doi.org/10.1155/2020/9512680.

Anyiam, C. K., Ogbobe, O., Oguzie, E. E., Madufor, I. C., Nwanonenyi, S. C., Onuegbu, G. C., Obasi, H. C., & Chidiebere, M. A. (2020). Corrosion inhibition of galvanized steel in hydrochloric acid medium by a physically modified starch. *SN Applied Sciences*, *2*(4), 520. https://doi.org/10.1007/s42452-020-2322-2.

Avwiri, G. O., & Igho, F. O. (2003). Inhibitive action of Vernonia amygdalina on the corrosion of aluminium alloys in acidic media. *Materials Letters*, *57*(22–23), 3705–3711. https://doi.org/10.1016/S0167-577X(03)00167-8.

Azmir, J., Zaidul, I. S. M., Rahman, M. M., Sharif, K. M., Mohamed, A., Sahena, F., Jahurul, M. H. A., Ghafoor, K., Norulaini, N. A. N., & Omar, A. K. M. (2013). Techniques for extraction of bioactive compounds from plant materials: A review. *Journal of Food Engineering*, *117*(4), 426–436. https://doi.org/10.1016/j.jfoodeng.2013.01.014.

Behpour, M., Ghoreishi, S. M., Khayatkashani, M., & Soltani, N. (2012). Green approach to corrosion inhibition of mild steel in two acidic solutions by the extract of Punica granatum peel and main constituents. *Materials Chemistry and Physics*, *131*(3), 621–633. https://doi.org/10.1016/j.matchemphys.2011.10.027.

Bhardwaj, N., Sharma, P., & Kumar, V. (2021). Phytochemicals as steel corrosion inhibitor: An insight into mechanism. *Corrosion Reviews*, *39*(1), 27–41. https://doi.org/10.1515/corrrev-2020-0046.

Bhawsar, J., Jain, P. K., & Jain, P. (2015). Experimental and computational studies of Nicotiana tabacum leaves extract as green corrosion inhibitor for mild steel in acidic medium. *Alexandria Engineering Journal*, *54*(3), 769–775. https://doi.org/10.1016/j.aej.2015.03.022.

Biotech, S., Engineering, B., Columbia, B., Technology, P., Wood, W., & Centre, S. (2021). Developing fibrillated cellulose as a sustainable technological material. *Nature*, *590*(February), 47–56.

Bouanis, M., Tourabi, M., Nyassi, A., Zarrouk, A., Jama, C., & Bentiss, F. (2016). Corrosion inhibition performance of 2,5-bis(4-dimethylaminophenyl)-1,3,4-oxadiazole for carbon steel in HCl solution: Gravimetric, electrochemical and XPS studies. *Applied Surface Science*, *389*, 952–966. https://doi.org/10.1016/j.apsusc.2016.07.115.

Charitha, B. P., & Rao, P. (2017). Starch as an ecofriendly green inhibitor for corrosion control of 6061-Al alloy. *Journal of Materials and Environmental Science*, *8*(1), 78–89.

Chauhan, L. R., & Gunasekaran, G. (2007). Corrosion inhibition of mild steel by plant extract in dilute HCl medium. *Corrosion Science*, *49*(3), 1143–1161. https://doi.org/10.1016/j.corsci.2006.08.012.

Chen, Z., Fadhil, A. A., Chen, T., Khadom, A. A., Fu, C., & Fadhil, N. A. (2021). Green synthesis of corrosion inhibitor with biomass platform molecule: Gravimetrical, electrochemical, morphological, and theoretical investigations. *Journal of Molecular Liquids*, *332*, 115852. https://doi.org/10.1016/j.molliq.2021.115852.

Chigondo, M., & Chigondo, F. (2016). Recent natural corrosion inhibitors for mild steel: An overview. *Journal of Chemistry*, *2016*, 6208937. https://doi.org/10.1155/2016/6208937.

Deivanayagam, P., Malarvizhi, I., Selvaraj, S., & Deepa, P. (2015). Corrosion inhibition efficacy of ethanolic extract of mimusops elengi leaves (MEL) on copper in natural sea water. *International Journal of Multidisciplinary Research and Development 2015*, *2*(4), 100–107. www.allsubjectjournal.com.

Duan, H., Wang, D., & Li, Y. (2015). Green chemistry for nanoparticle synthesis. *Chemical Society Reviews*, *44*(16), 5778–5792. https://doi.org/10.1039/c4cs00363b.

Eddy, N. O., Ebenso, E. E., & Ibok, U. J. (2010). Adsorption, synergistic inhibitive effect and quantum chemical studies of ampicillin (AMP) and halides for the corrosion of mild steel in H_2SO_4. *Journal of Applied Electrochemistry*, *40*(2), 445–456. https://doi.org/10.1007/s10800-009-0015-z.

El-Etre, A. Y. (2006). Khillah extract as inhibitor for acid corrosion of SX 316 steel. *Applied Surface Science*, *252*(24), 8521–8525. https://doi.org/10.1016/j.apsusc.2005.11.066.

El-Etre, A. Y., Abdallah, M., & El-Tantawy, Z. E. (2005). Corrosion inhibition of some metals using lawsonia extract. *Corrosion Science*, *47*(2), 385–395. https://doi.org/10.1016/j.corsci.2004.06.006.

El-Housseiny, S. (2017). Inhibition of the acid corrosion of zinc by cannabis plant extract. *American Journal of Chemistry*, *2017*(1), 6–15. https://doi.org/10.5923/j.chemistry.20170701.02.

El Bribri, A., Tabyaoui, M., Tabyaoui, B., El Attari, H., & Bentiss, F. (2013). The use of Euphorbia falcata extract as eco-friendly corrosion inhibitor of carbon steel in hydrochloric acid solution. *Materials Chemistry and Physics*, *141*(1), 240–247. https://doi.org/10.1016/j.matchemphys.2013.05.006.

Emran, K. M., Ali, S. M., & Lehaibi, H. A. A. (2018). Green methods for corrosion control. *Corrosion Inhibitors, Principles and Recent Applications*, 61–78. https://doi.org/10.5772/intechopen.72762.

Esmailzadeh, S., Aliofkhazraei, M., & Sarlak, H. (2018). Interpretation of cyclic potentiodynamic polarization test results for study of corrosion behavior of metals: A review. *Protection of Metals and Physical Chemistry of Surfaces*, *54*(5), 976–989. https://doi.org/10.1134/S207020511805026X.

Finšgar, M. (2020). Electrochemical, 3D topography, XPS, and ToF-SIMS analyses of 4-methyl-2-phenylimidazole as a corrosion inhibitor for brass. *Corrosion Science*, *169*(January), 108632. https://doi.org/10.1016/j.corsci.2020.108632.

Gaya, H. S., Shawai, S. A. A., Yusuf, G. A., & Abubakar, I. H. (2019). Plants extract for corrosion control of mild steel in acidic medium. *World Academics Journal of Engineering Sciences*, *6*(1), 47–51.

Gerengi, H., Ugras, H. I., Solomon, M. M., Umoren, S. A., Kurtay, M., & Atar, N. (2016). Synergistic corrosion inhibition effect of 1-ethyl-1-methylpyrrolidinium tetrafluoroborate and iodide ions for low carbon steel in HCl solution. *Journal of Adhesion Science and Technology*, *30*(21), 2383–2403. https://doi.org/10.1080/01694243.2016.1183407.

Gharby, S., Oubannin, S., Ait Bouzid, H., Bijla, L., Ibourki, M., Gagour, J., Koubachi, J., Sakar, E. H., Majourhat, K., Lee, L. H., Harhar, H., & Bouyahya, A. (2022). An overview on the use of extracts from medicinal and aromatic plants to improve nutritional value and oxidative stability of vegetable oils. *Foods*, *11*(20), 3258. https://doi.org/10.3390/foods11203258.

Gunasekaran, G., & Chauhan, L. R. (2004). Eco friendly inhibitor for corrosion inhibition of mild steel in phosphoric acid medium. *Electrochimica Acta*, *49*(25), 4387–4395. https:// doi.org/10.1016/j.electacta.2004.04.030.

Haldhar, R., Prasad, D., & Saxena, A. (2018). Armoracia rusticana as sustainable and eco-friendly corrosion inhibitor for mild steel in 0.5M sulphuric acid: Experimental and theoretical investigations. *Journal of Environmental Chemical Engineering*, *6*(4), 5230–5238. https://doi.org/10.1016/j.jece.2018.08.025.

Hebbar, N., Praveen, B. M., Prasanna, B. M., & Venkatesha, T. V. (2015). Anticorrosion potential of a pharmaceutical intermediate floctafenine for zinc in 0.1 M HCl solution. *International Journal of Industrial Chemistry*, *6*(3), 221–231. https://doi.org/10.1007/ s40090-015-0049-5.

Honarmand, E., Mostaanzadeh, H., Motaghedifard, M. H., Hadi, M., & Khayadkashani, M. (2017). Inhibition effect of opuntia stem extract on corrosion of mild steel: A quantum computational assisted electrochemical study to determine the most effective components in inhibition. *Protection of Metals and Physical Chemistry of Surfaces*, *53*(3), 560–572. https://doi.org/10.1134/S207020511703008X.

Hosary, A. A. E. L. (1972). Corrosion inhibition by naturally occurring substances--I. The effect of Hibiscus subdariffa (karkade) extract on the dissolution. *Corrosion Science*, *12*(March), 897–904.

Huang, L., Chen, W. Q., Wang, S. S., Zhao, Q., Li, H. J., & Wu, Y. C. (2022). Starch, cellulose and plant extracts as green inhibitors of metal corrosion: A review. *Environmental Chemistry Letters*, *20*(5), 3235–3264. https://doi.org/10.1007/s10311-022-01400-5.

Ismail, A. (2016). A review of green corrosion inhibitor for mild steel in seawater. *ARPN Journal of Engineering and Applied Sciences*, *11*(14), 8710–8714.

James, A. O., & Akaranta, O. (2011). Inhibition of corrosion of zinc in hydrochloric acid solution by red onion skin acetone extract. *Research Journal of Chemical Sciences*, *1*(1), 31–37. https://citeseerx.ist.psu.edu/viewdoc/download?doi=10.1.1.454.8682&rep=rep1 &type=pdf.

Jiang, T., Duan, Q., Zhu, J., Liu, H., Yu, L., Industrial, A., Duan, Q., Zhu, J., Liu, H., Yu, L., & Biodegradable, S. (2019). Starch-based biodegradable materials: Challenges and opportunities. *Advanced Industrial and Engineering Polymer Research*, *3*, 8–18. https://doi. org/10.1016/j.aiepr.2019.11.003.

Kaur, J., Daksh, N., & Saxena, A. (2022). Corrosion inhibition applications of natural and eco-friendly corrosion inhibitors on steel in the acidic environment: An overview. *Arabian Journal for Science and Engineering*, *47*(1), 57–74. https://doi.org/10.1007/ s13369-021-05699-0.

Khadom, A. A., & Yaro, A. S. (2011). Protection of low carbon steel in phosphoric acid by potassium iodide. *Protection of Metals and Physical Chemistry of Surfaces*, *47*(5), 662–669. https://doi.org/10.1134/S2070205111050078.

Khadraoui, A., Khelifa, A., Boutoumi, H., Mettai, B., Karzazi, Y., Karzazi, Y., & Hammouti, B. (2014). Corrosion inhibition of carbon steel in hydrochloric acid solution by mentha pulegium extract. *Portugaliae Electrochimica Acta*, *32*(4), 271–280. https://doi. org/10.4152/pea.201404271.

Khaled, K. F. (2008). Guanidine derivative as a new corrosion inhibitor for copper in 3% NaCl solution. *Materials Chemistry and Physics*, *112*(1), 104–111. https://doi.org/10.1016/j. matchemphys.2008.05.052.

Khorrami, M. S., Ali Mostafaei, M., Pouraliakbar, H., & Kokabi, A. H. (2014). Study on microstructure and mechanical characteristics of low-carbon steel and ferritic stainless steel joints. *Prostaglandins, Leukotrienes and Essential Fatty Acids*, *608*, 35–45. https://doi.org/10.1016/j.msea.2014.04.065.

Krishnaveni, K., Ravichandran, J., & Selvaraj, A. (2013). Effect of Morinda tinctoria leaves extract on the corrosion inhibition of mild steel in acid medium. *Acta Metallurgica Sinica (English Letters)*, *26*(3), 321–327. https://doi.org/10.1007/s40195-012-0219-9.

Krishnegowda, P. M., Venkatesha, V. T., Krishnegowda, P. K. M., & Shivayogiraju, S. B. (2013). Acalypha torta leaf extract as green corrosion inhibitor for mild steel in hydrochloric acid solution. *Industrial and Engineering Chemistry Research*, *52*(2), 722–728. https://doi.org/10.1021/ie3018862.

Lahhit, N., Bouyanzer, A., Desjobert, J., Hammouti, B., & Salghi, R. (2011). Fennel (Foeniculum vulgare) essential oil as green corrosion inhibitor of carbon steel in hydrochloric acid solution. *Portugaliae Electrochimica Acta*, *29*(2), 127–138. https://doi.org/10.4152/pea.201102127.

Lebrini, M., Robert, F., Lecante, A., & Roos, C. (2011). Corrosion inhibition of C38 steel in 1M hydrochloric acid medium by alkaloids extract from Oxandra asbeckii plant. *Corrosion Science*, *53*(2), 687–695. https://doi.org/10.1016/j.corsci.2010.10.006.

Loto, C. A., Loto, R. T., & Popoola, A. P. I. (2011). Inhibition effect of extracts of Carica papaya and Camellia sinensis leaves on the corrosion of duplex (α β) brass in 1M nitric acid. *International Journal of Electrochemical Science*, *6*(10), 4900–4914.

Mansfeld, F. (2009). Fundamental aspects of the polarization resistance technique-the early days. *Journal of Solid State Electrochemistry*, *13*(4), 515–520. https://doi.org/10.1007/s10008-008-0652-x.

Miralrio, A., & Vázquez, A. E. (2020). Plant extracts as green corrosion inhibitors for different metal surfaces and corrosive media: A review. *Processes*, *8*(8), 942. https://doi.org/10.3390/PR8080942.

Neffati, N., Aloui, Z., Karoui, H., Guizani, I., Boussaid, M., & Zaouali, Y. (2017). Phytochemical composition and antioxidant activity of medicinal plants collected from the Tunisian flora. *Natural Product Research*, *31*(13), 1583–1588. https://doi.org/10.1080/14786419.2017.1280490.

Nn, A. (2015). A review on the extraction methods use in medicinal plants, principle, strength and limitation. *Medicinal & Aromatic Plants*, *04*(03), 3–8. https://doi.org/10.4172/2167-0412.1000196.

Obot, I. B., & Obi-Egbedi, N. O. (2010). An interesting and efficient green corrosion inhibitor for aluminium from extracts of Chlomolaena odorata L. in acidic solution. *Journal of Applied Electrochemistry*, *40*(11), 1977–1984. https://doi.org/10.1007/s10800-010-0175-x.

Odusote, J. K., Owalude, D. O., Olusegun, S. J., & Yahya, R. A. (2015). Inhibition efficiency of moringa oleifera leaf extract on the corrosion of reinforced steel bar in HCl Solution. *The West Indian Journal of Engineering*, *38*(2), 64–70. https://sta.uwi.edu/eng/wije/.

Oguzie, E. E. (2006). Studies on the inhibitive effect of Occimum viridis extract on the acid corrosion of mild steel. *Materials Chemistry and Physics*, *99*(2–3), 441–446. https://doi.org/10.1016/j.matchemphys.2005.11.018.

Oguzie, E. E. (2007). Corrosion inhibition of aluminium in acidic and alkaline media by Sansevieria trifasciata extract. *Corrosion Science*, *49*(3), 1527–1539. https://doi.org/10.1016/j.corsci.2006.08.009.

Oke, G. O., Aluko, A. O., & Sanya, O. T. (2018). Inhibitive potential of Datura stramonium leaf extract on the corrosion behavior of mild steel in 1M HCl acidic solution. *Leonardo Journal of Sciences*, *32*, 76–92. https://ljs.academicdirect.org/A32/076_092.pdf.

Okoronkwo, A. E., Olusegun, S. J., & Oluwasina, O. O. (2015). The inhibitive action of chitosan extracted from archachatina marginata shells on the corrosion of plain carbon steel in acid media. *Anti-Corrosion Methods and Materials*, *62*(1), 13–18. https://doi.org/10.1108/ACMM-10-2013-1307.

Onen, P. M., Maitera, A. I., & Ushahemba, O. (2015). Corrosion inhibition of zinc in acid medium by Moringa oleifera and Mangifera indica leaves extracts. *International Journal of Development and Sustainability*, *4*(9), 940–950. www.isdsnet.com/ijds.

Onwumelu, H. A., Aralu, C. C., & Egwuat C. I. (2018). Inhibition of mild steel corrosion in HCl solution by plant extract of Biden pilosa. *IOSR Journal of Applied Chemistry (IOSR-JAC)*, *16*(1), 35–44.

Orubite, K. O., & Oforka, N. C. (2004). Inhibition of the corrosion of mild steel in hydrochloric acid solutions by the extracts of leaves of Nypa fruticans Wurmb. *Materials Letters*, *58*(11), 1768–1772. https://doi.org/10.1016/j.matlet.2003.11.030.

Orubite-Okorosaye, K., & Oforka, N. C. (2005). Corrosion inhibition of zinc on HCl using Nypa fruticans Wurmb extract and 1,5 diphenyl carbazone. *Journal of Applied Sciences and Environmental Management*, *8*(1), 17228. https://doi.org/10.4314/jasem.v8i1.17228.

Othman, N. K., Salleh, E. M., Dasuki, Z., & Lazim, A. M. (2018). Dimethyl sulfoxide-treated starch of Dioescorea hispida as a green corrosion inhibitor for low carbon steel in sodium chloride medium. *Corrosion Inhibitors, Principles and Recent Applications*. IntechOpen. https://doi.org/10.5772/intechopen.73552.

Oukhrib, R., Issami, E., El Ibrahimi, B., El Mouaden, K., Bazzi, L., Bammou, L., Chaouay, A., Salghi, R., Jodeh, S., Hammouti, B., & Amin-Alami, A. (2017). Ziziphus lotus as green inhibitor of copper corrosion in natural sea water. *Portugaliae Electrochimica Acta*, *35*(4), 187–200. https://doi.org/10.4152/pea.201704187.

Pais, M., & Rao, P. (2019). Biomolecules for corrosion mitigation of zinc: A short review. *Journal of Bio- and Tribo-Corrosion*, *5*, 1–11. https://doi.org/10.1007/s40735-019-0286-9.

Patni, N., Agarwal, S., & Shah, P. (2013). Greener approach towards corrosion inhibition. *Chinese Journal of Engineering*, *2013*, 1–10. https://doi.org/10.1155/2013/784186.

Pham, H., Nguyen, V., Vuong, Q., Bowyer, M., & Scarlett, C. (2015). Effect of extraction solvents and drying methods on the physicochemical and antioxidant properties of Helicteres hirsuta Lour. leaves. *Technologies*, *3*(4), 285–301. https://doi.org/10.3390/technologies3040285.

Popoola, L. T. (2019). Progress on pharmaceutical drugs, plant extracts and ionic liquids as corrosion inhibitors. *Heliyon*, *5*(2), e01143. https://doi.org/10.1016/j.heliyon.2019.e01143.

Prabhu, D., & Rao, P. (2013). Coriandrum sativum L.- A novel green inhibitor for the corrosion inhibition of aluminium in 1. 0 M phosphoric acid solution. *Journal of Environmental Chemical Engineering*, *1*, 1–8. https://doi.org/10.1016/j.jece.2013.07.004.

Quraishi, M. A., Yadav, D. K., & Ahamad, I. (2015). Green approach to corrosion inhibition by black pepper extract in hydrochloric green approach to corrosion inhibition by black pepper extract in hydrochloric acid solution. *The Open Corrosion Journal*, *2*(1), April. https://doi.org/10.2174/1876503300902010056.

Rani, B. E. A., & Basu, B. B. J. (2012). Green inhibitors for corrosion protection of metals and alloys: An overview. *Green Approaches to Corrosion Mitigation*, *2012*(i), 380217. https://doi.org/10.1155/2012/380217.

Rosliza, R., & Wan Nik, W. B. (2010). Improvement of corrosion resistance of AA6061 alloy by tapioca starch in seawater. *Current Applied Physics*, *10*(1), 221–229. https://doi.org/10.1016/j.cap.2009.05.027.

Roy, P., Karfa, P., Adhikari, U., & Sukul, D. (2014). Corrosion inhibition of mild steel in acidic medium by polyacrylamide grafted Guar gum with various grafting percentage: Effect of intramolecular synergism. *Corrosion Science*, *88*, 246–253. https://doi.org/10.1016/j.corsci.2014.07.039.

Saadouni, M., Larouj, M., Salghi, R., Lgaz, H., Jodeh, S., Zougagh, M., & Souizi, A. (2016). Corrosion control of carbon steel in hydrochloric acid by Sulfaguandine: Weight loss, electrochemical and theoretical studies. *Der Pharmacia Lettre*, *8*(4), 65–76.

Salim, R., Ech-chihbi, E., Oudda, H., El Hajjaji, F., Taleb, M., & Jodeh, S. (2019). A review on the assessment of imidazo[1,2-a]pyridines as corrosion inhibitor of metals. *Journal of Bio- and Tribo-Corrosion*, *5*(1), 1–9. https://doi.org/10.1007/s40735-018-0207-3.

Sangeetha, M., Rajendran, S., Sathiyabama, J., Krishnaveni, A., Shanthy, P., Manimaran, N., & Shyamaladevi, B. (2011). Corrosion inhibition by an aqueous extract of Phyllanthus amarus. *Portugaliae Electrochimica Acta*, *29*(6), 429–444. https://doi.org/10.4152/pea.201106429.

Sangeetha, T. V., & Fredimoses, M. (2011). Inhibition of mild copper metal corrosion in HNO_3 medium by acid extract of Azadirachta indica seed. *E-Journal of Chemistry*, *8*(Suppl. 1), 1–7. https://doi.org/10.1155/2011/135952.

Satapathy, A. K., Gunasekaran, G., Sahoo, S. C., Amit, K., & Rodrigues, P. V. (2009). Corrosion inhibition by Justicia gendarussa plant extract in hydrochloric acid solution. *Corrosion Science*, *51*(12), 2848–2856. https://doi.org/10.1016/j.corsci.2009.08.016.

Saxena, A., Thakur, K. K., Saxena, K. K., Chambyal, S., & Sharma, A. (2019). Electrochemical studies and surface examination of low carbon steel by applying the extract of Terminalia chebula. *Materials Today: Proceedings*, *26*(January), 1360–1367. https://doi.org/10.1016/j.matpr.2020.02.276.

Seal, T. (2016). Quantitative HPLC analysis of phenolic acids, flavonoids and ascorbic acid in four different solvent extracts of two wild edible leaves, Sonchus arvensis and Oenanthe linearis of North-Eastern region in India. *Journal of Applied Pharmaceutical Science*, *6*(2), 157–166. https://doi.org/10.7324/JAPS.2016.60225.

Shabani-Nooshabadi, M., Hoseiny, F. S., & Jafari, Y. (2015). Green approach to corrosion inhibition of copper by the extract of calligonum comosum in strong acidic medium. *Metallurgical and Materials Transactions A: Physical Metallurgy and Materials Science*, *46*(1), 293–299. https://doi.org/10.1007/s11661-014-2634-1.

Shah, A. M., Rahim, A. A., Hamid, S. A., & Yahya, S. (2013). Green inhibitors for copper corrosion by Mangrove tannin. *International Journal of Electrochemical Science*, *8*(2), 2140–2153.

Shahini, M. H., Ramezanzadeh, B., & Eivaz Mohammadloo, H. (2020). Recent advances in biopolymers/carbohydrate polymers as effective corrosion inhibitive macro-molecules: A review study from experimental and theoretical views. *Journal of Molecular Liquids*, *325*, 115110. https://doi.org/10.1016/j.molliq.2020.115110.

Sharma, S. K., Mudhoo, A., Khamis, E., & Jain, G. (2008). Green corrosion inhibitors: An overview of recent research. *Journal of Corrosion Science and Engineering*, *11*, 1–14.

Sharma, S. K., Mudhoo, A., Gargi, J., & Sharma, J. (2015). Inhibitory effects of ocimum tenuiflorum (Tulsi) on the corrosion of zinc in sulphuric acid: A green approach. *RASAYAN Journal of Chemistry*, *2*, 332–339.

Srivastava, M., Tiwari, P., Srivastava, S. K., Kumar, A., Ji, G., Prakash, R., Tiwari, P., Srivastava, S. K., Ji, G., & Prakash, R. (2017). Low cost aqueous extract of Pisum sativum peels for inhibition of mild steel corrosion. *Journal of Molecular Liquids*, *254*, 357–368. https://doi.org/10.1016/j.molliq.2018.01.137.

Stephen, J. T., & Adebayo, A. (2018). Inhibition of corrosion of mild steel in hydrochloric acid solution using akee apple seed extract. *Journal of Failure Analysis and Prevention*, *18*(2), 350–355. https://doi.org/10.1007/s11668-018-0431-7.

Suedile, F., Robert, F., Roos, C., & Lebrini, M. (2014). Corrosion inhibition of zinc by Mansoa alliacea plant extract in sodium chloride media: Extraction, characterization and electrochemical studies. *Electrochimica Acta*, *133*, 631–638. https://doi.org/10.1016/j.electacta.2013.12.070.

Suhaimi, N. F., Jalaludin, J., & Latif, M. T. (2019). Physicochemical properties, structures, bio-activities and future prospective for polysaccharides from Plantago L. (Plantaginaceae): A review. *International Journal of Biological Macromolecules*, *135*, 1–3. https://doi.org/10.1016/j.ijbiomac.2019.05.211.

Tuaweri, T. J., Ogbonnaya, E. A., & Onyemaobi, O. O. (2015). Corrosion inhibition of heat treated mild steel with neem leave extract in a chloride medium. *International Journal of Research in Engineering and Technology*, *04*(06), 404–409. https://doi.org/10.15623/ijret.2015.0406069.

Ugi, B., Ekerete, J., Ikeuba, I., & Uwah, I. (2015). *Mangifera indica* leave extracts as organic inhibitors on the corrosion of zinc sheet in 5 M H_2SO_4 solution. *Journal of Applied Sciences and Environmental Management*, *19*(1), 145. https://doi.org/10.4314/jasem.v19i1.19.

Umoren, S. A., Eduok, U. M., Solomon, M. M., & Udoh, A. P. (2016). Corrosion inhibition by leaves and stem extracts of Sida acuta for mild steel in 1 M H_2SO_4 solutions investigated by chemical and spectroscopic techniques. *Arabian Journal of Chemistry*, *9*, S209–S224. https://doi.org/10.1016/j.arabjc.2011.03.008.

Umoren, S. A., Solomon, M. M., Obot, I. B., & Suleiman, R. K. (2019). A critical review on the recent studies on plant biomaterials as corrosion inhibitors for industrial metals. *Journal of Industrial and Engineering Chemistry, 76,* 91–115. https://doi.org/10.1016/j.jiec.2019.03.057.

Vaidya, N. R., Aklujkar, P., & Rao, A. R. (2022). Modification of natural gums for application as corrosion inhibitor: A review. *Journal of Coatings Technology and Research, 19*(1), 223–239. https://doi.org/10.1007/s11998-021-00510-z.

Varma, R. S. (2016). Greener and sustainable trends in synthesis of organics and nanomaterials. *ACS Sustainable Chemistry and Engineering, 4*(11), 5866–5878. https://doi.org/10.1021/acssuschemeng.6b01623.

Vengatesh, G., Sundaravadivelu, M., & Sundaravadivelu, M. (2019). Non-toxic bisacodyl as an effective corrosion inhibitor for mild steel in 1M HCl: Thermodynamic, electrochemical, SEM, EDX, AFM, FT-IR, DFT and molecular dynamics simulation studies. *Journal of Molecular Liquids, 287,* 110906. https://doi.org/10.1016/j.molliq.2019.110906.

Verma, C., Ebenso, E. E., & Quraishi, M. A. (2017). Ionic liquids as green and sustainable corrosion inhibitors for metals and alloys: An overview. *Journal of Molecular Liquids, 233*(2016), 403–414. https://doi.org/10.1016/j.molliq.2017.02.111.

Verma, C., Ebenso, E. E., & Quraishi, M. A. (2020). Molecular structural aspects of organic corrosion inhibitors: Influence of -CN and -NO_2 substituents on designing of potential corrosion. *Decision Support Systems, January,* 113260. https://doi.org/10.1016/j.molliq.2020.113874.

Walia, M., & Singh, G. (2005). Corrosion inhibition of pure nickel by some phosphonium compounds in acid medium. *Surface Engineering, 21*(3), 176–179. https://doi.org/10.1179/174329405X49976.

Wang, S., Zhang, J., Gharbi, O., Vivier, V., Gao, M., & Orazem, M. E. (2021). Electrochemical impedance spectroscopy. *Nature Reviews Methods Primers, 1*(1), 2000–2008. https://doi.org/10.1038/s43586-021-00039-w.

Wedian, F., Al-Qudah, M. A., & Al-Mazaideh, G. M. (2017). Corrosion inhibition of copper by Capparis spinosa L. extract in strong acidic medium: Experimental and density functional theory. *International Journal of Electrochemical Science, 12*(6), 4664–4676. https://doi.org/10.20964/2017.06.47.

Xiang, L., Chen, S., & Yuan, J. (2014). Research on the properties of controlling corrosion and scale inhibition of the modified starch with microwave loading. *Advanced Materials Research, 997,* 9–15. https://doi.org/10.4028/www.scientific.net/AMR.997.9.

Yahaya, L. E., Royeun, S. O. A., Ogunwolu, S. O., Jayeola, C. O., and Igbinadolor, R. O. (2017). Green and black tea (Camellia sinendis) extract as corrosion inhibitor for mild steel in acid medium. *American-Eurasian Journal of Agricultural & Environmental Sciences, 17*(4), 273. https://doi.org/10.5829/idosi.aejaes.2017.273.279.

Zhang, Q. W., Lin, L. G., & Ye, W. C. (2018). Techniques for extraction and isolation of natural products: A comprehensive review. *Chinese Medicine (United Kingdom), 13*(1), 1–26. https://doi.org/10.1186/s13020-018-0177-x.

Zinola, C. F., & Luna, A. M. C. (1995). The inhibition of Ni corrosion in H_2SO_4 solutions containing simple non-saturated substances. *Corrosion Science, 37*(12), 1919–1929. https://doi.org/10.1016/0010-938X(95)00074-T.

Znini, M., Majidi, L., Bouyanzer, A., Paolini, J., & Desjobert, J. (2012). Essential oil of Salvia aucheri mesatlantica as a green inhibitor for the corrosion of steel in 0. 5 M H_2SO_4. *Arabian Journal of Chemistry, 5,* 467–474. https://doi.org/10.1016/j.arabjc.2010.09.017.

15 Phytochemicals/Plant Extracts as Corrosion Inhibitors for Magnesium in H$_2$SO$_4$ Solutions

Khasan Berdimuradov
Shahrisabz Branch of Tashkent Institute
of Chemical Technology

Abduvali Kholikov and Khamdam Akbarov
National University of Uzbekistan

Elyor Berdimurodov
New Uzbekistan University
Central Asian University
National University of Uzbekistan

Brahim El Ibrahimi
Ibn Zohr University

Burak Tuzun
Sivas Cumhuriyet University

Hicham Es-Soufi
Moulay Ismail University

Sadhucharan Mallick
Indira Gandhi National Tribal University
(Central University)

Oybek Mikhliev
Karshi Engineering Economics Institute

DOI: 10.1201/9781003394631-15

15.1 INTRODUCTION

Magnesium and its alloys have found wide application across many industries due to their high strength-to-weight ratio, machinability, and corrosion resistance. However, magnesium still experiences severe corrosion issues when exposed to acidic environments containing sulfuric acid (H_2SO_4). The corrosion of magnesium in H_2SO_4 solutions presents major challenges in applications ranging from acid pickling to nuclear waste storage [1–3].

Traditional corrosion inhibition methods rely heavily on toxic chromates or other inorganic inhibitors that come with environmental and health concerns. These conventional inhibitors also have limitations in efficacy and corrosion protection. There has been rising interest in exploring natural phytochemicals from plant sources as an alternative approach to inhibiting magnesium corrosion in H_2SO_4 conditions [4–6].

Phytochemicals are chemical compounds produced by plants that feature heterogeneous structures including alkaloids, phenolics, flavonoids, and terpenoids. Studies have shown promising corrosion inhibition abilities for a variety of phytochemicals. This has spurred research on extracting and isolating these compounds from plants as "green" corrosion inhibitors [7,8].

This chapter delves into the exploration of phytochemicals and plant extracts as potential corrosion inhibitors for magnesium in sulfuric acid (H_2SO_4) solutions. The discussion will encompass the background of corrosion, the specific challenges posed by H_2SO_4 to magnesium, and the traditional methods of corrosion inhibition. The main focus will be on the role of phytochemicals and plant extracts as corrosion inhibitors, their extraction methods, mechanisms of action, and comparative advantages over traditional inhibitors. The chapter will also present case studies and future perspectives in this research area [9–11].

15.1.1 IMPORTANCE OF CORROSION INHIBITION FOR MAGNESIUM IN H_2SO_4 SOLUTIONS

Magnesium, despite its advantageous properties such as light weight and high strength, is highly susceptible to corrosion, especially in H_2SO_4 solutions. This corrosion significantly reduces the lifespan and effectiveness of magnesium and its alloys in various industrial applications. Therefore, effective corrosion inhibition is crucial not only for the longevity of magnesium-based materials but also for the safety and efficiency of many industrial processes [12,13].

15.1.2 ROLE OF PHYTOCHEMICALS/PLANT EXTRACTS IN CORROSION INHIBITION

In the quest for environmentally friendly and effective corrosion inhibitors, researchers have turned their attention to phytochemicals and plant extracts. These natural compounds, due to their diverse chemical structures and functional groups, have shown promising results in inhibiting corrosion, particularly for metals like magnesium. This chapter will delve into the specifics of how these phytochemicals function as corrosion inhibitors, their advantages, and potential for future use.

15.2 PHYTOCHEMICALS AS CORROSION INHIBITORS

15.2.1 INTRODUCTION TO PHYTOCHEMICALS

Phytochemicals, also known as plant secondary metabolites, are naturally occurring compounds found in plant materials. These compounds are not directly involved in the growth, development, or reproduction of plants but play a significant role in the plant's defense mechanisms against pests, diseases, and environmental stressors [14,15].

Phytochemicals are categorized into several classes based on their chemical structures and biosynthetic origins. The major classes include phenolics, flavonoids, terpenoids, alkaloids, and saponins. Each class of phytochemicals possesses unique chemical structures and properties, which contribute to their diverse biological activities, including antioxidant, anti-inflammatory, antimicrobial, and anticancer activities.

In recent years, the corrosion inhibition potential of phytochemicals has been explored. Their rich chemical diversity, coupled with their eco-friendly nature, makes them promising candidates as corrosion inhibitors. The following sections will delve into the mechanisms of how these phytochemicals inhibit corrosion, their interaction with magnesium in H_2SO_4 solutions, and their advantages over traditional corrosion inhibitors.

15.2.2 DEFINITION AND TYPES OF PHYTOCHEMICALS

Phytochemicals, also known as plant secondary metabolites, are bioactive compounds found in plants that are not essential for the plant's basic metabolic functions but play a crucial role in the plant's defense mechanisms against various environmental stressors. These compounds are synthesized in plants as a response to various biotic and abiotic stress conditions and are known for their health-promoting properties in humans [16–18].

Phytochemicals are broadly classified into several types based on their chemical structure and biosynthetic pathways. It was found the following main types:

a. **Phenolics:** These are compounds characterized by one or more aromatic rings with one or more hydroxyl groups. They are further divided into subcategories such as phenolic acids, flavonoids, stilbenes, and lignans.

b. **Alkaloids:** These are a diverse group of nitrogen-containing compounds with complex structures. They are known for their potent biological activities and are often used in pharmaceuticals.

c. **Terpenoids:** Also known as isoprenoids, these are a large and diverse class of naturally occurring organic chemicals derived from five-carbon isoprene units. They include monoterpenes, sesquiterpenes, diterpenes, triterpenes, and tetraterpenes.

d. **Saponins:** These are glycosides with a distinctive foaming characteristic. They have a wide range of pharmacological activities including anti-inflammatory, cholesterol-lowering, and anticancer effects.
e. **Organosulfur compounds:** These are organic compounds that contain sulfur. They are found in cruciferous vegetables like garlic and onions and are known for their anticancer and antimicrobial properties.

Each type of phytochemical has unique properties that can contribute to their potential as corrosion inhibitors. The following sections will delve into the mechanisms of how these phytochemicals inhibit corrosion and their interaction with magnesium in H_2SO_4 solutions.

15.2.3 EXTRACTION OF PHYTOCHEMICALS FROM PLANTS

The extraction of phytochemicals from plants is a crucial step in utilizing them as corrosion inhibitors. The process involves separating the desired phytochemicals from the plant matrix and purifying them for further use. It was found a general outline of the extraction process [19–21]:

A. **Selection and preparation of plant material:** The first step involves choosing the appropriate plant material known to contain the desired phytochemicals. The plant material is then cleaned, dried, and ground into a fine powder to increase the surface area for extraction.
B. **Extraction method:** The extraction method depends on the type of phytochemicals to be extracted. Common methods include:
   i. **Solvent extraction:** This is the most common method used for extracting phytochemicals. It involves soaking the plant material in a suitable solvent that can dissolve the desired phytochemicals. The choice of solvent depends on the polarity of the phytochemicals. Common solvents include ethanol, methanol, acetone, and water.
   ii. **Steam distillation:** This method is commonly used for extracting volatile phytochemicals, such as essential oils. The plant material is subjected to steam, and the volatile phytochemicals are evaporated, condensed, and collected.
   iii. **Supercritical fluid extraction:** This is a more advanced method that uses supercritical fluids (usually carbon dioxide) as the extraction solvent. It is particularly useful for extracting thermally unstable or easily oxidized phytochemicals.
C. **Purification:** After extraction, the extract is often a complex mixture of different phytochemicals. Further purification steps, such as filtration, evaporation, and chromatographic techniques, are used to isolate the desired phytochemicals.
D. **Characterization:** The final step involves characterizing the extracted phytochemicals using various analytical techniques, such as spectroscopy and chromatography, to confirm their identity and purity.

The extraction of phytochemicals is a critical step in their use as corrosion inhibitors, as the effectiveness of the inhibitors depends on the purity and concentration of the phytochemicals.

15.3 MECHANISM OF PHYTOCHEMICALS AS CORROSION INHIBITORS FOR MAGNESIUM IN H_2SO_4 SOLUTIONS

15.3.1 EXPLANATION OF HOW PHYTOCHEMICALS INHIBIT CORROSION

Phytochemicals inhibit corrosion primarily through adsorption onto the metal surface, forming a protective barrier that prevents the corrosive medium from attacking the metal. The effectiveness of a phytochemical as a corrosion inhibitor depends on its ability to adsorb onto the metal surface, which is influenced by the nature of the phytochemical, the metal, and the corrosive medium [22–24].

The adsorption of phytochemicals onto the metal surface can occur through physical or chemical adsorption. Physical adsorption, also known as physisorption, involves weak van der Waals forces, while chemical adsorption, or chemisorption, involves the formation of covalent or ionic bonds between the phytochemical and the metal surface.

Phytochemicals, due to their complex structures, often contain functional groups such as hydroxyl (-OH), carbonyl (C=O), carboxyl (-COOH), and aromatic rings that can interact with the metal surface. These functional groups can donate electrons to form a coordinate bond with the metal, or they can accept protons from the acidic medium, reducing the concentration of H^+ ions and thus the acidity of the medium.

In the case of magnesium in H_2SO_4 solutions, the phytochemicals can adsorb onto the magnesium surface, forming a protective layer that prevents the direct contact of the magnesium with the sulfuric acid. This inhibits the corrosion process, which involves the oxidation of magnesium and the reduction of H^+ ions. Furthermore, some phytochemicals can chelate with the magnesium ions formed during the initial stages of corrosion, forming a stable complex that further strengthens the protective layer. It's important to note that the effectiveness of phytochemicals as corrosion inhibitors can be influenced by several factors, including the concentration of the phytochemical, the temperature, and the pH of the medium.

The research [25] investigates the effects of decyl glucoside (DG), a non-ionic surfactant, on the corrosion inhibition and battery performance of a magnesium-air (Mg-air) battery. The study uses both chemical and electrochemical techniques to evaluate the corrosion rate and the efficiency of the inhibitor. The corrosion rate of magnesium (Mg) in the battery electrolyte, which is a 3.5% NaCl solution, is significantly reduced in the presence of the DG surfactant. This is a crucial finding as corrosion can degrade the performance of a battery over time, reducing its lifespan and efficiency. The maximum inhibition efficiency, which is over 94%, is achieved at the critical micelle concentration (CMC) of the DG surfactant, which is 2.5 mM. The CMC is the concentration at which surfactant molecules begin to form micelles, or small clusters, in a solution. This suggests that the DG surfactant is most effective at inhibiting corrosion when it forms micelles. The study also found that the

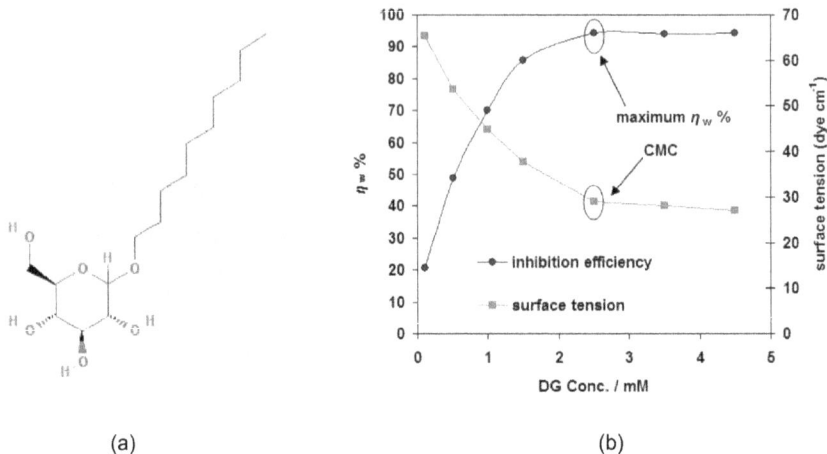

(a)   (b)

FIGURE 15.1 (a) Chemical structure of decyl glucoside (DG) and (b) dependence between the inhibitor concentration and protection degree [25].

presence of the DG surfactant increases the activation energy of the corrosion reaction. Activation energy is the minimum amount of energy required for a chemical reaction to occur. By increasing the activation energy, the DG surfactant makes the corrosion reaction less likely to occur, further inhibiting corrosion. The inhibition action of the DG surfactant is suggested to be due to a physisorption mechanism. Physisorption is a type of adsorption where a substance adheres to a surface due to weak van der Waals forces. This suggests that the DG surfactant adheres to the Mg surface, forming a protective layer that inhibits corrosion. Finally, the study found that the Mg-air battery containing the DG surfactant offers higher operating voltage, discharge capacity, and anodic utilization than in its absence. This suggests that the DG surfactant not only inhibits corrosion but also enhances the overall performance of the Mg-air battery (Figure 15.1).

15.3.2 SPECIFIC INTERACTION OF PHYTOCHEMICALS WITH MAGNESIUM AND H_2SO_4

The interaction of phytochemicals with magnesium in H_2SO_4 solutions is a complex process that involves several steps. It was concluded the following basics of the process [26–28]:

A. **Adsorption:** The first step in the corrosion inhibition process is the adsorption of the phytochemical onto the magnesium surface. This can occur through physical adsorption, where the phytochemical is held onto the surface by weak van der Waals forces, or chemical adsorption, where a chemical bond is formed between the phytochemical and the magnesium surface. The presence of functional groups such as hydroxyl, carbonyl, carboxyl, and aromatic rings in the phytochemical structure facilitates this adsorption process.

B. **Formation of protective layer:** Once adsorbed, the phytochemical forms a protective layer on the magnesium surface, preventing the H_2SO_4 from coming into direct contact with the magnesium. This layer acts as a barrier, reducing the rate of magnesium dissolution and hydrogen evolution, the two primary reactions in the corrosion process.

C. **Interaction with H_2SO_4:** The phytochemicals can also interact with the H_2SO_4 in the solution. Some phytochemicals can accept protons (H^+ ions) from the acid, reducing the acidity of the solution and thus the rate of corrosion. Others can chelate with the magnesium ions formed during the initial stages of corrosion, forming a stable complex that further strengthens the protective layer.

D. **Reinforcement of protective layer:** Over time, the protective layer can be reinforced through the continuous adsorption of phytochemicals and the formation of more stable complexes with magnesium ions. This leads to a decrease in the corrosion rate over time, a phenomenon known as the "time-dependent" or "self-healing" property of phytochemical inhibitors.

The specific interaction of phytochemicals with magnesium and H_2SO_4 can vary depending on the type of phytochemical, its concentration, and the conditions of the solution, such as temperature and pH. Further research is needed to fully understand these interactions and optimize the use of phytochemicals as corrosion inhibitors for magnesium in H_2SO_4 solutions.

15.4 ADVANTAGES OF PHYTOCHEMICALS OVER TRADITIONAL INHIBITORS FOR MAGNESIUM IN H_2SO_4 SOLUTIONS

15.4.1 ENVIRONMENTAL BENEFITS

Phytochemicals, as corrosion inhibitors, offer several environmental benefits over traditional inhibitors [29–31]:

A. **Biodegradability:** Phytochemicals are naturally occurring substances that are biodegradable. This means they can be broken down into simpler substances by biological processes, reducing the risk of environmental accumulation and contamination.

B. **Non-toxicity:** Many traditional corrosion inhibitors are toxic to both humans and the environment. In contrast, phytochemicals (Figure 15.2) are generally non-toxic and safe to handle. This reduces the risk of health hazards for workers involved in their production and application and minimizes the environmental impact in case of spills or leaks [32].

C. **Renewable source:** Phytochemicals are derived from plants, which are renewable resources. This contrasts with many traditional inhibitors that are derived from non-renewable resources, such as petroleum. The use of plant-based inhibitors thus contributes to the sustainability of corrosion protection strategies.

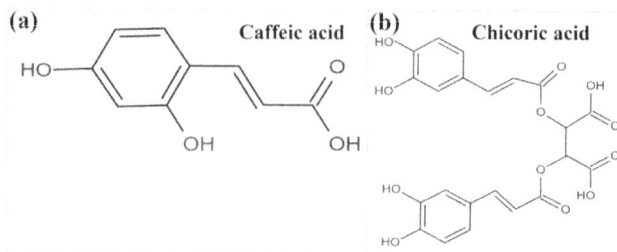

FIGURE 15.2 Chemical molecular structures of CA extractions mainly including (a) caffeic acid (Caf) and (b) chicoric acid (Chi) as corrosion inhibitor for Mg alloys [32].

  D. **Reduction in hazardous waste:** The use of phytochemicals as corrosion inhibitors can reduce the generation of hazardous waste. Traditional inhibitors often produce toxic by-products or residues that require special handling and disposal. In contrast, the by-products of phytochemicals are generally less harmful and easier to manage.
  E. **Carbon sequestration:** The cultivation of plants for the extraction of phytochemicals can contribute to carbon sequestration, helping to mitigate climate change.

These environmental benefits make phytochemicals an attractive alternative to traditional inhibitors for the protection of magnesium in H_2SO_4 solutions. However, further research is needed to fully understand their corrosion inhibition mechanisms and to optimize their use in different conditions.

15.4.2 Cost-Effectiveness

Phytochemicals also offer potential cost advantages over traditional inhibitors [33,34]:

  A. **Abundance of raw materials:** Phytochemicals are derived from plants, many of which are abundant and widely distributed. This makes the raw materials for phytochemical extraction readily available and often less expensive than the raw materials for traditional inhibitors.
  B. **Lower processing costs:** The extraction and processing of phytochemicals can be less complex and less energy-intensive than the production of traditional inhibitors, leading to lower processing costs.
  C. **Reduced waste disposal costs:** As phytochemicals are biodegradable and generally non-toxic, the costs associated with waste disposal can be significantly lower than for traditional inhibitors, which often require special handling and disposal due to their toxicity.
  D. **Potential for dual-purpose cultivation:** In some cases, the plants used for phytochemical extraction can also be used for other purposes, such as food production or landscaping. This dual-purpose cultivation can further improve the cost-effectiveness of phytochemical production.

E. **Lower regulatory costs:** Due to their environmental and health benefits, the use of phytochemicals may be subject to fewer regulatory restrictions and lower compliance costs than traditional inhibitors.

While these factors suggest that phytochemicals can be a cost-effective alternative to traditional inhibitors, it's important to note that the actual cost-effectiveness can depend on various factors, including the specific type of phytochemical, the local availability of the plant source, and the scale of production. Further research and development are needed to optimize the cost-effectiveness of phytochemicals as corrosion inhibitors for magnesium in H_2SO_4 solutions.

15.4.3 Efficiency and Effectiveness

Phytochemicals also demonstrate significant efficiency and effectiveness as corrosion inhibitors, offering several advantages over traditional inhibitors [35–37]:

A. **Broad spectrum of activity:** Phytochemicals, due to their diverse structures and functional groups, can inhibit a wide range of corrosive agents. This broad-spectrum activity can make them more effective than traditional inhibitors, which may only target specific types of corrosion.

B. **Self-healing properties:** Some phytochemicals have the ability to adsorb onto the metal surface and form a protective layer that can repair itself over time. This self-healing property can enhance the long-term effectiveness of the corrosion protection.

C. **Synergistic effects:** Phytochemicals often contain multiple active components that can work together to inhibit corrosion. These synergistic effects can enhance the overall efficiency and effectiveness of the corrosion inhibition.

D. **High adsorption efficiency:** Many phytochemicals have high adsorption efficiency, meaning they can form a dense and stable protective layer on the metal surface with a relatively small amount of inhibitor. This high adsorption efficiency can make them more cost-effective than traditional inhibitors.

E. **Resistance to harsh conditions:** Some phytochemicals can resist harsh conditions, such as high temperatures and extreme pH levels, that can reduce the effectiveness of traditional inhibitors. This makes them suitable for use in a wide range of environments.

While these advantages suggest that phytochemicals can be highly efficient and effective as corrosion inhibitors, it's important to note that their performance can depend on various factors, including the specific type of phytochemical, the concentration of the inhibitor, and the conditions of the corrosive environment. Further research is needed to fully understand their mechanisms of action and to optimize their use in different conditions.

15.5 MECHANISMS OF PHYTOCHEMICAL INHIBITION FOR MAGNESIUM IN H₂SO₄ SOLUTIONS

15.5.1 EXPLANATION OF ADSORPTION AND FILM FORMATION ON METAL SURFACE

The primary mechanism by which phytochemicals inhibit corrosion is through adsorption onto the metal surface, forming a protective film that prevents the corrosive medium from interacting with the metal. It was revealed a detailed explanation [1,32,38]:

A. **Adsorption:** Phytochemicals, due to their diverse structures and functional groups, have the ability to adsorb onto the metal surface. This adsorption can occur through physical or chemical means. Physical adsorption, or physisorption, involves weak forces such as van der Waals forces, while chemical adsorption, or chemisorption, involves the formation of covalent or ionic bonds between the phytochemical and the metal surface. The type and extent of adsorption depend on the nature of the phytochemical, the metal, and the corrosive medium.

B. **Film formation:** Once adsorbed, the phytochemicals form a protective film on the metal surface. This film acts as a barrier, preventing the corrosive medium (in this case, H_2SO_4) from coming into direct contact with the magnesium. The film is often composed of the phytochemical itself, its reaction products with the metal or the corrosive medium, or a combination of both (Figure 15.3).

C. **Reinforcement of film:** Over time, the protective film can be reinforced through the continuous adsorption of phytochemicals and the formation of more stable complexes with the metal ions. This leads to a decrease in the corrosion rate over time, a phenomenon known as the "time-dependent" or "self-healing" property of phytochemical inhibitors.

The effectiveness of this adsorption and film formation process in inhibiting corrosion depends on several factors, including the concentration of the phytochemical, the temperature, and the pH of the medium. Understanding these factors is crucial for optimizing the use of phytochemicals as corrosion inhibitors for magnesium in H_2SO_4 solutions.

15.5.2 DISCUSSION OF SPECIFIC FUNCTIONAL GROUPS INVOLVED IN INHIBITION

Phytochemicals are complex organic compounds that contain a variety of functional groups. These functional groups play a crucial role in the adsorption of the phytochemical onto the metal surface and the subsequent inhibition of corrosion. It was found some of the key functional groups involved (Figure 15.4) [5–7]:

A. **Hydroxyl groups (-OH):** Hydroxyl groups are polar, allowing them to form hydrogen bonds with the metal surface. They can also donate electrons to the metal, forming a coordinate bond. These properties make hydroxyl

(a)   Mg-Caf E_{ad} = -12.44 eV   (b)   Mg-CaCaf1 E_{ad} = -29.06 eV

(c)   Mg-CaCaf2 E_{ad} = -27.29 eV   (d)   Mg-CaCaf3 E_{ad} = -28.30 eV

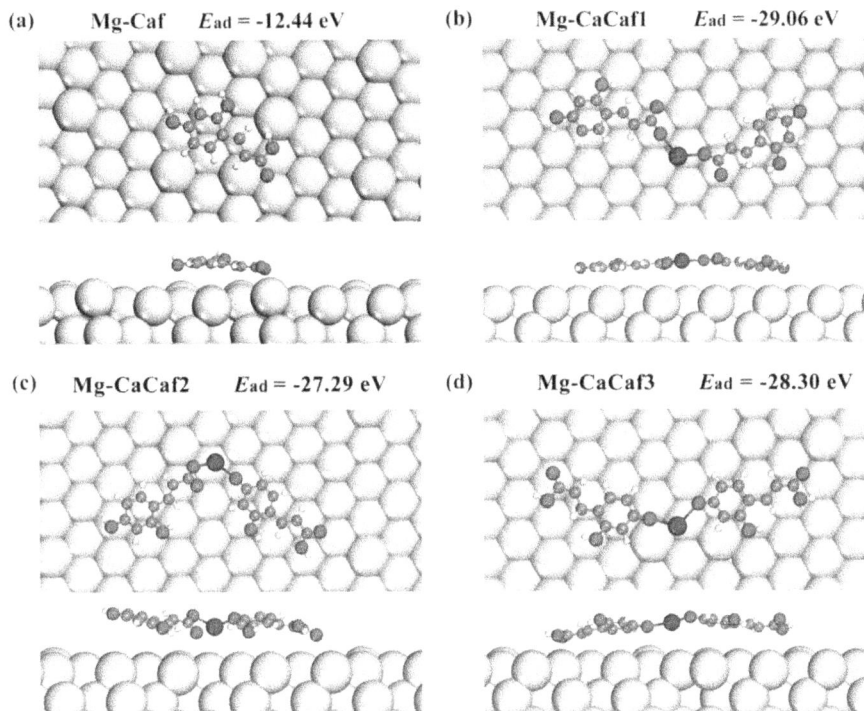

FIGURE 15.3 Optimized adsorption arrangements of various inhibitors on the Mg (0001) surface are depicted from both top and side perspectives: (a) Mg-Caf, (b) Mg-CaCaf1, (c) Mg-CaCaf2, and (d) Mg-CaCaf3 with depiction of several atoms of C, Ca, O, H, and Mg [32].

groups particularly effective in adsorbing onto the metal surface and inhibiting corrosion.

B. **Carbonyl groups (C=O):** Carbonyl groups can interact with the metal surface through the oxygen atom, which can form a coordinate bond with the metal. This can facilitate the adsorption of the phytochemical onto the metal surface.

C. **Carboxyl groups (-COOH):** Carboxyl groups can donate two electrons to the metal, forming a coordinate bond. They can also accept protons from the acidic medium, reducing the concentration of H^+ ions and thus the acidity of the medium.

D. **Aromatic rings:** Aromatic rings, due to their delocalized π electrons, can interact with the metal surface through π-bonding. This can facilitate the adsorption of the phytochemical onto the metal surface and enhance the stability of the protective film.

E. **Nitrogenous groups:** Nitrogenous groups, such as those found in alkaloids, can donate electrons to the metal, forming a coordinate bond. They can also form hydrogen bonds with the metal surface, enhancing the adsorption of the phytochemical.

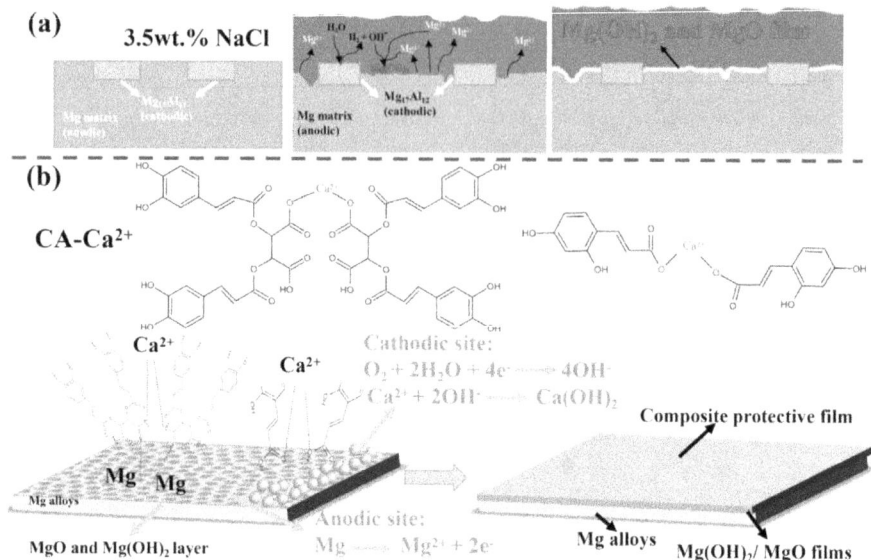

FIGURE 15.4 A schematic diagram depicting the mechanism of corrosion inhibition for
AZ91D alloy when immersed in a 3.5 wt.% NaCl solution. (a) shows the alloy with CA-Ca^{2+}
inhibitors present, while (b) shows the alloy without inhibitors. The inhibitors are proposed
to form a protective film on the alloy surface that prevents chloride ions from initiating corrosion. In the absence of inhibitors, the alloy undergoes corrosion when exposed to the chloride
solution [32].

These functional groups, through their interactions with the metal surface and the
corrosive medium, play a crucial role in the corrosion inhibition process. The specific functional groups present in a phytochemical, and their arrangement within
the molecule, can significantly influence the effectiveness of the phytochemical as a
corrosion inhibitor. Understanding these interactions is key to optimizing the use of
phytochemicals as corrosion inhibitors for magnesium in H$_2$SO$_4$ solutions.

15.6 EFFECTS OF MOLECULAR STRUCTURE ON PHYTOCHEMICAL INHIBITORS FOR MAGNESIUM IN H$_2$SO$_4$ SOLUTIONS

15.6.1 How Structural Properties Like Aromaticity Affect Inhibition

The molecular structure of phytochemicals, including properties like aromaticity,
plays a significant role in their effectiveness as corrosion inhibitors. It was generally
explained [12–14]:

A. **Aromaticity:** Aromatic compounds contain a ring of atoms with delocalized π electrons, which can interact with the metal surface through
   π-bonding. This interaction facilitates the adsorption of the phytochemical
   onto the metal surface, forming a protective layer that inhibits corrosion.

Additionally, the stability of aromatic compounds can enhance the durability of the protective layer.

B. **Presence of polar functional groups:** Functional groups like hydroxyl (-OH), carbonyl (C=O), and carboxyl (-COOH) are polar, allowing them to form bonds with the metal surface. These groups can also participate in acid–base reactions with the corrosive medium, reducing its acidity and thus the rate of corrosion.

C. **Size and shape of the molecule:** The size and shape of the phytochemical molecule can influence its ability to adsorb onto the metal surface and form a protective layer. Larger molecules with flat shapes can cover more surface area, providing better protection against corrosion.

D. **Solubility:** The solubility of the phytochemical in the corrosive medium can affect its availability for adsorption onto the metal surface. Phytochemicals with good solubility can reach the metal surface more easily, enhancing their effectiveness as corrosion inhibitors.

E. **Electron density:** The electron density of the phytochemical can influence its ability to donate electrons to the metal, forming a coordinate bond. Phytochemicals with high electron density can form stronger bonds with the metal, leading to a more stable protective layer.

Understanding the effects of these structural properties on the corrosion inhibition process is crucial for the design and selection of effective phytochemical inhibitors for magnesium in H_2SO_4 solutions. Further research is needed to fully elucidate these effects and to develop predictive models for the performance of phytochemical inhibitors based on their molecular structure.

15.6.2 QUANTITATIVE STRUCTURE–ACTIVITY RELATIONSHIP (QSAR) MODELING

Quantitative structure–activity relationship (QSAR) modeling is a method used in chemistry that relates a set of predictor variables (the structures of molecules) to the potency of the response variable (the biological activity), using a mathematical equation. In the context of corrosion inhibition, QSAR modeling can be used to predict the effectiveness of phytochemicals as corrosion inhibitors based on their molecular structure. It was indicated the general QSAR modeling [16–18]:

A. **Selection of descriptors:** The first step in QSAR modeling is the selection of descriptors, which are numerical values that represent the properties of the molecules. These can include structural properties (like the number of aromatic rings), physicochemical properties (like solubility or molecular weight), and electronic properties (like electron density or charge distribution).

B. **Data collection:** The next step is to collect data on the effectiveness of a set of phytochemicals as corrosion inhibitors. This can be done through experimental testing or by gathering data from the literature.

C. **Model building:** The collected data is then used to build a QSAR model, which is a mathematical equation that relates the descriptors to the effectiveness of the phytochemicals as corrosion inhibitors. This is usually done using statistical methods like regression analysis.

D. **Model validation:** The QSAR model is then validated by testing its predictions against a separate set of data not used in the model building. A good QSAR model should be able to accurately predict the effectiveness of phytochemicals as corrosion inhibitors based on their molecular structure.

E. **Application of the model:** Once validated, the QSAR model can be used to predict the effectiveness of new phytochemicals as corrosion inhibitors, guiding the selection and design of new inhibitors.

QSAR modeling can be a powerful tool for understanding the effects of molecular structure on the effectiveness of phytochemicals as corrosion inhibitors for magnesium in H_2SO_4 solutions. However, it requires a large amount of high-quality data and sophisticated statistical methods, and its predictions should always be validated through experimental testing.

15.7   FUTURE PERSPECTIVES

15.7.1   POTENTIAL FOR FURTHER RESEARCH IN THIS AREA

The use of phytochemicals as corrosion inhibitors for magnesium in H_2SO_4 solutions is a promising field with ample opportunities for further research. Future studies could focus on exploring more types of phytochemicals, understanding their mechanisms of action at a molecular level, and optimizing their extraction and application processes. Additionally, it is indicated developing predictive models, such as QSAR models, to guide the design and selection of effective phytochemical inhibitors.

15.7.2   POSSIBLE IMPROVEMENTS AND ADVANCEMENTS IN THE USE OF PHYTOCHEMICALS AS CORROSION INHIBITORS

Advancements in this field could lead to the development of more effective and sustainable corrosion inhibitors. Improvements could be made in the extraction and purification processes to increase the yield and purity of the phytochemicals. The formulation of the inhibitors could also be optimized to enhance their performance and ease of application. Furthermore, advancements in analytical techniques could enable more accurate and detailed characterization of the phytochemicals and their interactions with the metal and the corrosive medium.

15.7.3   IMPACT ON INDUSTRIES AND THE ENVIRONMENT

The use of phytochemicals as corrosion inhibitors could have a significant impact on various industries, including the automotive, aerospace, and energy industries, where magnesium and its alloys are commonly used. It could lead to improved corrosion protection, longer lifespan of materials, and cost savings. Moreover, as phytochemicals are derived from renewable resources and are generally non-toxic and biodegradable, their use could contribute to the sustainability of corrosion protection strategies and reduce the environmental impact of corrosion management. This aligns with the global trend towards green chemistry and sustainable industrial practices.

15.7.4 Impact on Economic Development

The development and commercialization of phytochemicals as corrosion inhibitors could stimulate economic growth by creating new markets and job opportunities. This could be particularly beneficial for regions with abundant plant resources, where the cultivation and processing of plants for phytochemical extraction could contribute to local economic development. Furthermore, the cost savings from improved corrosion protection could enhance the competitiveness of industries that rely heavily on magnesium and its alloys.

15.7.5 Challenges and Limitations

While the use of phytochemicals as corrosion inhibitors is promising, it also comes with challenges and limitations that need to be addressed. These include the variability in the composition of plant extracts, the potential for bioavailability issues, and the need for large-scale production methods. Further research is needed to overcome these challenges and to fully realize the potential of phytochemicals as corrosion inhibitors.

15.8 CONCLUSION

15.8.1 Chapter's Key Points

In this chapter, it was explored the use of phytochemicals and plant extracts as an alternative, environmentally friendly approach to inhibiting magnesium corrosion in sulfuric acid solutions.

The key points covered in this chapter include:

i. The adsorption mechanisms of phytochemicals onto magnesium surfaces to form protective films against corrosion
ii. The role of functional groups like hydroxyl, carbonyl, and aromatic rings in facilitating the adsorption and inhibition process
iii. Methods to extract and isolate phytochemicals from plant sources
iv. The effects of molecular structure on the corrosion inhibition efficacy of phytochemicals
v. Quantitative structure–activity relationship modeling to predict inhibitor performance
vi. Comparative advantages of phytochemicals over traditional inhibitors in terms of environmental impact, cost-effectiveness, and efficiency
vii. The potential for further research and development in this field to advance the use of phytochemicals as sustainable corrosion inhibitors

15.8.2 Final Thoughts on the Importance and Potential of Phytochemicals/Plant Extracts as Corrosion Inhibitors for Magnesium in H₂SO₄ Solutions

The research conducted so far clearly demonstrates the promise of phytochemicals and plant extracts as an alternative to traditional corrosion inhibitors for protecting magnesium alloys in acidic conditions. These natural compounds offer a renewable, environmentally friendly, and often more efficient approach to corrosion inhibition.

Further development in the extraction, purification, and optimization of phytochemicals could lead to effective commercial corrosion inhibitors that are safer, greener, and more sustainable. Their implementation on an industrial scale could significantly impact industries relying on magnesium alloys, while also reducing environmental hazards associated with traditional inhibitors. There is tremendous scope for ongoing research in this field to fully realize the potential of plant-derived corrosion inhibitors.

REFERENCES

[1] Abdulrahman AS, Ismail M. Green plant extract as a passivation-promoting inhibitor for reinforced concrete. *Internation Journal of Engineering Science and Technology*, 2011;3(8):6484–90.

[2] Cao Y, Jiang S, Zhang Y, Xu J, Qiu L, Wang L. Investigation into adsorption characteristics and mechanism of atrazine on nano-MgO modified fallen leaf biochar. *Journal of Environmental Chemical Engineering*, 2021;9(4):105727.

[3] Fan X-L, Li C-Y, Wang Y-B, Huo Y-F, Li S-Q, Zeng R-C. Corrosion resistance of an amino acid-bioinspired calcium phosphate coating on magnesium alloy AZ31. *Journal of Materials Science & Technology*, 2020;49:224–35.

[4] Feliu Jr S. Electrochemical impedance spectroscopy for the measurement of the corrosion rate of magnesium alloys: Brief review and challenges. *Metals*, 2020;10(6):775.

[5] Filgueiras AV, Capelo JL, Lavilla I, Bendicho C. Comparison of ultrasound-assisted extraction and microwave-assisted digestion for determination of magnesium, manganese and zinc in plant samples by flame atomic absorption spectrometry. *Talanta*, 2000;53(2):433–41.

[6] Galicia G, Pébère N, Tribollet B, Vivier V. Local and global electrochemical impedances applied to the corrosion behaviour of an AZ91 magnesium alloy. *Corrosion Science*, 2009;51(8):1789–94.

[7] Gao H, Li Q, Dai Y, Luo F, Zhang HX. High efficiency corrosion inhibitor 8-hydroxyquinoline and its synergistic effect with sodium dodecylbenzenesulphonate on AZ91D magnesium alloy. *Corrosion Science*, 2010;52(5):1603–9.

[8] Gomes MP, Costa I, Pébère N, Rossi JL, Tribollet B, Vivier V. On the corrosion mechanism of Mg investigated by electrochemical impedance spectroscopy. *Electrochimica Acta*, 2019;306:61–70.

[9] Gu X, Zheng Y, Cheng Y, Zhong S, Xi T. In vitro corrosion and biocompatibility of binary magnesium alloys. *Biomaterials*, 2009;30(4):484–98.

[10] Hassan RM, Ibrahim SM. Performance and efficiency of methyl-cellulose polysaccharide as a green promising inhibitor for inhibition of corrosion of magnesium in acidic solutions. *Journal of Molecular Structure*, 2021;1246:131180.

[11] Hu J, Zeng D, Zhang Z, Shi T, Song G-L, Guo X. 2-Hydroxy-4-methoxy-acetophenone as an environment-friendly corrosion inhibitor for AZ91D magnesium alloy. *Corrosion Science*, 2013;74:35–43.

[12] Kannan S, Madhu K, Nallaiyan R. Formulation of magnesium conversion coating with herbal extracts for biomedical applications. *Journal of Bio- and Tribo-Corrosion*, 2022;8(4):114.

[13] Lei L, Shi J, Wang X, Liu D, Xu H. Microstructure and electrochemical behavior of cerium conversion coating modified with silane agent on magnesium substrates. *Applied Surface Science*, 2016;376:161–71.

[14] Li H, Fan M, Wang K, Bian X, Jiang H, Ding W. Traditional Chinese medicine extracts as novel corrosion inhibitors for AZ91 magnesium alloy in saline environment. *Scientific Reports*, 2022;12(1):7367.

[15] Li L, Zhang M, Li Y, Zhao J, Qin L, Lai Y. Corrosion and biocompatibility improvement of magnesium-based alloys as bone implant materials: A review. *Regenerative Biomaterials*, 2017;4(2):129–37.

[16] Liu L, Yang Q, Huang L, Liu X, Liang Y, Cui Z, et al. The effects of a phytic acid/calcium ion conversion coating on the corrosion behavior and osteoinductivity of a magnesium-strontium alloy. *Applied Surface Science*, 2019;484:511–23.

[17] Ma L, Li W, Zhu S, Wang L, Guan S. Corrosion inhibition of Schiff bases for Mg-Zn-Y-Nd alloy in normal saline: Experimental and theoretical investigations. *Corrosion Science*, 2021;184:109268.

[18] Musdalslien UI, Standal NA, Johansen JG, Oehme M. Pilot plant tests with a wet electrostatic precipitator for reducing PCDD/PCDF in corrosive off-gas from magnesium production. *Chemosphere*, 1991;23(8–10):1097–108.

[19] Qian Y, Wu Y, Guo X, Wang L. A synergistic anti-corrosion effect of longan residue extract and sodium dodecylbenzenesulfonate composition on AZ91D magnesium alloy in NaCl solution. *Corrosion Engineering, Science and Technology*, 2022;57(4):322–30.

[20] Quraishi MA, Chauhan DS, Saji VS. Heterocyclic biomolecules as green corrosion inhibitors. *Journal of Molecular Liquids*, 2021;341:117265.

[21] Samadianfard R, Seifzadeh D, Habibi-Yangjeh A, Jafari-Tarzanagh Y. Oxidized fullerene/sol-gel nanocomposite for corrosion protection of AM60B magnesium alloy. *Surface and Coatings Technology*, 2020;385:125400.

[22] Shanab SMM, Ameer MA, Fekry AM, Ghoneim AA, Shalaby EA. Corrosion resistance of magnesium alloy (AZ31E) as orthopaedic biomaterials in sodium chloride containing antioxidantly active compounds from Eichhornia crassipes. *International Journal of Electrochemical Science*, 2011;6:3017–35.

[23] Toorani M, Aliofkhazraei M, Rouhaghdam AS. Microstructural, protective, inhibitory and semiconducting properties of PEO coatings containing CeO2 nanoparticles formed on AZ31 Mg alloy. *Surface and Coatings Technology*, 2018;352:561–80.

[24] Umoren SA, Solomon MM, Madhankumar A, Obot IB. Exploration of natural polymers for use as green corrosion inhibitors for AZ31 magnesium alloy in saline environment. *Carbohydrate Polymers*, 2020;230:115466.

[25] Deyab MA. Decyl glucoside as a corrosion inhibitor for magnesium-air battery. *Journal of Power Sources*, 2016;325:98–103.

[26] Verma C, Ebenso EE, Bahadur I, Quraishi MA. An overview on plant extracts as environmental sustainable and green corrosion inhibitors for metals and alloys in aggressive corrosive media. *Journal of Molecular Liquids*, 2018;266:577–90.

[27] Walker J, Shadanbaz S, Woodfield TBF, Staiger MP, Dias GJ. Magnesium biomaterials for orthopedic application: A review from a biological perspective. *Journal of Biomedical Materials Research Part B: Applied Biomaterials*, 2014;102(6):1316–31.

[28] Wu Y, Zhang Y, Jiang Y, Li N, Zhang Y, Wang L, et al. Exploration of walnut green husk extract as a renewable biomass source to develop highly effective corrosion inhibitors for magnesium alloys in sodium chloride solution: Integrated experimental and theoretical studies. *Colloids and Surfaces A: Physicochemical and Engineering Aspects*, 2021;626:126969.

[29] Wu Y, Zhang Y, Jiang Y, Qian Y, Guo X, Wang L, et al. Orange peel extracts as bio-degradable corrosion inhibitor for magnesium alloy in NaCl solution: Experimental and theoretical studies. *Journal of the Taiwan Institute of Chemical Engineers*, 2020;115:35–46.

[30] Yamamoto A, Terawaki T, Tsubakino H. Microstructures and corrosion properties on fluoride treated magnesium alloy. *Materials Transactions*, 2008;49(5):1042–7.

[31] Yang L, Shi X, Tian X, Han X, Mu J, Qi L. Effect of sealing treatment on corrosion behavior of plasma sprayed ZrO2 coated Cf/Mg composites. *Surface and Coatings Technology*, 2021;423:127627.

[32] Li P, Shao Z, Fu W, Ma W, Yang K, Zhou H, et al. Enhancing corrosion resistance of magnesium alloys via combining green chicory extracts and metal cations as organic-inorganic composite inhibitor. *Corrosion Communications*, 2023;9:44–56.

[33] Ye CH, Zheng YF, Wang SQ, Xi TF, Li YD. In vitro corrosion and biocompatibility study of phytic acid modified WE43 magnesium alloy. *Applied Surface Science*, 2012;258(8):3420–7.

[34] Ye C-H, Xi T-F, Zheng Y-F, Wang S-Q, Yang-De LI. In vitro corrosion and biocompatibility of phosphating modified WE43 magnesium alloy. *Transactions of Nonferrous Metals Society of China*, 2013;23(4):996–1001.

[35] Zhang S, Li Q, Chen B, Yang X. Preparation and corrosion resistance studies of nano-metric sol-gel-based CeO2 film with a chromium-free pretreatment on AZ91D magnesium alloy. *Electrochimica Acta*, 2010;55(3):870–7.

[36] Zhang Y, Shao Y, Zhang T, Meng G, Wang F. High corrosion protection of a polyaniline/organophilic montmorillonite coating for magnesium alloys. *Progress in Organic Coatings*, 2013;76(5):804–11.

[37] Zhou H, Li J, Li J, Ruan Q, Peng X, Li S, et al. A composite coating with physical interlocking and chemical bonding on WE43 magnesium alloy for corrosion protection and cytocompatibility enhancement. *Surface and Coatings Technology*, 2021;412:127078.

[38] Zhu J, Jia H, Liao K, Li X. Improvement on corrosion resistance of micro-arc oxidized AZ91D magnesium alloy by a pore-sealing coating. *Journal of Alloys and Compounds*, 2021;889:161460.

16 Phytochemicals/Plant Extracts as Corrosion Inhibitors for Copper in Biodiesel

Flávia Dias Fernandes and
Maria Lúcia Caetano Pinto da Silva
Universidade de São Paulo

16.1 INTRODUCTION

Biodiesel has emerged as a viable and sustainable alternative to conventional fossil fuels, offering a promising solution to reduce greenhouse gas emissions and decrease our dependence on non-renewable energy sources. Derived from vegetable oils, animal fats, or waste cooking oils through the transesterification process, biodiesel has gained significant popularity due to its environmental benefits and potential to promote greener transportation. However, like any new fuel, biodiesel comes with its unique set of challenges, one of which is its potential to induce corrosion in engine components, particularly in copper and its alloys used in the automotive and transportation sectors. Copper, a widely used material in fuel injection systems, heat exchangers, and connectors, exhibits excellent thermal conductivity and malleability, making it a preferred choice for various engine components. However, its susceptibility to corrosion in the presence of biodiesel raises concerns about the integrity and performance of the engine, necessitating the development of effective corrosion mitigation strategies.

In recent years, researchers and engineers have turned to nature for inspiration, exploring the potential of plant extracts as corrosion inhibitors for copper in biodiesel. Plant extracts are rich sources of bioactive compounds, such as alkaloids, flavonoids, tannins, and essential oils, known for their diverse functionalities and potential for inhibiting corrosion. Harnessing the power of these natural compounds, researchers seek to develop green corrosion inhibitors that can protect copper surfaces from degradation and extend the lifespan of engine components in biodiesel applications. Before delving into the potential of plant extracts as corrosion inhibitors, it is crucial to comprehend the underlying mechanisms of corrosion in biodiesel. Biodiesel is primarily composed of fatty acid methyl esters (FAMEs) resulting from the reaction between triglycerides and methanol during the transesterification process. The chemical structure of FAMEs varies depending on the

DOI: 10.1201/9781003394631-16

feedstock used for biodiesel production, and the presence of unsaturated fatty acid chains makes biodiesel more susceptible to oxidation and degradation. The degradation of biodiesel can lead to the formation of acidic compounds and peroxides, which, when in contact with metals like copper, can initiate the corrosion process. Furthermore, the presence of impurities in biodiesel, such as water, free fatty acids, and monoglycerides, can exacerbate the corrosive potential of the fuel. Water, for instance, can act as a corrosive agent, accelerating the breakdown of FAMEs and promoting the formation of corrosive species on metal surfaces. The combination of these factors makes understanding and mitigating corrosion in biodiesel a critical aspect of ensuring the reliable and safe operation of diesel engines.

Traditional corrosion inhibitors, often composed of toxic and environmentally harmful chemicals, may raise concerns about their impact on human health and the ecosystem. In contrast, plant extracts present an attractive and sustainable alternative due to their green and eco-friendly nature. Extracted from various plant sources, these natural compounds offer several advantages, including biodegradability, low toxicity, and renewable availability. Phytochemicals present in plant extracts possess diverse functionalities that contribute to their inhibitive properties. For instance, alkaloids can form complexation compounds with metal ions, hindering their interaction with corrosive species. Flavonoids, on the contrary, can chelate metal surfaces, leading to the formation of a protective film that retards the corrosion process. Tannins, known for their antioxidant properties, can scavenge free radicals and inhibit oxidative degradation of biodiesel, thus reducing the formation of corrosive compounds. To harness the potential of plant extracts as corrosion inhibitors for copper in biodiesel, researchers employ various evaluation methods to determine their effectiveness. Electrochemical techniques, such as potentiodynamic polarization and impedance spectroscopy, provide insights into the corrosion kinetics and the behavior of the inhibitor on the metal surface. Weight loss studies offer a straightforward approach to quantify the corrosion rate and assess the protective efficiency of the plant extract. Additionally, surface analysis techniques, such as scanning electron microscopy and X-ray diffraction, allow researchers to explore the morphology and composition of the protective film formed on the copper surface.

16.2   PRINCIPLES

16.2.1   BIODIESEL-BASED ELECTROLYTES

Biodiesel is constituted of fatty acid alkyl esters that can be saturated and unsaturated. The raw material chosen for the production of biodiesel is directly related to its oxidative stability (Monieta, Szmukata, and Adamczyk, 2022). This occurs because the increase in unsaturation in the fatty acid chain and the presence of double bonds in biodiesel make it more prone to oxidation due to its reactivity with O_2 and the onset of oxidation at the allylic and bis-allylic positions of the double bonds (Kumar, Kumar, and Sham, 2016). The bis-allylic positions are more reactive, and their protons are highly susceptible to free radical attack. The instability at these points occurs due to the attraction of electrons that are present in the vicinity of the double bonds on either side of the methylene group. Thus, the hydrogen that is in this position becomes more acidic and can be easily removed when exposed to a free radical (Kumar, 2017).

The presence of unsaturated fatty acids in biodiesel favors the oxidation mechanism by auto-oxidation that affects properties such as viscosity, acidity number, iodine value, peroxide value and induction time (Annisa, Darwis, and Hadinugrahaningsih, 2022). Initially, in the allylic and bis-allylic positions of the double bonds, hydrogen is removed from a carbon atom with the formation of a free radical which in the presence of oxygen tends to form peroxide. The peroxide formed removes hydrogen from carbon again to form another free radical and hydroperoxide. The final stage of propagation consists of the reaction of oxygen with this newly formed free radical. The reaction will end when two free radicals react with each other to form stable products such as aldehydes, carboxylic acids and sedimentary gum (Jain and Sharma, 2010). Therefore, this phenomenon is directly associated with the presence of oxygen and can be described by three main steps which are initiation (Equation 16.1), propagation (Equations 16.2 and 16.3) and termination (Equations 16.4–16.6). In the equations, RH is an unsaturated fatty acid, R· is a free radical, ROO· is a peroxide radical and ROOH is a hydroperoxide (Ramalho and Jorge, 2006). These oxidation reactions form acidic species that can lead to corrosive processes in automotive materials and structures that come into contact with biodiesel (Wu, Ge, and Choi, 2020).

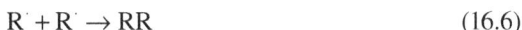

$$RH \rightarrow R^{\cdot} + H^{\cdot} \tag{16.1}$$

$$R^{\cdot} + O_2 \rightarrow ROO\cdot \tag{16.2}$$

$$ROO^{\cdot} + RH \rightarrow ROOH + R\cdot \tag{16.3}$$

$$ROO^{\cdot} + R^{\cdot} \rightarrow ROOR \tag{16.4}$$

$$ROO^{\cdot} + ROO^{\cdot} \rightarrow ROOR + O_2 \tag{16.5}$$

$$R^{\cdot} + R^{\cdot} \rightarrow RR \tag{16.6}$$

Impurities present in biodiesel can also act negatively in relation to its degradation and corrosivity and depend on the quality and type of raw material and the process for obtaining biodiesel (Fazal, Rubaiee, and Al-Zahrani, 2019). It can be mentioned as residual impurities of the catalyst used in the production of biodiesel, free fatty acids, glycerol, alcohol and water (Hoang, Tabatabaei, and Aghbashlo, 2020). The fact that this fuel is hygroscopic by nature accelerates the contamination of biodiesel by fungi and bacteria and leads to a reduction in its stability, biofouling and corrosive processes (Fazal, Rubaiee, and Al-Zahrani, 2019).

16.2.2 CORROSION MECHANISM OF COPPER EXPOSED TO BIODIESEL

One of the main materials used in the manufacture of diesel engine components is copper and its alloys. Its employment can be observed in injector pump, feed pump (Hoang, Tabatabaei, and Aghbashlo, 2020; Hosseinabadi, Moheimani, and

Javaherdashti, 2022), fuel pump, gasket and fitting, housing, bearing, low and high pressure (Chandran, 2020; Kugelmeier et al., 2021).

In an acidic medium, the copper corrosion starts with oxygen and occurs in two steps as shown in Equations 16.7 and 16.8. The formation of Cu(II) metallic copper ions can further enhance the corrosion process, as observed in Equation 16.9 (Fazal, Haseeb, and Masjuki, 2013).

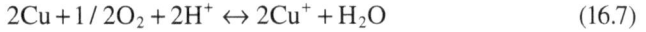

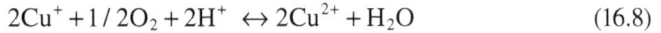

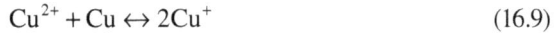

$$2Cu + 1/2O_2 + 2H^+ \leftrightarrow 2Cu^+ + H_2O \qquad (16.7)$$

$$2Cu^+ + 1/2O_2 + 2H^+ \leftrightarrow 2Cu^{2+} + H_2O \qquad (16.8)$$

$$Cu^{2+} + Cu \leftrightarrow 2Cu^+ \qquad (16.9)$$

After the exposure of copper to biodiesel, the formation of corrosion products in the form of oxides, carbonates and hydroxides has been reported by several authors (Fazal, Rubaiee, and Al-Zahrani, 2019). The presence of O_2, H_2O, the ROO^- radical and CO_2 dissolved in the biodiesel that interact with the copper surface justifies the formation of such products. The presence of oxygen can cause the formation of Cu_2O as observed in Equation 16.10. As this oxide is unstable, it transforms into a stable species in the form of CuO following Equation 16.11. The reaction between copper and the ROO^- radical formed in the biodiesel auto-oxidation (Equation 16.2) can lead to the formation of $CuCO_3$ (Equation 16.12). In addition, this corrosion product can also be formed from the reaction of Cu_2O and CuO with CO_2 from the atmosphere (Equations 16.13 and 16.14, respectively). The presence of dissolved H_2O, O_2 and CO_2 in biodiesel causes the formation of copper compounds based on carbonate and hydroxyl according to Equations 16.15 to 16.18 (Fazal, Haseeb, and Masjuki, 2013; Hoang, Tabatabaei, and Aghbashlo, 2020).

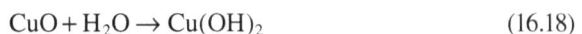

$$2Cu + 1/2O_2 \rightarrow Cu_2O \qquad (16.10)$$

$$Cu_2O + 1/2O_2 \rightarrow 2CuO \qquad (16.11)$$

$$Cu^{2+} + 2ROO^- \rightarrow CuCO_3 + RR + CO \qquad (16.12)$$

$$Cu_2O + 2CO_2 + 1/2O_2 \rightarrow 2CuCO_3 \qquad (16.13)$$

$$CuO + CO_2 \rightarrow CuCO_3 \qquad (16.14)$$

$$2Cu + H_2O + CO_2 + O_2 \rightarrow Cu(OH)_2 \cdot CuCO_3 \qquad (16.15)$$

$$2Cu(OH)_2 + CO_2 \rightarrow Cu(OH)_2 \cdot CuCO_3 + H_2O \qquad (16.16)$$

$$2Cu + O_2 + 2H_2O \rightarrow 2Cu(OH)_2 \qquad (16.17)$$

$$CuO + H_2O \rightarrow Cu(OH)_2 \qquad (16.18)$$

16.2.3 PLANT EXTRACT AS GREEN CORROSION INHIBITOR

Plant extract obtained from different parts of the plants has several phytochemical compounds that contain electron-rich polar functional groups such as hydroxyl (–OH), amine (–NH$_2$), amide (–CONH$_2$), ester (–COOC$_2$H$_5$) and aromatic rings that are easily adsorbed acting as strong anticorrosive materials. In the literature, studies are found evaluating different parts such as fruits, root, bark, flower, seed and sometimes entire plants used as effective corrosion inhibitors for various metals and alloys in different electrolytic media (Verma, 2022a).

The preparation of plant extracts involves these steps. Initially, the selection and washing of the part of the plant are made. Then, the collected material should be dried, and this is usually performed at room temperature. This step is an obligation for all parts of the plants, except for the fruit juice extract. After this step, the leaf is ground and sieved in order to transform the material into powder. Finally, various methods can be used to isolate and extract the plant extract such as solvent extraction, distillation method, pressing and sublimation according to the principle of extraction. In short, the extraction methods are based on heating, cooling and separation of active compounds in the presence of the solvent (Salleh et al., 2021).

It should be noted that plant extracts are ecologically correct and effective in replacing traditional inhibitors due to a series of benefits given as follows (Fernandes, Ferreira, and Da Silva, 2021a; Verma, 2022a; Thakur et al., 2022):

(i) Ease of fabrication and application.
(ii) Non-toxic, non-bioaccumulative and biodegradable.
(iii) High corrosion inhibition efficiency due to the presence of several complex phytochemicals that have heteroatoms such as O, N, P and S and exhibit strong bonding with the metallic surface.
(iv) Useful for different metal/electrolyte systems.

16.2.4 PLANT EXTRACT INHIBITION MECHANISM

Metallic materials are inherently unstable and react chemically or electrochemically with aggressive agents in the environment forming more stable substances in the form of corrosion products (Zakeri, Bahmani, and Aghdam, 2022). In this way, the use of corrosion inhibitors is necessary in that they use the principle of direct or indirect adsorption of inhibitor molecules on the surface of the metal in order to reduce its contact with the aggressive environment (Wei et al., 2020). Phytochemicals are complex molecules that contain multiple electron-rich centers that can be adsorbed in metal surfaces. The mechanism of adsorption of these phytochemicals can be by physisorption, by chemisorption or by physio-chemisorption. In the physisorption mechanism, electrostatic interactions occur between inhibitor molecules (phytochemicals) and charged metallic surfaces. In the chemisorption mechanism, charge transfer reactions occur between the vacant orbitals on the metal surface and free electron pairs in the inhibitor (Miralrio and Vázquez, 2020; Alrefaee et al., 2021).

The adsorption mechanism directly depends on the substrate surface charge during the interaction between the phytochemical compounds present in the plant extract and the metal surface. In the paper by Wei et al. (2020), some configurations are addressed to interpret the interaction of green corrosion inhibitors with a metallic surface:

For physisorption, two mechanisms can be observed:
   i.   The electrostatic interaction of the plant extract with the metallic surface of opposite charge leads to the physisorption of the phytochemicals on the metallic surface in a direct way.
   ii. Another form of physisorption can occur indirectly when there is an electrostatic interaction of pre-adsorbed ions and the positively charged surface. These anions adsorbed through electrostatic interaction lead to a negatively charged metal surface increasing the ability to adsorb protonated inhibitors. This adsorption mechanism occurs specifically in acidic medium.

For chemisorption, two mechanisms can be observed:
   iii. Phytochemical compounds are usually electron donors establishing donor–acceptor interaction, so that the unshared electron pairs of heteroatoms such as O and N or π-electrons of the aromatic ring of inhibitors interact with d-orbitals of the atomic surface of the substrate of metal to form a protective film.
   iv. Another form of chemisorption results from the interaction between metal ions and functional groups such as –OH and NH_2 from the plant extract leading to the formation of insoluble complex compounds that protect the metal surface from corrosion.

In order to better understand the processes around the adsorption of phytochemicals on the metallic surface, mathematical models known as adsorption isotherms can be applied to estimate the amount of adsorbate (inhibitor molecules/ions) in the absorbent (surface of the metallic substrate) at a constant temperature such as Langmuir, Temkin, Freundlich, Flory-Huggins and El-Awady (Adejo et al., 2012).

16.3   METHODS

Corrosion caused by biodiesel can be evaluated in two ways, taking into account the deterioration of biodiesel and the deterioration of materials that come into contact with this fuel (Yeşilyurt, Öner, and Yilmaz, 2019). Most existing studies in the literature to assess the compatibility of certain materials in biodiesel are carried out based on the durability of engines and laboratory tests such as immersion and electrochemical tests (Fazal, Rubaiee, and Al-Zahrani, 2019).

For engine durability tests parts are normally disassembled. In addition, lubricating oils are collected and analyzed after performing the static engine test or field test to assess failure, corrosion, biodiesel degradation and wear of fuel system parts. During this type of test, the main technical problems found are fuel pump failure, filter clogging, injector coking and moving parts locking (Dunn, 2009). This type of test is very important for the field of biodiesel use. However, few researchers perform it due to the fact that it is more time-consuming and more expensive than engine power, economy and emissions (Mofijur et al., 2013).

In immersion tests, the first step consists of cutting samples and inserting a hole if possible so that they can be hung during the experiment immersed in fuel. The samples are then sanded to a specific grain size, washed with deionized water, dried, and then dipped in acetone or alcohol to degrease. The samples are initially weighed on a digital scale with precision of up to four decimal places. Afterward, they are immersed in fuel for a specific time. After removal, the materials are washed and degreased with acetone and then weighed again. It is usually carried out with longer durations and causes a change in the color of the biodiesel and on the surface of the samples, which helps in the identification of changes in the composition of the fuel due to corrosive processes (Shehzad et al., 2021). According to the ASTM G1-03 – *Standard Practice for Preparing, Cleaning and Evaluation Corrosion Test Specimens* (2015) and with the weighing of the samples before and after immersion in biodiesel, it is possible to obtain the difference in the weight of the initial and final samples and calculate the rate of corrosion (*Cr*) of materials according to the Equation 16.19. In Equation 16.19, *K* is a constant, *T* is the exposure time, *A* is the area, *w* is the lost mass and *D* is the density. The *K* value mentioned can be found in the referred standard. The ASTM G31 – *Standard Guide for Laboratory Immersion Corrosion Testing of Metals* (2011) provides a standardized procedure to directly measure the weight loss of metals. The scope of the standard addresses how the temperature, pH, flow and composition of the solution should be controlled and how the test duration should be stipulated. Important factors such as sample preparation, test conditions, sample cleaning methods and care regarding test result interpretation are also discussed (Ebert and Gattu, 2020). The inhibition efficiency (*I_e*) in % of the use of the plant extract can be calculated in relation to the corrosion rate in the absence and presence of the additive in the biodiesel as shown in Equation 16.20 (Deyab, 2014). In Equation 16.20, C_r^0 is the corrosion rate in the absence of plant extract and *Cr* is the corrosion rate in the presence of plant extract:

$$Cr = \frac{K \cdot w}{A \cdot T \cdot D} \tag{16.19}$$

$$I_e(\%) = \frac{Cr0 - Cr}{Cr0} \times 100 \tag{16.20}$$

Electrochemical tests use reduction and oxidation reactions, and when the corrosive process occurs, the metal oxidizes in solution. This type of reaction involves the flow of electrons and current and therefore can be measured and calculated electronically. The duration for the electrochemical method is only a few hours (Shehzad et al., 2021). The most used methods for these types of tests are the open circuit potential, polarization and extrapolation of Tafel lines and electrochemical impedance spectroscopy. The open circuit potential shows the direct measure of the equilibrium potential in relation to a reference electrode (Wolynec, 2003). With the polarization and extrapolation of the Tafel lines, we can find corrosion rates, corrosion potential, corrosion current density and the evaluation of the regions present in the anodic region verifying whether the material is passivated in the medium or not (Kruger and Hopkins, 2003; Kakaei, Esrafili, and Ehsani, 2019). Furthermore, the corrosion current density values of the green corrosion inhibitor can be estimated according to Equation 16.21 where *Iblank* is the corrosion density current without inhibitor and *Iinhibitor* is the corrosion density current with inhibitor (Thakur et al., 2022). The electrochemical impedance technique provides the study of intrinsic properties of the material or specific processes that can influence the conductance, resistance or capacitance of an electrochemical system (Magar, Hassan, and Mulchandani, 2021). The data obtained by this technique can be modeled in order to convert the response data that are obtained in frequency into corrosion properties (Feliu, 2020). Thus, with the values of charge transfer resistance and capacitance, the efficiency of the green corrosion inhibitor can be calculated according to Equation 16.22 where R'_{ct} is the charge transfer resistance of the medium without inhibitor and R_{ct} is the charge transfer resistance of the medium with inhibitor (Thakur et al., 2022):

$$I_e(\%) = \frac{I_{blank} - I_{inhibitor}}{I_{blank}} \times 100 \tag{16.21}$$

$$I_e(\%) = \frac{R'_{ct} - R_{ct}}{R_{ct}} \times 100 \tag{16.22}$$

After the immersion test and the electrochemical test, other investigations can be carried out in order to better understand and evaluate the corrosive process. Table 16.1 lists some of the main techniques that can be performed after the experimental tests described above for the surface of copper and for biodiesel (Fazal, Rubaiee, and Al-Zahrani, 2019; Yeşilyurt, Öner, and Yilmaz, 2019).

As observed in Table 16.1, surface characterization is usually performed using spectroscopy and microscopy techniques. The scanning electron microscope provides morphological information and a comparison between the metal surface with and without the green corrosion inhibitor. Atomic force microscopy allows imaging around surface topography, and X-ray photoelectron spectroscopy is used for

---

TABLE 16.1

General Methods for Evaluating the Surface of Copper and Biodiesel after Corrosion Tests

Material	Techniques
Copper	Optical microscopy, scanning electron microscopy/energy dispersive spectroscopy (SEM/EDS), atomic force microscopy (AFM), X-ray diffraction (XRD), X-ray photoelectron spectroscopy (XPS)
Biodiesel	Acidity index (Ai), Fourier-transform infrared spectroscopy (FTIR), viscosity, density, compositional changes, oxidation stability, water content, cetane number

oxidation states, stoichiometry and electronic state determination. As for the bio-diesel, Fourier transform infrared spectroscopy can be used to obtain information about the functional groups and vibrational modes of green corrosion inhibitors (Miralrio and Vázquez, 2020). Information regarding the composition of corrosion products in the sample without and with the presence of the green corrosion inhibitor can be obtained using the X-ray diffraction technique (Zhang, Tan, and Li, 2022).

Acids are produced with the auto-oxidation of biodiesel where the high acid value can corrode the engine. To assess the degree of biodiesel auto-oxidation, the acid value is traditionally used (Chen et al., 2020). The acidity index (A_i) analysis car-ried out for biodiesel provides us with important information regarding its oxida-tion. One methodology applied for its determination consists of initially weighing 2 g of the biodiesel sample and then adding 25 mL of ether-alcohol solution (2:1 v/v) adding drops of phenolphthalein. Titration can be performed with standard 0.01 mol L^{-1} potassium hydroxide standardized to a pinkish solution. The calculation of the acidity index (A_i) can be done according to Equation 16.23. In this equation, A is the volume of the titration solution obtained for titration of the sample, B is the volume of the titration solution for titration of the blank sample, C is the concentration of the titration solution and M is the mass of the biodiesel sample (Fernandes, Ferreira, and Da Silva, 2021a):

$$A_i = \frac{(A-B)\times C\times 56.1}{M} \tag{16.23}$$

With current technological advances, computational methods have been widely used to evaluate the anticorrosive performance of plant extracts. These methods sup-port the data obtained from mass loss and electrochemical results, in addition to describing the interaction between metallic surface and corrosion inhibitors (Verma, 2022a). Molecular simulations can be used to accurately predict properties and then design a new green corrosion inhibitor. The use of these methods is also advanta-geous in reducing the time and cost to evaluate the corrosive performance of the extract. Quantum chemical approaches are accurate in identifying molecular struc-ture, knowing the reactivity of corrosion inhibitors and saving in terms of explora-tion costs due to blind screening checks. Molecular dynamics simulation and density functional theory are usually employed in quantum chemistry investigations to deepen knowledge about the inhibition mechanism of green corrosion inhibitors at the molecular level. The use of these techniques also brings important parameters regarding the structure such as energy and distribution of the molecular orbital, frac-tion of electrons transferred from the corrosion inhibitor to the substrate and absolute values of electronegativity (Zakeri, Bahmani, and Aghdam, 2022).

16.4 CALCULATIONS/WORKED EXAMPLES/ CURRENT APPLICATIONS

Table 16.2 lists research found in the literature regarding the use of plant extracts as green corrosion inhibitors in biodiesel in contact with copper. As can be seen, there is a diversity in relation to the raw materials used for the production of biodiesel.

TABLE 16.2

Use of Some Plant Extracts as a Green Corrosion Inhibitor for Copper in Biodiesel Medium

Type	Biodiesel (%)	Temperature (°C)	Plant Name	Part of Plant	References
Waste cooking oil	B10, B100	25–27	*Tinospora cordifolia*	Stem	Amgain et al. (2022)
Neem oil	B100	30, 50, 70, 90	*Piper nigrum*	Fruit	Iyappan et al. (2022)
Soybean oil	B100	20–28, 60	*Psidium guajava* L.	Leaves	Fernandes, Ferreira, and Da Silva (2021a)
Neem oil	B100	30, 40, 50, 60	*Ricinus communis*	Seed	Priyatharesini, Kumar, and Kumari (2021)
Palm biodiesel	B100	25–27	*Pyrola incarnata* Fisch.	Leaves	Chen et al. (2020)
Waste cooking oil	B10, B100	25–27	*Vitex negundo*	Leaves	Subedi et al. (2019)
Waste cooking oil	B10, B100	25–27	*Tinospora cordifolia*	Leaves	Amgain et al. (2018)

The percentage of biodiesel in the medium under study also varies as observed in studies with pure biodiesel and in mixtures with diesel. There are also some studies evaluating the effect of different temperatures and their effect on corrosive processes. Regarding the plant extracts used, the use of different parts of plants such as the stem, fruit, seed and leaves is noted.

The effect of *Tinospora cordifolia* plant leaf extract with different concentrations (500, 1,000, 1,500 and 2,000 ppm) on the corrosive behavior of copper in pure biodiesel and in mixtures with diesel was evaluated in the study by Amgain et al. (2018). The study was carried out at room temperature. For the evaluation of corrosion, the authors used the immersion test and the potentiodynamic polarization curves. Concentrations of 500–2,000 ppm of plant extract in pure biodiesel appear to be the most effective green corrosion inhibitor for copper to significantly increase the corrosion resistance property up to 90%. The results of anodic polarization tests showed that the plant extract is a mixed-type green corrosion inhibitor. The corrosion inhibition mechanism was explained by the adsorption process which obeyed the Langmuir adsorption isotherm model.

Vitex negundo leaf extract on copper passivation behavior was studied by Subedi et al. (2019) using immersion tests, inhibition efficiency and anodic polarization curves. The effect of different plant extract concentrations at 500, 1,000, 1,500 and 2,000 ppm was studied. The copper corrosion rate decreased as the extract concentration increased finding an inhibition efficiency of 96% and 60% in pure biodiesel and the mixture of 10% biodiesel, respectively. This can be explained mainly by the adsorption process of the plant extract on the metallic surface, in this case obeying the Langmuir adsorption isotherm.

Chen et al. (2020) studied how the *Pyrola* leaf extract can act on the corrosiveness of biodiesel, and for this purpose, an immersion test was performed. The values found for the acidity index show a value closer to the biodiesel as freshly received for the biodiesel with the plant extract. It is observed in this study that the corrosion rate decreases for the biodiesel with the plant extract. SEM micrographs were used to observe the morphological change of coppers before and after immersion testing. The SEM micrographs and the EDS data proved that the extract protected the copper surface more than the other conditions since the image shows the sanding scratches clearly and with an appearance closer to the image of the copper as received. The EDS data point to lower carbon and oxygen contents and the highest copper content on the analyzed surface for copper in contact with biodiesel with the plant extract. XPS spectra point to peaks corresponding to oxides (Cu_2O, CuO) and peaks that may indicate organic deposition of biodiesel is caused by fatty acids decomposed as C–C and –COOH. The atomic content of C(COOH) was lower for copper in biodiesel with the plant extract which means that the extract was effective in protecting the biodiesel.

Ricinus communis seed extract at different concentrations such as 0.5, 1, 1.5, 2 and 2.5 ppm was studied by Priyatharesini, Kumar and Kumari (2021) as a corrosion inhibitor for copper in neem oil-based biodiesel. The immersion test was carried out at the following temperatures 303, 313, 323 and 333 K. An inhibition efficiency greater than 95% was found for copper in biodiesel with the plant extract. The increase in temperature led to increased weight loss. The adsorption of the plant extract on the metallic copper surface obeyed the Langmuir adsorption isotherm. This phenomenon was attributed to the presence of an available pair of electrons present in the heteroatoms of the inhibitor molecules confirmed by FTIR spectral studies. The surface morphology of metallic copper after immersion test was studied by SEM, and for the sample in contact with biodiesel with plant extract, it was observed that the surface was more protected from corrosive processes avoiding damage to the metal such as cracks and the formation of pits.

In the paper by Fernandes, Ferreira and Da Silva (2021a), the *Psidium guajava* L. extract was added to biodiesel. In this study, the authors used the immersion test to evaluate the corrosive process and the effect of temperature. An inhibition efficiency was found for the plant extract used as 55.37% and 21.11% at room temperature and 60°C, respectively. The authors used acidity index and FTIR analyzes to assess biodiesel oxidation, and in this study, biodiesel oxidized under both temperature conditions. At 60°C, the acidity index showed higher values and was proportional to the corrosion rate values. The additive had better anticorrosive performance at room temperature, and its protective effect is observed with the naked eye where the copper plate immersed in pure biodiesel appeared darker. Through SEM micrographs for this condition, it was possible to observe which type of corrosion occurred on the surface pointing to alveolar corrosion for copper exposed to pure biodiesel, while in biodiesel with additives, the copper surface did not present this type of corrosion. The XRD analysis was possible to identify the formation of corrosion products in the form of oxides suggesting a lower formation of these for copper in additive biodiesel. The anticorrosive performance observed for the plant extract is due to the physisorption of antioxidant compounds present in the leaf on the metal surface acting as a protective biofilm.

Amgain et al. (2022) analyzed the effect of *Tinospora cordifolia* stem extract for Cu in pure biodiesel and its 10% mixture with petrodiesel at 25°C through immersion test and electrochemical tests. Different extract concentrations at 500, 1,000, 1,500 and 2,000 ppm in solution were investigated. For pure biodiesel, an inhibition efficiency above 90% was found for all concentrations studied. The results of the adsorption studies also support the argument of formation of a protective layer against corrosion on the metal surface due to the use of plant extract. The adsorption of the plant extract on the metallic surface was done by the Langmuir adsorption model.

Iyyappan et al. (2022) studied *Piper nigrum* fruit extract as a corrosion inhibitor for copper metal surface in biodiesel through an immersion test. The authors evaluated different temperatures (303, 323, 343 and 363 K for a period of 24 hours) and concentrations (from 0.5 to 2.5 ppm with 0.5 ppm increment). It was observed in the study that as the temperature increased, weight loss also increased. Furthermore, corrosion increases when the acidity value of biodiesel increases and decreases the inhibition efficiency. Black pepper extract significantly reduced the corrosion reaction and weight loss where it was found that the activation energy for the corrosion reaction increased in the presence of the black pepper inhibitor. Inhibition efficiency is very close for 2 and 2.5 ppm inhibitor dosages, and the inhibition efficiency reached 98% in one of the studied cases. The SEM images obtained clearly demonstrate the protective nature of the black pepper inhibitor on the copper surface in biodiesel medium preventing metal damage such as cracks and holes. Regarding the analysis of fuel performance with brake-specific fuel consumption data, it can be noted that biodiesel and engine performance were not affected by the addition of plant extract to biodiesel.

16.5   FUTURE PREDICTIONS

One of the limiting factors for not using pure biodiesel commercially is due to its corrosivity when in contact with metals. In this way, understanding the corrosive mechanism of biodiesel and protecting the components that come into contact with this fuel are necessary and a great challenge.

To enhance the application of this biofuel, the use of plant extracts as anti-corrosive additives is indicated in order to delay corrosive processes. With the elaboration of the chapter, it has been noticed that the addition of these additives to biodiesel in contact with copper is widely studied and there are still some potent points that need further scientific investigation.

Thus, the need for more studies involving different mixtures in the biodiesel and diesel composition, different concentrations of plant extract and longer periods of immersion tests is evident. The use of the electrochemical impedance technique, tests involving engine durability and fuel performance, and computational methods in investigations should also be encouraged. In addition, it is up to future studies to better detail the mechanism of action of plant extracts in relation to the copper surface and biodiesel.

As the plant extract contains multicomponents, a future research focused on the isolation of specific active molecules is also interesting in order to address an evaluation of its anticorrosive property.

16.6 SUMMARY

Biodiesel auto-oxidation is a process that leads to corrosion due to the formation of radicals and acids that can interact with the copper surface. In addition to this phenomenon, the presence of H_2O, O_2 and CO_2 leads to the formation of corrosion products when in contact with the surface of this metal. The corrosion products found can be in the form of oxides, carbonates and hydroxides.

In addition to the use of biodiesel already bringing more sustainability to the processes in which it is currently used due to ecological awareness and strict environmental regulations, the use of green corrosion inhibitors from plant extracts is highly encouraged and studied.

As can be seen in the cited studies, the use of plant extracts is satisfactory. Thus, these plant extracts can be satisfactorily applied to biodiesel in contact with copper in order to mitigate corrosive processes.

Most authors justify the inhibition of corrosive processes by the plant extract as a result of the adsorption of phytochemical compounds present in the plant extract on the copper surface acting as a protective film.

The main tests observed in the works for the evaluation of the anticorrosive performance of plant extracts consist of immersion and electrochemical tests in which corrosion rates and inhibition efficiency can be obtained. Furthermore, SEM analysis for metallic surface and acidity number for biodiesel is used in most studies.

In most papers, the corrosion inhibition mechanism by the plant extract is explained by the adsorption process that obeys the Langmuir adsorption isotherm model.

REFERENCES

Adejo, S. O.; Ekwenchi, M. M.; Momoh, F.; Odiniya, E. Adsorption characterization of ethanol extract of leaves of Portulaca oleracea as green corrosion inhibitor for corrosion of mild steel in sulphuric acid medium. *International Journal of Modern Chemistry*. v. 1, 125–134, 2012.

Alrefaee, S. H.; Rhee, K. Y.; Verma, C.; Quraishi, M.A.; Ebenso, E. E. Challenges and advantages of using plant extract as inhibitors in modern corrosion inhibition systems: Recent advancements. *Journal of Molecular Liquids*. v. 321, 114666, 2021.

Amgain, K.; Subedi, B. N.; Joshi, S.; Bhattarai, J. Investigation on the effect of Tinospora cordifolia plant extract asagreen corrosion inhibitor to aluminum and copper in biodiesel and its blend. *Corcon*. v. 19, 1–11, 2018.

Amgain, K.; Subedi, B. N.; Joshi, S.; Bhattarai, J. A comparative study of the anticorrosive response of Tinospora cordifolia stem extract for Al and Cu in biodiesel-based fuels. *Web of Conferences*. v. 355, 01005, 2022.

Annisa, S. F.; Darwis, Z.; Hadinugrahaningsih, T. The effect of tertiary butylhydroquinone antioxidant on the stability of rubber seed biodiesel. *Chemistry and Materials*. v. 2, 34–39, 2022.

ASTM G1-03. *Standard Praticing for Preparing, Cleaning and Evaluating Corrosion Test Specimens*. ASTM International, 2011.

ASTM G31. *Standard Guide for Laboratory Immersion Corrosion Testing of Metals*. ASTM International, 2015.

Chandran, D. Compatibility of diesel engine materials with biodiesel fuel. *Renewable Energy*. v. 147, 89–99, 2020.

Chen, T.; Hu, R.; Yao, X.; Yang, Q.; Shuai, S.; Wang, J.; Xu, M.; Zhang, D.; Fu, Y.; Li, L.; Zhao, W. Effect of Pyrola extract on the stability of palm biodiesel upon exposure to copper. *Renewable Energy*. v. 149, 1282–1289, 2020.

Deyab, M. A. Adsorption and inhibition effect of Ascorbyl palmitate on corrosion of carbon steel in ethanol blended gasoline containing water as a contaminant. *Corrosion Science*. v. 80, 359–365, 2014.

Dunn, R. O. Effects of minor constituents on cold flow properties and performance of biodiesel. *Progress in Energy and Combustion Science*. v. 35, 481–489, 2009.

Ebert, W. L.; Gattu, V. K. Metal Waste Forms. *Comprehensive Nuclear Materials* (Second Edition). v. 6, 467–482, 2020.

Fazal, M. A.; Haseeb, A. S. M. A.; Masjuki, H. H. Corrosion mechanism of copper in palm biodiesel. *Corrosion Science*. v. 67, 50–59, 2013.

Fazal, M. A.; Rubaiee, S.; Al-Zahrani, A. Overview of the interactions between automotive materials and biodiesel obtained from different feedstocks. *Fuel Processing Technology*. v. 196, 106178, 2019.

Feliu Jr., S. Electrochemical impedance spectroscopy for the measurement of the corrosion rate of magnesium alloys: Brief review and challenges. *Metals*. v. 10, 775, 2020.

Fernandes, F. D.; Ferreira, L. M.; Da Silva, M. L. C. P. Evaluation of the corrosion inhibitory effect of the ecofriendly additive of Terminalia catappa leaf extract added to soybean oil biodiesel in contact with zinc and carbon steel 1020. *Journal of Cleaner Production*, v. 321, 128863, 2021a.

Fernandes, F. D.; Ferreira, L. M.; Da Silva, M. L. C. P. Application of Psidium guajava L. leaf extract as a green corrosion inhibitor in biodiesel: Biofilm formation and encrustation. *Applied Surface Science Advances*. v. 6, 100185, 2021b.

Hoang, A. T.; Tabatabaei, M.; Aghbashlo, M. A review of the effect of biodiesel on the corrosion behavior of metals/alloys in diesel engines. *Energy Sources, Part A: Recovery, Utilization, and Environmental Effects*. v. 42:23, 2923–2943, 2020.

Hosseinabadi, N.; Moheimani, N. R.; Javaherdashti, R. Biofuels-related materials deterioration in biorefineries, transportation and internal combustion engines: A technical review. *Corrosion Engineering Science and Technology*. v. 57, 178–194, 2022.

Iyyappan, S.; Kannan, P.; Kumar, K. P. V.; Nagarajan, V. A. Enriched neem biodiesel with piper nigrum ethyl acetate extract for corrosion inhibition of copper. *Periodico di Mineralogia*. v. 91, 61–20, 2022.

Jain, S.; Sharma, M. P. Prospects of biodiesel from Jatropha in India: A review. *Renewable and Sustainable Energy Reviews*. v. 14, 763–771, 2010.

Kakaei, K.; Esrafili, M. D.; Ehsani, A. Chapter 8 - Graphene and Anticorrosive Properties. *Interface Science and Technology*. v. 27, 303–337, 2019.

Kruger, J.; Hopkins, J. Passivity. *ASM Handbook, Volume 13A: Corrosion: Fundamentals, Testing, and Protection*. 61–67, 2003.

Kugelmeier, C. L.; Monteiro, M. R.; Da Silva, R.; Kuri, S. E.; Sordi, V. L.; Rovere, C. A. D. Corrosion behavior of carbon steel, stainless steel, aluminum and copper upon exposure to biodiesel blended with petrodiesel. *Energy*, v. 226, 120344, 2021.

Kumar, N. Oxidative stability of biodiesel: Causes, effects and prevention. *Fuel*. v. 190, 328–350, 2017.

Kumar, R.; Kumar, V.; Sham, R. Stability of biodiesel- A review. *Renewable and Sustainable Energy Reviews*. v. 62, 866–881, 2016.

Magar, H. S.; Hassan, R. Y. A.; Mulchandani, A. Electrochemical impedance spectroscopy (EIS): Principles, construction, and biosensing applications. *Sensors*. v. 21, 6578, 2021.

Miralrio, A.; Vázquez, A. E. Plant extracts as green corrosion inhibitors for different metal surfaces and corrosive media: A review. *Processes*. v. 8, 942, 2020.

Mofijur, M.; Masjuki, H. H.; Kalam, M. A.; Atabani, A. E.; Shahabuddin, M.; Palash, S. M.; Hazrat, M. A. Effect of biodiesel from various feedstocks on combustion characteristics, engine durability and materials compatibility: A review. *Renewable and Sustainable Energy Reviews.* v. 28, 441–455, 2013.

Monieta, J.; Szmukata, M.; Adamczyk, F. The effect of natural deterioration on selected properties of rapeseed oil methyl esters. *Fuel.* v. 330, 125606, 2022.

Priyatharesini, P. I.; Kumar, K. P. V.; Kumari, S. S. Studies of the anticorrosive nature of green Ricinus seed extract with nem biodiesel in copper metal. *Biofuels.* v. 12, 559–568, 2021.

Ramalho, V. C.; Jorge, N. Antioxidantes utilizados em óleos, gorduras e alimentos gordurosos. *Química Nova.* v. 29, 755–760, 2006.

Salleh, S. Z.; Yusoff, A. H.; Zakaria, S. K.; Taib, M. A. A.; Seman, A. A.; Najmi, M. Plant extracts as green corrosion inhibitor for ferrous metal alloys: A review. *Journal of Cleaner Production.* v. 304, 127030, 2021.

Shehzad, A.; Ahmed, A.; Quazi, M. M.; Jamshaid, M.; Ashrafur, R. S. M.; Hassan, M. H.; Javed, H. M. A. Current research and development status of corrosion behavior of automotive materials in biofuels. *Energies.* v. 14, 1440, 2021.

Subedi, B. N.; Amgain, K.; Joshi, S.; Bhattarai, J. Green approach to corrosion inhibition effect of Vitex negundo leaf extract on aluminum and copper metals in biodiesel and its blend. *International Journal of Corrosion and Scale Inhibition.* v. 8, 744–759, 2019.

Thakur, A.; Assad, H.; Kaya, S.; Kumar, A. Chapter 17 - Plant Extracts as Environmentally Sustainable Corrosion Inhibitors II. *Eco-Friendly Corrosion Inhibitors.* Elsevier, 283–310, 2022. Elsevier, 283–310, 2022.

Verma, C. Chapter12 - Computational Methods of Corrosion Assessment. *Handbook of Science & Engineering of Green Corrosion Inhibitors.* Elsevier, 103–114, 2022a.

Verma, C. Chapter17 - Plant Extracts as Green Corrosion Inhibitors. *Handbook of Science & Engineering of Green Corrosion Inhibitors.* Elsevier, 173–192, 2022b.

Wei, H.; Heidarshenas, B.; Zhou, L.; Hussain, G.; Li, Q.; Ostrikov, K. Green inhibitors for steel corrosion in acidic environment: State of art. *Materials Today Sustainability.* v. 10, 100044, 2020.

Wolynec, S. Técnicas eletroquímicas em corrosão. [S.l: s.n.], 2003.

Wu, G.; Ge, J. C.; Choi, N. J. A comprehensive review of the application characteristics of biodiesel blends in diesel engines. *Applied Sciences.* v. 10, 8015, 2020.

Yeşilyurt, M. K.; Öner, İ. V.; Yilmaz, E. Ç. Biodiesel induced corrosion and degradation: Review. *Pamukkale Univ Muh Bilim Derg.* v. 25, 60–70, 2019.

Zakeri, A.; Bahmani, E.; Aghdam, A. S. R. Plant extracts as sustainable and green corrosion inhibitors for protection of ferrous metals in corrosive media: A mini review. *Corrosion Communications.* v. 5, 25–38, 2022.

Zhang, X.; Tan, B.; Li, W. Chapter16 - Plant Extracts as Environmentally Sustainable Corrosion Inhibitors I. *Eco-Friendly Corrosion Inhibitors.* Elsevier, 263–282, 2022.

IMPORTANT WEBSITES

https://www.intechopen.com/books/corrosion-inhibitors-principles-and-recent-applications/green-corrosion-inhibitors-past-present-and-future
https://www.intechopen.com/books/corrosion-inhibitors/green-corrosion-inhibitors
https://www.intechopen.com/books/plant-extracts/introductory-chapter-plant-extracts
https://www.sciencedirect.com/book/9780323854054/environmentally-sustainable-corrosion-inhibitors
https://www.sciencedirect.com/book/9780444627223/corrosion-engineering
https://www.sciencedirect.com/topics/engineering/biodiesel
https://www.sciencedirect.com/topics/biochemistry-genetics-and-molecular-biology/corrosion

https://www.sciencedirect.com/topics/medicine-and-dentistry/plant-extract
https://www.sciencedirect.com/topics/medicine-and-dentistry/phytochemical#:~:text=Phytoc
    hemicals%20are%20bioactive%20nonnutrient%20components,sesquiterpenes%2C%
    20capsaicinoids%2C%20and%20capsinoids

SUGGESTED READING MATERIALS

Guo, L.; Verma, C.; Zhang, D. Eco-Friendly Corrosion Inhibitors. *Principles, Designing and Applications*. Elsevier, 283–310, 2022.

Knothe, G.; Krahl, J.; Gerpen, J. V. *The Biodiesel Handbook*. Elsevier, 2010.

Verma, C. *Handbook of Science & Engineering of Green Corrosion Inhibitors. Modern Theory, Fundamentals & Practical Applications*. Elsevier, 2022.

17 Phytochemicals as Corrosion Inhibitors for Different Metals in Acidic Medium

Priya Vashishth, Himanshi Bairagi, Rashmi, Rajni Narang, and Bindu Mangla
J.C. Bose University of Science and Technology

Sudhish K. Shukla
Manav Rachna University

17.1 INTRODUCTION

Corrosion is generally perceived in terms of the degradation of a metal due to the consequences of a corrosive medium [1] and is attracting considerable interest in the destructive result of a chemical reaction between a metal or metal alloy and its environment. The Indian government spends around 3.5% lakh crores of the nation's GDP to overcome corrosion losses [2]. According to Brasunas Anton de and Uhlig, "Corrosion is the destructive attack on a metal by chemical or electrochemical reaction with its environment" and "Corrosion may be defined as an unintentional attack on a material through reaction with a surrounding medium." respectively. Henthrone Michael (2003) said, "The secret of effective engineering lies in controlling corrosion rather than preventing corrosion because it is impossible to eliminate it." According to Sastri, aggressive acid solutions are extensively used in industries for manufacturing processes and other applications like acid pickling, acid cleaning, acid de-scaling, and oil-well cleaning. Recent studies estimate that India and other countries are raising their funds for the demand for corrosion inhibitors.

Various factors affect corrosion, such as electrode potential, overvoltage, purity, the physical state of the metal, and the nature of the film formed on the metal. Temperature, concentration of oxygen, pH, and flow velocity of the medium are the essential factors that influence the corrosion rate of the metal [3]. Corrosion of metal is never preferred because it diminishes and reduces metal properties, increasing the risk of tragedies. It has become essential to prevent corrosion since it results in a considerable worldwide economic loss ($2.5 trillion) and produces toxic corrosion products that pose serious environmental threats. Corrosion causes annual losses of millions of dollars. The degradation of iron, aluminium, copper, and steel is a

DOI: 10.1201/9781003394631-17

significant cause of this loss. Even though many other metals may also corrode, the issue with iron and many other metals is that the oxide developed by oxidation does not firmly bind to the surface of the metal and flakes off rapidly, leading to "pitting." Large-scale pitting eventually weakens the metal's structural integrity and causes it to disintegrate. The importance of corrosion studies is threefold. The first section includes the economic consideration with the main aim of reducing the material losses resulting from the corrosion of pipes, tanks, metal components of machines, bridges, Passovers, etc. The second section includes improved safety of operating equipment, which may lead to drastic consequences because of corrosion. The third section comprises conservation, mainly of the metal sources; currently, the prime motive for corrosion research is provided by the economic factor.

Corrosion has affected the surfaces of various infrastructures like highways, bridges, buildings, chemical processing units, wastewater treatment, and virtually all metallic objects in our day-to-day life use. There are multiple types of corrosion in aqueous environments [4]. The reactivity of metal, presence of impurities, air, moisture, gases like sulphur dioxide and carbon dioxide, presence of electrolytes, metal's grain structure, composition during alloying, or the temperature during fabrication make a significant effect on the phenomenon of corrosion [5]. In 1824, Sir Humphrey Davy presented a series of papers to the Royal Society in London. He described how zinc and iron anodes could prevent copper sheathing corrosion on the wooden hulls of British naval vessels [6].

Because of the variety of corrosion mechanisms and material failures, corrosion types are classified according to their main characteristics. In the case of water-cooled systems with electrochemical deterioration. Some of the most common types of corrosion are shown in Figure 17.1.

Carbon steel and its alloys are the most valuable metals in various fields that suffer localized and uniform corrosion in different aggressive media, such as

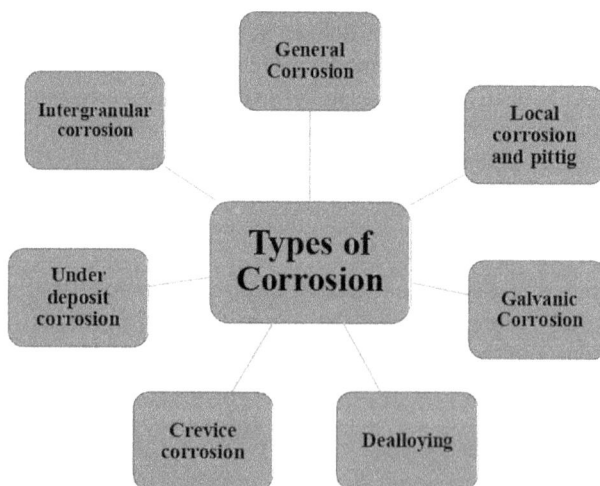

FIGURE 17.1 Various types of corrosion.

aqueous, acidic, basic, alkali, and alkaline solutions, containing chlorides with the passive oxide layer's destruction, forming pits. Mild steel (low carbon steel) is employed in conventional fields such as automobile, infrastructure, laboratories, pipeline, petroleum industries, and other related sectors in which corrosion resistance is required. The process of corrosion challenges emerges in tunnels owing to the abrasive nature of the substances moved by them. These substances are often utilized in a variety of processes, including petroleum-containing sulphur and water, high saltwater expansion, and the application of those tunnels for both cooling and heating purposes. However, all of the water that passes through these tunnels has exceptionally elevated concentrations of chloride as well as a significant quantity of sulphates and other metal ions. The application of corrosion-inhibiting substances via different sections of tunnels is critical for this purpose. This problem is causing a long list of destructions in our daily lives, beginning with the waste of precious materials, closures of plants, destruction of goods contaminating things, expensive repairs, decreasing effectiveness, and overpriced overdesigns. Acids are used frequently in testing facilities, enterprises, and other environments for activities like the purpose of pickling maintenance, and descaling. Corrosion is a chemical reaction that occurs on the surfaces of metals primarily a consequence of the acidic solutions' hostile nature. In industry activities involving acid cleaning up, acid pickling, acid descaling, and oil well acidizing, mineral acids, particularly hydrochloric acid, are frequently used [7]. Since these acids have such strong corrosive effects, the subject of corrosion avoidance is always both theoretically and practically interesting [8].

The use of corrosion-preventative techniques may avert deterioration to a certain level (Figure 17.2). Metals can be shielded against corrosion in several ways. Numerous corrosive evaluation techniques and preventive strategies have been developed in recent years and placed to use in an assortment of industry and laboratory

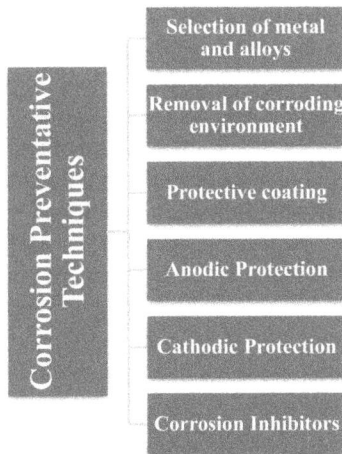

FIGURE 17.2 Different corrosion protection.

environments [9]. Compared to some recent reviews, articles, and prevention techniques, the present study offers up-to-date knowledge of the different types of protective coating on metals.

By avoiding dissolution of metals and acid intake inhibition of corrosion is the simplest feasible, affordable, and achievable method of controlling deterioration on the surfaces of metal in a number of media [10]. Based on the mechanism of action, inhibitors are of various types: (i) Anodic inhibitors reduce the anode area by acting on the anodic sites, polarizing the anodic reaction, and displacing the corrosion potential in the positive direction. Chromates, nitrates, tungstate, and molybdates are examples of anodic inhibitors. (ii) Cathodic inhibitors reduce the cathode area by acting on the cathodic sites, polarizing the cathode reaction, and displacing the corrosion potential in the negative direction. They reduce corrosion current and thereby retard the cathodic reaction and suppress the corrosion rate. Examples of cathodic inhibitors are phosphates, silicates, and borates. (iii) Substances that affect both cathodic and anodic reactions are called mixed inhibitors. Generally, organic compounds are also known as organic or adsorption-type inhibitors, which absorb on the metal surface and suppress metal dissolution and reduction reaction [11]. Towards this end, various corrosion inhibitors have been investigated for corrosion prevention. The effectiveness of the inhibition depends on the inhibitor molecule's capacity to interact with the material through physical or chemical adsorption. The stability of the adsorbed layer formed serves as a barrier to keep the corroding medium away from the metal surface. Physisorption occurs when inhibitor molecules contact electrostatically with metallic surfaces, whereas chemisorption occurs when electron pairs interact as donors and acceptors. Finally, film creation creates a physical barrier between the metal surface and the corrosive medium, shielding it from corrosive assaults [12]. Applying various methods, inhibitors can be deposited as a protective film or as a solution. Substances referred to as corrosion-inhibiting substances serve to minimize the negative effect of corrosive conditions on materials. They interact with the outside surface of the metal or with surrounding gases to hinder the electrochemical processes that lead to dissolution.

Organic or inorganic chemicals may be utilized as chemical compounds that hinder reactions. Commonly used as inhibitors to slow down the corrosion rate of metals in acidic media are compounds with polar functional groups with heteroatoms, atom having a lone pair of electrons, and a moiety with aromatic ring and conjugated multiple bonds that could associate with vacant d-orbital of metals [11–18]. Both the production of these chemicals and their usage may illustrate the risk involved. Concerns about corrosion inhibitor's effects on human health and surroundings have existed for a long time. It is vital to identify inexpensive, accessible, and environmentally friendly inhibitors for corrosion prevention is therefore crucial. For the reduction of the corrosion process, weight loss study, electrochemical study, characterization, and surface study are the most popular techniques, by which the analysis of corrosion rate, inhibition efficiency, and other such parameters can be determined [19].

17.2   CORROSION INHIBITION ANALYSIS

The weight loss methodology is adopted to determine the inhibition efficiency and corrosion rate since it is the most appropriate and effective method. Before being submerged in a corrosive solution, a sample of metal or alloy is weighed. It is then removed from the corrosive medium after a specified duration of time. After cleaning the metal specimen of all corrosion agents, it is weighed again [20]. The inhibition efficiency (*IE%*), surface coverage (*θ*), and corrosion rate (*C_r*) were used to quantify the inhibitory performance using the following formulas [21]:

$$IE\% = \frac{W_o - W_i}{W_o} \times 100$$

$$\theta = \frac{W_o - W_i}{W_o}$$

$$C_r = \frac{534 \times W}{A \times t \times d}$$

where *IE* is the inhibition efficiency, C_r is the corrosion rate, *θ* is the surface coverage, W_o is the weight loss of metal specimen without inhibitor, W_i is the weight loss of metal specimen with inhibitor, *W* indicates the weight loss in mg, *A* is the surface area of the coupon (cm²), *t* is the time of immersion (hours), and *d* is the density of the specimen (g cm⁻³).

In order to evaluate the efficacy of a corrosion inhibitor, it is helpful to apply a variety of computational physical-chemical applications, such as the kinetic, thermodynamic, adsorption, and computational chemistry methods. The assessment of the kinetic and thermodynamic variables of the inhibition process in corrosion studies depends heavily on temperature. As temperature rises, all electrochemical processes accelerate, and thus, adsorption isotherms and kinetic parameters are affected. Kinetic and thermodynamic parameters were evaluated by applying the Arrhenius equation given by Svante Arrhenius and the transition state equation given by Henry Eyring. These equations have the following expressions:

$$\log C_r = \frac{-E_a}{2.303 \ RT} + \log A$$

$$C_r = \frac{RT}{Nh} \exp\left(\frac{\Delta S}{R}\right) \exp\left(\frac{-\Delta H}{RT}\right)$$

where C_r represents the corrosion rate of low-carbon steel; E_a and *A* denote the activation energy and pre-exponential factor, respectively; *R* and *T* stand for molar gas constant and thermodynamic temperature, respectively; $Nh = 39.90 \times 10^{-11}$ J sec·mol⁻¹ (product of Avogadro's number and Planck's constant).

The nature and charge, electronic properties of the metal surface, test temperature of the reaction mixture, solvent molecule adsorption, and other ionic species are the primary determinants of the adsorption process and its parameters. Physical and chemical adsorption of the inhibitor molecules on the surfaces of metal can be distinguished. A corrosion inhibitor molecule interacts electrostatically with a charged metal to establish the physical adsorption, whereas a corrosion inhibitor molecule provides a lone pair of heteroatoms and forms a co-ordination bond with the metal's empty orbitals to produce chemical adsorption.

To extensively examine the adsorption mechanism of tomato stem extract at the mild steel/solution interface, a number of adsorption isotherms were used. The isotherms Langmuir, Frumkin, Temkin, Freundlich, Flory-Huggins, Hill deBor, Parsons, and El-Awady et al.'s kinetic/thermodynamics model are the most frequently used isotherms [22]. They have the following equation expressions [23]:

- **Langmuir isotherm equation:-** $\dfrac{C}{\theta} = \dfrac{1}{K_{ads}} + C$

- **Temkin isotherm equation:-** $e^{-2\alpha\theta} = KC$

- **El-Awady isotherm equation:-** $\ln\dfrac{\theta}{1-\theta} = y\ln C + \ln K'$

- **Flory-Huggins isotherm equation:-** $\ln\dfrac{\theta}{C} = x\ln(1-\theta) + \ln(K_{ads})$

- **Frumkin isotherm equation:-** $\ln\dfrac{\theta}{(1-\theta)C} = \ln K + 2\alpha\theta$

The linear polarization resistance (LPR) method is the most used electrochemical approach since it is a rapid, non-intrusive, and accessible equipment used to test in the field [24]. The basic idea underlying LPR is to disrupt the corrosion equilibrium on a metal specimen's surface by applying a modest perturbative direct current (DC) electrical pulse [25]. A standard half-cell is used to measure how the equilibrium reacts to this disruption. The slope of a potential-current density curve at the free-corroding potential, or the polarization resistance (R_p) of a material, is measured as E/i. The Stern-Geary approximation, which can be used to get the anodic (β_a) and cathodic (β_c) Tafel slopes empirically from actual polarization plots, can be used to connect the polarization resistance to the corrosion current (i_{corr}). Electrochemical impedance spectroscopy (EIS) is a highly effective electrochemical technique with numerous applications in corrosion research. This method yields useful information regarding the corrosion protection provided by an inhibitor. In this technique, an alternating current (AC) voltage (in the case of potentiostatic EIS) or current (in the case of galvanostatic EIS) is applied to the system under research to obtain a response in the form of an alternating current (voltage) or voltage (current) as a function of frequency. This procedure can be carried out in a two- or three-electrode setup using a potentiostat-galvanostat and a frequency response analyser (FRA) [26]. An AC voltage with tiny variations ranging from 5 to 10 mV is commonly employed in the system over a frequency range ranging from 100 kHz to 10 mHz. The electrochemical cell comprising the metal sample, adsorbed inhibitors, and the electrolyte medium is reflected by an equivalent circuit that shows provisions about the solution resistance R_s, charge transfer resistance R_{ct}, and the double-layer capacitance C_{dl} based on the shape of the Nyquist plot produced by the experiment [27]. A high R_{ct} value and decreasing C_{dl} values as inhibitor concentrations increase imply improved corrosion prevention [28].

The corrosion potential (E_{corr}) will be quantifiable on the metal surface at equilibrium; since oxidation of metal and reduction of metal have equal current densities (I_{corr}) at E_{corr}, there is no directly measurable current observed. However, potentiodynamic polarization, the limiting form of the Butler-Volmer equation, Tafel plots, and other methods can be used to deduce I_{corr} and subsequently the rate of corrosion per unit area. The Butler-Volmer equation's limiting form is shown below:

$$I = I_o \left\{ e^{(1-\alpha)f\eta} - e^{-\alpha f\eta} \right\}$$

where $e^{(1-\alpha)f\eta}$ = anodic component
   $e^{-\alpha f\eta}$ = cathodic component

$$f = \frac{nF}{RT}$$

$$\eta = E_{applied} - E_{corr}$$

$$\alpha = \text{transfer coefficient}$$

If the overpotential (η) is positive and large and corresponds to the working electrode being the anode during electrolysis, the cathodic component is much smaller than the anodic component and may be neglected, and the equation for anodic process can be written as:

$$\ln I = \ln I_o + (1-\alpha)f\eta$$

And if the overpotential (η) is negative and large and corresponds to the working electrode being the cathode during electrolysis, the anodic component is much smaller than the cathodic component and may be neglected and the equation for cathodic process can be written as:

$$\ln I = \ln I_o + \alpha f\eta$$

Electrochemical-based corrosion monitoring tests offer useful information regarding the rate of corrosion and mechanisms of corrosion protection since corrosion is a phenomenon involving electrochemistry [29]. The fundamental of polarization techniques is altering the current or potential on a sample under study and logging the resulting potential or current change. Either a DC or an AC source can be used to enable this [30]. Several significant and popular methods including Tafel extrapolation, electrochemical impedance, and LPR have been discussed. The anodic and cathodic reactions in the electrochemical cell are depicted by a Tafel curve, which is a current-potential diagram. In this procedure, the working electrode's potential was adjusted over a range at a predetermined pace, and the response in current that resulted was recorded. A curving line is used to indicate the total current created by the simultaneous anodic and cathodic reactions. A point that represents an estimate of the corrosion current (i_{corr}) and the corrosion potential is created by intersecting the linear parts of the logarithmic Tafel plot (E_{corr}). The inhibitory performance ($IE\%$) of the inhibitor was calculated using current densities (I_{corr}) by the formula:

$$IE\% = \frac{\left(I^{\circ}_{corr} - I'_{corr} \right)}{I^{\circ}_{corr}} \times 100$$

where I°_{corr} is the corrosion current density in blank solution and I'_{corr} is the corrosion current density at various concentrations of inhibitor.

Atomic force microscopy (AFM), scanning electron microscopy (SEM), and transmission electron microscopy are used to study the morphology of the metal's outer surface before and after inhibition. FTIR and FTIR-ATR detect the type of functional group and bonding between metal surface and inhibitor. Kinetic and thermodynamic parameters evaluate the feasibility of corrosion. Quantum chemical calculation method and density functional theory (DFT) with high accuracy are used for the calculation of the molecular structure parameters of inhibitors including electronegativity, frontier orbitals (HOMO and LUMO), chemical hardness, chemical softness, and X-ray fluorescence (XRF), and they are helpful in finding the composition analysis of steel. Energy dispersive X-ray analysis (EDAX) is employed to identify the elemental composition of the material. UV-Vis spectroscopy helps to determine the type of transition in inhibitor molecule and to ensure the formation of protective layer.

17.3   CORROSION INHIBITOR

Corrosion inhibition is the most practical, economical, and convenient technique to control corrosion on metal surfaces in different mediums by preventing metal dissolution and acid consumption [32]. These inhibitors work on the mechanism of molecular adsorption, i.e., the compound inhibits corrosion by controlling both anodic and cathodic reactions. The protonated species adsorb on the cathodic sites of the surface and decrease the evolution of hydrogen gas. Hence the rate of corrosion also decreases. Inhibitors can be applied as a solution or as a protective coating using different techniques. Corrosion inhibitors are very widely used to reduce corrosion processes in the environment and are substances that, when added in small amounts to aggressive environments, reduce the rate of corrosion. The adsorption of an inhibitor on a metal surface depends on different factors such as the nature of the metal, surface charge of the metal, adsorption mode and its chemical structure, and electrolyte solution [33].

Most organic molecules tend to adsorb to metal surfaces, but in general, the most desirable properties when seeking corrosion inhibition relate to the interaction of a double/triple bond or lone pair of electrons (N, S, P, or O) [34]. Most organic molecules tend to adsorb to metal surfaces, but in general, the most desirable properties when seeking corrosion inhibition relate to the interaction of a double/triple bond or lone pair of electrons (N, S, P, or O) with an empty metal d orbital or a combination of some of the previous mechanisms [31].

Corrosion inhibitors are classified based on how they prevent corrosion: cathodic, anodic, or mixed. Cathodic corrosion inhibitors lessen corrosion potential by blocking cathodic processes including oxygen reduction and hydrogen evolution [35]. Anodic corrosion inhibitors enhance corrosion potential by interacting with reactive areas on the metal surface and passivating them. Inhibitors that are neither cathodic nor anodic are classified as mixed-type inhibitors [36]. By physisorption, chemisorption, or film formation, these inhibitors can preserve the metal surface. Physisorption is driven by the electrostatic interaction of inhibitor molecules with the metal surface. Chemisorption, on the contrary, is caused by donor–acceptor interactions between vacant metal surface orbitals and free electron pairs in the inhibitor.

17.3.1 PHYTOCHEMICALS AS CORROSION INHIBITOR

One of the areas of green chemistry, where an eco-friendly chemical frequently applied to natural products, offers innovations in terms of sinking environmental collision, is related to metal surface protection. With the advent of green chemistry, the concept of green corrosion prevention is getting more and more attention. Plants are an excellent reservoir of endogenous products in the form of extracts or pure sustainably sourced compounds (plant molecules) that are environmentally acceptable, easily accessible, and eco-friendly in nature. The primary component that enables plants and plant wastes to function as corrosion inhibitor are the classes of phytochemicals that block the active site on metal surfaces by creating a protective film on it, causing inhibition effects due to inhibitory molecule adsorption above the metal surface. Plant extracts are made up of a very complex combination of plant components from many chemical families such as anthocyanins [37], flavonoids [38], tannins [39], terpenoids [20], alkaloids, carotenoids [40], saponins [41], etc. (Figure 17.3). These plant molecules are rich in π electrons (benzene ring, double/triple bond) and contain electrically negative functional groups in their chemical structure and heteroatom such as oxygen, sulphur, and nitrogen [42].

Plants have been identified as a source of naturally occurring chemicals, some with complex molecular structures and diverse physical, chemical, and biological characteristics. Most plant-derived chemicals are used in conventional applications such as medicines and biofuels. Furthermore, the usage of naturally occurring substances is appealing because of their low cost, vast availability, and, most significantly, environmental acceptability. Because of these benefits, extracts of various

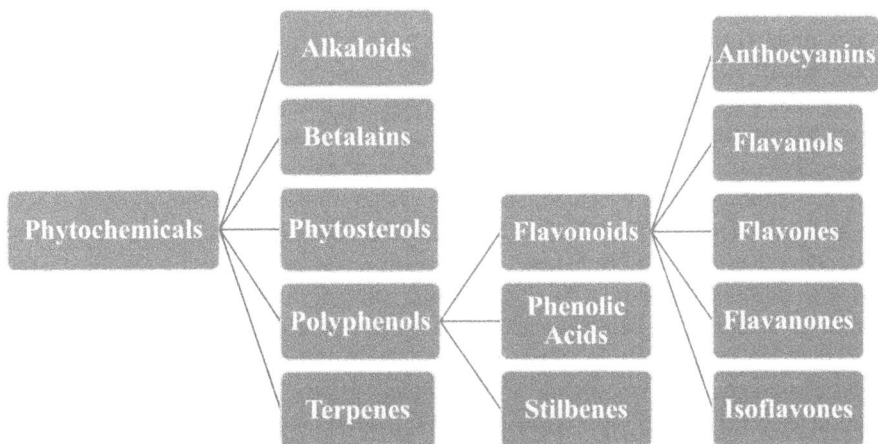

FIGURE 17.3 Various categories of phytochemicals.

common plants and plant products have been tested as corrosion inhibitors for metals and alloys in a variety of environments.

The use of phytochemicals as corrosion inhibitors may be dated back to the 1960s, when tannins and their derivatives were employed to preserve steel, iron, and other equipment against corrosion. El Hosary et al. reported on the use of common plant extracts as corrosion inhibitors in 1972. The food industry produces a lot of waste and waste products, which creates disposal and sustainability issues. These agricultural and industrial wastes may be used in biological processes that produce a broad assortment of alternative substrates, helping to address problems with environmental contamination brought on by their disposal. By-products of the food manufacturing process, however, may be recycled and reprocessed to provide a net gain. The valorization of by-products opens up a plethora of opportunities for additional revenue generation by providing a solution to the waste disposal issue facing food processing industries. For example, tomato processing waste, which includes the cull, damaged fruits, and pomace, among other things, might be used to recover a range of chemicals such as colours, oil, protein, polysaccharides, and animal feed. Not only in industries but also in fields large amount of plant residues, such as the stem, are being thrown out after harvesting the fruits carelessly without knowing the fact of their application in generating useful biomass for other specific uses such as developing corrosion, rather than creating another issue of dumping out them as waste. Tomato (*Lycopersicum esculentum*) is one of the most popular vegetables in the world, whether fresh or in processed forms such as canned tomatoes, sauce, juice, ketchup, and soup. A significant quantity of waste is created during tomato processing, consisting of peel, seed, and a minor bit of pulp. These by-products are often disposed of in landfills and only partially repurposed by composting or drying for animal feed. Nonetheless, contemporary eco-friendly technologies provide more efficient ways for recycling these wastes in order to utilize them as a sustainable supply of value-added compounds such as carotenoids, ascorbic acid, tocopherols, and polyphenols. Plant extracts are wonder extracts that are rich in phytochemicals that promote health and aid in the prevention of serious chronic degenerative diseases, and these compounds are high in phenolic compounds (phenolic acids and flavonoids), carotenoids (lycopene and carotenoids), vitamins (ascorbic acid and vitamin A), and glycoalkaloids (tomatine). Their bioactive components have antioxidant, anti-mutagenic, anti-proliferative, anti-inflammatory, and anti-atherogenic properties [43]. Choosing the antioxidant properties, it can be easily extracted as a corrosion inhibitor. Phytochemicals like lycopene, pectin, and flavonoids act as super antioxidant, which help in the inhibition of the corrosion rate of various metals and other substances like tin and mild steel. Phenolic acids (hydroxybenzoic and hydroxycinnamic acids) and flavanols (kaempferol 3-*O*-glucoside, (+)-catechin and quercetin) are obtained from the stem and pomace extract of plants, which perform tremendous job in protecting the metal surface from being corroded. Among the various categories of such inhibitors, here we focused on the phytochemical obtained from various parts of plants. Tobacco plant extracts from stems, twigs, and leaves have been shown to significantly prevent aluminium and steel corrosion in both saline solutions and powerful pickling acids. In 2 M HCl solutions, leaf extracts were proven to be efficient corrosion inhibitors for mild steel. The scientists observed a maximal inhibitory efficacy of 96%

with a tobacco concentration of only 0.01% (100 ppm). Tobacco extracts have been shown to contain large amounts of chemical substances such as terpenes, alcohols, polyphenols, carboxylic acids, nitrogen-containing compounds, and alkaloids, which have been shown to have electrochemical action such as corrosion inhibition [44]. Leguminous seeds, which are high in amino acids, have also been researched for their ability to suppress corrosion. In chloride ion-containing solutions, *Opuntia* extract protects metal against pitting corrosion [45]. Pomegranate alkaloids are effective at low temperatures. Because of the presence of the carbonyl group in *Eugenia caryophyllata*, acetyleugenol is more active at the surface than eugenol [46]. Rajesh Haldhar et al. studied the corrosion inhibition effects of the aerial parts of *Solanum surattense* (*S. surattense*) which showed 93% efficiency at an inhibitor concentration of 500 mg L^{-1} [47]. A few publications have reported on the use of pectin present in tomato derived from food industry waste as a corrosion inhibitor. Pectin from citrus peel was utilized as a corrosion inhibitor for mild steel in HCl solutions by Fiori-Bimbi, Alvarez, Vaca, and Gervasi, while commercially available pectin was used as a corrosion inhibitor for aluminium in hydrochloric acid by Fares, Maayta, and AlQudah. Mild steel, according to Zaher and Bouraoui, is extremely vulnerable to acid attack. Nitric acid (HNO$_3$), sulphuric acid (H$_2$SO$_4$) and hydrochloric acid are the most common acids used for acid cleaning and acid pickling (HCl). Rodriguez Torres et al. and El-Haddad et al. verified this, adding that the most often utilized acids are HCl and H$_2$SO$_4$. Furthermore, Umoren et al. proposed that HCl is preferable than H$_2$SO$_4$. They speculated that this discovery may be due to a variety of factors, including HCl's shorter pickling time and lower working temperature. In exchange, an excellent metal surface quality may be attained while utilizing less energy. As a result, industry participants may operate on a lesser budget. El Haddad et al. investigated mild steel corrosion in an HCl atmosphere. The authors discovered that an acidic atmosphere can hasten the corrosion of mild steel. As a result, they found that choosing the right corrosion prevention strategy in an acidic environment is critical. This was supported by Ogunleye et al., who said that knowing corrosion behaviour is critical for determining corrosion rate. To fight corrosion in an acidic environment, corrosion inhibitors have been widely used. In his original analysis, Javed predicted that the global corrosion inhibitor industry will be worth $8.7 billion in 2021.

Victoria Vorobyova et al. studied the anti-corrosive properties of tomato pomace extract on steel in sodium chloride (NaCl) solution. Gravimetric measurements, Tafel, and AC impedance (EIS) techniques were used. The obtained results revealed that tomato pomace extract was an efficient inhibitor for steel in sodium chloride media exhibiting a maximum inhibition efficiency of 98% at 500 ppm in 48 hours of immersion time [48]. Rajesh Haldhar et al. studied the corrosion inhibition effects of the aerial parts of *S. surattense* which showed 93% efficiency at 500 mg L^{-1} concentration of inhibitor [49]. Harish Kumar et al. studied the corrosion inhibition of mild steel in 0.5 M HCl by *Morus nigra* (mulberry leaves). The techniques employed for the study were FTIR, UV-visible, NMR, weight loss, EIS, potentiodynamic polarization (PDP) and DFT. The obtained results revealed that the *M. nigra* proved to be an effective inhibitor with an inhibition efficiency of 91.67% and follows both physical and chemical modes of adsorption [50]. *Allium sativum* L. acted as a corrosion inhibitor on mild steel studied by Harish Kumar et al. They did the gravimetric, electrochemical, and

DFT analyses to check the inhibition performance in 5 M HCl. The inhibition efficiency was found to be 94.76% and followed the physio-chemisorption mechanism [51]. Numerous investigations have been executed to examine the inhibition effect of phytochemical compounds obtained from plant extracts.

Sanni et al. studied the corrosion inhibition on stainless steel UNS N08904 (SS) corrosion in 0.5 M H_2SO_4 medium at 30°C was examined via gravimetric, potentiodynamic polarization, and SEM techniques. The obtained result shows that CSP is an excellent inhibitor for SS against corrosion, CSP adsorption on the surface of SS obeys Langmuir and Freundlich adsorption isotherms. The inhibition efficiency of 98% was obtained using 10 g CSP after 140 hours of exposure. The potentiodynamic polarization result describes CSP as a mixed-type inhibitor owing to changes in the lateral repulsion effect between the inhibitor molecules [52].

The *Eucalyptus* leaves extract (ELE) inhibition's impact towards mild steel corrosion in the HCl solution was examined by combined experimental and computational studies. The degree of inhibition was investigated by EIS and polarization test. The EIS analysis results showed that the increase in ELE concentration led to a significant increment of charge transfer resistance. An inhibition efficiency of ~88% was obtained using 800 ppm ELE after 5 hours of exposure. Polarization test results indicated the mixed inhibition effects of ELE with slight cathodic prevalence. The i_{corr} values for the uninhibited and inhibited (800 ppm ELE) samples were 0.93 and 0.25 $\mu A\ cm^{-2}$, respectively. The ELE molecules' adsorption on the surface of mild steel followed a Langmuir isotherm. Furthermore, the molecular simulation results evidenced the adsorption of ELE compounds on the iron surface [53].

The inhibition effect of *Atropa belladonna* extract (ABE) on the corrosion of carbon steel in 1 M HCl solution was investigated by potentiodynamic polarization, EIS, and electrochemical frequency modulation (EFM) techniques. The adsorption process obeyed a Langmuir adsorption isotherm. ABE acts as a mixed-type but mainly anodic inhibitor in 1 M HCl. The calculated adsorption thermodynamic parameters indicated that the adsorption was a spontaneous, exothermic process accompanied by an increase in entropy. The maximum inhibition approached 96.6% in the presence of 500 ppm ABE using Tafel polarization technique. Quantum chemical parameters were also calculated, which provided a reasonable theoretical explanation for the adsorption and inhibition behaviour of ABE on C-steel. ABE constituents have been simulated as adsorbate on Fe (1 1 0) substrate, and the adsorption energy has been identified on the iron surface [54].

The anticorrosion properties of natural extracts, as orange peel extract, cannot be explained only by the antioxidant properties of the polyphenol molecules, as flavonoids, contained in these extracts. As shown in this study, the presence of ene-diol or catechol functions or other derived functions are responsible for antioxidant activity of molecules and could partly explain the inhibition of cathodic reaction. The same kinds of functions could also be responsible for the anodic inhibition because of their chelating properties regarding metallic cations. However, the significant inhibition efficiency obtained with the extract is also due to the precipitation of a covering film on steel surface, which could be formed by other macromolecules acting in synergy with the antioxidant molecules [55].

Almond seeds (*Prunus dulcis*) extract was obtained with the aid of Soxhlet apparatus. The corrosion inhibition experiment was performed by setting up reactors containing mild steel coupon with variable concentrations of plant extract and 200 mL of 1.5 M HCl solution. The study revealed that the extract was an efficient inhibitor and was most effective as the concentration increased from 0.81% at 0.01 g mL^{-1} to 69.95% at 0.15 g mL^{-1}, respectively. Adsorption study on mild steel surface showed that the experimental data fitted better into the Temkin isotherm with regression R_2 closer to unity. Arrhenius constant and activation energy estimated at temperatures 308 to 328 K revealed that activation energy E_a increased with increasing inhibitor concentration from 5,348.23 J mol^{-1} at 0.01 g mL^{-1} to 6,151.44 J mol^{-1} at 0.05 g mL^{-1}. The outcome of the study revealed that mild steel is susceptible to corrosion, which is capable of destroying the material and increasing inhibitor concentration, and temperature has a significant influence on the corrosion [56].

The anticorrosive potential of *Ficus hispida* leaf extract (FHLE) as a corrosion inhibitor in 1 M HCl was investigated using weight loss measurement as well as potentiodynamic polarization and EIS techniques. Stigmasterol as the major constituent of *F. hispida* was confirmed by gas chromatography–mass spectrometry (GC–MS). Inhibition efficiency of 90% was achieved with 250 ppm of FHLE at 308 K. Temperature studies revealed an increase in inhibition efficiency with a decrease in temperature and activation energies increased in the presence of the extract. Cathodic and anodic polarization curves revealed that FHLE acts as mixed-type inhibitor, but the cathodic effect was more pronounced. Impedance diagrams showed that increasing FHLE concentration increased charge transfer resistance and decreased double-layer capacitance. The adsorption of FHLE on mild steel surface obeyed Langmuir adsorption isotherm. The morphology of the surface was examined by SEM and the surface composition was evaluated using energy-dispersive X-ray (EDX) spectroscopy to verify the presence of inhibitor on the mild steel surface. The adsorbed film on the mild steel surface containing the FHLE inhibitor was also characterized by diffuse reflectance Fourier transform infrared spectroscopy (DRFT-IR) and X-ray diffraction (XRD) studies [57].

The inhibition performance of natural inhibitor, i.e., black pepper extract on mild steel surfaces in hydrochloric acid (HCl), was inspected, and for this purpose, weight loss measurements, EIS measurements, and potentiodynamic polarization study at different concentrations and at different temperatures were recruited. The best inhibition efficiency was shown by black pepper extract, i.e., 97% in 1 M HCl. In the prolongation of our work, entropy (ΔS), enthalpy (ΔH), Gibbs free energy (ΔG), activation energy (E_a), and pre-exponential factor (A) of the reaction were calculated, and it is concluded that the inhibition process follows physical adsorption on the surface and the adsorption process follows Langmuir and Tempkin isotherms also. For surface analysis, AFM was done [58].

The corrosion inhibition of X70 steel in 1 M HCl by *Ginkgo* leaf extract (GLE) was investigated by conducting electrochemical measurements. The inhibition efficiency exceeded 90% in the presence of 200 mg L^{-1} GLE at all of the tested temperatures. The excellent inhibition capacity, which was attributed to the formation of inhibitor–adsorption films on the surface of the X70 steel, was confirmed by field emission SEM and AFM. The adsorption of GLE on steel surface followed the Langmuir adsorption model. The potential of zero charge measurement and quantum chemical calculation were adopted to elucidate the inhibition mechanism [59].

Gum Arabic (GA) is a plant exudate, consisting of glycoproteins (proteins with carbohydrate co-factor or prosthetic group) and polysaccharides (mainly consisting of galactose and arabinose). Because of its polymeric nature and tendency to dissolve in water, GA is widely used as anticorrosive material, especially in aqueous electrolytes. GA contains various electron-rich polar sites through which they easily get adsorbed on the metallic surface and behave as effective anticorrosive materials. Because of its natural and biological origin, GA is regarded as one of the environmentally sustainable and edible alternative to traditional toxic corrosion inhibitors. The present review piece of writing aims to illustrate the assortment of studies on GA as a corrosion inhibitor. Limitations of traditional organic corrosion inhibitors and advantages of using GA as an environmentally sustainable alternative have also been described along with the mechanism of corrosion inhibition [60].

The inhibitive action of natural honey on the corrosion of C-steel, which is used in the manufacture of petroleum pipelines, in high saline water was evaluated. The inhibition efficiency was calculated using weight loss measurements and potentiostatic polarization technique. It was found that natural honey exhibited a very good performance as an inhibitor for steel corrosion in high saline water. The inhibition efficiency increases with an increase in natural honey concentration. After some time, the inhibition efficiency decreased due to the growth of fungi in the medium. The adsorption of natural honey on the C-steel was found to follow the Langmuir adsorption isotherm [61].

The corrosion inhibition effect of pectin (a biopolymer) for X60 pipeline steel in the HCl medium was investigated using weight loss, electrochemical, water contact angle measurements, and SEM techniques. The results obtained show that pectin acts as a good corrosion inhibitor for X60 steel. Inhibition efficiency increased with an increase in pectin concentration and temperature. Potentiodynamic polarization results reveal that pectin could be classified as a mixed-type corrosion inhibitor with predominant control of the cathodic reaction. The effective corrosion inhibition potential of pectin could be related to the adsorption of pectin molecules at the metal/solution interface which is found to accord with the Langmuir adsorption isotherm model and a protective film formation. Quantum chemical calculations provided insights into the active sites and reactivity parameters governing pectin activity as a good corrosion inhibitor for X60 steel [62].

The corrosion inhibition of *Thevetia peruviana* (Kaner) flower extract (TPFE) is evaluated for mild steel in 1 M HCl solution using electrochemical, surface, and computational demonstrations. Results showed that TPFE acted as a potential corrosion inhibitor and showed the highest inhibition efficiency of 91.24% at 200 mg L^{-1} concentration. Electrochemical studies suggest that TPFE acted as a mixed- and interface-type corrosion inhibitor. The TPFE inhibits metallic corrosion through an adsorption mechanism that follows the Temkin adsorption isotherm model. Adsorption mechanism of corrosion inhibition was further supported using AFM, SEM-EDX, FT-IR, and UV-visible surface studies. Furthermore, the anticorrosive mechanism of major phytochemicals of the TPFE was studied using computational techniques: DFT and molecular dynamics (MD) simulations. Results showed that

TPFE interacts with donor-acceptor interactions, and its phytochemicals acquire the flat or horizontal orientations over the metallic surface [63].

In this study, leaves extract of *Arbutus unedo* L. plant was used as a green corrosion inhibitor of mild steel in hydrochloric acid. The aqueous extract is rich in polyphenols and in particular flavonoid compounds as determined by the quantitative aluminium chloride colorimetric method. The n-butanol extract shows the highest concentration of total phenol and total flavonoid contents with 219.46 GAE mg g^{-1} extract and 174.66 mg QE g^{-1} of dry extract, respectively. EIS and potentiodynamic polarization were used as experimental measurements to confirm the anti-corrosion performances of the extract of plant leaves at different temperatures. The addition of extract plant even at low concentration increases significantly the inhibition efficiency with a maximum of 91.72% at 298 K for 0.5 g L^{-1} of leaves extract. Surface analysis was performed to attest the presence of the protected film on the surface of mild steel in HCl solution. The calculated ΔG_{ads}^0 value (-24.441 kJ mol^{-1}) confirmed a spontaneous mixed chemical and physical adsorption of leaves extract compounds on the metal surface. The pKa analysis attested that the two leaves extract compounds, quercetin and catechin, exist in the neutral form in an acidic environment. Moreover, DFT and molecular dynamic simulation were also conducted to support the high inhibition efficiency results obtained by the experimental tests and to propose the adsorption mechanism [64].

The viability of using the leaves of *Crotalaria pallida*, a well-known medicinal plant, as an efficient corrosion inhibitor for mild steel in HCl media has been studied, and the results are summarized in this chapter. Methods including weight loss monitoring, electrochemical impedance analysis, and polarization techniques have been used, and the mechanism of action was further investigated using Biovia material studio software. The theoretical background of the inhibitor and the probable methods of interaction were carried out using Gaussian 09 packages. Ethanol extracts are rich in fatty acids (which are absent in the water extracts), and they make them more anticorrosive. The presence of fatty acids was confirmed using spectral analysis. Linolenic acid, oleic acid, and linoleic acids are the major fatty acids in the alcohol extract. The 4 (V/V %) alcohol extract has 95% efficiency, while the same volume percentage of water extract has only 87% efficiency. The efficiency decreases with an increase in acid concentration and temperature, suggesting the physisorption of plant constituents on the metal surface. The AFM and XPS results support the protection of mild steel by the adsorption of constituent phytochemicals in the leaves extract. Adsorption of these phytochemicals follows Langmuir unimolecular model. The polarization technique confirms the mixed-type behaviour of the active species of the extract. Inhibition efficiency is higher for linoleic acid than the others, which is confirmed by both DFT and material studio calculations [65].

To elucidate the mechanism of adsorption of inhibitors, the adsorption mechanisms of the preventing species must be pre-determined whether it is ionic or molecular in nature. The dominant mode of adsorption would be determined by some factors including the composition of the extract, type of acid anion, etc. The chemical compounds present in the inhibitor contain a conjugated double bond along with an oxygen heteroatom, which satisfies the general properties of a corrosion inhibitor. Parameters obtained from thermodynamic and kinetic studies justify the physical adsorption of the extract on the surface of mild steel in acid solutions. The chemical compounds are possibly protonated in acid solutions, and protonated species are expected to be improperly adsorbed since corroding mild steel samples have a positive surface charge in both acid solutions. The first step would be to adsorb the acid anions of sulphate and chloride, and these anions create an excess negative charge to the solution, which stimulates the cations to absorb more entirely. As a consequence of van der Waals' interaction, the protonated inhibitor may then accumulate on the negatively charged surface of the metal and form coordinate bonds by partial transfer of electrons from the O-heteroatom and multiple bonds to the unfilled d-orbitals of Fe. The compound present in the inhibitor might be combined with metal ions formed on a metal surface due to an unshared pair of electrons of O-atom to create the metal inhibitor complexes. Through the action of van der Waals' forces, these complexes adhere to the mild steel surface and create a protective layer that prevents corrosion on mild steel.

17.4 CONCLUSION

Metal corrosion in an acidic media is inescapable, and it has a significant impact on the global economy. Metal corrosion control is becoming increasingly significant from a technical, aesthetic, environmental, and economic perspective. This review study is primarily concerned with the corrosion inhibition of mild steel, as many important items are constructed of mild steel, which disintegrates when exposed to powerful acids. It can be deduced from the preceding observations that plant extracts are effective corrosion inhibitors in acidic media for preventing metal dissolution. Numerous synthetic, organic, and inorganic corrosion inhibitors are available, but green corrosion inhibitors should be used due to climate concerns. Because of the progress of the concept of green chemistry and economic development, the adoption of environmentally friendly alternatives is gaining importance in the field of engineering and research. Because of their eco-friendliness, natural plant extracts efficiently replace widely used corrosion inhibitors. The extract is made from several plant parts, including roots, bark, leaves, and stems. The extract of the plant can be made in either water or any organic solvent; however, aqueous extract is recommended. The adsorption of a substantial fraction of the plant components on metallic surfaces followed the Langmuir adsorption isotherm; nevertheless, Temkin and Freundlich's adsorption was also reported in a number of observations. Weight loss measurements, as well as several electrochemical techniques such as potentiodynamic polarization, EIS, quantum chemical analysis, gravimetric analysis, and the surface analysis approach, are the most useful strategies in investigating corrosion inhibition.

REFERENCES

[1] Z. Tasić, M. Mihajlović, A. T. Simonović, M. B. Radovanović, and M. M. Antonijević, "Ibuprofen as a corrosion inhibitor for copper in synthetic acid rain solution," *Scientific Reports*, vol. 9, p. 14710, 2019. [Online]. Available: https://www.nature.com/articles/s41598-019-51299-2 (Accessed: Dec. 08, 2022).
[2] I. A. Adejoro, F. K. Ojo, and S. K. Obafemi, "Corrosion inhibition potentials of ampicillin for mild steel in hydrochloric acid solution," *Journal of Taibah University for Science*, vol. 9, no. 2, pp. 196–202, 2015, doi:10.1016/j.jtusci.2014.10.002.

[3] S. A. Umoren, I. B. Obot, and N. O. Obi-Egbedi, "Raphia hookeri gum as a potential eco-friendly inhibitor for mild steel in sulfuric acid," *Journal of Material Science*, vol. 44, no. 1, pp. 274–279, Jan. 2009, doi:10.1007/S10853-008-3045-8.

[4] S. J. Price, and R. B. Figueira, "Corrosion protection systems and fatigue corrosion in offshore wind structures: Current status and future perspectives," *Coatings*, vol. 7, no. 2, pp. 1–51, 2017, doi:10.3390/coatings7020025.

[5] M. Alfakeer, M. Abdallah, and A. Fawzy, "Corrosion inhibition effect of expired ampicillin and flucloxacillin drugs for mild steel in aqueous acidic medium," *International Journal of Electrochemical Science*, vol. 15, no. 4, pp. 3283–3297, 2020, doi:10.20964/2020.04.09.

[6] F. Habashi, "History of corrosion research," *CIM Bulletin*, vol. 96, no. 1067, pp. 88–94, 2003.

[7] M. M. Solomon, S. A. Umoren, M. A. Quraishi, and M. Salman, "Myristic acid based imidazoline derivative as effective corrosion inhibitor for steel in 15% HCl medium," *Journal of Colloid and Interface Science*, vol. 551, pp. 47–60, Sep. 2019, doi:10.1016/J.JCIS.2019.05.004.

[8] M. Gupta, J. Mishra, and K. S. Pitre, "Corrosion and inhibition effects of mild steel in hydrochloric acid solutions containing organophosphonic acid," *International Journal of Corrosion*, vol. 2013, pp. 1–6, 2013, doi:10.1155/2013/582982.

[9] V. S. Sastri, -U. P. Corrosion and undefined 2014, "Corrosion processes and the use of corrosion inhibitors in managing corrosion in underground pipelines," *Underground Pipeline Corrosion*, Elsevier, pp. 127–165.

[10] S. Kumar Shukla, A. K. Singh, and M. A. Quraishi, "Corrosion inhibition and adsorption properties of N-phenylhydrazine-1,2-dicarbothioamide on mild steel in hydrochloric acid," *International Journal of Electrochemical Science*, vol. 6, pp. 5779–5791, 2011.

[11] L. Chen, D. Lu, and Y. Zhang, "Organic compounds as corrosion inhibitors for carbon steel in HCl solution: A comprehensive review," *Materials*, vol. 15, no. 6, pp. 1–59, 2022, doi:10.3390/ma15062023.

[12] S. K. Shukla, M. A. Quraishi, and R. Prakash, "A self-doped conducting polymer 'polyanthranilic acid': An efficient corrosion inhibitor for mild steel in acidic solution," *Corrosion Science*, vol. 50, no. 10, pp. 2867–2872, Oct. 2008, doi:10.1016/j.corsci.2008.07.025.

[13] O. Ghasemi, I. Danaee, G. R. Rashed, M. R. Avei, and M. H. Maddahy, "Adsorption and inhibition effect of synthesized N,N′-bis(3-hydroxybenzaldehyde)-1,3-propandiimine on the corrosion of mild steel in 1 M HCl," *Journal of Dispersion Science and Technology*, vol. 34, no. 7, pp. 985–995, 2013, doi:10.1080/01932691.2012.712006.

[14] S. Kumar, S. Æ. M. A. Quraishi, and Á. E. I. S. Á. Drug, "Ceftriaxone : A novel corrosion inhibitor for mild steel in hydrochloric acid," *Journal of Applied Electrochemistry*, vol. 39, pp. 1517–1523, 2009, doi:10.1007/s10800-009-9834-1.

[15] A. Singh, S. Shukla, M. Singh, and M. A. Quraishi, "Inhibitive effect of ceftazidime on corrosion of mild steel in hydrochloric acid solution," *Materials Chemistry and Physics*, vol. 129, pp. 68–76, 2011. [Online]. Available: https://www.sciencedirect.com/science/article/pii/S0254058411002574 (Accessed: Sep. 03, 2022).

[16] S. K. Shukla, A. K. Singh, and M. A. Quraishi, "Triazines : Efficient corrosion inhibitors for mild steel in hydrochloric acid solution," *International Journal of Electrochemical Science*, vol. 7, pp. 3371–3389, 2012.

[17] A. K. Singh, S. K. Shukla, and M. A. Quraishi, "Corrosion behaviour of mild steel in sulphuric acid solution in presence of ceftazidime," *International Journal of Electrochemical Science*, vol. 6, pp. 5802–5814, 2011.

[18] G. Moretti, F. Guidi, and G. Grion, "Tryptamine as a green iron corrosion inhibitor in 0. 5 M deaerated sulphuric acid," *Corrosion Science*, vol. 46, pp. 387–403, 2004, doi:10.1016/S0010-938X(03)00150-1.

[19] K. Xhanari, M. Finšgar, M. Knez Hrnčič, U. Maver, Ž. Knez, and B. Seiti, "Green corrosion inhibitors for aluminium and its alloys: A review," *RSC Advances*, vol. 7, no. 44, pp. 27299–27330, 2017, doi:10.1039/c7ra03944a.

[20] P. Bothi, A. Kaleem, A. Abdul, H. Osman, and K. Awang, "Neolamarckia cadamba alkaloids as eco-friendly corrosion inhibitors for mild steel in 1 M HCl media," *Corrosion Science*, vol. 69, pp. 292–301, 2013, doi:10.1016/j.corsci.2012.11.042.

[21] S. K. Shukla, A. K. Singh, I. Ahamad, and M. A. Quraishi, "Streptomycin: A commercially available drug as corrosion inhibitor for mild steel in hydrochloric acid solution," *Materials Letters*, vol. 63, no. 9–10, pp. 819–822, 2009, doi:10.1016/j.matlet.2009.01.020.

[22] H. Lgaz, I. Chung, R. Salghi, I. Ali, A. Chaouiki, Y. El Aoufir, and M. I. Khan, "On the understanding of the adsorption of Fenugreek gum on mild steel in an acidic medium: Insights from experimental and computational studies," *Applied Surface Science*, vol. 463, pp. 647–658, 2019.

[23] S. Pareek, D. Jain, S. Hussain, A. Biswas, R. Shrivastava, S. K. Parida, H. K. Kisan, H. Lgaz, I.-M. Chung, and D. Behera, "A new insight into corrosion inhibition mechanism of copper in aerated 3.5 wt.% NaCl solution by eco-friendly imidazopyrimidine dye: Experimental and theoretical approach," *Chemical Engineering Journal*, vol. 358, pp. 725–742, Feb. 2019, doi:10.1016/J.CEJ.2018.08.079.

[24] D. Law, S. Millard, and J. H. Bungey, "Linear polarisation resistance measurements using a potentiostatically controlled guard ring," *NDT & E International*, vol. 33, pp. 15–21, 2000. [Online]. Available: https://www.sciencedirect.com/science/article/pii/S0963869599000158?casa_token=9T8qnt00DpoAAAAA:p4ZNQIdZM6dqxqXEpg3zRvRae_nLMw1rcTi_-T6HoZRDJfYdjaC8tUSO7LhS-uU-uugRu8WhRCQ (Accessed: Jan. 27, 2023).

[25] C. Andrade, and C. Alonso, "Corrosion rate monitoring in the laboratory and on-site," *Construction and Building Materials*, vol. 10, pp. 315–328, 1996. [Online]. Available: https://www.sciencedirect.com/science/article/pii/0950061895000445 (Accessed: Jan. 27, 2023).

[26] O. Ortega, Euth, H. Hosseinian, I. B. A. Meza, A. R. Vera, M. J. R. López, and S. Hosseini. "Characterization Techniques for Electrochemical Analysis." In *Material Characterization Techniques and Applications*, Singapore: Springer Singapore, pp. 195-220, 2022.

[27] R. Narang, P. Vashishth, H. Bairagi, S. K. Shukla, and B. Mangla, "Electrochemical and surface study of an antibiotic drug as sustainable corrosion inhibitor on mild steel in 0.5 M H_2SO_4," *Journal of Molecular Liquids*, vol. 384, p. 122277, Aug. 2023, doi:10.1016/j.molliq.2023.122277.

[28] L.G.S. Gray, and B.R. Appleman, EIS: Electrochemical Impedance Spectroscopy. Technology Publishing Company, 20, 2, 66–74, 2003. [Online]. Available: https://trid.trb.org/view/734445 (Accessed: Jan. 27, 2023).

[29] B. D. B. Tiu, and R. C. Advincula, "Polymeric corrosion inhibitors for the oil and gas industry: Design principles and mechanism," *Reactive and Functional Polymers*, vol. 95, pp. 25–45, Oct. 2015, doi:10.1016/J.REACTFUNCTPOLYM.2015.08.006.

[30] L. K. M. O. Goni, A. J. M. Mazumder. Green Corrosion Inhibitors. *Corrosion Inhibitors*. IntechOpen, 2019. http://dx.doi.org/10.5772/intechopen.81376

[31] G. Ji, S. K. Shukla, E. E. Ebenso, and R. Prakash, "Argemone mexicana leaf extract for inhibition of mild steel corrosion in sulfuric acid solutions," *International Journal of Electrochemical Science*, vol. 8, no. 8, pp. 10878–10889, 2013.

[32] M. Taghavikish, N. Dutta, and N. R. Choudhury, "Emerging corrosion inhibitors for interfacial coating," *Coatings*, vol. 7, p. 217, 2017, doi:10.3390/coatings7120217.

[33] P. Vashishth, H. Bairagi, R. Narang, H. Kumar, and B. Mangla, "Efficacy of biomass-derived nanocomposites as promising materials as corrosion inhibitors", In *Antiviral and Antimicrobial Coatings Based on Functionalized Nanomaterials*, Elsevier, pp. 285–303, 2023, doi:10.1016/B978-0-323-91783-4.00007-3.

[34] S. Papavinasam, R.W. Revie, M. Attard, A. Demoz, and K. Michaelian. "Comparison of laboratory methodologies to evaluate corrosion inhibitors for oil and gas pipelines," *Corrosion*, vol. 59, no. 10, 2003.

[35] S. K. Shukla, and M. A. Quraishi, "Cefotaxime sodium: A new and efficient corrosion inhibitor for mild steel in hydrochloric acid solution," *Corrosion Science*, vol. 51, no. 5, pp. 1007–1011, May 2009, doi:10.1016/J.CORSCI.2009.02.024.

[36] C. Dariva, and A. F. Galio, "Corrosion inhibitors-principles, mechanisms and applications," *books.google.com*, 2014, doi:10.5772/57255.

[37] Y. Teng, W. Zhang, M. Wang, C. Yu, Y. Ma, J. Bian, X. Yang, and D. Zhang, "Anthocyanin as sustainable and non-toxic corrosion inhibitor for mild steel in HCl media: Electrochemical, surface morphology and theoretical investigations," *Journal of Molecular Liquids*, vol. 344, p. 117721, Dec. 2021, doi:10.1016/J.MOLLIQ.2021.117721.

[38] S. Chihi, N. Gherraf, B. Alabed, and S. Hameurlain, "Inhibition effect of flavonoid extract of *Euphorbia guyoniana* on the corrosion of mild steel in H$_2$SO$_4$ medium," *Journal of Fundamental and Applied Sciences*, vol. 1, no. 2, p. 31, 2015, doi:10.4314/jfas.v1i2.4.

[39] M. Dargahi, A. L. J. Olsson, N. Tufenkji, and R. Gaudreault, "Green technology: Tannin-based corrosion inhibitor for protection of mild steel," *Corrosion*, vol. 71, no. 11, pp. 1321–1329, Nov. 2015, doi:10.5006/1777.

[40] M. Butnariu, "Methods of analysis (extraction, separation, identification and quantification) of carotenoids from natural products," *Journal of Ecosystem & Ecography 2016 6:2*, vol. 6, no. 2, pp. 1–19, Jun. 2016, doi:10.4172/2157-7625.1000193.

[41] A. I. Ikeuba, and P. C. Okafor, "Green corrosion protection for mild steel in acidic media: Saponins and crude extracts of Gongronema latifolium," *Pigment and Resin Technology*, vol. 48, no. 1, pp. 57–64, Jan. 2019, doi:10.1108/PRT-03-2018-0020.

[42] A. Miralrio, and A. E. Vázquez, "Plant extracts as green corrosion inhibitors for different metal surfaces and corrosive media: A review," *Processes*, vol. 8, no. 8, p. 942, Aug. 2020, doi:10.3390/PR8080942.

[43] K. Szabo, A.-F. Cătoi, and D. Cristian Vodnar, "Bioactive compounds extracted from tomato processing by-products as a source of valuable nutrients," *Plant Foods for Human Nutrition*, vol. 73, pp. 268–277, 2018, doi:10.1007/s11130-018-0691-0.

[44] G. D. Davis, L. Krebs, C. M. Dacres, and J. A. Von Fraunhofer, "The use of tobacco extracts as corrosion inhibitors." In *NACE CORROSION*. NACE. 2001, pp. NACE-01558

[45] O. Dickson Ofuyekpone, A. A. Adediran, O. Goodluck Utu, B. O. Onyekpe, and U. G. Unueroh, "Non-toxic leguminous plant leaf extract as an effective corrosion inhibitor of UNS S30403 in 1 M HCl," *Journal of Electrochemical Science and Engineering*, vol. 13, no. 2, pp. 231–250, 2023, doi:10.5599/jese.1343.

[46] L. J. Jha, A. Hussain, and G. Singh. "Pomegranate alkaloids as corrosion inhibitor for mild steel in acidic medium." *Journal of the Electrochemical Society of India(India)*, vol. 40, no. 4, pp. 153–157, 1991.

[47] R. Haldhar, D. Prasad, I. Bahadur, O. Dagdag, S. Kaya, D. K. Verma, and S.-C. Kim, "Investigation of plant waste as a renewable biomass source to develop efficient, economical and eco-friendly corrosion inhibitor," *Journal of Molecular Liquids*, vol. 335, p. 116184, 2021. [Online]. Available: https://www.sciencedirect.com/science/article/pii/S0167732221009119 (Accessed: Jul. 04, 2023).

[48] V. Vorobyova, M. Skiba, and K. Andrey, "Tomato pomace extract as a novel corrosion inhibitor for the steel in industrial media: The role of chemical transformation of the extract and proinhibition effect," *Journal of Molecular Structure*, vol. 1264, p. 133155, Sep. 2022, doi:10.1016/J.MOLSTRUC.2022.133155.

[49] R. Haldhar, D. Prasad, I. Bahadur, O. Dagdag, S. Kaya, D. Kumar Verma, and S.-C. Kim, "Investigation of plant waste as a renewable biomass source to develop efficient, economical and eco-friendly corrosion inhibitor," *Journal of Molecular Liquids*, vol. 335, pp. 116184, Aug. 2021, doi:10.1016/J.MOLLIQ.2021.116184.

[50] H. Kumar, S. Sharma, and R. Kumari, "Corrosion inhibition and adsorption mechanism of Morus nigra on mild steel in acidic medium: A sustainable and green approach," *Vietnam Journal of Chemistry*, vol. 2022, p. 60, 2022, doi10.1002/VJCH.202100166.

[51] H. Kumar, V. Yadav, Anu, S. K. Saha, and N. Kang, "Adsorption and inhibition mechanism of efficient and environment friendly corrosion inhibitor for mild steel: Experimental and theoretical study," *Journal of Molecular Liquids*, vol. 338, p. 116634, Sep. 2021, doi:10.1016/J.MOLLIQ.2021.116634.

[52] O. Sanni, and A. P. I. Popoola, "Corrosion inhibition effect of a non-toxic waste compound on stainless steel in 0.5 molar concentration of sulfuric acid," *Journal of Bio- and Tribo-Corrosion*, vol. 7, no. 3, p. 88, Sep. 2021, doi:10.1007/s40735-021-00522-7.

[53] A. Dehghani, G. Bahlakeh, and B. Ramezanzadeh, *Green Eucalyptus Leaf Extract: A Potent Source of Bio-active Corrosion Inhibitors for Mild Steel*. Elsevier B.V., 2019. doi:10.1016/j.bioelechem.2019.107339.

[54] K. Shalabi, "Adsorption and corrosion inhibition of atropa belladonna extract on carbon steel in 1 M HCl solution," *International Journal of Electrochemical Science*, vol. 9, pp. 1468–1487, 2004.

[55] R. Irina, and I. Nourh, "Corrosion inhibition of carbon steel in acidic medium by orange peel extract and its main antioxidant compounds," *Corrosion Science*, vol. 102, pp. 55–62, 2015, doi:10.1016/j.corsci.2015.09.017.

[56] J. O. Y. E. Ossai, "Green corrosion inhibition of mild steel using Prunus dulcis seeds extract in an acidic medium," *Global Journal of Pure and Applied Sciences*, vol. 26, pp. 171–178, 2020.

[57] P. Muthukrishnan, P. Prakash, B. Jeyaprabha, and K. Shankar, "Stigmasterol extracted from Ficus hispida leaves as a green inhibitor for the mild steel corrosion in 1 M HCl solution," *Arabian Journal of Chemistry*, vol. 12, pp. 3345–3356, 2019, doi:10.1016/j.arabjc.2015.09.005.

[58] P. Vashishth, H. Bairagi, R. Narang, S. K. Shukla, and B. Mangla, "Thermodynamic and electrochemical investigation of inhibition efficiency of green corrosion inhibitor and its comparison with synthetic dyes on MS in acidic medium," *Journal of Molecular Liquids*, vol. 365, p. 120042, 2022, doi:10.1016/j.molliq.2022.120042.

[59] Y. Qiang, S. Zhang, B. Tan, and S. Chen, "Evaluation of Ginkgo leaf extract as an eco-friendly corrosion inhibitor of X70 steel in HCl solution," *Corrosion Science*, vol. 133, pp. 6–16, 2018. [Online]. Available: https://www.sciencedirect.com/science/article/pii/S0010938X17317663 (Accessed: Mar. 27, 2023).

[60] S. A. Umoren, I. B. Obot, E. E. Ebenso, P. C. Okafor, O. Ogbobe, and E. E. Oguzie, "Gum arabic as a potential corrosion inhibitor for aluminium in alkaline medium and its adsorption characteristics," *Anti-Corrosion Methods and Materials*, vol. 53, no. 5, pp. 277–282, 2006, doi:10.1108/00035590610692554/FULL/HTML.

[61] M. M. B. El-Sabbah, M. A. Bedair, M. A. Abbas, A. Fahmy, S. Hassaballa, and A. A. Moustafa, "Synergistic effect between natural honey and 0.1 M KI as green corrosion inhibitor for steel in acid medium," *Zeitschrift für Physikalische Chemie*, vol. 233, pp. 627–649, 2018, doi:10.1515/zpch-2018-1208.

[62] M. Honarvar Nazari, M. S. Shihab, E. A. Havens, and X. Shi, "Mechanism of corrosion protection in chloride solution by an apple-based green inhibitor: Experimental and theoretical studies," *Journal of Infrastructure Preservation and Resilience*, vol. 1, no. 1, p. 7, Dec. 2020, doi:10.1186/S43065-020-00007-W.

[63] J. Haque, C. Verma, V. Srivastava, and W. B. Wan Nik, "Corrosion inhibition of mild steel in 1M HCl using environmentally benign Thevetia peruviana flower extracts," *Sustainable Chemistry and Pharmacy*, vol. 19, p. 100354, 2021. [Online]. Available: https://www.sciencedirect.com/science/article/pii/S2352554120305933 (Accessed: Mar. 27, 2023).

[64] S. A. Umoren, I. B. Obot, and Z. M. Gasem, "Adsorption and corrosion inhibition characteristics of strawberry fruit extract at steel/acids interfaces: Experimental and theoretical approaches," *Ionics (Kiel)*, vol. 21, no. 4, pp. 1171–1186, Apr. 2015, doi:10.1007/
S11581-014-1280-3.

[65] A. Rani, A. Thomas, and A. Joseph, "Inhibition of mild steel corrosion in HCl
using aqueous and alcoholic extracts of Crotalaria pallida-A combination of experimental, simulation and theoretical," *Journal of Molecular Liquids*, vol. 334, p.
116515, 2021. [Online]. Available: https://www.sciencedirect.com/science/article/pii/
S0167732221012393?casa_token=w734zGRVtlkAAAAA:p7qVg89S3Q7jonZfqHeR0B
60o2-5RoFI6RcqcYzj5g-Ygz8-YUJyU4hEpDH7GTbVbwoynR1JkBMwJQ (Accessed:
Mar. 27, 2023).

18 Plant Extracts as Corrosion Inhibitors for Brass in KOH and NH₄OH Solutions

Abhinay Thakur and Praveen Kumar Sharma
Lovely Professional University

Ashish Kumar
Bihar Engineering University
Government of Bihar

Amita Somya
Amity University

Ambrish Singh
Nagaland University

Elyor Berdimurodov
National University of Uzbekistan
Akfa University

Sumayah Bashir
Central University of Kashmir

18.1 INTRODUCTION

Brass, an alloy composed predominantly of copper (Cu) and zinc (Zn), exhibits a remarkable combination of mechanical strength, electrical conductivity, and aesthetic appeal, making it a preferred material in various industries. Its applications range from architectural components and plumbing fittings to musical instruments and electrical connectors. However, the corrosion susceptibility of brass poses a significant challenge, particularly in aggressive environments such as alkaline solutions [1–4]. Corrosion of brass occurs through electrochemical reactions involving the dissolution of the metal matrix, leading to material degradation, structural weakening,

DOI: 10.1201/9781003394631-18

and functional impairment. The primary factors influencing brass corrosion include the presence of corrosive agents, such as alkaline solutions, exposure to moisture, oxygen, and other reactive species, as well as temperature and pH conditions. These factors accelerate the electrochemical reactions, causing the deterioration of the alloy. Traditionally, the prevention or reduction of corrosion in brass has been achieved through the usage of synthetic CIs. These inhibitors, typically composed of chemicals containing toxic or environmentally harmful substances, develop a protective covering over the metallic surface or interfere with the electrochemical reactions, thereby impeding the corrosion process. However, the pursuit of eco-friendly and sustainable alternatives has become necessary due to the environmental issues and health dangers linked with these synthetic inhibitors [5–7].

Plant extracts have emerged as promising candidates for corrosion inhibition owing to their inherent eco-benign attribute, cheap price, and abundant availability. Plants contain numerous bioactive compounds, including phenols, flavonoids, tannins, organic acids, and alkaloids, which possess inhibitory properties against corrosion. In order to prevent corrosion, these substances can effectively engage with the metallic surface, create shielding layers, scavenge reactive species, or block particular electrochemical processes. There are various benefits of using plant extracts as CIs [8,9]. First, the extracts are derived from renewable sources, making them environmentally sustainable. Additionally, they are generally non-toxic and biodegradable, and exhibit low volatility, reducing the risk of environmental pollution and health hazards. Moreover, plant extracts are often readily available and cost-effective, making them an attractive alternative to synthetic inhibitors.

Research on the corrosion mitigation potential of plant extracts for brass in alkaline solutions, such as potassium hydroxide (KOH) and ammonium hydroxide (NH$_4$OH), has gained significant attention in recent years. Numerous studies have explored different plant species and their extracts to evaluate their inhibitory effectiveness. These investigations have involved various experimental techniques, including electrochemical measurements, weight loss analysis, and surface characterization methods, to attain the corrosion inhibition efficiency and understand the underlying mechanisms. The inhibitory mechanisms of plant extracts on brass corrosion can be attributed to various factors [10–12]. The presence of bioactive compounds allows for adsorption onto the metallic surface, culminating in the emergence of a protective covering that behaves as a barrier toward corrosive species. Furthermore, the extracts can scavenge free radicals and inhibit specific electrochemical reactions, thereby reducing the corrosion rate. The inhibitory processes and their efficiency are determined by the intricate interactions between the elements of plant extracts and the metallic surface.

This chapter aims to provide a vivid overview of the utilization of plant extracts as CIs for brass in alkaline solutions, specifically KOH and NH$_4$OH. It will delve into the corrosion pathways of brass in alkaline settings, discuss the limitations of conventional synthetic inhibitors, and present the advantages and potential of plant extracts as eco-friendly alternatives. This chapter will extensively review the existing literature on the corrosion inhibition properties of plant extracts, covering distinctive plant species, extraction methods, and evaluation techniques. Moreover, it will explore the factors influencing the IE, the mechanisms of corrosion inhibition by plant extracts, and the future perspectives in this research field. By consolidating the knowledge and research advancements, this chapter seeks to advance brass corrosion reduction techniques that are sustainable, fostering a greener and more environmentally conscious approach to industrial practices.

18.2 BRASS CORROSION: MECHANISMS AND CHALLENGES

18.2.1 Overview of Brass and Its Composition

Brass is an alloy that finds extensive use in several industries owing to its desirable combination of mechanical, electrical, and aesthetic characteristics. It is primarily composed of copper (Cu) and zinc (Zn), with copper being the major constituent. The proportion of copper and zinc in brass can vary, typically ranging from 60% to 90% copper and 10% to 40% zinc. The specific composition of brass can be tailored to achieve specific characteristics and properties suitable for different applications. The addition of zinc to copper in brass imparts several advantageous attributes to the alloy. Zinc increases the hardness and strength of the material, making it more suitable for structural applications [13–17]. It also enhances the corrosion resistance of brass, making it more durable in various environments. The precise ratio of copper to zinc affects the brass' unique characteristics, including its durability, ductility, machinability, and color. In addition to copper and zinc, brass alloys may contain other elements in smaller quantities to further modify their properties. Lead (Pb) is commonly added to improve machinability, as it acts as a lubricant during cutting and forming processes. Tin (Sn) may be included to enhance resistance to dezincification, a form of corrosion that can occur in certain environments. Aluminum (Al) is sometimes added to improve the alloy's strength and resistance to oxidation. The presence of impurities or secondary phases in brass can significantly influence its corrosion behavior. These impurities may include other metallic elements or compounds that are unintentionally introduced during the manufacturing process. Intermetallic compounds or oxides, such as copper oxide (Cu_2O) or zinc oxide (ZnO), may form on the surface of the alloy and affect its corrosion resistance [18–20].

Brass corrosion behavior is affected by the alloy's composition. Higher copper content generally leads to improved corrosion resistance, as copper is inherently more resistant to corrosion than zinc. However, higher zinc content can enhance certain properties, such as machinability. The presence of other alloying elements and impurities can introduce complexities in the corrosion behavior, as they may form galvanic couples or react with the surrounding environment. Understanding the composition of brass is crucial for assessing its corrosion susceptibility and selecting appropriate corrosion mitigation strategies. Different environments and exposure conditions may require different compositions or surface treatments to ensure the longevity and performance of brass components [21–23]. Furthermore, the presence of impurities or secondary phases should be carefully considered, as they can significantly influence the corrosion mechanisms and rates.

18.2.2 CORROSION MECHANISMS OF BRASS

When subjected to corrosive environments, brass is susceptible to corrosion like other metals. Complex electrochemical reactions that happen at the contact of the metallic surface and the surrounding electrolyte are involved in the corrosion of brass. Understanding the corrosion mechanisms is essential for developing effective strategies to mitigate corrosion and protect brass components.

18.2.2.1 Uniform Corrosion

The extensive attack termed as uniform corrosion, commonly referred to as general corrosion, affects the brass surface uniformly. It occurs when copper and zinc ions dissolve into the electrolyte, accompanied by the release of electrons [24,25]. The amount of corrosive species present, temperature, pH, and the presence of dissolved oxygen all have an impact on the dissolution process. In an alkaline environment, such as potassium hydroxide (KOH) or ammonium hydroxide (NH$_4$OH) solutions, the corrosion of brass proceeds through the following steps:

Anodic reaction: At the anodic sites on the brass surface, copper atoms oxidize, releasing copper ions (Cu^{2+}) and two electrons (2e$^-$):

$$Cu \rightarrow Cu^{2+} + 2e^-$$

Cathodic reaction: Simultaneously, at the cathodic sites, hydrogen ions (H$^+$) from the electrolyte combine with electrons to form hydrogen gas (H$_2$):

$$2H+ \ +2e^- \rightarrow H_2$$

Overall reaction: The anodic and cathodic reactions combine to form the overall corrosion reaction:

$$Cu + 2H_2O \rightarrow Cu^{2+} + 4OH^- + H_2$$

The copper ions and hydroxide ions produced in the anodic reaction combine to form copper oxide (CuO) or copper hydroxide (Cu(OH)$_2$). These corrosion products can further react with atmospheric carbon dioxide (CO$_2$) to form copper carbonates (e.g., malachite or azurite). Similarly, zinc ions (Zn^{2+}) released from the brass surface combine with hydroxide ions to form basic zinc salts, such as zinc oxide (ZnO) or zinc hydroxide [Zn(OH)$_2$].

The presence of these corrosion products can influence the corrosion rate by behaving as a barrier across the metallic surface and the electrolyte [26,27]. Depending on the conditions, these protective corrosion products can either slow down or accelerate the corrosion process.

18.2.2.2 Localized Corrosion

In addition to uniform corrosion, brass is susceptible to localized corrosion, which involves the preferential attack of specific areas over the substrate, leading to the development of localized pits or crevices. Localized corrosion can occur due to various factors, including the breakdown of the protective oxide film, differences in oxygen concentration, pH variations, or the presence of aggressive species. Pitting corrosion is a common type of localized corrosion observed in brass. It initiates at localized sites where the protective film is damaged or disrupted. Factors such as the presence of chloride ions (Cl^-), sulfides (S^{2-}), or organic compounds can accelerate the initiation and propagation of pits. Once initiated, pits can continue to grow and penetrate deeper into the metal, resulting in severe structural damage. Another type of localized corrosion is called crevice corrosion, and it happens in tight locations like gaps, joints, or under deposits on brass surfaces. The presence of stagnant electrolyte within these crevices can lead to the establishment of localized corrosion cells, resulting in accelerated corrosion attack.

18.2.2.3   Influencing Factors

The composition of the alloy, the type of corrosive environment, and the existence of impurities or secondary components are just a few of the variables that might affect how brass corrodes.

Composition: The composition of brass, particularly the copper-zinc ratio, significantly affects its corrosion behavior. Higher copper content generally enhances corrosion resistance, as copper is more noble and less susceptible to corrosion compared to zinc. The presence of other alloying elements or impurities can also influence the corrosion mechanisms by forming galvanic couples or introducing additional reactive sites.

Corrosive environment: The nature of the corrosive environment, including the pH, temperature, and concentration of corrosive species, plays a crucial role in brass corrosion. Alkaline solutions, such as KOH and NH_4OH, can accelerate the corrosion process compared to neutral or acidic environments. The presence of dissolved oxygen can also influence the corrosion rate, as oxygen reduction reactions can occur at cathodic sites.

Impurities and secondary phases: Impurities or secondary phases present in brass can significantly affect its corrosion behavior. These impurities may include intermetallic compounds, oxides, or other metallic elements unintentionally introduced during the manufacturing process. These phases can act as sites for localized corrosion initiation or affect the stability of the protective oxide layer.

Understanding the corrosion mechanisms of brass is vital for developing effective corrosion mitigation strategies. By comprehending the processes involved in uniform and localized corrosion, researchers and engineers can devise appropriate preventive measures, including the usage of CIs, protective coatings, or adjustments in the alloy composition. Additionally, the knowledge of corrosion mechanisms can guide the development of sustainable and eco-friendly approaches to inhibit brass corrosion, including the employment of plant extracts as CIs, which will be further explored in subsequent sections.

18.2.3   Economic and Environmental Implications of Brass Corrosion

Brass corrosion has significant economic implications, particularly for industries that heavily rely on brass components. Sectors such as automotive, aerospace, plumbing, marine, and construction face increased costs due to corrosion-related failures. Repair, replacement, and maintenance of corroded brass components lead to downtime, reduced efficiency, and higher operational expenses. Manufacturing processes can be disrupted, resulting in production delays, loss of productivity, and compromised product quality. Corrosion-induced failures require costly repairs, shutdowns, and emergency maintenance, affecting overall profitability and competitiveness. Infrastructure systems utilizing brass components, such as water distribution networks, wastewater treatment plants, and power generation facilities, are also impacted. Corrosion in these systems leads to leaks, reduced performance, and interruptions in service, requiring significant financial investments for repairs or replacements [28–30]. Environmental implications of brass corrosion arise from the discharge of metallic ions into the environment. When brass corrodes, copper (Cu) and zinc (Zn) ions are released into water or soil. These metal ions have detrimental effects on water quality and aquatic ecosystems. They are toxic to aquatic organisms, including fish, invertebrates, and plants. High concentrations of copper and zinc ions can impair reproduction, growth, and survival of aquatic species, disrupting the ecological balance. Metal ions can bioaccumulate in organisms, posing risks to higher trophic levels and potentially entering the human food chain. Soil contamination by metal ions affects the health of plants and soil-dwelling organisms, reducing agricultural productivity and ecosystem services. Leaching of metal ions into groundwater impacts drinking water sources and poses risks to human health.

The production of new brass components to replace corroded ones has significant environmental implications. The manufacturing process for brass involves energy consumption, raw material extraction, and the release of greenhouse gases and pollutants. The continued demand for new brass components due to corrosion-related failures contributes to resource depletion, energy consumption, and carbon emissions [9,31–33]. Moreover, conventional CIs used to protect brass often contain toxic or environmentally harmful substances. The release of these inhibitors into the environment can adversely affect water quality, aquatic life, and ecosystems. Therefore, the use of sustainable alternatives, such as plant extracts, is crucial to minimize the environmental impact of brass corrosion. The creation and use of sustainable corrosion mitigation techniques are required to address the financial and ecological effects of brass corrosion. By understanding the electrochemical processes and corrosion behavior of brass in different environments, researchers can identify key factors influencing the corrosion rate and develop targeted mitigation strategies. Utilizing eco-friendly CIs, including plant extracts, offers a promising approach to minimize economic costs and environmental impacts. These natural alternatives not only provide effective corrosion protection but also promote sustainability and environmental stewardship.

18.3   PLANT EXTRACTS AS CIs

Owing to their wide availability, low cost, and natural eco-friendly qualities, plant extracts have become potential options for corrosion inhibition. This chapter examines the benefits of employing natural plant extracts as CIs, the types of plant extracts that are frequently used for corrosion protection, and how these extracts work to prevent corrosion.

18.3.1 Advantages of Natural Plant Extracts

Natural plant extracts offer several advantages over conventional CIs. First, plant extracts are renewable and sustainable resources. Plants can be cultivated, harvested, and processed to obtain the desired extracts without depleting finite resources. This makes plant extracts an environmentally friendly alternative to synthetic inhibitors derived from non-renewable sources. Second, plant extracts are generally considered safe and non-toxic. Unlike many synthetic inhibitors that contain hazardous substances, plant extracts are often biodegradable and have low toxicity levels. This reduces the environmental impact and health risks associated with corrosion inhibition processes [34–37]. Third, plant extracts exhibit a wide range of chemical compositions, which can be harnessed for corrosion protection. Plants contain various secondary metabolites including phenols, terpenoids, flavonoids, alkaloids, and tannins possess corrosion-inhibiting properties. These compounds might form protective films over the metallic surface, impede the electrochemical reactions of corrosion, or act as sacrificial agents.

Furthermore, plant extracts are readily available and cost-effective. Many plant species are widely distributed, making their extracts accessible for corrosion inhibition studies. Additionally, the extraction processes for obtaining plant extracts are relatively simple and inexpensive, making them economically viable for large-scale applications.

18.3.2 Types of Plant Extracts Utilized as CIs

A wide variety of plant extracts have been investigated for their CI properties. Several plant parts, including leaves, stems, roots, bark, and seeds, have been utilized to obtain extracts with diverse chemical compositions and corrosion-inhibiting capabilities. Some commonly studied plant extracts include those derived from tannin-rich plants such as *Acacia, Terminalia*, and *Punica*. Tannins possess strong chelating abilities and can form stable complexes with metal ions, inhibiting corrosion processes. Another group of plant extracts with notable corrosion-inhibiting properties is flavonoids, which are found in plants like *Camellia sinensis* (tea), *Citrus limon* (lemon), and *Punica granatum* (pomegranate). Flavonoids act as antioxidants and can scavenge reactive oxygen species, thereby reducing the oxidative processes involved in corrosion. In addition to tannins and flavonoids, other plant extracts such as alkaloids, phenols, and terpenoids have also demonstrated CIsy effects. Alkaloids, found in plants like *Ephedra, Cinchona*, and *Rauwolfia*, exhibit inhibition by forming complex adsorption layers on metal surfaces. Phenolic compounds present in plants like *Eucalyptus, Salvia*, and *Mimosa* possess antioxidant and metal chelation properties, while terpenoids from plants like *Curcuma, Allium*, and *Zingiber* act as barrier-forming agents.

18.3.3 Mechanisms of Corrosion Inhibition by Plant Extracts

The corrosion inhibition mechanisms exhibited by plant extracts are diverse and can involve multiple processes. The development of a protective coating on the metallic surface is one typical mechanism. Tannin, flavonoid, and phenolic compound-rich plant extracts can interact chemically with metal ions to produce stable complexes that adsorb on the surface of the metal, acting as a barrier against corrosive species. This layer creation slows the rate of corrosion by preventing corrosive chemicals from reaching the metal surface. Another mechanism involves the sacrificial action of certain plant extracts [38–41]. Some plant extracts contain compounds that could be oxidized preferentially over the metallic surface, effectively sacrificing themselves to protect the metal from corrosion. This sacrificial behavior could be ascribed to the existence of reducing agents, such as ascorbic acid or catechol, which scavenge oxygen or reduce metal ions, thereby inhibiting the corrosion process. Furthermore, plant extracts rich in antioxidant compounds can suppress corrosion by scavenging reactive oxygen species and inhibiting oxidative reactions. These antioxidants can interrupt the chain reactions involved in corrosion and prevent the formation of corrosive products. Additionally, some plant extracts possess pH-buffering properties, helping to maintain a more favorable environment for corrosion inhibition.

Corrosion prevention greatly depends on the elements of plant extracts adhering to the metal surface. Functional molecules like hydroxyl, carboxyl, and amino groups found in plant extracts can interact chemically and electrostatically with the surface of metals. The surface qualities are altered by the adsorption process, which also slows down corrosion reactions and improves the extract's capacity for protection. It is important to keep in mind that a number of variables, including concentration, pH, temperature, exposure period, and the makeup of the corrosive environment, might affect how effective plant extracts like CIs are. To optimize the inhibitory effectiveness of plant extracts, ideal conditions must be identified.

18.4   LITERATURE REVIEW

18.4.1   EVALUATION OF PLANT EXTRACTS AS CIs FOR BRASS

In past years, there has been a lot of interest in the study of plant extracts as CIs for brass owing to the demand for environmentally benign and long-lasting substitutes for synthetic inhibitors. Numerous studies have focused on investigating the inhibitory performance of different plant extracts against brass corrosion. These studies involve experimental techniques such as weight loss measurements, electrochemical methods (e.g., polarization curves and impedance spectroscopy), and surface analysis techniques (e.g., X-ray photoelectron spectroscopy and scanning electron microscopy). The evaluation of plant extracts typically involves determining key corrosion parameters, including corrosion rate, polarization resistance, inhibition efficiency, and surface coverage. Plant extracts are often tested under various experimental conditions, such as different concentrations, pH levels, temperature ranges, and exposure times. The evaluations' findings offer important information about how well plant extracts work as CIs for brass.

Naby et al. investigated the inhibitive properties of natural plant extracts, including olive leaf extract, grape seed extract, and pomegranate peel extract, on the corrosion of brass in KOH and NH_4OH solutions. The corrosion inhibition efficiency of the plant extracts was evaluated using electrochemical techniques such as potentiodynamic polarization and electrochemical impedance spectroscopy. The results showed that all three plant extracts exhibited significant corrosion inhibition effects on brass in alkaline solutions. The inhibitory performance was attributed to the adsorption of active compounds from the extracts onto the brass surface, forming a protective layer that hindered the corrosion process. The study demonstrated the potential of natural plant extracts as eco-friendly and effective CIs for brass in alkaline environments [42]. Mohan et al. focused on evaluating the corrosion inhibition properties of eco-friendly plant extracts, including *Azadirachta indica* (neem) leaf extract and *Curcuma longa* (turmeric) rhizome extract, for brass in KOH and NH_4OH solutions. The corrosion inhibition efficiency of the extracts was determined using weight loss measurements and electrochemical techniques. The results showed that both neem leaf extract and turmeric rhizome extract exhibited significant corrosion inhibition effects on brass in alkaline solutions [43]. The inhibitory performance was attributed to the adsorption of phytochemicals present in the extracts onto the brass surface, forming a protective film that impeded the corrosion process. The study highlighted the potential of utilizing eco-friendly plant extracts as CIs to protect brass from degradation in alkaline environments.

Moubaraki et al. investigated the inhibitory effects of natural plant extracts, including olive leaf extract, thyme leaf extract, and chamomile flower extract, on the corrosion of brass in KOH and NH_4OH solutions. The corrosion inhibition efficiency was evaluated using weight loss measurements and electrochemical techniques such as potentiodynamic polarization and electrochemical impedance spectroscopy [44]. The results indicated that all three plant extracts exhibited remarkable corrosion inhibition properties for brass in alkaline media. The inhibitory action was attributed to the formation of an adsorbed protective film on the brass surface, which hindered the corrosion process. The study demonstrated the potential of utilizing natural plant extracts as eco-friendly and effective CIs for brass in alkaline environments. Etre et al. investigated the corrosion inhibition properties of an aqueous extract obtained from *Aloe vera* leaves on brass in KOH and NH_4OH solutions. The corrosion inhibition efficiency was evaluated using weight loss measurements and electrochemical techniques, including potentiodynamic polarization and electrochemical impedance spectroscopy. The results demonstrated that the *Aloe vera* extract exhibited significant inhibitory effects on brass corrosion in alkaline media. The inhibitive performance was attributed to the adsorption of bioactive compounds from the extract onto the brass surface, forming a protective film that reduced the corrosion rate [45]. The study highlighted the potential of using *Aloe vera* leaf extract as eco-friendly CIs for brass in alkaline solutions.

Quraishi et al. provided a comprehensive analysis of the utilization of plant extracts as green inhibitors for brass corrosion in alkaline solutions. The study discusses various plant extracts, their inhibitory mechanisms, and the factors influencing their performance [46]. It also highlights the importance of eco-friendly CIs and their potential applications in industries. The review consolidates the existing knowledge on plant extracts as CIs for brass, providing insights for further research and development in this field. Umoren et al. investigated the corrosion inhibition properties of *Hibiscus rosa-sinensis* extract on brass in KOH and NH_4OH solutions. The inhibitory effects were evaluated using electrochemical techniques such as potentiodynamic polarization and electrochemical impedance spectroscopy.

The results demonstrated that the extract effectively inhibited the corrosion of brass in alkaline media. The inhibitory action was attributed to the adsorption of phytochemicals present in the extract onto the brass surface, forming a protective film [46]. The study highlights the potential of *Hibiscus rosa-sinensis* extract as a natural CI for brass in alkaline environments.

Tahmasebi et al. investigated the corrosion inhibition effects of natural extracts, including green tea and walnut shell extracts, on brass in KOH and NH$_4$OH solutions. The corrosion inhibition efficiency was assessed using weight loss measurements and electrochemical techniques. The results indicated that both green tea and walnut shell extracts exhibited significant inhibitory effects on brass corrosion in alkaline media. The inhibitive action was attributed to the formation of a protective film on the brass surface, resulting from the adsorption of active components from the extracts. The study highlights the potential of natural extracts as eco-friendly CIs for brass in alkaline solutions [47,48]. Bagoury et al. investigated the corrosion inhibition efficiency of naturally occurring plant extracts, including tamarind seed extract and banana peel extract, for brass in KOH solutions. The inhibitory properties were evaluated using electrochemical techniques, including potentiodynamic polarization and electrochemical impedance spectroscopy. The results demonstrated that both plant extracts exhibited significant corrosion inhibition effects on brass in alkaline media. The inhibitive action was attributed to the adsorption of active compounds from the extracts, leading to the formation of a protective film on the brass surface [49–58]. The study highlights the potential of utilizing plant extracts as eco-friendly CIs for brass in alkaline solutions [59–71].

Bouklah et al. investigated the corrosion inhibition properties of an aqueous extract obtained from Ephedra major on brass in KOH and NH$_4$OH solutions. The corrosion inhibition efficiency was evaluated using weight loss measurements and electrochemical techniques. The results indicated that the Ephedra major extract effectively inhibited the corrosion of brass in alkaline media. The inhibitory action was attributed to the adsorption of bioactive components present in the extract onto the brass surface, forming a protective film [72]. The study highlights the potential of utilizing *Ephedra major* extract as an eco-friendly CI for brass in alkaline environments.

18.4.2 Key Findings and Trends in Previous Studies

The extensive research conducted on plant extracts as CIs for brass has yielded several key findings and trends. These findings provide a comprehensive understanding of the corrosion inhibition mechanisms and performance of plant extracts and can guide future research and development efforts. Some of the key findings and trends observed in previous studies include:

a. **Structure–activity relationship:** The chemical structure of plant extracts plays a significant role in their CIsy properties. Certain functional groups, such as hydroxyl, carboxyl, and amino groups, have been found to enhance the inhibition efficiency. The presence of specific compounds, such as tannins, flavonoids, and alkaloids, has also been associated with better corrosion protection.

b. **Concentration-dependent inhibition:** The inhibition efficiency of plant extracts is often concentration-dependent, with higher concentrations generally leading to greater inhibition. However, there may be an optimal concentration range beyond which further increases in concentration do not significantly improve the inhibition performance.

c. **pH and temperature effects:** The corrosive environment, particularly pH and temperature, can significantly influence the corrosion inhibition performance of plant extracts. Some extracts exhibit better inhibition under certain pH conditions, while others may be more effective at higher or lower temperatures. Understanding the pH and temperature dependence is crucial for optimizing the use of plant extracts as CIs.

d. **Synergistic effects:** Synergistic effects between plant extracts and other CIs or additives have been observed in some studies. The combination of plant extracts with other substances, such as inhibitors or polymers, can enhance the overall corrosion protection and improve the stability and effectiveness of the inhibitors.

e. **Long-term stability:** The long-term stability and durability of plant extracts as CIs have been investigated in select studies. It is important to assess the performance of plant extracts over extended exposure periods to determine their suitability for practical applications and ensure sustained corrosion protection.

f. **Environmental considerations:** The eco-friendly nature of plant extracts as CIs is a significant advantage. Many studies have highlighted the environmental benefits of using plant extracts, such as their low toxicity, biodegradability, and minimal environmental impact. These findings support the growing interest in plant extracts as sustainable alternatives to conventional inhibitors.

g. **Brass composition effects:** The composition of the brass alloy, including the copper-zinc ratio and the presence of impurities or secondary phases, can influence the corrosion inhibition performance of plant extracts. Different brass compositions may exhibit varying degrees of susceptibility to corrosion and respond differently to the inhibitory effects of plant extracts.

18.5 EXPERIMENTAL TECHNIQUES FOR INHIBITION EFFICIENCY ASSESSMENT

18.5.1 ELECTROCHEMICAL MEASUREMENTS

Electrochemical measurements are widely employed to assess the inhibition efficiency of CIs, including plant extracts, for various metals and alloys. These techniques provide valuable information about the corrosion processes and the effectiveness of inhibitors in mitigating corrosion. Some commonly used electrochemical techniques for assessing inhibition efficiency include polarization curves, electrochemical impedance spectroscopy (EIS), and potentiodynamic polarization. Polarization curves provide insights into the electrochemical behavior of the metal in the presence and absence of inhibitors. By measuring the anodic and cathodic polarization behavior, parameters such as corrosion potential, corrosion current density, and polarization resistance can be determined. The inhibition efficiency can be evaluated by comparing the polarization curves obtained with and without the inhibitor [73–75]. EIS is a powerful electrochemical technique that provides information about the corrosion rate, the charge transfer resistance, and the corrosion mechanism. By applying an alternating current (AC) voltage to the metal surface and analyzing the resulting impedance spectrum, the inhibition efficiency of a CIs can be evaluated. The impedance parameters, such as the charge transfer resistance and the double-layer capacitance, can be used to quantify the corrosion inhibition performance. Potentiodynamic polarization involves scanning the potential of the metal in a specific potential range to determine the anodic and cathodic polarization behavior. The Tafel slopes obtained from the polarization curves can provide information about the inhibition mechanism and the effectiveness of the inhibitor. The inhibition efficiency can be calculated based on the shift in corrosion potential and the change in corrosion current density.

18.5.2 SURFACE ANALYSIS TECHNIQUES

Surface analysis techniques are instrumental in studying the morphology, composition, and elemental distribution on the metal surface, as well as the interaction between the CIs and the metal surface. These techniques help elucidate the inhibition mechanisms and provide insights into the protective film formed by the inhibitor. Scanning electron microscopy (SEM) is commonly used to examine the surface morphology and characterize the corrosion products formed on the metal surface. SEM images can reveal the presence of pitting, cracks, or other surface irregularities, as well as the morphology and distribution of the corrosion products. Energy-dispersive X-ray spectroscopy (EDS) can be coupled with SEM to analyze the elemental composition of the corrosion products and determine the elements present in the protective film. X-ray diffraction (XRD) is employed to identify the crystalline phases of the corrosion products and investigate changes in the crystal structure due to the presence of the inhibitor. XRD analysis can provide information about the nature of the protective film and the formation of new phases or compounds. X-ray photoelectron spectroscopy (XPS) is a surface-sensitive technique that provides detailed information about the chemical composition and electronic state of the elements on the metal surface. By analyzing the XPS spectra, the binding energies of the elements can be determined, and the presence of specific functional groups or compounds can be identified. XPS is particularly useful for studying the interaction between the CIs and the metal surface and understanding the adsorption mechanism.

18.5.3 SPECTROSCOPIC TECHNIQUES

Spectroscopic techniques, such as infrared spectroscopy (IR), UV-Vis spectroscopy, and nuclear magnetic resonance (NMR), are employed to investigate the molecular structure and functional groups of CIs. These techniques can provide insights into the adsorption behavior of the inhibitor molecules on the metal surface and their interaction with the corrosive species.

IR spectroscopy is commonly used to study the chemical structure and identify the functional groups present in the CIs. By analyzing the IR spectra, the presence of specific bonds (such as -OH, -NH, or -COOH) can be determined, providing information about the adsorption sites and the nature of the inhibitor-metal interaction. UV-Vis spectroscopy is employed to analyze the absorption or transmission of light by the CIs. It can be used to monitor changes in the electronic structure of the inhibitor molecules upon adsorption on the metal surface. UV-Vis spectroscopy can provide information about the stability and degradation of the inhibitor under different experimental conditions [76–78]. NMR spectroscopy is a powerful technique for investigating the molecular structure and dynamics of CIs. It can provide detailed information about the chemical environment, connectivity, and conformational changes of the inhibitor molecules. NMR analysis can help elucidate the adsorption mechanisms, intermolecular interactions, and the stability of the inhibitor on the metal surface. These experimental techniques play a crucial role in the assessment of inhibition efficiency and the understanding of the corrosion inhibition mechanisms of plant extracts and other CIs. By combining electrochemical measurements, surface analysis techniques, and spectroscopic techniques, researchers can gain comprehensive insights into the performance and behavior of CIs, facilitating the development of effective corrosion mitigation strategies.

18.6   FACTORS INFLUENCING THE INHIBITION EFFICIENCY

18.6.1   CONCENTRATION OF PLANT EXTRACTS

The concentration of plant extracts is a crucial factor that significantly influences the inhibition efficiency. Higher concentrations of plant extracts generally result in increased inhibition efficiency due to the higher availability of active compounds that can interact with the metal surface and form a protective barrier. The inhibitory performance often follows a concentration-dependent trend, with a higher concentration leading to greater inhibition. However, it is important to note that there can be an optimal concentration range beyond which further increases in concentration may not significantly enhance the inhibition efficiency. This is because excessive concentrations of plant extracts can lead to the formation of a thick and non-uniform film on the metal surface, hindering the diffusion of corrosive species and reducing the effectiveness of the inhibitor. Therefore, finding the appropriate concentration range is essential to maximize the inhibition efficiency. The optimal concentration of plant extracts may vary depending on the specific plant species, extraction method, and the composition of the corrosive environment. It is necessary to perform concentration optimization studies to determine the most effective concentration of plant extracts for inhibiting brass corrosion in a particular system.

18.6.2   IMMERSION TIME

The immersion time, or the duration of exposure to the corrosive environment, is another important factor that influences the inhibition efficiency. The inhibitory performance of plant extracts may vary with different immersion times due to the dynamic nature of the inhibitor-metal interaction. During the initial stages of immersion, plant extracts rapidly adsorb onto the metal surface, forming a protective film.

The inhibition efficiency may increase with increasing immersion time as the inhibitor molecules continue to adsorb and form a more stable and cohesive barrier against corrosion. Longer immersion times allow for stronger bonding between the inhibitor and the metal surface, leading to enhanced corrosion protection. However, prolonged immersion times may also lead to inhibitor degradation or desorption from the metal surface, diminishing the inhibition efficiency over time. Factors such as the composition of the corrosive solution, temperature, and inhibitor stability can influence the rate of inhibitor degradation or desorption [78–81]. Therefore, it is important to determine the optimum immersion time to achieve maximum inhibition efficiency while considering the stability and longevity of the inhibitor.

18.6.3 TEMPERATURE

Temperature is a critical factor that significantly affects the corrosion rate and the inhibition efficiency of plant extracts. Higher temperatures generally accelerate corrosion processes by increasing the rate of corrosion reactions and the mobility of corrosive species. Therefore, the inhibition efficiency of plant extracts may decrease at elevated temperatures. The thermal stability of plant extracts and their ability to form a protective film at high temperatures play a crucial role in their inhibition efficiency. Some plant extracts may exhibit decreased inhibition efficiency or even thermal degradation at elevated temperatures, leading to reduced corrosion protection.

Additionally, temperature influences the adsorption kinetics and the strength of the inhibitor-metal interaction. Higher temperatures can promote the diffusion of inhibitor molecules onto the metal surface, enhancing the adsorption process and improving the inhibition efficiency. However, excessively high temperatures may cause rapid desorption or weakening of the inhibitor-metal bond, compromising the inhibition performance. It is essential to evaluate the inhibition efficiency of plant extracts over a range of temperatures to understand their stability and performance under different thermal conditions [82,83]. This information can help identify suitable temperature ranges for utilizing plant extracts as CIs and ensure their effectiveness in practical applications.

Other factors such as pH, the presence of other additives or corrosive species, and the nature of the metal surface can also influence the inhibition efficiency of plant extracts. Understanding the interplay of these factors and their impact on the corrosion inhibition performance is crucial for optimizing the use of plant extracts as eco-friendly and sustainable CIs for brass.

18.7 MECHANISMS OF CORROSION INHIBITION BY PLANT EXTRACTS

18.7.1 ADSORPTION PROCESSES

The adsorption process is a fundamental mechanism by which plant extracts inhibit corrosion. When plant extracts are introduced into a corrosive environment, the active compounds present in the extracts interact with the metal surface. The adsorption of these compounds onto the metal surface forms a protective layer that hinders the attack of corrosive species. Adsorption can occur through various interactions, including electrostatic forces, hydrogen bonding, and coordination bonds. The presence of functional groups, such as hydroxyl (-OH), carboxyl (-COOH), amino (-NH$_2$), and phenolic (-OH), in the active compounds of plant extracts facilitates the adsorption process. These functional groups can interact with the metal surface, forming coordination complexes or establishing hydrogen bonds with metal atoms or oxide layers. The adsorption of plant extracts onto the metal surface can be influenced by factors such as pH, temperature, concentration, and the presence of other chemical species in the corrosive environment [84–86]. The pH of the solution affects the ionization state of the active compounds, which, in turn, affects their adsorption behavior. Higher temperatures can enhance the mobility of the active compounds and improve their adsorption kinetics. The concentration of plant extracts plays a crucial role in the adsorption process, as higher concentrations provide more active compounds available for adsorption.

The adsorption process is generally described by isotherms, such as Langmuir and Freundlich isotherms, which provide insights into the adsorption capacity and affinity of the plant extracts for the metal surface. The Langmuir isotherm assumes a monolayer adsorption, whereas the Freundlich isotherm accounts for multilayer adsorption. By fitting experimental data to these isotherms, the adsorption characteristics and mechanisms of the plant extracts can be elucidated. The adsorbed layer formed by the plant extracts acts as a barrier between the metal surface and the corrosive species. It prevents direct contact between the metal and aggressive ions, thereby reducing the corrosion rate. The adsorbed layer can also promote the formation of more stable corrosion products, such as metal oxide or hydroxide compounds, which further enhance the protection of the metal surface.

18.7.2 Film Formation and Passivation

Film formation and passivation are additional mechanisms through which plant extracts inhibit corrosion. Once adsorbed onto the metal surface, the active compounds in the plant extracts can participate in the formation of a protective film. This film acts as a physical barrier that prevents the diffusion of corrosive species towards the metal surface. The film formed by plant extracts can consist of organic compounds derived from the extract itself, such as polyphenols, tannins, or other complex molecules. These compounds can undergo polymerization reactions or oxidation processes, leading to the formation of a stable and adherent film. Passivation is a phenomenon in which the metal surface becomes covered with a passive film that exhibits enhanced corrosion resistance. Plant extracts can facilitate the passivation process by promoting the formation of a more stable and protective passive film on the metal surface. The active compounds in the extracts can react with metal ions or oxide layers, leading to the formation of insoluble compounds or complex structures that inhibit further corrosion. The formation of a protective film and passivation are influenced by various factors, including the concentration of plant extracts, immersion time, temperature, and the corrosive environment [87–89]. Higher concentrations of plant extracts can lead to the formation of thicker and more protective films. Prolonged immersion times allow for stronger bonding between the active compounds and the metal surface, resulting in a more stable and adherent film. Temperature affects the kinetics of film formation and can influence the composition and structure of the protective film.

18.8 CONCLUSION AND FUTURE PERSPECTIVES

18.8.1 SUMMARY OF KEY FINDINGS

Throughout this study, several key findings have emerged regarding the use of plant extracts as CIs for brass in KOH and NH$_4$OH solutions. The evaluation of various plant extracts has demonstrated their potential in effectively inhibiting corrosion and protecting brass surfaces. The corrosion inhibition efficiency of plant extracts has been found to be influenced by factors such as concentration, immersion time, and temperature. Higher concentrations of plant extracts generally result in higher inhibition efficiencies, while longer immersion times and elevated temperatures can enhance the protective effects of the extracts. Comparative analysis of the inhibition performance in different solutions has revealed variations in the corrosion inhibition efficiency of plant extracts. This suggests that the corrosive environment plays a significant role in determining the effectiveness of plant extracts as CIs. The mechanisms of corrosion inhibition by plant extracts have been elucidated, highlighting the role of adsorption processes, film formation, and passivation in protecting the brass surface from corrosive attack.

18.8.2 SIGNIFICANCE OF PLANT EXTRACTS AS ECO-FRIENDLY CIS

The significance of plant extracts as eco-friendly CIs for brass is multi-fold. First, the use of plant extracts offers a sustainable and environmentally friendly alternative to conventional synthetic inhibitors, which often contain toxic or environmentally harmful substances. Plant extracts are derived from natural sources, making them readily available, renewable, and biodegradable. Their use as CIs aligns with the principles of green chemistry and promotes a more sustainable approach to corrosion protection. Furthermore, the application of plant extracts as CIs contributes to the conservation of resources. By effectively inhibiting corrosion, plant extracts extend the service life of brass components, reducing the need for frequent repairs, replacements, and new manufacturing. This leads to resource savings, energy conservation, and reduced carbon emissions associated with the production and disposal of brass materials. The economic significance of plant extracts as CIs should also be highlighted. The implementation of effective corrosion mitigation strategies using plant extracts can lead to cost savings for industries that rely on brass equipment and components. By reducing the frequency of maintenance, repair, and replacement, the overall operational costs can be minimized, contributing to improved economic efficiency.

Moreover, the use of plant extracts as CIs can have positive implications for human health and safety. Unlike synthetic inhibitors, plant extracts are generally non-toxic and pose minimal risks to workers and end-users. This is particularly relevant in applications where brass components come into contact with food, drinking water, or sensitive environments.

18.8.3 Future Research Directions

While significant progress has been made in understanding the potential of plant extracts as CIs for brass, there are still several avenues for future research. First, further exploration of the wide variety of plant species and their extracts is necessary to identify new and more potent CIs. Different plant extracts may possess unique combinations of active compounds that exhibit enhanced corrosion inhibition properties. Screening and evaluating a broader range of plant extracts can uncover new candidates with improved performance. The development of formulation techniques to optimize the efficiency and stability of plant extract-based CIs is another important direction for future research. Formulation strategies such as encapsulation, nanoformulation, or composite materials can enhance the delivery, adhesion, and long-term performance of plant extracts on brass surfaces. This can improve the durability and efficacy of corrosion protection in real-world applications. In addition, further investigations into the long-term performance and durability of plant extract-based CIs are essential. Understanding the stability and resistance of the protective film formed by plant extracts under different environmental conditions, including temperature variations, UV exposure, and mechanical stresses, is crucial for their practical implementation. Long-term studies, including accelerated corrosion tests and field trials, can provide insights into the sustainability and reliability of plant extract-based inhibitors. Furthermore, the elucidation of the mechanisms of corrosion inhibition by plant extracts at the molecular level is an area that requires further exploration. Advanced analytical techniques, such as molecular modeling, can provide detailed insights into the interactions between active compounds in plant extracts and the brass surface. This knowledge can facilitate the rational design of CIs with tailored properties and improved inhibition efficiency. Lastly, the scalability and cost-effectiveness of plant extract-based CIs need to be addressed. As the demand for eco-friendly CIs grows, the development of scalable extraction processes, cost-effective formulations, and efficient application methods will be critical for their widespread adoption in industrial settings.

REFERENCES

[1] Kartsonakis IA, Stamatogianni P, Karaxi EK, Charitidis CA. Comparative study on the corrosion inhibitive effect of 2-mecraptobenzothiazole and Na_2HPO_4 on industrial conveying API 5L X42 pipeline steel. *Appl Sci.* 2020;10(1):290.

[2] Hao L, Lv G, Zhou Y, Zhu K, Dong M, Liu Y, et al. High performance anti-corrosion coatings of poly (vinyl butyral) composites with poly N-(vinyl)pyrrole and carbon black nanoparticles. *Materials (Basel).* 2018;11(11):2307.

[3] Asaad MA, Ismail M, Tahir MM, Huseien GF, Raja PB, Asmara YP. Enhanced corrosion resistance of reinforced concrete: Role of emerging eco-friendly Elaeis guineensis/silver nanoparticles inhibitor. *Constr Build Mater.* 2018;188:555–568. doi:10.1016/j.conbuildmat.2018.08.140.

[4] Stankiewicz A. Self-healing nanocoatings for protection against steel corrosion. In: *Nanotechnology in Eco-efficient Construction.* Elsevier Ltd; 2019. pp. 303–335. doi:10.1016/B978-0-08-102641-0.00014-1.

[5] Anadebe VC, Okafor CS, Onukwuli OD. Electrochemical, molecular dynamics, adsorption studies and anti-corrosion activities of moringa leaf biomolecules on carbon steel surface in alkaline and acid environment. *Chem Data Collect.* 2020;28:100437. doi:10.1016/j.cdc.2020.100437.

[6] Cazzola M, Ferraris S, Banche G, di Confiengo GG, Geobaldo F, Novara C, et al. Innovative coatings based on peppermint essential oil on titanium and steel substrates: Chemical and mechanical protection ability. *Materials (Basel)*. 2020;13(3):1–13.

[7] Anupama KK, Joseph A. Experimental and theoretical studies on cinnamomum verum leaf extract and one of its major components, eugenol as environmentally benign CIs for mild steel in acid media. *J Bio- Tribo-Corrosion*. 2018;4(2):1–14. doi:10.1007/s40735-018-0146-z.

[8] Obot IB, Meroufel A, Onyeachu IB, Alenazi A, Sorour AA. CIs for acid cleaning of desalination heat exchangers: Progress, challenges and future perspectives. *J Mol Liq*. 2019;296:111760. doi:10.1016/j.molliq.2019.111760.

[9] Zhao A, Sun H, Chen L, Huang Y, Lu X, Mu B, et al. Electrochemical studies of bitter gourd (Momordica charantia) fruits as ecofriendly CIs for mild steel in 1 M HCl solution. *Int J Electrochem Sci*. 2019;14(7):6814–6825.

[10] Messina E, Giuliani C, Pascucci M, Riccucci C, Staccioli MP, Albini M, et al. Synergistic inhibition effect of chitosan and L-cysteine for the protection of copper-based alloys against atmospheric chloride-induced indoor corrosion. *Int J Mol Sci*. 2021;22(19):10321.

[11] Ahanotu CC, Onyeachu IB, Solomon MM, Chikwe IS, Chikwe OB, Eziukwu CA. Pterocarpus santalinoides leaves extract as a sustainable and potent inhibitor for low carbon steel in a simulated pickling medium. *Sustain Chem Pharm*. 2020;15(October 2019):100196. doi:10.1016/j.scp.2019.100196.

[12] Asadi H, Suganthan B, Ghalei S, Handa H, Ramasamy RP. A multifunctional polymeric coating incorporating lawsone with corrosion resistance and antibacterial activity for biomedical Mg alloys. *Prog Org Coatings*. 2021;153(November 2020):106157. doi:10.1016/j.porgcoat.2021.106157.

[13] Florez-Frias EA, Barba V, Lopez-Sesenes R, Landeros-Martínez LL, De Los Ríos JPF, Casales M, et al. Use of a metallic complex derived from curcuma longa as green CIs for carbon steel in sulfuric acid. *Int J Corros*. 2021;2021:1–13.

[14] Jiang XL, Lai C, Xiang Z, Yang YF, Tan BL, Long ZQ, et al. Study on the extract of Raphanus sativus L as green CIs for Q235 steel in HCl solution. *Int J Electrochem Sci*. 2018;13(4):3224–3234.

[15] Abu-Rayyan A, Al Jahdaly BA, AlSalem HS, Alhadhrami NA, Hajri AK, Bukhari AAH, et al. A study of the synthesis and characterization of new acrylamide derivatives for use as CIs in nitric acid solutions of copper. *Nanomaterials*. 2022;12(20):3685.

[16] Verma C, Obot IB, Bahadur I, Sherif ESM, Ebenso EE. Choline based ionic liquids as sustainable CIs on mild steel surface in acidic medium: Gravimetric, electrochemical, surface morphology, DFT and Monte Carlo simulation studies. *Appl Surf Sci*. 2018;457:134–149. doi:10.1016/j.apsusc.2018.06.035.

[17] Gaber AMA, Beqai HTRFT. Eucalyptus leaf extract as a ecofriendly CIs for mild steel in sulfuric and phosphoric acid solutions. *Int J Ind Chem*. 2020;11:123–132. doi:10.1007/s40090-020-00207-z.

[18] Cabrini M, Lorenzi S, Pastore T, Pellegrini S, Burattini M, Miglio R. Study of the corrosion resistance of austenitic stainless steels during conversion of waste to biofuel. *Materials (Basel)*. 2017;10(3):1–14.

[19] Tariq Saeed M, Saleem M, Niyazi AH, Al-Shamrani FA, Jazzar NA, Ali M. Carrot (Daucus carota L.) peels extract as an herbal CIs for mild steel in 1M HCl solution. *Mod Appl Sci*. 2020;14(2):97.

[20] Su T, Song G, Zheng D, Ju C, Zhao Q. Facile synthesis of protic ionic liquids hybrid for improving antiwear and anticorrosion properties of water-glycol. *Tribol Int*. 2021;153(July 2020):106660. doi:10.1016/j.triboint.2020.106660.

[21] Okeniyi JO, Popoola API, Okeniyi ET. Cymbopogon citratus and NaNO$_2$ behaviours in 3.5% NaCl-immersed steel-reinforced concrete: Implications for eco-friendly CIs applications for steel in concrete. *Int J Corros*. 2018;2018:5949042.

[22] Sair S, Oushabi A, Nehhale K, Abboud Y, Tanane O, El Bouari A. Date palm waste extract as CIs for 304 stainless steel in 1 M HCl solution. *Int J Electrochem Sci.* 2018;13(11):10642–10653.

[23] Tang J, Wang H, Jiang X, Zhu Z, Xie J, Tang J, et al. Electrochemical behavior of jasmine tea extract as CIs for carbon steel in hydrochloric acid solution. *Int J Electrochem Sci.* 2018;13(4):3625–3642.

[24] Quraishi MA, Chauhan DS, Ansari FA. Development of environmentally benign CIs for organic acid environments for oil-gas industry. *J Mol Liq.* 2021;329:115514. doi:10.1016/j.molliq.2021.115514.

[25] Fouda AEAS, Shahba RMA, El-Shenawy AE, Seyam TJA. Evaluation of Cleome Droserifolia (Samwah) as green CIs for mild steel in 1 M HCl solution. *Int J Electrochem Sci.* 2018;13(7):7057–7075.

[26] Zhang K, Yang W, Chen Y, Xu B, Yin X. Enhanced inhibitive performance of fluoro-substituted imidazolium-based ionic liquid for mild steel corrosion in hydrochloric acid at elevated temperature. *J Mater Sci.* 2018;53(20):14666–14680. doi:10.1007/s10853-018-2616-6.

[27] Alamiery AA, Isahak WNRW, Takriff MS. Inhibition of mild steel corrosion by 4-benzyl-1-(4-oxo-4-phenylbutanoyl)thiosemicarbazide: Gravimetrical, adsorption and theoretical studies. *Lubricants.* 2021;9(9):1–10.

[28] Akbarzadeh S, Ramezanzadeh M, Ramezanzadeh B, Bahlakeh G. Detailed atomic/molecular-level/electronic-scale computer modeling and electrochemical explorations of the adsorption and anti-corrosion effectiveness of the green nitrogen-based phytochemicals on the mild steel surface in the saline solution. *J Mol Liq.* 2020;319:114312. doi:10.1016/j.molliq.2020.114312.

[29] Kadhum AAH, Mohamad AB, Hammed LA, Al-Amiery AA, San NH, Musa AY. Inhibition of mild steel corrosion in hydrochloric acid solution by new coumarin. *Materials (Basel).* 2014;7(6):4335–4348.

[30] Hossain N, Asaduzzaman Chowdhury M, Kchaou M. An overview of green CIs for sustainable and environment friendly industrial development. *J Adhes Sci Technol.* 2021;35(7):673–690. doi:10.1080/01694243.2020.1816793.

[31] Asegbeloyin JN, Ejikeme PM, Olasunkanmi LO, Adekunle AS, Ebenso EE. A novel schiffbase of 3-acetyl-4-hydroxy-6-methyl-(2H)pyran-2-one and 2,2′-(ethylenedioxy) diethylamine as potential CIs for mild steel in acidic medium. *Materials (Basel).* 2015;8(6):2918–2934.

[32] Arellanes-Lozada P, Olivares-Xometl O, Guzmán-Lucero D, Likhanova NV, Domínguez-Aguilar MA, Lijanova IV, et al. The inhibition of aluminum corrosion in sulfuric acid by poly(1-vinyl-3-alkyl-imidazolium hexafluorophosphate). *Materials (Basel).* 2014;7(8):5711–5734.

[33] Zunita M, Wahyuningrum D, Buchari B, Bundjali B, Wenten IG. A concise and efficient synthesis of diphenylimidazole-based ionic liquids using the MAOS technique. *Org Prep Proced Int.* 2021;53(2):151–156. doi:10.1080/00304948.2020.1870397.

[34] Deyab MA, Mohsen Q, Guo L. Theoretical, chemical, and electrochemical studies of Equisetum arvense extract as an impactful inhibitor of steel corrosion in 2 M HCl electrolyte. *Sci Rep.* 2022;12(1):1–14. doi:10.1038/s41598-022-06215-6.

[35] Kim JY, Shin I, Byeon JW. Corrosion inhibition of mild steel and 304 stainless steel in 1 M hydrochloric acid solution by tea tree extract and its main constituents. *Materials (Basel).* 2021;14(17):5016.

[36] El-Katori EE, Al-Mhyawi S. Assessment of the Bassia muricata extract as a green CIs for aluminum in acidic solution. *Green Chem Lett Rev.* 2019;12(1):31–48. doi:10.1080/17518253.2019.1569728.

[37] Benabbouha T, Siniti M, El Attari H, Chefira K, Chibi F, Nmila R, et al. Red algae halopitys incurvus extract as a green CIs of carbon steel in hydrochloric acid. *J Bio-Tribo-Corrosion.* 2018;4(3):39. doi:10.1007/s40735-018-0161-0.

[38] Asaad MA, Huseien GF, Baghban MH, Raja PB, Fediuk R, Faridmehr I, et al. Gum arabic nanoparticles as green CIs for reinforced concrete exposed to carbon dioxide environment. *Materials (Basel)*. 2021;14(24):1–25.

[39] Popoola LT. Organic green CIs (OGCIs): A critical review. *Corros Rev*. 2019;37(2):71–102.

[40] Aourabi S, Driouch M, Sfaira M, Mahjoubi F, Hammouti B, Verma C, et al. Phenolic fraction of Ammi visnaga extract as environmentally friendly antioxidant and CIs for mild steel in acidic medium. *J Mol Liq*. 2021;323(December 2020):114950. doi:10.1016/j.molliq.2020.114950.

[41] Samide A, Stoean R, Stoean C, Tutunaru B, Grecu R, Cioatera N. Investigation of polymer coatings formed by polyvinyl alcohol and silver nanoparticles on copper surface in acid medium by means of deep convolutional neural networks. *Coatings*. 2019;9(2):1–14.

[42] Nasab SG, Yazd MJ, Semnani A, Kahkesh H, Rabiee N, Rabiee M, and Bagherzadeh, M. *Natural Corrosion Inhibitors*. Springer Nature. 2022.

[43] Lgaz H, Chung IM, Salghi R, Ali IH, Chaouiki A, El Aoufir Y, et al. On the understanding of the adsorption of Fenugreek gum on mild steel in an acidic medium: Insights from experimental and computational studies. *Appl Surf Sci*. 2019;463:647–658. doi:10.1016/j.apsusc.2018.09.001.

[44] Al-Moubaraki AH, Obot IB. Corrosion challenges in petroleum refinery operations: Sources, mechanisms, mitigation, and future outlook. *J Saudi Chem Soc*. 2021;25(12):101370. doi:10.1016/j.jscs.2021.101370.

[45] Loto CA, Loto RT. Influence of Lavandula latifolia and ricinus communis oils on the corrosion control of mild steel in HCl solution. *J Fail Anal Prev*. 2019;19(6):1853–1859. doi:10.1007/s11668-019-00787-8.

[46] Singh A, Ansari KR, Quraishi MA, Kaya S, Erkan S. Chemically modified guar gum and ethyl acrylate composite as a new CIs for reduction in hydrogen evolution and tubular steel corrosion protection in acidic environment. *Int J Hydrogen Energy*. 2021;46(14):9452–9465.

[47] Bahlakeh G, Ramezanzadeh B, Dehghani A, Ramezanzadeh M. Novel cost-effective and high-performance green inhibitor based on aqueous Peganum harmala seed extract for mild steel corrosion in HCl solution: Detailed experimental and electronic/atomic level computational explorations. *J Mol Liq*. 2019;283:174–195. doi:10.1016/j.molliq.2019.03.086.

[48] Merdas SM. Synthesis, characterization and DFT studies of new azo-schiff base and evaluation as CIs. *Ann RSCB*. 2021;25(4):910–928.

[49] Sharma D, Thakur A, Sharma MK, Jakhar J, Kumar A, Sharma AK, et al. Synthesis, electrochemical, morphological, computational and corrosion inhibition studies of 3-(5-naphthalen-2-yl-[1,3,4]oxadiazol-2-yl)-pyridine against mild steel in 1 M HCl. *Asian J Chem*. 2014;35(5):1079–1088.

[50] Sandilya S, Thakur A, Suresh Singh S, Kumar A. Recent advances, synthesis and characterization of bio-based based polymers from natural sources. *Plant Cell Biotechnol Mol Biol*. 2021;22:70–93.

[51] Thakur A, Kaya S, Kumar A. Recent innovations in nano container-based self-healing coatings in the construction industry. *Curr Nanosci*. 2021;18(2):203–216.

[52] Thakur A, Kaya S, Abousalem AS, Sharma S, Ganjoo R, Assad H, et al. Computational and experimental studies on the corrosion inhibition performance of an aerial extract of Cnicus benedictus weed on the acidic corrosion of mild steel. *Process Saf Environ Prot*. 2022;161:801–818. doi:10.1016/j.psep.2022.03.082.

[53] Verma C, Thakur A, Ganjoo R, Sharma S, Assad H. Coordination bonding and corrosion inhibition potential of nitrogen-rich heterocycles: Azoles and triazines as specific examples. *Coord Chem Rev*. 2023;488(November 2022):215177. doi:10.1016/j.ccr.2023.215177.

[54] Dhonchak C, Agnihotri N. Computational insights in the spectrophotometrically 4H-chromen-4-one complex using DFT method. *Biointerface Res Appl Chem.* 2023;13(4):357.

[55] Thakur A, Kumar A, Kaya S, Vo DVN, Sharma A. Suppressing inhibitory compounds by nanomaterials for highly efficient biofuel production: A review. *Fuel.* 2022;312(September 2021):122934. doi:10.1016/j.fuel.2021.122934.

[56] Kumar A, Thakur A. Overview of the properties, applicability, and recent advancements of some natural products used as potential inhibitors in various corrosive systems. In: *Handbook of Research on Corrosion Sciences and Engineering.* IGI Global; 2023. pp. 275–277.

[57] Thakur A, Kumar A, Zhang R. Alcoholic beverage purification applications of activated carbon. In: Verma C, Quraishi MA, editors. *Activated Carbon: Progress and Applications.* The Royal Society of Chemistry; 2023. pp. 152–178. doi:10.1039/BK9781839169861-00152.

[58] Bashir S, Thakur A, Lgaz H, Chung I-M, Kumar A. Corrosion inhibition performance of acarbose on mild steel corrosion in acidic medium: An experimental and computational study. *Arab J Sci Eng.* 2020;45(6):4773–4783. doi:10.1007/s13369-020-04514-6.

[59] Kaya S, Thakur A, Kumar A. The role of in silico/DFT investigations in analyzing dye molecules for enhanced solar cell efficiency and reduced toxicity. *J Mol Graph Model.* 2023;124(June):108536.

[60] Thakur A, Kumar A, Kaya S, Marzouki R, Zhang F, Guo L. Recent advancements in surface modification, characterization and functionalization for enhancing the biocompatibility and corrosion resistance of biomedical implants. *Coatings.* 2022;12:1459.

[61] Thakur A, Kumar A. Recent trends in nanostructured carbon-based electrochemical sensors for the detection and remediation of persistent toxic substances in real-time analysis. *Mater Res Express.* 2023;10:034001.

[62] Parveen G, Bashir S, Thakur A, Saha SK, Banerjee P, Kumar A. Experimental and computational studies of imidazolium based ionic liquid 1-methyl-3-propylimidazolium iodide on mild steel corrosion in acidic solution experimental and computational studies of imidazolium based ionic liquid 1-methyl-3-propylimidazolium. *Mater Res Express.* 2020;7(1):016510.

[63] Bashir S, Thakur A, Lgaz H, Chung I-M, Kumar A. Computational and experimental studies on phenylephrine as anti-corrosion substance of mild steel in acidic medium. *J Mol Liq.* 2019;293:111539.

[64] Thakur A, Sharma S, Ganjoo R, Assad H, Kumar A. Anti-corrosive potential of the sustainable CIs based on biomass waste: A review on preceding and perspective research. *J Phys Conf Ser.* 2022;2267(1):012079.

[65] Bashir S, Thakur A, Lgaz H, Chung IM, Kumar A. Corrosion inhibition efficiency of bronopol on aluminium in 0.5 M HCl solution: Insights from experimental and quantum chemical studies. *Surf Interfaces.* 2020;20(April):100542. doi:10.1016/j.surfin.2020.100542.

[66] Thakur A, Kaya S, Abousalem AS, Kumar A. Experimental, DFT and MC simulation analysis of Vicia sativa weed aerial extract as sustainable and eco-benign CIs for mild steel in acidic environment. *Sustain Chem Pharm.* 2022;29(July):100785. doi:10.1016/j.scp.2022.100785.

[67] Thakur A, Kumar A. Recent advances on rapid detection and remediation of environmental pollutants utilizing nanomaterials-based (bio)sensors. *Sci Total Environ.* 2022;834(January):155219. doi:10.1016/j.scitotenv.2022.155219.

[68] Thakur A, Kumar A, Sharma S, Ganjoo R, Assad H. Materials today: Proceedings computational and experimental studies on the efficiency of Sonchus arvensis as green CIs for mild steel in 0.5 M HCl solution. *Mater Today Proc.* 2022;66:609–621. doi:10.1016/j.matpr.2022.06.479.

[69] Thakur A, Kumar A. Sustainable inhibitors for corrosion mitigation in aggressive corrosive media: A comprehensive study. *J Bio- Tribo-Corrosion*. 2021;7(2):1–48. doi:10.1007/s40735-021-00501-y.

[70] Bashir S, Lgaz H, Chung IM, Kumar A. Effective green corrosion inhibition of aluminium using analgin in acidic medium: An experimental and theoretical study. *Chem Eng Commun*. 2020;208(8):1121–1130. doi:10.1080/00986445.2020.1752680.

[71] Thakur A, Savaş K, Kumar A. Recent trends in the characterization and application progress of nano-modified coatings in corrosion mitigation of metals and alloys. *Appl Sci*. 2023;13:730.

[72] Ouachikh O, Bouyanzer A, Bouklah M, Desjobert JM, Costa J, Hammouti B, et al. Application of essential oil of Artemisia herba alba as green CIs for steel in 0.5 M H_2SO_4. *Surf Rev Lett*. 2009;16(1):49–54.

[73] Enyinnaya NP, James AO. Inhibitive potential of benign Caesar-weed leaves extract (CWLE) as CIs of aluminium in H_2SO_4 phase. *Interdiscip J Appl Basic Subj*. 2022;2:1–13.

[74] Berdimurodov E, Kholikov A, Akbarov K, Obot IB, Guo L. Thioglycoluril derivative as a new and effective CIs for low carbon steel in a 1 M HCl medium: Experimental and theoretical investigation. *J Mol Struct*. 2021;1234:130165. doi:10.1016/j.molstruc.2021.130165.

[75] Alibakhshi E, Ramezanzadeh M, Haddadi SA, Bahlakeh G, Ramezanzadeh B, Mahdavian M. Persian liquorice extract as a highly efficient sustainable CIs for mild steel in sodium chloride solution. *J Clean Prod*. 2019;210:660–672.

[76] Berisha A. Experimental, Monte Carlo and molecular dynamic study on corrosion inhibition of mild steel by pyridine derivatives in aqueous perchloric acid. *Electrochem*. 2020;1(2):188–199.

[77] Belarbi N, Dergal F, Chikhi I, Merah S, Lerari D, Bachari K. Study of anti-corrosion activity of Algerian L. stoechas oil on C38 carbon steel in 1 M HCl medium. *Int J Ind Chem*. 2018;9(2):115–125. doi:10.1007/s40090-018-0143-6.

[78] Madani A, Kaabi I, Sibous L, Bentouhami E. Synthesis, characterization and evaluation of the corrosion inhibition on mild steel of two new Schiff bases derived from 4,4′-diaminobiphenyl: Density functional theory investigation. *J Iran Chem Soc*. 2021;18(11):3077–3095. doi:10.1007/s13738-021-02252-6.

[79] Shahen S, Abdel-karim A, Gaber G. Eco-friendly roselle (Hibiscus sabdariffa) leaf extract as naturally CIs for Cu-Zn alloy in 1M HNO_3. *Egypt J Chem*. 2021;65(4):351–361.

[80] Jafar Mazumder MA. A review of green scale inhibitors: Process, types, mechanism and properties. *Coatings*. 2020;10(10):1–29.

[81] Alamri AH. Experimental and theoretical insights into the synergistic effect of iodide ions and 1-acetyl-3-thiosemicarbazide on the corrosion protection of c1018 carbon steel in 1 M HCL. *Materials (Basel)*. 2020;13(21):1–17.

[82] Verma C, Ebenso EE, Quraishi MA, Hussain CM. Recent developments in sustainable CIs: Design, performance and industrial scale applications. *Mater Adv*. 2021;2(12):3806–3850.

[83] Ogunleye OO, Eletta OA, Arinkoola AO, Agbede OO. Gravimetric and quantitative surface morphological studies of Mangifera indica peel extract as a CIs for mild steel in 1 M HCl solution. *Asia-Pacific J Chem Eng*. 2018;13(6):e2257.

[84] Motamedi M, Ramezanzadeh B, Mahdavian M. Corrosion inhibition properties of a green hybrid pigment based on Pr-Urtica Dioica plant extract. *J Ind Eng Chem*. 2018;66(May 2019):116–125. doi:10.1016/j.jiec.2018.05.021.

[85] Zaher A, Aslam R, Lee HS, Khafouri A, Boufellous M, Alrashdi AA, et al. A combined computational & electrochemical exploration of the Ammi visnaga L. extract as a green CIs for carbon steel in HCl solution. *Arab J Chem*. 2022;15(2):103573. doi:10.1016/j.arabjc.2021.103573.

[86] Zadeh MK, Yeganeh M, Shoushtari MT, Esmaeilkhanian A. Corrosion performance of polypyrrole-coated metals: A review of perspectives and recent advances. *Synth Met.* 2021;274(January):116723.

[87] Iroha NB, Dueke-Eze CU. Experimental studies on two isonicotinohydrazide-based schiff bases as new and efficient inhibitors for pipeline steel erosion corrosion in acidic cleaning solution. *Chem Africa.* 2021;4(3):635–646. doi:10.1007/s42250-021-00252-w.

[88] Mary RN, Nazareth R, Adimule SP, Potla K. Investigation of corrosion inhibition property of triazole-based schiff bases on maraging steel in acid mixtures. *J Fail Anal Prev.* 2021;21(2):547–562. doi:10.1007/s11668-020-01099-y.

[89] Saxena A, Prasad D, Haldhar R. Investigation of corrosion inhibition effect and adsorption activities of Cuscuta reflexa extract for mild steel in 0.5 M H_2SO_4. *Bioelectrochemistry.* 2018;124:156–164.

19 Plant Extracts as Corrosion Inhibitors for Steel in NaCl Solutions

Shveta Sharma, Humira Assad,
Abhinay Thakur, and Richika Ganjoo
Lovely Professional University

Ambrish Singh
Central University of Nagaland

Ashish Kumar
Bihar Engineering University
Government of Bihar

19.1 INTRODUCTION

Corrosion is the progressive degradation of metals caused by the chemical or electrochemical (EC) processes in the surrounding environment and is a pervasive and costly problem in numerous industries [1,2]. Among the affected metals, steel is particularly susceptible to corrosion, especially in chloride-containing environments [3–5]. Chloride ions have a highly corrosive nature, accelerating the corrosion process and posing significant challenges for industries that rely on steel-based infrastructure and equipment. To mitigate the detrimental effects of corrosion, traditional methods of corrosion protection have been employed, with synthetic inhibitors being a commonly used approach. These inhibitors work by either forming a protective barrier on the metal surface or altering the EC processes involved in corrosion [6]. While synthetic inhibitors have shown effectiveness in reducing corrosion rates, they are not without limitations. One of the main concerns with synthetic inhibitors is their high cost. The production and application of these inhibitors can be expensive, particularly when used on a large scale in industries with extensive steel infrastructure. Additionally, environmental considerations have come to the forefront, as many synthetic CIs are associated with harmful effects on ecosystems and human health [7]. The use of toxic chemicals raises concerns about their impact on the environment and potential risks to workers handling these inhibitors [8–11].

In light of these drawbacks, there has been a growing interest in exploring alternative CIs that are derived from natural sources and are environmentally friendly [12]. PE have emerged as a promising candidate in this regard. They offer several

DOI: 10.1201/9781003394631-19

advantages over synthetic inhibitors, rendering them a promising option for eco-friendly corrosion fortification [8]. Firstly, PE are abundant in nature and readily available, making them a sustainable choice for corrosion inhibition. Plants can be cultivated or sourced from existing agricultural practices, ensuring a constant supply of raw materials for extraction. This contrasts with synthetic inhibitors that rely on complex chemical synthesis processes and limited resources. Secondly, PE offer a renewable and environmentally friendly approach. Unlike synthetic CIs, which often persevere in the atmosphere and can have long-term ecological effects, PE are bio-degradable and do not contribute to the accumulation of harmful substances. They provide a greener alternative for industries seeking to reduce their environmental footprint and comply with stringent environmental regulations [9]. Moreover, the cost of PE is generally lower compared to synthetic inhibitors. Plant materials can be obtained at a fraction of the cost of synthesizing complex chemical compounds. This affordability makes PE economically viable for large-scale corrosion protection applications, potentially reducing the financial burden on industries [7].

The utilization of plant excerpts as CIs also aligns with the growing demand for sustainable and eco-friendly practices across industries. Governments, regulatory bodies, and consumers are increasingly prioritizing environmentally conscious approaches, making the adoption of PE a favourable choice for corrosion prevention. Inhibitory mechanisms play a vital role in the efficiency of PE as CIs [10]. Understanding these mechanisms is essential for elucidating the inhibitory processes and optimizing their performance.

This chapter discusses the various mechanisms through which PE interact with the metallic outer-layer and inhibit corrosion. This includes the formation of defensive coatings, adsorption processes, and modification of the EC properties of the metal surface. This chapter provides a comprehensive overview of the utilization of PE as CIs for steel in sodium chloride (NaCl) solutions. It explores the potential of PE as proficient alternatives to synthetic CIs and delves into their inhibitory mechanisms. This chapter's goal is to offer a thorough knowledge of the significance of PE as CIs for steel in saline mediums. It aims to showcase the potential of these natural compounds to address the limitations of traditional methods of corrosion protection. By examining the inhibitory mechanisms and factors affecting their performance, researchers and engineers can gain insights into optimizing the utilization of PE for corrosion prevention.

19.1.1 CORROSION IN CHLORIDE-CONTAINING ENVIRONMENTS

To understand the challenges of steel corrosion in Cl^--comprising mediums, it is essential to delve into the fundamentals of corrosion. Metal surfaces gradually deteriorate due to corrosion, a natural process that happens when metals contact with their surroundings [13]. The EC nature of corrosion involves anodic and cathodic reactions that take place simultaneously. In the context of steel corrosion, the primary EC reaction is the oxidation of iron (Fe) to form iron ions (Fe^{2+}). This anodic reaction occurs at localized areas on the metallic outer face, often signified as corrosion pits or anodic areas. Meanwhile, a cathodic reaction takes place, typically involving the reduction of oxygen from the surrounding environment. Chloride ions play a substantial role in accelerating the corrosion phenomenon. When chloride ions are present, they can create an aggressive environment by facilitating the formation of a highly corrosive species called the chloride ion complex. This complex can increase the mobility of metal ions, allowing them to migrate away from the anodic sites and exacerbating the corrosion process.

Along with the EC reactions, other factors can influence the corrosion rate. These include temperature, pH, and the presence of other corrosive species in the environment. Higher temperatures can accelerate corrosion rates by increasing the kinetics of the EC reactions. pH levels also play a crucial role, as highly acidic or alkaline conditions can enhance the corrosion of steel [14–18]. Understanding the fundamentals of corrosion provides a basis for comprehending the challenges specific to steel corrosion in chloride environments. These challenges are explored in the following section.

19.1.2 Challenges of Steel Corrosion in Chloride Environments

Steel, a widely used metal in various industries, is particularly vulnerable to weathering in chloride-containing environments. The manifestation of Cl^- ions introduces specific challenges and accelerates the corrosion process, posing significant risks to infrastructure, equipment, and the economy. Several factors contribute to the challenges of steel corrosion in chloride environments:

- *High Corrosion Rates*: Chloride ions have a corrosive effect on steel, leading to accelerated corrosion rates. The aggressive nature of chloride ions enhances the anodic dissolution of iron, resulting in rapid metal loss. This elevated corrosion rate can compromise the structural integrity of steel components and lead to catastrophic failures if left unchecked [19–23].
- *Pitting Corrosion*: Chloride ions are notorious for promoting pitting corrosion, a localized form of corrosion characterized by the formation of small pits or holes on the metal surface. Pitting corrosion can occur even in the presence of a protective oxide layer on the steel surface. The aggressive nature of chloride ions can break down the passivating film, allowing for localized attack and the initiation of corrosion pits.
- *Corrosion under Deposits*: In chloride-containing environments, the accumulation of deposits on the metal surface can exacerbate corrosion. Deposits, such as scale, sediments, or biofilms, can trap chloride ions and other corrosive species, creating a localized corrosive microenvironment. This phenomenon, known as corrosion under deposits, can result in accelerated corrosion rates and increased damage to the metal surface.
- *Crevice Corrosion*: Chloride-containing environments often have crevices or gaps where water can accumulate, creating a stagnant, oxygen-depleted environment. This environment is conducive to crevice corrosion, which occurs in confined spaces such as gaps between metal surfaces or underseals. Chloride ions can concentrate in these crevices, leading to localized corrosion and the potential for rapid deterioration.

• *Sensitization and Intergranular Corrosion*: In certain grades of stainless steel, the presence of chloride ions can lead to sensitization and intergranular corrosion. Sensitization occurs when chromium carbides precipitate along grain boundaries, depleting the area adjacent to the grain boundaries of chromium. This chromium-depleted region becomes susceptible to intergranular corrosion in the incidence of Cl^-, leading to accelerated degradation along the grain boundaries.

• *Environmental Factors*: The challenges of steel corrosion in chloride environments are also influenced by various environmental factors. Temperature, humidity, pH, and the presence of other corrosive species can all impact corrosion rates [24–31]. Higher temperatures and humidity levels can accelerate the corrosion process, while low pH levels or the presence of other aggressive species can further enhance the corrosive environment.

Understanding the specific challenges associated with steel corrosion in chloride-containing environments is crucial for developing effective corrosion mitigation strategies. By comprehending the mechanisms and factors that contribute to accelerated corrosion rates, researchers and engineers can devise targeted approaches to mitigate corrosion and prolong the service life of steel infrastructure and equipment.

19.2   TRADITIONAL METHODS OF CORROSION PROTECTION

19.2.1   SYNTHETIC INHIBITORS

Synthetic inhibitors have been widely employed as traditional methods of corrosion protection. These inhibitors are chemical compounds specifically designed to reduce or prevent corrosion by interfering with the EC reactions occurring on the metallic outer-layer. Anodic inhibitors and cathodic inhibitors are the two primary categories. Anodic inhibitors function by coating the metal surface with a shielding layer. By serving as a physical barrier, this coating keeps the metal from Anodic inhibitors typically include compounds such as chromates, molybdates, and phosphates. These compounds are capable of forming stable and insoluble films that adhere to the metal surface, providing corrosion resistance. Cathodic inhibitors, on the other hand, function by reducing the oxygen concentration at the cathodic sites, thereby impeding the reduction reaction. Common cathodic inhibitors include nitrites, azoles, and amines [32–34]. These compounds act as oxygen scavengers, reducing the availability of oxygen for the cathodic reaction and inhibiting the corrosion process. Synthetic inhibitors offer several advantages that have contributed to their widespread use in corrosion protection. They are generally effective in reducing corrosion rates and extending the lifespan of metal structures and equipment. Synthetic inhibitors can be easily applied through various methods such as immersion, spraying, or coating, making them versatile in different industrial settings. Moreover, they can be tailored to specific applications by adjusting their concentration or formulation.

19.2.2 DRAWBACKS OF TRADITIONAL METHODS

Despite their effectiveness, traditional methods of corrosion protection using synthetic inhibitors come with a range of drawbacks that have led to increased interest in alternative approaches.

- *High Cost*: One significant drawback of synthetic inhibitors is their high cost. The production and formulation of these inhibitors often involve complex chemical processes and specialized equipment, resulting in elevated production costs. Additionally, the cost of acquiring and maintaining the necessary equipment and expertise for their application can further contribute to the overall expense. This high cost is a significant barrier, particularly in industries with extensive steel infrastructure that require large-scale corrosion protection.
- *Environmental Concerns*: Synthetic inhibitors can pose environmental concerns due to their composition and potential ecological impact. Many traditional inhibitors contain toxic or hazardous substances, such as chromates, which are known carcinogens. The release of these compounds into the environment can have adverse effects on ecosystems and human health. Concerns over the environmental impact of synthetic inhibitors have led to regulatory restrictions and a push for greener alternatives.
- *Toxicity and Health Hazards*: Synthetic inhibitors can present health hazards to workers involved in their production, handling, and application. The toxic nature of certain compounds used as inhibitors poses risks to human health, particularly during exposure or accidents. Inhalation, skin contact, or ingestion of these substances can lead to adverse health effects. As a result, stringent safety protocols and protective measures are necessary when working with synthetic inhibitors.
- *Sustainability Concerns*: In recent years, the demand for sustainable and environmentally friendly practices has grown significantly. Synthetic inhibitors, with their potential ecological impact and reliance on non-renewable resources, may not align with the principles of sustainability. Industries are increasingly seeking greener alternatives that minimize environmental footprints and promote long-term sustainability.
- *Limited Long-Term Performance*: While synthetic inhibitors can provide effective corrosion protection in the short term, their long-term performance may be limited. Factors such as exposure to harsh environments, degradation of the inhibitor film, and depletion of the inhibitor concentration over time can diminish their effectiveness. This limitation necessitates regular reapplication or maintenance, leading to additional costs and logistical challenges.

The drawbacks associated with traditional methods of corrosion protection using synthetic inhibitors have driven the exploration of alternative approaches. The need for eco-friendly, cost-effective, and sustainable CIs has propelled research into natural sources, such as PE, that offer promising alternatives for corrosion control.

These alternatives aim to address the limitations of synthetic inhibitors while providing effective, long-lasting corrosion protection for steel in chloride-containing environments.

19.3 PLANT EXTRACTS AS ALTERNATIVE CORROSION INHIBITORS FOR STEEL

19.3.1 ECO-FRIENDLY CORROSION PROTECTION OF STEEL

Corrosion of steel in sodium chloride (NaCl) solutions is a common and significant challenge across various industries. The occurrence of Cl^- ions exacerbates the corrosion process, leading to the degradation of steel structures, equipment, and infrastructure. In recent years, there has been a growing emphasis on eco-friendly corrosion protection strategies for steel in NaCl solutions. Natural CIs stemmed from renewable resources, like PE, have gained significant attention as eco-friendly alternatives for the corrosion protection of steel in NaCl solutions. PE are obtained from several parts of plants, like roots, stems, leaves, and fruits, and they contain a rich array of organic compounds with corrosion-inhibitory properties as shown in Figure 19.1.

These compounds include phenols, alkaloids, flavonoids, tannins, and terpenoids, among others. Phenolic compounds are one of the most abundant and effective classes of CIs found in PE. They possess aromatic rings with hydroxyl groups that can form complexes with metal ions, inhibiting the corrosion progression. Alkaloids, another class of compounds present in PE, exhibit inhibitory effects by materializing a defensive coating on the metallic outer-layer or altering the EC reactions involved in corrosion. Flavonoids, known for their antioxidant properties, scavenge free radicals and inhibit the oxidative reaction that results in corrosion. Tannins, with their high molecular weight and complex structure, can bind to metallic exteriors, putting up a barrier against species that cause corrosion. Terpenoids, which are widely distributed in plants, have shown promising inhibitory effects by modifying the EC performance of steel. In an experiment, Shyamvarnan et al. [36] also studied

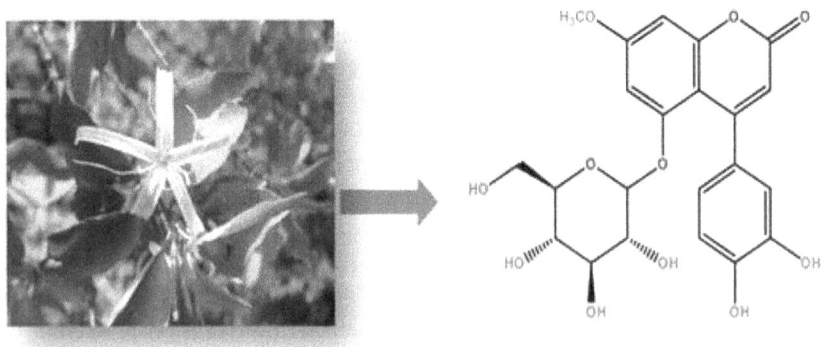

FIGURE 19.1   *Exostema caribaeum* and chemical structure of 4-PC employed as CI for AISI 1018 steel in 3% NaCl [35].

the anti-corrosive effects of the ethanol extract obtained from *Elettaria cardamo-mum* (commonly known as cardamom) on MS in a 3.5% NaCl solution. The dose range for the *E. cardamomum* extract was 50–250 ppm, and its impact on corrosion rate (C_R) and corrosion inhibition efficiency (IE) was evaluated using the weight loss (W_L) approach over a period of 1 and 7 days. The findings revealed that the IE augmented with higher doses of the extract and reduced as the immersion period extended. The maximum IE of 73.5% was achieved after 1 day of immersion with a 250-ppm concentration of *E. cardamomum* extract, while the IE decreased to 54.1% after 7 days. The mitigating influence was accredited to the adsorption of the CI onto the outer-layer of the MS. Thermodynamic analysis revealed that the adsorption obeyed Langmuir adsorption isotherm (LAI) and was characterized as physisorption. Ben et al.'s [37] study of *Olea europaea* (olive leaf) extract, evaluated on MS in a "0.1 M NaOH + 0.5 M NaCl" corrosive medium, revealed that IE rises as the extraction solvent's polarity does. They discovered that dichloromethane ethyl, hexane, methanol, and acetoacetate all had lower IEs. The excerpt of CH_3OH that had the highest level of inhibitory effectiveness was around 91.9%. By using GCMS analysis, the scientists hypothesized that the defensive function might be brought on by the existence of heteroatoms like O, N, and π-ēs. It was discovered that the extract had high levels of phenol and flavonoids. The PP investigations indicate that olive leaf extract is a mixed-type (MT) CI. The findings of EIS and PP were in good arrangement. Similarly, Othman et al. [38] investigated the mitigation of steel corrosion in a 3.5% NaCl media using rice straw extract (RSE) as the inhibitor. The IE was assessed over a period of 42 days using the W_L technique. The findings consistently displayed a high corrosion inhibition efficiency (IE) of 92% and a reduced C_R of 13×10^{-3} mm year^{-1}, indicating the strong inhibitory effect of RSE on steel corrosion. Cyclic polarization measurements revealed that RSE effectively reduced pitting corrosion by exhibiting a small hysteresis loop, indicating the suppression of localized corrosion. The surface analysis revealed the generation of a ferric-RSE covering, suggesting the complexation between the inhibitor and the metal surface. This layer served as a corrosion-resistant barrier, further enhancing the inhibitory performance of RSE. Furthermore, immersion of the steel in RSE resulted in a significant reduction in the existence of corrosive moieties on the metal outer-layer. This observation indicated that RSE not only inhibited the initiation of corrosion but also hindered the growth of caustic species, thereby preventing the progression of the corrosion process. Based on the experimental findings, interphase inhibition was suggested as the means through which RSE could prevent steel corrosion. RSE formed a protective interfacial coating on the metallic exterior, effectively separating it from the hostile environment. This interphase inhibition prevented the direct interaction between the metal and the saline medium, thereby reducing the C_R and inhibiting the formation of corrosion products.

Devikala et al. [39] explored the influence of aqueous *Allium sativum* (AS) extract in 3.5% NaCl on the prevention of MS corrosion using a variety of techniques, including potentiodynamic polarization, EC impedance spectroscopy, SEM, FTIR, and XRD. The findings demonstrated that AS is a superior Green inhibitor, with a 95% maximum inhibition efficiency. According to the EIS studies, the creation of a protective exterior coating including chemicals found in garlic extract was what caused the AS extract to lessen the C_R. Additionally, the PDP study demonstrates that garlic extract functions as an MT-CI and forms a defensive coating on metal surfaces.

Haddadi and colleagues [40] discovered that increasing the immersion duration up to 48 hours improved the inhibition capacity of the *Junglans regia* green fruit shell extract used as a CI on MS in 3.5 wt.% NaCl medium. With 1,000 ppm, the highest level of inhibitory efficacy of about 94% was attained. Both anodic and cathodic processes were delayed, according to PP experiments. Functional elements of the extract's phenolic chemicals, including carboxyl, hydroxyl, and carbonyl groups, were intended to physically and chemically adsorb on the metal surface. Therefore, a mixed physisorption-chemisorption process was brought about through electrostatic interactions and covalent bonding.

Motamedi et al. [41] focused on investigating the anti-corrosive characteristics of a hybrid complex composed of praseodymium nitrate (PrN) and *Urtica dioica* (UrDi) on MS in a NaCl medium. The objective of the study was to assess the corrosion resistance provided by this synthesized CI on the MS outer layer in a neutral NaCl medium. The study revealed significant results, demonstrating the efficiency of the hybrid complex as a CI. After immersing the MS in the solution for 24 hours, the synthesized inhibitor exhibited the highest corrosion resistance with an impedance value of 4,437 Ω cm^{-2}. This impedance value indicates the potency of the CI to hinder the flow of corrosive species and lessen the C_R. Furthermore, the IE of the synthesized inhibitor was approximately 94%. This high inhibition efficiency suggests that the hybrid complex effectively prevents or slows down the corrosion process on the MS outer-layer. Additionally, the surface coverage of the CI was also around 94%, indicating that a substantial portion of the MS outer-layer was protected by the inhibitor. The amalgamation of PrN and UrDi in the synthesized inhibitor contributed to its excellent corrosion inhibition properties. The specific chemical interactions and synergistic effects between these components likely enhance the inhibitory performance. The presence of PrN and UrDi on the MS outer-layer forms a protective coating that impedes the interaction between the metallic specimen and the corrosive environment. To gain insight into the surface morphology, the treated and untreated mild steel specimens were examined using SEM. This analysis provided a visual comparison of the morphological differences between the inhibited and uninhibited surfaces, offering further evidence of the protective effect of the *E. cardamomum* extract.

Singh et al. [42] used mass spectrometry (MS), FTIR, EIS, PDP, contact angle measurements, and SEM to investigate how the fruit extract of *Gingko biloba* (GFE) inhibits the deterioration of J55 steel in CO_2-saturated NaCl (3.5 wt.%) solution. It was discovered that the extract's dosage augmented the effectiveness of inhibition. The LAI was obeyed by the extract's adsorption on the J55 steel surface. Contact angle and SEM data elucidated the occurrence of an adsorbed coating comprising an inhibitor on the surface of J55 steel. The outcomes demonstrated that the GFE might be a powerful inhibitor of metal corrosion in CO_2-saturated saline mediums.

The *Camellia sinensis* (aqueous green tea) extract was evaluated by Pradipta et al. [43] as a CI on CS in a 3.5% NaCl caustic environment. The natural antioxidants in green tea were chosen because these substances, like green tea itself, include many polar atoms and electron-rich connections, which may act as MT-CIs. According to experiments on linear polarization resistance, green tea extract exhibited an identical anti-corrosion efficiency, ranging from 51% to 70%, as commercial calcium nitrite CI at equivalent doses. Green tea extract had a higher corrosion inhibition efficiency of about 75%–80% at equal levels as compared to commercial CI. Catechin or (-)-epicatechin, (-)-epigallocatechin gallate, and (-)-epicatechin gallate are expected to combine to form a mixed-type corrosion-inhibiting layer by liquid chromatography-mass spectrometry.

Kusumastuti et al. [44] investigated the effectiveness and mechanism of a green CI derived from *Morinda citrifolia* (Noni) on low CS. The motivation behind this study is to develop a cost-effective and environmentally friendly CI utilizing components obtained from tropical plants found in Indonesia. The main goal of this study is to assess Noni extracts' effectiveness as a green CI for carbon steel in aggressive environments. The experiment was conducted using a 3.5% NaCl solution, which represents a challenging corrosion environment. The dosage of the Noni extract and the immersion time were varied as experimental parameters, with all tests performed at room temperature. The C_R was determined using EC polarization measurements with a CMS 600-Gamry instrument and W_L measurements. Fourier transform infrared (FTIR) spectroscopy was used to examine how the CI adhered to the metallic outer layer and formed a defensive coating. Polarization curve exploration and fitting to the LAI were used to explore the inhibitory mechanism. According to the experimental findings, the Noni inhibitor's IE was strengthened when its concentration was increased. After an immersion time of 288 hours, the best inhibition was obtained at a concentration of 3 ppm of Noni fruit extract. With an IE of 76.92%, the C_R determined under these circumstances was 1.385 mpy (million pounds per year). The development of a monolayer film on the material's surface revealed Noni's MT- anti-corrosive behaviour.

Further, the effectiveness of a green CI made from *Tamarindus indiaca* (TAM) extract combined with zinc nitrate (ZS) was also examined by Akbarzadeh et al. MS was the material investigated, and the hostile environment used was 3.5% NaCl. After a 24-hour immersion, the results of EIS demonstrated the synergistic behaviour and 96% corrosion IE in TAM with 300 ppm and ZS 700 ppm. Results from the polarization spectrum showed that the anodic depression dominated the behaviour of the TAM and ZS mixture. FE-SEM and grazing incidence XRD images proved that a uniform protective layer had been formed [45].

Similarly, Ibrahim et al. [46] focused on investigating the potential of *Calotropis procera* leaves extract (CPLE) as a CI for MS in a CO_2-saturated 3.5 wt.% NaCl solution, employing different EC approaches. The objective is to evaluate the inhibitory efficiency of CPLE at various doses or concentrations. The findings indicate that CPLE exhibits significant inhibitory properties, with an IE of 80% at a dosage of 50 ppm. Polarization studies further reveal that CPLE acts as an MT, influencing both the anodic and cathodic reactions involved in the corrosion process. The adsorption behaviour of CPLE on the surface of the MS follows the LAI, indicating monolayer formation. This adsorption process is a crucial factor in the inhibitory action of CPLE.

By using PDP, EIS, and SEM techniques, Shabani et al. [47] examined the inhibitory effect of *Santolina chamaecyparissus* (SC) extract as a natural CI on oxidation of 304 SS in 3.5% NaCl medium. The results demonstrate that the extract exhibits MT inhibitory behaviour, and the addition of 1.0 g L^{-1} extract developed in an IE of 86.9%. Increasing SC concentration results in higher R_{ct} and lower C_{dl}, as per the impedance curves. The rate of corrosion rose with medium temperature increases and decreased with extract concentration increases. The extract's adsorption adhered to the Langmuir isotherm model.

To acquire additional comprehensions into the corrosion inhibition mechanism, molecular dynamics simulations (MDS) were conducted. These simulations provide a molecular-level understanding of the interactions between CPLE and the steel surface. It was observed that the existence of nitrogen (N) and oxygen (O) atoms in CPLE, along with the presence of π-electron systems, contributes to the high adsorption energies and strong interactions between CPLE and the steel surface. These interactions play an important role in the corrosion inhibition performance of CPLE.

Furthermore, Alibakshi et al. [48] proposed Persian Liquorice, derived from the root of *Glycyrrhiza glabra*, as a sustainable CI for MS in NaCl mediums. This natural extract contains various active compounds, including Licochalcone E (LCE), Glycyrrhizin (GL), Licochalcone A (LCA), Liquritigenin (LTG), 18βGlycyrrhetinic acid (GA), and Glabridin (GLD), which contribute to its excellent inhibition properties. To identify and analyse these active components, FT-IR spectroscopy was employed. The corrosion IE of Persian Liquorice (PL) was investigated using EIS, PDP, and EC current noise measurements at different concentrations of the extract. The EC experiments revealed a remarkable maximum IE of 98.8%, which demonstrates the effectiveness of Persian Liquorice in protecting mild steel against corrosion. This high level of inhibition power has not been reported for any other sustainable CI in chloride solutions, even after extended immersion periods of up to 72 hours. Surface analysis techniques such as SEM and AFM confirmed the materialization of a defensive covering on the MS outer-layer in the existence of PL. These results visually demonstrate the barrier formed by the inhibitor, which prevents the diffusion of harsh ions and further corrosion. To gain a deeper understanding of the inhibitory mechanism, MDS and quantum mechanics calculations were conducted. These computational approached demonstrated the accumulation of PL moieties onto the metallic outer layer through donor-acceptor interactions. This adsorption plays a crucial role in the formation of the protective layer and the subsequent inhibition of corrosion.

19.3.2 ADVANTAGES AND POTENTIAL OF PE

PE, derived from various parts of plants, offer numerous advantages and exhibit significant potential as CIs. These natural compounds have gained consideration as eco-friendly substitutes for synthetic inhibitors due to their inherent properties and sustainable nature. Here, we explore the advantages and potential of PE in corrosion inhibition.

- *Abundance and Renewability*: One of the major advantages of PE is their abundance in nature. Plants are widely available and can be cultivated or sourced from existing agricultural practices. This ensures a constant and renewable supply of raw materials for the extraction of corrosion-inhibiting compounds. Unlike synthetic inhibitors, which often rely on non-renewable resources, PE provide a sustainable solution for corrosion prevention.

- *Biodegradability and Environmental Safety*: PE are biodegradable and pose minimal risk to the environment. They break down naturally into harmless substances, reducing the risk of environmental contamination and accumulation. This aligns with the principles of green chemistry and promotes environmental safety. The biodegradability of PE minimizes their impact on ecosystems and contributes to sustainable practices.

- *Low Toxicity and Human Safety*: PE are generally considered safe for human health. The compounds present in PE are often recognized as safe and have a long history of use in traditional medicine and food industries. Unlike synthetic inhibitors, which may contain toxic substances, PE offer a safer alternative, reducing health risks for workers involved in their production, handling, and application.

- *Diverse Chemical Composition*: PE contain a rich diversity of organic compounds, including phenols, alkaloids, flavonoids, tannins, saponins, and terpenoids, among others. This diverse chemical composition provides PE with a broad spectrum of inhibitory properties. Different compounds can target specific corrosion mechanisms, leading to a multi-functional corrosion protection effect.

- *Multifaceted Inhibitory Mechanisms*: The inhibitory mechanisms of PE are often multifaceted, contributing to their effectiveness as CIs. PE can interact with the metallic exterior through adsorption, generating a defensive barrier that inhibits the access of corrosive species to the metal. They can also scavenge free radicals and inhibit the EC reactions involved in corrosion. Moreover, certain compounds in PE possess passivating properties, forming a stable oxide layer on the metallic exterior and augmenting its endurance to corrosion.

- *Tailorability and Customization*: PE offer the advantage of tailorability and customization for specific corrosion protection applications. The composition and concentration of active compounds can be adjusted to target the specific corrosion mechanisms and conditions. By optimizing the formulation of plant extract-based inhibitors, their inhibitory performance can be enhanced, leading to more effective corrosion protection.

- *Cost-Effectiveness*: PE provide a cost-effective alternative to synthetic inhibitors. Plant materials are generally affordable and can be obtained at a fraction of the cost of synthesizing complex chemical inhibitors. The extraction processes are relatively simple and can be scaled up for large-scale applications, reducing the overall cost of corrosion inhibition.

- *Compatibility with Coatings and Paints*: PE can be easily incorporated into protective coatings and paints. They are compatible with various coating systems, such as epoxy-based or polyurethane-based coatings, without compromising their inhibitory properties. This compatibility allows for the integration of PE into existing corrosion protection systems, offering an additional layer of defence against corrosion.

- *Promising Research Findings*: Numerous studies have demonstrated the efficacy of PE as CIs. These studies have evaluated the inhibitory functioning of various PE through corrosion tests, such as EC techniques, W_L approaches, and surface analysis. The studies consistently indicate the potential of PE in reducing corrosion rates, inhibiting the development of corrosion products, and extending the service life of metals.
- *Wide Range of Applications*: PE have shown promise in corrosion protection across various industries and applications. They have been successfully applied in oil and gas pipelines, marine environments, chemical processing plants, and many other settings. The versatility of PE allows for their use in different corrosive environments, making them a valuable option for corrosion prevention.

19.4   INHIBITIVE MECHANISMS OF PE ON STEEL

Corrosion of steel is a complex EC process influenced by several factors, including the corrosive environment and the composition of the metal surface. PE, derived from various plant sources, have been recognized for their potential as eco-friendly CIs for steel. These natural compounds contain a diverse range of organic compounds, such as phenols, alkaloids, flavonoids, tannins, saponins, and terpenoids, which exhibit inhibitory properties against corrosion. Understanding the inhibitive mechanisms of PE on steel is crucial for optimizing their performance and developing effective corrosion protection strategies. In this chapter, we delve into the various inhibitive mechanisms of PE on steel.

19.4.1   FILM FORMATION MECHANISM

PE exhibit the ability to form a protective film on the surface of steel, which plays a crucial role in inhibiting corrosion. This film acts as a physical barrier that prevents direct contact between the metal and the corrosive environment. By inhibiting the transport of corrosive species, such as oxygen and chloride ions, to the metal surface, the film reduces the corrosion rate and enhances the durability of the steel. The formation of the protective film involves the adsorption of plant extract compounds onto the metal surface. When the PE come into contact with the steel surface, the active compounds present in the extracts interact with the metal atoms, leading to adsorption. The adsorbed molecules can physically block the active sites on the metal surface, limiting the access of corrosive species to these sites. This physical barrier impedes the diffusion of ions and other corrosive agents, reducing the likelihood of EC reactions that initiate corrosion.

In addition to physically blocking the active sites, the adsorbed molecules from PE can also undergo chemical reactions with the metal surface. These reactions can result in the formation of a passivating layer, which further enhances the corrosion resistance of the steel. The passivating layer acts as a chemical barrier, preventing the corrosion-promoting reactions from occurring on the metal surface. The protective film formed by PE can be composed of organic compounds, such as phenols, flavonoids, and tannins, present in the extracts. These compounds possess functional groups, such as hydroxyl groups, that can interact with the metal surface. Through adsorption and subsequent reactions, the protective film is formed, providing a stable and durable barrier against corrosive species. By hindering the diffusion of corrosive species, the protective film inhibits both anodic and cathodic reactions involved in corrosion. At the anodic sites, where metal oxidation occurs, the film acts as a physical barrier, preventing the flow of electrons and the release of metal ions. At the cathodic sites, where reduction reactions take place, the film blocks the access of oxygen and other species, thereby impeding the cathodic reactions.

Furthermore, the protective film formed by PE can exhibit self-healing properties. In the event of minor damage or local disruption to the film, the active compounds in the PE can migrate and re-adsorb onto the exposed metal surface, repairing and re-establishing the protective film. This self-healing ability contributes to the long-term effectiveness of plant extract-based corrosion inhibition. It is important to note that the formation and stability of the protective film can be influenced by various factors, including the concentration of the plant extract, the nature of the corrosive environment, the pH level, and the temperature. Optimizing these parameters can enhance the inhibitory performance of the film and maximize the corrosion protection provided by PE.

19.4.2 Adsorption Mechanism

PE have the ability to adsorb onto the surface of metals, including steel, resulting in the inhibition of the corrosion process. This adsorption mechanism plays a crucial role in forming a protective layer that hinders the interaction between the metal surface and corrosive species. The adsorbed molecules from the PE can physically block the active sites on the metal surface or undergo chemical reactions with the metal, leading to the formation of a passivating layer. The adsorption of PE onto the metal surface is influenced by various factors, including the chemical composition of the extract, the concentration of the extract, and the properties of the metal surface. These factors contribute to the strength and stability of the adsorption process and ultimately determine the effectiveness of corrosion inhibition. The chemical composition of the plant extract plays a vital role in its adsorption behaviour. Organic compounds present in the extract, such as phenols, alkaloids, flavonoids, tannins, and other functional groups, can interact with the metal surface. Electrostatic interactions, hydrogen bonding, van der Waals' forces, and other intermolecular forces facilitate the adsorption process. The specific functional groups present in the plant extract compounds can form strong bonds with the metal surface, enhancing the adsorption capacity and stability.

The concentration of the plant extract also influences the adsorption mechanism. Higher concentrations of PE typically result in increased adsorption onto the metal surface. The higher concentration provides a greater number of active compounds available for adsorption, leading to more effective corrosion inhibition. However, there is an optimum concentration range for adsorption, and exceeding this range may lead to saturation or desorption of the adsorbed molecules. The properties of the metal surface, such as its roughness, morphology, and surface charge, also affect the adsorption mechanism. A rough or irregular metal surface provides more sites for adsorption, increasing the surface coverage of the protective layer. The morphology of the metal surface, including its grain boundaries and defects, can influence the adsorption kinetics and the distribution of adsorbed molecules. The surface charge of the metal affects the electrostatic interactions between the metal surface and the charged functional groups in the plant extract compounds. The adsorption process can result in the physical blocking of active sites on the metal surface. The adsorbed molecules act as a barrier, preventing corrosive species from reaching the metal surface and initiating corrosion reactions. This physical blocking mechanism reduces the diffusion of corrosive species, such as oxygen, water, or chloride ions, to the metal surface, thereby inhibiting the corrosion process.

Additionally, the adsorbed molecules can undergo chemical reactions with the metal surface, leading to the formation of a passivating layer. The functional groups present in the plant extract compounds can react with the metal ions or with the metal surface itself, resulting in the formation of stable compounds or oxides. This passivating layer acts as a barrier that inhibits further corrosion reactions and protects the metal surface from the corrosive environment. The adsorption mechanism of PE on steel provides a versatile and effective means of corrosion inhibition. The combination of physical blocking and passivation effects contributes to the overall inhibitory performance. By tailoring the chemical composition of the plant extract, optimizing the extract concentration, and considering the properties of the metal surface, researchers can enhance the adsorption mechanism and develop plant extract-based CIs with improved effectiveness and stability.

19.4.3 Passivation Mechanism

Some PE possess the remarkable ability to induce passivation of the metal surface, leading to the formation of a stable oxide layer. This passivation layer acts as both a physical and chemical barrier, effectively shielding the metal from further corrosion attacks. The passivation mechanism is particularly effective in environments where the formation of a stable oxide layer is critical for corrosion protection. The passivation process involves the formation of an oxide layer on the metal surface, which significantly reduces the corrosion rate. Plant extract compounds play a crucial role in promoting the formation of a stable and protective oxide layer through various mechanisms. One mechanism by which PE facilitate passivation is through the reduction of metal cations. The active compounds present in the PE possess reducing properties, allowing them to interact with metal cations present on the metal surface. By undergoing redox reactions, the plant extract compounds effectively reduce the metal cations to their lower oxidation states. This reduction process promotes the formation of metal oxides, which act as the passivation layer. The metal oxides exhibit greater stability and resistance to corrosion, thereby enhancing the overall corrosion protection.

Another mechanism employed by PE to induce passivation is the coordination of metal ions. The organic compounds present in the extracts can coordinate with metal ions present on the metal surface, forming coordination complexes. These coordination complexes can effectively promote the formation of a stable oxide layer by facilitating the deposition of metal ions as oxides. The coordination of metal ions by the plant extract compounds enhances the adhesion and stability of the passivation layer, providing robust protection against corrosion. Furthermore, the formation of complexes with metal ions is another mechanism by which PE induce passivation. The active compounds in the extracts can form complex structures with metal ions, leading to the deposition of metal complexes on the metal surface. These metal complexes act as a protective layer, effectively preventing the interaction of corrosive species with the metal surface. The formation of metal complexes provides an additional level of corrosion resistance, enhancing the longevity of the passivation layer and the overall corrosion protection. The ability of PE to induce passivation and promote the formation of a stable oxide layer is highly advantageous for corrosion protection. The passivation layer acts as a physical barrier that impedes the diffusion of corrosive species, such as ions and oxygen, to the metal surface. Additionally, it acts as a chemical barrier by altering the EC reactions involved in corrosion. The formation of a stable oxide layer effectively inhibits both the anodic and cathodic reactions, thereby significantly reducing the corrosion rate of the metal. The passivation mechanism employed by PE is influenced by various factors, including the chemical composition of the extract, the concentration of the extract, and the nature of the metal surface. By understanding and optimizing these factors, researchers can tailor the properties of plant extract-based CIs to enhance their passivation capabilities and overall corrosion protection performance.

19.4.4 OXYGEN-SCAVENGING MECHANISM

Certain PE possess remarkable oxygen-scavenging properties, making them effective inhibitors of the corrosion process. Oxygen reduction is a significant cathodic reaction in the corrosion process, and by scavenging oxygen, PE can impede this reaction and effectively inhibit corrosion. The oxygen scavenging mechanism involves the ability of plant extract compounds to react with dissolved oxygen, preventing its reduction at the cathodic sites. In the corrosion process, oxygen reduction at the cathodic sites plays a crucial role in sustaining the corrosion reactions. Oxygen serves as the primary electron acceptor, participating in the reduction reactions that occur at the cathodic regions of the metal surface. By scavenging oxygen, PE reduce the availability of oxygen for these cathodic reactions, thus hindering the corrosion process. Plant extract compounds possess inherent chemical properties that enable them to react with dissolved oxygen. These compounds can undergo various chemical reactions, such as oxidation or complexation, with oxygen molecules present in the corrosive environment. As a result, the plant extract compounds effectively consume or sequester oxygen, preventing its reduction at the cathodic sites.

By reducing the availability of oxygen, PE disrupt the EC reactions involved in the corrosion process. The reduction reactions that rely on oxygen as the electron acceptor are impeded, thereby inhibiting the overall corrosion process. The oxygen scavenging mechanism acts as a potent defence mechanism against corrosion, as it directly interferes with one of the key reactions that sustain the corrosion process. The ability of PE to scavenge oxygen is attributed to the presence of specific organic compounds within the extracts. Phenols, flavonoids, and other active compounds in the PE possess chemical functionalities that enable them to react with oxygen. These compounds can donate electrons or hydrogen atoms to oxygen molecules, effectively neutralizing their reactivity and preventing their participation in corrosive reactions. The oxygen scavenging mechanism can be further influenced by factors such as the chemical composition and concentration of the plant extract, as well as the environmental conditions. The specific chemical constituents of the plant extract determine its oxygen-scavenging activity, with different compounds exhibiting varying degrees of effectiveness. The concentration of the plant extract plays a role in the scavenging process, as higher concentrations can provide a greater number of active scavenging compounds. Additionally, environmental factors such as temperature, pH, and the presence of other corrosive species can affect the oxygen scavenging mechanism.

The oxygen scavenging mechanism of PE offers a unique and effective approach to inhibiting corrosion. By reducing the availability of oxygen at the cathodic sites, PE disrupt the EC reactions involved in corrosion, effectively impeding the progression of the corrosion process. This mechanism complements other inhibitive mechanisms of PE, such as film formation, adsorption, and passivation, providing comprehensive corrosion protection.

19.4.5 RADICAL-SCAVENGING MECHANISM

Many PE are rich in antioxidant compounds that exhibit the remarkable ability to scavenge free radicals. Free radicals, such as hydroxyl radicals (OH), are highly reactive species that can initiate and propagate corrosion processes. By scavenging free radicals, PE can effectively interrupt the chain reactions involved in corrosion and inhibit the corrosion process. The radical scavenging mechanism of PE involves the ability of their compounds to neutralize free radicals by donating electrons or hydrogen atoms. These active compounds act as sacrificial agents, readily giving up their electrons or hydrogen atoms to the free radicals, thereby effectively reducing their reactivity. By neutralizing free radicals, PE disrupt the chain reactions that lead to the propagation of corrosion, thereby inhibiting the overall corrosion process. The scavenging action of plant extract compounds reduces the reactivity of free radicals, preventing their participation in corrosive reactions. Free radicals are highly reactive and can initiate oxidation reactions, leading to the degradation and corrosion of materials. By donating electrons or hydrogen atoms, the antioxidant compounds in PE effectively stabilize the free radicals, rendering them less reactive and preventing them from initiating or propagating corrosion reactions.

The radical scavenging mechanism employed by PE is crucial for inhibiting corrosion as it interrupts the chain reactions that sustain the corrosion process. By neutralizing free radicals, PE effectively break the chain of reactions, preventing further damage and corrosion. This mechanism complements other inhibitive mechanisms of PE, such as film formation, adsorption, passivation, and oxygen scavenging, providing comprehensive protection against corrosion. The antioxidant compounds present in PE can vary, including phenols, flavonoids, tannins, and other active compounds. These compounds possess chemical functionalities that allow them to readily donate electrons or hydrogen atoms to free radicals. The structure and composition of these compounds influence their radical scavenging activity, with different compounds exhibiting varying degrees of effectiveness in inhibiting corrosion. Furthermore, the radical scavenging mechanism of PE is influenced by factors such as the concentration of the extract and the specific environmental conditions. Higher concentrations of PE can provide a greater number of active scavenging compounds, enhancing their effectiveness in neutralizing free radicals. Environmental factors, such as temperature, pH, and the presence of other corrosive species, can also influence the radical scavenging mechanism. The radical scavenging mechanism employed by PE offers a powerful and effective approach to inhibiting corrosion. By neutralizing free radicals, PE effectively disrupt the chain reactions involved in corrosion, preventing further degradation and damage to materials. This mechanism, in combination with other inhibitive mechanisms of PE, contributes to comprehensive and sustainable corrosion protection.

19.4.6 Synergistic Effects

The inhibitive mechanisms of PE on steel are not limited to a single mechanism but often involve a combination of different mechanisms. The simultaneous action of multiple inhibitive mechanisms can lead to synergistic effects, resulting in enhanced corrosion protection. This combination of mechanisms offers a more comprehensive and effective approach to corrosion inhibition compared to relying on a single mechanism alone. Understanding the various inhibitive mechanisms of PE on steel is crucial for optimizing their performance and developing effective corrosion protection strategies. One of the primary mechanisms of corrosion inhibition by PE is the formation of a protective film on the metal surface. This film acts as a physical barrier, preventing direct contact between the metal and the corrosive environment. The film can inhibit the transport of corrosive species, such as oxygen and chloride ions, to the metal surface, reducing the corrosion rate. The formation of the protective film involves the adsorption of plant extract compounds onto the metal surface, leading to the formation of a passivating layer. The adsorbed molecules can physically block the active sites on the metal surface or undergo chemical reactions with the metal, forming a stable and protective layer. Another important inhibitive mechanism is the adsorption of plant extract compounds onto the metal surface. The adsorbed molecules can physically block the active sites, limiting the access of corrosive species to these sites. Additionally, the adsorbed molecules can undergo chemical reactions with the metal, leading to the formation of a passivating layer. This adsorption mechanism is influenced by factors such as the chemical composition of the extract, the concentration of the extract, and the properties of the metal surface. Electrostatic interactions, hydrogen bonding, van der Waals' forces, and other intermolecular forces play a role in the adsorption process.

PE also have the ability to induce passivation of the metal surface, resulting in the formation of a stable oxide layer. This passivation layer acts as a physical and chemical barrier, protecting the metal against further corrosion. The passivation mechanism involves the formation of an oxide layer on the metal surface, which reduces the corrosion rate. The plant extract compounds can promote the formation of a stable and protective oxide layer through various mechanisms, including the reduction of metal cations, the coordination of metal ions, and the formation of complexes with metal ions. Furthermore, PE can exhibit oxygen-scavenging properties, which inhibit the corrosion process by reducing the availability of oxygen at the cathodic sites. Oxygen reduction is a significant cathodic reaction in the corrosion process, and by scavenging oxygen, PE can impede the cathodic reaction and inhibit corrosion. The oxygen-scavenging mechanism involves the ability of plant extract compounds to react with dissolved oxygen, preventing its reduction at the cathodic sites. This mechanism reduces the availability of oxygen for the cathodic reaction, thereby inhibiting the corrosion process.

In addition to oxygen scavenging, certain PE possess antioxidant properties and can scavenge free radicals. Free radicals, such as hydroxyl radicals, are highly reactive species that can initiate and propagate corrosion processes. By scavenging free radicals, PE can interrupt the chain reactions involved in corrosion and inhibit the corrosion process. The radical scavenging mechanism involves the ability of plant extract compounds to neutralize free radicals by donating electrons or hydrogen atoms. This scavenging action reduces the reactivity of free radicals, preventing their participation in corrosive reactions and inhibiting the corrosion process. The inhibitive mechanisms discussed above, including film formation, adsorption, passivation, oxygen scavenging, radical scavenging, and synergistic effects, highlight the diverse ways in which PE can inhibit corrosion. These mechanisms can be influenced by various factors, such as the chemical composition of the extract, concentration, surface characteristics of the metal, and the corrosive environment. Optimizing these factors is crucial for enhancing the inhibitive performance of PE and tailoring their corrosion protection capabilities to specific applications.

By gaining a comprehensive understanding of the inhibitive mechanisms of PE on steel, researchers can further explore the potential of these natural inhibitors and develop tailored corrosion protection strategies for different industrial applications. This knowledge can contribute to the development of sustainable and eco-friendly CIs that effectively mitigate the damaging effects of corrosion on steel structures and equipment.

19.5 FACTORS INFLUENCING INHIBITORY PERFORMANCE

Factors influencing the inhibitory performance of PE as CIs for steel in NaCl solutions are of great importance for understanding and optimizing their effectiveness. These factors include the concentration of PE, temperature effects, pH considerations, and the impact of immersion time on long-term performance.

19.5.1 Concentration of PE

The concentration of PE is a crucial factor that influences their inhibitory performance as CIs for steel in NaCl solutions. Generally, increasing the concentration of PE results in a higher concentration of active inhibitive compounds available for adsorption onto the metal surface. This increased concentration of active compounds enhances the inhibitory effectiveness against corrosion. When PE are present in higher concentrations, a greater number of inhibitive molecules are available to interact with the metal surface. This leads to increased coverage and a higher probability of adsorption onto the active sites of the metal, effectively forming a protective layer. The adsorbed molecules act as a physical barrier, preventing the corrosive species, such as chloride ions or oxygen, from accessing the metal surface and initiating corrosion reactions. However, it is important to note that there is an optimal concentration range for PE, beyond which further increases in concentration may not significantly improve the inhibitory performance. This can be attributed to several factors. Firstly, as the concentration increases, the number of available active compounds for adsorption reaches a saturation point. Beyond this saturation point, additional plant extract molecules may not have a significant impact on the inhibitory performance, as the active sites on the metal surface become fully occupied.

Furthermore, excessive concentrations of PE can lead to desorption of the previously adsorbed molecules. This desorption can occur due to overcrowding or repulsive forces between the molecules on the metal surface. As a result, the protective film formed by the PE may become less stable, reducing its inhibitory effectiveness. Finding the appropriate concentration of PE is crucial to maximize their inhibitory efficiency. This requires careful experimentation and optimization to determine the concentration range that provides the best balance between coverage and stability of the protective film. It is important to consider the specific properties of the plant extract compounds, the nature of the metal surface, and the corrosive environment when determining the optimal concentration. Moreover, it is worth noting that the concentration of PE should be optimized based on the intended application. Different environments and corrosion conditions may require different concentrations to achieve optimal corrosion protection. Therefore, it is essential to conduct systematic studies and evaluate the inhibitory performance of PE at various concentrations to identify the most effective concentration range for a specific application.

19.5.2 TEMPERATURE EFFECTS

Temperature is a significant factor that can have a profound impact on the inhibitory performance of PE as CIs for steel in NaCl solutions. Higher temperatures often accelerate the corrosion rate, necessitating the evaluation of the effectiveness of PE under elevated temperature conditions. Understanding the influence of temperature is crucial for assessing the long-term stability and performance of PE as CIs. As temperatures rise, the properties and behaviour of plant extract compounds can undergo changes, affecting their inhibitory performance. One key aspect to consider is the adsorption capability of PE onto the metal surface. Higher temperatures can alter the molecular structure and solubility of plant extract compounds, potentially impacting their adsorption behaviour. The adsorption process is fundamental for the formation of a protective film on the metal surface, and any temperature-induced changes can affect the stability and coverage of the film.

Additionally, the film-forming abilities of plant extract compounds may be affected by elevated temperatures. The thermal stability of the film, as well as its ability to withstand the corrosive environment, can be influenced by temperature. Changes in the molecular structure or the interaction between the plant extract compounds and the metal surface may occur, leading to variations in the film's protective properties. The stability of plant extract compounds at higher temperatures is another crucial consideration. Elevated temperatures can induce chemical reactions within the plant extract compounds, potentially affecting their inhibitory properties. Some compounds may undergo degradation or structural changes that can diminish their inhibitory effectiveness. Therefore, it is essential to study the stability of PE under specific temperature conditions encountered in the corrosive environment to ensure their long-term effectiveness. Moreover, the solubility of plant extract compounds can also be influenced by temperature. Some compounds may exhibit decreased solubility at higher temperatures, leading to reduced availability and inhibitory activity. It is important to evaluate the solubility behaviour of PE under different temperature conditions to understand their performance and ensure an appropriate concentration of active inhibitive compounds. To fully understand the impact of temperature on the inhibitory performance of PE, it is necessary to conduct systematic studies that evaluate their effectiveness over a range of temperatures. This enables researchers to identify the temperature range within which the inhibitory performance of PE is optimal.

19.5.3 pH Considerations

The pH of the NaCl solution or corrosive environment is a crucial factor that significantly influences the inhibitory performance of PE as CIs for steel. The pH level affects the chemical composition, charge distribution, adsorption behaviour, and reactivity of the plant extract compounds, ultimately influencing their inhibitory effectiveness. Understanding the pH dependence of PE is essential for determining their optimal pH range to achieve maximum inhibitory efficiency. The chemical composition of plant extract compounds can undergo changes with variations in pH. These compounds may contain functional groups that can ionize or exhibit pH-dependent charge distribution. As the pH of the solution changes, the ionization states of the functional groups within the plant extract compounds can be affected. This, in turn, influences their adsorption behaviour onto the metal surface. The adsorption process is crucial for the formation of a protective film, and the pH-dependent charge distribution can affect the electrostatic interactions between the plant extract compounds and the metal surface, influencing the stability and coverage of the film. Moreover, the reactivity of plant extract compounds can also be influenced by pH. Changes in pH can alter the availability and activity of specific functional groups within the compounds, affecting their inhibitory mechanisms. For example, pH changes may modify the ability of plant extract compounds to donate electrons or hydrogen atoms, which are crucial for passivation or radical scavenging mechanisms. Understanding the pH dependence of these inhibitory mechanisms is essential for optimizing the inhibitory performance of PE.

The stability and solubility of plant extract compounds can also be pH-dependent. Some compounds may exhibit variations in solubility and stability as the pH of the solution changes. pH-induced changes in solubility can impact the availability and concentration of active inhibitive compounds in the solution, ultimately affecting their inhibitory effectiveness. Additionally, pH-dependent stability can influence the durability and performance of the protective film formed by PE. Studying the pH dependence of PE is crucial to determine their optimal pH range for maximum inhibitory efficiency. This involves conducting systematic studies to evaluate the inhibitory performance of PE under different pH conditions. By examining the adsorption behaviour, reactivity, stability, and solubility of plant extract compounds at various pH levels, researchers can identify the pH range that provides the most effective corrosion protection. Furthermore, it is important to consider the specific pH conditions encountered in the corrosive environment to ensure the practical applicability of PE as CIs. The pH of the corrosive environment can vary depending on the industry, application, or specific location. By understanding the pH dependence of PE, researchers can tailor their application and optimize their inhibitory performance in specific pH conditions.

19.5.4 IMMERSION TIME AND LONG-TERM PERFORMANCE

The immersion time of steel in a corrosive solution is a critical factor that significantly influences the inhibitory performance of PE as CIs. The effectiveness of PE may vary over time, and their inhibitory efficiency can change with extended exposure to the corrosive environment. Therefore, it is essential to consider the immersion time and evaluate the durability and stability of the inhibitory film formed by PE to ensure their long-term corrosion protection capabilities. In short-term evaluations, PE may exhibit promising inhibitory performance, effectively reducing the corrosion rate and forming a protective film on the metal surface. However, the long-term behaviour of PE as CIs cannot be overlooked. The inhibitory efficiency may decrease over time due to several factors, including the degradation or desorption of the plant extract compounds, changes in the corrosive environment, or alteration of the inhibitory film. Assessing the durability and stability of the inhibitory film is crucial for determining the long-term corrosion protection capabilities of PE. The film formed by PE should be able to withstand the continuous exposure to the corrosive environment without undergoing significant degradation or detachment. Factors such as the mechanical strength, adhesion properties, and chemical stability of the film should be evaluated to ensure its long-term effectiveness. Additionally, the immersion time can influence the interactions between the plant extract compounds and the metal surface. Over time, the adsorbed molecules may undergo modifications, such as rearrangements or chemical reactions with the metal. These changes can impact the coverage and stability of the inhibitory film, affecting its ability to protect the metal against corrosion.

The corrosive environment itself may also change with prolonged immersion time, potentially affecting the inhibitory performance of PE. For example, the concentration of corrosive species, such as chloride ions, may increase due to the gradual depletion of plant extract compounds or the accumulation of corrosion products. These changes can alter the kinetics and mechanisms of the corrosion process, subsequently impacting the inhibitory efficiency of PE. To ensure the long-term corrosion protection capabilities of PE, researchers should conduct comprehensive evaluations that consider extended immersion times. These evaluations can involve monitoring the corrosion rate, film stability, and changes in the surface morphology of the metal over time. By assessing the inhibitory performance at different immersion periods, researchers can gain insights into the durability and stability of plant extract-based CIs. Furthermore, it is important to investigate the potential for reapplication or replenishment of PE during extended immersion periods. Reapplication of PE can restore the inhibitory film or enhance its protective properties, thereby maintaining the long-term corrosion protection capabilities.

19.6   CONCLUSION AND FUTURE PERSPECTIVES

The utilization of PE as CIs for steel in NaCl solutions offers a promising and eco-friendly approach to combat the widespread and costly problem of corrosion. The inhibitory mechanisms exhibited by PE, such as film formation, adsorption, passivation, oxygen scavenging, radical scavenging, and synergistic effects, provide multiple lines of defence against corrosion. These mechanisms act in concert to inhibit the corrosion process, protecting the metal surface from degradation and extending its service life. The advantages of PE as CIs are evident. They are derived from renewable resources and possess properties that make them sustainable and environmentally friendly alternatives to synthetic inhibitors. PE are biodegradable and pose minimal risks to human health and the environment. Additionally, they can be extracted using environmentally friendly methods, such as water or ethanol extraction, further enhancing their eco-friendliness.

However, there are still several areas that warrant further exploration and research in the field of plant extract-based corrosion inhibition. Firstly, the identification and isolation of the active compounds responsible for the inhibitory effects in PE are essential. Understanding the structure-activity relationships of these compounds will enable researchers to design and develop more potent and selective inhibitors. This knowledge can be further utilized to optimize extraction techniques and develop standardized protocols for extracting the active compounds. Another area of future research is the investigation of the long-term performance and durability of plant extract-based CIs. Extended exposure to corrosive environments, variations in temperature, and other environmental factors may affect the stability and effectiveness of the inhibitory film formed by PE. Studies focused on the long-term behaviour, including film stability, degradation mechanisms, and the impact of environmental factors, will provide valuable insights for the practical application of PE as CIs. Additionally, the development of efficient and sustainable methods for the large-scale production of PE is crucial. Scaling up the production process while maintaining the quality and consistency of the extracts is a challenge that needs to be addressed. This will ensure the availability of PE as viable alternatives for industrial corrosion protection applications. Lastly, further exploration of the inhibitory performance of PE under different corrosive environments, such as acidic or alkaline conditions, and in the presence of other corrosive species is needed. Understanding how PE perform under these diverse conditions will expand their applicability and broaden their potential use in various industries.

REFERENCES

[1] Assad H, Kumar A. Understanding functional group effect on corrosion inhibition efficiency of selected organic compounds. *Journal of Molecular Liquids*. 2021;344:117755.
[2] Assad H, Kumar S, Saha SK, Kang N, Fatma I, Dahiya H, et al. Evaluating the adsorption and corrosion inhibition capabilities of Pyridinium-P-Toluene Sulphonate on MS in 1M HCl medium: An experimental and theoretical study. *Inorganic Chemistry Communications*. 2023;153:110817.

[3] Yadav M, Goel G, Hatton FL, Bhagat M, Mehta SK, Mishra RK, et al. A review on biomass-derived materials and their applications as corrosion inhibitors, catalysts, food and drug delivery agents. *Current Research in Green and Sustainable Chemistry*. 2021;4:100153.

[4] Youssef AM, Hasanin MS, El-Aziz MEA, Turky GM. Conducting chitosan/hydroxyle-thyl cellulose/polyaniline bionanocomposites hydrogel based on graphene oxide doped with Ag-NPs. *International Journal of Biological Macromolecules*. 2021;167:1435–44.

[5] Zhang W, Li H-J, Chen L, Sun J, Ma X, Li Y, et al. Performance and mechanism of a composite scaling-corrosion inhibitor used in seawater: 10-Methylacridinium iodide and sodium citrate. *Desalination*. 2020;486:114482.

[6] Assad H, Kumar A. Carbon nanotubes (SWCNTs/MWCNTs) and functionalized carbon nanotubes in corrosion prevention. *Corrosion Prevention Nanoscience: Nanoengineering Materials and Technologies*. 2022;85:133287.

[7] Assad H, Thakur A, Sharma AK, Kumar A. Density functional theory-based molecu-lar modeling. *Computational Modelling and Simulations for Designing of Corrosion Inhibitors*: Elsevier; 2023. pp. 95–113.

[8] Thakur A, Assad H, Kaya S, Kumar A. Plant extracts as environmentally sustainable corrosion inhibitors II. *Eco-Friendly Corrosion Inhibitors*: Elsevier; 2022. pp. 283–310.

[9] Thakur A, Assad H, Kaya S, Kumar A. Plant extracts as bio-based anticorrosive materi-als. In *Handbook of Biomolecules*: Elsevier; 2023. pp. 591–618.

[10] Thakur A, Kumar A, Sharma S, Ganjoo R, Assad H. Computational and experimental studies on the efficiency of Sonchus arvensis as green corrosion inhibitor for mild steel in 0.5 M HCl solution. *Materials Today: Proceedings*. 2022;66:609–21.

[11] Verma C, Thakur A, Ganjoo R, Sharma S, Assad H, Kumar A, et al. Coordination bonding and corrosion inhibition potential of nitrogen-rich heterocycles: Azoles and triazines as specific examples. *Coordination Chemistry Reviews*. 2023;488:215177.

[12] Assad H, Ganjoo R, Sharma S, editors. A theoretical insight to understand the struc-tures and dynamics of thiazole derivatives. *Journal of Physics: Conference Series*; 2022: IOP Publishing.

[13] Thakur A, Assad H, Sharma S, Ganjoo R, Kaya S, Kumar A. *Coordination Polymers as Corrosion Inhibitors. Functionalized Nanomaterials for Corrosion Mitigation: Synthesis, Characterization, and Applications*: ACS Publications; 2022. pp. 231–54.

[14] Bashir S, Thakur A, Lgaz H, Chung I-M, Kumar A. Computational and experimental studies on Phenylephrine as anti-corrosion substance of mild steel in acidic medium. *Journal of Molecular Liquids*. 2019;293:111539.

[15] Bashir S, Thakur A, Lgaz H, Chung I-M, Kumar A. Corrosion inhibition performance of acarbose on mild steel corrosion in acidic medium: An experimental and computa-tional study. *Arabian Journal for Science and Engineering*. 2020;45(6):4773–83.

[16] Bashir S, Thakur A, Lgaz H, Chung I-M, Kumar A. Corrosion inhibition efficiency of bronopol on aluminium in 0.5 M HCl solution: Insights from experimental and quantum chemical studies. *Surfaces and Interfaces*. 2020;20:100542.

[17] Ganjoo R, Bharmal A, Sharma S, Thakur A, Assad H, Kumar A. Imidazolium based ionic liquids as green corrosion inhibitors against corrosion of mild steel in acidic media. *Journal of Physics: Conference Series*. 2022;2267(1):012023.

[18] Ganjoo R, Sharma S, Thakur A, Kumar A. Thermodynamic study of corrosion inhibi-tion of dioctylsulfosuccinate sodium salt as corrosion inhibitor against mild steel in 1 M HCl. *Materials Today: Proceedings*. 2022;66:529–33.

[19] Parveen G, Bashir S, Thakur A, Saha SK, Banerjee P, Kumar A. Experimental and compu-tational studies of imidazolium based ionic liquid 1-methyl-3-propylimidazolium iodide on mild steel corrosion in acidic solution. *Materials Research Express*. 2020;7(1):016510.

[20] Sharma D, Thakur A, Sharma MK, Jakhar K, Kumar A, Sharma AK, et al. Synthesis, electrochemical, morphological, computational and corrosion inhibition studies of 3-(5-naphthalen-2-yl-[1,3,4]oxadiazol-2-yl)-pyridine against mild steel in 1 M HCl. *Asian Journal of Chemistry*. 2023;35(5):1079–88.

[21] Sharma S, Ganjoo R, Kr. Saha S, Kang N, Thakur A, Assad H, et al. Investigation of inhibitive performance of Betahistine dihydrochloride on mild steel in 1 M HCl solution. *Journal of Molecular Liquids.* 2022;347:118383.

[22] Sharma S, Ganjoo R, Kr. Saha S, Kang N, Thakur A, Assad H, et al. Experimental and theoretical analysis of baclofen as a potential corrosion inhibitor for mild steel surface in HCl medium. *Journal of Adhesion Science and Technology.* 2021;36(19):2067–92.

[23] Sharma S, Ganjoo R, Thakur A, Kumar A. Investigation of corrosion performance of expired Irnocam on the mild steel in acidic medium. *Materials Today: Proceedings.* 2022;66:540–3.

[24] Thakur A, Kaya S, Abousalem AS, Kumar A. Experimental, DFT and MC simulation analysis of Vicia Sativa weed aerial extract as sustainable and eco-benign corrosion inhibitor for mild steel in acidic environment. *Sustainable Chemistry and Pharmacy.* 2022;29:100785.

[25] Thakur A, Kaya S, Abousalem AS, Sharma S, Ganjoo R, Assad H, et al. Computational and experimental studies on the corrosion inhibition performance of an aerial extract of Cnicus Benedictus weed on the acidic corrosion of mild steel. *Process Safety and Environmental Protection.* 2022;161:801–18.

[26] Thakur A, Kaya S, Kumar A. Recent innovations in nano container-based self-healing coatings in the construction industry. *Current Nanoscience.* 2022;18(2):203–16.

[27] Thakur A, Kumar A. Sustainable inhibitors for corrosion mitigation in aggressive corrosive media: A comprehensive study. *Journal of Bio- and Tribo-Corrosion.* 2021;7(2). DOI:10.1007/s40735-021-00501-y.

[28] Thakur A, Kumar A. Recent advances on rapid detection and remediation of environmental pollutants utilizing nanomaterials-based (bio)sensors. *Science of the Total Environment.* 2022;834:155219.

[29] Thakur A, Kumar A, Kaya S, Vo D-VN, Sharma A. Suppressing inhibitory compounds by nanomaterials for highly efficient biofuel production: A review. *Fuel.* 2022;312:122934.

[30] Thakur A, Kumar A, Sharma S, Ganjoo R, Assad H. Computational and experimental studies on the efficiency of Sonchus arvensis as green corrosion inhibitor for mild steel in 0.5 M HCl solution. *Materials Today: Proceedings.* 2022;66:609–21.

[31] Thakur A, Sharma S, Ganjoo R, Assad H, Kumar A. Anti-corrosive potential of the sustainable corrosion inhibitors based on biomass waste: A review on preceding and perspective research. *Journal of Physics: Conference Series.* 2022;2267(1):012079.

[32] Verma C, Quraishi MA. Chelation capability of chitosan and chitosan derivatives: Recent developments in sustainable corrosion inhibition and metal decontamination applications. *Current Research in Green and Sustainable Chemistry.* 2021;4:100184.

[33] Verma C, Thakur A, Ganjoo R, Sharma S, Assad H, Kumar A, et al. Coordination bonding and corrosion inhibition potential of nitrogen-rich heterocycles: Azoles and triazines as specific examples. *Coordination Chemistry Reviews.* 2023;488:215177.

[34] Wu Y, Zhang Y, Jiang Y, Li N, Zhang Y, Wang L, et al. Exploration of walnut green husk extract as a renewable biomass source to develop highly effective corrosion inhibitors for magnesium alloys in sodium chloride solution: Integrated experimental and theoretical studies. *Colloids and Surfaces A: Physicochemical and Engineering Aspects.* 2021;626:126969.

[35] Espinoza-Vázquez A, Rodríguez-Gómez FJ, Figueroa-Vargas IA, Pérez-Vásquez A, Mata R, Miralrio A, et al. 4-Phenylcoumarin (4-PC) glucoside from Exostema caribaeum as corrosion inhibitor in 3% NaCl saturated with CO2 in AISI 1018 steel: Experimental and theoretical study. *International Journal of Molecular Sciences.* 2022;23(6):3130.

[36] Shyamvarnan B, Shanmugapriya S, Arockia Selvi J, Kamaraj P, Mohankumar R. Corrosion inhibition effect of Elettaria cardamomum extract on mild steel in 3.5% NaCl medium. *Materials Today: Proceedings.* 2021;40:S192–S7.

[37] Ben Harb M, Abubshait S, Etteyeb N, Kamoun M, Dhouib A. Olive leaf extract as a green corrosion inhibitor of reinforced concrete contaminated with seawater. *Arabian Journal of Chemistry*. 2020;13(3):4846–56.

[38] Othman NK, Yahya S, Ismail MC. Corrosion inhibition of steel in 3.5% NaCl by rice straw extract. *Journal of Industrial and Engineering Chemistry*. 2019;70:299–310.

[39] Devikala S, Kamaraj P, Arthanareeswari M, Patel MB. Green corrosion inhibition of mild steel by aqueous Allium sativum extract in 3.5% NaCl. *Materials Today: Proceedings*. 2019;14:580–9.

[40] Haddadi SA, Alibakhshi E, Bahlakeh G, Ramezanzadeh B, Mahdavian M. A detailed atomic level computational and electrochemical exploration of the Juglans regia green fruit shell extract as a sustainable and highly efficient green corrosion inhibitor for mild steel in 3.5 wt% NaCl solution. *Journal of Molecular Liquids*. 2019;284:682–99.

[41] Motamedi M, Ramezanzadeh B, Mahdavian M. Corrosion inhibition properties of a green hybrid pigment based on Pr-Urtica Dioica plant extract. *Journal of Industrial and Engineering Chemistry*. 2018;66:116–25.

[42] Singh A, Lin Y, Ebenso EE, Liu W, Pan J, Huang B. Gingko biloba fruit extract as an eco-friendly corrosion inhibitor for J55 steel in CO2 saturated 3.5% NaCl solution. *Journal of Industrial and Engineering Chemistry*. 2015;24:219–28.

[43] Pradipta I, Kong D, Tan JBL. Natural organic antioxidants from green tea inhibit corrosion of steel reinforcing bars embedded in mortar. *Construction and Building Materials*. 2019;227:117058.

[44] Kusumastuti R, Pramana RI, Soedarsono JW. The use of morinda citrifolia as a green corrosion inhibitor for low carbon steel in 3.5% NaCl solution. *AIP Conference Proceedings:* AIP Publishing; 2017. vol. 1823, no. 1.

[45] Akbarzadeh S, Ramezanzadeh B, Bahlakeh G, Ramezanzadeh M. Molecular/electronic/atomic-level simulation and experimental exploration of the corrosion inhibiting molecules attraction at the steel/chloride-containing solution interface. *Journal of Molecular Liquids*. 2019;296:111809.

[46] Ibrahim T, Gomes E, Obot IB, Khamis M, Abou Zour M. Corrosion inhibition of mild steel byCalotropisproceraleaves extract in a CO2 saturated sodium chloride solution. *Journal of Adhesion Science and Technology*. 2016;30(23):2523–43.

[47] Shabani-Nooshabadi M, Ghandchi M. Santolina chamaecyparissus extract as a natural source inhibitor for 304 stainless steel corrosion in 3.5% NaCl. *Journal of Industrial and Engineering Chemistry*. 2015;31:231–7.

[48] Alibakhshi E, Ramezanzadeh M, Haddadi SA, Bahlakeh G, Ramezanzadeh B, Mahdavian M. Persian Liquorice extract as a highly efficient sustainable corrosion inhibitor for mild steel in sodium chloride solution. *Journal of Cleaner Production*. 2019;210:660–72.

20 Phytochemicals/Plant Extracts as Corrosion Inhibitors for Aluminium in NaCl Solutions

Kanika Cial and M. Mobin
Aligarh Muslim University

20.1 INTRODUCTION

20.1.1 PLANT EXTRACTS

Plants have always been a part of human history as a source of health, and they continue to do so now. However, over the course of history, man has developed a curiosity about the origin of the characteristics that are possessed by plants. The study of plants dates back hundreds of years, and the research has made it possible for them to be examined not only for obtaining food but also in the field of medicine (for instance, extracts have been investigated as a potential alternative treatment for cancer, treatment for diabetes, and dermatological disorders, among other things), agriculture (repellent effects and antifeedant, control of diseases), and corrosion inhibitors.

The Greek term phyto, meaning "plant," is where we got the modern word "phytochemical"; these are the natural chemical compounds with biological activity that enhance human health in the same ways as macronutrients and micronutrients do. They are abundant in whole plant foods including vegetables, fruits, cereal grains, and drinks made from these foods like tea and wine. Plants rich in beneficial phytochemicals may provide a complement to the human body by functioning as natural antioxidants. They can be found in the plant's roots, stems, leaves, flowers, fruits, or seeds. The outermost layers of plants' tissues tend to have high concentrations of phytochemicals. Almost 2,000 different phytochemicals have been isolated from plants which include bioactive chemicals such as alkaloids, tannins, flavonoids, and phenolics (Veiga et al., 2020). Different phytochemicals derived from plant extracts have been listed in Figure 20.1.

DOI: 10.1201/9781003394631-20

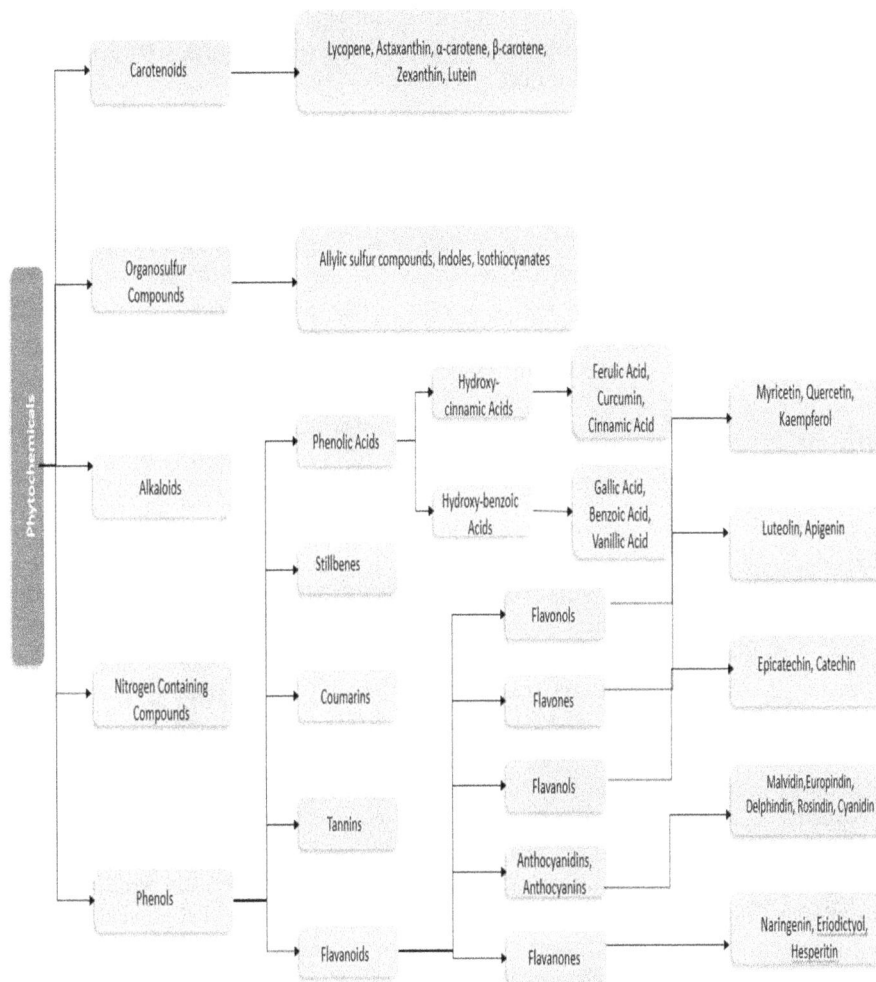

FIGURE 20.1   Classification of phytochemicals.

20.2   PLANT EXTRACTS AS CORROSION INHIBITOR

In the 1930s, the natural products were used as anti-corrosives for the first time when *Chelidonium majus* (commonly referred as celandine) and a few other plants were used in a sulphuric acid pickling bath. Many plant extracts were later discovered to be efficient as corrosion inhibitors in a wide range of environments, and there was a subsequent surge in interest in the use of natural chemicals for this purpose. A study published by Hosary et al. (1972) claimed to prevent corrosion of aluminium and steel when exposed to neutral and acidic environments using tobacco leaves, stems, and twigs.

The available literature indicates that extracts derived from a wide range of plant components, such as fruits, leaves, bark, peel, flowers, root, seed, and even entire plant, have frequently been used as corrosion inhibitors. It has been shown that the extracts derived from different parts of the plants, the leaf extracts have the greatest overall protective effectiveness at quite low doses. Plant extracts have the potential to limit the corrosion of metals due to the presence of high concentration of elements such as polyphenols, flavonoids, tannins, alkaloids, and polysaccharides (Alrefaee et al., 2021). Some examples of plant extracts that have been reported in literature to mitigate corrosion are flavonoids, glycosides, alkaloids, saponins, phytosterol, tannins, anthraquinones, phenolic compounds, and triterpenes. These phytochemicals mostly comprise of hydroxyl, ester, and carboxylic acid functional groups that facilitate the absorption on the metal substrate.

20.3 EXTRACTION METHODS

The biologically active chemicals that are found in plants are difficult to gain access to, and hence, to acquire the desired chemicals from a variety of sources like plant or seed samples, extraction is a necessary operation. Humans have been figuring out how to extract useful chemicals from their natural surroundings for an immeasurably long time. Thyme, saffron, marjoram, cumin, and peppermint are only some of the fragrant food plants whose essential oils have been documented as having been extracted in Ancient Greek approximately 500 and 400 BC (Geow et al., 2021). The ancient technique that was employed to extract volatile chemicals was time-consuming, labour-intensive, and expensive, and therefore, these techniques have fallen out of favour in recent years (Danlami et al., 2014). The classical method for recovering bioactive chemicals from plant materials was maceration, which is based on the leaching of compounds from solid material. For better mass transfer and chemical solubility, maceration makes use of solvents in conjunction with heat and/or agitation. Increased public awareness of the usage of natural substances, as well as rising public awareness of sustainable development and environmental preservation, has fuelled intensive research over the past few decades to enhance extraction processes (Rasul, 2018).

The phytochemical extraction from various plant parts requires some basic steps such as authentication and extraction of plant material, separation and isolation of the constituent of interest, characterisation of the isolated compound, and quantitative evaluation. Further, these steps are important as various techniques require different approaches, and to get higher efficiency, along with choice of solvents, is an important criterion for extracting techniques. The pressing and solvent extraction methods are considered to be traditional extraction techniques, while the supercritical fluid extraction, microwave-assisted extraction, and ultrasound-assisted extraction techniques are considered to be advanced extraction techniques (Zhang et al., 2018). Different extraction methods have been employed for extracting phytochemicals. Each method is associated with certain advantages and disadvantages that are reported in Table 20.1.

TABLE 20.1
Advantages and Disadvantages of Extraction Method

Methods	Time Duration	Temperature	Advantages	Disadvantages
Supercritical fluid extraction	Short	Near room temperature	High extraction efficiency	Volatile analytes may get lost
Microwave assisted extraction	Short	Room temperature	Highly efficient, with low use of solvents and much cleaner process. Extraction time is fast	Filtration after extraction is required. Instrument setup is costly as it requires microwave. Extraction volume is limited
Decoction	Moderate	Under heat	Very easy to operate and no need for expensive instrument.	Not recommended for heat-sensitive compounds
Hydro distillation and steam distillation	Long	Under heat	Inexpensive method	Complete extraction is not possible, requires more fuel and space and thus becomes non-economical
Enzyme assisted extraction	Moderate	Room temperature, or heated after enzyme treatment	Process is environment friendly. Yield is higher than conventional methods.	The process becomes disadvantageous as it requires enzymes that can change their stability and actions with change in pH, temperature, presence of other ions, etc.
Ultrasound assisted extraction	Moderate	Room temperature, or under heat	Requires less energy and also investment cost is low	Requires large solvent volume of solvent and many extraction steps may also be required.
Pressurised liquid extraction	Fast	Under heat	High extraction efficiency	Instrument very costly
Maceration	Long	Room temperature	High yield with low energy consumption	It is not recommended for the extraction of compounds that are sensitive to heat
Soxhlet extraction	Long	Under heat	Simple	Depends on labour and the extraction time is lengthy.

(Continued)

TABLE 20.1 (*Continued*)
Advantages and Disadvantages of Extraction Method

Methods	Time Duration	Temperature	Advantages	Disadvantages
Reflux extraction	Long	Under heat	Process is useful as these solvents can be recovered, set-up is easy, and the process has an advantage over Soxhlet as no further filtration is needed because simultaneous filtration occurs in the complete process	Hazardous as toxic solvents are used. Requires large volume of solvent and is a very slow process.
Percolation	Long	Room temperature, occasionally under heat	Requires less time than maceration and provides complete extraction	Requirement of skilled labour, solvent, and more time than Soxhlet extraction. Particle size of material should be kept in mind during the entire process.

20.3.1 Type of Solvents Used in Extraction

Choosing a solvent system is heavily reliant on the type of bioactive chemical being studied. Many different solvent solutions can be used to isolate the bioactive component from plants. Polar solvents, such as ethyl acetate, ethanol, or methanol, are used for the extraction of hydrophilic substances. Lipophilic substances are best extracted using dichloromethane or a dichloromethane/methanol combination of 1:1 (Cos et al., 2006). The solvents used to extract the different substances along with the factors that need to be considered in selecting solvents for the different extraction processes are given in Figures 20.2 and 20.3.

20.4 ALUMINIUM

Aluminium is one of the most important nonferrous metals used in the metallurgy sector having an annual market of 25 million tonnes, making it one of the most important metals in the business. It is important to note that aluminium does not refer to a single material; rather, it refers to a group of alloys whose mechanical characteristics vary greatly (Xhanari & Finšgar, 2016). Typical alloying elements such as copper, magnesium, tin, silicon, manganese, and zinc are some of them that are used

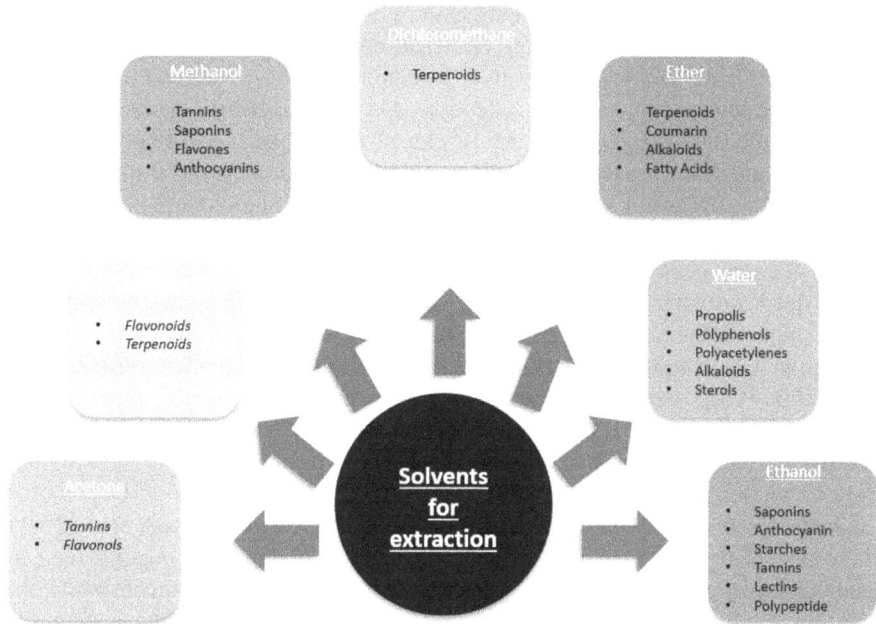

FIGURE 20.2 Different solvents for extraction.

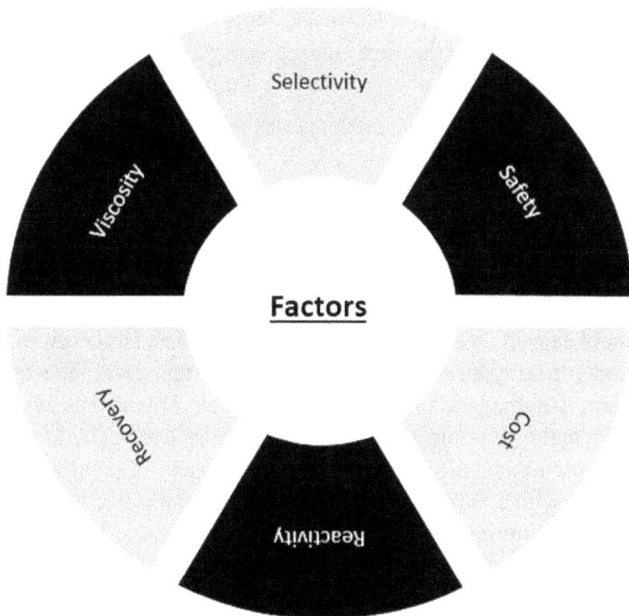

FIGURE 20.3 Factors that need to be considered in selecting solvents for the different extraction processes.

in aluminium alloys. Aluminium along with its alloys found usage in a wide variety of industries including construction, electrical engineering, packaging, transportation, and the home because of its great formability, strong electrical and thermal conductivity, low weight, and high reflectivity (Ashkenazi, 2019). The low density of aluminium (density: 2.7 g cm⁻³), which is somewhat less than one-third of the density of ordinary steel, is directly connected to the exceptional mechanical qualities that aluminium has (Winston Revie, 2011). As a result, aluminium and the alloys it can form provide an excellent balance of strength and weight. Also, they are quite inexpensive and may be found in approximately double the quantity that iron can.

The quick formation of an oxide layer on aluminium, which does not permit further oxidation of the metal contributes to its exceptional resistance to corrosion both in air as well as in aqueous environments. However, the coating of aluminium oxide is very thin. If it is exposed to specific conditions, it may be readily broken down by either physical or chemical processes. On the contrary, due to the presence of intermetallic particles at the surface, this metal along with its different alloys is extremely prone to corrosion. As a consequence, there is a requirement for the development of a suitable approach in order to increase its resistance to corrosion (Radi et al., 2021). There have been efforts made to preserve the surface of aluminium against the corrosion that may occur in a variety of settings that are corrosive. In recent years, the inclusion of inhibitors has come to be regarded as one of the most frequent strategies for preventing the corrosion of aluminium (Alrefaee et al., 2021). It has been extensively documented that aliphatic amines, aromatic amines, and nitrogen heterocyclic molecular compounds are effective in inhibiting the corrosion of aluminium in hostile conditions (Verma et al., 2017). However, some of these substances are not only expensive but also harmful to the surrounding ecosystem. Therefore, effort is being done to find an alternative form of protection for aluminium alloy that is less harmful to the environment.

20.4.1 ALUMINIUM CORROSION AND ITS MECHANISM

Some types of aluminium alloys even have a high strength-to-weight ratio that is equivalent to that of high-strength structural steels. In addition, due to the creation of a resistant oxide layer, aluminium exhibits high corrosion resistance when exposed to the atmosphere as well as many other aquatic conditions. However, in the presence of solutions containing chloride, the oxide layer that has developed on the surface of the aluminium is damaged, and the metal corrodes. This is true even if the oxide layer shields the metal in some situations. In order to lessen the amount of corrosion that aluminium and its alloys experience in a variety of conditions, researchers have experimented with a number of approaches and combinations of these methods. Anodising, coatings (inorganic, organic, or organic-inorganic kinds), and inhibitors (inorganic and organic types) are some of the most often tried-out forms of corrosion prevention (Xhanari & Finšgar, 2016).

Corrosion of aluminium in neutral medium is caused by the presence of chloride ions, which causes pitting corrosion, that is, the most destructive form of corrosion, and can be distinguished by the presence of pits and irregular deformities on the surface of aluminium. The diameter and depth of pits depend on the type of material, on the environment in which aluminium and its different alloys are exposed to, and on corrosive medium which is in use. Aluminium alloys are susceptible to pitting attack in chloride-containing solutions, despite the fact that the corrosion rate of such alloys is often fairly modest. Aluminium alloys with a high purity level have a greater resistance to pitting. Among aluminium alloys, the purest metal (1xxx grade) is the most durable option. In terms of corrosion resistance in saltwater and aqueous chloride-containing solutions, the commercial Al-Mg alloys (5xxx grades) perform well. The Al-Si (4xxx grade) and Al-Si-Mg (6xxx grade) aluminium alloys, on the contrary, have less resistance to localised corrosion. The pitting corrosion resistance of Al-Cu alloys (2xxx grades) is the lowest of all the aluminium alloys (Ashkenazi, 2019).

From the Pourbaix diagram of aluminium, it can be inferred that the aluminium metal is thermodynamically stable in the pH range of 5–8.5 due to the formation of passive, amorphous layer formed on the metal itself (Winston Revie, 2011). When exposed to acidic and alkaline solutions, and especially solutions containing chloride content, the oxide layer that has developed on the surface of the aluminium is damaged. Aluminium corrosion occurs in different steps:

At first, chloride ions are attracted to the surface of the aluminium-containing aluminium oxide, and then, after that, aluminium ions form a complex with the chloride ions to produce soluble species. Further diffusion of soluble species from the metal surface leads to the consequent thinning of the protective film from the surface which exposes the metal to further react with the aggressive medium. As a result of the inhibitor being adsorbed onto the surface of the aluminium, a thin layer is formed on the surface of the metals, which helps to slow down the corrosion (Winston Revie, 2011; Xhanari & Finšgar, 2016).

The anodic processes (Equations 20.1 and 20.2) and the cathodic processes (Equations 20.3, 20.4, and 20.5) of aluminium corrosion in a solution containing 3.5% NaCl demonstrate the dissolution of aluminium and the reduction of dissolved oxygen:

$$Al \rightarrow Al^{3+} + 3e^- \qquad (20.1)$$

$$Al^{3+} + 3H_2O \rightarrow Al(OH)_3 + 3H^+ \qquad (20.2)$$

$$AlCl_3 + 3H_2O \rightarrow Al(OH)_3 + 3HCl \qquad (20.3)$$

$$3H^+ + 3e^- \rightarrow \frac{3}{2}H_2 \qquad (20.4)$$

$$\frac{1}{2}O_2 + H_2O + 2e^- \rightarrow 2OH^- \qquad (20.5)$$

20.4.2 Literature Review on different Phytochemicals/Plant Extracts Employed for Mitigation of Corrosion of Aluminium or Its Alloy in Neutral Medium (NaCl Solution)

In a study by Liu et al. (2014), a natural inhibitor (*Coptis chinensis*) was chosen as effective corrosion inhibitor for aluminium 7075 alloy in neutral medium. The plant extract was extracted from rhizome (root) part of the plant, and corrosion inhibition techniques such as electrochemical impedance spectroscopy and surface analysis were considered in order to establish the extract as an effective corrosion inhibitor. From the experimental data, it can be inferred that CCR extract has an optimum inhibition efficiency of 94% at 1,000 ppm. Electrochemical results showed that root extract acts as mixed-type inhibitor, and adsorption was assumed to obey Temkin isotherm. Also, surface morphological studies such as scanning electrochemical microscopy (SECM) and UV-Vis spectroscopy were employed to further study the corrosion inhibition of the extract.

The anti-corrosive properties of an ethanolic solution of *Laurus nobilis* L. oil in a 3% NaCl solution were tested using weight loss, potentiodynamic polarisation, and linear polarisation methods. By using polarisation techniques, scientists have shown that this oil, when added at quantities between 10 and 50 ppm, weakens the density of cathodic currents. The AA5754 alloy was shown to have improved corrosion resistance in 3% NaCl solution compared to pure aluminium, whereas the studied oil had a more significant inhibitory influence on the corrosion process of pure aluminium. *Laurus nobilis* L. oil's active components considerably reduce pitting corrosion on specimen surfaces, as shown by scanning electron microscopy (Halambek et al., 2013).

Another study examined how garlic extract can prevent corrosion of the 1050 aluminium alloy in an aerated 3.5 wt.% NaCl solution. Different techniques such as linear polarisation, electrochemical impedance spectroscopy (EIS), and electrochemical noise analysis (EN) were utilised to assess the corrosion behaviour of the chosen alloy in the presence of an inhibitor. From EIS and polarisation data, significant deduction in corrosion rate can be seen. To provide a thorough knowledge of the interfacial interactions and adsorption process, theoretical methods based on molecular dynamics simulation were also applied. EIS findings show that garlic extract produces the most effective inhibition by boosting corrosion resistance at an optimal dose of 8 mL L^{-1} (Hajsafari et al., 2021).

Ocimum basilicum seed extract, also known as OBSE, was utilised on 2024 aluminium alloy in sodium chloride solution at a concentration of 3 wt.% as an environmentally friendly corrosion inhibitor. The findings of the antioxidant assay that was conducted using 2, 2-diphenyl-1-picryl-hydrazyl-hydrate (DPPH), which is a convenient and affordable method, demonstrated that the extract possesses remarkable antioxidant activity. At a concentration of 1 g L^{-1}, OBSE provided the highest level of corrosion prevention (95.5% efficiency). The polarisation graphs made it clear that the corrosion control was accomplished by a process known as mixed inhibition control. In addition, the Langmuir isotherm is followed by the adsorption of the inhibitor on the surface (Fernine et al., 2021).

An extract of *Ambrosia maritima* L. was employed as a corrosion inhibitor to study aluminium corrosion in 2 M sodium hydroxide solution and 0.5 M sodium chloride solution. Electrochemical techniques were used in the study to establish inhibition effect of the extract. The addition of chloride ions and *A. maritima* extract slowed the hydrogen evolution, as measured by chemical gasometry. The *A. maritima* extract functioned as a mixed-type inhibitor since it affected both the anodic and cathodic processes in the presence and absence of Cl⁻ ions. The inhibitor extract was most potent in the presence of chloride ions, although it was still effective in the absence of Cl- ions. There was an increase in the protection provided by the extract as the temperature and the concentration rose (Abdel-Gaber et al., 2008).

The use of commercial henna as a potential environmentally friendly corrosion inhibitor for 5083 aluminium alloy in seawater was described by Zulkifli et al. (2017). For the purpose of the analysis, gravimetric and electrochemical methods were utilised. Both ethyl acetate and methanol were used as solvents throughout the extraction process. The formation of a layer on the surface of the alloy by lawsone, which is the primary component of henna, led to the suppression of corrosion. A Langmuir adsorption isotherm was followed by the adsorption process. The findings suggested and illustrated that *Lawsonia inermis* extracted in ethyl acetate provides superior protection against corrosion.

The use of a biodegradable and physiologically active extract from *Treculia africana* (TA) leaves for corrosion mitigation of aluminium alloy AA7075-T7351 in NaCl (2.86%) medium was assessed using potentiodynamic polarisation, EIS, and gravimetric methods. Different adsorption models such as Langmuir, Freundlich, and Temkin isotherms were all satisfied by the spontaneous adsorption of TA on the AA7075-T7351 surface. Bode phase and magnitude graphs showed that when inhibitor concentration increased, phase angles and slopes increased. Micrographs taken with a scanning electron microscope confirmed that the extract had an inhibitory effect by showing an improvement in surface smoothness (Udensi et al., 2021).

An investigation was conducted to determine whether an extract from the roots of the Udi plant (*Terminalia glaucescens* Planch) could inhibit the corrosion of AA 6063 in solutions of acid (0.3 M H_2SO_4) and salt (3.5 wt.% NaCl). Extract of *T. glaucescens* Planch was utilised as a replacement for chemical inhibitors. The presence of bioactive components in the extract, such as tannins, flavanoids, and keller killani, has been hypothesised to be the cause of the extract's potential for inhibition. After adding the extract, the corrosion rates were lowered in both the H_2SO_4 and the 3.5 wt.% NaCl conditions (Olakolegan et al., 2020).

Berberine was employed in an experiment to evaluate how effectively it protected 7075 AA against the corrosive effects of 3.5 wt.% NaCl. It was revealed that increasing the concentration of the inhibitor increased the efficacy of protection. The adsorption of the inhibitor on the alloy's surface followed the Langmuir isotherm. Berberine, according to the findings, can function as an inhibitor in the test media, preventing the 7075-aluminium alloy from degradation (Singh et al., 2014).

In another study, the effectiveness of pumpkin seed (PS) in inhibiting corrosion of AA 5075-T6 in 3.5% NaCl was evaluated. Electrochemical and theoretical studies indicated that pumpkin seeds are an effective natural inhibitor for the corrosion of aluminium alloy. The PS showed its best level of protective efficiency at room temperature (298 K), which was 95%. In addition to this, PS worked as an inhibitor on the cathodic part. The Langmuir isotherm describes how the inhibitor is adsorbed onto the surface of AA. Surface investigations demonstrated the efficiency of PS by demonstrating its ability to adsorb on the alloy surface (Radi et al., 2021).

The effect of the fruit juice of *Phoenix dactylifera* L. (PDL) on the corrosion of 7075 aluminium alloy in a sodium chloride medium (3.5%) was investigated. Studies using EIS and potentiodynamic polarisation (PDP) were carried out in order to gain knowledge of the corrosion efficiency and behaviour of the alloy while the inhibitor was present. The potency of the PDL improved in proportion to the degree to which its concentration was raised, and it demonstrated activity as a cathodic inhibitor. It was discovered that physisorption was the adsorption mechanism, and the Temkin model accurately described the adsorption pattern (Gerengi, 2012).

An investigation into the potential inhibitory effects of an extract of *Linum usitatissimum* seeds was carried out by Elgahawi et al. (2017). For the purpose of this investigation, AA2024 was selected as the metal and 3.5% NaCl as the corrosive medium. From different techniques such as electrochemical noise measurement (EN), EIS, and PDP, it was observed that, on increasing extract concentration from 80 to 1,200 ppm, inhibition efficiency increased from 65% to 82%. In a solution containing 3.5% sodium chloride, it was established by PDP that the seeds of *L. usitatissimum* function as a cathodic inhibitor for the alloy. The deterioration of the material can be prevented by either developing an insoluble salt on the surface of the AA2024 alloy or by forming a barrier layer that restricts electrolyte diffusion towards the surface of the alloy.

20.5   CONCLUDING REMARKS AND FUTURE OUTLOOK

Extensive studies have been conducted on the usage of plant extracts as corrosion inhibitor for the mitigation of aluminium corrosion in neutral environment. Plant extracts have enough potential to stop the corrosion of aluminium in neutral environment with little or no impact on the environment. The plant extracts include a wide variety of simple and complex phytochemicals that have the ability to adsorb efficiently across the surface of the metal and offer a protective shield. Plant extracts can be obtained using either aqueous or organic solvents; the use of aqueous extracts is preferentially recommended because of the considerably more sustainable features they possess. However, plant extracts often contain only trace amounts of active ingredients; hence, a substantial quantity of the plants themselves must be consumed in order to achieve the desired level of inhibition, which makes the plant extract an expensive product. In addition, the method of extraction is far too complicated to be suitable for use in the industrial sector on a big scale. The identification and isolation of important active components in the plant extracts that inhibit corrosion is one of the choices that may be undertaken in the future. The procedure of obtaining plant extract is a very important step. The various methods available for extraction have their own distinct qualities. Organic solvents contribute to the emission of greenhouse gases into the atmosphere, which endangers both people and the natural world. In addition, the use of an excessive amount of solvent may result in the development of a substantial amount of waste by-products. Therefore, further studies are needed to be undertaken in order to propose procedures that are both effective and kind to the environment.

REFERENCES

Abdel-Gaber, A. M., Khamis, E., Abo-ElDahab, H., & Adeel, S. (2008), "Inhibition of aluminium corrosion in alkaline solutions using natural compound", *Materials Chemistry and Physics*, Vol. 109, pp. 297–305.

Alrefaee, S. H., Rhee, K. Y., Verma, C., Quraishi, M. A., & Ebenso, E. E. (2021), "Challenges and advantages of using plant extract as inhibitors in modern corrosion inhibition systems: Recent advancements", *Journal of Molecular Liquids*, Vol. 321, pp. 114666.

Ashkenazi, D. (2019), "How aluminum changed the world: A metallurgical revolution through technological and cultural perspectives", *Technological Forecasting and Social Change*, Vol. 143, pp. 101–113.

Cos, P., Vlietinck, A. J., Berghe, D. vanden, & Maes, L. (2006), "Anti-infective potential of natural products: How to develop a stronger in vitro "proof-of-concept"", *Journal of Ethnopharmacology*, Vol. 106, pp. 290–302.

Danlami, J. M., Arsad, A., Zaini, M. A. A., & Sulaiman, H. (2014), "A comparative study of various oil extraction techniques from plants", *Reviews in Chemical Engineering*, Vol. 30, pp. 605–626.

Elgahawi, H., Gobara, M., Baraka, A., & Elthalabawy, W. (2017), "Eco-friendly corrosion inhibition of AA2024 in 3.5% NaCl using the extract of Linum usitatissimum seeds", *Journal of Bio- and Tribo-Corrosion*, Vol. 3, pp. 55.

Fernine, Y., Ech-chihbi, E., Arrousse, N., el Hajjaji, F., Bousraf, F., Ebn Touhami, M., Rais, Z., & Taleb, M. (2021), "Ocimum basilicium seeds extract as an environmentally friendly antioxidant and corrosion inhibitor for aluminium alloy 2024-T3 corrosion in 3 wt% NaCl medium", *Colloids and Surfaces A: Physicochemical and Engineering Aspects*, Vol. 627, pp. 127232.

Geow, C. H., Tan, M. C., Yeap, S. P., & Chin, N. L. (2021), "A review on extraction techniques and its future applications in industry", *European Journal of Lipid Science and Technology*, Vol. 123, pp. 2000302.

Gerengi, H. (2012), "Anticorrosive properties of date palm (Phoenix dactylifera L.) fruit juice on 7075 type aluminum alloy in 3.5% NaCl solution", *Industrial and Engineering Chemistry Research*, Vol. 51, pp. 12835–12843.

Hajsafari, N., Razaghi, Z., & Tabaian, S. H. (2021), "Electrochemical study and molecular dynamics (MD) simulation of aluminum in the presence of garlic extract as a green inhibitor", *Journal of Molecular Liquids*, Vol. 336, pp. 116386.

Halambek, J., Berković, K., & Vorkapić-Furač, J. (2013), "Laurus nobilis L. oil as green corrosion inhibitor for aluminium and AA5754 aluminium alloy in 3% NaCl solution", *Materials Chemistry and Physics*, Vol. 137, pp. 788–795.

Hosary, A. A. el, Saleh, R. M., Shams, A. M., & Din, E. L. (1972), "Corrosion inhibition by naturally occurring substances--I. The effect of Hibiscus subdariffa (Karkade) extract on the dissolution of A1 AND Zn*", *Corrosion Science*, Vol. 12, pp. 897–904.

Liu, W., Singh, A., Lin, Y., Ebenso, E. E., Tianhan, G., & Ren, C. (2014), "Corrosion inhibition of Al-alloy in 3.5% NaCl solution by a natural inhibitor: An electrochemical and surface study", *International Journal of Electrochemical Science*, Vol. 9, pp. 5560–5573.

Olakolegan, O. D., Owoeye, S. S., Oladimeji, E. A., & Sanya, O. T. (2020), "Green synthesis of Terminalia Glaucescens Planch (Udi plant roots) extracts as green inhibitor for aluminum (6063) alloy in acidic and marine environment", *Journal of King Saud University - Science*, Vol. 32, pp. 1278–1285.

Radi, M., Melian, R., Galai, M., Dkhirche, N., Makha, M., Verma, C., Fernandez, C., & EbnTouhami, M. (2021), "Pumpkin seeds as an eco-friendly corrosion inhibitor for 7075-T6 alloy in 3.5% NaCl solution: Electrochemical, surface and computational studies", *Journal of Molecular Liquids*, Vol. 3 (37), pp. 116547.

Rasul, M. G. (2018), "Conventional extraction methods use in medicinal plants, their advantages and disadvantages", *International Journal of Basic Sciences and Applied Computing*, Vol. 2, pp. 10–14.

Singh, A., Lin, Y., Liu, W., Yu, S., Pan, J., Ren, C., & Kuanhai, D. (2014), "Plant derived cationic dye as an effective corrosion inhibitor for 7075 aluminum alloy in 3.5% NaCl solution", *Journal of Industrial and Engineering Chemistry*, Vol. 20, pp. 4276–4285.

Udensi, S. C., Ekpe, O. E., & Nnanna, L. A. (2021), "Corrosion inhibition performance of low cost and eco-friendly Treculia africana leaves extract on aluminium alloy AA7075-T7351 in 2.86% NaCl solutions", *Scientific African*, Vol. 12, pp. e00791.

Veiga, M., Costa, E. M., Silva, S., & Pintado, M. (2020), "Impact of plant extracts upon human health: A review", *Critical Reviews in Food Science and Nutrition*, Vol. 60, pp. 873–886.

Verma, C., Ebenso, E. E., & Quraishi, M. A. (2017), "Corrosion inhibitors for ferrous and non-ferrous metals and alloys in ionic sodium chloride solutions: A review", *Journal of Molecular Liquids*, Vol. 248, pp. 927–942.

Winston Revie (2011), *Uhlig's Corrosion Handbook* (third ed.), John Willey & Sons Inc., New York.

Xhanari, K., & Finšgar, M. (2016), "Organic corrosion inhibitors for aluminium and its alloys in acid solutions: A review", *RSC Advances*, Vol. 6, pp. 62833–62857.

Zhang, Q. W., Lin, L. G., & Ye, W. C. (2018), "Techniques for extraction and isolation of natural products: A comprehensive review", *Chinese Medicine*, Vol. 13, pp. 20.

Zulkifli, F., Ali, N., Yusof, M. S. M., Khairul, W. M., Rahamathullah, R., Isa, M. I. N., & Wan Nik, W. B. (2017), "The effect of concentration of Lawsonia inermis as a corrosion inhibitor for aluminum alloy in seawater," *Advances in Physical Chemistry*, Vol. 2017, Article ID 8521623.

21 Phytochemicals/Plant Extracts as Corrosion Inhibitors for Copper in NaCl Solutions

Elyor Berdimurodov
New Uzbekistan University
Central Asian University
National University of Uzbekistan

Khasan Berdimuradov
Shahrisabz branch of Tashkent Institute
of Chemical Technology

Ashish Kumar
Bihar Engineering University
Government of Bihar

Omar Dagdag
Gachon University

Mohamed Rbaa
Ibn Tofail University

Bhawana Jain
Siddhachalam Laboratory

Oybek Mikhliev
Karshi Engineering Economics Institute

Abduvali Kholikov and Khamdam Akbarov
National University of Uzbekistan

DOI: 10.1201/9781003394631-21

21.1 INTRODUCTION

21.1.1 Background on Corrosion and Its Effects on Metal Materials

Corrosion is a natural process that involves the deterioration of metal materials as a result of electrochemical reactions with their surrounding environment. This degradation occurs when the metal atoms lose electrons and form positively charged ions, which then react with other substances in the environment to create undesirable compounds. The formation of these compounds, such as oxides, hydroxides, and salts, leads to a weakening of the metal's structural integrity and a reduction in its mechanical properties [1–3].

Corrosion can manifest in various forms, such as uniform corrosion, pitting corrosion, galvanic corrosion, crevice corrosion, and stress corrosion cracking, among others. Each type of corrosion has its unique characteristics, but they all ultimately contribute to the degradation of metal materials [4,5].

The effects of corrosion on metal materials can have significant implications in various industries, such as:

Construction: Corrosion can weaken the structural integrity of buildings and infrastructure, leading to safety hazards and costly repairs or replacements.

Transportation: Corroded components in vehicles, ships, and aircraft can reduce performance, increase maintenance costs, and pose potential safety risks.

Electronics: Corrosion of electronic components can lead to a loss of functionality, reduced efficiency, and premature failure of devices.

Oil and gas: Corrosion in pipelines and other equipment can result in leaks, environmental damage, and increased operational costs.

The economic impact of corrosion is substantial, with estimates suggesting that the global cost of corrosion is approximately 3%–4% of a country's gross domestic product (GDP). Consequently, efforts to prevent or mitigate corrosion are crucial in preserving the performance and longevity of metal materials, reducing maintenance costs, and safeguarding against potential safety hazards [6,7].

21.1.2 Importance of Copper and Its Susceptibility to Corrosion

Copper is a versatile and widely used metal due to its unique combination of properties, including high electrical and thermal conductivity, ductility, corrosion resistance, and antimicrobial characteristics. These properties make copper a valuable material in various industries and applications, such as electrical wiring, plumbing, electronics, heat exchangers, and architectural elements [8,9].

Despite its inherent corrosion resistance, copper is not immune to corrosion, particularly in specific environments. Copper is susceptible to corrosion in the presence of aggressive ions like chloride (Cl^-), which can be found in saline environments such as coastal areas or in contact with deicing salts. Copper can also corrode when exposed to certain acids, ammonia, and sulfur compounds [10,11].

The corrosion of copper typically results in the formation of a protective oxide layer on the metal surface, known as patina. This layer, which consists of copper oxide, copper hydroxide, and basic copper salts, can provide some protection against further corrosion. However, in environments with high concentrations of aggressive ions such as chloride, the protective oxide layer can break down, leading to accelerated corrosion rates and localized forms of corrosion like pitting.

When copper corrodes, its performance, appearance, and service life can be significantly affected. In electrical systems, for example, corrosion can increase electrical resistance, leading to a loss of efficiency and potential failures. In plumbing systems, corrosion can cause leaks, blockages, and reduced water quality. Furthermore, the corrosion of copper in architectural elements can result in unsightly staining and a loss of aesthetic appeal.

Given the importance of copper in numerous applications and its susceptibility to corrosion in certain environments, it is crucial to implement effective corrosion prevention and mitigation strategies. These strategies may include the use of corrosion inhibitors, proper material selection, and appropriate design considerations to minimize the risk of corrosion and prolong the service life of copper-based materials [12–14].

21.1.3 NEED FOR CORROSION INHIBITORS IN NaCl SOLUTIONS

Corrosion inhibitors are substances that, when added to a corrosive environment, can significantly reduce the rate of corrosion of a metal. These inhibitors can act through various mechanisms, such as adsorption on the metal surface, formation of a protective film, or modification of the corrosive environment. In the context of NaCl solutions, the need for corrosion inhibitors is particularly important due to the presence of chloride ions, which can have a detrimental effect on the corrosion resistance of metals, including copper [15–17].

In NaCl solutions, chloride ions can accelerate the corrosion process by disrupting the protective oxide layers formed on the metal surface and promoting localized forms of corrosion, such as pitting and crevice corrosion. These localized forms of corrosion can lead to rapid material degradation and potentially catastrophic failures in certain applications.

The use of corrosion inhibitors in NaCl solutions can significantly reduce the corrosion rate of metals like copper and help maintain their performance and service life. By inhibiting the corrosive action of chloride ions, corrosion inhibitors can minimize the risk of structural damage, reduce maintenance costs, and improve safety in various applications, such as [18–20]:

Marine and coastal infrastructure: Structures and equipment exposed to seawater or salt-laden air, such as bridges, piers, and offshore platforms, can benefit from the use of corrosion inhibitors to minimize chloride-induced corrosion.

Transportation: Vehicles and equipment operating in regions where deicing salts are used, such as automobiles, aircraft, and railroad systems, may experience accelerated corrosion due to the presence of chloride ions in the environment. Corrosion inhibitors can help protect these systems and prolong their service life.

Electronics: Electronic devices and components exposed to saline environments or contaminated with chloride-based substances can experience increased corrosion rates, resulting in reduced performance and premature failure. Corrosion inhibitors can help maintain the reliability and longevity of these devices.

Given the detrimental effects of chloride ions on the corrosion resistance of metals and the widespread use of NaCl solutions in various applications, the use of corrosion inhibitors is essential in preserving the performance, integrity, and service life of metal materials, particularly copper, in such environments [21].

21.1.4 Overview of Phytochemicals and Plant Extracts as Corrosion Inhibitors

Phytochemicals and plant extracts have emerged as promising eco-friendly alternatives to conventional corrosion inhibitors, which are often associated with environmental and health concerns due to their toxicity and non-biodegradability. These naturally occurring compounds possess various properties that make them effective at inhibiting the corrosion of metals, including copper, in different environments like NaCl solutions [22,23].

Phytochemicals, such as alkaloids, flavonoids, tannins, terpenoids, and others, can interact with the metal surface and form a protective barrier, preventing corrosive species from reaching the metal. They can also modify the corrosive environment, reducing the aggressiveness of the solution and the corrosion rate. Plant extracts, on the contrary, are complex mixtures of various phytochemicals that can work synergistically to provide effective corrosion inhibition [24,25].

The use of phytochemicals and plant extracts as corrosion inhibitors offers several advantages over conventional inhibitors:

Low toxicity: Phytochemicals and plant extracts are generally less toxic than synthetic inhibitors, reducing the risk of adverse health and environmental effects.

Biodegradability: Being derived from natural sources, phytochemicals and plant extracts are more biodegradable and environmentally friendly compared to their synthetic counterparts.

Sustainability: Plant-based inhibitors can be obtained from renewable sources, contributing to a more sustainable approach to corrosion protection.

Cost-effectiveness: Some phytochemicals and plant extracts can be extracted from agricultural waste or low-cost plant materials, providing an economical source of corrosion inhibitors.

Numerous studies have demonstrated the effectiveness of various phytochemicals and plant extracts as corrosion inhibitors for different metals, including copper, in various environments such as NaCl solutions. However, further research is needed to optimize their performance, understand their mode of action, and develop efficient extraction and application methods to fully exploit their potential as eco-friendly and sustainable corrosion inhibitors.

21.2 PHYTOCHEMICALS AND PLANT EXTRACTS

21.2.1 Definition and Classification of Phytochemicals

Phytochemicals, also known as secondary metabolites, are biologically active compounds produced by plants. These compounds are not directly involved in the primary processes of growth, reproduction, or development, but they play important roles in the plant's defense mechanisms, communication with other organisms, and adaptation to their environment [26–28].

Phytochemicals can be classified into several major categories based on their chemical structure and functional groups. Some of the most common classes of phytochemicals include:

Alkaloids: Nitrogen-containing organic compounds that are primarily derived from plants and fungi. Alkaloids often have pronounced pharmacological effects on humans and other animals. Examples include caffeine, nicotine, and morphine.

Flavonoids: A large group of polyphenolic compounds that are responsible for the color and flavor of many fruits, vegetables, and flowers. Flavonoids are known for their antioxidant, anti-inflammatory, and anticancer properties. Examples include quercetin, rutin, and catechins.

Tannins: Polyphenolic compounds found in various plant tissues, particularly in the bark, seeds, and leaves. Tannins have astringent properties and can bind and precipitate proteins. Examples include gallic acid, ellagic acid, and condensed tannins.

Terpenoids: A diverse class of naturally occurring organic compounds derived from the isoprene unit. Terpenoids are responsible for the aroma and flavor of many plants and are often used in the perfume and flavor industries. Examples include limonene, menthol, and carotenoids.

Saponins: A class of amphiphilic glycosides that can form stable foams in aqueous solutions. Saponins are found in various plant species and have been used for their detergent and emulsifying properties. Examples include glycyrrhizin, digitonin, and ginsenosides.

Phenolic acids: A group of aromatic acids that possess one or more hydroxyl groups attached to the aromatic ring. Phenolic acids are found in various plant tissues and are known for their antioxidant and antimicrobial properties. Examples include ferulic acid, caffeic acid, and vanillic acid.

Quinones: A class of organic compounds characterized by the presence of fully conjugated cyclic diones. Quinones are involved in various biological processes such as electron transport, photosynthesis, and respiration. Examples include ubiquinone (coenzyme Q10), plastoquinone, and anthraquinone.

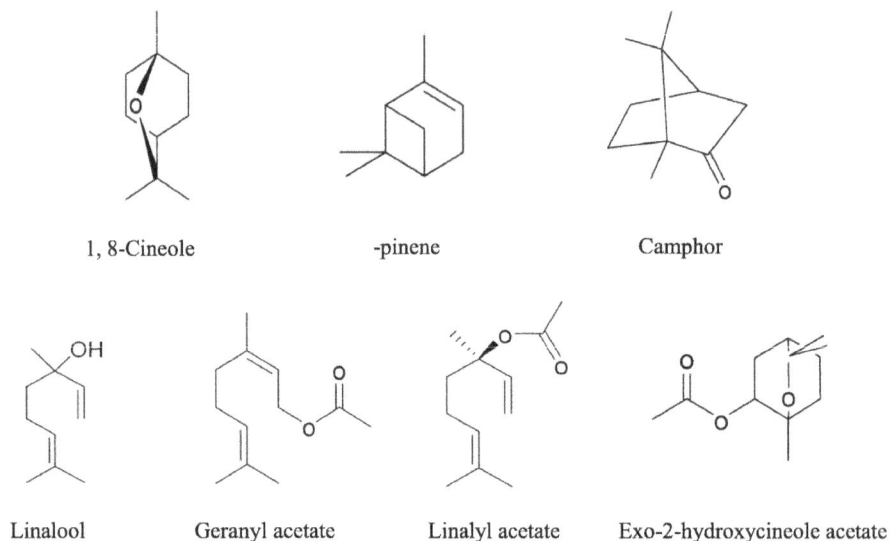

1, 8-Cineole   -pinene   Camphor

Linalool   Geranyl acetate   Linalyl acetate   Exo-2-hydroxycineole acetate

FIGURE 21.1   Basic chemicals in essential oils, which are mainly responsible for corrosion inhibition of copper in saline solution [4].

For example, the essential oils (Figure 21.1) containing oxygenated monoterpenes like 1,8-cineole and camphor can act as effective corrosion inhibitors for copper in saline solutions by adsorbing onto the metal surface, forming a protective film, inhibiting electrochemical reactions, and benefiting from synergistic effects between their various components. This environmentally friendly approach can help extend the service life of copper and reduce the negative impacts of corrosion on structures and equipment exposed to saline environments [4].

21.3   MECHANISMS OF ACTION

21.3.1   ADSORPTION OF PHYTOCHEMICALS ON COPPER SURFACE

The primary mechanism by which phytochemicals inhibit corrosion is the adsorption of the active compounds onto the metal surface, forming a protective barrier that prevents corrosive species from reacting with the metal. This adsorption process is influenced by the chemical structure and functional groups of the phytochemicals, as well as the properties of the metal surface and the surrounding environment [29,30].

There are several ways in which phytochemicals can adsorb onto the copper surface [22,23]:

Physisorption: Also known as physical adsorption, physisorption involves the formation of weak van der Waals forces between the phytochemical molecules and the copper surface. This type of adsorption is generally reversible and can be influenced by factors such as temperature, concentration, and the nature of the phytochemical.

Chemisorption: This type of adsorption involves the formation of chemical bonds between the phytochemical molecules and the copper surface, resulting in a more stable and stronger interaction. Chemisorption can involve the donation or acceptance of electrons, leading to the formation of coordination bonds, covalent bonds, or ionic bonds. The functional groups of the phytochemicals, such as hydroxyl, carboxyl, or amino groups, play a crucial role in chemisorption.

Mixed adsorption: In some cases, phytochemicals can interact with the copper surface through a combination of both physisorption and chemisorption processes.

The adsorption of phytochemicals on the copper surface generally follows well-established isotherms, such as the Langmuir, Freundlich, or Temkin isotherms, which describe the relationship between the amount of adsorbed phytochemical and its concentration in the solution.

The effectiveness of the adsorption process depends on several factors, including the concentration of the phytochemical, the temperature and pH of the environment, and the presence of other species in the solution. Some phytochemicals can also form complexes with the corrosive species in the solution, reducing their reactivity and further inhibiting the corrosion process.

In summary, the adsorption of phytochemicals onto the copper surface is a crucial mechanism by which these compounds inhibit corrosion. The nature of the adsorption process depends on the chemical structure and functional groups of the phytochemicals, as well as the properties of the copper surface and the surrounding environment.

Based on the provided description of Figure 21.2, the study investigates the corrosion inhibition performance of *Aloe saponaria* (syn. *Aloe maculata*) extract, referred to as AST, on bronze B66 in a 3% NaCl solution. The electrochemical measurements, including Tafel polarization and impedance spectroscopy, were conducted to analyze the influence of AST on the corrosion behavior of bronze B66 [31].

Tafel polarization: In the cathodic domain, the addition of AST displaces the corrosion potential (E_corr) in the negative direction, indicating a shift in the electrochemical equilibrium towards a more negative potential. Moreover, the cathodic current density decreases, with the reduction being more pronounced at higher concentrations of the inhibitor. This suggests that the AST inhibitor affects the cathode reaction mechanism, hindering the corrosion process. In the anodic domain, there is a slight decrease in the anodic current density, although less significant compared to the cathodic domain.

Impedance plots: The impedance diagrams show that the polarization resistance (R_p) increases as the inhibitor concentration increases, indicating that the AST extract forms a protective layer on the bronze surface, slowing down the charge transfer at the metal-electrolyte interface. In the low-frequency range, the charge transfer resistance (R_{ct}) values increase with the inhibitor concentration, reaching 11,188 Ωcm^2 at 150 ppm. The associated capacitance is approximately 2.89 μF cm^{-2}. The significant increase in polarization resistance with concentration highlights the effectiveness of the inhibitor in decreasing charge transfer at the studied interface.

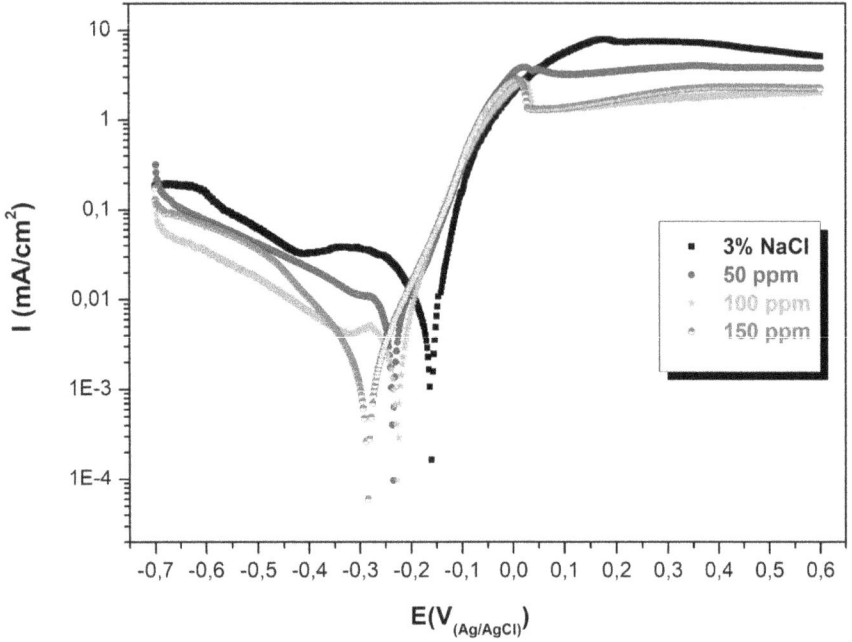

(a)

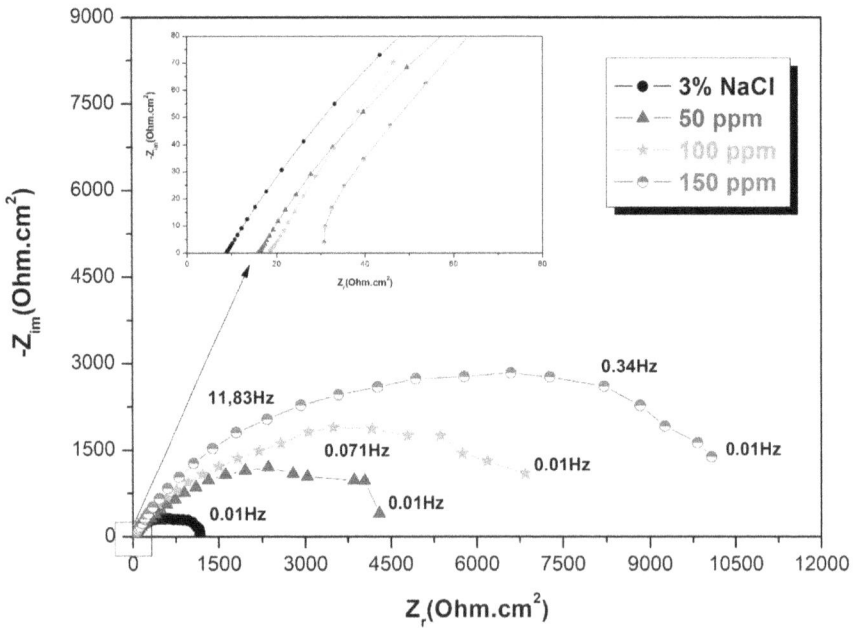

(b)

FIGURE 21.2   (a) Tafel and (b) impedance plots of bronze B66 in 3% NaCl by *Aloe saponaria* (syn. *Aloe maculata*) extract [31].

21.3.2 FORMATION OF PROTECTIVE LAYERS BY PHYTOCHEMICALS

The formation of protective layers on the metal surface is another important mechanism by which phytochemicals can inhibit corrosion. Upon adsorption, some phytochemicals can create a physical or chemical barrier that prevents the access of corrosive species to the metal surface, thus reducing the rate of the electrochemical reactions that cause corrosion [19,20].

Two main types of protective layers can be formed by phytochemicals on the copper surface:

Passive films: Some phytochemicals can promote the formation of a passive film on the copper surface, which consists of a thin, dense, and stable oxide layer. This film acts as a barrier, separating the metal from the corrosive environment and reducing the rate of electrochemical reactions. The presence of functional groups, such as hydroxyl or carboxyl groups, in the phytochemicals can facilitate the formation of these passive films by interacting with the metal ions on the surface or by modifying the local environment.

Organic layers: Phytochemicals can also form organic layers on the copper surface through adsorption. These layers consist of a network of phytochemical molecules that are either physically or chemically bonded to the metal surface. The organic layers can provide a barrier against the diffusion of corrosive species, thus reducing the rate of corrosion. The effectiveness of these organic layers depends on the nature of the phytochemicals, their concentration, and the conditions of the environment.

In some cases, the protective layers formed by phytochemicals can also have a self-healing property. If the protective layer is damaged or partially removed, the phytochemicals present in the solution can re-adsorb onto the exposed metal surface, repairing the protective layer and maintaining corrosion inhibition.

The formation of protective layers by phytochemicals is an essential mechanism for inhibiting corrosion, particularly in copper-based materials. By creating a physical or chemical barrier on the metal surface, phytochemicals can effectively reduce the interaction between the metal and the corrosive environment, thus preserving the structural integrity and performance of the material.

The SEM images and EDS data (Figure 21.3) comparison helps to understand the corrosion behavior of copper in a 1 M HCl solution and the effectiveness of PJ extract as a corrosion inhibitor. By analyzing the surface morphology and elemental composition, researchers can identify the mechanisms by which the inhibitor protects the copper surface from corrosion, leading to potential applications in the development of environmentally friendly and effective corrosion protection solutions [16].

The comparison of two- and three-dimensional AFM images (Figure 21.4) helps to understand the corrosion behavior of copper in a 1 M HCl solution and the effectiveness of PJ inhibitor as a corrosion inhibitor. By analyzing the surface topography and observing the differences in surface features, researchers can identify the mechanisms by which the inhibitor protects the copper surface from corrosion. This information can be useful in the development of environmentally friendly and effective corrosion protection solutions [16].

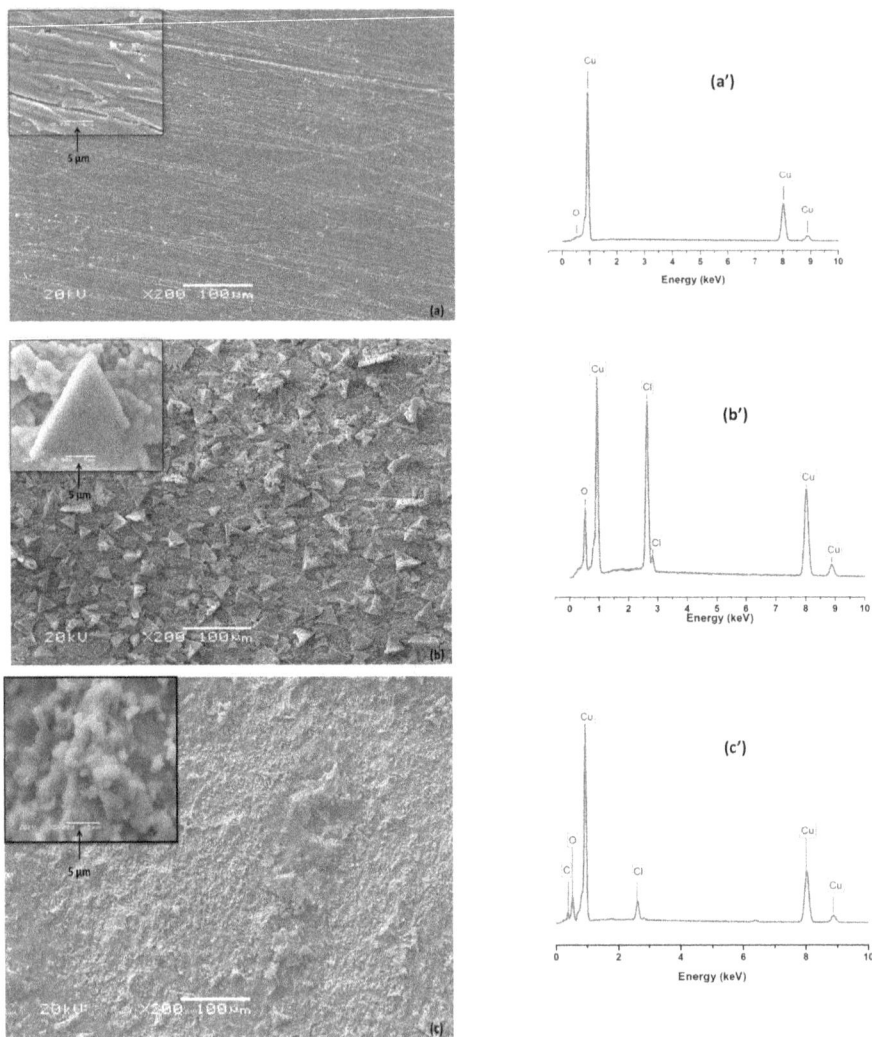

FIGURE 21.3 SEM pictures (left) and EDS data (right) for unprocessed polished copper (a, a′) and following a 24-hour immersion at ambient temperature in 1 M HCl solution without an inhibiting agent (b, b′) and with 1 g L^{-1} of PJ extract (c, c′) [16].

21.3.3 INTERACTION OF PHYTOCHEMICALS WITH CORROSION REACTION INTERMEDIATES

Another mechanism by which phytochemicals can inhibit corrosion is through their interaction with the intermediates of the corrosion reaction. Corrosion is an electrochemical process involving anodic and cathodic reactions that occur simultaneously on the metal surface. By interacting with the intermediates of these reactions, phytochemicals can disrupt the electrochemical process and reduce the overall corrosion rate [17–19].

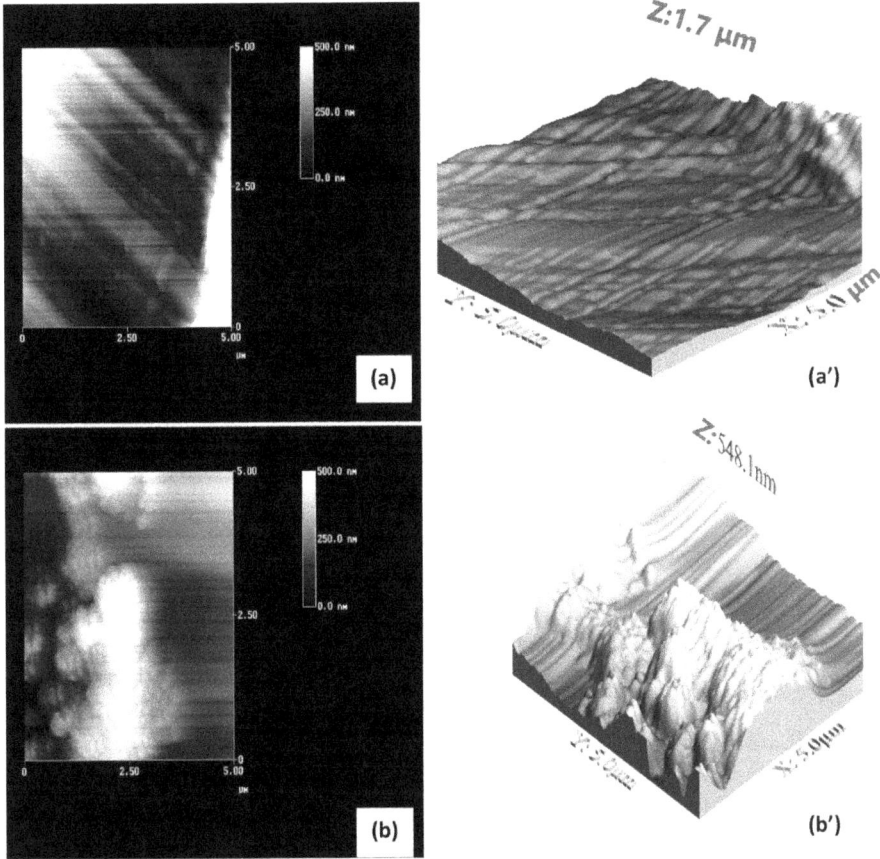

FIGURE 21.4 AFM pictures of copper samples prior to and following a 24-hour submersion in 1 M HCl solution: (a, a′) in the absence of a corrosion-preventing agent and (b, b′) containing 1 g L^{-1} of PJ inhibiting substance at ambient temperature [16].

The interaction of phytochemicals with corrosion reaction intermediates can occur in various ways:

Inhibition of anodic reactions: Anodic reactions involve the dissolution of the metal and the generation of metal ions (e.g., Cu^{2+} for copper). Phytochemicals can interact with these metal ions, forming complexes that are less soluble and less reactive. This can slow down the anodic reactions and decrease the rate of metal dissolution. The functional groups present in the phytochemicals, such as hydroxyl or carboxyl groups, play an essential role in the formation of these complexes.

Inhibition of cathodic reactions: Cathodic reactions involve the reduction of a species in the environment, such as oxygen or hydrogen ions. Phytochemicals can interfere with these reactions by scavenging the reactive species or by competing for the available reaction sites on the metal surface. For example, some phytochemicals can act as oxygen scavengers, reducing the amount of dissolved oxygen in the environment and limiting the rate of cathodic reactions.

Modulation of the electrochemical potential: Phytochemicals can also affect the electrochemical potential of the metal surface, shifting the potential either toward more positive or toward more negative values. This can alter the rates of the anodic and cathodic reactions, leading to a reduction in the overall corrosion rate. The specific effect of the phytochemicals on the electrochemical potential depends on their chemical structure, concentration, and the environment.

By interacting with the intermediates of the corrosion reaction, phytochemicals can alter the balance between the anodic and cathodic reactions, effectively reducing the overall corrosion rate. This mechanism, combined with the adsorption of phytochemicals onto the metal surface and the formation of protective layers, contributes to the corrosion inhibition properties of phytochemicals and their potential as eco-friendly and sustainable corrosion inhibitors.

21.4 TYPES IN PHYTOCHEMICALS/PLANT EXTRACTS AS CORROSION INHIBITORS FOR COPPER IN NACL SOLUTIONS

21.4.1 Alkaloids

Alkaloids are a diverse group of naturally occurring organic compounds that primarily contain basic nitrogen atoms. They are derived from plant and animal sources and exhibit a wide range of biological activities, including anti-inflammatory, antibacterial, and anticancer properties. Due to their complex molecular structures and the presence of various functional groups, alkaloids have demonstrated potential as corrosion inhibitors for copper in NaCl solutions [8–10].

Several alkaloids have been studied for their corrosion inhibition properties on copper, including [5–7]:

Quinoline alkaloids: Quinoline and its derivatives, such as 8-hydroxyquinoline, have been reported to show strong corrosion-inhibitive properties on copper in NaCl solutions. The presence of nitrogen and oxygen atoms in the quinoline ring allows for the formation of coordinate bonds with the copper surface, leading to the formation of a protective film that reduces the corrosion rate.

Piperidine alkaloids: Piperidine and its derivatives, such as piperine, have been reported to exhibit corrosion inhibition properties on copper in NaCl solutions. The nitrogen atom in the piperidine ring can form coordinate bonds with the copper surface, while the presence of other functional groups, such as methoxy or amide groups, can enhance the adsorption of the alkaloid onto the metal surface.

Indole alkaloids: Indole alkaloids, such as tryptamine and serotonin, have also been studied for their corrosion inhibition properties on copper in NaCl solutions. The presence of nitrogen and oxygen atoms in the indole ring allows for the formation of coordinate bonds with the copper surface, while other functional groups, such as hydroxyl or amine groups, can further enhance the adsorption and the formation of protective films.

Isoquinoline alkaloids: Isoquinoline alkaloids, such as berberine, have been reported to exhibit corrosion inhibition properties on copper in NaCl solutions. The nitrogen atom in the isoquinoline ring and the presence of other functional groups, such as methoxy or hydroxyl groups, can contribute to the adsorption of the alkaloid onto the copper surface and the formation of protective films.

The efficacy of alkaloids as corrosion inhibitors for copper in NaCl solutions depends on their chemical structure, concentration, and environmental conditions, such as temperature and pH. The presence of various functional groups, such as nitrogen and oxygen atoms, enables alkaloids to form coordinate bonds with the copper surface, leading to the formation of protective films that reduce the corrosion rate. Further research is needed to optimize the application of alkaloids as corrosion inhibitors for copper and to explore the synergistic effects of combining different alkaloids or other phytochemicals for enhanced corrosion protection [31,32].

21.4.2 Flavonoids

Flavonoids are a large group of naturally occurring polyphenolic compounds found in various plant species. They are well known for their antioxidant, anti-inflammatory, and antimicrobial properties. Due to their diverse chemical structures and the presence of multiple functional groups, flavonoids have also demonstrated potential as corrosion inhibitors for copper in NaCl solutions.

Some examples of flavonoids that have been studied for their corrosion inhibition properties on copper include:

Quercetin: Quercetin is a common flavonoid found in many fruits, vegetables, and plant-derived beverages. Its molecular structure includes hydroxyl groups that can form hydrogen bonds or coordinate bonds with the copper surface. Quercetin has been reported to exhibit good corrosion inhibition properties on copper in NaCl solutions, forming a protective film on the metal surface that reduces the corrosion rate.

Rutin: Rutin is a flavonoid glycoside found in various plants, including buckwheat, asparagus, and citrus fruits. The presence of hydroxyl and glycosidic groups in its structure allows for the formation of hydrogen bonds or coordinate bonds with the copper surface. Rutin has been reported to show corrosion inhibition properties on copper in NaCl solutions.

Catechin: Catechin is a type of flavonoid commonly found in tea, cocoa, and various fruits. Its molecular structure includes multiple hydroxyl groups that can interact with the copper surface, forming hydrogen bonds or coordinate bonds. Catechin has been reported to exhibit corrosion inhibition properties on copper in NaCl solutions.

Naringin: Naringin is a flavonoid glycoside found in grapefruit and other citrus fruits. Its molecular structure includes hydroxyl and glycosidic groups
that can form hydrogen bonds or coordinate bonds with the copper surface.
Naringin has been reported to show corrosion inhibition properties on copper in NaCl solutions.

The effectiveness of flavonoids as corrosion inhibitors for copper in NaCl solutions
depends on their chemical structure, concentration, and environmental conditions,
such as temperature and pH. The presence of functional groups, such as hydroxyl
and glycosidic groups, allows flavonoids to form hydrogen bonds or coordinate bonds
with the copper surface, leading to the formation of protective films that reduce the
corrosion rate. Further research is needed to optimize the application of flavonoids
as corrosion inhibitors for copper and to explore the synergistic effects of combining
different flavonoids or other phytochemicals for enhanced corrosion protection.

21.4.3 TANNINS

Tannins are a class of naturally occurring polyphenolic compounds found in various
plant species, particularly in the bark, leaves, and fruits. They are widely known for
their astringent properties and have been used traditionally for various applications,
such as leather tanning, dyeing, and medicine. Due to their complex molecular structures and the presence of multiple functional groups, tannins have also demonstrated
potential as corrosion inhibitors for copper in NaCl solutions [3,4,31].

Tannins can be generally classified into two main groups: hydrolyzable tannins
and condensed tannins.

Hydrolyzable tannins: Hydrolyzable tannins are esters of gallic acid or ellagic
acid and a sugar, usually glucose. Examples of hydrolyzable tannins include
tannic acid and ellagi tannins. These tannins have been found to exhibit
corrosion inhibition properties on copper in NaCl solutions. The presence
of multiple hydroxyl and carboxyl groups in their structure allows hydrolyzable tannins to form hydrogen bonds or coordinate bonds with the copper
surface, leading to the formation of protective films that reduce the corrosion rate.
Condensed tannins: Condensed tannins, also known as proanthocyanidins, are
polymers of flavonoid units, such as catechin and epicatechin. These tannins have also been reported to exhibit corrosion inhibition properties on
copper in NaCl solutions. The presence of multiple hydroxyl groups in their
structure enables condensed tannins to form hydrogen bonds or coordinate
bonds with the copper surface, resulting in the formation of protective films
that decrease the corrosion rate.

The effectiveness of tannins as corrosion inhibitors for copper in NaCl solutions
depends on their chemical structure, concentration, and environmental conditions,
such as temperature and pH. The presence of functional groups, such as hydroxyl

and carboxyl groups, allows tannins to form hydrogen bonds or coordinate bonds with the copper surface, leading to the formation of protective films that reduce the corrosion rate. Further research is needed to optimize the application of tannins as corrosion inhibitors for copper and to explore the synergistic effects of combining different tannins or other phytochemicals for enhanced corrosion protection.

The presence of functional groups, such as hydroxyl or carbonyl groups, in their structure allows monoterpenoids to form hydrogen bonds or coordinate bonds with the copper surface, leading to the formation of protective films that reduce the corrosion rate.

Sesquiterpenoids: Sesquiterpenoids are a class of terpenoids that consist of three isoprene units. Examples include farnesol and artemisinin. These terpenoids have also been reported to exhibit corrosion inhibition properties on copper in NaCl solutions. The presence of functional groups, such as hydroxyl or carbonyl groups, enables sesquiterpenoids to interact with the copper surface, forming protective films that reduce the corrosion rate.

21.4.4 SAPONINS

Saponins are a class of naturally occurring glycosides found in various plant species. They are characterized by the presence of a hydrophobic aglycone (sapogenin) and one or more hydrophilic sugar moieties. Saponins are known for their ability to form foams in aqueous solutions and exhibit various biological activities, including antimicrobial and anti-inflammatory properties. Due to their amphiphilic nature and the presence of multiple functional groups, saponins have demonstrated potential as corrosion inhibitors for copper in NaCl solutions.

21.4.5 PHENOLIC ACIDS

Phenolic acids are a group of naturally occurring aromatic compounds that primarily contain a carboxylic acid functional group and a phenol ring. Examples of phenolic acids include gallic acid, caffeic acid, and ferulic acid. These compounds have been reported to exhibit corrosion inhibition properties on copper in NaCl solutions. The presence of hydroxyl and carboxyl groups in their structure allows phenolic acids to form hydrogen bonds or coordinate bonds with the copper surface, leading to the formation of protective films that reduce the corrosion rate.

21.4.6 QUINONES

Quinones are a class of naturally occurring aromatic compounds that consist of a fully conjugated cyclic dione structure. Examples of quinones include anthraquinone, benzoquinone, and naphthoquinone. Quinones have been reported to exhibit corrosion inhibition properties on copper in NaCl solutions. The presence of carbonyl and conjugated π-electron systems in their structure allows quinones to form coordinate bonds with the copper surface, leading to the formation of protective films that reduce the corrosion rate.

21.4.7 COMPARISON OF DIFFERENT TYPES OF PHYTOCHEMICALS AS CORROSION INHIBITORS

The effectiveness of different types of phytochemicals as corrosion inhibitors for copper in NaCl solutions depends on their chemical structure, concentration, and environmental conditions such as temperature and pH. Each class of phytochemicals possesses unique structural features and functional groups that contribute to their corrosion inhibition properties [4,8].

In general, phytochemicals with multiple functional groups capable of forming hydrogen bonds or coordinate bonds with the metal surface, such as hydroxyl, carboxyl, or nitrogen-containing groups, tend to exhibit better corrosion inhibition properties. Additionally, the presence of π-electron systems in some phytochemicals can enhance their adsorption onto the metal surface and contribute to the formation of protective films.

It is difficult to directly compare the effectiveness of different types of phytochemicals as corrosion inhibitors, as their performance may vary depending on the specific environmental conditions and the experimental setup. However, by understanding the structural features and mechanisms of action of different phytochemicals, researchers can design more effective and environmentally friendly corrosion inhibitors based on natural compounds. Further research is needed to optimize the application of phytochemicals as corrosion inhibitors and to explore the synergistic effects of combining different phytochemicals or other eco-friendly additives for enhanced corrosion protection.

The provided description presents the most stable configurations of various adsorbed molecules on a copper (111) surface, as obtained from molecular dynamics (MD) simulations (Figure 21.5). The molecules considered include neutral palmitic acid, phytol, 17-octadecynoic acid, linolenic acid, diisooctyl phthalate, and chrysophanol. The results are presented as side (a) and top (b) views.

Adsorption configurations: The side and top views of these configurations provide insights into how each molecule interacts with the copper surface. These interactions could involve various types of bonding, such as covalent, ionic, or van der Waals forces, depending on the specific molecule and surface chemistry. The most stable configuration for each molecule represents the lowest energy state, which is indicative of the strongest and most favorable interaction between the molecule and the copper surface.

Molecular dynamics simulations: MD simulations are a powerful computational tool used to model the behavior of molecules and materials at the atomic scale. By simulating the motion of atoms and molecules over time, researchers can obtain detailed information about the structural, mechanical, and thermodynamic properties of the system under study. In this case, MD simulations are used to identify the most stable adsorption configurations for each molecule on the Cu (111) surface, providing valuable information about the nature and strength of the molecule-surface interactions.

Potential applications: Understanding the adsorption of these molecules on the copper surface can have various implications and applications, such as in the development

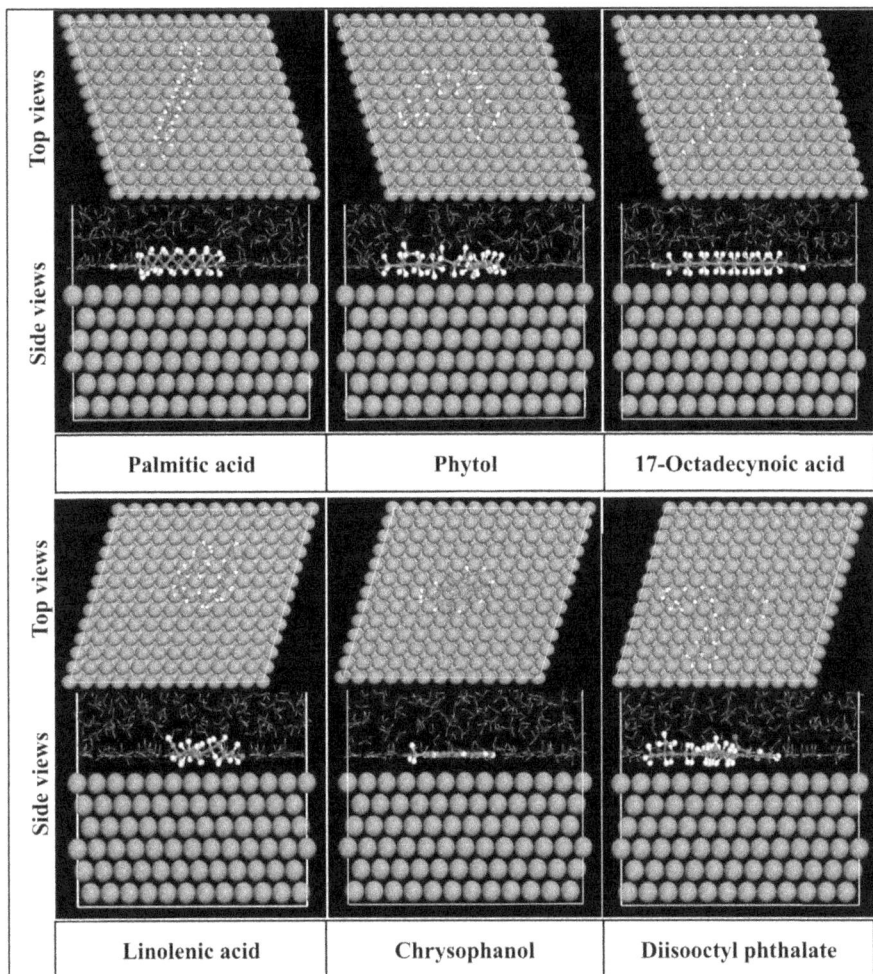

FIGURE 21.5 Side (a) and top (b) perspectives of the most stable arrangements for the adsorption of neutral palmitic acid, phytol, 17-octadecynoic acid, linolenic acid, diisooctyl phthalate, and chrysophanol molecules on the Cu (111) surface, as determined from molecular dynamics (MD) simulations [31].

of corrosion inhibitors, self-assembled monolayers, or surface modifications to tailor specific properties. By studying the most stable configurations, researchers can identify the key factors that contribute to effective molecule-surface interactions, leading to potential advancements in materials science and surface engineering. In the context of corrosion prevention, for example, insights into the adsorption behavior of these molecules could be used to design inhibitors that form strong, protective layers on the copper surface, thereby reducing corrosion rates and prolonging the lifetime of copper-based materials.

21.5 CHALLENGES AND FUTURE RESEARCH DIRECTIONS

The use of phytochemicals as corrosion inhibitors for copper in NaCl solutions presents several challenges and opportunities for future research. Some of the key challenges and research directions include:

Structure-activity relationship: Understanding the structure-activity relationship of phytochemicals is essential for the development of effective corrosion inhibitors. Determining which functional groups and structural features contribute to the corrosion inhibition properties of a specific phytochemical can guide the selection and design of new inhibitors. Further research is needed to identify the key structural features and mechanisms of action responsible for the corrosion inhibition properties of phytochemicals [2,6,7,11].

Synergistic effects: Combining different phytochemicals or other eco-friendly additives could lead to synergistic effects, resulting in improved corrosion protection. Investigations into the interactions between different phytochemicals and their combined effects on corrosion inhibition can help optimize the use of natural compounds for corrosion protection.

Optimization of concentration and formulation: Research is needed to determine the optimal concentration and formulation of phytochemicals for effective corrosion inhibition. Factors such as environmental conditions (e.g., temperature and pH), metal surface properties, and the presence of other additives need to be considered when optimizing the application of phytochemicals as corrosion inhibitors.

Long-term stability and performance: The long-term stability and performance of phytochemicals as corrosion inhibitors need to be assessed to ensure their effectiveness over time. This includes evaluating the durability of the protective films formed by phytochemicals and their resistance to environmental factors such as UV radiation, temperature fluctuations, and exposure to aggressive ions.

Scale-up and commercialization: The scale-up and commercialization of phytochemical-based corrosion inhibitors present challenges related to the sustainable sourcing of raw materials, large-scale extraction and purification processes, and the development of cost-effective and eco-friendly industrial formulations. Research is needed to address these challenges and to establish viable pathways for the commercialization of phytochemical-based corrosion inhibitors.

Environmental impact and toxicity: While phytochemicals are generally considered environmentally friendly and non-toxic, the potential environmental impact and toxicity of specific phytochemicals used as corrosion inhibitors need to be assessed. This includes understanding the bioavailability, biodegradability, and potential ecological effects of these compounds.

By addressing these challenges and exploring future research directions, the development of effective and environmentally friendly phytochemical-based corrosion inhibitors can contribute to the sustainable protection of copper and other metals from corrosion in a wide range of applications and industries.

21.6 CONCLUSION

Phytochemicals, derived from plant sources, have demonstrated potential as corrosion inhibitors for copper in NaCl solutions. Different classes of phytochemicals, such as alkaloids, flavonoids, tannins, terpenoids, saponins, phenolic acids, and quinones, exhibit unique structural features and functional groups that contribute to their corrosion inhibition properties. The effectiveness of these phytochemicals as corrosion inhibitors depends on their chemical structure, concentration, and environmental conditions such as temperature and pH.

Understanding the structure-activity relationship and mechanisms of action of phytochemicals is essential for the development of effective corrosion inhibitors. Future research directions include exploring the synergistic effects of combining different phytochemicals or other eco-friendly additives, optimizing the concentration and formulation of phytochemicals, and evaluating their long-term stability and performance.

By addressing the challenges and exploring future research directions, the development of effective and environmentally friendly phytochemical-based corrosion inhibitors can contribute to the sustainable protection of copper and other metals from corrosion in various applications and industries.

REFERENCES

[1] Chaubey N, Qurashi A, Chauhan DS, Quraishi MA. Frontiers and advances in green and sustainable inhibitors for corrosion applications: A critical review. *Journal of Molecular Liquids*. 2021;321:114385.

[2] Dahmani K, Galai M, Ouakki M, Cherkaoui M, Touir R, Erkan S, et al. Quantum chemical and molecular dynamic simulation studies for the identification of the extracted cinnamon essential oil constituent responsible for copper corrosion inhibition in acidified 3.0 áwt% NaCl medium. *Inorganic Chemistry Communications*. 2021;124:108409.

[3] Deyab MA, Mohsen Q, Guo L. Aesculus hippocastanum seeds extract as eco-friendly corrosion inhibitor for desalination plants: Experimental and theoretical studies. *Journal of Molecular Liquids*. 2022;361:119594.

[4] Dhouibi I, Masmoudi F, Bouaziz M, Masmoudi M. A study of the anti-corrosive effects of essential oils of rosemary and myrtle for copper corrosion in chloride media. *Arabian Journal of Chemistry*. 2021;14(2):102961.

[5] El Ibrahimi B. Sustainable corrosion inhibitors for copper and its alloys. *Sustainable Corrosion Inhibitors*. 2021;107:175.

[6] El-Asri A, Rguiti MM, Jmiai A, Oukhrib R, Bourzi H, Lin Y, et al. Carissa macrocarpa extract (ECM) as a new efficient and ecologically friendly corrosion inhibitor for copper in nitric acid: Experimental and theoretical approach. *Journal of the Taiwan Institute of Chemical Engineers*. 2023;142:104633.

[7] Faraj MW, Alamiery A. Palm oil as green corrosion inhibitors for different metal surfaces and corrosive media: A review. *International Journal of Corrosion and Scale Inhibition*. 2022;11:465–77.

[8] Fekri MH, Omidali F, Alemnezhad MM, Ghaffarinejad A. Turnip peel extract as green corrosion bio-inhibitor for copper in 3.5% NaCl solution. *Materials Chemistry and Physics*. 2022;286:126150.

[9] Feng Y, He J, Zhan Y, An J, Tan B. Insight into the anti-corrosion mechanism Veratrum root extract as a green corrosion inhibitor. *Journal of Molecular Liquids*. 2021;334:116110.

[10] Fernine Y, Ech-Chihbi E, Arrousse N, El Hajjaji F, Bousraf F, Touhami ME, et al. Ocimum basilicum seeds extract as an environmentally friendly antioxidant and corrosion inhibitor for aluminium alloy 2024-T3 corrosion in 3 wt% NaCl medium. *Colloids and Surfaces A: Physicochemical and Engineering Aspects.* 2021;627:127232.

[11] Grekulović V, Mitovski A, Vujasinović MR, Štrbac N, Zdravković M, Gorgievski M, et al. Electrochemical behavior of copper in chloride medium in the presence of walnut shell macerate.

[12] Guo X, Wu F, Cheng T, Huang H. Extraction of a high efficiency and long-acting green corrosion inhibitor from silkworm excrement and its adsorption behavior and inhibition mechanism on copper. *Colloids and Surfaces A: Physicochemical and Engineering Aspects.* 2021;631:127679.

[13] Hasanin MS, Al Kiey SA. Environmentally benign corrosion inhibitors based on cellulose niacin nano-composite for corrosion of copper in sodium chloride solutions. *International Journal of Biological Macromolecules.* 2020;161:345–54.

[14] Huang L, Chen W-Q, Wang S-S, Zhao Q, Li H-J, Wu Y-C. Starch, cellulose and plant extracts as green inhibitors of metal corrosion: A review. *Environmental Chemistry Letters.* 2022;20(5):3235–64.

[15] Ituen E, Singh A, Li R, Yuanhua L, Guo C. Nanostructure, surface and anticorrosion properties of phyto-fabricated copper nanocomposite in simulated oilfield descaling fluid. *Surfaces and Interfaces.* 2020;19:100514.

[16] Jmiai A, El Ibrahimi B, Tara A, Chadili M, El Issami S, Jbara O, et al. Application of Zizyphus Lotuse - pulp of Jujube extract as green and promising corrosion inhibitor for copper in acidic medium. *Journal of Molecular Liquids.* 2018;268:102–13.

[17] Kasapović D, Klepo L, Korać F. Investigation of inhibitory effect of the Rubus idealis L. extract on corrosion of copper. *Bulletin of the Chemists & Technologists of Bosnia & Herzegovina.* 2022;(59).

[18] Kasapović D, Korać F, Bikić F. Testing the effectiveness of raspberry flower extract as an inhibitor of copper's corrosion in 3% NaCl. *Zaštita materijala.* 2022;63(2):115–21.

[19] Kusumaningrum I, Soenoko R, Siswanto E, Gapsari F. Investigation of Artocarpus Heteropyllus peel extract as non-toxic corrosion inhibitor for pure copper protection in nitric acid. *Case Studies in Chemical and Environmental Engineering.* 2022;6:100223.

[20] Martinović I, Pilić Z, Zlatić G, Soldo V, Šego M. N-Acetyl cysteine and D-penicillamine as green corrosion inhibitors for copper in 3% NaCl. *International Journal of Electrochemical Science.* 2023;18:100238.

[21] Martinović I, Zlatić G, Pilić Z, Šušak M, Falak F. Antioxidant capacity and corrosion inhibition efficiency of Sambucus nigra L. extract. *Chemical and Biochemical Engineering Quarterly.* 2023;37(2):79–87.

[22] Messaoudi H, Djazi F, Litim M, Keskin B, Slimane M, Bekhiti D. Surface analysis and adsorption behavior of caffeine as an environmentally friendly corrosion inhibitor at the copper/aqueous chloride solution interface. *Journal of Adhesion Science and Technology.* 2020;34(20):2216–44.

[23] Pourzarghan V, Fazeli-Nasab B. The use of Robinia pseudoacacia L fruit extract as a green corrosion inhibitor in the protection of copper-based objects. *Heritage Science.* 2021;9(1):1–14.

[24] Pourzarghan V, Fazeli-Nasab B. The use of Robinia pseudoacania L fruit extract as a natural corrosion inhibitor in the protection of historical bronze objects. 2021;1:1–32.

[25] Raghavendra N. Latest exploration on natural corrosion inhibitors for industrial important metals in hostile fluid environments: A comprehensive overview. *Journal of Bio- and Tribo-Corrosion.* 2019;5(3):54.

[26] Shen J, Yang D, Ma L, Gao Z, Yan A, Liao Q. Exploration of neonicotinoids as novel corrosion inhibitors for copper in a NaCl solution: Experimental and theoretical studies. *Colloids and Surfaces A: Physicochemical and Engineering Aspects.* 2022;636:128058.

[27] Shinato KW, Zewde AA, Jin Y. Corrosion protection of copper and copper alloys in different corrosive medium using environmentally friendly corrosion inhibitors. *Corrosion Reviews*. 2020;38(2):101–9.

[28] Tan B, Xiang B, Zhang S, Qiang Y, Xu L, Chen S, et al. Papaya leaves extract as a novel eco-friendly corrosion inhibitor for Cu in H_2SO_4 medium. *Journal of Colloid and Interface Science*. 2021;582:918–31.

[29] Tasić ŽZ, Petrović Mihajlović MB, Radovanović MB, Antonijević MM. New trends in corrosion protection of copper. *Chemical Papers*. 2019;73:2103–32.

[30] Umoren SA, Solomon MM, Obot IB, Suleiman RK. A critical review on the recent studies on plant biomaterials as corrosion inhibitors for industrial metals. *Journal of Industrial and Engineering Chemistry*. 2019;76:91–115.

[31] Benzidia B, Barbouchi M, Hsissou R, Zouarhi M, Erramli H, Hajjaji N. A combined experimental and theoretical study of green corrosion inhibition of bronze B66 in 3% NaCl solution by Aloe saponaria (syn. Aloe maculata) tannin extract. *Current Research in Green and Sustainable Chemistry*. 2022;5:100299.

[32] Ammouchi N, Allal H, Zouaoui E, Dob K, Zouied D, Bououdina M. Extracts of Ruta chalepensis as green corrosion inhibitor for copper CDA 110 in 3% NaCl medium: Experimental and theoretical studies. *Analytical and Bioanalytical Electrochemistry*. 2019;11(07):830–50.

22 Phytochemicals/Plant Extracts as Corrosion Inhibitors for Zinc in NaCl Solutions

Mosarrat Parveen, Mohammad Mobin,
Saman Zehra, Ruby Aslam, and Kanika Cial
Aligarh Muslim University

22.1 INTRODUCTION

Metal corrosion is a thermodynamically predictable process that causes metals to react chemically or electrochemically with the environment to achieve a more stable state. Corrosion's adverse effects on the metallic structure, reconstruction costs, and public welfare are significant. Corrosion causes significant negative impacts on a material's health as a result of environmental contacts. Consistent corrosion causes plant shutdowns, costly maintenance, and resource waste, and large amounts are expended in expensive designs to prevent these losses each year. Numerous sectors face serious corrosion-related issues. It causes loss of massive amounts of metallic materials.

According to estimates, corrosion costs the economy trillions of dollars each year and damages a wide range of sectors, such as water transportation, petroleum and gas, automotive, and pipelines. The National Association of Corrosion Engineers estimates that in 2016, corrosion cost the world economy \$2.5 trillion, or nearly 3.5% of GDP. However, the expense of corrosion can be decreased by up to 15%–35% by properly using existing corrosion-preventing technologies (375–875 billion US dollars). Corrosion is a critical challenge that needs to be handled by scientists and engineers operating globally in the corrosion field and engineering throughout the world because of its association with extremely high economic and safety damages [1]. Due to its low cost and good mechanical qualities, mild steel is the most crucial engineering material, particularly for construction, and automotive applications. However, in many different processes, such as etching, acid pickling, and oil-well acidification, steel comes into contact with corrosive conditions which can cause the steel to rust and deteriorate. It is usual practice to employ inhibited acid solutions to minimize acid's corrosive impact on metals. Since the usage of inhibitors varies depending on the system, it needs to be properly examined.

Due to their simplicity in synthesis and application as well as their excellent efficiency at relatively low concentrations, the use of synthetic corrosion inhibitors is

DOI: 10.1201/9781003394631-22

among the most popular used corrosion prevention techniques. These organic substances provide a surface barrier that protects metals from corrosive degradation that affects alloys and metals by sticking to their surfaces *via* their heteroatoms and electrons. The majority of synthetic organic inhibitors are harmful to human health and have a detrimental effect on the environment [2–4]. Green and environmentally safe corrosion inhibitors have emerged as a result of rising ecological consciousness and stringent environmental restrictions [5–7].

In the field of chemistry, the idea of "green chemistry" has become increasingly significant. The creation of economical and environmentally friendly corrosion inhibitors is necessary for the practical application of this idea. A corrosion inhibitor can be defined as a chemical compound or mixture of chemicals that, when present in an optimum concentration, hinders or lowers the corrosion rate without appreciably modifying the amount of any harsh agent. Corrosion inhibitors can be designated as either organic or inorganic based on their chemical makeup. According to the properties that they possess, the inhibitors can be categorized as either liquid-phase inhibitors or gaseous-phase inhibitors. According to the nature of the protective layer that develops on the surface of the metal, the inhibitors can be divided into one of three categories: an adsorption type, an oxidation film type, or a precipitation type. On the basis of the electrochemical processes that they engage in, the inhibitors can be divided into three categories: cathode-type, anode-type, and mixed-type corrosion inhibitors. Under acidic situations, organic corrosion inhibitors are frequently employed. These inhibitors primarily use adsorption (physical adsorption or chemical adsorption) to produce protective layers on metal surfaces.

Conventional inhibitors have been created over many years to reduce corrosive environment challenges posed by CO_2, H_2S, etc. Eco-friendly inhibitors, particularly plant extracts, have replaced conventional metal corrosion inhibitors in recent years. These cost-effective, environmentally friendly alternatives have gained popularity since they can be used in a variety of industries and cause no harm to the environment [8]. The phytochemicals and secondary metabolites are substances that can be retrieved from many different plant species and can be employed. The environmentally benign, ecologically acceptable, affordable, widely accessible, and highly effective corrosion inhibition of the naturally extracted phytochemicals offers the use of plant extracts as effective substitutes for the pricy and potentially hazardous conventional synthetic corrosion preventives.

When metal components or their alloys are utilized to create various components or materials for engineering and other industrial purposes, corrosion has significant negative consequences on their metallurgical. For instance, zinc metal is employed in a wide range of applications, some of which include reaction vessels, pipes, tanks, etc., all of which are known to corrode when exposed to various solvents. There has been a lot of interest in the study of zinc corrosion since it is a topic of great theoretical and practical relevance. Corrosion inhibitors are necessary when using acid solutions for de-scaling, acid pickling, and acidizing oil wells in order to stop the corrosion attack on metallic components. Chemicals called inhibitors interact with a material's surface to slow down corrosion or interact with the operating environment to make a material less corrosive [9].

Zn is a non-ferrous metal that happens naturally in the earth's crust and is extensively employed in metallic coating. In addition to rock and soil, this substance is also found in water, air, and the biosphere, as well as in plants, animals, and human beings. It is placed fourth among the metals in global manufacturing and usage. Zinc is a metal that is used in a wide variety of industrial items, including rubber, pharmaceuticals, and metal products. Zinc is employed as metal, mostly as a corrosion-resistant coating for steel and iron (galvanized metal), an alloying component of bronze and brass, a component of die-casting alloys, and rolled zinc [7]. In addition, zinc plays a significant role in a vast array of capital uses, including paints, cosmetics, medications, battery systems, and electrical equipment. Zinc, on the contrary, is a type of active metal and is rapidly corroded in an acidic environment. For protection during zinc machining, finding an efficient zinc inhibitor is crucial [10]. Aqueous acids are among the most hazardous corroding agents for zinc [11]. Acids, especially sulfuric and hydrochloric acids, are quite likely to attack zinc metal. Therefore, it is necessary to employ inhibitors while cleaning zinc surfaces with acidic solutions to remove scale [12].

Extracts from plants have been used by several researchers as inhibitors to lessen the solubility of metals in various common industrial solutions [13–18]. Many biodegradable phytochemicals can be found in plant extracts. The phytochemicals found in plant extracts, such as alkaloids, tannins, flavonoids, carbohydrates, phenolics, and proteins, also provide their inhibitory properties. These metabolites typically contain triple or double conjugate bonds that serve as important adsorption locations, as well as polar functional groups and hetero elements like nitrogen, sulfur, or oxygen. Plants include a variety of chemical substances, including alkaloids, glycosides, organic acids, and resins. Only a few examples include volatile oils, sugars (including sugars, inulin, gums, and sputum), amino acids, proteins, and enzymes, tannins, plant colors (including chlorophyll, carotenoids, flavonoids), oils and waxes, and inorganic elements (trace elements). There are many different types of oil-soluble plant extracts, some of which include essential oils, spray-dried powder, oxidase vegetable protein powder, the pure active substance, and powder form, liposome-encapsulated microcapsules, polysaccharides, or other porous polymer-encapsulated microcapsules, and a variety of vegetable oils such as sunflower oil, coconut oil, and olive extract oil. A review of the literature indicates that different extracts, comprising leaf, root, stem, bark, pulp, fruit, etc., have been effectively used as long-lasting inhibitors for the corrosion of various metals and alloys. The collection of published research on the issue of "phytochemicals as corrosion inhibitors for zinc in demanding NaCl solutions" is discussed in the current review article.

22.2   MECHANISM OF CORROSION OF ZINC IN NaCl MEDIA

The type of corrosion that takes place in a sodium chloride aqueous solution varies according to the metal and alloys. Understanding zinc corrosion in NaCl solution is challenging due to the complex nature of the film that forms on zinc samples exposed to NaCl media. In terms of the zinc oxidation reaction's mechanism, zinc undergoes an oxidation reaction at the anode to produce the metallic ion (Zn^{2+}) and release electrons. By interacting with the chloride ions, these metallic ions generate intermediate ions.

It develops a complicated layer composed of zinc oxide, ZnO; zinc hydroxide, Zn$(OH)_2$; and zinc hydroxide chloride or simonkolleite, $Zn_5(OH)_8Cl_2.2H_2O$; the two primary corrosion-related substances are zinc oxide and zinc hydroxide [19].

There have been numerous studies on the behavior of zinc corrosion in NaCl solution, when there are two separate reactions that result in zinc deterioration.

(i) Oxygen reduction is depicted by the cathodic reaction.

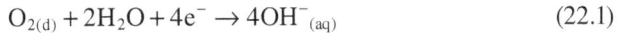

$$O_{2(d)} + 2H_2O + 4e^- \rightarrow 4OH^-_{(aq)} \tag{22.1}$$

The zinc dissolves during the anodic process.

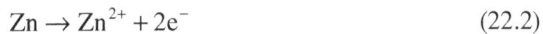

$$Zn \rightarrow Zn^{2+} + 2e^- \tag{22.2}$$

Zinc hydroxide is predicted to form when a reaction between zinc cation and the hydroxide ions takes place.

$$Zn^{2+} + 2OH^- \rightarrow Zn(OH)_2 \tag{22.3}$$

$$Zn(OH)_2 + 2OH^- \rightarrow 5Zn(OH)_4^{2-} \tag{22.4}$$

On the cathodic sites, the pH is supposed to be high for the formation of zincate ions $Zn(OH)_4^{2-}$. Further, formation of zinc hydroxide chloride occurs when zinc hydroxide reacts with chloride ion (Cl⁻).

$$5Zn(OH)_2 + 2Cl^- + H_2O \rightarrow Zn_5(OH)_8 Cl_2.H_2O + 2OH^- \tag{22.5}$$

22.3 PHYTOCHEMICALS/PLANT EXTRACTS AS CORROSION INHIBITORS FOR ZINC IN NaCl SOLUTIONS: AN OVERVIEW

The inhibitive impact of natural extract from *Bagassa guianensis* plant (EEBGP) was investigated by M. Lebrini and his colleagues using various techniques such as PDP, EIS, isothermal adsorption model, and X-ray photoelectron spectroscopy (XPS) for zinc corrosion in 3% NaCl [20]. It was shown that the inhibitory efficiency was around 97% at 100 ppm. The green inhibitor affected both electrochemical reactions, as seen by the polarization curves, and the presence of the extract in 3% NaCl caused a shift toward positive potentials (Figure 22.1).

The results showed that a protective film had been established around the EEBGP extract and the flavonoids component. The impedance study revealed two capacitive loops with a charge-transfer and a passive layer. The electrochemical studies also demonstrated that the all-diverse chemical groups showed identical activity as the whole extract. But, the inhibitory effect is mostly controlled by the flavonoid family. The single-layer adsorption property of the Langmuir isotherm inhibits interactions between the molecules that have been deposited on the surface of the zinc. Yet, this

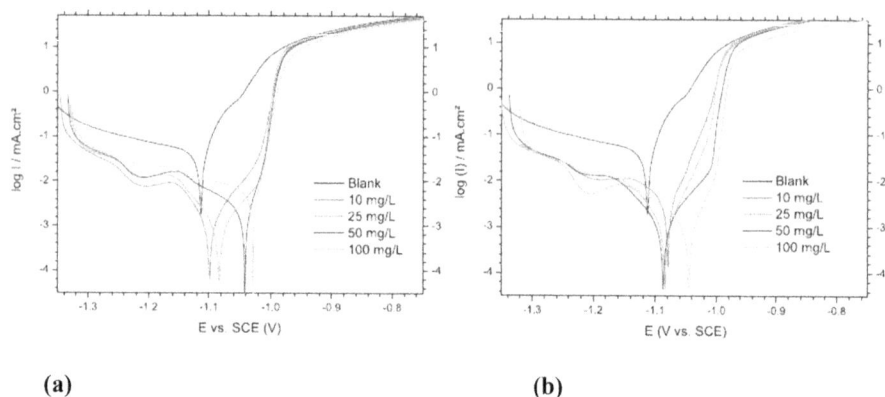

(a)   (b)

FIGURE 22.1 Polarization curves for zinc in 3% NaCl in the presence of (a) EEBGP obtained by UAE and (b) EEBGP obtained under reflux.

response is remarkable in the case of plant extracts, and the reason for this is simply due to the intermolecular interactions (both electrical and steric) of a large number of powerful components that are present in the extract that was analyzed. Investigations into plant chemistry as well as XPS were carried out in order to evaluate the adsorption of molecules from the entire extract (EEBGP) onto the surface of the zinc. These compounds included flavonoids. In order to understand how zinc in the saline solution is adsorbed by the whole extract and flavonoids extract, the XPS analyses were carried out on the EEBGP and the zinc substrate after immersing it in 3% NaCl medium with $100 \, mg \, L^{-1}$ concentration for 4 hours shown to validate the chemical compositions of the film that was deposited on the material's interface (Figure 22.2).

It was reported by R. Tambun et al. that zinc corrosion can be controlled by using powdered, extracted, or tannic forms of Petai peel [21]. The type of inhibitor, its concentration, and the duration of immersion are the key factors in this investigation. The inhibitors were used in concentrations of 2%, 4%, and 6% (w/v), and the immersion period in 3% NaCl solution was adjusted up to 12 days. The corrosion rate and inhibition effectiveness are the variables that were measured. According to the findings, zinc corrosion in a NaCl solution can be slowed down by using one of three types of inhibitors at a concentration of 2%. Nevertheless, as inhibitor concentration increases, zinc corrosion decreases (Figure 22.3). Petai peel has a high potential for use as corrosion protection in the coating industries with an inhibitory efficacy of up to 90%. By covering the zinc's surface and forming a passive layer on the cathodic side, tannin serves as a corrosion inhibitor and effects the cathode reduction reaction. If the process at the cathode is blocked, the anode's zinc oxidation reaction will also be inhibited. The zinc corrosion rate will reduce because the passive layer will stop corrosion ions from reaching the zinc surface.

B.A. Abd-El-Naby et al. [22] studied corrosion inhibition by using potentiodynamic polarization and electrochemical impedance spectroscopy, also known as EIS technique at 30 °C; it was found that some natural extracts, such as those from Lupine, Hlfabar, and Damsisa, had an inhibiting impact on the corrosion of zinc in a solution of 0.5 M sodium chloride. The inhibition efficiency of the Lupine, Hlfabar, and

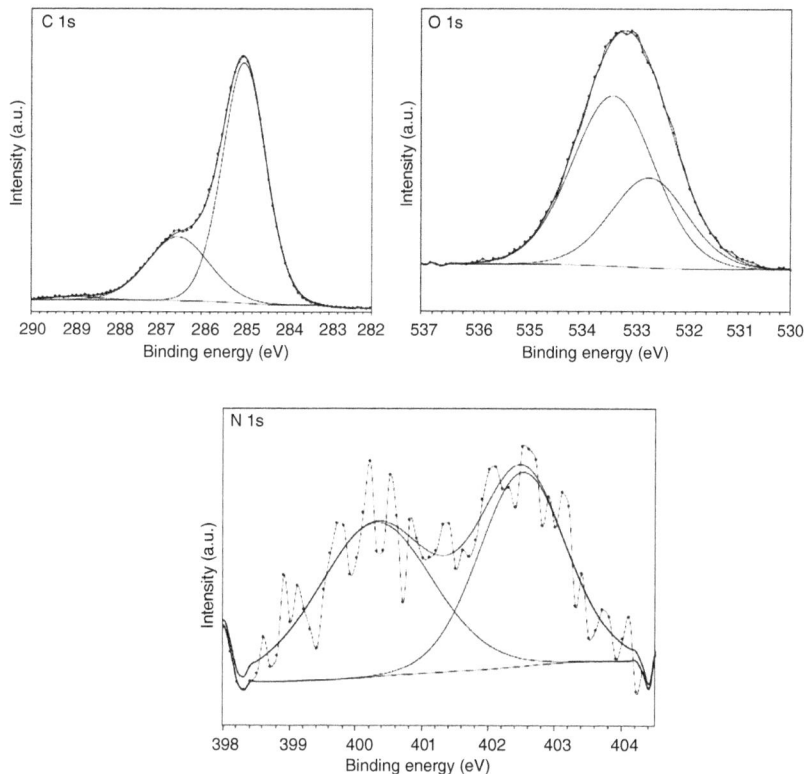

FIGURE 22.2 C 1s, O 1s, and N 1s XPS deconvoluted profiles for *Bagassa guianensis* plant extract (EEBGP).

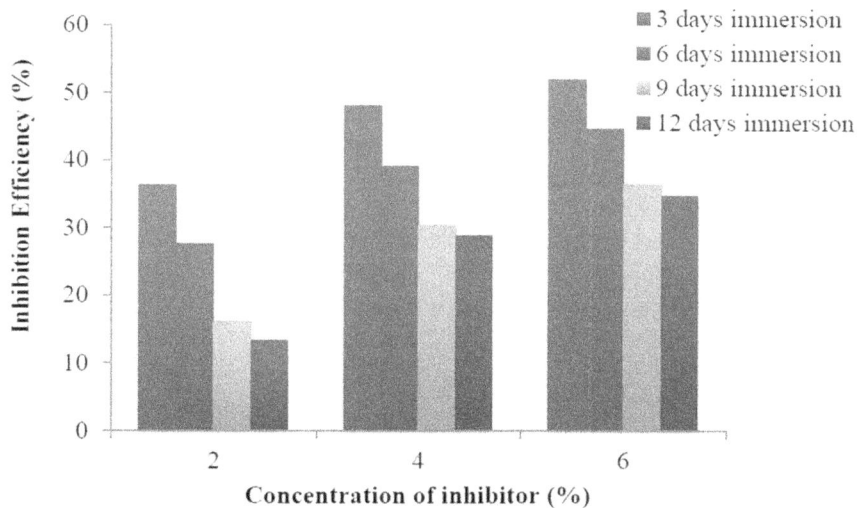

FIGURE 22.3 Effect of Petai peel powder as inhibitor on the efficiency of zinc corrosion.

Damsisa was thought to be 89.1%, 94.7%, and 90.7% at concentrations of 40, 40, and 15 ppm, respectively. The mixed-type inhibitory effects of inhibitors were revealed by potentiodynamic polarization curves. The data show that these inhibitor extracts prevent zinc from corroding in neutral media containing chloride ions by raising R_{ct} and lowering double-layer capacitance. In EIS measurements, data observed the total resistance, and consequently, with higher inhibitor doses, the effectiveness of the inhibition increases, demonstrating that these extracts have an inhibitory impact on zinc corrosion.

The naturally occurring biological substances in *Bagassa guianensis* that exhibit corrosion inhibition of zinc in an ASTM medium studied by M. Lebrini et al. investigated that this plant extract could operate as a potent inhibitor for the zinc dissolution in sodium chloride media [23]. The resultant extract has 85% inhibition rate which was investigated by potentiodynamic polarization curves. Also, from PDP curves, it can be inferred that the presence of the basic extract in the ASTM media caused the corrosion potential values to shift to the positive potentials, affecting both the anodic and cathodic reactions. The experimental findings from the EIS approach reveal a pattern of frequencies, and therefore, a modeling feature with frequency dispersion tendency and a constant phase component (CPEα,Q) has been used. In both the uninhibited and inhibited solutions of zinc, all of the Nyquist graphs reveal a single capacitive loop, and the associated Bode diagrams display a one-time constant signifying a charge-transfer phenomenon. However, at an amount of $10 \, mg \, L^{-1}$ of EEBGP, the Nyquist diagram merely shows one depressed capacitance semicircle. The occurrence of this depressed capacitance loop can be explained by dispersion during the spectra recording due to a quick shift in potential, and moreover, it is also not significant. Indeed, at low frequencies, the data collection points last few minutes.

By using polarization and electrochemical impedance spectroscopy, F. Suedile et al. investigated the ethanol extract of *Mansoa alliacea* as a corrosion preventive method for zinc in NaCl 3% medium [24]. One of the most notable plants native to the northern part of South America is called *M. alliacea*. The plant is well-known for its garlic aroma and its molecules that contain sulfur. Additionally, it is used to treat aches. Potentiodynamic polarization curves demonstrated the mixed-type nature of inhibitor. The resulting extract yields 90% inhibition. The experimental data that were gathered using the EIS method display a frequency distribution, and as a result, a modeling element that has frequency dispersion properties and a constant phase component (CPE,Q) has been applied.

Lawsonia leaf extract, which contains tannin, can function as a selective inhibitor for the corrosion of zinc in NaCl and NaOH solutions, according to El-Etre and colleagues [25]. The aqueous extract of *Lawsonia* leaves is tested using C-steel, nickel, and zinc in acidic, neutral, and alkaline solutions using the polarization technique. By enhancing the inhibition efficiency, the extract acts as a mixed inhibitor by changing the anodic and cathodic Tafel constants. The characteristics of the metal and the kind of media affect how much inhibition is present. The *Lawsonia* leaf extract's ability to inhibit zinc corrosion is affected by the media's composition, and its efficacy decreases from neutral to alkaline to acid in that sequence. The response to media dependence that has been observed here is very different from the one that has previously been observed for the corrosion of nickel and C-steel. The extract's inhibitory effect demonstrated that, in all systems examined, the Langmuir adsorption isotherm is followed by the adsorption of lawsonia molecules on the metal surface. *Lawsonia* extract mostly consists of hydroxy aromatic substances like tannin and lawsone. It was believed that the tannin's inhibitory action was caused by the creation of a protective film made up of tannates on the surface of the metal. Tannins, particularly in basic media, have been found to react with a variety of metallic ions to generate complex compounds. The ligand lawsone may chelate with multiple metal cations to produce intricate complexes. As a result, one of the likely reasons for the observed inhibitory effect of lawsone is the creation of insoluble complex compounds as a result of the interaction between the metal's cations and the molecules of lawsone that have been adsorbed on its surface. The inhibitory activity of zinc did not display the same tendency because zinc complexes are not very stable [26].

22.4 CONCLUSION AND FUTURE PERSPECTIVE

It is possible to draw the conclusion that plant extracts are excellent replacements for currently employed, expensive, and dangerous inorganic and synthetic organic corrosion inhibitors. Many phytochemicals and other compounds found in plant extracts have excellent adsorption and corrosion-inhibiting properties for metallic zinc. The effectiveness of several sorts of extracts, including leaves, roots, stems, and other types, as corrosion inhibitors for zinc metals in neutral (NaCl) conditions has been investigated. Future study on corrosion inhibition is advised to employ plant extracts on metallic zinc corrosion inhibitors since they are affordable, accessible, recyclable, biocompatible, benign, highly efficient, and environmentally secure. The majority of the phytochemicals extracted from plants are heterocyclic compounds that can interact strongly with metal surfaces to prevent corrosion. Prior to its employment in actual testing situations, a number of factors must be considered.

REFERENCES

[1] C. Verma, E.E. Ebenso, I.B.M.A. Quraishi, An overview on plant extracts as environmental sustainable and green corrosion inhibitors for metals and alloys in aggressive corrosive media, *J. Mol. Liq.*, 266 (2018) 577–590.

[2] C. Verma, A. Singh, G. Pallikonda, M. Chakravarty, M.A. Quraishi, I. Bahadur, E.E. Ebenso, Aryl sulfonamidomethylphosphonates as new class of green corrosion inhibitors for mild steel in 1 M HCl: electrochemical, surface and quantum chemical investigation, *J. Mol. Liq.*, 209 (2015) 306–319.

[3] C. Verma, P. Singh, I. Bahadur, E.E. Ebenso, M.A. Quraishi, Electrochemical, thermodynamic, surface and theoretical investigation of 2-aminobenzene-1,3-dicarbonitriles as green corrosion inhibitor for aluminum in 0.5 M NaOH, *J. Mol.Liq.*, 209 (2015) 767–778.

[4] Y. Tang, X. Yang, W. Yang, Y. Chen, R. Wan, Experimental and molecular dynamics studies on corrosion inhibition of mild steel by 2-amino-5-phenyl-1,3,4-thiadiazole, *Corros. Sci.*, 52 (2010) 242–249.

[5] K.M. Ismail, Evaluation of cysteine as environmentally friendly corrosion inhibitor for copper in neutral and acidic chloride solutions, *Electrochim. Acta*, 52 (2007) 7811–7819.

[6] M. Bobina, A. Kellenberger, J.-P.C. Millet, N. Muntean, Vaszilcsin, corrosion resistance of carbon steel in weak acid solutions in the presence of l-histidine as corrosion inhibitor, *Corros. Sci.*, 69 (2013) 389–395.

[7] A.M. Abdel-Gaber, H.H. Abdel-Rahman, A.M. Ahmed, M.H. Fathalla, Corrosion behaviour of zinc in alcohol-water solvents, *Anti-Corros. Methods Mater.*, 53 (2006) 218–223.

[8] Z. Shang, J. Zhu, Overview on plant extracts as green corrosion inhibitors in the oil and gas fields, *J. Mater. Res. Technol.*, 15 (2021) 5078–5094.

[9] S. Cao, D. Liu, H. Ding, J. Wang, H. Lu, J. Gui, Task-specific ionic liquids as corrosion inhibitors on carbon steel in 0.5M HCl solution: an experimental and theoretical study, *Corros. Sci.*, 153 (2019) 301–313.

[10] W. Huang, J. Zhao, Adsorption of quaternary ammonium gemini surfactants on zinc and the inhibitive effect on zinc corrosion in vitriolic solution, *Colloids Surf. A: Physicochem. Eng. Aspects*, 278 (2006) 246–251.

[11] S. Manov, F. Noli, A.M. Lamazouere, L. Aries, Surface treatment for zinc corrosion protection by a new organic chelating reagent, *J. Appl. Electrochem.*, 29 (1999) 995–1003.

[12] G. Wranglen, An Introduction to Corrosion and Protection of Metals; Springer, 1985.

[13] A.S. Fouda, K. Shalabi, A.M. Nofal, M.A. Elzekred, Methanol extract of Rumex Vesicarius L. as ecofriendly corrosion inhibitor for carbon steel in sulfuric acid solution, *Chem. Sci. Trans.*, 7 (2018) 101–111.

[14] A.S. Fouda, E. Abdel Haleem, Berry leaves extract as green effective corrosion inhibitor for Cu in nitric acid solutions, *Surf. Eng. Appl. Electrochem.*, 54 (2018) 498–507.

[15] A.S. Fouda, S.M. Rashwan, M.M.K. Darwish, N.M. Arman, Corrosion inhibition of Zn in a 0.5M HCl solution by Ailanthus altissima extract, *Port. Electrochim. Acta*, 36 (2018) 309–323.

[16] H.M. Elabbasy, A.S. Fouda, Olive leaf as green corrosion inhibitor for C-steel in Sulfamic acid solution, *Green Chem. Lett. Rev.*, 12 (2019) 332–342.

[17] A.S. Fouda, H.M. Elabbasy, Corrosion inhibition effect of methanol extract of nerium oleander on copper in nitric acid solutions. *Int. J. Electrochem. Sci.*, 14 (2019) 6884–6901.

[18] H.M. Elabbasy, Investigation of Withania Somnifera extract as corrosion inhibitor for copper in nitric acid solutions, *Int. J. Electrochem. Sci.*, 14 (2019) 5355–5372.

[19] M. Mouanga, P. Bercot, J.Y. Rauch, Comparison of corrosion behaviour of zinc in NaCl and in NaOH solutions. Part I: corrosion layer characterization, *Corros. Sci.*, 52 (2010) 3984–3992.

[20] M. Lebrini, F. Suedile, P. Salvin, C. Roos, A. Zarrouk, C. Jama, F. Bentiss, Bagassa guianensis ethanol extract used as sustainable eco-friendly inhibitor for zinc corrosion in 3% NaCl: electrochemical and XPS studies, *Surf. Interfaces*, 20 (2020) 100588.

[21] R. Tambun, D.H. Sidabutar, V. Alexander, The potential of Petai peel as a zinc corrosion inhibitor in sodium chloride solution, *Int. J. Corros. Scale Inhib.*, 9 (2020) 929–940.

[22] B.A. Abd-El-Naby, O.A. Abdullatef, A.M. Abd-El-Gabr, M.A. Shaker, G. Esmail, Effect of some natural extracts on the corrosion of zinc in 0.5M NaCl, *Int. J. Electrochem. Sci.*, 7 (2012) 5864–5879.

[23] M. Lebrini, F. Suedile, C. Roos, Corrosion inhibitory action of ethanol extract from Bagassa guianensis on the corrosion of zinc in ASTM medium, *J. Mater. Environ. Sci.*, 9 (2018) 414–423.

[24] F. Suedile, F. Robert, C. Roos, M. Lebrini, Corrosion inhibition of zinc by Mansoa alliacea plant extract in sodium chloride media: extraction, characterization and electrochemical studies, *Electrochim. Acta*, 133 (2014) 631–638.

[25] A.Y. El-Etre, M. Abdallah, Z.E. El-Tantawy, Corrosion inhibition of some metals using lawsonia extract, *Corros. Sci.*, 47 (2005) 385–395.

[26] C.E. Mortimer, *Chemistry, a Conceptual Approach*, fifth ed., D. Van Nostrand Co., 1983.

23 Phytochemicals/Plant Extracts as Corrosion Inhibitors for Magnesium in NaCl Solutions

Xin Liu, Yifan Lv, Xuexue Xu, and Sheng Wu
Harbin Engineering University

Xuerong Zheng
Hainan University

23.1 INTRODUCTION

Magnesium is the lightest metal structural material (the density is about $1.74\,g\,cm^{-3}$, which is 25% of steel, 66% of aluminium, and 40% of titanium alloy). Then, magnesium resources are quite abundant, ranking eighth in the reserves of the earth's crust. Therefore, magnesium has a great development prospect in the fields of industry, biological application, and marine engineering.

As we all know, pure magnesium is very easy to oxidize in industrial or marine environments, and its mechanical properties are not very good compared with other metals, which leads to the fact that pure magnesium cannot be widely used in industrial production or other fields. Therefore, magnesium will often combine with Mn, Al, Zn, Zr, and rare earth elements to form alloys in practical applications. Compared with pure magnesium, Mg alloys have a high strength ratio and stiffness ratio, and the machinability, impact resistance, casting performance, and corrosion resistance of Mg alloys have been greatly improved. In addition, the thermal conductivity, electromagnetic shielding performance, and damping performance have also received good feedback. These excellent characteristics promote Mg alloys to play more and more important application roles in many fields, such as aerospace, aeroplanes, electronic products, medical equipment, automobile industry, and marine engineering.

However, the electrode potential of magnesium is as low as −2.34V, and its chemical activity is very high, so oxidation reaction is easy to occur in air or other corrosive media. Generally, an oxide film with porous defects is formed on the surface of magnesium and Mg alloys, which cannot protect the substrate for a long time. Therefore, in a corrosive environment, magnesium and Mg alloys easily corrode, especially when there are chloride ions, sulfur ions, and other corrosive ions in the

DOI: 10.1201/9781003394631-23

475

environment, the corrosion will be faster and more serious, which greatly limits the application of magnesium and Mg alloys.

In neutral NaCl solution, generally, the anodic reaction of magnesium is to convert Mg into Mg^{2+} through an electrochemical process. The anodic reaction formula is as follows:

$$Mg \rightarrow Mg^{2+} + 2e^-$$

In NaCl solution, the cathodic reaction usually involves the semi-reduction reaction of oxygen, which is as illustrated:

$$O_2 + 4H^+ + 4e^- \rightarrow 2H_2O$$

The overall reaction is therefore:

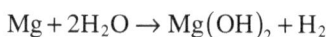

$$Mg + 2H_2O \rightarrow Mg(OH)_2 + H_2$$

For Mg alloys, the added alloying elements can change the reaction rate or reaction steps, but the corrosion mechanism remains unchanged. Besides, the Mg alloy not only suffers uniform corrosion but also suffers serious local corrosion due to the formation of an internal electric couple between the Mg alloy matrix and internal impurities or the second phase. Compared with the passivation film of aluminium alloy and stainless steel, the quasi-passivated hydroxide film on the surface of Mg alloy is very unstable and has a very limited effect on improving the corrosion resistance of Mg alloy. Therefore, it is usually necessary to use anti-corrosion methods to inhibit the failure of Mg alloys.

Several methods are usually applied to improve the corrosion resistance of magnesium and its alloys, including surface coating, electroless plating, and the use of corrosion inhibitors. Among all these technologies, adding a corrosion inhibitor to the corrosion medium is considered to be the simplest and most practical method to slow down the corrosion of magnesium and its alloys. Commonly used Mg alloy corrosion inhibitors include inorganic corrosion inhibitors and organic corrosion inhibitors. The common inorganic corrosion inhibitor is F^-, which reacts with Mg^{2+} to form a magnesium fluoride protective layer on the surface of Mg or Mg alloys, effectively reducing corrosion. Adding chromate, phosphate, molybdate, vanadate, tungstate, and cerium salts can also form a protective layer to slow down the corrosion rate. The commonly used organic corrosion inhibitors for Mg and Mg alloys contain polar groups (heterocycle or N, S, O, P, and other heteroatoms). The organic corrosion inhibitor will be adsorbed on the metal surface through the polar group to form a protective film, and the non-polar group will be arranged in the corrosion medium to achieve the purpose of corrosion inhibition. Common organic corrosion inhibitors include carboxylates (glycine, stearate, myristic acid, etc.), surfactants (sodium dodecyl benzene sulfonate [SDBS], sodium dodecyl sulfate, sodium lignosulfonate, etc.), ionic liquids (imidazolium and long-chain phosphonium compounds), organic polymers (sodium alginate, carboxymethylcellulose, polyaspartic acid, etc.), and other organic corrosion inhibitors (benzotriazole, 2-mercaptobenzothiazole, 8-hydroxyquinoline [8-HQ], sodium benzoate, etc.) [1].

In addition, mixed corrosion inhibitors are increasingly used to produce synergistic effects to achieve more efficient protection. At present, inorganic-organic corrosion inhibitors (such as Na_3PO_4 and SDBS [2], NaF and 3-amino-1,2,4-triazole-5-thiol [3], $Zn(NO_3)_2$ and L-phenylalanine [4] form a precipitation-adsorption film) and multiple organic corrosion inhibitors (such as 8-HQ and SDBS [5], 8-HQ and 1.10-phenanthroline [6] form a complementary, uniform and dense adsorption film) are commonly used. The combination of two precipitation-type corrosion inhibitors can also be applied, such as aminopropyltriethoxysilicate (APTS-Na) and $Zn(NO_3)_2$, which can simultaneously generate $Mg(OH)_2$ and $Zn(OH)_2$ deposition film [7].

Although adding a corrosion inhibitor is the simplest and most efficient anti-corrosion method, it should be noted that many corrosion inhibitors have certain toxicity to the environment, especially organic corrosion inhibitors. In addition, it is reported that most compounds that can effectively inhibit the corrosion of other metals in water are ineffective for magnesium and its alloys. For example, Lamaka et al. [8] selected 151 compounds (organic and inorganic) as corrosion inhibitors for pure magnesium and Mg alloys in NaCl solution. The study found that only 15 compounds have corrosion inhibition, and more than 60% of them are marked as toxic, carcinogenic, and harmful to the natural environment.

Therefore, natural phytochemicals or chemicals extracted from plants can be selected as corrosion inhibitors for Mg alloys. On the one hand, they are specially developed for corrosion resistance of Mg alloys, and on the other hand, they are less toxic to the environment. Natural corrosion inhibitor refers to the extracts from leaves, bark, seeds, fruits, and roots of plants. Its structure is similar to that of organic corrosion inhibitors and usually consists of compounds or mixtures containing heteroatoms. Therefore, these compounds can be used as metal corrosion inhibitors. However, only a few studies have been carried out using plant extracts as corrosion inhibitors for magnesium and its alloys.

23.2 CORROSION INHIBITORS FOR MG AND ITS ALLOY

23.2.1 Plant Extracts as Corrosion Inhibitors

Natural or extracted corrosion inhibitors refer to organic compounds extracted from plants such as garlic, kelp, olive leaves, *Coptis chinensis*, persimmon peel, green tea, carrots, and mango leaves. Their structure is similar to organic corrosion inhibitors, usually containing heteroatoms, which can be adsorbed on the surface of Mg alloy to provide a protection effect. The current research mainly focuses on using weight-loss experiment, hydrogen evolution measurement, electrochemical impedance spectroscopy, and polarization curve to test the corrosion inhibition effect of the extracted corrosion inhibitor on magnesium and Mg alloys and employing computational chemistry to simulate the adsorption behavior of molecules to explore the corrosion inhibition mechanism.

Orange peel extracts (OPE) have been proven to have a corrosion inhibition effect on carbon steel in the acid environment [9]. However, its effect on Mg alloys is rarely explored. Since there are O atoms, conjugated rings, and hydroxyl groups in OPE, they may easily be adsorbed on the surface of Mg alloy, thus delaying the corrosion of magnesium. Wu et al. [10] used OPE as a corrosion inhibitor for Mg alloy, explored the corrosion inhibition performance of OPE on AZ91D Mg alloy in NaCl solution by means of experiment and computational chemistry, and compared OPE with three main pure components in OPE. The results show that when Mg alloy is immersed in OPE, the ring structure, O atom and OH group in its structure would interact with the magnesium surface to form a protective self-assembled film, which plays a more important role in the inhibition effect. In addition, the acidic conditions formed by corrosion promote the protonation of OPE, and the protonated composition is conducive to the interaction with $Mg/MgO/Mg(OH)_2$ to form a more solid interaction. The absorption behavior of OPE follows the Langmuir absorption model. At a very low concentration of $0.030\,g\,L^{-1}$, its efficiency is up to 85.7%, and the inhibition effect is better than the three pure components. Theoretical calculation results show that the absorption energy of ascorbic acid@Mg, neohesperidin@Mg, naringin@Mg, and mixture@Mg is $-20.56\,kJ\,mol^{-1}$, $-57.45\,kJ\,mol^{-1}$, $-74.12\,kJ\,mol^{-1}$, and $-100.00\,kJ\,mol^{-1}$. The negative adsorption energy indicates that the adsorption is spontaneous. Moreover, the adsorption energy of mixture@Mg is much larger than that of single-component@Mg, suggesting stronger adsorption. The stronger adsorption energy is beneficial for the formation of protective film (Figures 23.1 and 23.2).

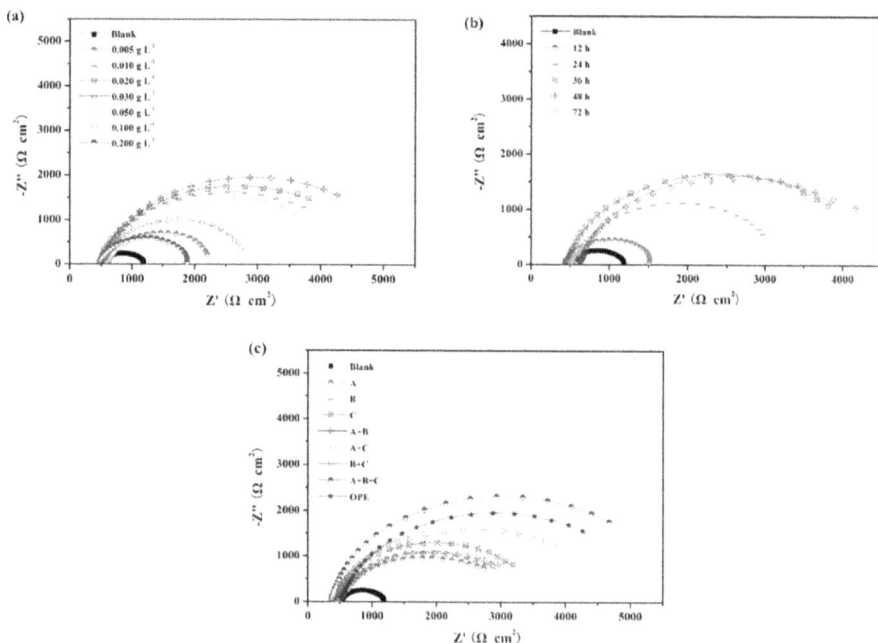

FIGURE 23.1 Nyquist plot for AZ91D Mg alloy electrode in 0.05 wt% NaCl solution (a) without and with different concentrations of orange peel extracts, (b) after different immersion times in $0.030\,g\,L^{-1}$ orange peel extract-ethanol solution, and (c) in the absence and presence of differently mixed inhibitors with the same concentrations of $0.030\,g\,L^{-1}$. (A is ascorbic acid, B is neohesperidin, and C is naringin) [10].

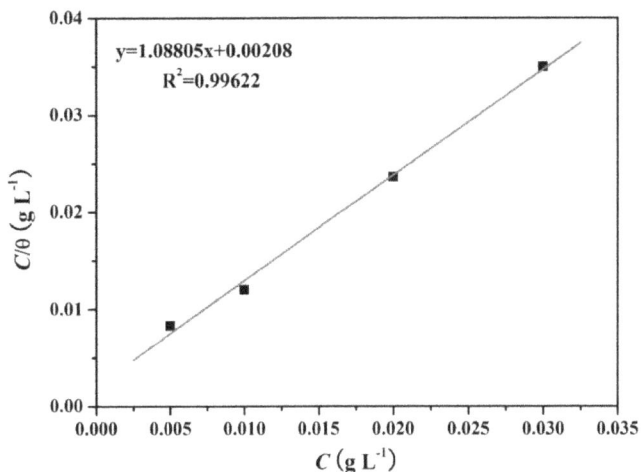

FIGURE 23.2 Langmuir adsorption isotherm for orange peel extracts on AZ91D Mg alloy in 0.05 wt% NaCl solution at 298K [10].

Similarly, walnut green husk extract (WGHE) is explored to be an inhibitor for Mg alloys in NaCl solution [11]. According to the electrochemical measurement, the best inhibition efficiency of $1.0 \, \text{g L}^{-1}$ WGHE is only 44.8%. In addition, the corrosion inhibition efficiency will not be further improved even if the concentration of WGHE is increased. While immersion can effectively improve the corrosion inhibition efficiency. Combined with the calculation result, the corrosion resistance is attributed to the adsorption of WGHE, the rupture of WGHE, and other reactions.

The corrosion inhibition behavior of 2-hydroxy-4-methoxy-acetophenone (paeonol) as a natural green corrosion inhibitor on AZ91D Mg alloy in 0.05 wt% NaCl solution was studied by electrochemical method and computational chemical simulation [12]. The results showed that paeonol could inhibit the corrosion of AZ91D. When the concentration of paeonol is 50 ppm, the maximum inhibition efficiency is 90%, which is caused by the formation of a paeonol-magnesium complex by the chelation of paeonol and magnesium, which is mixed with the original $Mg(OH)_2$ film on the surface to inhibit the anodic dissolution of AZ91D (Figures 23.3 and 23.4).

Dinodi et al. [13] modified stearic acid, natural palmitic acid, and myristic acid to produce corresponding carboxylates, then studied and compared the corrosion inhibition performance of stearate, palmitate, and myristate on ZE41 in dilute brine medium. The results showed that carboxylate was a mixed inhibitor. The anodic inhibition is achieved through densification of porous surface film by the precipitates of magnesium carboxylate salts, formed from the physisorbed carboxylates. The authors also determined that the inhibition efficiency is related to its aliphatic chain length (Figures 23.5–23.7).

Tannic acid is a chemical substance extracted from a Chinese medicine called gallnut. Wu et al. [14] reported that sodium tannate has a certain corrosion inhibition effect on AZ61 Mg alloy, and when its concentration is $1.0 \times 10^{-6} \text{mol L}^{-1}$, the corrosion inhibition efficiency reached 80.3%. The corrosion inhibition mechanism

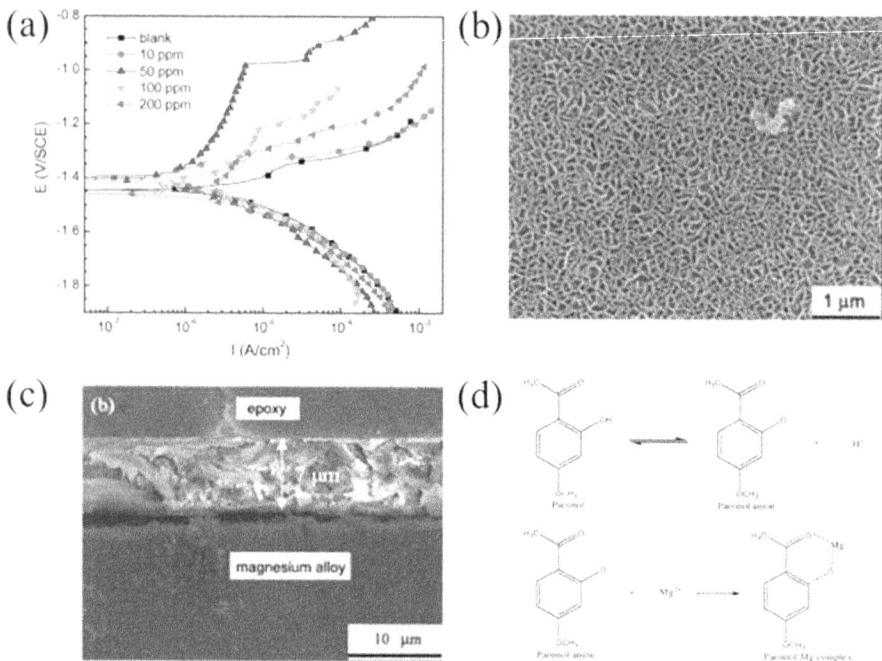

FIGURE 23.3 Inhibition performance and mechanism of paeonol for AZ91D Mg alloy in 0.05 wt% NaCl solution; (a) potentiodynamic polarization curves of AZ91D Mg alloy in 0.05 wt% NaCl blank solution and the solutions containing various concentrations of paeonol, (b) SEM images of AZ91D Mg alloy after 3 days of immersion in the solution containing 50 wt.ppm paeonol, (c) cross-section morphologies of AZ91D Mg alloy after 3 days of immersion in the solution containing 50 wt.ppm paeonol, and (d) formation mechanism of paeonol-Mg complex [12].

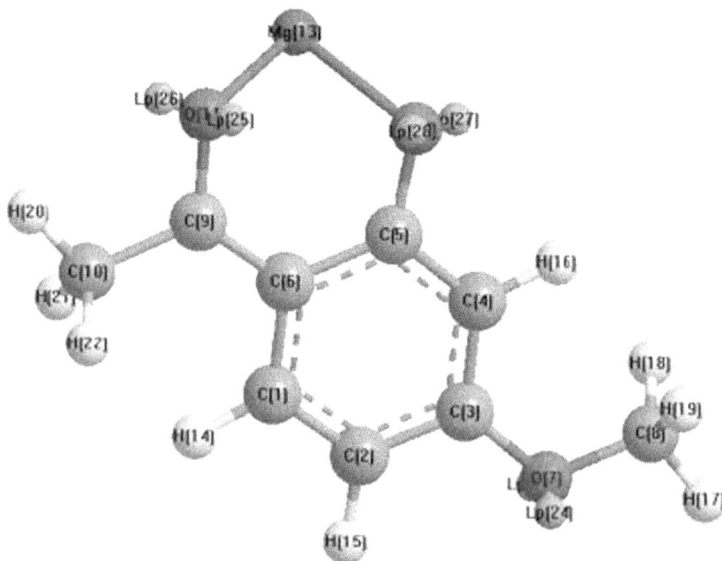

FIGURE 23.4 Molecular structure of paeonol-Mg complex [12].

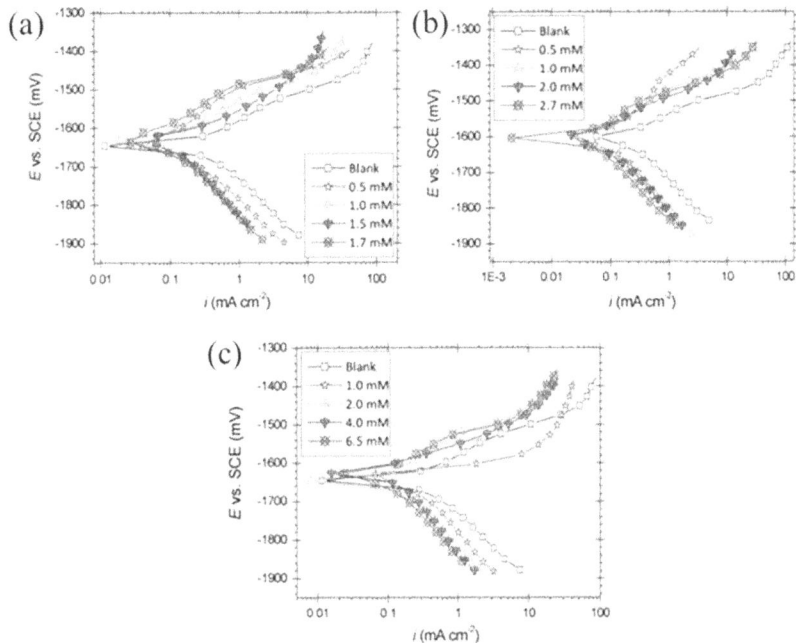

FIGURE 23.5 Potentiodynamic polarization curves for the corrosion of ZE41 alloy in 0.2M Na$_2$SO$_4$-0.1M NaCl solutions containing different concentrations of (a) stearate at 40°C, (b) palmitate at 30°C, and (c) myristate at 40°C [13].

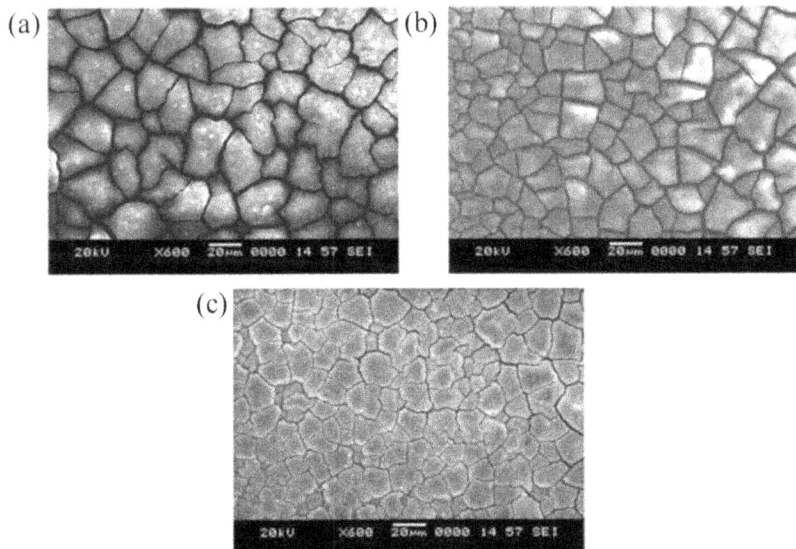

FIGURE 23.6 SEM image of ZE41 specimen surface after 1 hour immersion in 0.2M Na$_2$SO$_4$-0.1M NaCl solutions containing (a) 1.7 mM of stearate, (b) 2.7 mM of palmitate, and (c) 6.5 mM of myristate [13].

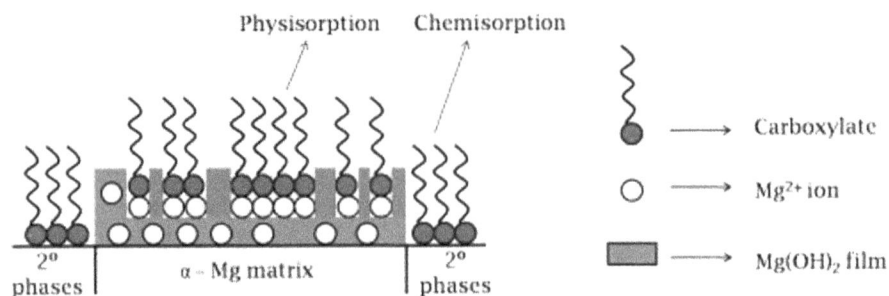

FIGURE 23.7 Schematic representation for the surface adsorption of alkyl carboxylates over different phases of ZE41 [13].

is that the hydroxyl group in the tannin molecule and the corrosion product $Mg(OH)_2$ on the surface of Mg alloy form a dense protective film, thus inhibiting the corrosion of Mg alloy.

Sodium alginate (ALG) is a polysaccharide biopolymer extracted and refined from kelp. Dang et al. [15] reported that when the concentration of ALG was 500 ppm, the corrosion inhibition efficiency of AZ31 alloy in 3.5% NaCl solution reached 90%, which could effectively inhibit the corrosion of AZ31 alloy. The corrosion inhibition efficiency began to decrease with increasing concentration. This inhibition is due to the adsorption of ALG on the exposed Mg alloy surface, where ALG and $Mg(OH)_2$ mix to form a more continuous and compact film, thus delaying the corrosion of the alloy.

Carrageenan (KLJ) is also a polysaccharide polymer compound extracted from natural algae. Zhang et al. [16] tested the corrosion inhibition of sodium dodecyl sulfonate (SDS), sodium dodecyl benzene sulfonate (SDBS), carrageenan (KLJ), and sodium alginate (SA) on AZ91D Mg alloy in 3.5% NaCl solution. The results showed that the greater the steric resistance of the molecular monomer of the corrosion inhibitor, the lower the corrosion inhibition efficiency. The four molecular corrosion inhibition efficiencies were ranked as SDS>SDBS>KLJ>SA.

Umoren et al. [17] screened the inhibition properties of seven natural polymers including chitosan (CHI), dextran (Dex), carboxymethyl cellulose (CMC), sodium alginate (ALG), pectin (PEC), hydroxylethyl cellulose (HEC), and gum Arabic (GA) on Mg alloys in NaCl medium. CHI, Dex, CMC, PEC, and GA accelerated the corrosion, while ALG and HEC moderately inhibited the corrosion of the alloy. The corrosion inhibition efficiency of HEC and ALG is 64.13% and 58.27% respectively.

23.2.2   SYNERGISTIC INHIBITION EFFECT

Although the use of phytochemicals/plant extracts as corrosion inhibitors for magnesium and Mg alloys can slow down corrosion, it still faces problems such as limited corrosion inhibition efficiency and difficulty in meeting high corrosion protection requirements. Using two or more corrosion inhibitors in combination can effectively improve the corrosion inhibition efficiency of corrosion inhibitors by virtue of synergistic effect. High-efficiency composite corrosion inhibitors can also be developed by using the synergistic effect between natural corrosion inhibitors or between natural corrosion inhibitors and common corrosion inhibitors. Common composite corrosion inhibitors include the synergy of precipitation-adsorption type corrosion inhibitors, cathodic-anodic type corrosion inhibitors, inorganic-organic type inhibitors, and multiple organic corrosion inhibitors.

DL-malic acid (DMA) is widely found in immature apples, grapes, peaches, etc. Qiu et al. [18] used DL-malic acid and NaF as synergistic corrosion inhibitors for Mg alloy. Based on high-throughput screening (HTS) of visual recognition, electrochemical method, and surface characterization technology, the effect of their interaction and ratio on the corrosion performance of Mg-Al-Mn alloy (AM50) in NaCl solution was studied. The results showed that the highest corrosion inhibition efficiency was 94.1% and the synergistic factor was 3.8 when containing equimolar (0.05 M) NaF and DMA components. The specific protective mechanism is that the $Mg(OH)_2$ film formed is less protective due to the presence of chloride ions. In the

---

TABLE 23.1

Corrosion Inhibition Effect of Some Phytochemicals/Plant Extracts for Mg Alloy in NaCl Solution

S/N	Substrate	Corrosive Medium	Molecular Structure	Inhibition Efficiency	Reference
1	AZ91D	0.05% NaCl	OPE:	85.7%	[10]

component 1: ascorbic acid

component 2: naringin

component 3: neohesperidin

(Continued)

TABLE 23.1 (*Continued*)
Corrosion Inhibition Effect of Some Phytochemicals/Plant Extracts for Mg Alloy in NaCl Solution

S/N	Substrate	Corrosive Medium	Molecular Structure	Inhibition Efficiency	Reference
2	AZ91D	0.05% NaCl	WGHE		44.8%	[11]

component 1: menadione

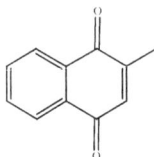

component 2: ferulic acid

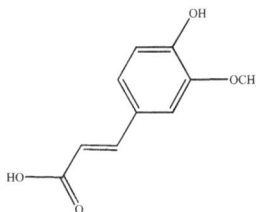

component 3: juglone

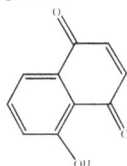

component 4: myricetin

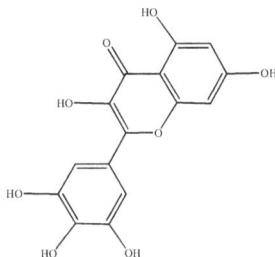

(*Continued*)

TABLE 23.1 (*Continued*)
Corrosion Inhibition Effect of Some Phytochemicals/Plant Extracts for Mg Alloy in NaCl Solution

S/N	Substrate	Corrosive Medium	Molecular Structure	Inhibition Efficiency	Reference
3	AZ91D	0.05% NaCl	paeonol 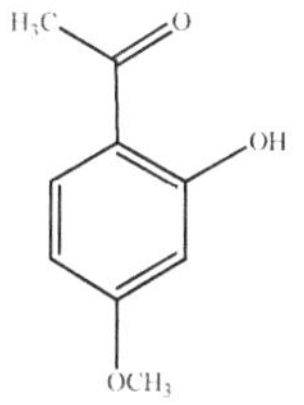	90%	[12]
4	ZE41	0.2M Na$_2$SO$_4$-0.1M NaCl	stearic acid 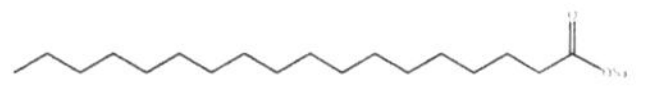 palmitic acid 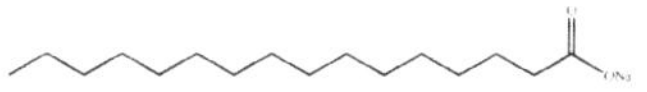 myristic acid 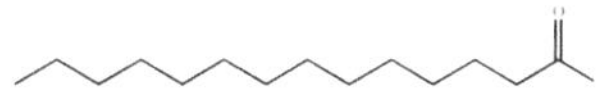	88.0% 80.5% 79.6%	[13]
5	AZ61	0.5% NaCl	sodium tannins 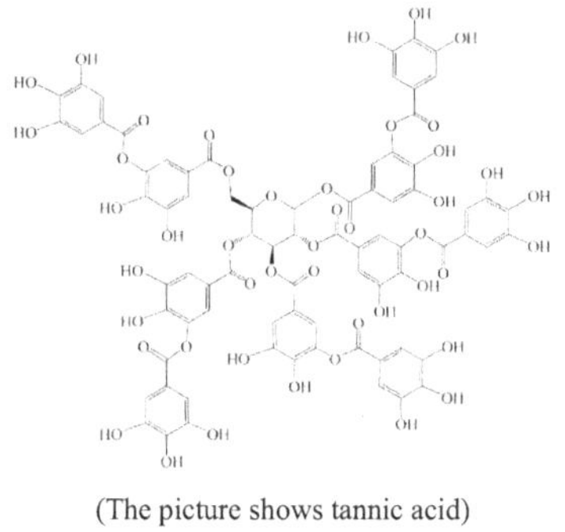 (The picture shows tannic acid)	80.3%	[14]
6	AZ31	3.5% NaCl	sodium alginate 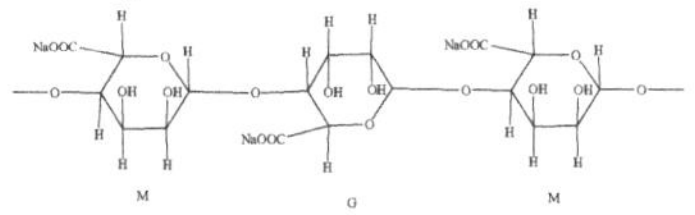	90.00%	[15]
7	AZ91D	3.5% NaCl	KLJ 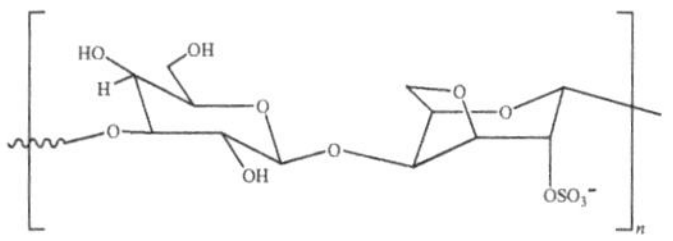	-	[16]

(*Continued*)

TABLE 23.1 (*Continued*)

Corrosion Inhibition Effect of Some Phytochemicals/Plant Extracts for Mg Alloy in NaCl Solution

S/N	Substrate	Corrosive Medium	Molecular Structure	Inhibition Efficiency	Reference
8	AZ31	3.5% NaCl	sodium alginate (ALG)	58.27%	[17]
				64.13%	

hydroxylethyl cellulose (HEC)

$R = H$ or

presence of corrosion inhibitors F^- and Na^+, $NaMgF_3$ precipitation will be formed on the surface of the Mg alloy. When the DMA molecule is introduced, on the one hand, it forms a coordination compound Mg(DMA) through the combination of a carboxyl group and the active site of the alloy surface, which is preferentially chemically adsorbed onto the $Mg(OH)_2$/MgO layer. On the other hand, due to the high electronic density of DMA, Cl^- and F^- are repelled from the metal/electrolyte interface, which will reduce the concentration of F^- and the ratio of $[F: Mg^{2+}]$ at the interface, thus changing the larger cubic $NaMgF_3$ particles formed on the surface of Mg alloy into smaller spherical $NaMgF_3$ particles, realizing the synergistic effect of corrosion inhibition (Figures 23.8 and 23.9).

The powder of chicory was extracted from chicory leaves (CA) and mixed with $Ca(NO_3)_2 \cdot 4H_2O$, $Fe(NO_3)_3 \cdot 9H_2O$, $Fe(SO_4)_2 \cdot 7H_2O$, and $NiSO_4 \cdot 6H_2O$, respectively, and then added into the corrosion solution to obtain mixed inhibitors of CA, CA-Ca^{2+}, CA-Fe^{3+}, CA-Fe^{2+}, and CA-Ni^{2+} [19]. The added concentrations of chicory and metal cations are 1,000 and 100 ppm, respectively. The test results showed that the corrosion potential of all mixed inhibitor samples shifted in the positive direction due to the inhibition of anodic and cathodic reactions during the corrosion process. The corrosion inhibition efficiency of CA, CA-Fe^{2+}, CA-Fe^{3+}, and CA-Ni^{2+} is 69%, 54%, 77%, and 46%, respectively, while that of CA-Ca^{2+} is 92%, indicating that CA-Ca^{2+} has the best corrosion inhibition efficiency for Mg alloys. This is because the organic group contained in CA can form a chelate with metal ions to fill the pores of the porous $MgO/Mg(OH)_2$ film, thus forming a relatively dense protective layer to protect the Mg alloy from corrosion. In addition, Ca ions will hydrolyze to produce precipitate $Ca(OH)_2$, which will be further deposited on the porous $MgO/Mg(OH)_2$ layer

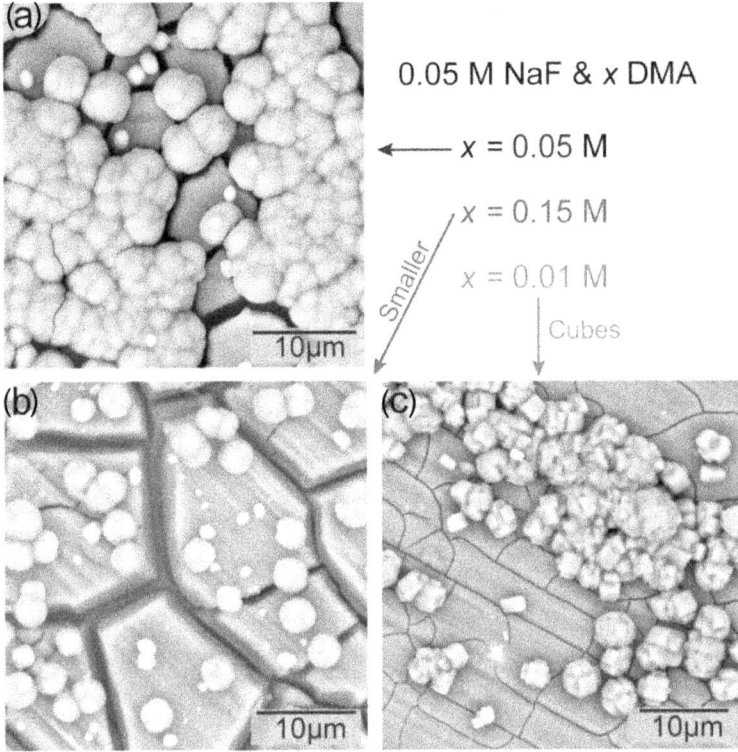

0.05 M NaF & x DMA

← x = 0.05 M

x = 0.15 M

x = 0.01 M

Cubes

Smaller

FIGURE 23.8 Morphology of NaMgF$_3$ particles obtained from mix solutions containing 0.05 M NaF and different molar concentrations of DMA: (a) 0.05 M, (b) 0.15 M, and (c) 0.01 M [18].

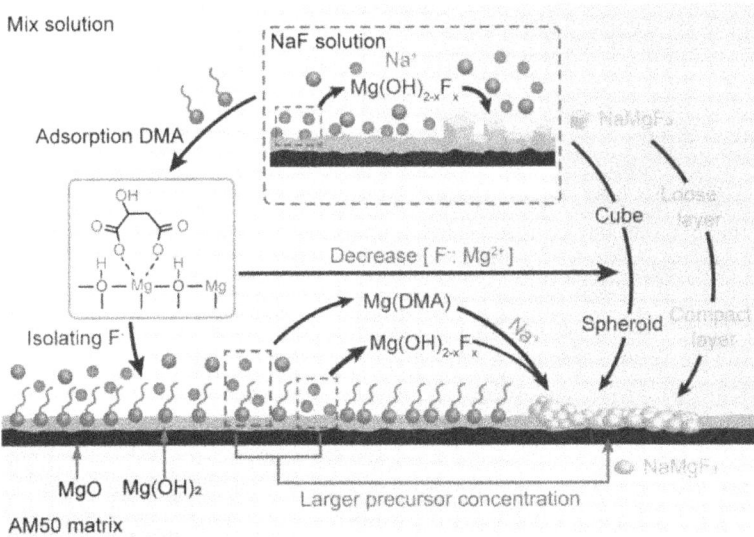

FIGURE 23.9 Schematic illustration of the synergistic corrosion inhibition mechanism of NaF and DMA hybrid inhibitor [18].

to form a dense protective layer. The possible molecules and their adsorption energy, charge density difference, and Bader charge were calculated by calculation software, and the good corrosion inhibition behavior of CA-Ca^{2+} was verified.

Zhang et al. [20] studied the synergistic corrosion inhibition effect of sodium alginate (SA) and sodium silicate (SS) on AZ91D Mg alloy in 3.5 wt% NaCl solution. The corrosion inhibition efficiency can reach 98.56%, and the protection effect on Mg alloy is better than that of SA or SS alone. Through the above research, a synergistic inhibition mechanism is proposed. At the beginning of immersion, a large amount of Mg^{2+} and H^+ are produced at the surface defects of the sample. $Si_2O_3^{2-}$ and Mg^{2+} combine to form magnesium silicate, which is deposited at the defect to provide limited protection. With the addition of SA, Mg^{2+} and H^+ can act as bridges so that SA negative ions can be absorbed on the alloy surface, thus producing effective synergistic protection.

Hou et al. [21] studied the inhibition effect of sodium alginate (SA) and sodium phosphate (SP) on the corrosion of AZ31 Mg alloy in 3.5% NaCl solution at the same time. The results showed that the presence of SA and SP had a synergistic inhibition effect on AZ31. In 3.5% NaCl solution containing 0.05% SA and 0.15% SP, the inhibition efficiency reached the peak, which was attributed to the Mg-SA and $Mg_3(PO_4)_2$ formed by the interaction of the two corrosion inhibitors and Mg alloy, and the two interact to form a dense, uniform, and protective composite film (Figure 23.10).

Longan is a popular fruit. Its seeds and peel are extracted and used as a friendly corrosion inhibitor for mild steel in HCl solution [22]. However, the research shows that the corrosion inhibition efficiency of longan residue extract (LRE) for AZ91D Mg alloy in NaCl solution is only 79.3% [23]. However, when the longan residue extract and SDBS were used as the composite corrosion inhibitor of AZ91D Mg alloy, the corrosion inhibition efficiency reached 95.67%, which confirmed the synergistic protection of the two on Mg alloy, which was caused by the joint adsorption of the two to form a dense film [23].

23.2.3 COATINGS CONTAINING CORROSION INHIBITORS

It is an effective way to protect Mg alloy by coating corrosion inhibitor-based self-repairing coating on the surface of Mg alloy. Corrosion inhibitors can be directly added into the coating of Mg alloys as a pigment. In addition to the shielding property of the coating, the inhibitor in the coating can also react with the Mg alloy substrate to prevent further corrosion in case of damage. In addition, the corrosion inhibitor can also be added into the nanocontainer to prepare the composite corrosion inhibitor, and then, the composite inhibitor is added into the coating as a filler. When the coating is damaged, it can respond to certain stimuli (such as local pH change, corrosive ions, mechanical fracture, or corrosion potential) based on the physical and chemical properties of the surface of the nanocontainer to release the loaded corrosion inhibitor and realize the active protection of metal [24–26]. Similarly, some natural plant corrosion inhibitors or extracted corrosion inhibitors have been applied to prepare self-repairing coatings. However, there is little research on the coating containing phytochemicals/plant extracts corrosion inhibitor of Mg alloy.

Upadhyay et al. [27] added four natural organic corrosion inhibitors quinaldic acid (QDA), beta (BET), dopamine hydrochloride (DOP), and diazolidinyl urea (DZU) to the sol-gel coating and applied them to the Mg AZ31B substrate. The potentiodynamic polarization scan and electrochemical impedance spectroscopy show that the four kinds of coatings provide a good barrier for the Mg alloy substrate. In addition, the four inhibitors containing coatings were tested using the scanning vibrating electrode technique (SVET) under defect conditions, and it shows that only the QDA-enhanced coating can maintain the passivation of the coating under the condition of defects in the coating and achieve self-repairing performance.

Li et al. [28] synthesized paeonol condensation tyrosine (PCTyr) Schiff (PCTyr Schiff) from tyrosine and paeonol, then added it to the sol-gel coating and applied it to the ZE21B alloy matrix. The performance of the prepared coating was tested in the simulated body fluid (SBF), and the results showed that the coating reduced the corrosion rate, showed the self-repairing effect, and improved the compatibility, providing a new clue for the surface modification of cardiovascular biomaterials.

Asadi et al. [29] prepared polycaprolactone (PCL) polymer coating on AZ31 Mg alloy and added Lawson to it to improve its corrosion resistance. Lawson (2-hydroxy-1,4-naphthoquinone) is a natural red-orange dye extracted from the leaves of the *Lawsonia inermis* plant from "Henna", which has a strong chelating effect with metal cations. The corrosion resistance of the coating was evaluated by electrochemical technology and *in vitro* immersion test. The results showed that the corrosion resistance of the coating to bare Mg alloy was significantly improved. In addition to corrosion inhibition, the coating also showed strong antibacterial activity and cell compatibility.

Wang et al. [30] *in situ* grew corrosion inhibitor (VO_4^{3-} and MoO_4^{2-}) intercalated magnesium aluminium-layered hydroxide (Mg-Al-LDH) on the surface of AZ31 Mg alloy and then modified by sodium stearate (SS), lauric acid (LA), and myristic acid (MA), which can bring down the surface energy. Thus, the superhydrophobic anti-corrosion coating was prepared. Through potentiodynamic polarization curve and electrochemical impedance spectroscopy, it was found that Mg-Al-LDH intercalated by MoO_4^{2-} intercalation and LA-modified coating had excellent corrosion inhibition performance (99.99%), which was caused by the synergistic action of Mg-Al-LDH physical protection, MoO_4^{2-} corrosion inhibition, and LA superhydrophobic property (Figures 23.11–23.13).

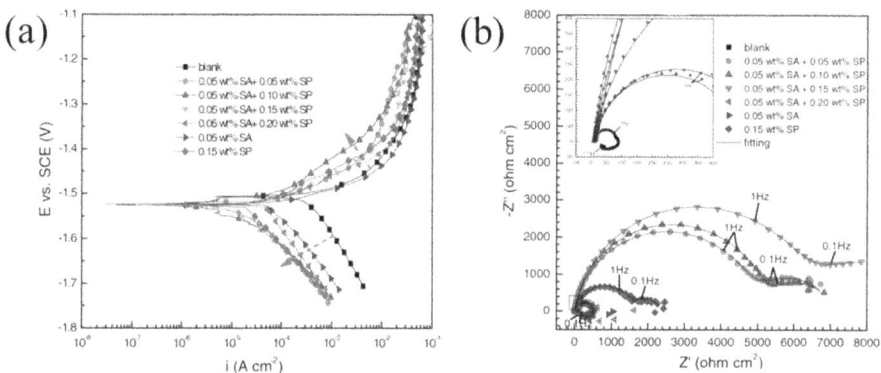

FIGURE 23.10 Potentiodynamic polarization curves (a) and Nyquist plots (b) of Mg alloy AZ31 specimens in contact with 3.5 wt% NaCl solutions containing various concentrations of SA and SP[21].

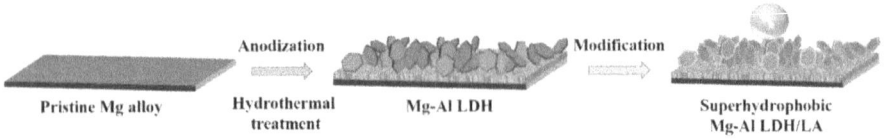

FIGURE 23.11   Schematic illustration of the fabrication process of the hydrophobic Mg-Al LDH coating on AZ31 [30].

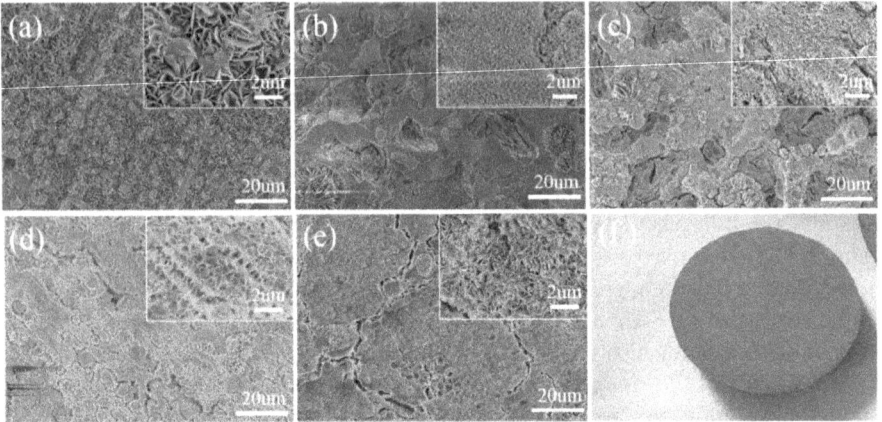

FIGURE 23.12   SEM micrographs of Mg-Al-NO$_3^-$ LDH coating modified by (a) SS, (b) LA, and (c) MA and lauric acid modified Mg-Al LDH coating intercalated with (d) vanadate and (e) molybdate; (f) digital graph [30].

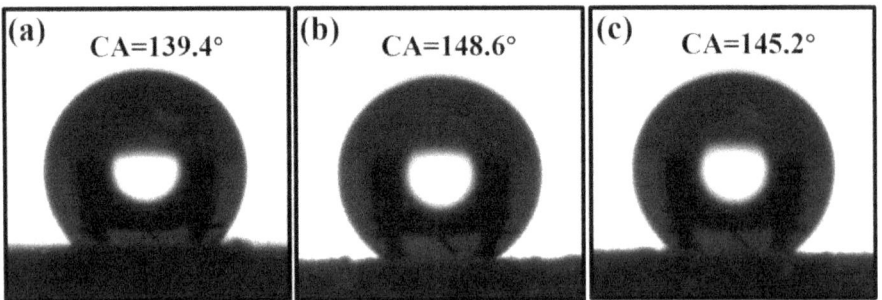

FIGURE 23.13   The water static contact angle of Mg-Al-NO$_3^-$ LDH coating modified by (a) SS, (b) LA, and (c) MA [30].

Plasma electrolytic oxidation (PEO) is an effective surface treatment process for magnesium and Mg alloys, which protects the substrate from corrosion by forming a ceramic coating [31,32]. However, relatively high porosity is the main concern of the PEO-coated Mg alloy to maintain long-term corrosion ability in a corrosive environment, so it is usually necessary to seal the pores of the PEO layer [33]. The common method is to coat the PEO surface with a sol-gel coating to form a composite coating

[34,35]. Lamaka et al. [36] proposed for the first time to directly load corrosion inhibitors (8-hydroxyquinoline and Ce^{3+}, etc.) into the pores generated by the PEO process to achieve corrosion resistance and self-healing performance. Phytic acid is a green corrosion inhibitor extracted from natural crops [37,38]. Guo et al. [39] deposited PEO film, a self-assembled nanoparticle film containing phytic acid (SANP-P) and a fluorocarbon coating layer on Mg-Gd-Y alloy to prepare a composite film (called PSPF). The results show that phytic acid improves the compactness of self-assembled nanoparticle films. PSPF film shows enhanced shielding performance and can provide good corrosion protection for Mg alloy. Phytic acid plays a key and important role in its good long-term performance in neutral salt environment (Figure 23.14).

Tannin is a natural, non-toxic water-soluble polyphenol, mainly extracted from the roots, branches, leaves, flowers, fruits and seeds of various trees, which can inhibit the corrosion of iron and other metals [40]. Zmozinski et al. [41] synthesized

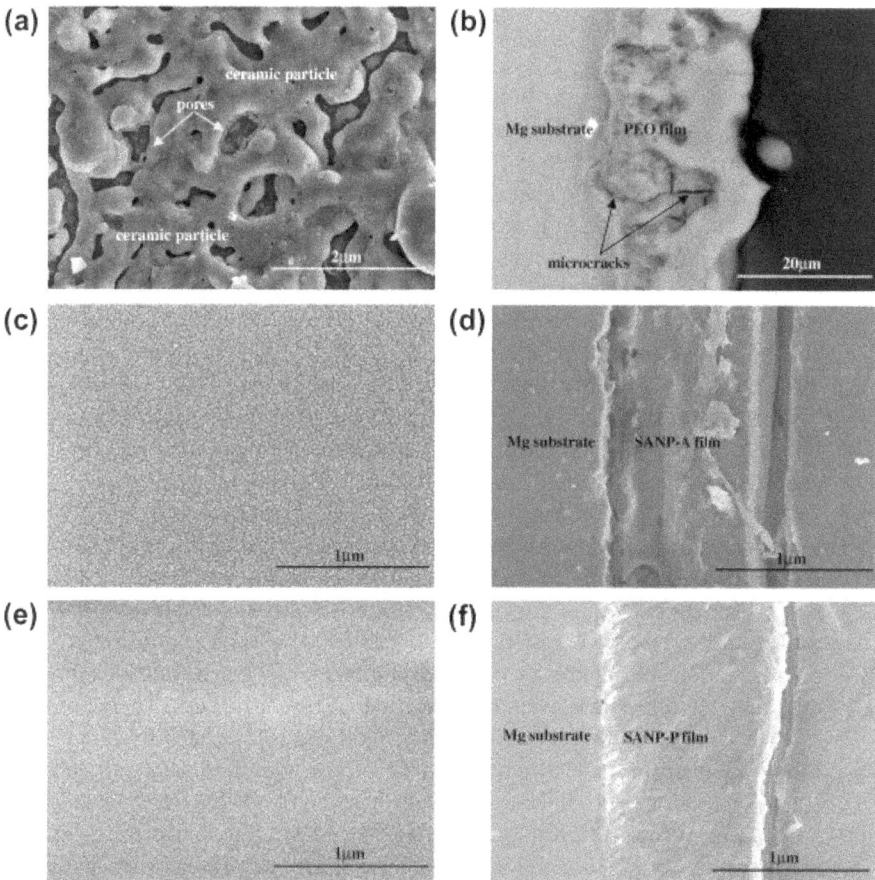

FIGURE 23.14 SEM images of surface and cross-section morphologies of the Mg-Gd-Y alloy substrates with (a and b) PEO film, (c and d) SANP-A (SANP-acetic acid) film, and (e and f) SANP-P film [39].

tannin-based anti-corrosive pigment, namely magnesium tannin through the reaction of tannin and magnesium salt. The epoxy coating containing synthetic pigment was further prepared and tested in the neutral saline environment. The results showed that magnesium tannin was a good pigment.

23.3 RESEARCH GAPS AND FUTURE PREDICTIONS

Natural corrosion inhibitors not only have excellent corrosion inhibition performance but also have advantages of low toxicity and biodegradability compared with other organic corrosion inhibitors, which meet the requirements of low carbon and green chemical industry. Thus, the development of natural plant corrosion inhibitors has become a new trend at present. However, there are still problems in the extraction and separation of natural plant inhibitors as well as the characterization and determination of their effective components, which limits their further research. The development of natural plant inhibitors for Mg and Mg alloys in the future mainly focuses on the following main directions:

1. The current extraction method has the problems of complex process and low purity. Therefore, it is necessary to explore new extraction methods to shorten the extraction time and improve the extraction efficiency. Moreover, the use of more accurate characterization means to analyze the active substances and corresponding molecular structure of plant extracts requires upgrading purification equipment and using modern advanced analytical and measuring instruments.

2. Extracting inhibitors from plants consumes a large amount of raw materials and has a low utilization rate, which is still a high-cost corrosion resistance method. In order to reduce costs, it is necessary to select resource-rich plants and use them or their purified products as corrosion inhibitors and increase their production on an industrial scale to achieve large-scale application to cut the cost.

3. Even natural plants or extracts may be toxic, so it is necessary to select nontoxic corrosion inhibitors according to the service environment of Mg alloys and carry out research on their advantages in terms of efficacy, residual toxicity, degradation and disposal, in order to obtain more stable and environmentally friendly corrosion inhibitors suitable for wide application.

4. Some compounds suitable for corrosion inhibition of other metals have little effect on Mg alloys and even accelerate the corrosion process. Therefore, a more suitable inhibitor, especially for Mg alloy, can be extracted from plants. Moreover, there are some differences in the alloy composition, microstructure, corrosion performance, and corrosion mechanism of four commonly used Mg alloys (Mg-Al-Si series, Mg-Al-Zn series, Mg-Al-Mn series, and Mg-Al-RE series). Suitable corrosion inhibitors can be developed according to the characteristics of Mg alloys. The development of natural plant inhibitors also needs to pay attention to the use of environmental differences: for example, the development of corrosion inhibitors suitable for oil/water/gas and gas/liquid/solid multiphase systems and high-temperature corrosion inhibitors.

5. The addition of single natural plant corrosion inhibition often presented a limited corrosion resistance effect, so it can be combined with other methods: for example, the design of self-healing coating enhanced by natural inhibitor, synergism of multiple corrosion inhibitors, synthesis of new nano-composite corrosion inhibitors to achieve controlled release of natural corrosion inhibitors, extraction natural compounds with anti-corrosion and antifouling functions to achieve the integration of anti-corrosion and antifouling functions.

6. In view of the great difficulty in understanding the corrosion resistance mechanism of plant extract as an inhibitor at the atomic-molecular level at present, the computational chemical method and high-throughput corrosion testing technology can be introduced to study the structure-activity relationship between the molecular structure and the corrosion inhibition efficiency of the corrosion inhibitor, which usually provide theoretical and data support for the development and screening of high-efficiency corrosion inhibitors for Mg alloys. Furthermore, it is a good strategy to use quantum chemical theory and molecular design to synthesize natural corrosion inhibitors with high efficiency, multi-function, and environmental friendliness.

7. The study of natural or extracted corrosion inhibitors should be guided by their practical application: compared with scientific research, the actual application requirements of corrosion inhibitors may be slightly different, such as the requirements for the melting point and boiling point of corrosion inhibitors, the requirements for non-toxic performance, and the solubility of corrosion inhibitors under specific circumstances, which are matters needing attention in addition to the consideration of corrosion inhibition performance.

23.4 SUMMARY

This chapter summarizes the corrosion inhibition effect of phytochemicals/plant extracts as corrosion inhibitors on magnesium and its alloys in neutral NaCl solution. Magnesium and its alloys are widely used in industrial applications, marine engineering, and biomedical fields because of their light weight, excellent mechanical properties, and strong biocompatibility. In this chapter, the research on phytochemicals/plant extracts as corrosion inhibitors for magnesium and its alloys is divided into three parts: plant extracts corrosion inhibitors, synergistic corrosion inhibitors, and coatings containing corrosion inhibitors. In most cases, phytochemicals/plant extracts and their modified substances moderately inhibit the corrosion of the alloy. It was also found that some organic plant extracts accelerated the corrosion of magnesium or its alloys. The main conclusions include:

1. Limited information about the direct use of phytochemicals/plant extracts or their modified substances as corrosion inhibitors to protect magnesium and its alloys in neutral NaCl solution. The corrosion inhibition efficiency varies greatly with the type of inhibitor, and even some compounds accelerate the corrosion process.

2. Limited information about the synergism of phytochemicals/plant extracts and other corrosion inhibitors as corrosion inhibitors for magnesium and its alloys. The study of synergistic corrosion inhibition mechanisms is not yet in-depth.
3. There is little research on the application of corrosion inhibitors extracted from natural plants on magnesium and Mg alloy coating.

Based on the knowledge gap found, it is suggested that future research should focus on filling the existing research gap.

REFERENCES

[1] S.A. Umoren, M.T. Abdullahi, M.M. Solomon, An overview on the use of corrosion inhibitors for the corrosion control of Mg and its alloys in diverse media, *J Mater Res Technol*, 20 (2022) 2060–2093.
[2] D.B. Huang, J.Y. Hu, G.L. Song, X.P. Guo, Inhibition effect of inorganic and organic inhibitors on the corrosion of Mg-10Gd-3Y-0.5Zr alloy in an ethylene glycol solution at ambient and elevated temperatures, *Electrochim Acta*, 56 (2011) 10166–10178.
[3] Y.M. Qiu, J. Li, Y.F. Bi, X.P. Lu, X.H. Tu, J.J. Yang, Insight into synergistic corrosion inhibition of 3-amino-1,2,4-triazole-5-thiol (ATT) and NaF on magnesium alloy: Experimental and theoretical approaches, *Corros Sci*, 208 (2022) 110618.
[4] Y.M. Zhang, Y.X. Wu, N. Li, Y.M. Jiang, Y.F. Qian, L. Wang, J.L. Zhang, Synergistic inhibition effect of L-Phenylalanine and zinc salts on chloride-induced corrosion of magnesium alloy: Experimental and theoretical investigation, *J Taiwan Inst Chem E*, 121 (2021) 48–60.
[5] H. Gao, Q. Li, Y. Dai, F. Luo, H.X. Zhang, High efficiency corrosion inhibitor 8-hydroxyquinoline and its synergistic effect with sodium dodecylbenzenesulphonate on AZ91D magnesium alloy, *Corros Sci*, 52 (2010) 1603–1609.
[6] W. Al Zoubi, M.E. Khan, Y.G. Ko, A simple method to functionalize the surface of plasma electrolysis produced inorganic coatings for growing different organic structure, *Prog Org Coat*, 171 (2022) 107008.
[7] J.Y. Hu, D.B. Huang, G.L. Song, X.P. Guo, The synergistic inhibition effect of organic silicate and inorganic Zn salt on corrosion of Mg-10Gd-3Y magnesium alloy, *Corros Sci*, 53 (2011) 4093–4101.
[8] S.V. Lamaka, B. Vaghefinazari, D. Mei, R.P. Petrauskas, D. Höche, M.L. Zheludkevich, Comprehensive screening of Mg corrosion inhibitors, *Corros Sci*, 128 (2017) 224–240.
[9] N. M'hiri, D. Veys-Renaux, E. Rocca, I. Ioannou, N.M. Boudhrioua, M. Ghoul, Corrosion inhibition of carbon steel in acidic medium by orange peel extract and its main antioxidant compounds, *Corros Sci*, 102 (2016) 55–62.
[10] Y.X. Wu, Y.M. Zhang, Y.M. Jiang, Y.F. Qian, X.G. Guo, L. Wang, J.L. Zhang, Orange peel extracts as biodegradable corrosion inhibitor for magnesium alloy in NaCl solution: Experimental and theoretical studies, *J Taiwan Inst Chem E*, 115 (2020) 35–46.
[11] Y.X. Wu, Y.M. Zhang, Y.M. Jiang, N. Li, Y.N. Zhang, L. Wang, J.L. Zhang, Exploration of walnut green husk extract as a renewable biomass source to develop highly effective corrosion inhibitors for magnesium alloys in sodium chloride solution: Integrated experimental and theoretical studies, *Colloid Surface A*, 626 (2021) 126969.
[12] J.Y. Hu, D. Zeng, Z. Zhang, T.H. Shi, G.L. Song, X.P. Guo, 2-Hydroxy-4-methoxy-acetophenone as an environment-friendly corrosion inhibitor for AZ91D magnesium alloy, *Corros Sci*, 74 (2013) 35–43.
[13] N. Dinodi, A.N. Shetty, Alkyl carboxylates as efficient and green inhibitors of magnesium alloy ZE41 corrosion in aqueous salt solution, *Corros Sci*, 85 (2014) 411–427.

[14] F.J. Wu, S.T. Zhang, Z.H. Tao, Corrosion inhibition of sodium tannate for AZ61 magnesium alloy in sodium chloride solution, *Mater Prot*, 43 (2010) 22–25. (in Chinese)

[15] N. Dang, Y.H. Wei, L.F. Hou, Y.G. Li, C.L. Guo, Investigation of the inhibition effect of the environmentally friendly inhibitor sodium alginate on magnesium alloy in sodium chloride solution, *Mater Corros*, 66 (2015) 1354–1362.

[16] P. Zhang, Q. Li, F.N. Chen, Study of the effect environmental and friendly corrosion inhibitors on magnesium alloy, *J Funct Mater*, 44 (2013) 2384–2388. (in Chinese)

[17] S.A. Umoren, M.M. Solomon, A. Madhankumar, I.B. Obot, Exploration of natural polymers for use as green corrosion inhibitors for AZ31 magnesium alloy in saline environment, *Carbohydr Ploym*, 230 (2020) 115466.

[18] Y.M. Qiu, X.H. Tu, X.P. Lu, J.J. Yang, A novel insight into synergistic corrosion inhibition of fluoride and DL-malate as a green hybrid inhibitor for magnesium alloy, *Corros Sci*, 199 (2022) 110177.

[19] Z. Shao, P. Li, C. Zhang, B. Wu, C. Tang, M. Gao, Enhancing the anti-corrosion performance and biocompatibility of AZ91D Mg alloy by applying roughness pretreatment and coating with in-situ Mg (OH)$_2$/Mg-Al LDH. *J Magnes Alloy*, 10 (2022) 14

[20] P. Zhang, Q. Li, L.Q. Li, X.X. Zhang, Z.W. Wang, A study of environment-friendly synergistic inhibitors for AZ91D magnesium alloy, *Mater Corros*, 66 (2015) 31–34.

[21] L.F. Hou, N. Dang, H.Y. Yang, B.S. Liu, Y.G. Li, Y.H. Wei, X.B. Chen, A combined inhibiting effect of sodium alginate and sodium phosphate on the corrosion of magnesium alloy AZ31 in NaCl solution, *J Electrochem Soc*, 163 (2016) C486–C494.

[22] L.L. Liao, S. Mo, H.Q. Luo, N.B. Li, Longan seed and peel as environmentally friendly corrosion inhibitor for mild steel in acid solution: Experimental and theoretical studies, *J Colloid Interf Sci*, 499 (2017) 110–119.

[23] Y.F. Qian, Y.X. Wu, X.G. Guo, L. Wang, A synergistic anti-corrosion effect of longan residue extract and sodium dodecylbenzenesulfonate composition on AZ91D magnesium alloy in NaCl solution, *Corros Eng Sci Techn*, 57 (2022) 322–330.

[24] L.W. Ma, J.K. Wang, Y.J. Wang, X. Guo, S.H. Wu, D.M. Fu, D.W. Zhang, Enhanced active corrosion protection coatings for aluminum alloys with two corrosion inhibitors co-incorporated in nanocontainers, *Corros Sci*, 208 (2022) 110663.

[25] Q.Q. Chen, X.P. Lu, M. Serdechnova, C. Wang, S. Lamaka, C. Blawert, M.L. Zheludkevich, F.H. Wang, Formation of self-healing PEO coatings on AM50 Mg by in-situ incorporation of zeolite micro-container, *Corros Sci*, 209 (2022) 110785.

[26] D.G. Shchukin, H. Mohwald, Smart nanocontainers as depot media for feedback active coatings, *Chem Commun*, 47 (2011) 8730–8739.

[27] V. Upadhyay, Z. Bergseth, B. Kelly, D. Battocchi, Silica-based sol-gel coating on magnesium alloy with green inhibitors, *Coatings*, 7 (2017) 86.

[28] W.J. Li, Y. Su, L. Ma, S.J. Zhu, Y.F. Zheng, S.K. Guan, Sol-gel coating loaded with inhibitor on ZE21B Mg alloy for improving corrosion resistance and endothelialization aiming at potential cardiovascular application, *Colloid Surface B*, 207 (2021) 111993.

[29] H. Asadi, B. Suganthan, S. Ghalei, H. Handa, R.P. Ramasamy, A multifunctional polymeric coating incorporating lawsone with corrosion resistance and antibacterial activity for biomedical Mg alloys, *Prog Org Coat*, 153 (2021) 106157.

[30] X. Wang, C. Jing, Y.X. Chen, X.S. Wang, G. Zhao, X. Zhang, L. Wu, X.Y. Liu, B.Q. Dong, Y.X. Zhang, Active corrosion protection of super-hydrophobic corrosion inhibitor intercalated Mg-Al layered double hydroxide coating on AZ31 magnesium alloy, *J Magnes Alloy*, 8 (2020) 291–300.

[31] L. Zhang, J.Q. Zhang, C.F. Chen, Y.H. Gu, Advances in microarc oxidation coated AZ31 Mg alloys for biomedical applications, *Corros Sci*, 91 (2015) 7–28.

[32] M. Esmaily, J.E. Svensson, S. Fajardo, N. Birbilis, G.S. Frankel, S. Virtanen, R. Arrabal, S. Thomas, L.G. Johansson, Fundamentals and advances in magnesium alloy corrosion, *Prog Mater Sci*, 89 (2017) 92–193.

[33] Y. Chen, X.P. Lu, S.V. Lamaka, P.F. Ju, C. Blawert, T. Zhang, F.H. Wang, M.L. Zheludkevich, Active protection of Mg alloy by composite PEO coating loaded with corrosion inhibitors, *Appl Surf Sci*, 504 (2020) 144462.

[34] Z.J. Li, X.Y. Jing, Y. Yuan, M.L. Zhang, Composite coatings on a Mg-Li alloy prepared by combined plasma electrolytic oxidation and sol-gel techniques, *Corros Sci*, 63 (2012) 358–366.

[35] H. Tang, W. Tao, C. Wang, H. Yu, Fabrication of hydroxyapatite coatings on AZ31 Mg alloy by micro-arc oxidation coupled with sol-gel treatment, *RSC Adv*, 8 (2018) 12368–12375.

[36] S.V. Lamaka, G. Knörnschild, D.V. Snihirova, M.G. Taryba, M.L. Zheludkevich, M.G.S. Ferreira, Complex anticorrosion coating for ZK30 magnesium alloy, *Electrochim Acta*, 55 (2009) 131–141.

[37] A.P.M. Bloot, D.L. Kalschne, J.A.S. Amaral, I.J. Baraldi, C. Canan, A review of phytic acid sources, obtention, and applications, *Food Rev Int*, (2021) 1–20.

[38] Y.J. Zhang, B.J. Dou, Y.W. Shao, X.J. Cui, Y.Q. Wang, G.Z. Meng, X.Z. Lin, Influence of phytic acid on the corrosion behavior of carbon steel with different surface treatments, *Anti-Corros Method M*, 65 (2018) 658–667.

[39] X.H. Guo, K.Q. Du, Q.Z. Guo, Y. Wang, R. Wang, F.H. Wang, Effect of phytic acid on the corrosion inhibition of composite film coated on Mg-Gd-Y alloy, *Corros Sci*, 76 (2013) 129–141.

[40] M. Dargahi, A.L.J. Olsson, N. Tufenkji, R. Gaudreault, Green technology: Tannin-based corrosion inhibitor for protection of mild steel, *Corrosion*, 71 (2015) 1321–1329.

[41] A.V. Zmozinski, R.S. Peres, K. Freiberger, C.A. Ferreira, S.M. Mesquita Tamborim, D.S. Azambuja, Zinc tannate and magnesium tannate as anticorrosion pigments in epoxy paint formulations, *Prog Org Coat*, 121 (2018) 23–29.

24 Sustainability/Greenness of Phytochemicals/ Plant Extracts

Sampad Ghosh
Maulana Abul Kalam Azad University of Technology

Nabakumar Pramanik
National Institute of Technology

Rajeev Kumar
Nalanda College of Engineering

Vinay Kumar Chaudhary
Government Engineering College

24.1 INTRODUCTION

Corrosion is a natural phenomenon where metals and alloys try to revert back to their more stable thermodynamic form due to reaction with the environment that surrounds them. Corrosion is very expensive due to loss of materials or their properties, which leads to loss of time during maintenance, the shutting down of systems, and severe failure of some structures, which in some cases may be hazardous and cause injury.

Many efforts were made to bring under control the onset and thereafter critical effects of corrosion using many preventive measures. To protect metals or alloys from corrosion, approaches such as isolating the structure from aggressive media (using coatings or film-forming chemicals) or compensating for the loss of electrons (corrosion is an oxidation process) from the corroded structure (e.g. cathodic protection by impressed current or by using active sacrificial anodes) are generally employed.

Out of several methods, inhibitors' use is one of the most realistic methods of controlling corrosion, particularly in aggressive media. In order to reduce the rate of corrosion generally, the inhibitors are used. A variety of compounds both inorganic and organic are used as inhibitors for metal corrosion in different aggressive media. There are several ways of preventing corrosion and the rates at which it can propagate with a view of improving the lifetime of metallic and alloy materials. The use of inhibitors for the control of corrosion of metals and alloys that are in contact with the aggressive environment is one among the acceptable practices used to reduce

DOI: 10.1201/9781003394631-24

and/or prevent corrosion. However, many researchers [1–5] reviewed the work on the mechanism of corrosion inhibition and other researchers [6,7] consolidated the work carried out on the effectiveness of green inhibitors. Some researchers [8–10] have reviewed the work on the utilization of plant sources in corrosion protection. Speller [11] is the first researcher who identified the inhibition performance of corroded scaled water pipes in the hydrochloric acid medium using organic inhibitor.

Corrosion inhibition may include organic or inorganic compounds that adsorb on the metallic structure to isolate it from its surrounding media to stop the oxidation-reduction process. Organic inhibitors create their inhibition by adsorbing their molecules on the metal or alloy surface to form a protective layer [12]. Alternatively, inorganic inhibitors act as anodic inhibitors, and their metallic atoms are enclosed in the film to improve their corrosion resistance. Most investigated corrosion inhibitors are toxic and cause severe environmental hazards upon disposal. Therefore, their use has been limited by environmental regulations. Nevertheless, inhibitors still play a critical role in corrosion prevention. Inhibitors are classified according to the following:

- Three types of electrode process, namely, anodic, cathodic, and mixed;
- The chemical nature of the environment to acid inhibitors (organic or inorganic), neutral inhibitors, alkaline inhibitors, and vapor-phase inhibitors.

The present trend in research on environmentally friendly corrosion inhibitors is taking us back to exploring the use of natural products as possible sources of cheap, nontoxic, and eco-friendly corrosion inhibitors. These natural products are either synthesized or extracted from aromatic herbs, spices, and medicinal plants. Of increasing interest is the use of medicinal plant extracts as corrosion inhibitors for metals in acid solutions. This is because these plants serve as incredibly rich sources of naturally synthesized chemical compounds that are environmentally acceptable, inexpensive, readily available, and renewable sources of materials [13,14]. This chapter is a brief compilation of the work done by various researchers for examination of the use of natural products or phytochemicals as corrosion inhibitors on metals.

24.2 CORROSION INHIBITORS

Corrosion inhibitors are compounds that are commonly added in small quantities to an environment to prevent corrosion of exposed metal in a corrosive environment [15]. Added in a small quantity in industrial processes, they are able to diminish the progression of corrosion, maintaining a protective, inhibitive film on metal superficial.

Corrosion inhibitors are classified as organic or inorganic compounds and over the last century, inorganic inhibitors such as chromates, vanadates, and various nitrates have been systematically used in various corrosive media [16] due to their remarkable inhibitory properties on different metals and alloys. However, their use is prohibited due to high toxicity, and thus, studies have been carried out on the utilization of different organic compounds, especially newly synthesized heterocyclic molecules [17]. At the moment when it is understood that their preparation is extremely expensive, and when a large amount of hazardous organic solvents is used for their synthesis, the application of such compounds is avoided. Although it has been proven that the vast majority of organic inhibitors protect the metal from the aggressive medium, mostly by adsorption, through atoms that possess free electron pairs (N, S, O and P) or through double and triple bonds in the molecule, their utilization is also avoided due to the risks to human life and its environment [17]. These corrosion inhibitors decrease the degree of corrosion through the following methods:

- Intended for substrates to the surface of metal, plummeting the degree of diffusion
- Onto the surface of the metal, sorption of molecules/ions
- Decreasing or increasing the cathodic or anodic reaction
- Inhibitors that have *in situ* submission benefits and are frequently easy to use

Many features such as easy obtainability, amount and cost, and vitally well-being to the atmosphere and its species are required for consideration when selecting an inhibitor.

The increasing awareness of "green principles," i.e., ecological, economic, and environmentally friendly (e^3 concept) [18], in all fields of science and technology has been influenced also in the corrosion discipline. In that connection, the use of traditional, synthetic organic compounds has been limited and replaced by green organic compounds, usually extracted by conventional or innovative processing technology, such as pulsed electric fields, ultrasound, microwave, supercritical fluid, and high hydrostatic pressure [19]. The green corrosion inhibitors are renewable, biodegradable, and nontoxic, which made them a superior anticorrosion candidate than their synthetic counterparts. They are employed as inhibitors for various metallic materials in different media, such as HCl, H_2SO_4, H_3PO_4, and HNO_3. Nowadays, plant extracts have acquired numerous examinations and applications as green corrosion inhibitor, due to their natural origin.

In order to evaluate the inhibition activity of green corrosion inhibitors in various corrosive medium, several standard analyses, including gravimetric test (weight loss), and potentiodynamic polarization and impedance spectroscopy analyses are employed [20,21]. The weight loss test is the simplest technique used to calculate the corrosion rate, and it is based on metal weight measurements before and after immersion in the corrosive medium for a certain period. In comparison with weight loss analysis, the potentiodynamic polarization technique [22] is able to provide a number of useful information regarding the rate of metal dissolution and its protection. Besides these methods, the atomic force microscope (AFM) and scanning electron microscopy (SEM) are also two of the powerful techniques applied to investigate the surface morphology, at nano- to micro-scale. Both of them are capable of providing valuable information regarding the generation and progression of corrosion at the metal-solution interface. Nowadays, computational studies and quantum chemical calculation are also in progress, and they are important for the prediction of anticorrosion mechanism.

24.3   PRINCIPLE OF GREEN CORROSION INHIBITORS

The corrosion process is directly related to Gibbs' free energy. The higher the Gibbs' free energy, the higher the corrosion rate. To lower down the free energy, green corrosion inhibitors can be used. Inhibitors accumulate on the surface of the active sites of the corroded metals and form a barrier against corrosion. The relative rate of corrosion is also regulated by the Pilling-Bedworth ratio, by which the status of the film surface can be identified [23]. When the ratio is less than 1, i.e., the volume of the product generated by corrosion is less than that of the volume of the metal, insufficient oxide to protect the metal is produced which means the metal is rendered nonprotective. When the ratio is greater than 1, i.e., the volume of the product generated by corrosion is greater than the volume of metal, the metal is protective; however, too large value of the ratio may create cracking in the oxides formed. When the ratio becomes 1, there is a good spatial match between the oxide and the metal, and the oxide is rendered protective [24].

Inhibition that occurs based on the adsorption technique is affected by many factors like type of the electrolytes used, temperature, chemical compositions of the inhibitors, etc. For chemical adsorption process, there is a chance of the formation of coordinate covalent bond. The green corrosion inhibition is generally done at room temperature, and the inhibition efficiency is generally inversely proportional to temperature [25].

The action of the green corrosion inhibitors is also influenced by the structure of the ingredients which can be summarized as follows [26]:

- The inhibitors are generally in onium ions form which are adsorbed on the metal surface having active cathodic sites.
- The plant extracts contain alkaloid bebeerines. It has O- and N- atoms carrying free electrons which facilitate to form bonds with the electrons that are freely available on the metal surfaces.
- Increasing electrical resistance of the metal surface by forming a film (coat) on it.
- The potential cathodic process of steel can be affected by S-containing unsaturated compounds such as allyl propyl disulfide exited from many plant extracts.
- In pyrrolidine, the strength of the bond between pyrrolidine and metals depends on the availability of the negative charge on the N- atoms which in turn denotes the basicity of the compounds.
- The $-OCH_3$ group in alkaloid ricimine favors the interaction with the active metal surfaces.

24.4   FEW REPORTS ON CORROSION PREVENTION USING PHYTOCHEMICALS OR GREEN CORROSION INHIBITORS

Nowadays, cleaner production, biodegradability, sustainable development, and pollution prevention are to be implemented in industrial unit operations, synthesis, and applications. Various researchers have recently reported the corrosion inhibitor effectiveness of metals by natural plant extracts like *Swertia aungustifolia* [27], Coriander [28], Thyme [29], Hibiscus [30], Anise [31], *Ricinus communis* [32], *Telfaria occidentalis* [33], *Prunus cerasus* [34], *Capparis deciduas* [35], *Piper guinensis* [36], *Azadirachta indica* [37], *Solanum tuberosum* [38], natural henna [39], *Zanthoxylum alatum* [40], and *Nicotiana tabacum* [41].

These inhibitors are also termed as green corrosion inhibitors which by a little dose can effectively reduce the problem of corrosion on the metal or alloy surface. During the process of adsorption of effective species, i.e., naturally synthesized chemical compounds from plant extracts, the rate of corrosion on metallic surface can be affected by:

- Altering the rate of reactions at the anode and/or cathode,
- Affecting the rate of diffusion of aggressive ions that interact with the metals,
- Enhancing the electrical resistance of the metallic surface by the formation of a protective film.

24.5 PLANT EXTRACT AS CORROSION INHIBITOR

The leaf extract of *Chromolaena odorata* L. has been studied as green inhibitor for the corrosion of aluminum in acidic medium by gasometric and thermometric techniques at different temperatures [42]. The feasibility study showed that this green inhibitor is biodegradable as well as cost-effective. It has a variety of applications including anodizing and surface coating in industries. Gum Arabic, a water-soluble, branched, neutral natural gum is very much popular in case of corrosion inhibition. As it contains a high concentration of organic components, it is used as adsorptive material in mild steel and aluminum corrosion in acidic environment.

Punica granatum extract as corrosion inhibitor was monitored by Deepa et al. [43] on brass, and they found that the efficiency of inhibition is directly proportional to the increase in the concentration of acid. The maximum efficiency achieved was 94.52% at 1,000 ppm concentration of bio-inhibitor. On the basis of these results, the physisorption mechanism was also studied.

Hitherto saps of *Accacia arabica*, *Annona squamosa*, *Ricimus communis*, *Eugenia jambolans*, *Azadirachta indica*, *Pongamia glabra,* etc. have been scrutinized for analysis of reluctance of corrosion in the acidic environment taking mild steel as a case study. In addition, it has been found that some herbs like anise, black cumin, and coriander can be used as potential, new, green corrosion inhibitors [44].

Piperanine was isolated from black pepper (BP) extract and studied as corrosion inhibitor. The inhibition efficiency of C38 steel in 1 M HCl solution was studied by weight loss method at a temperature range of 298K–353K. NMR technique was used for identification of piperanine. Results inferred that 97.5% corrosion inhibition was observed at 10^{-3}M inhibitor concentration. Piperanine adsorbs on the metal surface according to Langmuir isotherm [45].

Ervatinine, isolated from the leaves of *Ervatamia coronaria* plant, was tested as corrosion inhibitor on the mild steel in acidic medium. The corrosion inhibition efficiency was examined by gravimetric, electrochemical impedance, Tafel polarization, SEM, and XRD techniques. Results showed that the ervatinine alkaloid present in the plant extract acts as a good corrosion inhibitor. The adsorption of inhibitor on metal surface followed Langmuir adsorption isotherm and ervatinine physically adsorb on it. The surface morphological examination *via* SEM techniques indicated that the ervatinine retards corrosion on the specimen surfaces by forming a protective layer. The results suggest that ervatinine acts as a good corrosion inhibitor [46].

The corrosion inhibition potential of vasicine molecule isolated from *Adhatoda vasica* plant extract was studied on mild steel in acidic medium [47]. Corrosion measurement results showed that the inhibition efficiency of vasicine in acidic medium increases with an increase in inhibitor concentration and decreases with a rise in temperature.

Green synthesis of inhibitors has been investigated by several other researchers [48]. Table 24.1 lists some natural products that have anti-corrosive properties.

24.6   ESSENTIAL OILS AS SUSTAINABLE CORROSION INHIBITOR

It has been found that oils show inhibition efficiency up to 98%, so it is certain that oils are effective corrosion inhibitors. The inhibition efficiency of jojoba oil, on the corrosion of iron in acidic solution was studied by using weight loss measurement and electrochemical polarization methods. Studies revealed that jojoba oil was an excellent corrosion inhibitor and showed 100% inhibition at $0.515\,g\,L^{-1}$ concentrations of jojoba oil, indicating that jojoba oil was inhibited. The adsorption on the metal of jojoba oil, obeyed the Frumkin isotherm [79].

Mentha pulegium (Pennyroyal Mint) oil was tested as a corrosion inhibitor of steel in 1M HCl solution using weight loss measurements, electrochemical polarization, and electrochemical impedance spectroscopy (EIS) methods. Results showed that the inhibition efficiency was found to increase with oil content to attain 80% at $2.76\,g\,L^{-1}$ and oil acts as a cathodic inhibitor [79]. Natural oil extracted from *Athamanta sicula* was evaluated as a corrosion inhibitor of mild steel in molar hydrochloric acid [80]. The corrosion rate and inhibition efficiency were determined by using gravimetric, EIS, and Tafel polarization curve methods. The oil was a mixed-type inhibitor and retards the corrosion rate of mild steel in aggressive medium. The corrosion-inhibiting nature of *Artemisia* oil as steel in 2M H_3PO_4 was reported using gravimetric, electrochemical polarization, and electron impedance spectroscopy methods [81]. The oil reduces the corrosion rate by increasing the concentration, and the maximum inhibition efficiency attains 79% at $6\,g\,L^{-1}$ at different temperatures. They found that the inhibition efficiency of the oil decreases with the rise in temperature. The adsorption isotherm of natural product on the steel has also been determined.

The inhibiting effect of *Artemisia herba-alba* oil [82], essential oil of fennel (*Foeniculum vulgare*) [83], *Warionia saharae* [84], *Pulicaria mauritanica* [85], *Eucalyptus globulus* (Myrtaceae) [86], *Asteriscus graveolens* [87], *Glycine max* [88], *Argan* oil [89], garlic essential oil [90], and *Pistachio* essential oils [91] as a corrosion inhibitor was tested on metals in acidic solutions. The results obtained showed that the oils act as an eco-friendly corrosion inhibitor.

TABLE 24.1
List of Natural Products and Their Anti-Corrosive Properties

Sl. No.	Metal/Alloy	Aggregation Media	Source of Inhibitors	Observations	Ref.
1	Steel	1 M HCl	Cladodes of *Opuntia ficus-indica*	Microwave extraction method was used which led to 94% inhibitor efficiency (IE).	[49]
2	Mild steel	0.2 M HCl	Leaves and bark of mango plant	WL method was employed. 0.2 M sulfuric acid concentration gave a good IE.	[50]
3	Mild steel	0.1 M HCl	*Calotropis procera* plant extract	WL method was used and the polarization curves were analyzed.	[51]
4	Mild steel	1 M HCl	Bark and leaves of *Neolamarckia cadamba*	Acidic/alkaline extraction method was applied. IE was 91% at 30°C.	[52]
5	Mild steel	1 M HCl	Watermelon peel, seeds, and rind	Boiling in 1 M HCl solution.	[53]
6	Mild steel	1 M HCl	Leaves of henna	Boiling water was used as a solvent. IE was 90.34% at 30°C.	[54]
7	Mild steel	1 M HCl	Pennyroyal oil from *Mentha pulegium*	Weight loss measurements, electrochemical polarization, and electrochemical impedance spectroscopy (EIS) were used.	[55]
8	Mild steel	1 M HCl	*Justicia gendarussa* extract	WL method, atomic force microscopy, and electron spectroscopy for chemical analysis were used, which suggested that the inhibitor was a mixed-type inhibitor. The isotherm was that of Langmuir.	[56]
9	Mild steel	1 M HCl and H₂SO₄	Seeds and leaves extract of *Phyllanthus amarus*	WL and gasometric techniques were used. The Temkin isotherm fitted this process.	[57]
10	Mild steel	1 M HCl and 0.5 M H₂SO₄	*Murraya koenigii* leaves	Electrochemical measurements were done. The inhibitor was of mixed type. The adsorption isotherm was followed by Langmuir and Dubinin-Radushkevich isotherm.	[58]
11	Mild steel	1 M H₂SO₄	Coriander, black cumin	WL, EIS, and linear polarization were used.	[59]

(Continued)

TABLE 24.1 (Continued)
List of Natural Products and Their Anti-Corrosive Properties

Sl. No.	Metal/Alloy	Aggregation Media	Source of Inhibitors	Observations	Ref.
12	Mild steel	2 M H_2SO_4	Ethanolic extract of *Medicago sative*	The alternating current and direct current. Electrochemical techniques were used.	[60]
13	Mild steel	20%, 50%, and 88% H_3PO_4	*Zanthoxylum alatum* plant extract	WL, EIS, XPS, and FTIR analyses were carried out. The IE was 88%.	[61]
14	Mild steel	3% NaCl	*Ricinus communis*, Coumarins plants	Galvanostatic, anodic, and cathodic polarization measurements were done.	[62]
15	Aluminum	1 M HCl	Roots of Ginseng	WL method was followed. IE was 93.1%. The adsorption isotherm was Freundlich isotherm.	[63]
16	Aluminum	2 M HCl	Mucilage extract of *Opuntia*	Langmuir isotherm fitted the process. WL and hydrogen evolution techniques were used.	[64]
17	Aluminum	Highly basic (pH = 12)	*Hibiscus rosasinensis*	The additive was Zn^{2+}. The inhibitor was a cathodic type. WL, followed by AC impedance techniques, was used.	[65]
18	Aluminum	0.5 M NaOH	*Hibiscus sabdariffa* leaves	10% ethanol was used as an additive. Hydrogen evolution, followed by EIS, SEM, and WL techniques, was used.	[66]
19	Mild steel	0.5 M H_2SO_4	*Artemisia herba-alba*		[67]
20	Carbon steel	2 M HCl	Dry olive leaves	Boiling water was used.	[68]
21	Aluminum	0.5 to 2 M HCl	Citrus peel	Pectin was the major constituent. At 10°C, the IE was 91%.	[69]
22	Bronze	Simulated acid rain solution	*Salvia hispanica* seeds	Inhibition was carried out by soaking in methanol.	[70]

(Continued)

TABLE 24.1 (Continued)
List of Natural Products and Their Anti-Corrosive Properties

Sl. No.	Metal/Alloy	Aggregation Media	Source of Inhibitors	Observations	Ref.
23	Low-carbon steel	1 M HCl	Schinopsis lorentzii tree powder	TAPPI T204 OM-8 and ASTM 1110–96 standards were followed.	[71]
24	Tin	NaCl (2%), acetic acid (1%), and citric acid (5%) solution	Tomato peel wastes	Acidic/alkaline extraction method was applied. IE was 90% at 30°C.	[72]
25	Mild steel and copper	NaCl and SO_2 environment	Natural oil from Cassia auriculata, Strychnos nuxvomica		[73]
26	Tin	HNO_3	Ficus carica L., Glycyrrhiza glabra plants	Thermometric method was used.	[74]
27	316 stainless steel	HCl (5%)	Medicago polymorpha Roxb.	The process was carried out at ambient temperature. Open circuit potential studies were done.	[75]
28	X52 mild steel	20% H_2SO_4	Cotula cinerea, Retama raetam plant	WL and electrochemical methods were used.	[76]
29	Carbon steel	1 M HCl	Aqueous extract of passion fruits and cashew peels	WL and EIS were used. The adsorption isotherm was of Langmuir.	[77]
30	Concrete steel surface	10%–23% NaOH	Magraba banana stems	Banana plant juice was used as an inhibitor. Inhibition was monitored using the WL method.	[78]

24.7   COMPOUNDS ISOLATED FROM PLANTS AS SUSTAINABLE CORROSION INHIBITOR

The inhibition efficiencies of two Amazonian trees' (*Guatteria ouregou* and *Simira tinctoria*) alkaloid extracts on the corrosion of low-carbon steel in 0.1M HCl solution were investigated by using electrochemical techniques. The results obtained show that both extracts provide adequate inhibition of corrosion of low-carbon steel in acidic media. Authors found that harmane was an active component of *S. tinctoria* extract, and the anticorrosion activity in low-carbon steel is in aggressive environment [92]. Lawsone, an active principle, was isolated from henna (*Lawsonia inermis*) plant and was used as a corrosion inhibitor [93]. The inhibition efficiency of an active molecule was studied in 1M HCl solution on mild steel surface by weight loss method. Results revealed that the corrosion rate decreased with an increase in the concentration of Lawsone.

Piperanine was isolated from black pepper (BP) extract and studied as a corrosion inhibitor. The inhibition efficiency of C38 steel in 1 M HCl solution was studied by the weight loss method at a temperature range of 298K–353K. NMR technique was used for identification of piperanine. Results inferred that 97.5% corrosion inhibition was observed at 10^{-3}M inhibitor concentration. Piperanine adsorbs on the metal surface according to Langmuir isotherm [94].

Ervatinine, isolated from the leaves of *Ervatamia coronaria* plant, was tested as a corrosion inhibitor on mild steel in acidic medium. The corrosion inhibition efficiency was examined by gravimetric, electrochemical impedance, Tafel polarization, SEM, and XRD techniques. Results showed that the ervatinine alkaloid present in the plant extract acts as a good corrosion inhibitor. The adsorption of inhibitor on the metal surface follows Langmuir adsorption isotherm and ervatinine physically adsorbs on it. The surface morphological examination *via* SEM techniques indicated that the ervatinine retards the corrosion on the specimen surfaces by forming a protective layer. The results suggest that ervatinine acts as a good corrosion inhibitor [95].

The corrosion inhibition potential of vasicine molecule isolated from *Adhatoda vasica* plant extract was studied on mild steel in acidic medium [96]. Corrosion measurement results showed that the inhibition efficiency of vasicine in acidic medium increases with an increase in inhibitor concentration and decreases with a rise in temperature.

24.8   EXPIRED DRUGS AS SUSTAINABLE CORROSION INHIBITORS

Some drugs which are nontoxic in nature have been used as good corrosion inhibitors for different metals. Some expired drugs used as corrosion inhibitors such as Lumerax [97], Ciprofloxacin [98], Penicillin [99], Ketosulfone Drug [100], Biotin [101], Chloroquine phosphate [102], and Atorvastatin [103] for metal in acidic media. Four environmentally friendly penicillin derivatives, including penicillin G, oxacillin, penicillin V, and amoxicillin, are used as corrosion inhibitors [99]. In this study, corrosion inhibition of steel in hydrochloric acid was studied by weight loss measurement and Tafel polarization technique. The experimental results revealed that the derivatives are mixed type of inhibitors and adsorb on the metal surface.

24.9 STATISTICAL AND THEORETICAL APPROACH TO CORROSION INHIBITION

Simulation is an important analytical computational means for composite scientific and engineering issues. Simulation turns into standard statistical methods where the output can be analyzed using analysis of variance (ANOVA), which is a method of collecting all data and applying them to the statistical models. ANOVA is used to analyze the divergence between the group means and their related procedures. In recent decades, statistical approaches have been widely applied to corrosion problems. Many authors have investigated the statistical significance by using F-test. Rajendran and his team studied the inhibitive effect of *Phyllanthus amarus* extract Zn^{2+} [104] and henna leaves extract Zn^{2+} [105] on carbon steel by weight loss and electrochemical methods and proved that the given system was statistically significant. Yaro and his co-workers [106] have used ANOVA on the data for the corrosion prevention of mild steel (MS) in 1 M phosphoric acid by apricot juice obtained from weight loss data. ANOVA has shown that the corrosion rate is influenced by temperature, inhibitor concentration, and their combined interaction. Loto and his co-workers [107] studied the inhibition performance of *Vernonia amygdalina* extract on the MS corrosion in neutral medium. The outcome of the statistical test was verified with the results at 95% assurance. Similarly, another author investigated the performance of fenugreek (*Trigonella foenum graecum*) [108] seed extract as an inhibitor on mild steel under corrosive medium. A statistical view of their investigations revealed that the synergistic effect existing between fenugreek extract and potassium iodide is statistically significant.

During the extraction of plant fragments, it must be kept in mind that the solvent used for extraction must be less harmful. Also, some process requires long processing time and high temperatures, which disfavor the green corrosion inhibition. To overcome these problems, nowadays supercritical fluid extraction process is applied as an alternative technique. But, the effect of physicochemical parameters on its mechanism is currently under investigation. Moreover, the active ingredients of the inhibitors are very much less. Nowadays, some pharmaceutical drugs are used for this purpose. But, all the drugs are not biodegradable as it limits the processing of dry raw materials for green corrosion inhibitors. However, in spite of the above-mentioned limitations, green corrosion inhibitors are still the most promising alternatives because they are generally synthesized from natural products which in turn boost their ready availability and cost-effectiveness.

24.10 CONCLUSION

Metals, an essential part of our civilization, have the characteristic properties of degrading in the presence of harmful atmospheric conditions created by oxygen or acidic, basic, or even neutral conditions, which in turn compel us to think about preventing the phenomenon of corrosion. There are several techniques being applied to prevent corrosion. Among them, use of inhibitor is the best option. Many countries have already used these green inhibitors to prevent corrosion. Due to environmental concerns, some bio-based inhibitors nowadays are well pronounced.

These are also known as green corrosion inhibitors as they have no toxicity, are naturally available, and are very safe to handle. Moreover, the extraction process of the products is also generally cost-effective. As these products are easily available from an economical point of view, it is very feasible. In a nutshell, it can be said that though green corrosion inhibitors have some limitations, these inhibitors are very much needed to make a cleaner, safer, sustainable surrounding and most importantly to get rid of the problems caused by corrosion. A lot of potential is still untapped, and further research should also be focused on plant extraction methods and their active constituents as well as scale-up experiments for industrial applications that are needed to commercialize these natural extracts to effectively replace conventional chemicals.

CONFLICT OF INTEREST

The authors declare that the research was conducted in the absence of any commercial or financial relationships that could be construed as potential conflicts of interest.

REFERENCES

[1] B. Sanyal, Organic compounds as corrosion inhibitors in different environments, *Prog. Org. Coat.*, 9 (1981) 165–236.
[2] P. B. Raja, M. G. Sethuraman, Natural products as corrosion inhibitor for metals in corrosive media: A review, *Mater. Letters*, 62 (2008) 113–116.
[3] B. E. Amitha Rani, J. B. Bharathi Bai, Green inhibitors for corrosion protection of metals and alloys: An overview, *Int. J. Corros.*, 10 (2012) 1–15.
[4] M. Bethencourt, F. J. Botana, J. J. Calvino, M. Marcos, M. A. Rodriguesz Chacon, Lanthanide compounds as environmentally friendly corrosion inhibitors of aluminum alloys: A review, *Corr. Sci.*, 40 (1998) 1803–1819.
[5] D. M. Abdullah, A review: Plant extracts and oils as corrosion inhibitors in aggressive media, *Indus Lub. Trib.*, 63 (2011) 227–233.
[6] D. Kesavan, M. Gopiraman, N. Sulochana, Green inhibitors for corrosion of metals: A review, *Chem. Sci. Rev. Lett.*, 1 (2012) 1–8.
[7] N. Patni, S. Agarwal, P. Shah, Patni, N., Agarwal, S., P. Shah, Greener approach towards corrosion inhibition. *Chin. J. Eng.*, 78 (2013) 1–10.
[8] R. Rajilakshmi, A. Prithiba, S. Leelavathi, An overview of emerging scenario in the frontiers of eco-friendly corrosion inhibitors of plant origin for mild steel, *J. Chemica Acta*, 1 (2012) 6.
[9] M. Sangeetha, S. Rajendran, T. S. Muthumegala, A. Krishnaveni, Green corrosion inhibitors-an overview, *Zaštita Materijala*, 52 (2011) 3–19.
[10] P. Deepa Rani, S. Selvaraj, Inhibitive and adsorption properties of punica granatum extract on brass in acid media, *J. Phytology*, 2(11) (2010) 58–64.
[11] F.N. Speller, *Corrosion Causes and Prevention*, McGraw-Hill, New York, 1935.
[12] G.T. Hefter, N.A. North, S.H Tan, Organic corrosion inhibitors in neutral solutions; part 1-inhibition of steel, copper, and aluminum by straight chain carboxylates, *Corrosion*, 53(8) (1997) 657–667.
[13] R.M. Saleh, A.A. Ismail, A.H. El Hosary, Corrosion inhibition by naturally occurring substances. vii. The effect of aqueous extracts of some leaves and fruit peels on the corrosion of steel, Al, Zn and Cu in acids, *Br. Corros. J.*, 17(3) (1982) 131–135.
[14] O.K. Abiola, A.O. James, The effects of Aloe vera extract on corrosion and kinetics of corrosion process of zinc in HCl solution, *Corros. Sci.*, 52(2) (2010) 661–664.

[15] B.N. Popov, Corrosion inhibitors, in: B. N. Popov (Ed.), *Corrosion Engineering, Principles and Solved Problems*, Elsevier, Amsterdam, 2015, pp. 581–597.

[16] O. Gharbi, S. Thomas, C. Smith, N. Birbilis, Chromate replacement: What does the future hold?, *Mater. Degrad.*, 12 (2018) 1–8.

[17] J. Aljourani, K. Raeissi, M.A. Golozar, Benzimidazole and its derivatives as corrosion inhibitors for mild steel in 1 M HCl solution, *Corros. Sci.*, 51 (2009) 1836–1843.

[18] A. Režek Jambrak, Non-thermal and innovative processing technologies, in: P. Ferranti, E. M. Berry, J. R. Anderson (Eds.), *Encyclopedia of Food Security and Sustainability*, Elsevier, Oxford, 2019, pp. 477–483.

[19] J. Azmir, I.S.M. Zaidul, M.M. Rahman, K.M. Sharif, A. Mohamed, F. Sahena, M.H.A. Jahurul, K. Ghafoor, N.A.N. Norulaini, A.K.M. Omar, Techniques for extraction of bioactive compounds from plant materials: A review, *J. Food Eng.*, 117 (2013) 426–436.

[20] S. Marzorati, L. Verotta, S.P. Trasatti, Green corrosion inhibitors from natural sources and biomass wastes, *Molecules*, 48 (2019) 1–24.

[21] A. Ninčević Grassino, Plant extracts as a natural corrosion inhibitors of metals and its alloys used in food preserving industry, in A. Méndez-Vilas (Ed.), *Science within Food: Up to Date Advances on Research and Education Ideas*, Food Science Book Series 1, Formatex Research Center, Badajoz, 2017, pp. 185–193.

[22] N. Pramanik, R. Kumar, A. Ray, V.K. Chaudhary, S. Ghosh, Corrosion behavior of mild steel in presence of urea, sodium chloride, potassium chloride and glycine: A kinetic & potentiodynamic polarization study approach. *J. Bio- Tribo-Corros.*, 8 (2022) 112–119.

[23] C. Xu, W. Gao, Pilling-Bedworth ratio for oxidation of alloys. *Mater. Res. Innov.*, 4 (2000) 231–235.

[24] R.E. Bedworth, N.B. Pilling, The oxidation of metals at high temperatures, *J. Inst. Metals*, 29 (1923) 529–582.

[25] S. Shehata Omnia, A. Lobna, A.A. Korshed, *Green Corrosion Inhibitors, Past, Present, and Future*, IntechOpen, Raleka, Croatia, 2017.

[26] O.S. Shehata, L.A. Korshed, A. Attia, Green methods for corrosion control, Chapter 3, in: M. Aliofkhazraei (Ed.), *Corrosion Inhibitors, Principles and Recent Applications*, IntechOpen, 121 2017.

[27] S.J. Zakvi, G.N. Mehta, Acid corrosion of mild steel and its inhibition by swertia aungustifolia study by electrochemical techniques, *Trans. SAEST*, 23(4) (1988) 407–410.

[28] M.A. Al-Khaldi, K.Y. Al-qahtani, Corrosion inhibition of steel by Coriander extracts in hydrochloric acid solution, *J. Mater. Environ. Sci.*, 4(5) (2013) 593–600.

[29] A. Chetouani, B. Hammouti, Thyme as a natural corrosion inhibitor for iron in HCl solutions, *Bull. Electrochem.*, 20(8) (2004) 343–345.

[30] Z.V.P. Murthy, K. Vijayaragavan, Mild steel corrosion inhibition by acid extract of leaves of *Hibiscus sabdariffa* as a green corrosion inhibitor and sorption behaviour, *Green Chem. Lett. Rev.*, 7(3) (2014) 209–219.

[31] A.S. Fouda, G.Y. Elewady, K. Shalabi, S. Habouba, Anise extract as green corrosion inhibitor for carbon steel in hydrochloric acid solutions, *Int. J. Inno. Res. Sci., Engg. Tech.*, 3(4) (2014) 11210–11224.

[32] M. Abdulwahab, A.P.I. Popoola, O.S.I. Fayomi, Inhibitive effect by *Ricinus communis* on the HCl/H_3PO_4 acid corrosion of aluminium alloy, *Int. J. Electrochem. Sci.*, 7 (2012) 11706–11717.

[33] E.E. Ozugie, Inhibition of acid corrosion of mild steel by *Telfaria occidentalis* extract, *Pigm. Resin Technol.*, 34(6) (2005) 321–326.

[34] H. Ashassi-Sorkhabi, D. Seifzadeh, The inhibition of steel corrosion in hydrochloric acid solution by juice of *Prunus cerasus*, *Int. J. Electrochem. Sci.*, 1 (2006) 92–98.

[35] P.S. Pratihar, P. Monika, S. Verma, A. Sharma, *Capparis decidua* seeds: Potential green inhibitor to combat acid corrosion of copper, *Rasayan J. Chem.*, 8(4) (2015) 411–421.

[36] M.E. Ikpi, I.I. Udoh, P.C. Okafor, U.J. Ekpe, E.E. Ebenso, Corrosion inhibition and adsorption behaviour of extracts from piper guineensis on mild steel corrosion in acid media, *Int. J. Electrochem. Sci.*, 7 (2012) 12193–12206.

[37] P.C. Okafor, E.E. Ebenso, U.J. Ekpe, *Azadirachta Indica* extracts as corrosion inhibitor for mild steel in acid medium, *Int. J. Electrochem. Sci.*, 5 (2010) 978–993.

[38] P. Bothi Raja, G. S. Mathur, *Solanum tuberosum* as an inhibitor of mild steel corrosion in acid media, *Iran. J. Chem. Chem. Eng.*, 28(1) (2009) 77–84.

[39] A. Chetouani, B. Hammouti, Corrosion inhibition of iron in hydrochloric acid solutions by naturally henna, *Bull. Electrochem.*, 19(1) (2003) 23–25.

[40] G. Gunasekaran, L.R. Chauhan, Eco friendly inhibitor for corrosion inhibition of mild steel in phosphoric acid medium. *Electrochim. Acta*, 49(25) (2004) 4387–4395.

[41] J. Bhawsar, P.K. Jain, P. Jain, Experimental and computational studies of *Nicotiana tabacum* leaves extract as green corrosion inhibitor for mild steel in acidic medium, *Alexandria Eng. J.*, 54(3) (2015) 769–775.

[42] O. Ime, O.E. Nelson, An interesting and efficient green corrosion inhibitor for Aluminium from extracts of Chlomolaena odorata L. in acidic solution, *J. Appl. Electrochem.*, 40 (2010) 1977–1984.

[43] P.D. Rani, S. Selvaraj, Inhibitive and adsorption properties of punica granatum extract on brass in acid media, *J. Phytology*, 2 (2010) 58–64.

[44] E. Khamis, N. Alandis, Herbs as new type of green inhibitors for acidic corrosion of steel, *Material Wissenschaft und Werkstofftechnik*, 33 (2002) 550–554.

[45] M. Dahmani, S.S. Al-Deyab, A. Et-Touhami, B. Hammouti, A. Bouyanzer, R. Salghi, A. ElMejdoubi, Investigation of piperanine as HCl ecofriendly corrosion inhibitors for C38 Steel, *Int. J. Electrochem. Sci.*, 7 (2012) 2513–2522.

[46] M.G. Sethuraman, V. Aishwarya, C. Kamal, T. Jebakumar, I. Edison, Studies on ervatinine - the anticorrosive phytoconstituent of *Ervatamia coronaria*, *Arabian J. Chem.*, 10(S1) (2017) S522–S530.

[47] V. Thailan, K. Kannan, G. Sivaperumal, Effect of optimum concentration and temperature on inhibition of vasicine for mild steel corrosion in 1N hydrochloric acid, *J. Chem. Pharm. Res.*, 7(8) (2015) 975–986.

[48] D.G. Eyu, H. Esah, C. Chukwuekezie, J. Idris, I. Mohammad, Effect of green inhibitor on the corrosion behavior of reinforced carbon steel in concrete, *ARON J. Eng. Appl. Sci.*, 8 (2013) 326–332.

[49] N. Saidi, H. Elmsellem, M. Ramdani, A. Chetouani, K. Azzaoui, F. Yous, A. Aouniti, B. Hammouti, Using pectin extract as eco-friendly inhibitor for steel corrosion in 1 M HCl media, *Der. Pharma. Chem.*, 7 (2015) 87–94.

[50] C.A. Loto, The effect of mango bark and leaf extract solution additives on the corrosion inhibition of mild steel in dilute sulphuric acid-part I, *Corros. Prev. Control*, 48 (2001) 38–41.

[51] G.H. Awad, Effect of some plant extracts on the corrosion of mild steel in 0.1N hydrochloric acid solutions, In *6th European Symposium on Corrosion Inhibitors*, pp. 385–395, 1985.

[52] P.B. Raja, A.K. Qureshi, A.A. Rahim, H. Osman, K. Awang, Neolamarckiacadamba alkaloids as eco-friendly corrosion inhibitors for mild steel in 1 M HCl media, *Corros. Sci.*, 69 (2013) 292–301.

[53] N.A. Odewunmi, S.A. Umoren, Z.M. Gasem, Watermelon waste products as green corrosion inhibitors for mild steel in HCl solution, *J. Environ. Chem. Eng.*, 3 (2015) 286–296.

[54] A. Ostovari, S.M. Hoseinieh, M. Peikari, S.R. Shadizadeh, S.J. Hashemi, Corrosion inhibition of mild steel in 1M HCl solution by henna extract: A comparative study of the inhibition by henna and its constituents (Lawsone, Gallic acid, α-d-Glucose and Tannic acid), *Corros. Sci.*, 51 (2009) 1935–1949.

[55] A. Bouyanzer, B. Hammouti, L. Majidi, Pennyroyal oil from Menthapulegium as corrosion inhibitor for steel in 1M HCl, *Mater. Lett.*, 60 (2006) 2840–2843.

[56] A.K. Satapathy, G. Gunasekaran, S.C. Sahoo, K. Amit, P.V. Rodrigues, Corrosion inhibition by Justiciagendarussa plant extract in hydrochloric acid solution, *Corros. Sci.*, 51 (2009) 2848–2856.

[57] P.C. Okafor, M.E. Ikpi, I.E. Uwah, E.E. Ebenso, U.J. Ekpe, S.A. Umoren, Inhibitory action of Phyllanthusamarus extracts on the corrosion of mild steel in acidic media, *Corros. Sci.*, 50 (2008) 2310–2317.

[58] M.A. Quraishi, A. Singh, V.K. Singh, D.K. Yadav, A.K. Singh, Green approach to corrosion inhibition of mild steel in hydrochloric acid and sulphuric acid solutions by the extract of Murrayakoenigii leaves, *Mater. Chem. Phys.*, 122 (2010) 114–122.

[59] E. Khamis, N. Alandis, Herbs as new type of green inhibitors for acidic corrosion of steel, *Materialwissenschaftund Werkstoffiechnik*, 33 (2002) 550–554

[60] A.M. Al-Turkustani, S.T. Arab, L.S.S. Al-Qarni, MedicagoSative plant as safe inhibitor on the corrosion of steel in 2.0 M H_2SO_4 solution, *J. Saudi Chem. Soc.*, 15(1) (2011) 73–82.

[61] G. Gunasekaran, L.R. Chauhan, Eco-friendly inhibitor for corrosion inhibition of mild steel in phosphoric acid medium, *Electrochim. Acta*, 49(25) (2004) 4387–4395.

[62] Y.F. Barakat, A.M. Hassan, A.M Baraka, Corrosion inhibition of mild steel in aqueous solution containing H_2S by some naturally occurring substances, *Materialwissenchaft und Werkstofftechnik*, 29 (1998) 365–370.

[63] I.B. Obot, N.O. Obi-Egbedi, Ginseng, a new efficient and effective eco-friendly corrosion inhibitor for aluminium alloy of type AA 1060 in hydrochloric acid solution, *Int. J. Electrochem. Sci.*, 4 (2009) 1277–1288.

[64] A.Y. El-Etre, Inhibition of aluminum corrosion using Opuntia extract, *Corros. Sci.*, 45 (2003) 2485–2495.

[65] S. Rajenderan, J. Jeyasundari, P. Usha, J.A. Selvi, B. Narayanasamy, A.P.P. Regis, P. Renga, Corrosion behavior of aluminium in the presence of an aqueous extract of hibiscus rosasinensis, *Port. Electrochim. Acta*, 27 (2009) 153–164.

[66] E.A. Noor, Potential of aqueous extract of Hibiscus sabdariffa leaves for inhibiting the corrosion of aluminum in alkaline solutions, *J. Appl. Electrochem.*, 39 (2009) 1465–1475.

[67] K. Boumhara, H. Harhar, M. Tabyaoui, A. Bellaouchou, A. Guenbour, A. Zarrouk, Corrosion inhibition of mild steel in 0.5 M H_2SO_4 solution by artemisia herba-alba Oil, *J. Bio- Tribo-Corrosion*, 5(8) (2019) 1–9.

[68] A.Y. El-Etre, Inhibition of acid corrosion of carbon steel using aqueous extract of olive leaves, *J. Colloid Interface Sci.*, 314 (2007) 578–583.

[69] M.M. Fares, A.K. Maayta, M.M. Al-Qudah, Pectin as promising green corrosion inhibitor of aluminum in hydrochloric acid solution, *Corros. Sci.*, 60 (2012) 112–117.

[70] A.K. Larios-Galvez, J. Porcayo-Calderon, V.M. Salinas-Bravo, J.G. Chacon-Nava, J.G. Gonzalez-Rodriguez, L. Martinez-Gomez, Use of Salvia hispanica as an eco-friendly corrosion inhibitor for bronze in acid rain, Anti-Corros, *Methods Mater.*, 64 (2017) 654–663.

[71] H. Gerengi, H.I. Sahin, Schinopsis lorentzii extract as a green corrosion inhibitor for low carbon steel in 1 M HCl solution, *Ind. Eng. Chem. Res.*, 51 (2012) 780–787.

[72] A.N. Grassino, J. Halambek, S. Djakovi'c, S. Rimac Brnci'c, M. Dent, Z. Grabari'c, Utilization of tomato peel waste from canning factory as a potential source for pectin production and application as tin corrosion inhibitor, *Food Hydrocoll.*, 52 (2016) 265–274.

[73] Narayanasamy, Poongothai, P. Rajendran, M. Natesan, N. Palaniswamy, Wood bark oils as vapour phase corrosion inhibitors for metals in NaCl and SO2 environments, *Ind. J. Chem. Technol.*, 12 (2005) 641–647.

[74] M.E. Ibrahim, A.M. El-Khrisy, E.M.M. Al-Abdallah, A.Baraka, evaluation of the inhibitor action of certain natural substances used as corrosion inhibitors-1 in the dissolution of tin in nitric acid, *Mo Metalloberflanche Beschichten won Metall und Kunstsoff* 35, no. 4 (1981) 134–136.

[75] T.K. Soror, New naturally occurring product extract as corrosion inhibitor for 316 stainless steel in 5% HCl, *J. Mater. Sci. Technol.*, 20 (2004) 463–466.

[76] M. Dakmouche, S. Ladjel, N. Gherraf, M. Saidi, M. Hadjaj, M.R. Ouahrani, Inhibition effect of some plant extracts on the corrosion of mild steel in H₂SO₄ medium, *Asian J. Chem.*, 21 (2009) 6176–6180.

[77] J.C. Rocha, J.A. Cunha, P. Gomes, E.D. 'Elia, Corrosion inhibition of carbon steel in hydrochloric acid solution by fruit peel aqueous extract, *Corros. Sci.*, 52 (2010) 2341–2348.

[78] M.E.I- Sayed, O.Y. Mansour, I.Z. Selim, M.M. Ibrahim, Identification and utilization of banana plant juice and its liquor as anti-corrosive mate-rials, *J. Sci. Ind. Res.*, 60 (2001) 738–747.

[79] A. Bouyanzer, B. Hammouti, L. Majidi, Pennyroyal oil from *Mentha pulegium* as corrosion inhibitor for steel in 1M HCl, *Mater. Lett.*, 60(23) (2006) 2840–2843.

[80] Y. EL- Ouadi, A. Bouratoua, A. Bouyenzer, Z. Kabouche, R. Touzani, H. EL- Msellem, B. Hammouti, A. Chetouani, Effect of Athamanta sicula oil on inhibition of mild steel corrosion in 1M HCl, *Der Pharma Chemica*, 7(2) (2015) 103–111.

[81] M. Benabdellah, M. Benkaddour, B. Hammouti, M. Bendahhou, A. Aouniti, Inhibition of steel corrosion in 2M H₃PO₄ by artemisia oil, *Appl. Surf. Sci.*, 252(18) (2006) 6212–6217.

[82] O. Ouachikh, A. Bouyanzer, J.M. Desjobert, J. Costa, B. Hammouti, L. Majidi, Application of essential oil of Artemisia herba alba as green corrosion inhibitor for steel in 0.5 M H₂SO₄, *Surf. Rev. Lett.*, 16 (2009) 49–54.

[83] N. Lahhit, A. Bouyanzer, J.-M. Desjobert, B. Hammouti, R. Salghi, J. Costa, C. Jama, F. Bentiss, L. Majidi, Fennel (*Foeniculum Vulgare*) essential oil as green corrosion inhibitor of carbon steel in hydrochloric acid solution, *Portugaliae Electrochimica Acta*, 29(2) (2011) 127–138.

[84] M. Znini, L. Majidi, A. Laghchimi, J. Paolini, B. Hammouti, J. Costa, A. Bouyanzer, S.S. Al-Deyab, Chemical composition and anticorrosive activity of *warionia saharea* essential oil against the corrosion of mild steel in 0.5 M H₂SO₄, *Int. J. Electrochem. Sci.*, 6 (2011) 5940–5955.

[85] G. Cristofari, M. Znini, L. Majidi, A. Bouyanzer, S.S. Al-Deyab, J. Paolini, B. Hammouti, J. Costa, Chemical composition and anti-corrosive activity of *pulicaria mauritanica* essential oil against the corrosion of mild steel in 0.5 M H₂SO₄, *Int. J. Electrochem. Sci.*, 6 (2011) 6699–6717.

[86] S. Rekkab, H. Zarrok, R. Salghi, A. Zarrouk, LH. Bazzi, B. Hammouti, Z. Kabouche, R. Touzani, M. Zougagh, Green corrosion inhibitor from essential oil of eucalyptus globulus (myrtaceae) for C38 steel in sulfuric acid solution, *J. Mater. Environ. Sci.*, 3(4) (2012) 613–627.

[87] M. Znini, G. Cristofari, L. Majidi, A. Ansari, A. Bouyanzer, J. Paolini, J. Costa, B. Hammouti, Green approach to corrosion inhibition of mild steel by essential oil leaves of asteriscus graveolens (forssk.) in sulphuric acid medium, *Int. J. Electrochem. Sci.*, 7 (2012) 3959–3981.

[88] J. Bhawsar, P.K. Jain, P. Jain, A. Soni, Anticorrosive activity of *Glycine max* (L) oil against the corrosion of mild steel in acidic medium, *Int. J. Res. Chem. Environ.*, 3(4) (2013) 68–74.

[89] L. Afia, R. Salghi, L. Bammou, El. Bazzi, B. Hammouti, L. Bazzi, A. Bouyanzer, Anti-corrosive properties of Argan oil on C38 steel in molar HCl solution, *J. Saudi Chem. Soc.*, 18 (2014) 19–25.

[90] L. Afia, O. Benali, R. Salghi, E.E. Ebenso, S. Jodeh, M. Zougagh, B. Hammouti, Steel corrosion inhibition by acid garlic essential oil as a green corrosion inhibitor and sorption behavior, *Int. J. Electrochem. Sci.*, 9 (2014) 8392–8406.

[91] R. Salghi, D.B. Hmamou, O. Benali, S. Jodeh, I. Warad, O. Hamed, E.E. Ebenso, A. Oukacha, S. Tahrouch, B. Hammouti, Study of the corrosion inhibition effect of pistachio essential oils in 0.5 M H₂SO₄, *Int. J. Electrochem. Sci.*, 10 (2015) 8403–8411.

[92] A. Lecante, F. Robert, P.A. Blandinières, C. Roos, Anti-corrosive properties of *S. tinctoria* and *G. ouregou* alkaloid extracts on low carbon steel, *Curr. Appl Phys.*, 11(3) (2011) 714–724.

[93] S.H.S. Dananjaya, M. Edussuriya, A.S. Dissanayake, Inhibition action of lawsone on corrosion of mild steel in acidic media, *TOJSAT: Online J. Sci. Technol.*, 2(2) (2012) 32–36.

[94] M. Dahmani, S.S. Al-Deyab, A. Et-Touhami, B. Hammouti, A. Bouyanzer, R. Salghi, A. ElMejdoubi, Investigation of piperanine as HCl ecofriendly corrosion inhibitors for C38 steel, *Int. J. Electrochem. Sci.*, 7 (2012) 2513–2522.

[95] M.G. Sethuraman, V. Aishwarya, C. Kamal, T. Jebakumar, I. Edison, Studies on ervatinine - the anticorrosive phytoconstituent of *Ervatamia coronaria*, *Arabian J. Chem.*, 10(S1) (2017) S522–S530.

[96] V. Thailan, K. Kannan, G. Sivaperumal, Effect of optimum concentration and temperature on inhibition of vasicine for mild steel corrosion in 1N hydrochloric acid, *J. Chem. Pharm. Res.*, 7(8) (2015) 975–986.

[97] P. Dohare, D.S. Chauhan, B. Hammouti, M.A. Quraishi, Experimental and DFT investigation on the corrosion inhibition behavior of expired drug lumerax on mild steel in hydrochloric acid, *Anal. Bioanal. Electrochem*, 9(6) (2017) 762–783.

[98] A.A. Inemesit, O.O. Nnanake-Abasi, Inhibition of mild steel corrosion in hydrochloric acid solution by ciprofloxacin drug. Hindawi Publishing Corporation, *Int. J. Corr.*, Article ID 301689 (2013) 1–5.

[99] Y. Liang, C. Wang, J. Li, L. Wang, J. Fu, The penicillin derivatives as corrosion inhibitors for mild steel in hydrochloric acid solution: Experimental and theoretical studies, *Int. J. Electrochem. Sci.*, 10 (2015) 8072–8086.

[100] B.M. Prasanna, P.B. Mokshanatha, N. Hebbar, T.V. Venkatarangaiah, H.C. Tandon, Ketosulfone drug as a green corrosion inhibitor for mild steel in acidic medium, *Ind. Eng. Chem. Res.*, 53 (2014) 8436–8444.

[101] X. Xu, A. Singh, Z. Sun, K.R. Ansari, Y. Lin, Theoretical, thermodynamic and electrochemical analysis of biotin drug as an impending corrosion inhibitor for mild steel in 15% hydrochloric acid, *R. Soc. Open Sci.*, 4 (2017) 170933.

[102] U.O. Stanley, U.O. Pearl, Corrosion inhibition of mild steel in 0.1M hydrochloric acid media by Chloroquine Diphosphate, *ARPN J. Eng. Appl. Sci.*, 7(3) (2012) 272–276.

[103] P. Singh, D.S. Chauhan, K. Srivastava, V. Srivastava, M.A. Quraishi, Expired atorvastatin drug as corrosion inhibitor for mild steel in hydrochloric acid solution, *Int. J. Ind. Chem.*, 8 (2017) 363–372.

[104] M. Sangeetha, S. Rajendran, J. Sathiyabama, A. Krishnaveni, P. Shanthy, N. Manimaran, B. Shyamaladevi, Corrosion inhibition by an aqueous extract of phyllanthus amarus, *Port. Electrochim. Acta.*, 29 (2011) 429–434.

[105] S. Rajendran, M. Agasta, R. Bama Devi, B. Shyamala Devi, K. Rajam, J. Jeyasundari, Corrosion inhibition by an aqueous extract of Henna leaves (Lawsonia Inermis L), *Mater. Prot.*, 50 (2009) 77–84.

[106] A.S. Yaro, A.A. Khadom, R.K. Wael, Apricot Juice as Green corrosion inhibitor of mild steel in phosphoric acid, *Alexandria Eng. J.*, 52 (2013) 129–135.

[107] C.A. Loto, O.O. Joseph, R.T. Loto, A.P.I. Popoola, Inhibition effect of vernonia amygdalina extract on the corrosion of mild steel reinforcement in concrete in 3.5M NaCl environment, *Int. J. Electrochem. Sci.*, 8 (2013) 11087–11100.

[108] S. Harikrishna, A. Begum, K. Roy, Performance of fenugreek (Trigonella foenum graecum) seed extract as inhibitor on mild steel under corrosive medium- A statistical view, *Int. J. Chem. Tech. Res.*, 5 (2013) 1829–1834.

25 Synergism in Anticorrosive Phenomenon of Phytochemicals/ Plant Extracts

Madhulata Shukla
Veer Kunwar Singh University

25.1 INTRODUCTION

The decay of materials by the chemical reactions occurring between the materials/ metals and their reactive atmosphere is termed corrosion. It is a dangerous incidence producing an overwhelming effect on technological and industrial applications, predominantly in the field of oil and gas manufacturing industries [1]. It is triggered mainly in an aqueous media and occurs due to reduction–oxidation (i.e. redox) reactions taking place in gas and oil production, pipeline structures, and management [2]. Redox reaction consists of both reduction–oxidation processes occurring simultaneously. Redox reaction mainly affects the outer surface of materials (mainly metals) that are in contact with their environment, influencing the discharge of electrons by dissolving metal and their sequential migration of electrons to other sites on the surface responsible for the hydrogen ions to be reduced, which results in deterioration of the metal surface. The process of corrosion involves three things, namely, an anode, a cathode, and an electrolyte. At the anode, free electrons are generated as a consequence of the corrosion of metals, which passes through the electrolytic toward the cathode. At the cathode, hydrogen ions are reduced to form hydrogen gas. In many industrialized countries, corrosion has become a matter of vast concern. Crude oil and natural gas generally comprise numerous contaminated materials that are naturally corrosive. Around the world, oil companies are spending a huge proportion of money on tackling corrosion [3]. Hence, monitoring the decomposition of metals in oil and gas refineries has become a very important issue to be taken care of concerning environmental, technical, and economic aspects to protect against enormous expenses on materials and equipment. The acidic environment in metals and alloys in industries promotes the easy corrosion reaction rate, as the acid can affect the metal's surface through the interfacial reaction, and it results in the dissolution of metals to produce the ions and corrodes the heavy instruments and machineries in industries. In industries, acid solutions are used

514

DOI: 10.1201/9781003394631-25

for various purposes such as acid cleaning, acid pickling, industrial acid descaling, and mill scale removal from metal surfaces. These processes involve extremely concentrated acid solutions, which results in the interruption of metallic apparatuses along with the formation of surface imperfections such as rust and scale. Hence, corrosion is considered to be a critical and worldwide problem that must be taken care of to keep the country's economy increasing and protect the environment for safe living. Various factors are responsible for creating corrosion problems. Some of the environmental causes responsible for huge corrosion reactions are shown in Figure 25.1. As shown in Figure 25.1, temperature, humidity, rainfall, nature of the electrolyte, pH, level of pollution, and dust particles in the atmosphere play an important role in deciding the reaction rate of corrosion in a particular field. Furthermore, along with the above-mentioned aspects, a few microorganisms such as bacteria and fungi inside a biofilm on steel are also responsible for promoting the corrosion rate.

Corrosion has a huge effect on the economy, environmental impact, and safety of a country. So, corrosion has been a serious problem for any country and must be taken care of to reduce the corrosion rate. Corrosion has a huge impact on infrastructure, economy, as well as health and safety issues. A few impacts of corrosion are shown in Figure 25.2.

Several approaches and techniques are applied in science and technology to tackle this corrosion problem. Variable material selection, deposition of coating, electrochemical methods, and application of different corrosion inhibitors are most common [4,5]. The use of corrosion inhibitors (CIs) is the cheapest and most convenient use to reduce the corrosion rate in industries and factories. CIs are substances that are used in small concentrations to corrosion media to reduce the metal degradation rate. Chromates and their derivatives have the highest corrosion inhibitor property but due to the huge toxicity in nature of these chemicals toward the human life and ecosystem, their use has been restricted. Inhibitors can either be obtained from natural resources or can be synthesized using the available synthetic methods or by modifying the available methods. Various factors are responsible for corrosion in metals and alloys. Recently, the practical use of carbohydrates and their derivatives

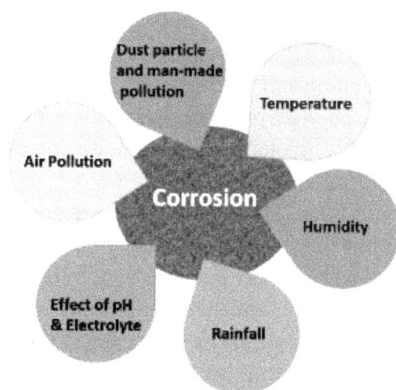

FIGURE 25.1 Environmental causes of corrosion.

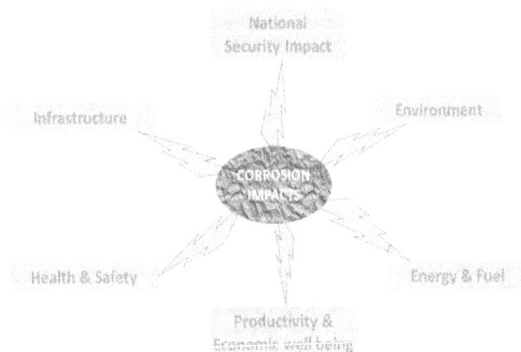

FIGURE 25.2 Impact of corrosion.

has been an emerging effort to be used as CIs that decrease environmental pollution [6,7]. Organic CIs having a hydrophobic nature show limited solubility due to the presence of aromatic rings and non-polar hydrocarbon chains, which unfavorably affect the shielding proficiency. Hence, a recent study on corrosion science is mainly focused on the production of different CIs that contain hydrophilic polar functional groups in their molecular structure. Although the practice of using CIs is the only option for monitoring metallic deterioration in numerous corrosive environments, organic corrosion inhibitor usage is the best and most profitable technique to avoid corrosion. Organic CIs are applied for various industrial utilization processes such as pickling process, oil, and gas field lines, boiler cleaner, oil well acidizing, vapor phase coating, paints and coating, cooling systems, and lubricants and are shown in Figure 25.3.

However, numerous problems are associated with these organic inhibitors. The biggest challenge is its very less solubility in polar solvents [8]. Consequently, present investigations in the field of corrosion science are focused on creating CIs that comprise mainly hydrophilic polar functional groups in their structure. But, several complications were provoked by the usage of small organic molecules and inorganic CIs, and hence, CIs containing polymeric chains came to the attention. Extracts of natural products which comprise alkaloids, polyphenols, quinine, nicotine, terpenes, and other functional groups such as carboxylic acids, and having atoms similar to C, N, O, S, etc., provoke adsorption by developing a thin film coating on metal's surface to protect it and hinder corrosion. Inorganic chemicals mainly chromates and their derivatives are known for their strong inhibitory properties but due to their toxic and harmful nature toward human life and the ecosystem, their application has been restricted. Instead of these harmful chemicals, natural products such as plant extracts are widely used, and it is cost-effective as well. Green chemistry research and its applications in CIs are being promoted in the twenty-first century. In the last few decades, the perception of "Green" has enormously stimulated almost all areas of science and technology, where the attraction for plant extracts as a metallic corrosion inhibitor has got significant attention. Recent research contributions in corrosion chemistry area are related to plant extract inhibitors that are eco-friendly, particularly based on natural product

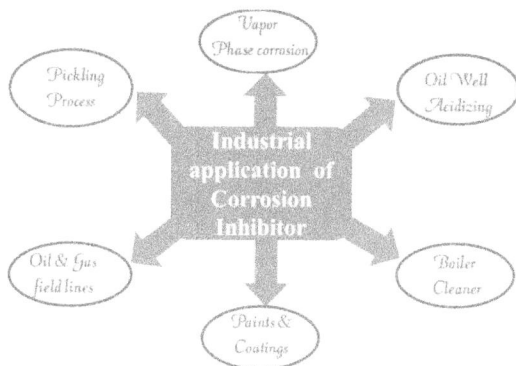

FIGURE 25.3 Industrial application of organic corrosion inhibitors.

materials, and biopolymers. Polymers, especially biopolymers such as polyethylene glycol (PEG), ionic liquids (ILs), natural rubbers, suberin, cutin, cutan, and melanin, are also becoming a point of interest due to the various derivatives they form. Owing to the low vapor pressure of ionic liquids (ILs) and polyethylene glycol (PEG), they are considered eco-friendly substitutes. The chemicals synthesized using water, ILs, or supercritical CO_2 (i.e. green solvents) are also regarded as eco-friendly chemicals. Many reviews report that these materials are extensively applied as metallic CIs for a large number of corrosive electrolytes. The growing requirement for the production of eco-friendly CIs forces many researchers to work with corrosion chemistry to produce the utmost demanded sustainable CIs. Scientists are also searching for a possible alternative in ILs as a green option to use them as green CIs apart from inhibitors attained from plants. Natural and eco materials used as CIs are shown in Figure 25.4.

This chapter deals with various types of green/sustainable plant extract CIs and their applications in numerous industries and environments. Plants are abundant with enormous active phytochemicals, and they are considered as an ideal substitute to overcome the use of traditional hazardous and toxic chemicals [9,10]. Various parts of plants such as roots, leaves, seeds, and fruits can be easily extracted through simple procedures and techniques and can be applied as CIs. It has been reported by Verma et al. that plant extracts have exceptionally high corrosion inhibitor efficiency, especially leaf extract, because of the presence of an abundant source of phytochemicals compared to other parts [9]. Owing to its easy availability, renewability, eco-friendly, and simple extraction techniques, it is suitable to use for several metals in different electrolytic media. The use of plant extract as a corrosion inhibitor reveals the significance of the go-green policy in the field of science and technology [11–16]. In addition to corrosion in oil and gas industries, several other industries are also promoting corrosion caused by harmful chemicals like sulfur and naphthenic acid. It must be avoided heavily to keep our environment suitable for the standard of living of humans. Hence, corrosion is considered to be a worldwide problem that must be taken care of as a serious problem and immediate action must be taken for safety purposes. On the contrary, natural products such as biopolymers and plant extracts are available as Corrosion

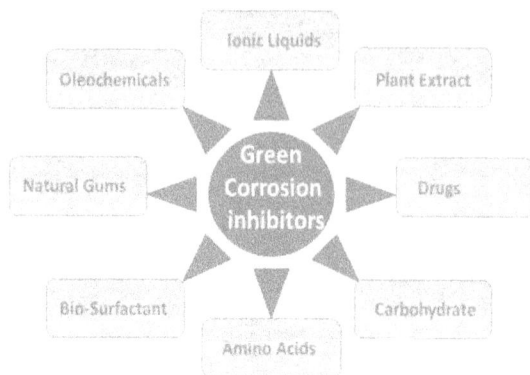

FIGURE 25.4 Natural materials used as green corrosion inhibitors.

inhibitors (CIs), which are widely used and also cost-effective. Green chemistry research and its application in the field of corrosion science are increasing day by day. Different parts of plants are used to tackle the corrosion problems originating due to the degradation of metal surfaces in industries. Several bioactive chemicals present in different parts of the plants act as CIs. This is attracting researchers to work in the field of corrosion chemistry and CIs. More than hundreds of research papers have been published in the last few years which motivate us to write a book chapter on plant extract as CIs [17].

25.2 CONVENTIONAL CIs

Several steps are being taken to prevent corrosion and to control it to its maximum extent. This will lead to minimizing the financial losses and increase the GDP of a country and will also help in improving the industrial safety and machines' durability. First and foremost, the option for minimizing the corrosion process is to choose a suitable material based on its design and application. Secondly, environmental conditions such as temperature, humidity, nature of electrolyte used, and pH should be controlled and a suitable corrosion inhibitor of the required amount must be added as per our requirement. Unraveling the machinery parts from corrosive media using protective coatings [18,19] and compensating for the electrons that have been lost through cathodic protection [20] are among the techniques to avoid and control the corrosion process. Using corrosion inhibitors to a minimum amount is also one of the most suitable techniques for reducing the corrosion rate. CIs slow down the rate of corrosion of exposed metals by getting adsorbed on the surface of metals, forming a protective layer and interacting with the anodic/cathodic reaction sites, and lowering the oxidation-reduction reactions [21]. Depending upon the type of electrode used and reactions taking place at the cathode/anode, the type of inhibitors is chosen for different types of materials. Some CIs are very specific to a particular material. It may happen that one CI used for a specific material may not work well for some other material. It also depends on the environmental conditions, types of electrolytes used, working temperature, and many more factors.

So, it is very important to choose the right CI for a specific material to prevent the corrosion rate efficiently. The use of CIs is the best and most effective method used for preventing or lowering the corrosion reaction rate. General organic and inorganic CIs are extensively used due to their easy availability, simple applicability, and outstanding efficiency at lower concentrations[9]. Inorganic CIs such as chromates, dichromate, phosphates, and arsenates are generally used in neutral environments, whereas organic CIs containing polar functional groups and heteroatoms like N, O, and S along with π electrons and hydrophobic groups are used in acidic conditions that prevent the corrosive moiety away from the material surfaces [22]. Fatty amides, polymers, pyridines, and imidazolines are the organic chemicals used as inhibitors. Organic chemicals get adsorbed to the surface of the material via one of the following mechanisms [23]:

a. Hydrolysis of organic compounds results in the formation of fatty dicarboxylic acids, which consequently form the insoluble ferrous salts on the metal surface
b. Donor-acceptor interactions arising among the π-electrons of CI molecules and the vacant d-orbital of metal substrate
c. Exchange of unshared electron pairs of the heteroatoms with the vacant d-orbital of metal substrate
d. Interaction of vacant orbital of the heteroatoms with the d-orbital electrons of iron substrate

Inorganic inhibitors can work at higher temperatures as well as compared to organic inhibitors [17]. But, these organic and inorganic chemicals are very much hazardous and toxic to the environment and human health. Heavy release of harmful chemicals during the synthesis of these types of inhibitors also causes several long-term diseases such as cancer and tuberculosis [24]. These chemicals also affect the soil and marine life. Hence due to the toxicity of organic and inorganic CIs, researchers are switching more toward green corrosion inhibitors (GCIs).

25.3 GREEN CORROSION INHIBITORS (GCIs)

As the concept of green chemistry and sustainable development is growing day by day, eco-friendly chemicals are getting attention in the field of science and engineering. Researchers are focusing mostly on green synthesis in every field. Also in the field of corrosion chemistry, GCIs are getting huge attention day by day. GCIs include several natural and biodegradable products such as plant extract, natural honey, oil, drugs, and herbs. A huge number of synthetic GCIs such as ionic liquids, biopolymers, and surfactant have been reported in the literature [25–28]. Some natural and GCIs have been shown above in Figure 25.4. Important criteria to describe a chemical as green corrosion inhibitor were set by legislative bodies of the Paris Commission (PARCOM) and North Sea (UK, Norway, Netherlands, Denmark). Green chemicals are those that are biodegradable, are non-bio-accumulative, and have zero or very less marine toxicity levels [29,30].

25.3.1 Plant Extract as Green CIs (GCIs)

The application of plant extract as a metallic corrosion inhibitor is growing promptly these days. It is due to easy availability, cost-effectiveness, renewability, and eco-friendly nature as compared to organic and inorganic inhibitors [31]. Several reviews are available explaining the different parts of plants such as root, leaf, stem, bark, pulp, and fruits used as sustainable corrosion inhibitors to prevent corrosion in different metals and alloys [32–38]. Recently, scientists are approaching the potential use of agro-industrial process byproducts and agricultural wastes as GCIs at a large scale [39].

Other than nutritive components present in different parts of plants, it carries several other types of chemicals that are responsible for plants' smell, color, and taste and are called phytochemicals. Terpenoids, organic acids, and polyphenols are different classes of phytochemicals present in plants. Different types of phytochemicals are present depending on the plant components selected for extraction. Earlier it has been stated in the literature that the phytochemical contents varied by plant part. Leaf extract contains flavonoids, alkaloids, saponins, organic acid, tannins, and anthraquinones, whereas stem extract is entirely rich in alkaloids, tannins, and anthraquinones [40]. Among different varieties of plant extracts, leaf extracts are reported to be the best and overall protective covering, even at lower concentrations also. It is due to the fact that phytochemicals are mainly manufactured in leaves, where they are produced in the presence of sunlight, water, and CO_2 [41,42]. Alkaloids, flavonoids, saponins, phytosterol, glycosides, tannins, anthraquinones, triterpenes, phenolic compounds, and phlobatannins are the few most common phytochemicals that have good corrosion-inhibiting effect. Many such phytochemicals contain polar functional groups like hydroxyl (-OH), sulfide (-S-), thiol (-SH), nitro (NO_2), ester ($-COOC_2H_5$), amide ($-CONH_2$), carboxylic acid (-COOH), and amino ($-NH_2$) that promote the absorption properties on materials and form the protective coating [43,44]. Molecular structure of a few phytochemicals is shown in Figure 25.5.

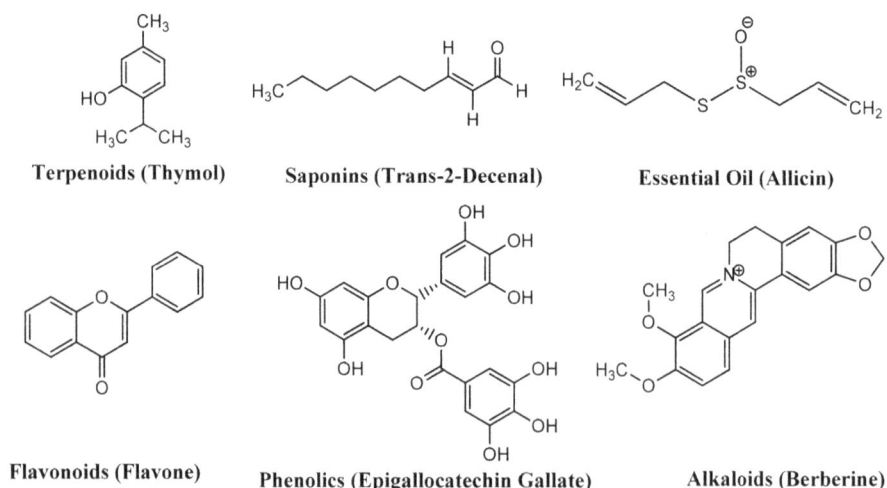

Terpenoids (Thymol)   Saponins (Trans-2-Decenal)   Essential Oil (Allicin)

Flavonoids (Flavone)   Phenolics (Epigallocatechin Gallate)   Alkaloids (Berberine)

FIGURE 25.5 Molecular structure of a few phytochemicals.

These plant extract GCIs, being non-toxic in nature and cost-effective, are used massively replacing the conventional organic and inorganic toxic chemicals.

Different functional groups in GCIs in solution state play an important role in corrosion reaction. Researchers analyze the chemical properties of GCIs using FTIR, UV-visible, or X-ray photoelectron spectroscopy. FTIR method is used to analyze the interaction of GCIs with different corrosive materials. Depending upon the different chemical properties of GCIs, a particular GCI is chosen for specific materials.

25.3.2 APPLICATION OF PLANT-BASED GCIs IN CORROSIVE MEDIA

25.3.2.1 HCl Medium

Hydrochloric acid (HCl) is a well-known acid used in the pickling and acidification of oil wells in industries. HCl is preferred for pickling over other acids due to its greater surface quality within shorter intervals of time and at low temperatures. 5–15% HCl at around 30°C is preferable to carry out the pickling process, but if HCl is used above this temperature, a huge amount of hydrogen and chlorine gas is released and it is dangerous to the environment as well as human safety also. The main aim of using plant extract in an HCl medium is to prevent metal corrosion during pickling. Several reports are available to explain the use of plant extract in an HCl medium in the range of 0.1–2 mol L^{-1} [45–47]. It was found by Raja et al. in 2013 that alkaloids extracted from *Alstonia angustifolia var. latifolia* leaves were an effective GCI for mild steel [46]. Alkaloids obtained from *Retama monosperma* (L.) Boiss. seeds extract was found to be useful for avoiding carbon steel dissolution in a 1 mol L^{-1} HCl environment [47]. The inhibition effect of *Neolamarckia cadamba* crude extract was investigated by Raja et al. for steel corrosion in a 1 mol L^{-1} HCl solution condition [48]. The temperature and concentration of *Elaeoselinum thapsioides* (BEET) butanolic extract affected its inhibition efficiency, as well as the corrosion rate of carbon steel in 1 mol L^{-1} HCl. It was also presumed that an increase in several BEET molecules leads to shielding the metal's active surface and hence reduces the metal's reactivity in the HCl medium. It has been observed by Benahmed et al. that the inhibition efficiency increases drastically and corrosion activity decreases rapidly with the increase in the concentration of BEET [45]. An Asian plant named *Esfand* was found to be rich in an amine functional group with abundant electron donor atoms, which were found to be very effective in providing suitable adsorption conditions on the steel surface in 1 mol L^{-1} HCl solution [49]. Recently, the inhibitive properties of *Nepeta pogonesperma* extract were examined by Shahini et al. in 2022 for mild steel in a 1 mol L^{-1} HCl solution [50]. It was also analyzed from SEM and AFM data that with an increase in GCI concentration, extra coverage of the steel surface was found, as clear from the smoother surface morphology obtained with fewer defects.

The mechanism of *Nepeta pogonesperma* extract inhibitor adsorption on the steel surface in the presence of HCl solution has been proposed and shown in reference by Zakeri et al. [17].

It has been clearly explained that interaction between the steel surface and HCl solution results in these areas being positively charged. Hence Cl⁻ ions from HCl get adsorbed on the steel surface resulting from positive-to-negative change occurring at the interface charge. Hence protonated moiety proceeds for physical adsorption via electrostatic interaction and chemical adsorption proceeds by the donation of electrons to the d-orbital of substrate moiety. *Aniba rosaeodora* alkaloidic extract has been studied in a 1 mol L^{-1} HCl medium by Chevalier et al. for a C38 steel substrate. This GCI was found to show an effective inhibiting effect to prevent corrosion on a steel substrate [51]. Several reports are available explaining the inhibiting effect of GCIs in HCl solution. Detailed evaluation of GCIs' performance in HCl medium on different substrates has been explained by Zakeri et al. through the bar diagram [17].

25.3.2.2 H$_2$SO$_4$ Medium

Compared with HCl, sulfuric acid (H$_2$SO$_4$) was found to be cheaper and used in many industries for acid pickling and cleaning during metal processing techniques. H$_2$SO$_4$ is also used for synthesizing copper and zinc products. But in oil well acidification, it is rarely used, as it forms insoluble sulfate sludges as byproducts. A large number of research papers are available explaining the application of plant extract as GCIs in the H$_2$SO$_4$ environment [52–55]. It has been shown by Odewunmi et al. using electrochemical techniques that watermelon rind extract (WMRE) acts as a corrosion inhibitor in HCl and H$_2$SO$_4$ solutions for mild steel [54]. Watermelon, *Citrullus lanatus* reported comprising of huge quantity of anti-nutrients namely saponin, alkaloids, phytate, hydrogen cyanide, tannins, phenol, oxalate, and flavonoids in fresh and dried watermelon peel (WMP), watermelon pulp (WMPu) and watermelon rind (WMR) [55]. The pulp of watermelon is the most useful and valuable part used in food industries. The effect of *Litchi chinesis* as GCIs has been investigated by Ramananda et al. in 0.5 mol L^{-1} H$_2$SO$_4$ solution using electrochemical techniques [56]. As per the Tafel polarization test, litchi peel extract was found to hamper the cathodic-anodic reaction and behaves as a mixed-type corrosion inhibitor. The inhibitor efficiency of *Salvia officinalis* L. in 0.5 mol L^{-1} H$_2$SO$_4$ solution was inspected on steel substrate by Khiya et al. and found to act as mixed-type GCI and hampers the reaction rate at both cathode and anode [53]. Muthukrishnan et al. in 2017 reported that *Lannea coromandelica* leaf extract (LCLE) contains polyphenols in a substantial amount and was tested to show good inhibitor properties in H$_2$SO$_4$ environment for mild steel [57]. Oguzie in 2008 investigated the leaf extracts of *Occimum viridis*, *Telferia occidentalis*, *Azadirachta indica*, and *Hibiscus sabdariffa* on mild steel in 2 M HCl and 1 M H$_2$SO$_4$ acidic solutions using a gasometric technique at 30°C and 60°C temperature [58]. Prasad et al. in 2022 explored *Cinnamoum tamala* leaves extract as a GCI for low carbon steel in a 0.5 mol L^{-1} H$_2$SO$_4$ medium and found that its inhibition efficiency is 96.76% as established by both the electrochemical tests as well as the gravimetric measurements [59].

It has been reported in many studies that though GCIs show sufficient results in H$_2$SO$_4$, much more effective corrosion inhibition property is observed in the HCl medium. It is predominantly due to the adsorption efficiency of chloride and sulfate ions in the adsorption of GCIs. Cl$^-$ ions are comparatively less hydrated than SO$_4^{2-}$ ions, and hence, Cl$^-$ ions display a higher capacity to adsorb on substrate surfaces [54,57]. More adsorption ability of Cl$^-$ leads to more negative charge on the surface as compared to SO$_4^{2-}$ and hence promotes the adsorption of Cl$^-$ containing species [54]. Temperature effects the rate of corrosion hugely in the case of H$_2$SO$_4$ as stated by Kobylin et al. The corrosion rate increases with the increase in temperature and corrodes the steel materials more aggressively when H_2SO_4 solution is used for the acid pickling method [60].

25.3.3 SYNERGISTIC ACTION WITH PLANT EXTRACT

Synergy means combination or coordination of the activity of two or more agents to produce a joint effect greater than the sum of their separate effects. It has been reported in many studies that the inhibition property of plant extracts can be upgraded by adding cationic or anionic salts to them. The efficiency of plant extract CIs has been improved using halide ions and has been reported to be a successful strategy [61,62]. Many studies are available explaining the positive effect of the addition of halogen and zinc ion-based salts to plant extract which increases the inhibition effect. This chapter represents a short review of the synergistic effect of different types of salt on the corrosion-inhibitive action of plant extract [61]. Rajeswari et al. have reported that *Eleusine aegyptiaca* and *Croton rottleri* leaf extracts were found to show more corrosion inhibition efficiency when leaf extract solutions are combined with halide ions [63]. Various active phytochemicals are present in plants which increases synergistic inhibition. Synergistic inhibition of halide ions was examined by Oguzie at al. in 2008 on the leaves extracts of *Telferia occidentalis*, *Occimum viridis*, *Azadirachta indica*, and *Hibiscus sabdariffa*, and seed extracts of *Garcinia kola* in acidic solutions containing steel [58]. Also, it has been reported by Eddy et al. that KCl synergistic interactions with *Lasianthera africana extract* increase the adsorption of the inhibitor while KBr and KI hamper the inhibition property [64]. Ethanol extract of *Lasianthera africana* was investigated for its inhibition and adsorption properties under H_2SO_4 acidic solution for mild steel using thermometric, gravimetric, gasometric, and infrared methods and was found to behave as a good inhibitor for corrosion of mild steel in H_2SO_4 condition. Alkaloids, tannin, saponin, flavonoid, and anthaquinone present in plant extracts are the reasons to enhance the inhibitive properties in acidic conditions. The inhibitory efficiency of steel in acidic solutions was found to be increased using the mixture of iodide ions with *Artemisia halodendron* [65], *Cassia italic* [66], *rice husk* [67], *bamboo leaf extract* [68]. Cang et al. in 2017 studied using electrochemical potentiometric polarization methods that the *A. halodendron* extract when mixed with iodide ion shows increased efficiency of the inhibition effect and it increases with an increase in the concentration of the iodide ion [65]. Pramudita et al. in 2019 studied the synergic effect of rice husk extract combined with potassium iodide for bio-corrosion in 1 M H_2SO_4 solution using weight loss method, and it has been reported that the inhibition effect increases with an increase in the concentration of the iodide ion [67]. These days the conjugation of natural plant extract with inorganic CIs has been an interesting topic of research. Several reports are available explaining the use of different plant root and leaves extracts along with inorganic CIs in acidic medium. Recently, a mixture of *Cardaria draba* leaves (CDL) extract along with potassium iodide solution was found to show an inhibitor efficiency of 94% [69] which was found to show lower efficiency individually with *C. draba* leaves [70] and potassium iodide solution [71]. The inhibitor efficiency at 60°C for CDL was found to be maximum (94.3%) for 10 and 2.5 mL L^{-1} for KI solution (94%). *A. halodendron* leaves extract was observed to behave as an eco-friendly inhibitor in 1 M HCl solution for the corrosion of the mild steel. The addition of iodide ion to *A. halodendron* leaves extract increases the inhibitory performance to a larger amount [65]. Finally, it may be stated that the mixture of plant extract combined with inorganic salt shows greater corrosion inhibition efficiency due to the protective layer covering the metal surfaces [72].

25.4 CONCLUSION

This chapter provides important information regarding the roles of natural CIs mainly the plant extracts. Plant extracts are attracting enormous attention these days due to their natural, eco-friendly, and biological origin. Plant extracts containing phytochemical complexes interact strongly with the metallic surface through their electron-rich sites and protect the metal surfaces from corrosion. Phytochemicals combine with polar functional groups and conjugate with multiple bonds to protect the metals from corrosion in acidic medium. This chapter covers a descriptive collection of recent outcomes and new findings on plant extracts as CIs. Synergic CIs efficiency of different plant extracts with inorganic salts have been compiled and reported to show better results to protect the steel or metal surfaces in acidic condition as compared to plant extract alone.

ACKNOWLEDGMENT

MLS acknowledges the financial assistance from UGC, India [F.30–446/2018 (BSR)].

REFERENCES

[1] P. R. Roberge, and R. Pierre, *Handbook of Corrosion Engineering Library of Congress Cataloging-in-Publication Data*. McGraw-Hill Education
[2] L. T. Popoola, A. S. Grema, G. K. Latinwo, and B. Gutti, "Corrosion problems during oil and gas production and its mitigation," *Int. J. Ind. Chem.*, vol. 4, pp. 1–15, 2013.
[3] L. Zhao, H. K. Teng, Y. S. Yang, and X. Tan, "Corrosion inhibition approach of oil production systems in offshore oilfields," *Mater. Corros.*, vol. 55, no. 9, pp. 684–688, 2004, doi:10.1002/maco.200303789.
[4] S. H. Alrefaee, K. Y. Rhee, C. Verma, M. A. Quraishi, and E. E. Ebenso, "Challenges and advantages of using plant extract as inhibitors in modern corrosion inhibition systems: Recent advancements," *J. Mol. Liq.*, vol. 321, p. 114666, Jan. 2021, doi:10.1016/J.MOLLIQ.2020.114666.
[5] C. Verma, E. E. Ebenso, M. A. Quraishi, and C. M. Hussain, "Recent developments in sustainable corrosion inhibitors: Design, performance and industrial scale applications," *Mater. Adv.*, vol. 2, no. 12, pp. 3806–3850, 2021, doi:10.1039/D0MA00681E.
[6] C. Verma, M. A. Quraishi, K. Kluza, M. Makowska-Janusik, L. O. Olasunkanmi, and E. E. Ebenso, "Corrosion inhibition of mild steel in 1M HCl by D-glucose derivatives of dihydropyrido [2,3-d:6,5-d'] dipyrimidine-2,4,6,8(1H,3H, 5H,7H)-tetraone," *Sci. Rep.*, vol. 7, no. February, pp. 1–17, 2017, doi:10.1038/srep44432.
[7] C. Verma, L. O. Olasunkanmi, E. E. Ebenso, M. A. Quraishi, and I. B. Obot, "Adsorption behavior of glucosamine-based, pyrimidine-fused heterocycles as green corrosion inhibitors for mild steel: Experimental and theoretical studies," *J. Phys. Chem. C*, vol. 120, no. 21, pp. 11598–11611, Jun. 2016, doi:10.1021/acs.jpcc.6b04429.

[8] A. N. Khramov, N. N. Voevodin, V. N. Balbyshev, and M. S. Donley, "Hybrid organo-ceramic corrosion protection coatings with encapsulated organic corrosion inhibitors," *Thin Solid Films*, vol. 447–448, pp. 549–557, 2004, doi:10.1016/j.tsf.2003.07.016.

[9] C. Verma, E. E. Ebenso, I. Bahadur, and M. A. Quraishi, "An overview on plant extracts as environmental sustainable and green corrosion inhibitors for metals and alloys in aggressive corrosive media," *J. Mol. Liq.*, vol. 266, pp. 577–590, 2018, doi:10.1016/j.molliq.2018.06.110.

[10] C. Verma, E. E. Ebenso, and M. A. Quraishi, "Corrosion inhibitors for ferrous and non-ferrous metals and alloys in ionic sodium chloride solutions: A review," *J. Mol. Liq.*, vol. 248, pp. 927–942, 2017, doi:10.1016/j.molliq.2017.10.094.

[11] H. S. Gadow, and M. M. Motawea, "Investigation of the corrosion inhibition of carbon steel in hydrochloric acid solution by using ginger roots extract," *RSC Adv.*, vol. 7, no. 40, pp. 24576–24588, 2017, doi:10.1039/C6RA28636D.

[12] X. Wang, "Viburnum sargentii Koehne fruit extract as corrosion inhibitor for mild steel in acidic solution," *Int. J. Electrochem. Sci.*, vol. 13, pp. 5228–5242, Jun. 2018, doi:10.20964/2018.06.36.

[13] Y. Qiang, S. Zhang, B. Tan, and S. Chen, "Evaluation of Ginkgo leaf extract as an eco-friendly corrosion inhibitor of X70 steel in HCl solution," *Corros. Sci.*, vol. 133, pp. 6–16, Apr. 2018, doi:10.1016/j.corsci.2018.01.008.

[14] X. Zheng, M. Gong, Q. Li, and L. Guo, "Corrosion inhibition of mild steel in sulfuric acid solution by loquat (Eriobotrya japonica Lindl.) leaves extract," *Sci. Rep.*, vol. 8, no. 1, p. 9140, 2018, doi:10.1038/s41598-018-27257-9.

[15] M. P. Casaletto, V. Figà, A. Privitera, M. Bruno, A. Napolitano, and S. Piacente, "Inhibition of Cor-Ten steel corrosion by 'green' extracts of Brassica campestris," *Corros. Sci.*, vol. 136, pp. 91–105, 2018, doi:10.1016/j.corsci.2018.02.059.

[16] T. K. Bhuvaneswari, V. S. Vasantha, and C. Jeyaprabha, "Pongamia pinnata as a green corrosion inhibitor for mild steel in 1N sulfuric acid medium," *Silicon*, vol. 10, no. 5, pp. 1793–1807, 2018, doi:10.1007/s12633-017-9673-3.

[17] A. Zakeri, E. Bahmani, and A. S. R. Aghdam, "Plant extracts as sustainable and green corrosion inhibitors for protection of ferrous metals in corrosive media: A mini review," *Corros. Commun.*, vol. 5, pp. 25–38, 2022, doi:10.1016/j.corcom.2022.03.002.

[18] E. Bahmani, A. Zakeri, and A. Sabour Rouh Aghdam, "Protection of silver electro-deposit surfaces against accelerated and natural atmospheric corrosion by Ge incorporation in Ag structure: A microstructural investigation," *Surf. Interfaces*, vol. 26, p. 101288, 2021, doi:10.1016/j.surfin.2021.101288.

[19] E. Bahmani, A. Zakeri, and A. Sabour Rouh Aghdam, "Microstructural analysis and surface studies on Ag-Ge alloy coatings prepared by electrodeposition technique," *J. Mater. Sci.*, vol. 56, no. 10, pp. 6427–6447, 2021, doi:10.1007/s10853-020-05601-7.

[20] C. Googan, "The cathodic protection potential criteria: Evaluation of the evidence," *Mater. Corros.*, vol. 72, no. 3, pp. 446–464, 2021, doi:10.1002/maco.202011978.

[21] A. D. Arulraj, J. Prabha, R. Deepa, B. Neppolian, and V. S. Vasantha, "Effect of components of solanum trilobatum-L extract as corrosion inhibitor for mild steel in acid and neutral medium," *Mater. Res. Express*, vol. 6, no. 3, p. 36527, 2019, doi:10.1088/2053-1591/aaf267.

[22] B. El Ibrahimi, A. Jmiai, L. Bazzi, and S. El Issami, "Amino acids and their derivatives as corrosion inhibitors for metals and alloys," *Arab. J. Chem.*, vol. 13, no. 1, pp. 740–771, 2020, doi:10.1016/j.arabjc.2017.07.013.

[23] B. R. Fazal, T. Becker, B. Kinsella, and K. Lepkova, "A review of plant extracts as green corrosion inhibitors for CO_2 corrosion of carbon steel," *npj Mater. Degrad.*, vol. 6, no. 1, p. 5, 2022, doi:10.1038/s41529-021-00201-5.

[24] S. Z. Salleh, A. H. Yusoff, S. K. Zakaria, M. A. A. Taib, A. A. Seman, M. N. Masri, M. Mohamad, S. Mamat, S. A. Sobri, A. Ali, and P. T. Teo, "Plant extracts as green corrosion inhibitor for ferrous metal alloys: A review," *J. Clean. Prod.*, vol. 304, p. 127030, 2021, doi:10.1016/j.jclepro.2021.127030.

[25] S. Aribo, S. J. Olusegun, L. J. Ibhadiyi, A. Oyetunji, and D. O. Folorunso, "Green inhibitors for corrosion protection in acidizing oilfield environment," *J. Assoc. Arab Univ. Basic Appl. Sci.*, vol. 24, pp. 34–38, 2017, doi:10.1016/j.jaubas.2016.08.001.

[26] A. Zhang, Y. Wang, and H. Wang, "Preparation of inorganic-polymer nano-emulsion inhibitor for corrosion resistance of steel reinforcement for concrete," *Alexandria Eng. J.*, vol. 66, pp. 537–542, 2023, doi:10.1016/j.aej.2022.11.020.

[27] B. D. B. Tiu, and R. C. Advincula, "Polymeric corrosion inhibitors for the oil and gas industry: Design principles and mechanism," *React. Funct. Polym.*, vol. 95, pp. 25–45, 2015, doi:10.1016/j.reactfunctpolym.2015.08.006.

[28] M. Javidi, and S. Bekhrad, "Failure analysis of a wet gas pipeline due to localised CO_2 corrosion," *Eng. Fail. Anal.*, vol. 89, pp. 46–56, 2018.

[29] S. A. Umoren, M. M. Solomon, I. B. Obot, and R. K. Suleiman, "A critical review on the recent studies on plant biomaterials as corrosion inhibitors for industrial metals," *J. Ind. Eng. Chem.*, vol. 76, pp. 91–115, 2019, doi:10.1016/j.jiec.2019.03.057.

[30] S. Papavinasam, "The main environmental factors influencing corrosion," In: *Corrosion Control in the Oil and Gas Industry*. Boston: Gulf Professional Publishing, 2014, pp. 179–247.

[31] Y. Javadzadeh, and S. Hamedeyazdan, "Floating drug delivery systems for eradication of Helicobacter pylori in treatment of peptic ulcer disease," In: *Trends in Helicobacter pylori Infection*, B. M. Roesler, Ed. Rijeka: IntechOpen, 2014, p. 13, Ch. 11.

[32] A. Dehghani, G. Bahlakeh, and B. Ramezanzadeh, "Green Eucalyptus leaf extract: A potent source of bio-active corrosion inhibitors for mild steel," *Bioelectrochemistry*, vol. 130, p. 107339, 2019, doi:10.1016/j.bioelechem.2019.107339.

[33] H. Hassannejad, and A. Nouri, "Sunflower seed hull extract as a novel green corrosion inhibitor for mild steel in HCl solution," *J. Mol. Liq.*, vol. 254, no. Complete, pp. 377–382, 2018, doi:10.1016/j.molliq.2018.01.142.

[34] I.-M. Chung, R. Malathy, S.-H. Kim, K. Kalaiselvi, M. Prabakaran, and M. Gopiraman, "Ecofriendly green inhibitor from Hemerocallis fulva against aluminum corrosion in sulphuric acid medium," *J. Adhes. Sci. Technol.*, vol. 34, no. 14, pp. 1483–1506, Jul. 2020, doi:10.1080/01694243.2020.1712770.

[35] A. A. Khadom, A. N. Abd, and N. A. Ahmed, "Xanthium s trumarium leaves extracts as a friendly corrosion inhibitor of low carbon steel in hydrochloric acid: Kinetics and mathematical studies," *South African J. Chem. Eng.*, vol. 25, pp. 13–21, 2018.

[36] S. Pal, H. Lgaz, P. Tiwari, I.-M. Chung, G. Ji, and R. Prakash, "Experimental and theoretical investigation of aqueous and methanolic extracts of Prunus dulcis peels as green corrosion inhibitors of mild steel in aggressive chloride media," *J. Mol. Liq.*, vol. 276, pp. 347–361, 2019, doi:10.1016/j.molliq.2018.11.099.

[37] Y. Liu, Z. Song, W. Wang, L. Jiang, Y. Zhang, M. Guo, F. Song, and N. Xu, "Effect of ginger extract as green inhibitor on chloride-induced corrosion of carbon steel in simulated concrete pore solutions," *J. Clean. Prod.*, vol. 214, pp. 298–307, 2019, doi:10.1016/j.jclepro.2018.12.299.

[38] A. Ostovari, S. M. Hoseinieh, M. Peikari, S. R. Shadizadeh, and S. J. Hashemi, "Corrosion inhibition of mild steel in 1M HCl solution by henna extract: A comparative study of the inhibition by henna and its constituents (Lawsone, Gallic acid, α-D-glucose and tannic acid)," *Corros. Sci.*, vol. 51, no. 9, pp. 1935–1949, 2009, doi:10.1016/j.corsci.2009.05.024.

[39] R. Salim, E. Ech-chihbi, H. Oudda, F. El Hajjaji, M. Taleb, and S. Jodeh, "A review on the assessment of imidazo[1,2-a]pyridines as corrosion inhibitor of metals," *J. Bio Tribo Corros.*, vol. 5, no. 1, p. 14, Apr. 2018, doi:10.1007/s40735-018-0207-3.

[40] S. A. Umoren, U. M. Eduok, M. M. Solomon, and A. P. Udoh, "Corrosion inhibition by leaves and stem extracts of Sida acuta for mild steel in 1M H_2SO_4 solutions investigated by chemical and spectroscopic techniques," *Arab. J. Chem.*, vol. 9, pp. S209–S224, 2016, doi:10.1016/j.arabjc.2011.03.008.

[41] M. Schreiner, and S. Huyskens-Keil, "Phytochemicals in fruit and vegetables: Health promotion and postharvest elicitors," *CRC. Crit. Rev. Plant Sci.*, vol. 25, no. 3, pp. 267–278, Jul. 2006, doi:10.1080/07352680600671661.

[42] S. H. Alrefaee, K. Y. Rhee, C. Verma, M. A. Quraishi, and E. E. Ebenso, "Challenges and advantages of using plant extract as inhibitors in modern corrosion inhibition systems: Recent advancements," *J. Mol. Liq.*, vol. 321, p. 114666, 2021, doi:10.1016/j.molliq.2020.114666.

[43] P. M. Krishnegowda, V. T. Venkatesha, P. K. M. Krishnegowda, and S. B. Shivayogiraju, "Acalypha torta leaf extract as green corrosion inhibitor for mild steel in hydrochloric acid solution," *Ind. Eng. Chem. Res.*, vol. 52, no. 2, pp. 722–728, Jan. 2013, doi:10.1021/ie3018862.

[44] E. E. Oguzie, K. L. Oguzie, C. O. Akalezi, I. O. Udeze, J. N. Ogbulie, and V. O. Njoku, "Natural products for materials protection: Corrosion and microbial growth inhibition using Capsicum frutescens biomass extracts," *ACS Sustain. Chem. Eng.*, vol. 1, no. 2, pp. 214–225, Feb. 2013, doi:10.1021/sc300145k.

[45] M. Benahmed, I. Selatnia, N. Djeddi, S. Akkal, and H. Laouer, "Adsorption and corrosion inhibition properties of butanolic extract of Elaeoselinum thapsioides and its synergistic effect with Reutera lutea (Desf.) Maires (Apiaceae) on A283 carbon steel in hydrochloric acid solution," *Chem. Africa*, vol. 3, no. 1, pp. 251–261, 2020, doi:10.1007/s42250-019-00093-8.

[46] P. B. Raja, A. K. Qureshi, A. A. Rahim, K. Awang, M. R. Mukhtar, and H. Osman, "Indole alkaloids of Alstonia angustifolia var. latifolia as green inhibitor for mild steel corrosion in 1 M HCl media," *J. Mater. Eng. Perform.*, vol. 22, no. 4, pp. 1072–1078, 2013, doi:10.1007/s11665-012-0347-4.

[47] N. El Hamdani, R. Fdil, M. Tourabi, C. Jama, and F. Bentiss, "Alkaloids extract of Retama monosperma (L.) Boiss. seeds used as novel eco-friendly inhibitor for carbon steel corrosion in 1M HCl solution: Electrochemical and surface studies," *Appl. Surf. Sci.*, vol. 357, pp. 1294–1305, 2015, doi:10.1016/j.apsusc.2015.09.159.

[48] P. B. Raja, A. K. Qureshi, A. Abdul Rahim, H. Osman, and K. Awang, "Neolamarckia cadamba alkaloids as eco-friendly corrosion inhibitors for mild steel in 1M HCl media," *Corros. Sci.*, vol. 69, pp. 292–301, 2013, doi:10.1016/j.corsci.2012.11.042.

[49] G. Bahlakeh, B. Ramezanzadeh, A. Dehghani, and M. Ramezanzadeh, "Novel cost-effective and high-performance green inhibitor based on aqueous Peganum harmala seed extract for mild steel corrosion in HCl solution: Detailed experimental and electronic/atomic level computational explorations," *J. Mol. Liq.*, vol. 283, pp. 174–195, 2019, doi:10.1016/j.molliq.2019.03.086.

[50] M. H. Shahini, M. Ramezanzadeh, and B. Ramezanzadeh, "Effective steel alloy surface protection from HCl attacks using Nepeta pogonesperma plant stems extract," *Colloids Surf. A Physicochem. Eng. Asp.*, vol. 634, p. 127990, 2022, doi:10.1016/j.colsurfa.2021.127990.

[51] M. Chevalier, F. Robert, N. Amusant, M. Traisnel, C. Roos, and M. Lebrini, "Enhanced corrosion resistance of mild steel in 1M hydrochloric acid solution by alkaloids extract from Aniba rosaeodora plant: Electrochemical, phytochemical and XPS studies," *Electrochim. Acta*, vol. 131, pp. 96–105, 2014, doi:10.1016/j.electacta.2013.12.023.

[52] M. H. Hussin, M. Jain Kassim, N. N. Razali, N. H. Dahon, and D. Nasshorudin, "The effect of Tinospora crispa extracts as a natural mild steel corrosion inhibitor in 1M HCl solution," *Arab. J. Chem.*, vol. 9, pp. S616–S624, 2016, doi:10.1016/j.arabjc.2011.07.002.

[53] Z. Khiya, M. Hayani, A. Gamar, S. Kharchouf, S. Amine, F. Berrekhis, A. Bouzoubae, T. Zair, and F. El Hilali, "Valorization of the Salvia officinalis L. of the Morocco bioactive extracts: Phytochemistry, antioxidant activity and corrosion inhibition," *J. King Saud Univ. Sci.*, vol. 31, no. 3, pp. 322–335, 2019, doi:10.1016/j.jksus.2018.11.008.

[54] N. A. Odewunmi, S. A. Umoren, and Z. M. Gasem, "Utilization of watermelon rind extract as a green corrosion inhibitor for mild steel in acidic media," *J. Ind. Eng. Chem.*, vol. 21, pp. 239–247, 2015, doi:10.1016/j.jiec.2014.02.030.

[55] J. T. Johnson, E. U. Iwang, J. T. Hemen, M. O. Odey, and E. E. Efiong, "Evaluation of anti-nutrient contents of watermelon Citrallus lanatus," *An. Biol. Res.*, vol. 3, no. 11, pp. 5145–5150, 2012.

[56] M. Ramananda Singh, P. Gupta, and K. Gupta, "The litchi (Litchi chinensis) peels extract as a potential green inhibitor in prevention of corrosion of mild steel in 0.5M H_2SO_4 solution," *Arab. J. Chem.*, vol. 12, no. 7, pp. 1035–1041, 2019, doi:10.1016/j.arabjc.2015.01.002.

[57] P. Muthukrishnan, B. Jeyaprabha, and P. Prakash, "Adsorption and corrosion inhibiting behavior of Lannea coromandelica leaf extract on mild steel corrosion," *Arab. J. Chem.*, vol. 10, pp. S2343–S2354, 2017, doi:10.1016/j.arabjc.2013.08.011.

[58] E. E. Oguzie, "Evaluation of the inhibitive effect of some plant extracts on the acid corrosion of mild steel," *Corros. Sci.*, vol. 50, no. 11, pp. 2993–2998, 2008, doi:10.1016/j.corsci.2008.08.004.

[59] D. Prasad, O. Dagdag, Z. Safi, N. Wazzan, and L. Guo, "Cinnamoum tamala leaves extract highly efficient corrosion bio-inhibitor for low carbon steel: Applying computational and experimental studies," *J. Mol. Liq.*, vol. 347, p. 118218, 2022, doi:10.1016/j.molliq.2021.118218.

[60] P. Kobylin, T. Kaskiala, and J. Salminen, "Modeling of H_2SO_4–$FeSO_4$–H_2O and H_2SO_4–$Fe_2(SO_4)_3$–H_2O systems for metallurgical applications," *Ind. Eng. Chem. Res.*, vol. 46, no. 8, pp. 2601–2608, Apr. 2007, doi:10.1021/ie061302y.

[61] A. N. Abd, A. A. Khadom, and N. A. Ahmed, "Corrosion inhibition of low carbon steel in hydrochloric acid by cardaria draba leaves extracts," *J. Pet. Res. Stud.*, vol. 20, pp. 213–229, 2018.

[62] F. Bentiss, M. Lebrini, and M. Lagrenée, "Thermodynamic characterization of metal dissolution and inhibitor adsorption processes in mild steel/2,5-bis(N-thienyl)-1,3,4-thiadiazoles/hydrochloric acid system," *Corros. Sci.*, vol. 47, no. 12, pp. 2915–2931, 2005, doi:10.1016/j.corsci.2005.05.034.

[63] V. Rajeswari, D. Kesavan, M. Gopiraman, P. Viswanathamurthi, K. Poonkuzhali, and T. Palvannan, "Corrosion inhibition of Eleusine aegyptiaca and Croton rottleri leaf extracts on cast iron surface in 1M HCl medium," *Appl. Surf. Sci.*, vol. 314, pp. 537–545, 2014, doi:10.1016/j.apsusc.2014.07.017.

[64] N. O. Eddy, S. A. Odoemelam, and A. O. Odiongenyi, "Joint effect of halides and ethanol extract of Lasianthera africana on inhibition of corrosion of mild steel in H_2SO_4," *J. Appl. Electrochem.*, vol. 39, no. 6, pp. 849–857, 2009, doi:10.1007/s10800-008-9731-z.

[65] H. Cang, Z. Tang, H. Li, L. Li, J. Shao, and H. Zhang, "Study on the synergistic effect of iodide ion with the extract of Artemisia halodendron on the corrosion inhibition," *Int. J. Electrochem. Sci.*, vol. 12, pp. 10484–10492, 2017, doi:10.20964/2017.11.24.

[66] A. E.-A. S. Fouda, H. H. Al-Zehry, and M. Elsayed, "Synergistic effect of potassium iodide with Cassia italica extract on the corrosion inhibition of carbon steel used in cooling water systems in 0.5 M H_2SO_4," *J. Bio Tribo Corros.*, vol. 4, no. 2, p. 23, 2018, doi:10.1007/s40735-018-0138-z.

[67] M. Pramudita, S. Sukirno, and M. Nasikin, "Synergistic corrosion inhibition effect of rice husk extract and KI for mild steel in H_2SO_4 solution," *Bull. Chem. React. Eng. Catal.*, vol. 14, no. 3, pp. 697–704, 2019, doi:10.9767/bcrec.14.3.4249.697-704.

[68] X. Li, S. Deng, H. Fu, and X. Xie, "Synergistic inhibition effects of bamboo leaf extract/major components and iodide ion on the corrosion of steel in H_3PO_4 solution," *Corros. Sci.*, vol. 78, pp. 29–42, 2014, doi:10.1016/j.corsci.2013.08.025.

[69] A. A. Khadom, A. N. Abd, and N. A. Ahmed, "Synergistic effect of iodide ions on the corrosion inhibition of mild steel in 1 M HCl by Cardaria draba leaf extract," *Results Chem.*, vol. 4, no. November, p. 100668, 2022, doi:10.1016/j.rechem.2022.100668.

[70] A. N. Abd, A. A. Khadom, and N. A. Ahmed, "(JPR & S) Corrosion inhibition of low carbon steel in hydrochloric acid by Cardaria draba leaves extracts," *J. Pet. Res. Stud.*," vol. 8, no. 20, pp. 213–229, 2018.

[71] A. A. Khadom, A. N. Abd, and N. A. Ahmed, "Potassium iodide as a corrosion inhibitor of mild steel in hydrochloric acid: Kinetics and mathematical studies," *J. Bio Tribo Corros.*, vol. 4, pp. 1–10, 2018.

[72] S. Gk, J. M. Jacob, P. Rugma, and A. R. Jr, "Synergistic effect of salts on the corrosion inhibitive action of plant extract : A review," *J. Adhes. Sci. Technol.*, vol. 35, no. 2, pp. 1–31, 2020, doi:10.1080/01694243.2020.1797336.

Index

For Product Safety Concerns and Information please contact our EU
representative GPSR@taylorandfrancis.com
Taylor & Francis Verlag GmbH, Kaufingerstraße 24, 80331 München, Germany